Sustainable Polymer Composites and Nanocomposites

Inamuddin · Sabu Thomas
Raghvendra Kumar Mishra
Abdullah M. Asiri
Editors

Sustainable Polymer Composites and Nanocomposites

Volume 2

Editors
Inamuddin
Department of Applied Chemistry
Aligarh Muslim University
Aligarh, India

Sabu Thomas
School of Chemical Sciences
Mahatma Gandhi University
Kottayam, Kerala, India

Raghvendra Kumar Mishra
Mahatma Gandhi University
Kottayam, Kerala, India

Abdullah M. Asiri
Chemistry Department, Faculty of Science
King Abdulaziz University
Jeddah, Saudi Arabia

ISBN 978-3-030-05398-7   ISBN 978-3-030-05399-4   (eBook)
https://doi.org/10.1007/978-3-030-05399-4

Library of Congress Control Number: 2018963283

© Springer Nature Switzerland AG 2019, corrected publication 2019
This work is subject to copyright. All rights are reserved by the Publisher, whether the whole or part of the material is concerned, specifically the rights of translation, reprinting, reuse of illustrations, recitation, broadcasting, reproduction on microfilms or in any other physical way, and transmission or information storage and retrieval, electronic adaptation, computer software, or by similar or dissimilar methodology now known or hereafter developed.
The use of general descriptive names, registered names, trademarks, service marks, etc. in this publication does not imply, even in the absence of a specific statement, that such names are exempt from the relevant protective laws and regulations and therefore free for general use.
The publisher, the authors and the editors are safe to assume that the advice and information in this book are believed to be true and accurate at the date of publication. Neither the publisher nor the authors or the editors give a warranty, express or implied, with respect to the material contained herein or for any errors or omissions that may have been made. The publisher remains neutral with regard to jurisdictional claims in published maps and institutional affiliations.

This Springer imprint is published by the registered company Springer Nature Switzerland AG
The registered company address is: Gewerbestrasse 11, 6330 Cham, Switzerland

Contents

Processing, Characterization and Application of Micro and Nanocellulose Based Environmentally Friendly Polymer Composites 1
Adriana de Campos, Ana Carolina Corrêa, Pedro Ivo Cunha Claro, Eliangela de Morais Teixeira and José Manoel Marconcini

Extraction of Cellulose Nanofibers and Their Eco/Friendly Polymer Composites 37
Stephen C. Agwuncha, Chioma G. Anusionwu, Shesan J. Owonubi, E. Rotimi Sadiku, Usman A. Busuguma and I. David Ibrahim

Synthesis, Characterization and Applications of Polyolefin Based Eco-Friendly Polymer Composites 65
Akash Deep, Deepanshu Bhatt, Vishal Shrivastav, Sanjeev K. Bhardwaj and Poonma Malik

Spectroscopy and Microscopy of Eco-friendly Polymer Composites 105
Ashish K. Shukla, Chandni Sharma, Syed M. S. Abidi and Amitabha Acharya

Biocompatible and Biodegradable Chitosan Composites in Wound Healing Application: In Situ Novel Photo-Induced Skin Regeneration Approach 143
Amr A. Essawy, Hassan Hefni and A. M. El-Nggar

Mechanical, Thermal and Viscoelastic Properties of Polymer Composites Reinforced with Various Nanomaterials 185
T. H. Mokhothu, A. Mtibe, T. C. Mokhena, M. J. Mochane, O. Ofosu, S. Muniyasamy, C. A. Tshifularo and T. S. Motsoeneng

Preparation and Characterization of Antibacterial Sustainable Nanocomposites 215
T. C. Mokhena, M. J. Mochane, T. H. Mokhothu, A. Mtibe,
C. A. Tshifularo and T. S. Motsoeneng

Extraction of Nano Cellulose Fibres and Their Eco-friendly Polymer Composite 245
Bashiru Kayode Sodipo and Folahan Abdul Wahab Taiwo Owolabi

Static and Dynamic Mechanical Properties of Eco-friendly Polymer Composites 259
Bernardo Zuccarello

Synthesis, Characterization, and Applications of Hemicellulose Based Eco-friendly Polymer Composites 293
Busra Balli, Mehmet Harbi Calimli, Esra Kuyuldar and Fatih Sen

Impact of Nanoparticle Shape, Size, and Properties of the Sustainable Nanocomposites 313
Thandapani Gomathi, K. Rajeshwari, V. Kanchana, P. N. Sudha
and K. Parthasarathy

Polymeric Composites as Catalysts for Fine Chemistry 337
P. SundarRajan, K. GracePavithra, D. Balaji and K. P. Gopinath

Fabrication Methods of Sustainable Hydrogels 355
Cédric Delattre, Fiona Louis, Mitsuru Akashi, Michiya Matsusaki,
Philippe Michaud and Guillaume Pierre

Application of Sustainable Nanocomposites for Water Purification Process 387
Hayelom Dargo Beyene and Tekilt Gebregiorgs Ambaye

Sustainable Nanocomposites in Food Packaging 413
H. Anuar, F. B. Ali, Y. F. Buys, M. A. Siti Nur E'zzati,
A. R. Siti Munirah Salimah, M. S. Mahmud,
N. Mohd Nordin and S. A. Adli

Mechanical Techniques for Enhanced Dispersion of Cellulose Nanocrystals in Polymer Matrices 437
Jamileh Shojaeiarani, Dilpreet S. Bajwa and Kerry Hartman

Processing and Industrial Applications of Sustainable Nanocomposites Containing Nanofillers 451
Khadija Zadeh, Sadiya Waseem, Kishor Kumar Sadasivuni,
Kalim Deshmukh, Aqib Muzaffar, M. Basheer Ahamed
and Mariam Al-Ali AlMaadeed

Recent Advances in Paper-Based Analytical Devices: A Pivotal Step Forward in Building Next-Generation Sensor Technology 479
Charu Agarwal and Levente Csóka

Polymers and Polymer Composites for Adsorptive Removal of Dyes in Water Treatment 519
Weiya Huang, Shuhong Wang and Dan Li

Current Scenario of Nanocomposite Materials for Fuel Cell Applications 557
Raveendra M. Hegde, Mahaveer D. Kurkuri and Madhuprasad Kigga

Rubber Clay Nanocomposites 593
Mariajose Cova Sanchez, Alejandro Bacigalupe, Mariano Escobar and Marcela Mansilla

Organic/Silica Nanocomposite Membranes Applicable to Green Chemistry 629
Mashallah Rezakazemi, Amir Dashti, Nasibeh Hajilary and Saeed Shirazian

Extraction of Cellulose Nanofibers and Their Eco-friendly Polymer Composites 653
M. Hazwan Hussin, Djalal Trache, Caryn Tan Hui Chuin, M. R. Nurul Fazita, M. K. Mohamad Haafiz and Md. Sohrab Hossain

Recyclable and Eco-friendly Single Polymer Composite 693
Mohd Azmuddin Abdullah, Muhammad Afzaal, Safdar Ali Mirza, Sakinatu Almustapha and Hanaa Ali Hussein

Processing Aspects and Biomedical and Environmental Applications of Sustainable Nanocomposites Containing Nanofillers 727
Mohd Azmuddin Abdullah, Muhammad Shahid Nazir, Zaman Tahir, Yasir Abbas, Majid Niaz Akhtar, Muhammad Rafi Raza and Hanaa Ali Hussein

Smart Materials, Magnetic Graphene Oxide-Based Nanocomposites for Sustainable Water Purification 759
Janardhan Reddy Koduru, Rama Rao Karri and N. M. Mubarak

Functionalized Carbon Nanomaterial for Artificial Bone Replacement as Filler Material 783
Fahad Saleem Ahmed Khan, N. M. Mubarak, Mohammad Khalid and Ezzat Chan Abdullah

Inorganic Nanocomposite Hydrogels: Present Knowledge and Future Challenge 805
Nasrin Moini, Arash Jahandideh and Gary Anderson

Processing, Characterization and Application of Natural Rubber Based Environmentally Friendly Polymer Composites 855
Nayan Ranjan Singha, Manas Mahapatra, Mrinmoy Karmakar and Pijush Kanti Chattopadhyay

Electrical Properties of Sustainable Nano-Composites Containing Nano-Fillers: Dielectric Properties and Electrical Conductivity 899
Sabzoi Nizamuddin, Sabzoi Maryam, Humair Ahmed Baloch, M. T. H. Siddiqui, Pooja Takkalkar, N. M. Mubarak, Abdul Sattar Jatoi, Sadaf Aftab Abbasi, G. J. Griffin, Khadija Qureshi and Nhol Kao

Thermal Properties of Sustainable Thermoplastics Nanocomposites Containing Nanofillers and Its Recycling Perspective 915
Pooja Takkalkar, Sabzoi Nizamuddin, Gregory Griffin and Nhol Kao

Application of Sustainable Nanocomposites in Membrane Technology 935
Pravin G. Ingole

Reliable Natural-Fibre Augmented Biodegraded Polymer Composites 961
Ritu Payal

An Overview on Plant Fiber Technology: An Interdisciplinary Approach 977
Alan Miguel Brum da Silva, Sandra Maria da Luz, Irulappasamy Siva, Jebas Thangiah Winowlin Jappes and Sandro Campos Amico

Nanocellulose-Reinforced Adhesives for Wood-Based Panels 1001
Elaine Cristina Lengowski, Eraldo Antonio Bonfatti Júnior, Marina Mieko Nishidate Kumode, Mayara Elita Carneiro and Kestur Gundappa Satyanarayana

Nanocellulose in the Paper Making 1027
Elaine Cristina Lengowski, Eraldo Antonio Bonfatti Júnior, Marina Mieko Nishidate Kumode, Mayara Elita Carneiro and Kestur Gundappa Satyanarayana

Impact of Nanoparticle Shape, Size, and Properties of Silver Nanocomposites and Their Applications 1067
Arpita Hazra Chowdhury, Rinku Debnath, Sk. Manirul Islam and Tanima Saha

Toxicological Evaluations of Nanocomposites with Special Reference to Cancer Therapy 1093
Arpita Hazra Chowdhury, Arka Bagchi, Arunima Biswas and Sk. Manirul Islam

Synthesis, Characterization and Application of Bio-based Polyurethane Nanocomposites 1121
Sonalee Das, Sudheer Kumar, Smita Mohanty and Sanjay Kumar Nayak

Clay Based Biopolymer Nanocomposites and Their Applications in Environmental and Biomedical Fields 1159
K. Sangeetha, P. Angelin Vinodhini and P. N. Sudha

Thermal Behaviour and Crystallization of Green Biocomposites 1185
Vasile Cristian Grigoras

Eco-friendly Polymer Composite: State-of-Arts, Opportunities and Challenge 1233
V. S. Aigbodion, E. G. Okonkwo and E. T. Akinlabi

Synthesis, Characterization, and Applications of Hemicelluloses Based Eco-friendly Polymer Composites 1267
Xinwen Peng, Fan Du and Linxin Zhong

Self-healing Bio-composites: Concepts, Developments, and Perspective 1323
Zeinab Karami, Sara Maleki, Armaghan Moghaddam and Arash Jahandideh

Chemical Modification of Lignin and Its Environmental Application 1345
Zhili Li, Yuanyuan Ge, Jiubing Zhang, Duo Xiao and Zijun Wu

Synthesis and Characterization and Application of Chitin and Chitosan-Based Eco-friendly Polymer Composites 1365
Aneela Sabir, Faizah Altaf and Muhammad Shafiq

Nanocomposites for Environmental Pollution Remediation 1407
Anjali Bajpai, Maya Sharma and Laxmi Gond

Correction to: Extraction of Nano Cellulose Fibres and Their Eco-friendly Polymer Composite C1
Bashiru Kayode Sodipo and Folahan Abdul Wahab Taiwo Owolabi

Processing Aspects and Biomedical and Environmental Applications of Sustainable Nanocomposites Containing Nanofillers

Mohd Azmuddin Abdullah, Muhammad Shahid Nazir, Zaman Tahir, Yasir Abbas, Majid Niaz Akhtar, Muhammad Rafi Raza and Hanaa Ali Hussein

1 Introduction

Polymer matrices such as rubber, plastic, acrylic, ethylene are commonly available in the market. These materials have advantages due to their lightweight, straightforward processing, and low cost [1], with outstanding corrosion stability and ductility. The major disadvantages of the polymer components are the low thermal and environmental stability (against UV), low acid resistance, and conductivity [2]. To overcome the problems, the polymer matrices are reinforced with fillers (particles, fibers, or platelets, synthetic or natural, organic or inorganic) at macro, micro, or nanoscale [3–5]. The resultant composite materials have characteristics different from the individual constituent. The composites consisting of two or more constituents may have significantly different physical or chemical properties, but the

---

M. A. Abdullah (✉) · H. A. Hussein
Institute of Marine Biotechnology, Universiti Malaysia Terengganu,
21030 Kuala Nerus, Terengganu, Malaysia
e-mail: azmuddin@umt.edu.my; joule1602@gmail.com

M. S. Nazir (✉)
Department of Chemistry, COMSATS, Institute of Information Technology,
Defence Road, Off Raiwind Road, Lahore, Punjab, Pakistan
e-mail: shahid.nazir@cuilahore.edu.pk

Z. Tahir · Y. Abbas
Department of Chemical Engineering, COMSATS, Institute of Information Technology,
Defence Road, Off Raiwind Road, Lahore, Punjab, Pakistan

M. N. Akhtar
Department of Physics, Muhammad Nawaz Shareef (MNS) University of Engineering and Technology, Multan, Punjab, Pakistan

M. R. Raza
Department of Mechanical Engineering, COMSATS, Institute of Information Technology, Sahiwal 57000, Punjab, Pakistan

© Springer Nature Switzerland AG 2019
Inamuddin et al. (eds.), *Sustainable Polymer Composites and Nanocomposites*,
https://doi.org/10.1007/978-3-030-05399-4_25

constituent components remain separate and distinct within the finished structure. The characteristics of the filler and the host material will determine the performance, although, with time, this may still deteriorate upon exposure to UV, high temperature, pH, and humidity [6].

In the petroleum industry, epoxy resins are derived products possessing superior properties, dimensional stability, and good wettability with matrices for composite fillers and structural adhesives [7]. The industries based on epoxies in 2010 have reported 2 million tonnes of production with an annual turnover of USD20 billion [8]. Approximately 90% of epoxies are produced from diglycidal ether of bisphenol A (DGEBA) [8] which is an endocrine disruptor in humans and causing morbidity, decreased fertility, and cancer [9]. Polymer composites have therefore gained prominence in sciences, technology and engineering essentially to address the issue of global sustainability and as a replacement to the conventional materials such as the epoxies, metals, and ceramics. Engineered composite materials can be found as plywood and polymer matrix composites, ceramic matrix composites, metal matrix composites and smart composite materials which require greater strength to weight ratio and cost-effectiveness. The implications are seen in industries from construction, automobiles, and furniture, to common applications such as the landfill liner which uses waste clay from boron production, wood fillers in the thermoplastic matrix, and thermal insulation of paper cups, to the more sophisticated applications in aerospace, microelectronics, biomedicals and pharmaceuticals [10].

Fibre-reinforced polymers (FRPs) possess strong ecological appeal, low cost, and higher specific mechanical properties. The polymer matrix has been reinforced with synthetic or artificial fibers to improve the performance of the final products [11]. Renewable and sustainable materials from the industrial and agricultural residues and energy crops especially have bright future as reinforcement material to improve specific properties in composite or nanocomposites. The agro-based materials are attractive for sustainable technologies and safe production, whilst fulfilling the energy needs of the industries [12]. The total world biomass production of the lignocellulosic material is estimated at 10 Mg per hectare per year [13], and the total cellulose production reaches 10^{11}–10^{12} tonnes per year [14], with the annual consumption of 7.5×10^{10} tonnes [15]. In the light of indiscriminate dumping of residual plastics which has caused serious global environmental problems, and the free availability of biomass wastes from agricultural sectors, the development of biocomposite material based on agro-residues with synthetic polymer will pave the way for enhanced waste recycling and the conversion into value-added products [16].

This review article discusses the characteristics and fabrication of nano-fillers with a special focus on cellulose, chitosan, and magnetic nanocomposites, the composites development and the preparation of nanofibers with renewable polymers. The applications in drug delivery, tissue engineering, biosensor and electrical insulation, and the catalysis and environmental remediation are highlighted.

2 Characteristics and Fabrication of Nano-fillers

The range of the host materials used in nanohybrids include the organic polymers, silica or even liquid media. A standardization committee, ISO TC229 "Nanotechnologies" in 2005 in a joint group with IEC 113 "Nanotechnology Standardization for Electrical and Electronic Products and Systems" has defined nanocomposites as those containing fillers with at least one dimension of size less than 100 nm (ISO/TS27687:2008, ISO/TS11360:2010, ISO/TS88004:2011, ISO/TS 80004-2:2015) [17]. The properties (such as mechanical, electrical, optical or thermal) of a new generation of materials can be controlled with the nano-fillers which depend upon the chemical composition, shape, size, dispersion and orientation of the nanofillers. Nanocomposites are multi-phasic materials where the matrix material incorporates units with one, two or three dimension and may contain the nanofillers up to 10% of the total mass of the matrix. The fillers and the matrix are mixed at a moderate temperature, followed by the fabrication into the desired shape, but the functionalized fillers may result in better incorporation into the composites [18]. The advantages are the integration of several component materials and their properties in a single material. The combined properties of the matrix material and the nanosized filler yield novel functional materials, which can be tailor-made for specific applications [19, 20].

The nanofillers can be divided into the one-dimensional nanoplate, two-dimensional nanofibres and three-dimensional nanoparticles (Fig. 1). The nanoplate nanofillers have a thickness in the nanometer scale, though the lateral dimension could be several hundred nanometers to microns in the range [21]. Smectite clay minerals, particularly montmorillonites (MMTs) and graphene are examples of one-dimensional nanofillers. To improve the silicate compatibility with the polymer chains, organophilic modification of the internal and external cations of the clay minerals can be carried out. Graphene is a one-atom-thick planar, single sheet of sp2-bonded carbon atoms, synthesized from graphite by micromechanical cleavage [22–24], and its one-atom thickness is advantageous for ultrafast computational and sensor applications [25]. The nanofiber nanofillers have diameters in the nanometer scale such as the carbon nanofibers (CNFs) and nanotubes (CNTs), metallic nano-rods, and whiskers. The CNFs and CNTs are the most widely used

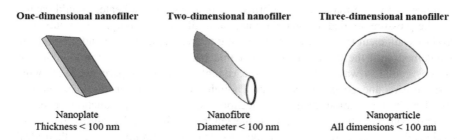

Fig. 1 Shapes of nanofillers used in the nanocomposites (adapted from [17])

nanofiber nanofillers in the polymer nanocomposites [26]. The CNTs, available in single wall or multi walls, with the lengths of tens of microns and the diversity in the synthesis routes and fabrication, may result in a different structure, diameter, aspect ratio, crystallinity, purity, and entanglement. The extraordinary properties of the CNTs such as density less than 2 g cm^{-3} with axial stiffness in the order of 1 TPa and the strength of 50 GPa [22], render the CNTs to be versatile. The nanoparticles (NPs) have the three dimensions at the nanometer scale such as the carbon black, silica, or quantum dots. The trend now is to individually disperse, control and modify the structure for enhanced properties of the composites [22].

Single biopolymer such as starch materials has poor mechanical properties and high water permeation which can be improved by reinforcing with the fibers, flakes, platelets, and particles [27–30], or the polymer matrix blending with other polymers [31]. Apart from the properties of the components, the bonding and the interfacial adhesion between the matrix and the filler also influence the final composite properties. With the nanofillers, the ratio between the area of the interface and the volume of the reinforcement is much higher than when the micro-sized or conventional reinforcements are used. NPs addition, therefore, reduces the retrogradation of the polymer matrix [32–34], as the large surface area per unit volume and the high aspect ratio of the nanofillers transfer their properties to the polymer matrix. A high fraction of the atoms in nanometric fillers is at the surface which enhances reactivity. With good dispersion, the interface region is maximized to allow the conformation of the interfacial percolation network, resulting in improved nanocomposite properties [35]. When the light reflection coefficient of the nanofiller is similar to the matrix, there will be optimal effect in terms of transparency, whereas the composites becoming opaque due to the light scattering effect of the microfillers. Nanofillers also improve the mechanical properties [29–31], as well as the electrical [36, 37] and thermal properties [36–39].

Different types of lignocellulosic fibers have been blended with biopolymers and synthetic polymers [40]. Compression and injection moulding for composites and material processing have been reported for various types of fibers and biopolymers (Table 1) [17, 41]. Compression moulding is widely used in the bakery industry, before being developed later as the processing methods in the plastics industry especially in the manufacturing of thermoset plastic parts, although also applicable to thermoplastics composite with unidirectional tapes, woven fabrics, randomly orientated fiber mat or chopped strand [42]. The raw materials, in the form of granules, putty-like masses, or preforms, are first placed in a heated mould cavity, and the pressure applied to force the material to fill up the cavity. A hydraulic ram produces sufficient force during the moulding process, and the heat and pressure are maintained until the plastic material is cured [42]. The compression moulding is low costs and low wastes, fast setup time, capable of large size parts whilst allowing intricate parts and good surface finish. Though producing fewer knit lines and less fibre-length degradation than the injection moulding, the production speed of compression moulding is lesser than the latter, and limited largely to the flat or moderately curved parts with no undercuts, and furthermore with less-than-ideal product consistency [42].

Table 1 Fabrication methods for lignocelluloses and biopolymers [41, 48]

Lignocelluloses/biopolymer		Fabrication methods
Long fibers	Jute yarn	Pultrusion and compression molding
	Flax and sisal	
	Lyocell	
	Kapok	
	Unidirectional/multidirectional flax	
Short fibers	Cellulose filament yarn	Fabricated by single/twin screw extruder and injection molding
	Jute yarn	
	Flax yarn	
	Cellulose fiber	Fabricated hydraulic/compression molding
	Sisal fiber	
	Kenaf fibers	
	Flax fibers	Twin screw extruder/internal mixture and injection molding
	Jute fibers	
	Vetiver grass	
	Bamboo fibers	
Particulate fibers	Cellulose pulp, Sisal, Coir, Luff sponge	Screw extruder/internal mixture and compression molding
	Cellulose whiskers	
	Soft wood, Avicel fiber, Alfa	
	Pine saw dust	
	Wood fiber	
	Hemp fiber	
	Saw dust	
	Luffa fiber	
	Paper slug	
	Rice husk, wood powder	Screw extruder/internal mixing and injection molding
	Lignocel fiber	
	Hard wood dust	
	Liquefied wood mill	
	Hemp	
	Pine soft wood fiber	
	Wheat straw	
Biopolymers from microorganisms	PLA Polyhydroxy alkanoates (PHB) Polyhydroxy fatty acids (PHF)	Extrusion Injection molding Injection molding

(continued)

Table 1 (continued)

Lignocelluloses/biopolymer		Fabrication methods
Biopolymers from renewable resources, biodegradable, of plant origin	Starch-based	Extrusion and injection molding
	Cellulose-based	Injection molding
	Lignin-based	Injection molding
Biopolymers from renewable resources, biodegradable, of animal origin	Chitin and chitosan	Dry molding
Biopolymers, biodegradable, of fossil origin	Polyester Ethanol (poly vinyl alcohol)	Extrusion

Conventional injection moulding process based on ceramic involves feedstock preparation, injection moulding, debinding process and sintering. In the case of composite granulate used as the feedstock, the key issues will be the particle sizes in microns, submicron or nano region, size distribution, shape and specific surface area and purity [43]. An optimum feedstock/thermoplastic binder content is mixed homogeneously to form a moderate viscosity mixture, free from agglomerates, with sufficient fluidity [44]. Injection moulding is used to mould the feedstock under concurrent heating and pressurization, with constant monitoring to minimize defects. This is followed by binding, where the soluble and insoluble components of the binder are removed, before sintering [43]. The use of conventional injection moulding machines on renewable and biopolymers is not easy due to a large number of products from compounds of the most varied shapes and variants. Flax fibers for example exhibit comparable elongation at break (3.3%) to the E-glass (3.4%), despite having a lower density of 1530 kg m^{-3}, E-modulus of 58 ± 15 GPa and tensile strength of 1339 ± 486 MPa, than the E-glass (2550, 71 and 3400, respectively). However, the flax/lactic acid resins (70:30) fabricated by the compression moulding show higher tensile strength of 62 MPa and tensile modulus of 9 GPa, as compared to the flax/PLA fabricated by the injection moulding (40–55 and 3–6, respectively) [45–47]. The injection moulding system, especially the plasticizing unit, must be adjusted to ensure correct feeding characteristics. The special requirements are to allow gentle plastification through the customized design, using coated systems to counter abrasion and chemical resistance, and high injection rates or pressures with adequate venting [48]. In the injection moulding process, maximum possible accuracy and precision of material, machine, and the mould must be streamlined to produce very small components especially in the field of miniaturisation such as the minimal connectors for automobile engineering, ball bearing retainers in nano-mechanics or micropipettes in medical technology or biotechnology. This may necessitate appropriate know-how in the field of tool engineering with customized systems based on the standard platform to handle shot weights of a few grams, and component weights of a few tenths or even a few centigrams [48].

3 Green Nanocomposites

Nanofillers or nanomaterials are mainly classified into organic nanofillers which include cellulose, chitosan, organic clays, fullerenes or CNTs; and the inorganic nanofillers which include magnetic NPs, metals-based NPs or inorganic clays. Renewable polymers such as starch, cellulose, and chitin are biocompatible, environmentally-friendly and biodegradable. The preparation of sustainable polymer nanocomposites reinforced with cellulose, chitin, and starch is discussed in the next section.

3.1 Cellulose Nanocomposites

Cellulose $(C_6H_{10}O_5)_n$ is a predominant constituent of lignocellulosic material (30%) and the most abundant biopolymer resource, employed for millennia in a wide variety of pre-industrial and industrial processes [12, 49]. It is a linear chain of a hundred to a thousand 1-4-β-d-anhydroglucose units, forming the structural component in prokaryotic and eukaryotic organisms including the green plants and several forms of algae, [12, 50]. Other natural substances in the plant fibers such as lignin, waxes, hemicelluloses and pectin are also found in large quantity [51–53]. Some bacterial species such as *Gluconacetobacter xylinum*, *Sarcina*, *Pseudomonas*, *Aerobacter*, *Alcaligenes*, *Acetobacter*, *Rhizobium*, *Achromobacter* and *Zoogloea* produce or secrete bacterial cellulose (BC) to make biofilms. Algal species such as *Pyrrophyta*, *Chrysophycease*, *Xanthophyceae*, *Phaeophyta*, and the *Chlorophyta* have also been reported to produce cellulose [54]. Cellulose can be synthesized in vitro by using specific cellulose activator and the synthesized microfibril particles and the subfibrils are arranged and assembled [55]. Cellulose shows the phenomenon of allotropy where the allomorphs have the same chemical substances but more than one crystalline form (Table 2). The crystalline structures of native cellulose I have two phases [56], where the Cellulose phases I_a for bacteria or algae have triclinic geometry, the tunicates or animal I_a and I_β cellulose have monoclinic chains and the high plant possesses both forms of cellulose [56]. Cellulose I can be transformed into Cellulose III_I by refluxing with an ammonical solution in

Table 2 Cellulose allomorphs with crystal geometry [56]

Cellulose allomorph	Crystal geometry
Alpha cellulose (I_a)	Triclinic, one parallel chain
Beta cellulose (I_β)	Monoclinic, two parallel chains
Cellulose II	Monoclinic, two antiparallel chains
Cellulose III_I	One monoclinic, parallel chain
Cellulose III_{II}	Two monoclinic, antiparallel chains
Cellulose IV_I	Two orthorhombic, parallel chains
Cellulose IV_{II}	Two orthorhombic, antiparallel chains

ethylenediamine, and Cellulose III_I can be converted into Cellulose IV_I via glycerol refluxing or from the hydrolysis of Cellulose I_a. Cellulose II can be produced by reacting with NaOH, resulting in the monoclinic structure [56]. Prokaryotic, *Eubacteria*, and *Sarcina* synthesize Cellulose II, but the purple bacteria *Acetobacter*, *Rhizobium* and *Agrobacterium* synthesize only Cellulose I [55, 56].

The unique properties of cellulose include biodegradability, capacity for broad chemical modification, and the formation of versatile semicrystalline morphologies [57, 58]. Microcrystalline cellulose (MCC) can be prepared from various plant fibers [59–61]. Green isolation of cellulose from oil palm empty fruit bunches (OPEFB) have been reported using autoclave-based and ultrasonication pretreatments to replace non-green chlorite method [16, 62]. The ultrasonic treatment yields 49% cellulose with 91.3% α-cellulose content and 68.7% crystallinity, while the autoclave treatment produces 64% cellulose, with 93.7% α-cellulose content and 70% crystallinity. The cellulose/PP composite fabrications with 25% cellulose loading via the injection-moulding technique attains a high tensile strength of 27 MPa, without any addition of coupling agents, with high thermal stability and low water and diesel sorption [16]. High crystallinity index of 87% cellulose I MCC from OPEFB has been achieved through the use of elaborate serial bleaching of oxygen-ozone and H_2O_2, followed by acid hydrolysis [63]. Generally, an acid hydrolysis process is used to make aqueous suspensions of nano-crystallites [64]. The crystalline rod-like cellulose nanocrystals (CNC) [65, 66], can be isolated from delaminated cellulosic fibers or MCC using controlled sulfuric acid treatment to remove the amorphous material. The esterification of hydroxyl groups by sulphate ions yields a negatively charged surface to form a stable colloidal dispersion of CNC, with diameter and length less than 100 nm [67–71]. As the CNCs tend to form aggregation and agglomeration [72] the mechanical or ultrasonification treatment allows uniform aggregate dispersion for a more stable colloidal suspension [69, 73]. Two methods—acid hydrolysis and chemical-swelling method, have been compared to produce CNC from the MCC extracted from OPEFB pulp. The swelling treatment where the MCCs are swelled and partly separated to whiskers by chemical and ultrasonification treatments [74, 75] results in typical cellulose I structure, similar to the MCC, with no cellulose II. The acid hydrolysis treatment using 64% H_2SO_4 (96% purity) and strong agitation exhibits the coexistence of cellulose I and cellulose II allomorphs [71], similar to that reported for the CNC isolated from waste sugarcane bagasse [76].

The higher or lower crystallinity will determine the intended use of final composite products. The crystal size and shape are influenced by the cellulose source, the plant species and age, the local growth conditions (climate, soil) and the acid hydrolysis conditions [66, 77]. The crystallinity of CNC from acid hydrolysis of OPEFB-MCC at 84%, for example, is lower than the MCC (87%) and the CNC from the swelling-treatment (88%) [71]. The reduced crystallinity from acid hydrolysis may have been a result of the esterification of the cellulose chains [71, 78], or also possibly due to the higher surface amorphous ratio of CNC [79]. With this high aspect ratio, lightweight and better thermal properties than the MCC, the CNCs are more cost-effective and suitable as reinforcing agents in the bio-renewable composite preparation [71]. The tensile strength of the CNC can be processed into the highest attainable composite strengths, far more than the high-volume content reinforcements [67, 80]. The CNCs from the bacterial cellulose fibrils are suitable as fillers for a transparent polymer composite due to their smaller lateral dimensions than the wavelength of the visible light [81]. The nanocomposite films of isotactic PP reinforced with CNCs dispersed with surfactant also exhibit far superior properties than the clean matrix or the composites containing other fillers [82].

3.2 Chitosan Nanocomposites

Chitosan (Cs) is the second most abundant biomaterial after cellulose. It is a naturally occurring amino polysaccharide, derived as a deacetylated form of chitin. The chemical structure of chitin and Cs is shown in Fig. 2 [83]. The primary amine group in Cs confers uniques properties such as the cationic nature, anti-microbial activity, biocompatibility, and biodegradability, which are favourable in the biomedical application for controlled drug release [84]. The X-ray results of chitin and Cs show nearly similar diffraction pattern between shrimp shell (α-chitin) and anhydrous squid pen (β-chitin). The crystallographic parameters of 'α' and 'β' chitin exhibit two antiparallel molecules per unit cell in α-chitin, but only one

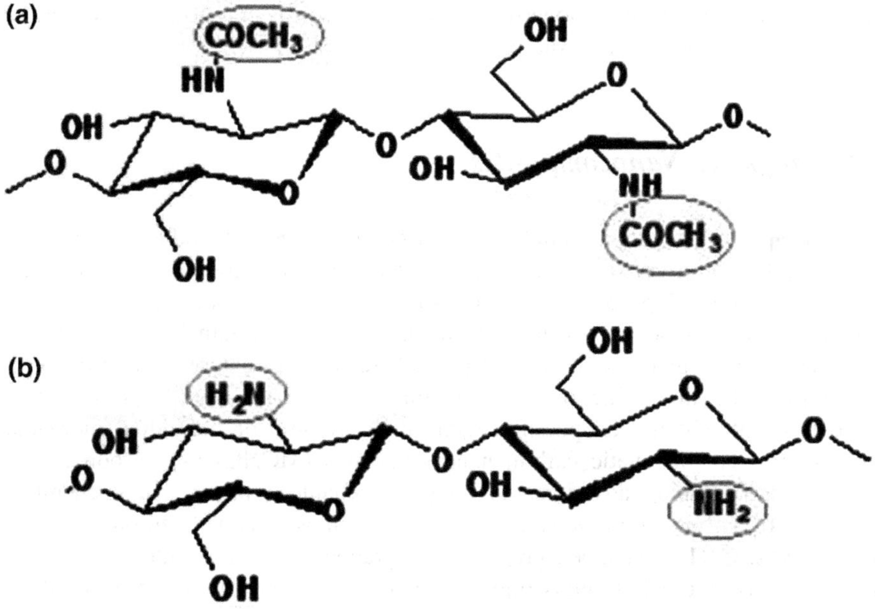

Fig. 2 Chemical structure of **a** chitin, and **b** chitosan (adapted from [84])

parallel arrangement in β-chitin. Despite this difference, it appears that N-acetyl glycosyl moiety is the independent crystallographic unit in both allomorphs [85, 86]. However, α-chitin samples suggest that the lobster-tendon chitin is observed at 'hkl' value of [001] [87], which is absent in the more crystalline sagitta chitin [88–90]. Further studies are needed to analyze the α-chitin crystal structure while the anhydrous β-chitin appears to be well defined and established [91, 92].

The incorporation of even 5–10% (w/w) of nanofillers into the Cs matrix could already improve the properties as compared to the 40–50% classical fillers in the conventional composites. The Cs composition in the composites can also be kept high for enhanced bioactivity and biocompatibility, thermal stability and transparency [93]. Electrospinning of Cs into nano-fibrous materials has been reported [94–98]. The high viscosity of Cs limits its spinnability and the application of an alkali treatment could hydrolyze the Cs chains and reduce its molecular weight (MW). Further treatment with aqueous 70–90% acetic acid for 48 h produces optimum nanofibers with moisture regain 74% greater than the treated and untreated chitosan powder. However, the diameter of 140 nm obtained, strongly affected by the electrospinning conditions and the solvent concentration [98], may be slightly higher than the 100 nm limit set for nanofiber [17]. Pure Cs nanofibers have been successfully produced, after the optimization of the electrospinning process parameters, from the blends of Cs and poly (ethylene oxide) (PEO) solubilized in the acetic acid. The Cs nanofibers exhibit better structural stability for at least six months in aqueous solutions (phosphate buffer (PBS) or water) [96]. Depending on the geometry and aspect ratio (particle length to thickness ratio), the Cs nanofillers can be fabricated into layer (nanoplate), rod-like (nanofiber) and spherical (nanoparticle) structure. The aspect ratio is the key factor in the enhancement of composite properties (Fig. 3) [99].

3.3 Magnetic Nanocomposites

Various types of magnetic particles have been combined with different materials such as gels, liquid crystals, renewable polymers, silica, carbon or metal-organic frameworks to develop novel nanocomposites. Magnetic polymer nanocomposites (MPNC) have the inorganic magnetic nanoparticles, fibers or lamellae embedded or dispersed in an organic polymer [100] which respond to external static or alternating magnetic field. The organic-inorganic synergies in MNPC possess new properties that will not be possible with single organic or individual inorganic components. The magnetic carbon nanocomposites (MCNC) use carbon as the hosting matrix, taking advantage of unique mechanical, physical and chemical properties of carbon, and diverse morphologies such as activated carbon, fullerene (C_{60}), CNFs, CNTs, expanded graphite, and graphene. Static or alternating magnetic fields as external stimulus offer relatively large penetration depth to induce

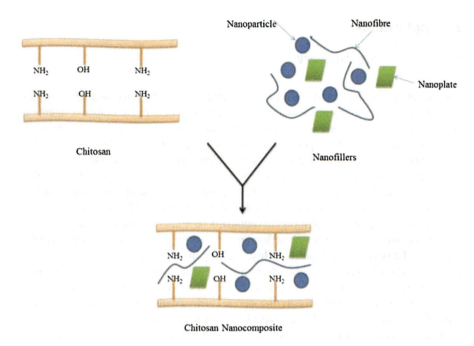

Fig. 3 Chitosan nanocomposite containing different types of nanofillers (adapted from [99])

magneto-mechanical forces, to change the shape or to move the host materials [20, 101]. These unique magnetic properties offer perspectives for fundamental understanding and applications in medical therapy and diagnosis, separations, actuation, or catalysis [101]. The common method of magnetic NPs synthesis is the co-precipitation of Fe^{2+} and Fe^{3+} ions by a base, to yield NPs of broad size distribution. To produce highly monodisperse magnetic NPs and the MPNCs with good particle distribution, thermal decomposition of metal precursors such as metal carbonyls ($Co_2(CO)_8$, $Fe(CO)_5$, $Ni(CO)_4$) and metal oleates, should be carried out in the presence of polymers. For dispersion in PP matrix, a specific amount of metal precursor such as the liquid of $Fe(CO)_5$ is injected into the dissolved PP solution, turning the transparent mixture into yellow, before gradually changing to black to indicate the NPs formation. Upon heating, $Fe(CO)_5$ is decomposed into $Fe_2(CO)_9$ and $Fe_3(CO)_{12}$ and the $Fe_3(CO)_{12}$ is then decomposed to finally form the metallic NPs [102, 103]. For raw OPEFB, cellulose and kapok fibers as base materials in the synthesis of magnetic biosorbents (MBSs), the fibers are first homogenized to the fine sizes of 0.005–0.02 mm, before mixing with the Fe_2O_3 NPs in deionized water, sonicated and shaken on an orbital shaker for the cycle of 3 times and the MBSs recovered using a permanent magnet (4000 G) [104].

4 Applications

4.1 Nano-drug Delivery

Nanomedicines confer several advantages in precision and targeted medicine such as concentrating the quantity and enhancing the absorption of chemotherapeutic agents at the specific site and tissue, increasing the retention time whilst reducing premature degradation for intracellular penetration, and improving the interaction with the biological environment to minimize their systemic distribution and reduce the adverse side effects [105–107]. Nano-vehicles/carriers including the nano-spheres/nanoshells, polymeric micelles, liposomes, polymer conjugates, dendrimers, and nucleic acid-based NPs have been the focus in the drug delivery system (DDS). The challenge is in ensuring that the encapsulating agent should not be immunogenic or at least not show any adverse immune response or react towards circulating proteins and molecules in the body. The DDS should exhibit solubility, stability in vivo, and bio-distribution, long circulation in the blood, passive or active targeting to the pathological or diseased sites, and responsive to the changes in the environmental conditions [105, 108]. Polymeric nano-DDS have shown potential as therapeutic carriers with remarkable results shown in preclinical studies especially for targeted delivery to tumour cells [109, 110]. Ligation of the anticancer drug to a biocompatible polymer such as hyaluronic acid (HA) could enhance drug target and controlled release [111–114]. HA is the main component of the extracellular matrix and ligand for CD44 and Receptor for Hyaluronan-Mediated Motility (RHAMM) which are over-expressed in a variety of tumour cell surfaces. HA-Cisplatin (Pt) NPs, formed through anionic polymer–metal complexation between Pt and HA, have shown no significant difference between Pt and HA-Pt cytotoxicity in vitro [108, 109]. The challenge, however, is that the hydrophobic drugs may be released too slowly from the DDS. To increase efficiency and efficacy, the water uptake into the NPs must be enhanced to speed up the release whilst at the same time sustain the release profile [115].

Nanofibers possess high surface to volume ratio and porosity to accommodate drug loading capacity, mass transfer and attachment to targeted cells [116]. Surface functionalization further may provide an avenue for the design of more precise drug and gene delivery [117]. However, to achieve this, the processing variables during fabrication of the nanofibers must be controlled precisely [118, 119]. Due to the unique size-dependent optical, electronic, magnetic and structural properties, NPs formulation with the therapeutic agents are being considered for target specific diagnostic applications based on imaging, diagnostics and targeted therapy to improve the anti-cancer therapeutics [120, 121]. The cellulose-based materials for example not only are biodegradable, and biocompatible, with no or low toxicity [122] but also can self-assemble or be designed into organized structures with specific properties and functions. Stimuli-responsive self-assembled cellulose materials based on temperature-, pH-, light- and redox-trigger have been developed [123]. A thermo-responsive hydroxylpropyl cellulose DDS has been found to be able to self-assemble into nano-particulate systems for curcumin encapsulation with satisfactory loading. The discharge of curcumin is totally dependant on the

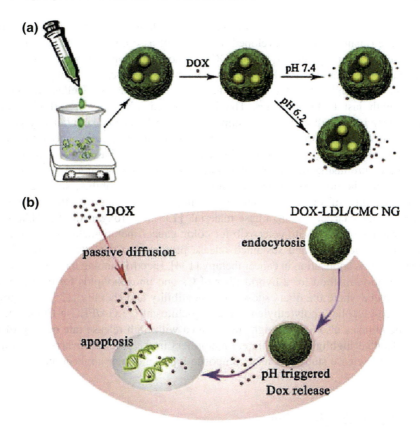

Fig. 4 a Illustration of the synthesis and structures of LDL/CMC nanogels, DOX loading, and pH-dependent drug release; **b** schematic diagram showing the proposed model for intracellular delivery processing of DOX-loaded LDL/CMC nanogels in tumor cells [125]

temperature in the physiologically relevant range [124]. A nano-gel obtained by the self-assembly of low-density lipoproteins (LDL) and sodium carboxymethyl cellulose (CMC) has demonstrated that the doxorubicin (DOX) release is pH-dependent, and the DOX-loaded LDL/CMC nanogels reduce endocytosis in HepG2 cells (Fig. 4) [125] has also drawn much attention due to the unique structure which leads to its application in anti-cancer drug delivery. Biodegradable CMC/Graphene oxide (GO) nanocomposite hydrogel beads, prepared via physical crosslinking with $FeCl_3·6H_2O$ for controlled release of DOX, also exhibit faster release rate in an acidic medium which is attributable to the instability of hydrogen bonding between GO and DOX [126]. The graft copolymer hydroxypropyl methylcellulose grafted with aminated-glycidyl methacrylate-grafted cellulose-grafted polymethacrylic acid-succinyl cyclodextrin (Cell-g-(GMA/en)-PMA-SeCD) has not shown any cytotoxicity but the anticancer drug 5-fluorouracil (5-FU)-DDS exhibits significant cytotoxicity, only to a lesser extent than the naked 5-FU. The study suggests the potential of Cell-g-(GMA/en)-PMA-SCD for drug loading and sustained release of 5-FU [127].

In an investigation on the potential use for live tumor cell imaging, the MCC-based nano-carriers developed by the self-assembly of MCC-graft-poly (p-dioxanone) copolymers (Cell-g-PPDO) in aqueous solution with fluorescent conjugated polymers (FCPs) loading have exhibited significantly stronger fluorescence in the cytoplasm of the cells treated with FCPs-loaded micelles than those treated with just FCPs in water. The FCPs-loaded micelles also exhibit lower cytotoxicity in the cell culture with improved intracellular uptake efficiency [128]. With applications in various industries such as food, biotechnology, agriculture and environmental protection [129–132], the nano-biocomposite based on Cs nanofillers are another promising materials for biomedical and pharmaceutical applications. Furthermore, the properties such as non-toxicity, biocompatibility, specificity and biodegradability of the Cs nanofillers are important to improve the biological properties of the target system or base-materials [133]. Hydroxyapatite-Cs (HAp-Cs) nanocomposite loaded with celecoxib for colon cancer therapy achieves high drug association efficiencies and sustained release profiles, but the serious side effects may hamper its application in cancer therapy [134]. Electrospinning has been used as the fabrication method for a hybrid fiber of Cs and phospholipids for transdermal drug delivery where the fibers show biocompatibility and stability [135]. The poly (lactide-co-glycolide)/poly(ethylene glycol)-g-chitosan (PLGA/PEG-g-Cs) electrospun membranes also exhibit high surface area with high release rate of ibuprofen [136]. Table 3 highlights the different nanofillers in the chitosan matrix, the drugs used for the release study, the nanocomposite types and their applications.

4.2 Tissue Engineering

In tissue engineering, the newly developed material provides scaffold for the tissue growth. The scaffold must have significant mechanical properties to bear the stresses and load, along with appropriate pore size, biocompatibility and biodegradability. BCs have high porosity which can be beneficial for medical application such as the mesoporous dressing to keep the wound moist for quick healing. *G. xylinum* secretes abundant 3D network of biocompatible fibrils which can be used to develop a scaffold for cells fabrication for artificial skin, artificial blood vessels and micro-vessels or wound dressing, or as template for cartilage regeneration. BC scaffold provides attachment as collagen type II substrate, and mechanical support for multiplication and differentiation of chondrocytes for the normal cartilage growth [56]. The properties and functionalities can be improved with the addition of NPs, nanotubes or nanofillers such as the nanocomposites BC/collagen, BC/gelatin, BC/Fibroin, BC/Chitosan [137]. Important factors to be optimized for reinforced bone scaffolds and composite dressings for dermal wounds are the surface area, porosity, density, rigidity and the morphology as thin mats or non-wovens, Pertinent among these are the safe and eco-friendly disassembly of base-materials as alternatives to acid hydrolysis, and to give higher yields and scaling-up possibilities [138]. In the electrospinning fabrication of Cs-based

Table 3 Different nanofillers in chitosan matrix, drug used for the release study, the nanocomposite type and their applications

Nanofiller	Drug used	Nanocomposite type	Applications	References
MCM-41 or MCM-41-APS	Metformin (MET) (an oral diabetes medicine)	Thin films, (Cs, polyethylene glycol)-block-poly(propylene glycol)-block-(polyethylene glycol)	Controlled DDS of MET	[166]
Fe_3O_4 NPs, CdTe@ZnS QDs	Doxorubicin	Surface modified magnetic NP (MNPs) with carboxymethyl-Cs	Controlled DDS and cellular imaging	[167]
Fe and Mn magnetic nanocrystals (MNCs)	Doxorubicin	N-naphthyl-Odimethylmaleoyl-drug loaded magnetic NPs (NCs-DMNPs)	MR-guided imaging for cancer therapy and pH-sensitive drug release	[168]
Fe_3O_4	Curcumin	Cs, PVP, PEG nanocomposite coated with Fe_3O_4 nanoparticles	pH sensitive drug delivery	[169]
Fe_3O_4	Gemcitabine	Magnetite/Cs (core/shell) nanocomposite	Cancer therapy via stimuli-sensitive nanomedicine	[170]
Fe_3O_4	2,7,12,18-tetramethyl-3,8-di (1-propoxyethyl)-13,17-bis-(3hydroxypropyl) porphyrin	Magnetic nanocomposite	DDS for targeting photodynamic therapy	[171]
Fe_3O_4	Bacillus Calmette-Guerin	Magnetic hydrogel	Intravesical drug delivery	[172]
Iron oxide	Methotrexate, mitomycin C	Iron oxide NPs decorated with Cs and functionalized with polyethylene glycolated methotrexate and cyanine dye	Fluorescence and MRI as self-targeted therapeutic delivery device for cancer treatment	[173, 174]
Zinc oxide (ZnO)	Ibuprofen	Hydrogel beads	Controlled DDS	[175]
ZnO	Gentamicin	ZnO-Cs gel	Wound care application	[176]
Gold	5-fluorouracil	Nanocomposite	Site specific and controlled DDS	[177]
Silver	Ibuprofen	Hydrogel beads	Antibacterial DDS	[178]
Graphene oxide (GO)	Camptothecin (CPT)	Covalent functionalized GO with Cs	Delivery of water insoluble CPT and genes	[179]

(continued)

Table 3 (continued)

Nanofiller	Drug used	Nanocomposite type	Applications	References
GO	Fluorescein sodium	Cs-GO nanocomposite	Transdermal drug delivery	[180]
Montmorillonite (MMT), silicate	Ofloxacin, 5-aminosalicylic acid, Ibuprofen, vitamin B_{12}	Hydrogel, solid-liquid interaction, nanohybrid films, nano hydrogel	Drug carrier for sustained release, electrically-induced DDS	[181–184]
MMT	5-fluorouracil	Cs coated alginate-MMT nanocomposite	Controlled release of drug to small intestine	[185]
MMT	Oxytetracycline	Nanocomposite	Enhanced drug permeation	[186]
Hydroxyapatite	Doxorubicin	Hydrogel	Drug delivery	[187]
Rectorite	Bovine serum albumin	Nanocomposite films	Controlled drug release carrier	[188]
Nanosilica	Carvedilol	Spherical nanosilica matrix	Sustained release of poor water soluble drug	[189]
Aluminosilicate	Doxorubicin	Cs-clay nanocomposite	Controlled and extended drug release	[190]

nanofibers, the spinnability and morphology are strongly dependant on the solution viscosity (in this case Cs, PEO and Triton-X) and the Cs-to-PEO ratio. The nanofibers can be designed into structures that promote the attachment of human osteoblasts and chondrocytes whilst maintaining the characteristic cell morphology and viability in tissue engineering and remodelling [94]. The PEO yields in the blend influence the degree of swelling and hydrophilicity of the films and the nanofibers. Better dispersion of the Cs in the presence of PEO will improve the mechanical properties of the composite which is when the electrospinning is performed at the optimal Cs/PEO weight ratio [139]. The composites of Cs, gelatin and polyurethane have been proposed for cardiac valves and for nerve conduits based on the fibers manufactured from the electrospun self-assembled particles. Gel drying with supercritical CO_2 leads to the structures similar to the extracellular matrix, which are of interest in orthopaedics [138].

4.3 Biosensor, Electrical Conductive Polymer and Insulator

Cellulose/HAp composites are effective for the removal of heavy metal ions from aqueous solution, taking advantage of the ion-exchange and sorption properties of the HAp with the highly accessible hydrophilic groups of the cellulose [140]. Chemically-modified carbon electrode (Cellulose-HAp-CME) composite has been successfully developed for trace Pb(II) ions detection in the physiologically relevant range of 10–60 ppb in digested and undigested blood serum (Fig. 5). This may be of great benefits for rapid environmental and clinical analyses of heavy metal ions [141]. A novel fluorescent amphiphilic cellulose nano-aggregates sensing system designed to detect the explosives in aqueous solution has shown the sensitivity significantly enhanced by up to 50-times, suggesting that the cellulose-based nano-aggregates maximize the interaction between the sensing material and the analyte [142]. A material based on cellulose acetate (CA) modified with the room temperature ionic liquid 1-butyl-3-methylimidazolium basified (trifluoro-methylsulfonyl) $(BMI·N(Tf)_2)$ biosensor shows a wide linear range of 34.8–370.3 µM and the detection limit of 5.5 µM for the methyldopa determination in pharmaceutical samples under optimized conditions [143].

The use of CNC fillers in the bulk cellulose improves the dielectric constant which is useful in energy storage applications [144]. Figure 6 illustrates the interaction between the CNC and the cellulose matrix and the dielectric property of cellulose and the cellulose/CNC hydrogels. The enhanced dielectric constant is attributed to the interfacial polarization effects occurring in the vicinity of the hydroxyl groups present on the CNC surface [144]. To observe a change in the dielectric properties, a polymer-filler interfacial interaction is necessary, according to the Maxwell-Wagner-Sillars (MWS) process. The large surface area of the CNCs in the cellulose reinforced matrix provides for strong interactions with the hydroxyl and carboxylic acid groups on the cellulose [145–147]. Cellulosic nanomaterials have been explored for the generation of electrically conductive materials and flexible

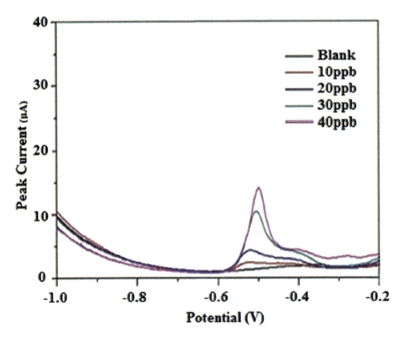

Fig. 5 Stripping voltammograms of Cellulose-HAp-CME for Pb(II) detection in undigested blood serum at pH 2, deposition potential: −1.2 V, deposition time: 240 s, frequency: 25 Hz, amplitude: 0.020 V, step potential: 0.005 V and rotation speed: 700 rpm [141]

films using polyaniline and cellulose nanofibers [148], polypyrrole and bacterial cellulose [149], single-walled CNTs with 2,2,6,6-tetramethylpiperidine-1-oxyl (TEMPO)-oxidized cellulose nanofibers [150], graphene and cellulose nanofibers [151, 152], graphene and CNCs [153], and CNCs modified with terpyridine moieties and crosslinked with Fe(II) [154]. Wood nanofibrillated cellulose (NFC) and algal *Cladophora* cellulose have potentials for electrical insulator applications [155]. The tand plots at different relative humidities (RHs) (Fig. 7) suggest that both samples exhibit a loss factor peak, where the peak is shifted to higher frequencies with increasing relative humidity (RH). These CNCs with sufficient loss factor values may be applicable in the formation of percolative composites with high dielectric constant and high dielectric loss factor for microwave receptor applications [155].

4.4 Catalysis and Environmental Remediation

There has been an increasing use of engineered magnetic nanoparticles as green catalysts and for remediation and water treatments. The cellulose-based core/shell composite with the core as catalyst solid particles have been applied for the synthesis

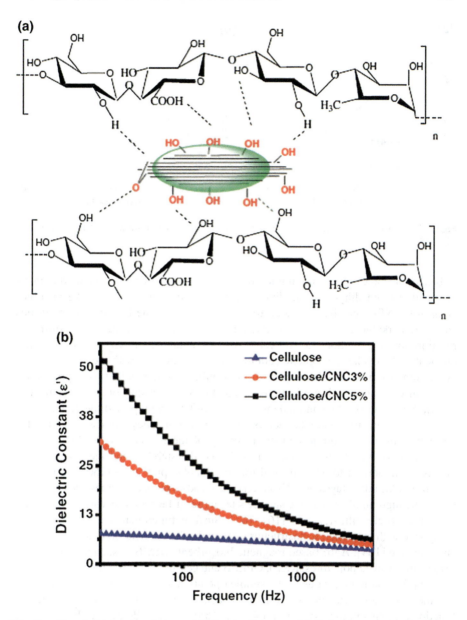

Fig. 6 a Chemical interaction between CNC and cellulose, and **b** dielectric property of cellulose and cellulose/CNC hydrogels [144]

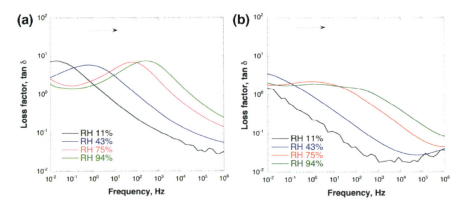

Fig. 7 Loss factor (tanδ) versus frequency for a *Cladophora* cellulose and b NFC, at different RHs [155]

of benzodiazepines [156] and multicomponent syntheses of polysubstituted tetrahydropyridines and dihydropyrimidinones [157]. A magnetic cellulose/Ag nanobiocomposite (NBC) catalyst prepared to synthesize chromene-linked nicotinonitriles shows remarkable magnetic property, and can easily be separated out from the reaction mixture without considerable loss of catalytic activity [158]. Another efficient and recyclable magnetic cellulose/Ag NBC catalyst with the sizes of about 20–25 nm size has been developed for the synthesis of tetrazolo[1,5-α]pyrimidines [159]. In environmental applications, magnetic NPs focusing on zero-valent iron (ZVI), magnetite (Fe_3O_4) and maghemite (γ-Fe_2O_3) have reported the contaminant removal mechanisms where the factors affecting the efficiency include the ability for the contaminant desorption, and the recovery of the magnetic NPs (MNPs). The aggregation of MNPs, the methods to enhance the stability and the toxicological effects are important to be addressed for sustainable application [160]. Apart from reusability, the advantages of MNPs include high adsorption capacities and efficiency due to the high number of active surface sites large surface area-to-volume ratio, and low intraparticle diffusion rate [161–163]. Surface functionalized MNPs and the composite with core-shell nanostructures can make the application more economical and effective [164]. Agro-based magnetic biosorbents (AMBs), such as from *Ceiba pentandra* (Raw kapok fibres, RKF), OPEFB and celluloses (CEL), are applicable for field application not only due to the simple operation and reusability for wastewater treatment, but also because of the high capacity and selectivity from the different functional groups in the structure [104, 164, 165]. The AMBs exhibit dispersion of magnetic NPs (Fig. 8) where the regeneration is successfully performed for 5 adsorption/desorption cycles. The highest Pb(II) removal efficiency of 99.4% and 49 mg/g adsorption capacity is achieved with the kapok AMBs and the AMBs overall exhibit 10.3% higher efficiency than the raw sorbents [104].

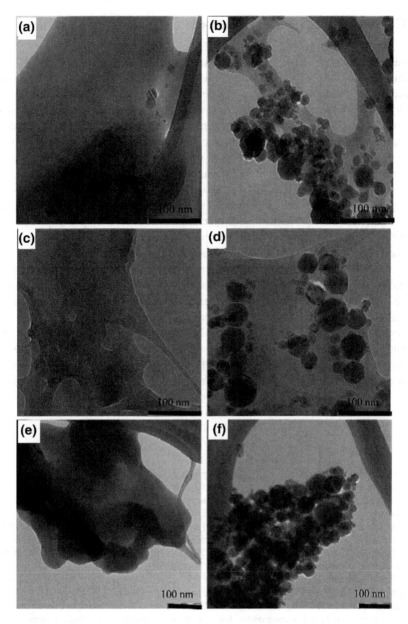

Fig. 8 TEM images of **a** raw OPEFB; **b** Fe_2O_3@EFB; **c** cellulose; **d** Fe_2O_3@ CEL; **e** RKF; and **f** Fe_2O_3@RKF nanocomposites [104]

5 Conclusion and Future Outlook

Global sustainable development necessitates relooking at the whole range of input and output processes in the product manufacturing to ensure effective utilization of resources, use of eco-friendly and cleaner processes, with optimal reaction efficiency, minimal toxic reagents, or zero-wastes. Product development based on composite materials has the advantages in the integration of different component materials and their properties in a single material. The combined properties of the matrix material and the filler will result in novel functional materials, which can be tailor-made for specific applications. Composite materials based on cellulose, chitosan and magnetic nanofillers have far-reaching impacts in meeting the sustainable development goals. The unique properties of cellulose and chitosan nanofillers such as non-toxicity, biodegradability, and biocompatibility have a special place in the area of biomedical, pharmaceutical and environmental applications. The capacity for broad chemical modification and the formation of versatile semicrystalline fiber morphologies confer improved and enhanced optical, mechanical, electrical and thermal properties to the base material matrix. The magnetic nanocomposite offers unique magnetic properties for fundamental understanding and applications in medical therapy and diagnosis, separations, actuation, or catalysis, with a high number of active surface sites, and reusability for environmental remediation. The fabrication aspect should address the issue of maximum possible accuracy and precision of the material, machine and the mould that must be streamlined for miniaturisation of the end-products, with higher yields and scaling-up possibilities.

References

1. Liu Q, Paavola J (2016) Lightweight design of composite laminated structures with frequency constraint. Compos Struct 156:356–360
2. Azwa ZN, Yousif BF, Manalo AC, Karunasena W (2013) A review on the degradability of polymeric composites based on natural fibres. Mater Des 47:424–442
3. Al-Oqla FM, Sapuan SM, Anwer T, Jawaid M, Hoque ME (2015) Natural fiber reinforced conductive polymer composites as functional materials: a review. Synth Met 206:42–54
4. Dhand V, Mittal G, Rhee KY, Park SJ, Hui D (2015) A short review on basalt fiber reinforced polymer composites. Compos B Eng 73:166–180
5. Mittal G, Dhand V, Rhee KY, Park SJ, Lee WR (2015) A review on carbon nanotubes and graphene as fillers in reinforced polymer nanocomposites. J Ind Eng Chem 21:11–25
6. Hung PY, Lau KT, Cheng LK, Leng J, Hui D (2018) Impact response of hybrid carbon/glass fibre reinforced polymer composites designed for engineering applications. Compos B Eng 133:86–90
7. Pillai SK, Ray SS (2011) Epoxy-based carbon nanotubes reinforced composites. In: Advances in nanocomposites-synthesis, characterization and industrial applications. InTech
8. Auvergne R, Caillol S, David G, Boutevin B, Pascault JP (2013) Biobased thermosetting epoxy: present and future. Chem Rev 114(2):1082–1115
9. Maffini MV, Rubin BS, Sonnenschein C, Soto AM (2006) Endocrine disruptors and reproductive health: the case of bisphenol-A. Mol Cell Endocrinol 254:179–186

10. Poletto M, Ornaghi Júnior HL, Visakh PM, Arao Y (2016) Composites and nanocomposites based on renewable and sustainable materials. Int J Polym Sci 2016
11. Böer P, Holliday L, Kang THK (2014) Interaction of environmental factors on fiber-reinforced polymer composites and their inspection and maintenance: a review. Constr Build Mater 50:209–218
12. Abdullah MA, Nazir MS, Ajab H, Daneshfozoun S, Almustapha S (2017) Advances in eco-friendly pre-treatment methods and utilization of agro-based lignocelluloses. Ind Biotechnol: Sustain Prod Biores Utilization 371–419
13. Perlack RD, Wright LL, Turhollow AF, Graham RL, Stokes BJ, Erbach DC (2005) Biomass as feedstock for a bioenergy and bioproducts industry: the technical feasibility of a billion-ton annual supply. Oak Ridge National Lab, Oak Ridge
14. Yuan Z, Cheng S, Leitch M, Xu CC (2010) Hydrolytic degradation of alkaline lignin in hot-compressed water and ethanol. Biores Technol 101(23):9308–9313
15. Jacqueline I, Kroschwitz A (2001) Encyclopedia of chemical technology, vol 20. Wiley-Interscience, New York
16. Abdullah MA, Nazir MS, Raza MR, Wahjoedi BA, Yussof AW (2016) Autoclave and ultra-sonication treatments of oil palm empty fruit bunch fibers for cellulose extraction and its polypropylene composite properties. J Clean Prod 126:686–697
17. ISO/TS 80004-2:(2015) Nanotechnologies—vocabulary—part 2, no. April 2015
18. Lamouroux E, Fort Y (2016) An overview of nanocomposite nanofillers and their functionalization. In: Spectroscopy of polymer nanocomposites, pp 15–64
19. Schrittwieser S, Ludwig F, Dieckhoff J, Tschoepe A, Guenther A, Richter M, Schotter J et al (2014) Direct protein detection in the sample solution by monitoring rotational dynamics of nickel nanorods. Small 10(2):407–411
20. Roeder L, Bender P, Tschöpe A, Birringer R, Schmidt AM (2012) Shear modulus determination in model hydrogels by means of elongated magnetic nanoprobes. J Polym Sci Part B: Polym Phys 50(24):1772–1781
21. Okada A, Usuki A (2006) Twenty years of polymer-clay nanocomposites. Macromol Mater Eng 291(12):1449–1476
22. Verdejo R, Bernal MM, Romasanta LJ, Tapiador FJ, Lopez-Manchado MA (2011) Reactive nanocomposite foams. Cell Polym 30(2):45
23. Novoselov KS, Geim AK, Morozov SV, Jiang D, Zhang Y, Dubonos SV, Firsov AA et al (2004) Electric field effect in atomically thin carbon films. Science 306(5696):666–669
24. Kotov NA (2006) Materials science: carbon sheet solutions. Nature 442(7100):254
25. George TF, Jelski D, Letfullin RL, Zhang GP (2011) Computational studies of new materials II
26. Popov VN (2004) Carbon nanotubes: properties and application. Mater Sci Eng R: Rep 43(3):61–102
27. Siqueira G, Bras J, Dufresne A (2010) Cellulosic bionanocomposites: a review of preparation, properties and applications. Polymers 2(4):728–765
28. García NL, Ribba L, Dufresne A, Aranguren M, Goyanes S (2011) Effect of glycerol on the morphology of nanocomposites made from thermoplastic starch and starch nanocrystals. Carbohyd Polym 84(1):203–210
29. Famá LM, Pettarin V, Goyanes SN, Bernal CR (2011) Starch/multi-walled carbon nanotubes composites with improved mechanical properties. Carbohyd Polym 83(3):1226–1231
30. Famá L, Rojo PG, Bernal C, Goyanes S (2012) Biodegradable starch based nanocomposites with low water vapor permeability and high storage modulus. Carbohyd Polym 87(3):1989–1993
31. García NL, Ribba L, Dufresne A, Aranguren MI, Goyanes S (2009) Physico-mechanical properties of biodegradable starch nanocomposites. Macromol Mater Eng 294(3):169–177
32. Angellier H, Molina-Boisseau S, Dole P, Dufresne A (2006) Thermoplastic starch—waxy maize starch nanocrystals nanocomposites. Biomacromolecules 7(2):531–539
33. Cao X, Chen Y, Chang PR, Huneault MA (2007) Preparation and properties of plasticized starch/multiwalled carbon nanotubes composites. J Appl Polym Sci 106(2):1431–1437

34. Ma X, Chang PR, Yu J, Lu P (2008) Characterizations of glycerol plasticized-starch (GPS)/carbon black (CB) membranes prepared by melt extrusion and microwave radiation. Carbohyd Polym 74(4):895–900
35. Qiao R, Brinson LC (2009) Simulation of interphase percolation and gradients in polymer nanocomposites. Compos Sci Technol 69(3–4):491–499
36. Kilbride BE, Coleman JN, Fraysse J, Fournet P, Cadek M, Drury A, Blau WJ et al (2002) Experimental observation of scaling laws for alternating current and direct current conductivity in polymer-carbon nanotube composite thin films. J Appl Phys 92(7):4024–4030
37. Sandler J, Kirk JE, Kinloch IA, Shaffer MSP, Windle AH (2003) Ultra-low electrical percolation threshold in carbon-nanotube-epoxy composites. Polymer 44(19):5893–5899
38. Biercuk MJ, Llaguno MC, Radosavljevic M, Hyun JK, Johnson AT, Fischer JE (2002) Carbon nanotube composites for thermal management. Appl Phys Lett 80(15):2767–2769
39. Wei C, Srivastava D, Cho K (2002) Thermal expansion and diffusion coefficients of carbon nanotube-polymer composites. Nano Lett 2(6):647–650
40. Yu L (2009) Biodegradable polymer blends and composites from renewable resources. Wiley
41. Malkapuram R, Kumar V, Negi YS (2009) Recent development in natural fiber reinforced polypropylene composites. J Reinf Plast Compos 28(10):1169–1189
42. http://www.efunda.com/processes/plastic_molding/molding_compression.cfm. Accessed 17 Apr 2018
43. Stanimirović Z, Stanimirović I (2012) Ceramic injection molding. In: Some critical issues for injection molding. InTech
44. Rak ZS (1999) New trends in powder injection moulding. Powder Metall Met Ceram 38(3–4):126–132
45. Åkesson D, Skrifvars M, Seppälä J, Turunen M (2011) Thermoset lactic acid-based resin as a matrix for flax fibers. J Appl Polym Sci 119(5):3004–3009
46. Bax B, Müssig J (2008) Impact and tensile properties of PLA/Cordenka and PLA/flax composites. Compos Sci Technol 68(7–8):1601–1607
47. Zhu J, Zhu H, Njuguna J, Abhyankar H (2013) Recent development of flax fibres and their reinforced composites based on different polymeric matrices. Materials 6(11):5171–5198
48. (SHI) Sumitomo (2017) Demag plastics machinery GmbH. https://www.sumitomo-shi-demag.eu/processes/biopolymers.html. Accessed 17 Apr 2018
49. Kalia S, Dufresne A, Cherian BM, Kaith BS, Avérous L, Njuguna J, Nassiopoulos E (2011) Cellulose-based bio-and nanocomposites: a review. Int J Polym Sci, 2011
50. Crawford RL (1982) Lignin biodegradation and transformation. Wiley, p 118
51. Bledzki AK, Reihmane S, Gassan J (1996) Properties and modification methods for vegetable fibers for natural fiber composites. J Appl Polym Sci 59(8):1329–1336
52. Bismarck A, Mishra S, Lampke T (2005) Plant fibers as reinforcement for green composites. Nat Fibers Biopolym Biocompos 10:9780203508206
53. Satyanarayana KG, Sukumaran K, Mukherjee PS, Pavithran C, Pillai SGK (1990) Natural fibre-polymer composites. Cement Concr Compos 12(2):117–136
54. Perasso R, Baroin A, Qu LH, Bachellerie JP, Adoutte A (1989) Origin of the algae. Nature 339(6220):142
55. Wertz JL, Mercier JP, Bédué O (2010) Cellulose science and technology. CRC Press
56. Lejeune A, Deprez T (2010) Cellulose: structure and properties, derivatives and industrial uses. Nova Science Publishers
57. Zhu S, Wu Y, Chen Q, Yu Z, Wang C, Jin S, Wu G et al (2006) Dissolution of cellulose with ionic liquids and its application: a mini-review. Green Chem 8(4):325–327
58. Klemm D, Heublein B, Fink HP, Bohn A (2005) Cellulose: fascinating biopolymer and sustainable raw material. Angew Chem Int Ed 44(22):3358–3393
59. Dalväg H, Klason C, Strömvall HE (1985) The efficiency of cellulosic fillers in common thermoplastics. Part II. Filling with processing aids and coupling agents. Int J Polym Mater 11(1):9–38

60. Maldas D, Kokta BV, Raj RG, Daneault C (1988) Improvement of the mechanical properties of sawdust wood fibre—polystyrene composites by chemical treatment. Polymer 29(7): 1255–1265
61. Zadorecki P, Michell AJ (1989) Future prospects for wood cellulose as reinforcement in organic polymer composites. Polym Compos 10(2):69–77
62. Nazir MS, Wahjoedi BA, Yussof AW, Abdullah MA (2013) Eco-friendly extraction and characterization of cellulose from oil palm empty fruit bunches. BioResources 8(2):2161–2172
63. Haafiz MM, Eichhorn SJ, Hassan A, Jawaid M (2013) Isolation and characterization of microcrystalline cellulose from oil palm biomass residue. Carbohyd Polym 93(2):628–634
64. Visakh PM, Thomas S (2010) Preparation of bionanomaterials and their polymer nanocomposites from waste and biomass. Waste Biomass Valorization 1(1):121–134
65. Bras J, Hassan ML, Bruzesse C, Hassan EA, El-Wakil NA, Dufresne A (2010) Mechanical, barrier, and biodegradability properties of bagasse cellulose whiskers reinforced natural rubber nanocomposites. Ind Crops Prod 32(3):627–633
66. Brito BS, Pereira FV, Putaux JL, Jean B (2012) Preparation, morphology and structure of cellulose nanocrystals from bamboo fibers. Cellulose 19(5):1527–1536
67. Bondeson D, Mathew A, Oksman K (2006) Optimization of the isolation of nanocrystals from microcrystalline cellulose by acid hydrolysis. Cellulose 13(2):171
68. Azizi Samir MAS, Alloin F, Dufresne A (2005) Review of recent research into cellulosic whiskers, their properties and their application in nanocomposite field. Biomacromolecules 6(2):612–626
69. Filson PB, Dawson-Andoh BE (2009) Sono-chemical preparation of cellulose nanocrystals from lignocellulose derived materials. Biores Technol 100(7):2259–2264
70. Petersson L, Kvien I, Oksman K (2007) Structure and thermal properties of poly (lactic acid)/cellulose whiskers nanocomposite materials. Compos Sci Technol 67(11–12):2535–2544
71. Haafiz MM, Hassan A, Zakaria Z, Inuwa IM (2014) Isolation and characterization of cellulose nanowhiskers from oil palm biomass microcrystalline cellulose. Carbohyd Polym 103:119–125
72. Liu D, Zhong T, Chang PR, Li K, Wu Q (2010) Starch composites reinforced by bamboo cellulosic crystals. Biores Technol 101(7):2529–2536
73. Pandey JK, Chu WS, Kim CS, Lee CS, Ahn SH (2009) Bio-nano reinforcement of environmentally degradable polymer matrix by cellulose whiskers from grass. Compos B Eng 40(7):676–680
74. Oksman K, Mathew AP, Bondeson D, Kvien I (2006) Manufacturing process of cellulose whiskers/polylactic acid nanocomposites. Compos Sci Technol 66(15):2776–2784
75. Pereda M, Amica G, Rácz I, Marcovich NE (2011) Structure and properties of nanocomposite films based on sodium caseinate and nanocellulose fibers. J Food Eng 103(1): 76–83
76. Mandal A, Chakrabarty D (2011) Isolation of nanocellulose from waste sugarcane bagasse (SCB) and its characterization. Carbohyd Polym 86(3):1291–1299
77. Valentini L, Bon SB, Cardinali M, Fortunati E, Kenny JM (2014) Cellulose nanocrystals thin films as gate dielectric for flexible organic field-effect transistors. Mater Lett 126:55–58
78. Tang L, Huang B, Lu Q, Wang S, Ou W, Lin W, Chen X (2013) Ultrasonication-assisted manufacture of cellulose nanocrystals esterified with acetic acid. Biores Technol 127:100–105
79. Wang N, Ding E, Cheng R (2007) Thermal degradation behaviors of spherical cellulose nanocrystals with sulfate groups. Polymer 48(12):3486–3493
80. Liu H, Liu D, Yao F, Wu Q (2010) Fabrication and properties of transparent polymethylmethacrylate/cellulose nanocrystals composites. Biores Technol 101(14):5685–5692
81. Yano H, Sugiyama J, Nakagaito AN, Nogi M, Matsuura T, Hikita M, Handa K (2005) Optically transparent composites reinforced with networks of bacterial nanofibers. Adv Mater 17(2):153–155
82. Ljungberg N, Cavaillé JY, Heux L (2006) Nanocomposites of isotactic polypropylene reinforced with rod-like cellulose whiskers. Polymer 47(18):6285–6292

83. Bornet A, Teissedre PL (2008) Chitosan, chitin-glucan and chitin effects on minerals (iron, lead, cadmium) and organic (ochratoxin A) contaminants in wines. Eur Food Res Technol 226(4):681–689
84. Ali A, Ahmed S (2017) A review on chitosan and its nanocomposites in drug delivery. Int J Biol Macromol
85. Rinaudo M (2006) Chitin and chitosan: properties and applications. Prog Polym Sci 31(7): 603–632
86. Gardner KH, Blackwell J (1975) Refinement of the structure of β-chitin. Biopolymers 14(8): 1581–1595
87. Minke RAM, Blackwell J (1978) The structure of α-chitin. J Mol Biol 120(2):167–181
88. Saito Y, Okano T, Chanzy H, Sugiyama J (1995) Structural study of α chitin from the grasping spines of the arrow worm (*Sagitta* spp.). J Struct Biol 114(3):218–228
89. Chanzy H (1998) Chitin crystals. Jacques Andre, Lyon
90. Chrétiennot-Dinet MJ, Giraud-Guille MM, Vaulot D, Putaux JL, Saito Y, Chanzy H (1997) The chitinous nature of filaments ejected by Phaeocystis (Prymnesiophyceae). J Phycol 33(4):666–672
91. Blackwell J (1969) Structure of β-chitin or parallel chain systems of poly-β-(1→4)-*N*-acetyl-D-glucosamine. Biopolymers 7(3):281–298
92. Gaill F, Persson J, Sugiyama J, Vuong R, Chanzy H (1992) The chitin system in the tubes of deep sea hydrothermal vent worms. J Struct Biol 109(2):116–128
93. Fernandes SC, Freire CS, Silvestre AJ, Neto CP, Gandini A, Berglund LA, Salmén L (2010) Transparent chitosan films reinforced with a high content of nanofibrillated cellulose. Carbohyd Polym 81(2):394–401
94. Bhattarai N, Edmondson D, Veiseh O, Matsen FA, Zhang M (2005) Electrospun chitosan-based nanofibers and their cellular compatibility. Biomaterials 26(31):6176–6184
95. Geng X, Kwon OH, Jang J (2005) Electrospinning of chitosan dissolved in concentrated acetic acid solution. Biomaterials 26(27):5427–5432
96. Mengistu Lemma S, Bossard F, Rinaudo M (2016) Preparation of pure and stable chitosan nanofibers by electrospinning in the presence of poly (ethylene oxide). Int J Mol Sci 17(11): 1790
97. Ojha SS, Stevens DR, Hoffman TJ, Stano K, Klossner R, Scott MC, Gorga RE et al (2008) Fabrication and characterization of electrospun chitosan nanofibers formed via templating with polyethylene oxide. Biomacromolecules 9(9):2523–2529
98. Homayoni H, Ravandi SAH, Valizadeh M (2009) Electrospinning of chitosan nanofibers: processing optimization. Carbohyd Polym 77(3):656–661
99. Pillai SK, Ray SS (2012) Chitosan-based nanocomposites. Nat Polym 2:33–68
100. Mai YW, Yu ZZ (2006) Polymer nanocomposites. Woodhead publishing
101. Behrens S, Appel I (2016) Magnetic nanocomposites. Curr Opin Biotechnol 39:89–96
102. Zhu J, Wei S, Li Y, Sun L, Haldolaarachchige N, Young DP, Guo Z et al (2011) Surfactant-free synthesized magnetic polypropylene nanocomposites: rheological, electrical, magnetic, and thermal properties. Macromolecules 44(11):4382–4391
103. Smith TW, Wychick D (1980) Colloidal iron dispersions prepared via the polymer-catalyzed decomposition of iron pentacarbonyl. J Phys Chem 84(12):1621–1629
104. Daneshfozoun S, Abdullah MA, Abdullah B (2017) Preparation and characterization of magnetic biosorbent based on oil palm empty fruit bunch fibers, cellulose and *Ceiba pentandra* for heavy metal ions removal. Ind Crops Prod 105:93–103
105. Gul-e-Saba, Abdullah MA (2015) Polymeric nanoparticle mediated targeted drug delivery to cancer cells. In: Biotechnology and Bioinformatics, pp 1–34
106. Farokhzad OC, Langer R (2009) Impact of nanotechnology on drug delivery. ACS Nano 3(1):16–20
107. Alexis F, Pridgen E, Molnar LK, Farokhzad OC (2008) Factors affecting the clearance and biodistribution of polymeric nanoparticles. Mol Pharm 5(4):505–515

108. Abdullah MA, Gul-e-Saba AA (2014) Cytotoxic effects of drug-loaded hyaluronan-glutaraldehyde cross-linked nanoparticles and the release kinetics modeling. J Adv Chem Eng 1(104):2
109. Van Vlerken LE, Amiji MM (2006) Multi-functional polymeric nanoparticles for tumour-targeted drug delivery. Expert Opin Drug Deliv 3(2):205–216
110. Parveen S, Sahoo SK (2006) Nanomedicine. Clin Pharmacokinet 45(10):965–988
111. Gul-e-Saba, Abdah A, Abdullah MA (2010) Hyaluronan-mediated CD44 receptor cancer cells progression and the application of controlled drug delivery system. Int J Curr Chem 1(4):245–265
112. Jin YJ, Ubonvan T, Kim DD (2010) Hyaluronic acid in drug delivery systems. J Pharm Invest 40(spc):33–43
113. Jaracz S, Chen J, Kuznetsova LV, Ojima I (2005) Recent advances in tumor-targeting anticancer drug conjugates. Bioorg Med Chem 13(17):5043–5054
114. Akima K, Ito H, Iwata Y, Matsuo K, Watari N, Yanagi M, Tatekawa I et al (1996) Evaluation of antitumor activities of hyaluronate binding antitumor drugs: synthesis, characterization and antitumor activity. J Drug Target 4(1):1–8
115. Marinich JA, Ferrero C, Jiménez-Castellanos MR (2012) Graft copolymers of ethyl methacrylate on waxy maize starch derivatives as novel excipients for matrix tablets: drug release and fronts movement kinetics. Eur J Pharm Biopharm 80(3):674–681
116. Hu X, Liu S, Zhou G, Huang Y, Xie Z, Jing X (2014) Electrospinning of polymeric nanofibers for drug delivery applications. J Controlled Release 185:12–21
117. Yoo HS, Kim TG, Park TG (2009) Surface-functionalized electrospun nanofibers for tissue engineering and drug delivery. Adv Drug Deliv Rev 61(12):1033–1042
118. Pillay V, Dott C, Choonara YE, Tyagi C, Tomar L, Kumar P, Ndesendo VM et al (2013) A review of the effect of processing variables on the fabrication of electrospun nanofibers for drug delivery applications. J Nanomater 2013
119. Sridhar R, Lakshminarayanan R, Madhaiyan K, Barathi VA, Lim KHC, Ramakrishna S (2015) Electrosprayed nanoparticles and electrospun nanofibers based on natural materials: applications in tissue regeneration, drug delivery and pharmaceuticals. Chem Soc Rev 44(3):790–814
120. Lim EK, Chung BH (2016) Preparation of pyrenyl-based multifunctional nanocomposites for biomedical applications. Nat Protoc 11(2):236
121. Siddiqui IA, Adhami VM, Christopher J (2012) Impact of nanotechnology in cancer: emphasis on nanochemoprevention. Int J Nanomed 7:591
122. Lin N, Dufresne A (2014) Nanocellulose in biomedicine: current status and future prospect. Eur Polymer J 59:302–325
123. Yang J, Li J (2017) Self-assembled cellulose materials for biomedicine: a review. Carbohydr Polym
124. Bielska D, Karewicz A, Kamiński K, Kiełkowicz I, Lachowicz T, Szczubiałka K, Nowakowska M (2013) Self-organized thermo-responsive hydroxypropyl cellulose nanoparticles for curcumin delivery. Eur Polymer J 49(9):2485–2494
125. He L, Liang H, Lin L, Shah BR, Li Y, Chen Y, Li B (2015) Green-step assembly of low density lipoprotein/sodium carboxymethyl cellulose nanogels for facile loading and pH-dependent release of doxorubicin. Colloids Surf B 126:288–296
126. Rasoulzadeh M, Namazi H (2017) Carboxymethyl cellulose/graphene oxide bio-nanocomposite hydrogel beads as anticancer drug carrier agent. Carbohyd Polym 168:320–326
127. Anirudhan TS, Nima J, Divya PL (2015) Synthesis, characterization and in vitro cytotoxicity analysis of a novel cellulose based drug carrier for the controlled delivery of 5-fluorouracil, an anticancer drug. Appl Surf Sci 355:64–73
128. Guo Y, Zhang J, Wang L, Ge W, Chen M, Wang X, Sun R (2015) Preparation of fluorescent core/shell nanoparticles from amphiphilic cellulose-based copolymers for tumor cell imaging. J Controlled Release: Official J Controlled Release Soc 213:e132

129. Coradin T, Allouche J, Boissière M, Livage J (2006) Sol-gel biopolymer/silica nanocomposites in biotechnology. Curr Nanosci 2(3):219–230
130. De Azeredo HM (2009) Nanocomposites for food packaging applications. Food Res Int 42(9):1240–1253
131. Liu X, Hu Q, Fang Z, Zhang X, Zhang B (2008) Magnetic chitosan nanocomposites: a useful recyclable tool for heavy metal ion removal. Langmuir 25(1):3–8
132. Rhim JW, Park HM, Ha CS (2013) Bio-nanocomposites for food packaging applications. Prog Polym Sci 38(10–11):1629–1652
133. Cheung RCF, Ng TB, Wong JH, Chan WY (2015) Chitosan: an update on potential biomedical and pharmaceutical applications. Marine drugs 13(8):5156–5186
134. Venkatesan P, Puvvada N, Dash R, Kumar BP, Sarkar D, Azab B, Mandal M et al (2011) The potential of celecoxib-loaded hydroxyapatite-chitosan nanocomposite for the treatment of colon cancer. Biomaterials 32(15):3794–3806
135. Mendes AC, Gorzelanny C, Halter N, Schneider SW, Chronakis IS (2016) Hybrid electrospun chitosan-phospholipids nanofibers for transdermal drug delivery. Int J Pharm 510(1):48–56
136. Jiang H, Fang D, Hsiao B, Chu B, Chen W (2004) Preparation and characterization of ibuprofen-loaded poly (lactide-co-glycolide)/poly (ethylene glycol)-g-chitosan electrospun membranes. J Biomater Sci Polym Ed 15(3):279–296
137. de Oliveira Barud HG, da Silva RR, da Silva Barud H, Tercjak A, Gutierrez J, Lustri WR, Ribeiro SJ et al (2016) A multipurpose natural and renewable polymer in medical applications: bacterial cellulose. Carbohyd Polym 153:406–420
138. Muzzarelli RA (2011) Biomedical exploitation of chitin and chitosan via mechano-chemical disassembly, electrospinning, dissolution in imidazolium ionic liquids, and supercritical drying. Marine Drugs 9(9):1510–1533
139. Garcia CEG, Martínez FAS, Bossard F, Rinaudo M (2018) Biomaterials based on electrospun chitosan. Relation between processing conditions and mechanical properties. Polymers 10(3):257
140. Choi S, Jeong Y (2008) The removal of heavy metals in aqueous solution by hydroxyapatite/cellulose composite. Fibers Polym 9(3):267–270
141. Ajab H, Dennis JO, Abdullah MA (2018) Synthesis and characterization of cellulose and hydroxyapatite-carbon electrode composite for trace plumbum ions detection and its validation in blood serum. Int J Biol Macromol 113:376–385
142. Wang X, Guo Y, Li D, Chen H, Sun RC (2012) Fluorescent amphiphilic cellulose nanoaggregates for sensing trace explosives in aqueous solution. Chem Commun 48(45):5569–5571
143. Moccelini SK, Franzoi AC, Vieira IC, Dupont J, Scheeren CW (2011) A novel support for laccase immobilization: cellulose acetate modified with ionic liquid and application in biosensor for methyldopa detection. Biosens Bioelectron 26(8):3549–3554
144. Gao X, Sadasivuni KK, Kim HC, Min SK, Kim J (2015) Designing pH-responsive and dielectric hydrogels from cellulose nanocrystals. J Chem Sci 127(6):1119–1125
145. Sadasivuni KK, Ponnamma D, Thomas S, Grohens Y (2014) Evolution from graphite to graphene elastomer composites. Prog Polym Sci 39(4):749–780
146. Yuan JK, Yao SH, Dang ZM, Sylvestre A, Genestoux M, Bai1 J (2011) Giant dielectric permittivity nanocomposites: realizing true potential of pristine carbon nanotubes in polyvinylidene fluoride matrix through an enhanced interfacial interaction. J Phys Chem C 115(13):5515–5521
147. Ponnamma D, Sadasivuni KK, Grohens Y, Guo Q, Thomas S (2014) Carbon nanotube based elastomer composites—an approach towards multifunctional materials. J Mater Chem C 2(40):8446–8485
148. Mattoso LHC, Medeiros ES, Baker DA, Avloni J, Wood DF, Orts WJ (2009) Electrically conductive nanocomposites made from cellulose nanofibrils and polyaniline. J Nanosci Nanotechnol 9(5):2917–2922

149. Xu J, Zhu L, Bai Z, Liang G, Liu L, Fang D, Xu W (2013) Conductive polypyrrole–bacterial cellulose nanocomposite membranes as flexible supercapacitor electrode. Org Electron 14(12):3331–3338
150. Koga H, Saito T, Kitaoka T, Nogi M, Suganuma K, Isogai A (2013) Transparent, conductive, and printable composites consisting of TEMPO-oxidized nanocellulose and carbon nanotube. Biomacromolecules 14(4):1160–1165
151. Patel MU, Luong ND, Seppälä J, Tchernychova E, Dominko R (2014) Low surface area graphene/cellulose composite as a host matrix for lithium sulphur batteries. J Power Sources 254:55–61
152. Yan C, Wang J, Kang W, Cui M, Wang X, Foo CY, Lee PS (2014) Highly stretchable piezoresistive graphene–nanocellulose nanopaper for strain sensors. Adv Mater 26(13): 2022–2027
153. Valentini L, Bon SB, Fortunati E, Kenny JM (2014) Preparation of transparent and conductive cellulose nanocrystals/graphene nanoplatelets films. J Mater Sci 49(3):1009–1013
154. Namazi H, Baghershiroudi M, Kabiri R (2017) Preparation of electrically conductive biocompatible nanocomposites of natural polymer nanocrystals with polyaniline via in situ chemical oxidative polymerization. Polym Compos 38(S1)
155. Le Bras D, Strømme M, Mihranyan A (2015) Characterization of dielectric properties of nanocellulose from wood and algae for electrical insulator applications. J Phys Chem B 119(18):5911–5917
156. Maleki A, Kamalzare M (2014) Fe_3O_4@ cellulose composite nanocatalyst: preparation, characterization and application in the synthesis of benzodiazepines. Catal Commun 53:67–71
157. Maleki A, Jafari AA, Yousefi S (2017) $MgFe_2O_4$/cellulose/SO_3H nanocomposite: a new biopolymer-based nanocatalyst for one-pot multicomponent syntheses of polysubstituted tetrahydropyridines and dihydropyrimidinones. J Iran Chem Soc 14(8):1801–1813
158. Maleki A, Movahed H, Ravaghi P (2017) Magnetic cellulose/Ag as a novel eco-friendly nanobiocomposite to catalyze synthesis of chromene-linked nicotinonitriles. Carbohyd Polym 156:259–267
159. Maleki A, Ravaghi P, Aghaei M, Movahed H (2017) A novel magnetically recyclable silver-loaded cellulose-based bionanocomposite catalyst for green synthesis of tetrazolo [1, 5-a] pyrimidines. Res Chem Intermed 43(10):5485–5494
160. Tang SC, Lo IM (2013) Magnetic nanoparticles: essential factors for sustainable environmental applications. Water Res 47(8):2613–2632
161. Nassar NN (2010) Kinetics, mechanistic, equilibrium, and thermodynamic studies on the adsorption of acid red dye from wastewater by γ-Fe_2O_3 nanoadsorbents. Sep Sci Technol 45(8):1092–1103
162. Tan Y, Chen M, Hao Y (2012) High efficient removal of Pb (II) by amino-functionalized Fe_3O_4 magnetic nano-particles. Chem Eng J 191:104–111
163. Feng Y, Gong JL, Zeng GM, Niu QY, Zhang HY, Niu CG, Yan M et al (2010) Adsorption of Cd (II) and Zn (II) from aqueous solutions using magnetic hydroxyapatite nanoparticles as adsorbents. Chem Eng J 162(2):487–494
164. Gómez-Pastora J, Bringas E, Ortiz I (2014) Recent progress and future challenges on the use of high performance magnetic nano-adsorbents in environmental applications. Chem Eng J 256:187–204
165. Nalbandian MJ, Zhang M, Sanchez J, Choa YH, Nam J, Cwiertny DM, Myung NV (2016) Synthesis and optimization of Fe_2O_3 nanofibers for chromate adsorption from contaminated water sources. Chemosphere 144:975–981
166. Shariatinia Z, Zahraee Z (2017) Controlled release of metformin from chitosan–based nanocomposite films containing mesoporous MCM-41 nanoparticles as novel drug delivery systems. J Colloid Interface Sci 501:60–76
167. Ding Y, Yin H, Shen S, Sun K, Liu F (2017) Chitosan-based magnetic/fluorescent nanocomposites for cell labelling and controlled drug release. New J Chem 41(4):1736–1743

168. Lim EK, Sajomsang W, Choi Y, Jang E, Lee H, Kang B, Huh YM et al (2013) Chitosan-based intelligent theragnosis nanocomposites enable pH-sensitive drug release with MR-guided imaging for cancer therapy. Nanoscale Res Lett 8(1):467
169. Prabha G, Raj V (2016) Preparation and characterization of polymer nanocomposites coated magnetic nanoparticles for drug delivery applications. J Magn Magn Mater 408:26–34
170. Arias JL, Reddy LH, Couvreur P (2012) Fe_3O_4/chitosan nanocomposite for magnetic drug targeting to cancer. J Mater Chem 22(15):7622–7632
171. Sun Y, Chen ZL, Yang XX, Huang P, Zhou XP, Du XX (2009) Magnetic chitosan nanoparticles as a drug delivery system for targeting photodynamic therapy. Nanotechnology 20(13):135102
172. Zhang D, Sun P, Li P, Xue A, Zhang X, Zhang H, Jin X (2013) A magnetic chitosan hydrogel for sustained and prolonged delivery of Bacillus Calmette-Guérin in the treatment of bladder cancer. Biomaterials 34(38):10258–10266
173. Lin J, Li Y, Li Y, Wu H, Yu F, Zhou S, Hou Z et al (2015) Drug/dye-loaded, multifunctional PEG–chitosan–iron oxide nanocomposites for methotraxate synergistically self-targeted cancer therapy and dual model imaging. ACS Appl Mater Interfaces 7(22):11908–11920
174. Jia M, Li Y, Yang X, Huang Y, Wu H, Huang Y, Zhang Q et al (2014) Development of both methotrexate and mitomycin C loaded PEGylated chitosan nanoparticles for targeted drug codelivery and synergistic anticancer effect. ACS Appl Mater Interfaces 6(14):11413–11423
175. Yadollahi M, Farhoudian S, Barkhordari S, Gholamali I, Farhadnejad H, Motasadizadeh H (2016) Facile synthesis of chitosan/ZnO bio-nanocomposite hydrogel beads as drug delivery systems. Int J Biol Macromol 82:273–278
176. Vasile BS, Oprea O, Voicu G, Ficai A, Andronescu E, Teodorescu A, Holban A (2014) Synthesis and characterization of a novel controlled release zinc oxide/gentamicin–chitosan composite with potential applications in wounds care. Int J Pharm 463(2):161–169
177. Chandran PR, Sandhyarani N (2014) An electric field responsive drug delivery system based on chitosan–gold nanocomposites for site specific and controlled delivery of 5-fluorouracil. RSC Adv 4(85):44922–44929
178. Yadollahi M, Farhoudian S, Namazi H (2015) One-pot synthesis of antibacterial chitosan/silver bio-nanocomposite hydrogel beads as drug delivery systems. Int J Biol Macromol 79:37–43
179. Bao H, Pan Y, Ping Y, Sahoo NG, Wu T, Li L, Gan LH et al (2011) Chitosan-functionalized graphene oxide as a nanocarrier for drug and gene delivery. Small 7(11):1569–1578
180. Justin R, Chen B (2014) Characterisation and drug release performance of biodegradable chitosan–graphene oxide nanocomposites. Carbohyd Polym 103:70–80
181. Depan D, Kumar AP, Singh RP (2009) Cell proliferation and controlled drug release studies of nanohybrids based on chitosan-g-lactic acid and montmorillonite. Acta Biomater 5(1):93–100
182. Aguzzi C, Capra P, Bonferoni C, Cerezo P, Salcedo I, Sánchez R, Viseras C et al (2010) Chitosan–silicate biocomposites to be used in modified drug release of 5-aminosalicylic acid (5-ASA). Appl Clay Sci 50(1):106–111
183. Hua S, Yang H, Wang W, Wang A (2010) Controlled release of ofloxacin from chitosan–montmorillonite hydrogel. Appl Clay Sci 50(1):112–117
184. Liu KH, Liu TY, Chen SY, Liu DM (2008) Drug release behavior of chitosan–montmorillonite nanocomposite hydrogels following electrostimulation. Acta Biomater 4(4):1038–1045
185. Azhar FF, Olad A (2014) A study on sustained release formulations for oral delivery of 5-fluorouracil based on alginate–chitosan/montmorillonite nanocomposite systems. Appl Clay Sci 101:288–296
186. Salcedo I, Sandri G, Aguzzi C, Bonferoni C, Cerezo P, Sánchez-Espejo R, Viseras C (2014) Intestinal permeability of oxytetracycline from chitosan-montmorillonite nanocomposites. Colloids Surf B 117:441–448

187. Taleb MFA, Alkahtani A, Mohamed SK (2015) Radiation synthesis and characterization of sodium alginate/chitosan/hydroxyapatite nanocomposite hydrogels: a drug delivery system for liver cancer. Polym Bull 72(4):725–742
188. Wang X, Du Y, Luo J, Lin B, Kennedy JF (2007) Chitosan/organic rectorite nanocomposite films: structure, characteristic and drug delivery behaviour. Carbohyd Polym 69(1):41–49
189. Sun L, Wang Y, Jiang T, Zheng X, Zhang J, Sun J, Wang S et al (2012) Novel chitosan-functionalized spherical nanosilica matrix as an oral sustained drug delivery system for poorly water-soluble drug carvedilol. ACS Appl Mater Interfaces 5(1):103–113
190. Yuan Q, Shah J, Hein SRDK, Misra RDK (2010) Controlled and extended drug release behavior of chitosan-based nanoparticle carrier. Acta Biomater 6(3):1140–1148

Smart Materials, Magnetic Graphene Oxide-Based Nanocomposites for Sustainable Water Purification

Janardhan Reddy Koduru, Rama Rao Karri and N. M. Mubarak

1 Introduction

Water pollution is a global environmental concern [61–64, 75]. Heavy metals are one of the primary contaminants in the aqueous environment. Continuous exposure to heavy metals leads to high-risk health problems for humans. Heavy metals are naturally occurring throughout the Earth's crust [2]. Anthropogenic activities, including mining operations, industry, and the use of metals and metal-containing compounds for the domestic and agricultural purpose, are the main sources of water pollution [83]. Hence, water is one of the major routes through which heavy metals and radionuclides may enter the human body [22]. Figure 1 shows the possible sources of water pollution. The real application of frequently used conventional wastewater purification techniques is limited to the removal of heavy metals at trace levels [22]. However, the low installation cost and easy operation of adsorption technique make it one of the preferred methods for water purification [26, 40, 25, 27]. Moreover, the use of activated carbon in the adsorption process is effective, but the use of it in a real application is limited, due to the complex installation process, coupled with the high-cost operation [41]. Hence, these drawbacks have

---

J. R. Koduru
Department of Environmental Engineering, Kwangwoon University,
Seoul 01897, Republic of Korea
e-mail: reddyjchem@gmail.com

R. R. Karri (✉)
Petroleum and Chemical Engineering, Faculty of Engineering,
Universiti Teknologi Brunei, Gadong, Brunei Darussalam
e-mail: kramarao.iitd@gmail.com

N. M. Mubarak
Department of Chemical Engineering, Faculty of Engineering and Science,
Curtin University, 98009 Miri, Sarawak, Malaysia
e-mail: mubarak.mujawar@curtin.edu.my

© Springer Nature Switzerland AG 2019
Inamuddin et al. (eds.), *Sustainable Polymer Composites and Nanocomposites*,
https://doi.org/10.1007/978-3-030-05399-4_26

necessitated the search for an alternative material that can be renewable and economic for water purification. The various potential applications of GO-based nanocomposites have been reported by different research groups [37]. Both the chemical stability of magnetic GO-based nanocomposites and literature survey, induce us to write a book chapter on magnetic GO (MGO's) based nanocomposites for the removal of heavy metals and radionuclides from water, with the purpose of reducing their environmental impact.

The numerous merits of graphene, which include high specific surface area, and thermal conductivity, high optical transmittance, and large Young's modulus have led to researchers paying great attention to it [78]. Similar to graphene, graphene-based material or graphene oxide (GO) shows the above significant properties. "However, GO is more easily dispersed than graphene, due to the presence epoxy, hydroxyl, and carboxyl functional groups, thus making its processing, synthesis, and application more convenient" (Fig. 2) [11]. Due to its imperishable hydrophilicity, GO found to be an efficient adsorbent and hence found many applications, including water purification [11]. Sreeprasad et al. [77] and Maaz et al. [47] have reported that nickel ferrite-GO composite is a better reaction media than iron ferrites, because of having higher catalytic and electron transfer efficiency through the Ni^{2+} in the nickel ferrite. Moreover, previous reports have proved the amazing removal response of magnetic nanoparticles/graphene or GO composites for pollutants like chromium [17, 67], copper [20], arsenic [105], cadmium [14], lead [100], and cobalt and an organic dye. Recently, Ligamdinne

Fig. 1 Schematic of the sources of water pollution

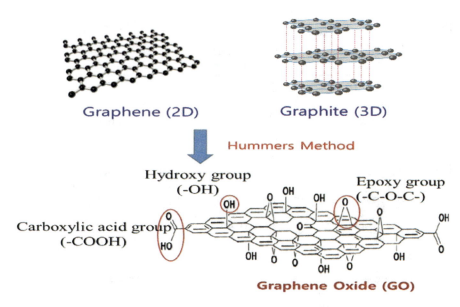

Fig. 2 Graphical representation of graphene oxide (GO) produced from graphite

et al. reported (Fig. 3) the removal of Co(II), Pb(II), Cr(III), As(III) and As(V), and radionuclides, U(VI) and Th(IV) from water, using the synthesis of "inverse spinel nickel ferrite incorporated-graphene oxide" based nanocomposites [35, 36, 39, 40]. The reported results demonstrated that the magnetic GO-based nanocomposites are promising, economic, could be separated by the external magnetic field.

Graphene can be extracted from graphite and it is merely a sheet of graphite [65]. It is defined as a single layer of sp^2 bonded carbon atoms in hexagonal lattice arrangement [97]. At the same time, graphene possesses few promising properties such as good electronic properties caused by the bonding and anti-bonding of the pi orbitals. Furthermore, graphene is clarified to be the strongest substance in terms of mechanical strength since it possesses high tensile strength and it is light in weight. For instance, it is more than 40 times stronger than diamond and more than 300 times stronger than A36 structural steel, at 130 GPa [81]. Meanwhile, for the optical properties, high absorption of white light up to 2.3% is capable to be observed from graphene.

Besides the impressive properties, appropriate method to produce graphene must be taken into consideration. There are two different type of methods to produce graphene which are exfoliation methods and direct growth of graphene layer [30]. Methods to generate graphene include "Scotch Tape Method", dispersion of graphite, exfoliation of graphite oxide, epitaxial growth and lastly CVD [23]. However, improved Hummers method which falls under the method of dispersion of graphite is used since it is an improved method which reduces the toxic gas emission and at the same time enhances the reaction performance [50].

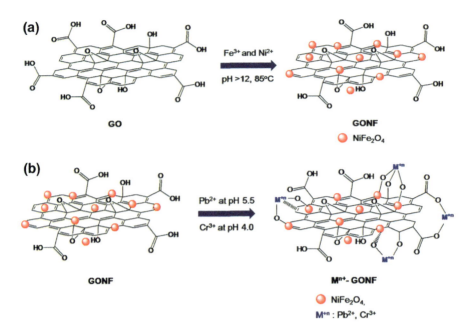

Fig. 3 Graphical representation of **a** nano-magnetic GONF composite preparation, **b** Pb(II) and Cr(III) adsorption onto GONF (reproduced from [39] with permission)

Generally, graphene generated via improved Hummers method are prepared to be further functionalized as chemical functionalization of graphene can be one of the best approaches for cadmium removal [101]. However, before functionalization takes place, the GO can be further transformed into GNs via acid treatment. Functionalization can be defined as the route where the addition of new properties, purposes, structures, or abilities to a substance via the alteration of the material in the aspect of surface chemistry. It is acknowledged that this is an essential method utilized throughout different fields such as biological engineering, chemistry, nanotechnology, materials science and the likes [93]. Functionalization can be done through the attachment of particles or nanoparticles to the surface of a substance, either via a chemical bond or via adsorption. For instance, the functionalized graphene can be produced via noncovalent and covalent alteration techniques. Both techniques share a similar process which is a superficial alteration of GO followed by reduction.

However, functionalization via ionic liquids (ILs) [48, 59] is known as a better covalent bonding technique [69]. The term of IL can be explained as poor coordination of the ions can be found in the salt below 100 °C or at room temperature. Ions in IL avoid the creation of a stable crystal lattice by having at least an ion which the charge is undergone delocalization and an organic component. Properties which include solubility of starting materials and other solvents, melting point, and viscosity are dependent on the counter ion and organic component [66]. For instance, implementation of ILs for synthetic difficulties is common and hence ILs can be known as "designer solvents". Furthermore, one of the promising advantages of IL is the zero presence of volatility. This condition has resulted in the solvents to have less toxicity compared to low-boiling-point solvents. For instance, by covalent bonding approach, GO obtained via modified Hummers method can be further functionalized via IL such as BF_4 [Bmim] with magnetic Fe_3O_4 nanoparticles to form core-shell structured Fe_3O_4@GO nanospheres to perform optimal extraction of cadmium [1].

2 Properties of Graphene

2.1 Electrical and Electronic Properties

The revolution of graphene can be initiated with the electronic and electrical properties of graphene [51]. Electrical conductivities of graphene, modified graphene and modified graphene/iron pentacarbonyl porous films with composites of chitosan (5, 10, 15, and 30 wt%) are shown in Fig. 4a. It is noted from these graphene derivatives that as the chitosan composite wt% is increased, it lowered the electrical conductivity. The number of layers existing in the graphene is the major factor to affect the properties. Hence, different materials are illustrated for monolayer, bilayer, and tri-layer of graphene. Former studies on graphene have proven that probability of altering charge carriers from holes to electron has led to the possible application in transistors [52]. However, merely monolayer graphene is valid for the electron-hole dependence. Yet, the dependence will get poorer with the disturbance of electrical field transmission by other layers once the number of layers is experiencing increment. The tremendously high mobility of electron at

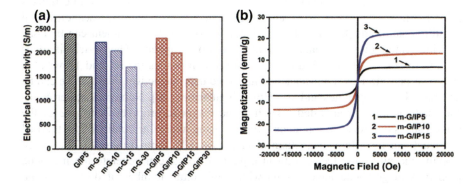

Fig. 4 a Electrical conductivities of graphene and its derivative films; **b** hysteresis loops of m-G/IP porous films [G—graphene; m-G—modified graphene, and m-G/IP modified graphene/iron pentacarbonyl porous films; m-GO-X, m-GO/IPX are composites of chitosan (5, 10, 15, and 30 wt%) with m-G and m-G/IP] (reproduced from [42] with permission from Elsevier)

different temperatures and exposure to magnetic fields result in the existence of quantum Hall effect in graphene for the hole and electron carriers [102]. For instance, mechanically generated graphene is found to exceed 2000 cm^2/V s at room temperature. Furthermore, in graphene, it only happens at only half integers of quantum Hall effect instead of happening in the classic integer which is at 4 e^2/h where the electron charge represents e while Planck's constant represents h. This circumstance results from the special band structure of graphene. Besides, utilized substrate and temperature are the major components to affect the performance of electron mobility within graphene. For instance, staggering mobility of suspended and annealed graphene onto Si/SiO_2 can reach more than 200,000 cm^2/V s which is considered as the highest recorded value among all the semiconductor substances [7].

2.2 Magnetic Properties

Magnetic properties of graphene might be affected by the presence of different types of defects [68]. The defects consist of topological defects, point defects, and extended defects. For instance, topological defects can be caused from the shapes such as heptagons and pentagons while the point defects are like adatoms, vacancies and the likes. However, extended defects include edges, voids, and cracks. Besides, defective parts like wrinkles, corrugations, and ripples can be found on the graphene surface. The defective lattice of the graphene such as voids and cracks will result in developing the local magnetic moments and forming interactions between the moments such as ferromagnetic [87]. The connection between the magnetic moments is ferromagnetic or antiferromagnetic if there is a detection of one Bohr magneton of magnetic moment formed by the vacancy or hydrogen chemisorption defects [95]. However, for the disorderly arranged graphene, induction of ferromagnetism can merely be done by monoatomic defects [94]. Furthermore, magnetism can be generated from adatoms, vacancies, and substitutional atoms [85]. In addition, induction of magnetic moment can be done by adding the monovalent and divalent adatom on graphene. The hysteresis loops of m-G/IP porous films with different chitosan composites (5, 10, and 15 wt%) is shown in Fig. 4b. Few studies also demonstrate that magnetism in graphene can be generated by zigzag edges and von Hove singularities [29]. Numerous experiments and methods have been tested regarding the magnetic properties and one of the studies states that reduced graphene oxide by using hydrazine continued by thermal treatment can form ferromagnetism in graphene [89]. Furthermore, exfoliation of graphite via ultrasonic technology in organic solvents can generate the magnetic properties in graphene nanocrystals with a minimum number of defects but no detection of ferromagnetism is observed at the temperature below than 2 K [72]. On the other hand, the presence of ferromagnetism can result from the high concentration of defects and it can be found mostly in the partly hydrogenated epitaxial graphene [91].

2.3 Chemical Properties

Pristine graphene sheets are regularly not reactive. Hence, graphene sheets should undergo superficial functionalization to activate its reactivity. This condition has illustrated that the domination of the surface is significant on the graphene chemistry while the graphene nanoribbons are dominated by the edges. Graphene reactivity also depends much on the thickness. As an example, monolayer graphene is found to have higher reactivity such as 10 times more than that of bilayer or multilayer graphene [73]. This statement is proved by utilizing the Raman spectroscopy in the peak measurement of relative disorder. Comparison of bulk graphene with graphene edges in terms of reactivity via spectroscopy examination is taken place and the discovery is that at least two times higher reactivity is found in the edges than that one of bulk monolayer graphene sheet [57].

2.4 Mechanical Properties

Performance and stability of the practical applications will be inevitably affected by the externally applied stress and unnecessary strain. The crystal-like graphene which covers an interatomic distance will eventually be affected by the externally applied stresses and hence it leads to the redistributed local charge. Indirectly, electronic transport will be varied significantly because of the developing band gap discovered in electronic structure. Anticipations from researchers can be witnessed once graphene is proven to have better performance than CNT in terms of its high stiffness and strength [12]. Hence, atomic force microscope has been utilized to make elastic properties measurement of the single layer graphene. As a result, 1 ± 0.1 TPa of Young's modulus and 130 ± 10 GPa of inherent strength are obtained [5]. Besides, measurement of strain with the applied tension and compression loads to the single layer graphene via Raman spectroscopy is recorded at the value of 1.3% for the strain and 0.7% for the compression and tension correspondingly [84]. There is a different degree of Young's modulus and fracture strength for a different layer of graphene. For instance, 1.02, 1.04 and 0.98 TPa of Young's modulus and 130, 126 and 101 GPa of fracture strength are owned by the single layer, bilayer, and tri-layer graphene correspondingly [31]. Therefore, it is obvious that increment of the layers will directly cause the increment in sliding tendency but weaken the properties. Furthermore, measurement of the alteration in 2D peaks and G with the presence of external stress can be made via Raman spectroscopy to record the measurement of the strains within the graphene sheets due to compression and tension. Potential to alter the band gap has been discovered by the introduction of measured strains because electric band structure can be varied by strain. Implementation of hydrogen plasma to carry out the reduction of graphene oxide has successfully led to the production of modified graphene with Young's modulus of 0.25 TPa [18]. Besides, the fracture toughness of pure graphene is recorded at 4 MPa since the potential formation of agglomerates and brittle nature due to imperfect graphene are present [99]. In short, reduction of the properties is highly depending on the increment of graphene nanosheet layers.

2.5 Thermal Properties

Phonon transport is the significant variable to affect the performance of graphene in terms of thermal conductivity [106]. Phonon transport can be explained as the ballistic and diffusive conduction at low and high temperature correspondingly. Yet, transport of electronic thermal can be ignorable since the non-doped graphene possesses carrier density which is low. The inherent thermal conduction of graphene can reach to approximately the range from 2000 to 6000 W/mK for the graphene sheets to undergo suspension at room temperature [3]. However, the value of 600 W/mK is recorded for the graphene which is undergoing suspension via silicon dioxide substrate [71]. On the other hand, localization and phonon scattering can take place due to the graphene defects which include isotopic doping [24], edge scattering and sample production deposits [56]. Thus, the guaranteed high quality of graphene sheets generated via micromechanical cleavage approach results in better thermal conduction. Besides, the thermal conductivity of the mechanically exfoliated graphene was recorded within the range of 4800–5300 W/mK [4]. The thermal conductivity is clearly outstanding than that of multi-wall, natural diamond and single wall CNTs which are 3000, 2200 and 3500 W/mK respectively [57]. This has proven that the outstanding thermal conductivity of graphene is most likely to replace the usage of copper.

3 Preparation Methods of MGO Nanocomposites

Most of the MGO nanocomposites are synthesized using the hydrothermal method. Although the hydrothermal process is generally carried out at high temperatures, this technique is an eco-friendly and economically feasible method [19]. Based on the synthetic approaches of MGOs, this hydrothermal method is performed in the presence of organic molecules as precursors and in the presence of alkaline media. The hydrothermal method used to perform at a temperature between 160 and 180 °C in a Teflon-line autoclave [80], is also known as the solvothermal method. Cheng et al. [10] reported: "one-step fabrication of magnetic GO composite gel" for the efficient adsorption removal of dye. The preparation of GO magnetic gel involved the hydrothermal method in the presence of alkaline (NH_3–H_2O) and polymer (polyvinyl alcohol). The gel exhibited both enhanced adsorption removal capacity towards cationic and anionic dyes, and magnetic separation capability, compared with the bare GO [86]. Generally, the ultra-sonochemical method is used to prevent re-aggregation, and improve the dispersion and reduction of the size of the material. It was used mostly before or after the synthesis of MGOs by the hydrothermal method. The main principle of the solvochemical method is the generation of ultrasound using a titanium horn that can serve to reduce the Van der Waals forces in the GO by ultra-sound irradiation of liquid [55]. Szabo et al. successfully prepared MGOs by sonication of a mixture of magnetic nanoparticles and GO solution [82].

Lately, microwave synthesis has become of great significance in the preparation of inorganic nanomaterials. In the synthesis of inorganic nanomaterials, compared to conventional heating technique, the microwave synthesis technique consumes less energy, environmentally friendly, and provides a homogeneous heating process for speedy reaction [74]. It can also offer rapid and selective heating of the reactant to a high temperature that leads to the production of "self-generated pressure in the sealed reaction vessel" [82]. Some of the previous works used the microwave synthesis method for the preparation of MGOs, including Mn_3O_4-rGO nanocomposites [74], and GO-NiO·4ZnO·4CoO·2Fe_2O_4 nanocomposites [45].

4 Structural Characterization and Properties of MGOs

The formation and structural functionalities of the prepared GO's and MGOs can be characterized using spectroscopic techniques that include XRD, XPS and FT-IR, and RS. The surface morphology, size, porosity, and dimensions of the prepared GO's and MGOs were evaluated using microscopic techniques, including AFM, TEM, and SEM. The surface area and surface primary adsorption characteristics were evaluated using BET theory analysis. Magnetic property measurements of GO's and MGOs were performed using a magnetometer. The SEM images of GO film, modified GO film, graphene porous film, and modified graphene porous film are shown in Fig. 5 [42]. The microstructure evolution of the unmodified GO films before (Fig. 5a) and after (Fig. 5c, e) the hydrazine-induced foaming process. Clearly, the GO film with highly-oriented GO sheets is converted to porous graphene film with random porous structures due to the excessive expansion derived from the weak interlayer interactions (Fig. 5c, e). The CS modified graphene (m-G) porous film (Fig. 5d, f) has distinct and continuous porous structures inherited from its precursor (Fig. 5b), which contrasts sharply the random structures of its unmodified counterpart (Fig. 5c, e). Additionally, some small pores are observed in the porous graphene film (Fig. 5f), which are beneficial for further decreasing the density of the porous film while retaining its reasonable strength.

XRD studies are mainly used to identify the formation, structure, and crystalline nature of MGOs. The XRD 2θ strong peak in the range 8°–12° indicates the formation of GO. By the magnetization, the crystalline property of GO is decreased by the increase of the mesoporous carbonaceous nature with alteration of the original position of the GO peak [47]. The XRD diffraction peaks are also used to identify the ferrite in MGOs. By decreasing the size of MGOs with increasing porosity, the positions of diffraction peaks shift to lower 2θ range [70].

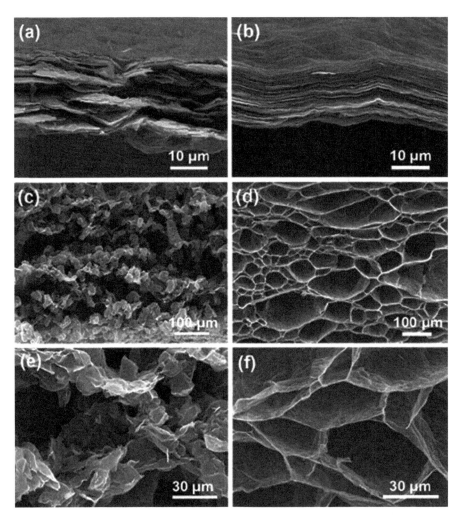

Fig. 5 SEM images of **a** GO film, **b** modified GO film, **c, e** graphene porous film, and **d, f** modified graphene porous film (reproduced from [42] with permission from Elsevier)

RS is an important technique to qualitatively identify the MGOs. As is known, the graphitic materials show two prominent Raman peaks around 1350 and 1600 cm^{-1} called the D and G bands. Here, the G band corresponds to the stretching vibrations of carbons at sp^2 hybridization, whereas the D band represents the vibrations of carbons at sp^3 hybridization, which can break the symmetry and selection rule [36]. By the magnetization of GO, these D and G bonds alter their positions, based on their principal interactions. But, in the case of nickel ferrite-rGO (rGONF), both the sp^2 domain (D) and sp^3 domain (G) carbons are shifted to lower range at ~1303–1591 cm^{-1}, which indicates that both D and G band carbons are involved in the formation of reduced GO-based magnetic nanocomposite. The XPS is used to

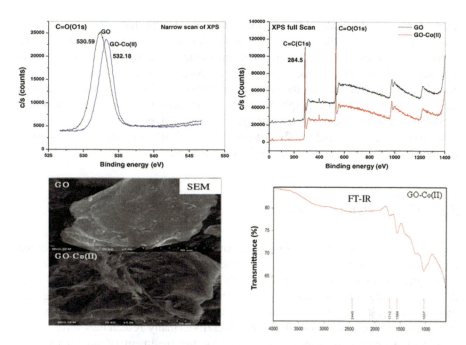

Fig. 6 XPS, SEM, FT-IR results of the Co(II)-loaded GO (reproduced from [40] with permission)

qualitative and quantitative identify the chemical composition of MGOs. The bonding energy peaks of 700–730 eV indicate the Fe peaks of magnetic materials.

The microscopic techniques, including SEM, TEM, and AFM, are used to measure the size of nanocomposites, MGOs, and their surface morphology, which is an important factor to know for the adsorption process. Their porous structure and surface area can be further evaluated by using (N_2) adsorption-desorption isotherms through BET analysis. The magnetic nature of MGOs is identified using the magnetic measurement system (MPMS). When the size of MGOs decreases to the nanoscale, it shows superparamagnetic nature. Lingamdinne et al. [36] confirmed the superparamagnetic property of magnetic nanocomposite by MT curves obtained at 1000 Oe magnetic field. They also observed the increase of superparamagnetic property by the reduction of nanocomposite [16]. The XPS, SEM, FT-IR of the Co(II)-GO were shown in Fig. 6.

5 Applications to Sustainable Water Purification

Graphene oxide is easily dispersed in water due to the plentiful hydrophilic (carboxylic, epoxide, and hydroxyl groups) groups on its surface. List of various GO based nanomaterials utilized for removal of heavy metal ions removal is given in

Table 1. Due to the hydrophilic nature of GO, it adsorbs the pollutants to stable complexes, causing difficulties for the separation and recovery of GO [96]. To overcome these difficulties of separation, magnetic functionalization of GO is an alternative solution [104]. Due to the magnetic particle has unique properties, they have been widely applied to the removal of various pollutants. Some researchers have developed magnetic graphene oxide composites for efficient applications, including water treatment, energy storage, and drug delivery [11]. Here, we review some research reports, and briefly critique the use of MGO based nanocomposites for the adsorption of heavy metals and radionuclides.

5.1 Heavy Metals Removal

Exposure to heavy metals can present serious health risks to human beings. For the purification of water, adsorption is an effective, economic, and easy operation technique, compared to conventional methods [54]. But it is limited, due to the difficulties in the filtration and regeneration of adsorbents. The use of magnetic materials for water purification can overcome the above difficulties, including the filtration and regeneration of adsorbents; therefore, many researchers have developed and widely used magnetic materials for the treatment of pollutants [60]. However, the nano metal ferrites show poor stability [47]. To overcome these difficulties, hybrid materials have been synthesized through magnetic ferrites and GO by the hydrothermal method. Due to the presence of epoxy, carboxylic, and hydroxyl functions at GO, they can enhance the adsorption of heavy metals [33]. The specific adsorption mechanisms of graphene oxide-based nanomaterials for metal ions removal are given in Table 2.

Chandra et al. [8] thru a chemical reaction developed 10 nm average size superparamagnetic magnetite reduced graphene oxide (M-RGO) composites. The M-RGO showed higher adsorption capacity over 99.9% for removal of both ionic states of Arsenic. Zhang et al. employed ferric hydroxide-GO composite for the magnificent adsorption of arsenate from water [98]. Here, high arsenate adsorption removal was observed over a pH range of 4–9 and reduced the arsenate concentration of contaminated water from 20 to 0.5 ppm. Water-soluble "magnetic graphene oxide nanocomposite" has been produced via a "copper catalyzed azide-alkyne cycloaddition", and was utilized for the adsorption removal of Pb (II), Cd(II), and Cu(II), from aqueous solutions [100]. The results found that the nanocomposite has the higher surface area and extraordinary removal capacity for heavy metals.

RGO–MnO_2 composites showed excellent Hg(II) uptake capacity [77]. The composite shows enhanced adsorption removal capacity compared to its base material. Liu et al. [43] employed an MGO for the successful removal of Co(II). The adsorption Co(II) on MGO was the rate-limiting kinetics, with "inner-sphere surface complexation" at low pH values. Meanwhile, at higher pH values, the removal mechanism of Co(II) was associated with inner-sphere surface

Table 1 List of various graphene oxide based nanomaterials utilized for removal of heavy metal ions removal

Metal ion	Adsorbent	Max. adsorption capacity (mg/g)	Conditions	Model (adsorption isotherm; kinetics)	Remarks
Cd	GO	1792.60	303 K; pH 4.0	Langmuir and Freundlich; pseudo second-order	• The equilibrium contact time is 120 min • The graphene oxide is generated by using amorphous graphite
	PAMAMs/ GO	253.81	298 K; pH 5.0	Langmuir; pseudo second-order	• The adsorption process achieves equilibrium within 60 min • The adsorbent dosage is 0.1 g
	Few-layered GO nanosheets	106.30	303 K; pH 6.0	Langmuir	• The dosage of adsorbent is 0.1 g/L • The adsorption capacity is strongly based on pH and humic acid
	GO/cellulose membranes	26.8	298 K; pH 4.5	Langmuir; pseudo second-order	• Good adsorption and no precipitation of metal hydroxides • It can be reused up to ten cycles
Pb	Few-layered GO	842.00	293 K; pH 6.0	Langmuir	• pH value strongly affects the adsorption capacity • The adsorption capacity is strongly independent of ionic strength
	Graphene nanosheet	476.19	298 K; pH 6.2	Langmuir	• The equilibrium contact time is 35 min • The dosage of adsorbent is 40 mg/L
	Ag/GO	312.57	298 K; pH 5.3	Langmuir; pseudo second-order	• 0.05 mg of adsorbents used showed the maximum adsorption performance • The equilibrium time for the lead adsorption is 50 min
Cu	Chitosan/ SH/GO	425.00	293 K; pH 5.0	Freundlich; pseudo second-order	• The dosage of adsorbents is 0.2 mg/mL • The adsorption efficiency is strongly dependent on pH, temperature and adsorbent dosage
	TiO_2/GO	45.20	293 K; pH 6.0	Langmuir	• The adsorption capacity is strongly based on the pH value
	GO aerogels	19.65	298 K; pH 6.2	Langmuir; pseudo second-order	• The dosage of adsorbents is 0.6 g/L • It involves ion exchange mechanism

(continued)

Table 1 (continued)

Metal ion	Adsorbent	Max. adsorption capacity (mg/g)	Conditions	Model (adsorption isotherm; kinetics)	Remarks
Cr	Chitosan/GO	310.40	318 K; pH 3.0	Redlich–Peterson/double exponential	• The adsorbent dosage is 0.5 g/L • Both internal and external diffusion take place effectively in the adsorption process
	Fe_3O_4/GO	32.33	293 K; pH 4.5	Langmuir; pseudo second-order	• pH value and ionic strength are the crucial factors to affect the adsorption capacities • The adsorbent dosage is 0.2 g/L

Adapted from [34] with permission from Elsevier

Table 2 Specific adsorption mechanisms of graphene oxide-based nanomaterials for metal ions removal

Graphene oxide-based nanomaterials	Adsorption mechanisms included for metal ions removal	Advantages	Drawbacks
Graphene oxide (GO)	• Electrostatic interactions • Ion exchange	• Good dispersion in water • Great colloidal constancy • Contains rich oxygenated functional groups	• A restricted number of sorption sites
Reduced graphene oxide (rGO)	• Electrostatic interactions • Lewis-base–acid mechanism	• Restoration of sp^2 domains • Better electron-transport properties	• Less oxygen-containing functional groups • Weaker colloidal stability
Magnetic graphene oxide nanocomposites	• Electrostatic interactions with graphene oxide • Interactions with the particles surface • Magnetic properties of the nanoparticles	• Bigger surface area compared to the pure GO • Increased number of binding sites compared to pure GO • Ease the recovery process from solutions	• Co-reduction of GO during the combination of the particles weakens the colloidal stability
Graphene oxide materials functionalized with organic molecules	• Electrostatic interactions • Complexation with organic molecules	• Bigger surface area compared to pure GO • Great colloidal stability • The greater number of functional groups ($-NH_2$, $-OH$)	• Large variations of the stability of the loaded molecules depending on the alteration approach physically or chemically

Reproduced from [34] with permission from Elsevier

complexation and precipitation. Co(II)-loaded MGO can be rapidly separated and recovered from aqueous solution by external magnetic field [43]. Liu et al. [44] reported a facile self-assembly of magnetic particles with GO through electrostatic interaction in the presence of 3-aminopropyltrimethoxysilane. The prepared porous MGO not only conventionally separated by a magnetic field but also enhanced the adsorption capacity of GO for Cr(VI) removal. They also concluded that the prepared MGO shows higher adsorption capacity than that of GO and Fe_3O_4. Bhunia et al. [6] developed a heterogeneous matrix of iron/iron oxide dispersed on RGO (RGO-FeO)/Fe_3O_4) that can be used for the effective adsorption of heavy metal ions, including Cr(VI), Hg(II), Pb(II), Cd(II), and As(III). Recently, Lingamdinne et al. employed porous inverse spinel composite (MGO) and porous inverse RMGO nanocomposites using nickel ferrite and GO, and applied them for the removal of Arsenic [38], Pb(II) and Cr(III) [35 36 37 39 40 41]. They also reported the as-prepared nanocomposites to show considerably enhanced adsorption capacity for Pb(II), Cr(III), As(III), and As(V), compared to that of GO. They stated that the adsorption efficiency of rMGO for As(III) and As(V) is 106.40 and 65.78 mg/g respectively was greater than that of MGO [38]. The adsorption process was thermodynamically favourable for the adsorption of Pb(II) and Cr(III) onto the nanocomposites and was spontaneous endothermic. But the As(III) and As(V) removal was enhanced with increased temperature up to 300 K, while it started decreasing with further increase of the temperature above 300 K. The unavailability of metal ions undergoing degradation process via bioprocesses and reactions chemically has led to the wider exposure to adsorption process and hence it is now considered as the most promising method to remove heavy metal ions [53].

Copper (II) ions can be removed effectively via the interaction between copper (II) ions and oxygen groups on GO which are positively charged and negatively charged respectively [92]. Besides, Pb(II), U(VI) and Co(II) ions can be adsorbed via the usage of GO as well [46]. For instance, powder X-ray diffraction (XRD), scanning electron microscopy (SEM), infrared spectroscopy (FT-IR) and X-ray photoelectron spectroscopy (XPS) are utilized to determine the adsorption characterization of GO towards copper, lead, zinc and cadmium. As a result, the largest adsorption capacities at pH 5 have been recorded as 294, 345, 530, 1119 mg/g for Cu(II), Zn(II), Cd(II) and Pb(II) respectively [76]. In addition, the adsorption capacities mentioned previously can be done in a wide range of pH values which are 3–7 for both Cu(II) and Pb(II), 4–8 for Cd(II) and 5–8 for Zn(II) [76]. It is studied that the removal of heavy metal ions via GO can be explained as the adsorption process is restricted chemically due to the participation of superficial complexity of metal ions with the presence of oxygen in the functional groups which lies on the surface of GO. As an example, cellulose hydrogel/GO possesses adsorption capacity of approximately 94 mg/g [9].

Furthermore, with the presence of metal oxides on GO and GNs, they are considered as high-performance adsorbents in the past [21]. For instance, the fabrication of GO–TiO_2 is implemented to remove Pb(II), Zn (II) and Cd(II) ions. Hence, the adsorption capacities of the GO–TiO_2 on the Pb(II), Zn (II) and Cd(II) ions at pH value of 5 are recorded at 65.8, 88.9 and 72.8 mg/g respectively [32].

The covalent bond which is securing the rGO and TiO_2 in the GO–TiO_2 has resulted in the poor electron-hole recombination and hence great amount of Cr(VI) can be reduced. Although different types of magnetic graphene composites have been utilized, to eradicate the separation difficulties, a combination of GO or GNs with magnetic materials is a good choice. The significant of the combination is to minimize the agglomeration and restacking of graphene sheets and at the same time increasing the surface area and adsorption efficiency [79].

Moreover, removal of cadmium ions and copper ions from wastewater can be done via fabrication of GNs through the method of modified Hummers [103]. For instance, at the condition of pH 6.0 ± 0.1 and temperature around 303 K, it is recorded that highest adsorption capacities of Co(II) and Cd(II) can be achieved with the value of 68.2 and 106.3 mg/g respectively. This has proven that the adsorption of heavy metal ions through GNs is depending on the ionic strength and pH. However, adsorption of chromium (VI) ions and chromium (III) ions can be done at the low range of pH levels but the adsorptions of copper (II) ions, lead (II) ions, and gold (III) ions are most likely to be occurring at higher range of pH levels [15].

Other than that, MMSP-GO to adsorb heavy metal ions such as Cd (II) and Pb(II) is worth studied as well. Once the synthesis of magnetic mesoporous silica comes with the functionalization with polyethyleneimine molecules, the conjugation of a great number of amine groups with carboxyl groups on the GO sheets could increase the affinity between the pollutants and mesoporous silica. Significant and effective data has been recorded for the MMSP-GO composites regarding its maximum adsorption capacities for Pb(II) and Cd(II) which are 333 and 167 mg/g respectively [90].

5.2 Organic Pollutants Removal

Presently, many process industries, including the paper, textile, paint, plastic, and leather industries, use pigments and dyes to colour their products, and excess of these colours end up in the discharge, which ultimately ends up as industrial effluent. These dyes are organic compounds, and the presence of dyes in an aquatic environment not only affects the aesthetics, it also inhibits the penetration of sunlight, and thus reduces photosynthesis for waterborne plants. Overall, the presence of dyes poses the threat of toxic materials, which are resistant to a chemical reaction in wastewaters, and which leads to cancer, mutagenesis, and other severe problems in human and aquatic creatures. The complicated chemical structures of dyes make these materials highly resistant to biodegradation [49]. Therefore, these dyes have to be removed efficiently before the effluent is discharged to the nearby aquatic environment.

In the last couple of years, magnetic nanoparticles have been extensively applied for the removal of toxic metal ions and organic pollutants. This is due to the features, like low toxicity, high chemical stability, and good magnetic properties.

Adsorbents based on magnetic nanoparticles are used in the removal of toxic dyes and heavy metals from aqueous solutions with precision and high accuracy [88]. However, the bare magnetic nanoparticles can easily be oxidized in atmospheric conditions. Therefore, to remedy these effects and increase the life of nanoparticles, researchers explored ways to coat or functionalize the magnetic nanoparticles and enhance the functional groups. In recent years, usage of GO-based magnetic nanoparticles as an adsorbent has increased. Association of graphene has resulted in high removal capacity of pollutants. Chandra et al. [8] reported that compared to free nanoparticles, RGO and GO embedded materials have shown higher binding capacity. Deng et al. [13] synthesized MGO and used it to investigate the simultaneous removal of Cd(II) and ionic dyes like orange gelb (OG) and methylene blue (MB). The maximum adsorption capacities reported for the removal of MB using graphene nanosheet/magnetite, magnetic rectorite/iron oxide, and multi-walled carbon nanotubes/Fe_2O_3, were 43.82, 31.18, and 42.30 mg/g, respectively. They also reported that the maximum sorption capacities of MGO (64.23 mg/g for MB, and 20.85 mg/g for OG) were much higher than those of "exfoliated graphene oxide" (17.3 mg/g for MB, and 5.98 for OG) used as adsorbent [58]. The maximum adsorption capacities reported for removal of dyes using bare iron oxide nanoparticles were 2.78 and 15.62 mg/g for MB and OG, respectively. Khurana et al. [28] investigated the "Eriochrome Black T" (EBT) adsorption from textile wastewater using MGO in a batch process. In this study, for the removal of toxic textile azo dye EBT, they synthesized MGO that was impregnated with α-Fe_2O_3. The maximum adsorption capacity for dye removal was reported to be 210.53 mg/g.

6 Conclusion and Future Perspective

This chapter indicates that adsorption using MGOs is becoming an alternative option to replace the conventional adsorbents used for water purification. It also shows that these MGO composites have comparable or even greater adsorption and regeneration capacity, compared to the available adsorbents and activated carbons. Moreover, the adsorbents coming out at long last with high adsorption efficiency for the handling of wastewater containing metal pollutants (heavy metals and radionuclides) and organic pollutants could be successfully implemented as beneficial to society.

The synthesis of graphene has been widely done by the latest and greatest method which is improved Hummers method. However, the experimental procedures to complete the fabrication of the graphene is time-consuming although the experimental complexity is considerably low. Therefore, replacement or removal of certain chemicals is required to be further discovered and studied to shorten the fabrication period and result in a better fabrication method.

The synthesis of graphene has been generally done by the enhanced Hummers technique. Applications are hindered due to the time-consuming methods for fabricating the Graphene. Accordingly, appropriate techniques need to be discovered to reduce the fabrication period and thus enhance the performance of graphene obtained by these fabrications method. Another major hurdle in these applications is the scalability of these methods. Most of the recently reported studies are limited to lab scale experiments. Therefore, studies should also focus on the scalability of these applications from lab scale experiments into commercial industrial scale applications. It should also be noted that, industries which produce effluents and wastewater at a larger scale will consume higher quantities of GO's and MGO's, thus increasing the cost of operation to many folds. Therefore, the commercialized MGOS should be prepared in such a way that they can be re-used or regenerate with low operating cost.

Acknowledgments This work has been supported by the National Research Foundation (NRF) of Korea funded by the Ministry of Science, ICT & Future Planning (MSIP) (2017R1C1B5016656) of the Korea Government, Seoul, Korea.

References

1. Alvand M, Shemirani F (2016) Fabrication of Fe_3O_4@graphene oxide core-shell nanospheres for ferrofluid-based dispersive solid phase extraction as exemplified for Cd (II) as a model analyte. Microchim Acta 183:1749–1757. https://doi.org/10.1007/s00604-016-1805-8
2. Azimi A, Azari A, Rezakazemi M, Ansarpour M (2017) Removal of heavy metals from industrial wastewaters: a review. ChemBioEng Rev 4:37–59. https://doi.org/10.1002/cben.201600010
3. Balandin AA (2011) Thermal properties of graphene and nanostructured carbon materials. Nat Mater 10:569. https://doi.org/10.1038/nmat3064
4. Balandin AA, Ghosh S, Bao W, Calizo I, Teweldebrhan D, Miao F, Lau CN (2008) Superior thermal conductivity of single-layer graphene. Nano Lett 8:902–907
5. Berger C et al (2004) Ultrathin epitaxial graphite: 2D electron gas properties and a route toward graphene-based nanoelectronics. J Phys Chem B 108:19912–19916. https://doi.org/10.1021/jp040650f
6. Bhunia P, Kim G, Baik C, Lee H (2012) A strategically designed porous iron–iron oxide matrix on graphene for heavy metal adsorption. Chem Commun 48:9888
7. Bolotin KI et al (2008) Ultrahigh electron mobility in suspended graphene. Solid State Commun 146:351–355. https://doi.org/10.1016/j.ssc.2008.02.024
8. Chandra V, Park J, Chun Y, Lee JW, Hwang I-C, Kim KS (2010) Water-dispersible magnetite-reduced graphene oxide composites for arsenic removal. ACS Nano 4:3979–3986. https://doi.org/10.1021/nn1008897
9. Chen X, Zhou S, Zhang L, You T, Xu F (2016) Adsorption of heavy metals by graphene oxide/cellulose hydrogel prepared from NaOH/urea aqueous solution. Mater 9:582
10. Cheng Z, Liao J, He B, Zhang F, Zhang F, Huang X, Zhou L (2015) One-step fabrication of graphene oxide enhanced magnetic composite gel for highly efficient dye adsorption and catalysis. ACS Sustain Chem Eng 3:1677–1685
11. Chung C, Kim Y-K, Shin D, Ryoo S-R, Hong BH, Min D-H (2013) Biomedical applications of graphene and graphene oxide. Acc Chem Res 46:2211–2224
12. Dasari BL, Nouri JM, Brabazon D, Naher S (2017) Graphene and derivatives—synthesis techniques, properties and their energy applications. Energy 140:766–778. https://doi.org/10.1016/j.energy.2017.08.048

13. Deng J-H, Zhang X-R, Zeng G-M, Gong J-L, Niu Q-Y, Liang J (2013) Simultaneous removal of Cd (II) and ionic dyes from aqueous solution using magnetic graphene oxide nanocomposite as an adsorbent. Chem Eng J 226:189–200
14. Deng J-H, Zhang X-R, Zeng G-M, Gong J-L, Niu Q-Y, Liang J (2013) Simultaneous removal of Cd(II) and ionic dyes from aqueous solution using magnetic graphene oxide nanocomposite as an adsorbent. Chem Eng J 226:189–200
15. Duru I, Ege D, Kamali AR (2016) Graphene oxides for removal of heavy and precious metals from wastewater. J Mater Sci 51:6097–6116
16. Fan Z, Wang K, Wei T, Yan J, Song L, Shao B (2010) An environmentally friendly and efficient route for the reduction of graphene oxide by aluminum powder. Carbon 48:1686–1689
17. Gollavelli G, Chang C-C, Ling Y-C (2013) Facile synthesis of smart magnetic graphene for safe drinking water: heavy metal removal and disinfection control. ACS Sustain Chem Eng 1:462–472
18. Gomez-Navarro C, Burghard M, Kern K (2008) Elastic properties of chemically derived single graphene sheets. Nano Lett 8:2045–2049. https://doi.org/10.1021/nl801384y
19. Hashim N et al (2016) A brief review on recent graphene oxide-based material nanocomposites: synthesis and applications. J Mater Environ Sci 7:3225–3243
20. Hu X-J et al (2013) Removal of Cu(II) ions from aqueous solution using sulfonated magnetic graphene oxide composite. Sep Purif Technol 108:189–195
21. Hur J, Shin J, Yoo J, Seo YS (2015) Competitive adsorption of metals onto magnetic graphene oxide: comparison with other carbonaceous adsorbents. The Sci World J 2015:1–11. https://doi.org/10.1155/2015/836287
22. Abbas A, Al-Amer AM, Laoui T, Al-Marri MJ, Nasser MS, Khraisheh M, Atieh MA (2016) Heavy metal removal from aqueous solution by advanced carbon nanotubes: critical review of adsorption applications. Sep Purif Technol 157:141–161
23. Ionita M, Vlăsceanu GM, Watzlawek AA, Voicu SI, Burns JS, Iovu H (2017) Graphene and functionalized graphene: extraordinary prospects for nanobiocomposite materials. Compos B Eng 121:34–57
24. Jiang J-W, Lan J, Wang J-S, Li B (2010) Isotopic effects on the thermal conductivity of graphene nanoribbons: localization mechanism. J Appl Phys 107:054314. https://doi.org/10.1063/1.3329541
25. Karri RR, Sahu JN (2018) Modeling and optimization by particle swarm embedded neural network for adsorption of zinc (II) by palm kernel shell based activated carbon from aqueous environment. J Environ Manage 206:178–191
26. Karri RR, Jayakumar N, Sahu J (2017) Modelling of fluidised-bed reactor by differential evolution optimization for phenol removal using coconut shells based activated carbon. J Mol Liq 231:249–262
27. Karri RR, Sahu JN, Jayakumar NS (2017) Optimal isotherm parameters for phenol adsorption from aqueous solutions onto coconut shell based activated carbon: error analysis of linear and non-linear methods. J Taiwan Inst Chem Eng 80:472–487
28. Khurana I, Shaw AK, Bharti, Khurana JM, Rai PK (2018) Batch and dynamic adsorption of Eriochrome Black T from water on magnetic graphene oxide: experimental and theoretical studies. J Environ Chem Eng 6:468–477
29. Kou L, Tang C, Guo W, Chen C (2011) Tunable magnetism in strained graphene with topological line defect. ACS Nano 5:1012–1017. https://doi.org/10.1021/nn1024175
30. Krane N (2011) Selected topics in physics: physics of nanoscale carbon. Freie Universität, Berlin
31. Lee C, Wei X, Li Q, Carpick R, Kysar Jeffrey W, Hone J (2009) Elastic and frictional properties of graphene. Physica Status Solidi (b) 246:2562–2567. https://doi.org/10.1002/pssb.200982329
32. Lee Y-C, Yang J-W (2012) Self-assembled flower-like TiO_2 on exfoliated graphite oxide for heavy metal removal. J Ind Eng Chem 18:1178–1185

33. Li J, Guo S, Zhai Y, Wang E (2009) Nafion–graphene nanocomposite film as enhanced sensing platform for ultrasensitive determination of cadmium. Electrochem Commun 11:1085–1088
34. Lim JY, Mubarak NM, Abdullah EC, Nizamuddin S, Khalid M, Inamuddin (2018) Recent trends in the synthesis of graphene and graphene oxide based nanomaterials for removal of heavy metals—a review. J Ind Eng Chem. https://doi.org/10.1016/j.jiec.2018.05.028
35. Lingamdinne L, Kim I-S, Ha J-H, Chang Y-Y, Koduru J, Yang J-K (2017) Enhanced adsorption removal of Pb(II) and Cr(III) by using nickel ferrite-reduced graphene oxide nanocomposite. Metals 7:225
36. Lingamdinne LP, Choi Y-L, Kim I-S, Yang J-K, Koduru JR, Chang Y-Y (2017) Preparation and characterization of porous reduced graphene oxide based inverse spinel nickel ferrite nanocomposite for adsorption removal of radionuclides. J Hazard Mater 326:145–156
37. Lingamdinne LP, Koduru JR, Chang Y-Y, Karri RR (2018) Process optimization and adsorption modeling of Pb(II) on nickel ferrite-reduced graphene oxide nano-composite. J Mol Liq 250:202–211
38. Lingamdinne LP, Choi Y-L, Kim I-S, Chang Y-Y, Koduru JR, Yang J-K (2016) Porous graphene oxide based inverse spinel nickel ferrite nanocomposites for the enhanced adsorption removal of arsenic. RSC Adv 6(77):73776–73789
39. Lingamdinne LP, Koduru JR, Choi Y-L, Chang Y-Y, Yang J-K (2016) Studies on removal of Pb(II) and Cr(III) using graphene oxide based inverse spinel nickel ferrite nano-composite as sorbent. Hydrometallurgy 165:64–72
40. Lingamdinne LP, Koduru JR, Roh H, Choi Y-L, Chang Y-Y, Yang J-K (2016) Adsorption removal of Co(II) from waste-water using graphene oxide. Hydrometallurgy 165:90–96
41. Lingamdinne LP, Roh H, Choi Y-L, Koduru JR, Yang J-K, Chang Y-Y (2015) Influencing factors on sorption of TNT and RDX using rice husk biochar. J Ind Eng Chem 32:178–186
42. Liu J, Zhang H-B, Liu Y, Wang Q, Liu Z, Mai Y-W, Yu Z-Z (2017) Magnetic, electrically conductive and lightweight graphene/iron pentacarbonyl porous films enhanced with chitosan for highly efficient broadband electromagnetic interference shielding. Compos Sci Technol 151:71–78. https://doi.org/10.1016/j.compscitech.2017.08.005
43. Liu M, Chen C, Hu J, Wu X, Wang X (2011) Synthesis of magnetite/graphene oxide composite and application for cobalt(ii) removal. J Phys Chem C 115:25234–25240
44. Liu M, Wen T, Wu X, Chen C, Hu J, Li J, Wang X (2013) Synthesis of porous Fe_3O_4 hollow microspheres/graphene oxide composite for Cr(vi) removal. Dalton Trans 42:14710
45. Liu P, Yao Z, Zhou J (2015) Preparation of reduced graphene oxide/ $NiO·4ZnO·4CoO·2Fe_2O_4$ nanocomposites and their excellent microwave absorption properties. Ceram Int 41:13409–13416
46. Liu, ZJ, Yang, JW, Li, CZ, Li, JX, Jiang, YJ, Dong, YH, Li, YY (2014) Adsorption of Co (II), Ni (II), Pb (II) and U (VI) from aqueous solutions using polyaniline/graphene oxide composites. Korean Chem Eng Res 52(6):781–788. https://doi.org/10.9713/kcer.2014.52.6.781
47. Maaz K, Karim S, Mumtaz A, Hasanain SK, Liu J, Duan JL (2009) Synthesis and magnetic characterization of nickel ferrite nanoparticles prepared by co-precipitation route. J Magn Magn Mater 321:1838–1842
48. Mesbah M, Shahsavari S, Soroush E, Rahaei N, Rezakazemi M (2018) Accurate prediction of miscibility of CO_2 and supercritical CO_2 in ionic liquids using machine learning. J CO_2 Utilization 25:99–107. https://doi.org/10.1016/j.jcou.2018.03.004
49. Mokhtari P, Ghaedi M, Dashtian K, Rahimi M, Purkait M (2016) Removal of methyl orange by copper sulfide nanoparticles loaded activated carbon: kinetic and isotherm investigation. J Mol Liq 219:299–305
50. Muzyka R, Kwoka M, Smędowski Ł, Díez N, Gryglewicz G (2017) Oxidation of graphite by different modified Hummers methods. New Carbon Mater 32:15–20. https://doi.org/10.1016/S1872-5805%5b17%5d60102-1
51. Novoselov KS et al (2004) Electric field effect in atomically thin carbon films. Science 306:666–669

52. Novoselov KS, Jiang D, Schedin F, Booth TJ, Khotkevich VV, Morozov SV, Geim AK (2005) Two-dimensional atomic crystals. Proc Natl Acad Sci U S A 102:10451
53. Nupearachchi CN, Mahatantila K, Vithanage M (2017) Application of graphene for decontamination of water implications for sorptive removal. Groundwater Sustain Dev 5:206–215. https://doi.org/10.1016/j.gsd.2017.06.006
54. Oraby EA, Eksteen JJ (2015) The leaching of gold, silver and their alloys in alkaline glycine–peroxide solutions and their adsorption on carbon. Hydrometallurgy 152:199–203
55. Peng Y, Ji J, Chen D (2015) Ultrasound assisted synthesis of ZnO/reduced graphene oxide composites with enhanced photocatalytic activity and anti-photocorrosion. Appl Surf Sci 356:762–768
56. Pettes MT, Jo I, Yao Z, Shi L (2011) Influence of polymeric residue on the thermal conductivity of suspended bilayer graphene. Nano Lett 11:1195–1200. https://doi.org/10.1021/nl104156y
57. Phiri J, Gane P, Maloney TC (2017) General overview of graphene: production, properties and application in polymer composites. Mater Sci Eng B 215:9–28. https://doi.org/10.1016/j.mseb.2016.10.004
58. Ramesha G, Kumara AV, Muralidhara H, Sampath S (2011) Graphene and graphene oxide as effective adsorbents toward anionic and cationic dyes. J Colloid Interface Sci 361:270–277
59. Razavi SMR, Rezakazemi M, Albadarin AB, Shirazian S (2016) Simulation of CO_2 absorption by solution of ammonium ionic liquid in hollow-fiber contactors. Chem Eng Process Process Intensification 108:27–34. https://doi.org/10.1016/j.cep.2016.07.001
60. Reddy DHK, Lee S-M (2013) Application of magnetic chitosan composites for the removal of toxic metal and dyes from aqueous solutions. Adv Coll Interface Sci 201–202:68–93
61. Rezakazemi M, Dashti A, Riasat Harami H, Hajilari N, Inamuddin (2018) Fouling-resistant membranes for water reuse. Environ Chem Lett 1–49. https://doi.org/10.1007/s10311-018-0717-8
62. Rezakazemi M, Ghafarinazari A, Shirazian S, Khoshsima A (2013) Numerical modeling and optimization of wastewater treatment using porous polymeric membranes. Polym Eng Sci 53:1272–1278. https://doi.org/10.1002/pen.23375
63. Rezakazemi M, Khajeh A, Mesbah M (2018) Membrane filtration of wastewater from gas and oil production. Environ Chem Lett 16:367–388. https://doi.org/10.1007/s10311-017-0693-4
64. Rezakazemi M, Shirazian S, Ashrafizadeh SN (2012) Simulation of ammonia removal from industrial wastewater streams by means of a hollow-fiber membrane contactor. Desalination 285:383–392. https://doi.org/10.1016/j.desal.2011.10.030
65. Rezakazemi M, Zhang Z (2018) 2.29 Desulfurization materials A2. In: Ibrahim D (ed) Comprehensive energy systems. Elsevier, Oxford, pp 944–979. https://doi.org/10.1016/B978-0-12-809597-3.00263-7
66. Sanes J, Avilés M-D, Saurín N, Espinosa T, Carrión F-J, Bermúdez M-D (2017) Synergy between graphene and ionic liquid lubricant additives. Tribol Int 116:371–382. https://doi.org/10.1016/j.triboint.2017.07.030
67. Sarı A, Tuzen M, Soylak M (2007) Adsorption of Pb(II) and Cr(III) from aqueous solution on Celtek clay. J Hazard Mater 144:41–46
68. Sarkar SK, Raul KK, Pradhan SS, Basu S, Nayak A (2014) Magnetic properties of graphite oxide and reduced graphene oxide. Physica E 64:78–82. https://doi.org/10.1016/j.physe.2014.07.014
69. Saurín N, Sanes J, Bermúdez M-D (2016) New graphene/ionic liquid nanolubricants. Mater Today Proc 3:S227–S232. https://doi.org/10.1016/j.matpr.2016.02.038
70. Senthilkumar B, Kalai Selvan R, Vinothbabu P, Perelshtein I, Gedanken A (2011) Structural, magnetic, electrical and electrochemical properties of $NiFe_2O_4$ synthesized by the molten salt technique. Mater Chem Phys 130:285–292
71. Seol JH et al (2010) Two-dimensional phonon transport in supported graphene. Science 328:213

72. Sepioni M et al (2010) Limits on intrinsic magnetism in graphene. Phys Rev Lett 105:207205
73. Sharma R, Baik JH, Perera CJ, Strano MS (2010) Anomalously large reactivity of single graphene layers and edges toward electron transfer chemistries. Nano Lett 10:398–405. https://doi.org/10.1021/nl902741x
74. She X, Zhang X, Liu J, Li L, Yu X, Huang Z, Shang S (2015) Microwave-assisted synthesis of Mn_3O_4 nanoparticles@reduced graphene oxide nanocomposites for high performance supercapacitors. Mater Res Bull 70:945–950
75. Shirazian S, Rezakazemi M, Marjani A, Moradi S (2012) Hydrodynamics and mass transfer simulation of wastewater treatment in membrane reactors. Desalination 286:290–295. https://doi.org/10.1016/j.desal.2011.11.039
76. Sitko R et al (2013) Adsorption of divalent metal ions from aqueous solutions using graphene oxide. Dalton Trans 42:5682–5689. https://doi.org/10.1039/C3DT33097D
77. Sreeprasad TS, Maliyekkal SM, Lisha KP, Pradeep T (2011) Reduced graphene oxide–metal/metal oxide composites: facile synthesis and application in water purification. J Hazard Mater 186:921–931
78. Stankovich S et al (2006) Graphene-based composite materials. Nature 442:282–286
79. Sun H, Cao L, Lu L (2011) Magnetite/reduced graphene oxide nanocomposites: one step solvothermal synthesis and use as a novel platform for removal of dye pollutants. Nano Res 4:550–562
80. Sun L, Wang G, Hao R, Han D, Cao S (2015) Solvothermal fabrication and enhanced visible light photocatalytic activity of Cu_2O-reduced graphene oxide composite microspheres for photodegradation of Rhodamine B. Appl Surf Sci 358:91–99
81. Sur UK (2012) Graphene: a rising star on the horizon of materials science. Int J Electrochem 2012: Article ID 237689, 12 pages. http://dx.doi.org/10.1155/2012/237689
82. Szabo T, Nánai L, Nesztor D, Barna B, Malina O, Tombácz E (2018) A simple and scalable method for the preparation of magnetite/graphene oxide nanocomposites under mild conditions. Adv Mater Sci Eng 2018:1–11
83. Tangahu BV, Sheikh Abdullah SR, Basri H, Idris M, Anuar N, Mukhlisin M (2011) A review on heavy metals (As, Pb, and Hg) uptake by plants through phytoremediation. Int J Chem Eng 2011:1–31
84. Tsoukleri G et al (2009) Subjecting a graphene monolayer to tension and compression. Small 5:2397–2402. https://doi.org/10.1002/smll.200900802
85. Ugeda MM, Brihuega I, Guinea F, Gómez-Rodríguez JM (2010) Missing atom as a source of carbon magnetism. Phys Rev Lett 104:096804
86. Urbas K, Aleksandrzak M, Jedrzejczak M, Jedrzejczak M, Rakoczy R, Chen X, Mijowska E (2014) Chemical and magnetic functionalization of graphene oxide as a route to enhance its biocompatibility. Nanoscale Res Lett 9:656
87. Vozmediano MAH, López-Sancho MP, Stauber T, Guinea F (2005) Local defects and ferromagnetism in graphene layers. Phys Rev B 72:155121
88. Wang H et al (2012) Fe nanoparticle-functionalized multi-walled carbon nanotubes: one-pot synthesis and their applications in magnetic removal of heavy metal ions. J Mater Chem 22:9230
89. Wang Y, Huang Y, Song Y, Zhang X, Ma Y, Liang J, Chen Y (2009) Room-temperature ferromagnetism of graphene. Nano Lett 9:220–224. https://doi.org/10.1021/nl802810g
90. Wang Y, Liang S, Chen B, Guo F, Yu S, Tang Y (2013) Synergistic removal of Pb (II), Cd (II) and humic acid by Fe_3O_4@ mesoporous silica-graphene oxide composites. PLoS One 8: e65634
91. Xie L et al (2011) Room temperature ferromagnetism in partially hydrogenated epitaxial graphene. Appl Phys Lett 98:193113. https://doi.org/10.1063/1.3589970
92. Yang S-T et al (2010) Folding/aggregation of graphene oxide and its application in Cu^{2+} removal. J Colloid Interface Sci 351:122–127. https://doi.org/10.1016/j.jcis.2010.07.042
93. Yang Y, Asiri AM, Tang Z, Du D, Lin Y (2013) Graphene based materials for biomedical applications. Mater Today 16:365–373. https://doi.org/10.1016/j.mattod.2013.09.004

94. Yazyev OV (2008) Magnetism in disordered graphene and irradiated graphite. Phys Rev Lett 101:037203
95. Yazyev OV, Helm L (2007) Defect-induced magnetism in graphene. Phys Rev B 75:125408
96. Yu S, Wang X, Tan X, Wang X (2015) Sorption of radionuclides from aqueous systems onto graphene oxide-based materials: a review. Inorg Chem Front 2:593–612
97. Zhang C, Shao Y, Zhu L, Wang J, Wang J, Guo Y (2017) Acute toxicity, biochemical toxicity and genotoxicity caused by 1-butyl-3-methylimidazolium chloride and 1-butyl-3-methylimidazolium tetrafluoroborate in zebrafish (Danio rerio) livers. Environ Toxicol Pharmacol 51:131–137
98. Zhang K, Dwivedi V, Chi C, Wu J (2010) Graphene oxide/ferric hydroxide composites for efficient arsenate removal from drinking water. J Hazard Mater 182:162–168
99. Zhang P, Ma L, Fan F, Zeng Z, Peng C, Loya PE, Liu Z, Gong Y, Zhang J, Zhang X Ajayan PM (2014) Fracture toughness of graphene. Nat Commun 5:3782
100. Zhang W, Shi X, Zhang Y, Gu W, Li B, Xian Y (2013) Synthesis of water-soluble magnetic graphene nanocomposites for recyclable removal of heavy metal ions. J Mater Chem A 1:1745–1753
101. Zhang Y-Y, Pei Q-X, Cheng Y, Zhang Y-W, Zhang X (2017) Thermal conductivity of penta-graphene: the role of chemical functionalization. Comput Mater Sci 137:195–200. https://doi.org/10.1016/j.commatsci.2017.05.042
102. Zhang Y, Small JP, Pontius WV, Kim P (2005) Fabrication and electric-field-dependent transport measurements of mesoscopic graphite devices. Appl Phys Lett 86:073104. https://doi.org/10.1063/1.1862334
103. Zhao G, Li J, Ren X, Chen C, Wang X (2011) Few-layered graphene oxide nanosheets as superior sorbents for heavy metal ion pollution management. Environ Sci Technol 45:10454–10462
104. Zhu J, He J, Du X, Lu R, Huang L, Ge X (2011) A facile and flexible process of β-cyclodextrin grafted on Fe_3O_4 magnetic nanoparticles and host–guest inclusion studies. Appl Surf Sci 257:9056–9062
105. Zhu J et al (2012) Magnetic graphene nanoplatelet composites toward arsenic removal. ECS J Solid State Sci Technol 1:M1–M5
106. Zhu Y, Murali S, Cai W, Li X, Suk Ji W, Potts Jeffrey R, Ruoff Rodney S (2010) Graphene and graphene oxide: synthesis, properties, and applications. Adv Mater 22:3906–3924. https://doi.org/10.1002/adma.201001068

Functionalized Carbon Nanomaterial for Artificial Bone Replacement as Filler Material

Fahad Saleem Ahmed Khan, N. M. Mubarak, Mohammad Khalid and Ezzat Chan Abdullah

1 Introduction

The human race is witnessing lots of remarkable advancement in the field of Science and Engineering. Nanotechnology is one of that remarkable advancement accomplished by mankind. According to Dr. Richard Smalley (Late), "Nanotechnology is the art and science of building stuff that does stuff at the scale of nanometer". In general, nanotechnology is specifically an engineering of human-made structures starting from a range of 1–100 μm [1]. It is an interdisciplinary field that comprises biomedical engineering, chemical engineering, chemistry, physics, and material and particle science. At present more than 600 products available in the market globally which uses nanomaterials for their products [2]. Furthermore, United States National Institute of Health has referred the assistance of nanotechnology for systems like diagnosis, treatment, monitoring, and control of biological as nanomedicine [3]. Nanotechnology is continuously promoting positive impacts on healthcare and making life more convenient than ever imagined

---

F. S. A. Khan · N. M. Mubarak (✉)
Department of Chemical Engineering, Faculty of Engineering and Science,
Curtin University, 98009 Sarawak, Malaysia
e-mail: mubarak.mujawar@curtin.edu.my

M. Khalid
Graphene & Advanced 2D Materials Research Group (GAMRG),
School of Science and Technology, Sunway University,
No. 5, Jalan Universiti, Bandar Sunway, 47500 Subang Jaya,
Selangor, Malaysia

E. C. Abdullah
Department of Chemical Process Engineering, Malaysia-Japan International
Institute of Technology (MJIIT), Universiti Teknologi Malaysia (UTM),
Jalan Sultan Yahya Petra, 54100 Kuala Lumpur, Malaysia

© Springer Nature Switzerland AG 2019
Inamuddin et al. (eds.), *Sustainable Polymer Composites and Nanocomposites*,
https://doi.org/10.1007/978-3-030-05399-4_27

before. In addition, nanomaterials are significantly getting attention for the applications related to bone engineering as it holds properties that have the strength similar to natural bone. However, construction of artificial bone with uniqueness, identical properties likewise natural bone is a strenuous task. Therefore, most of the researchers rely on nanotechnology and nanomaterials when it comes to constructing artificial bone.

Carbon nanomaterial is one of the remarkable outcomes from nanotechnology. Carbon-based nanomaterial considered more accentuated and promising when it comes to the application based on the bio-medical field. Carbon nanomaterials are categorized as low-dimensional materials, having sp^2 and sp^3 carbon atoms arranged into a continuous network [4]. Since the discovery of the first well-known carbon nanomaterial i.e. fullerene, have significantly aroused the interest around the globe. Carbon nanomaterial holds tunable physical, chemical, electrical, optical, thermal, and mechanical properties, and its related composite provides remarkable usage in sensors, biomedicine, electrodes, electrocatalysis, energy storage as well as conversion. One of the main features of carbon nanomaterial i.e. biocompatibility [5] increased its demand in the field of bio-medical. Carbon nanomaterials are one of the highlighted topics when it comes to the modern medical field. Out of all carbon-based nanomaterials, carbon nanotubes (CNTs) are considered more suitable and promising for applications related to bio-medical e.g. nano-electronic bio-sensing [6], drug delivery [7], and bone tissue engineering [8].

Carbon nanomaterials are dimensionally categorized as fullerene (0-D), carbon nanotubes (1-D) and graphene (2-D). The origin of carbon nanotubes is from the synthetic carbon allotropes and categorized as sp^2 hybridized network of carbon atoms. At first, Sumio Iijima discovered the CNTs structure as helical microtubules of graphitic carbon in the year 1991 through an arc discharge process that was initially designed for fullerene production [9]. Since the discovery, significant researches took place on this extraordinary nanomaterial. Theoretically, carbon nanotubes synthesized by rolling the sheets of graphene with connecting hexagonal rings seamlessly. Vapour phase growth, corona-discharge, catalyst-supported growth, hydrocarbons pyrolysis and laser ablation are the conventional techniques for carbon nanotubes synthesis [10]. At present, plasma-enhanced chemical vapour deposition and chemical vapour deposition are the most recent synthesis methods for carbon nanotubes [11] (Fig. 1).

Carbon nanotubes are categorized generally as single-walled carbon nanotubes (SWNTs), multi-walled carbon nanotubes (MWNTs) and double-walled carbon nanotubes (DWNTs) [12]. Major making of MWNTs and DWNTs were testified via arc discharge process. And SWNTs synthesis was primarily described independently by Iijima and Ichihashi (Tokyo, Japan), and Bethune's IBM group (California, USA) in the year 1993 [13]. Furthermore, compared to all other inorganic nanoparticles in which heavy toxic metals are present, for example, quantum dots (Qds); CNTs are mainly composed of pure atoms of carbon that are relatively non-toxic [14]. Thus, carbon nanotubes direct use is restricted due to its biological toxicity in application related to bio-medical. The toxicity of carbon nanotubes are affected due to the metal impurities; shape, structure and length of the

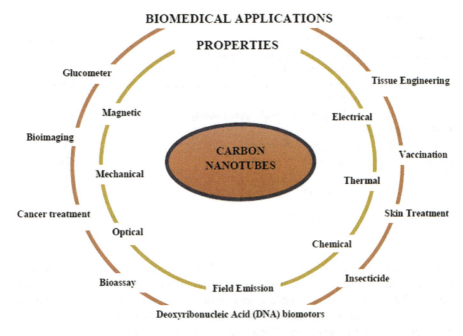

Fig. 1 Properties and biomedical applications of carbon nanotubes

tubes; layers thickness; aggregation degree; etc. Moreover, key factors for the toxicity of nanotubes include metal catalyst residual; length and hydrophobic surface of carbon nanotubes [15]. The metals catalyst (Fe, Ni etc.) containments in nanotubes support free radical generation in cells and are introduced through the process of synthesis and purification which lead to cytotoxicity [16]. In addition, experimental results as evidence clearly show that shorter length of nanotubes, i.e. less than 200 nm, compared to normal length (up to tens of micrometres) is more suitable to enter the cells [14].

Furthermore, synthesized carbon nanotubes hold hydrophobic surface, and are insoluble in many aqueous solutions. Due to the strong hydrophobic interactions, carbon nanotubes stick together and form as bundles that later make it difficult to apart. The formation of bundles trend makes the processability of carbon nanotubes complicated as well extremely difficult their integration into aqueous based biological media, which considered essential for the applications based on bio-medical [17]. Biological behaviours of the tubes precisely based on the surface chemistry. Furthermore, the experimental study proved that raw carbon nanotubes avoid the growth of ovary cells in Chinese hamster [14]. Table 1; listed the theoretical and experimentally calculated properties of carbon nanotubes.

Biomedical refers to the living system; therefore, modification of carbon-based nanomaterial is an important step, like biocompatibility and bioactivity, can be enhanced with chemically modified carbon nanotubes [18]. The modification of carbon nanomaterial includes surface modification as well as chemical alteration

Table 1 Properties of carbon nanotubes (CNTs)

Parameters	Carbon Nanotubes	References
Lattice structure	Nanotubes: ropes, tubes organized in the triangular lattice by parameters of a = 1.7 nm, tube-tube = 0.314	[78]
Elastic modulus	SWNTs and MWNTs $\sim$1 and 1.28 terapascal (TPa), respectively	[79]
Maximum tensile strength	100 gigapascal (GPa), approximately	[79]
Specific gravity	0.8–1.8 gcc^{-1} (theoretical)	[78]
Thermal expansion	Insignificant (theoretical)	[78]
Thermal conductivity	2000 W/m/K	[79]
Oxidation in air	Greater than 700 °C	[78]

steps. Furthermore, stable dispersion helpful for biomedical use, for example, functionalized graphene make it easier for chemotherapeutics and image agent delivery [19]. At present, a countless number of natural and synthetic biopolymers and bio-ceramics are being used for designing artificial bone prosthetic but their usage remained until laboratory. Therefore, designing an artificial bone prosthetic that holds similar properties to that of human bones is a challenging task for the researchers. Furthermore, researchers keeping in mind of getting the radical benefit of nano-technology, nano-materials scaffolds and cell-based biomaterials for constructing artificial bone [20].

2 Bone Structure and Mechanics

Bone is the primary component of the skeletal structure and varies from the convective tissues due to the properties of rigidity and hardness. These properties help the skeleton to keep the shape of the body, cover the main organs, as well as transfer force of muscular contraction at the time of movement [21]. Bones are actually made of the fibrous protein collagen and soak with calcium phosphate like mineral [22]. The content of mineral serves as a reservoir for ions, more generally calcium as human body stored approximately 99% of calcium, which helps in circulating extracellular fluid composition [21]. In addition, bone contains water too, and it is very vital according to mechanical perspective. The organic matrix of the bone composed of collagen and non-collagenous proteins, 90 and 10% respectively [23]. In fact, the collagen proteins especially form-I act as storage for hydroxyapatite, and the mineral phase helps to generate the rigidity of the skeletal structure. However, according to mechanical perspective, bone is considered as an aeolotropic and non-homogeneous material. Furthermore, spongy and cortical bones can be assumed as an orthotropic having 9 and 5 independent material constants, respectively. Bone is possibly imagined as a linear elastic material with significant viscoelastic effects, in terms of the physiological range of loading [24]. Bone is stronger in compression and retaining higher Young's moduli of elasticity [25].

State to the term composite and structure of the bone; it proves that bone is a composite material. Moreover, considering bone as bio-composite, the hierarchical structure is seen at various level. For example, at microstructural level, cortical bone shows osteon (large hollow fibers with an outer diameter of 200–250 μm) fibers which are made up of concentric lamellae and pores. The fibers are comprised of hydroxyapatite mineral and collagen at nano level [26].

Bone is an internal part of the body and is surrounded by cells entire life. Because of having no-expendable nature, the process of resorption and formation for every bone take place at the surface compared to soft biological tissue which can have growth at both interstitials as well as oppositional. Bone is holding varies values of porosity and it mainly depends on bones macrostructure as it has a porous structure. However at the level of macroscopic, cortical and cancellous are two types of bone structure. Cortical bone (compact) covers 80% of the mass in a mature human and is mainly in charge of the protective function of the skeleton [27]. Osteon is a large hollow cylinder and a primary structural unit of compact bone. However, the spongy bone which is a network of narrow rods and plates of calcified bone tissue. This calcified bone tissue is known as trabeculae, covered by the bone marrow. The function of bone marrow is to provide nutrients and discard disposal for the bone cells.

In comparison with cortical bone, the content of mineralization in spongy bone is less. However, experimental studies as evidence show that spongy bone is significantly active in remodelling compared to cortical bone [28]. Furthermore, the primary cellular elements of the bone are osteoclast, osteoblast, osteocytes and bone lining [29]. Osteoblasts and osteoclasts origin are different for nurturing and categorize as temporary cells because of short life length [30].

3 History of Artificial Organ

An artificial organ is a human developed device that is an alternative to natural organ and holding bit identical but not exactly similar properties. The function of developing such devices is to make the life of patient's as convenient as possible. Referred to artificial organ definition, the device is not required to attach to any sources like filling or chemical processing units. Any kind of stationary resources, chemical refilling, exchanging of fillers, attached to the device will not categorize under the artificial organ. For example, dialysis machine, an extraordinary technology to support kidney patient's life is not an artificial organ. However, developing and installing an artificial organ is significantly expensive, and required years of experimental researches compared to the natural organ. The experiments for artificial organs mostly conduct on animals or people who are close to death. As a matter of fact, the word artificial organ rarely heard because mostly it refers to as the replacement of human or animal bones and/or joints. At present, an immense number of artificial organs has been implanted with a high range of success. Some of the remarkable studies on different types of artificial organs have been done by scientist all around the world which include brain pacemakers, cardia, corporta cavernosa, ear (cochlear implant), eye (visual prosthetic), heart, limbs, liver, lungs, bladder, ovaries and many more [31].

The first-ever bone defect treatment was trephined prehistoric skull. However, in the year 1668, a Dutch surgeon Job van Meek'ren named himself first for operating a successful bone defect treatment. He used a dog skull to fill the defect of a soldier's cranium which was taken off after 2 years on soldier's wish. Furthermore, in the scientific era, the interest on osteogenesis and bone transplantation was begun in the year 1739 by Du Hammel. But significant numbers of patients with autogenous bone transplantation were already recorded 200 years earlier [32]. Some of the successful artificial organ transplantation is listed below (cited from 33):

- In the year 1857, Eduard Zirm conducted first successful cornea transplantation (Czech Republic).
- In the year 1954, Joseph Murray operated first successful kidney transplantation (Boston, Massachusetts).
- In the year 1966, Richard Lillehei and William Kelly named themselves first for successful pancreas transplant (Minneapolis, USA).
- In the year 1967, Christiaan Barnard operated first heart transplant successfully (Cape Town, SA).

Moreover, bone transplant referred to fix fractures and joints and to treat skeletal defects. However, there are possible side effects for the transplantation of autogenous bone material like the surgical cut on the skin, low strength bone of donors, and inclined postoperative morbidity. As a matter of fact, not only the amount of autogenous bone is limited but also bone graft has insignificant properties too, like mechanical, biological and physiological [32]. The bone transplants or implants are classified into four classes as:

- Autograft: Engraftment of organs within one individual [33].
- Isograft: Transplantation of organs between genetically similar individuals [34].
- Allograft: Engraftment of organs of genetical individuals of the same species [35]
- Xenograft: Engraftment of organs of individuals of dissimilar species [36].

For bone treatment, two well-known procedures are autograft and allografts in the field of orthopaedic. Due to the limited donors, disease transfer (hepatitis), structural problems, pain etc. [37], autograft and allograft methods are no longer preferred and replaced by artificial bone replacement. Table 2; illustrate the tensile strength, elastic modulus, and ultimate strain of the human bone tissue.

As a matter of fact, strength itself not considered enough for artificial bones to fully integrate with natural bones. In order to compete with natural bone, there are

Table 2 Human bone mechanical properties

Tissue	Tensile strength (MPa)	Elastic modulus (GPa)	Ultimate strain (%)	References
Cortical bone (longitudinally)	130	12.0	3	[80]
Cortical bone (transverse)	60	13.4	1	[80]
Cancellous bone	2	0.39	2.5	[80]

certain characteristics that an artificial bone requires which include good handling, suitable mechanical capability and excellent bio-degradability and bone- regeneration [38].

3.1 Artificial Bone Materials

Biomaterials mainly used to construct devices that are associated with the biological system to co-exist for long-lasting use with limited chance of failures. In 1981, Williams express biomaterials as *"non-viable materials with the application of making medical devices, intended to associate with the biological system"* [39]. For the past few years, the demand for biomaterials has rapidly increased. In the USA, biomaterial industry is generating more than $300 billion annually, and it kept inclining [40]. For many years' medical specialists have been working on finding a substitute for the treatment related to bones like bone repair or replacements. In past, a substance like leather, metals (gold, silver and platinum), bones from other species for direct transplant had utilized for bone repairing treatment. However, alternative materials for bones are categorized as natural and artificial (chemical composition) depending on the source of origin. Natural materials have some remarkable advantages, and due to that popularly used. Additionally, chitosan, collagen, fibrin, and chitin are named as natural materials [39]. And artificial materials are further divided into metal, ceramic, polymer, composites and biological origin substances [40]. Based on host reaction, biomaterials classified as:

- Bio-tolerant: It is in the body, and mainly surrounded by the fibrous membrane, for example, bone cement.
- Bio-inert: These biomaterials are not associated or interact when exposing to biological tissues, for example, titanium oxide.
- Bio-active: These biomaterials integrate with the bone, for example, hydroxyapatite.
- Bio-resorbable: These biomaterials indulge in the bone, for example, calcium.

Furthermore, features such as biocompatibility, elasticity, toughness, corrosion, fatigue resistance and allergic diathesis are vital in bone surgery [41].

4 Carbon Nanomaterials

Carbon nanomaterial is one of the remarkable outcomes from nanotechnology. Carbon nanomaterials are categorized as low-dimensional materials, having sp^2 and sp^3 carbon atoms arranged into a continuous network [4]. Since the discovery of the first well-known carbon nanomaterial in the year 1985 by Sean O'Brien, Richard Smalley, Robert Curl, Harry Kroto, and James Heath, i.e. fullerene, have significantly aroused the public's interest. Carbon nanomaterial tunable physical, chemical, electrical, optical, thermal, and mechanical properties and its related composite provide remarkable usage in sensors, biomedicine, electrodes, electrocatalysis, energy storage as well as conversion. One of the main feature of carbon nanomaterial i.e. biocompatibility [5] that increased its demand in the field of bio-medical. Out of all carbon-based nanomaterials, carbon nanotubes (CNT) consider more suitable and promising for applications related to bio-medical e.g. nano-electronic bio-sensing [6], drug delivery [7], and bone tissue engineering [8].

5 Carbon Nanotubes

1991, an exciting year for carbon science as Sumio Iijima of Nippon Electric Company (NEC) discovered a thin material identical as needle under an electron microscope while analyzing carbon nanomaterial, and named it carbon nanotubes [42]. In the same year, the scientist made the availability of fullerene as a compound too. However, fullerene was first-ever discovered in the year 1985 by researchers from the University of Houston and University of Sussex. Due to the identical shape of this element to football holding thirty-two faces, fullerene was first named as Buckminster-fullerene or Bucky-ball [43]. Carbon nanotubes display as one dimensional with the tubular structure, resemblance to rolled G nano-sheets with minute nanometer thickness. Initially, carbon nanotubes were discovered as multi-walled carbon nanotubes (MWNTs), and later as single-walled carbon nanotubes (SWNTs). Earlier applying electric discharge, laser ablation, and similar techniques used for the synthesis of fullerene but with the addition of some catalyst made the availability of carbon nanotubes in bulk form. CNTs generally have the diameter similar to fullerenes i.e. 1 nm but 1000 time longer than fullerene. These tubes even referred as macromolecules, poly disperses in size and having a molecular weight of 1,000,000 Daltons [44]. Table 3; list the discoveries and development of carbon nanomaterials, mainly carbon nanotubes:

In addition, carbon nanotubes have been studied extensively by researchers due to its promising properties [4] such as electrical, mechanical, conductivity, and chemical which help to assist in the application like bio-sensors, nano-oscillators, drug delivery systems or pure structural components in nano-devices [45].

Table 3 Discoveries and developments of carbon nanotubes (CNTs)

Year	Discovery	References
1889	Discovered carbon filaments (thermal decomposition of gaseous hydrocarbons)	[81]
1890 & 1903	Hydrocarbon (thermal decomposition), results in production of carbon-fibers	[82, 83]
1939	Transmission electron microscopy (TEM) became commercialized which help to research in-depth of carbon fibers	[83]
1952	TEM evidence was printed which display the nano sized diameter of hollow graphitic carbon fiber	[10, 84]
1958	Carbon fibers of bamboo texture were discovered with the help of electron diffraction	[83]
1976	Technique of vapor growth generated carbon fibers	[85]
1979	Arc discharge method synthesized hollow carbon fibers	[86, 87]
1987	In USA, graphitic (hollow carbon fibrils) patent were published	[83]
1991	Arc discharge synthesized carbon nanotubes, and brought into hot spot under scientific society.	[10]
1992	Discovered SWNTs	[46]
1993	Iron catalyst was used to manufacture single walled carbon nanotubes; Cobalt catalyst were used to produced SWNTs	[83]
2004	Templates were applied for carbon micro tubes synthesis	[88]
2007	Carbon nano-buds were manufactured	[83]

Carbon nanotubes properties lead them to applications related to bio-medical. SWNTs and MWNTs consider more suitable for bio-medical applications. Extensive researches have been particularly done on both SWNTs and MWNTs.

5.1 Structure and Properties of Carbon Nanotubes

5.1.1 Single-Walled Carbon Nanotubes (SWNTs)

Single-walled carbon nanotubes (SWNTs), manufactured in 1993 [46], having a diameter and length range from 1 to 7 nm and 20 to 40 nm, respectively [45]. This carbon nanotube type has generated some remarkable properties as well as the capability to apply successfully in a wide number of fields. SWNTs described as a single layer of a graphite crystal, rolled up into a continuous cylinder and capped with hemisphere (carbon rings in hexagonal and pentagonal) at both sides [46]. Since SWNTs hold some impressive properties, therefore, the synthesis of SWNTs has significantly become a matter of global studies. At present, many methods have been designed for the synthesis of SWNTs but electric arc discharge, laser ablation, and catalytic chemical vapour deposition are popular among all [47]. Catalytic chemical vapour deposition abbreviated as CCVD method generates a significant amount of SWNTs at the economic cost compared to all other methods [46]. Moreover, in the market, the vastest method used is laser ablation for SWNTs [48].

5.1.2 Multi-walled Carbon Nanotubes (MWNTs)

In 1952, Radushkevich and Lukyanovich were the original discoverers of MWNTs. Thus, the present carbon nanotube boom in material science without any doubt was instigated by Sumio Iijima in 1991 [49]. Multi-walled carbon nanotubes (MWNTs), extended hollow cylinder, are composed of sp^2 carbon with a diameter range from 2 to 100 nm and growing length up to 10 microns as well as aspect ratio ranges from 10 to 10 million. However, the thickness of MWNTs's wall along the axis remains constant and inner channel is straight. The ends of MWNTs covered by half fullerene spheres due to the channel which is not accessible directly from the outer side but by expanding the nanotube it can be accessed, for example, oxidation, milling or ion beam treatment [50]. However, based on larger diameter and Raman spectrum of MWNTs, clearly differentiate them from SWNTs and double walled CNTs [51]. The techniques to manufacture MWNTs and SWNTs are generally same. Likewise, SWNTs synthesis, a wide variety of methods has been discovered for MWNTs synthesis but the well-known are catalytic chemical vapour deposition, arc discharge, and laser ablation. Likewise, SWNTs, catalytic chemical vapour deposition is the most efficient and most widely used process (Table 4).

Table 4 Single-walled and multi walled CNTs comparison

Single-walled carbon nanotube (SWNTs)	Multi-walled carbon nanotubes (MWNTs)	References
Mono-layer of graphene	Several layers of graphene	[89]
The catalyst is not optional for synthesis	For the synthesis catalyst not required	[79]
Appropriate control over growth as well as on atmospheric condition is compulsory for bulk synthesis	Bulk synthesis is convenient	[90]
Lack of purity	Negligible impurity	[79,90]
Chance of defect is higher at the time of functionalization	Chance of defect is limited but if occurred, it is a challenging task to improve	[90]
Convenient for characterization and evaluation	Holds a very complex structure	[79, 90]
Comparatively flexible and easy to twist	Difficult to be twisted	[90]

5.2 Synthesis of Carbon Nanotubes

Carbon nanotubes can be manufactured by a range of methods but well-known include laser ablation, arc discharge, and chemical vapour deposition. Processes like laser ablation, arc discharge and chemical vapour discharge are fallen under the category of physical and chemical processes, respectively [52]. In the process of laser ablation, high power laser is used which helps to vaporize graphite source combined with a metal catalyst. Since graphite contains carbon that transforms large numbers of single-walled carbon nanotubes (SWNTs) on the metal catalyst [53]. And in arc discharge method, high quality but limited quantities of SWNTs and MWNTs are manufactured by introducing electric discharge from the electrode (carbon-based). However, in chemical vapour deposition (CVD), carbon nanotubes are produced in a chamber by reacting hydrocarbons (for example CH_4) with an appropriate metal catalyst [48]. All of these methods come with advantages and limitations. Some of them listed in Table 5 and Fig. 2.

6 Functionalization of Carbon Nanomaterials

Carbon nanomaterials, fullerenes, graphene and carbon nanotubes (SWNTs, MWNTs & DWNTs), in organic (especially in an aqueous solvent) are poorly soluble. Like, carbon nanotubes stick together and from as bundles due to strong

Table 5 Carbon nanotubes (CNTs) synthesis method

Synthesis methods	Advantages	Disadvantages	References
Chemical-vapor deposition (CVD)	Cost-efficient; scalability; continuous operation process; low operating temperature around 500–800 °C; diameter can be adjusted	Lack of quality i.e. Wall structure contains significant defects and deposits of carbonaceous contamination; a combination of SWNTs and MWNTs	[79, 91]
Arc-discharge (AR)	Simplicity and versatility on the basis of carbon-based material and catalyst; better quality nanotubes produce; minor defects; economical process	Significant consumption of energy; lack of capability for industrial up-scaling; high temperature required (>1700 °C); shorter length of CNTs produced	[10, 73]
Laser ablation (LA)	Synthesis at room temperature; generate high purity and yield of tubes	Purification of the crude product is must; method restricts till laboratory scale; expensive process	[79, 73]

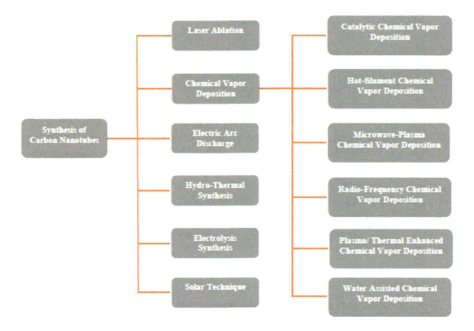

Fig. 2 Synthesis approaches of carbon nanotubes

hydrophobic interactions which later make it difficult to apart. The formation of bundles trend makes the processability of carbon nanomaterials complex and challenging their integration into aqueous based biological media, which considered essential for the applications related to bio-medical [17]. Therefore, functionalization of carbon nanomaterials is important in order to attain their remarkable strengths. As a result, a significant part of the current journals on carbon nanomaterials has an emphasis on improving their solubilization and dispersion using functionalization methods. In fact, massive range of applications in many fields has opened through functionalized carbon nanomaterials, particularly carbon nanotubes and graphene because of their extraordinary properties like lightweight, significant aspect ratio, and electrical conductivity, as well as mechanical, thermal strength. Furthermore, carbon nanomaterials uses have become applicable in many fields because of the modification of its surfaces. The exceptional physiochemical properties of functionalized carbon nanomaterials have been used for anti-viral drug development, energy, treatment for cancer, applications for biotechnological, and aerospace. Moreover, theoretical work has also been done to understand as well as optimize functionalization. On the other hand, non-functionalized carbon nanomaterials carry limitations like the capability to generate stable bundles/aggregates due to highly significant interactions, e.g. van der Waals force, strong π–π stacking. Consequences of the aggregation bring undesirable changes, like affect the aspect ratio plus declines the nanocomposites properties. Therefore, stable dispersion and functionalized role are vital in order to attain better nanomaterial support system.

The functionalization of nanomaterials with the mandatory moieties relies on the base material chemistry. And the functionalization mode on nano's surface mainly relies on the problem and proposed use of the material.

In the past, many procedures had discovered; among them, the well-known that still applies today include covalent and non-covalent functionalization. All these approaches are with organic and inorganic compounds help to attain improved solubility and dispersibility. For example, the use of pristine carbon nanotubes, particularly in living organism and cells bring toxic effects, therefore functionalization is essential to be concerned [17]. Furthermore, due to the availability of carboxylic, epoxy and hydroxyl groups, covalent functionalization is more preferable. In addition, covalent functionalization suited as better moieties for functional group conversion. Besides this, the existence of an sp^2 hybridized π network gives the chance for non-covalent interaction among host species and carbon nanomaterials [54]. Thus, modifying the surface of carbon nanomaterial in order to retain the remarkable properties is a challenging task. At present, different functionalization techniques for both covalent and non-covalent have developed for carbon nanomaterials.

6.1 Covalent Approach for CNTs

Lately, carbon nanotubes have received significant interest because of its astonishing unique electronic, optical, mechanical, thermal, and structural properties [55], especially for medical and material science fields. But the insolubility of carbon nanotubes is a chief barrier for its use in real world. The insolubility of carbon nanotubes is generally due to its hydrophobic structure, surface area, forces like Van der Waals, etc. There is only one solution for the carbon nanotubes issues (insolubility and poor dispersion) which scientist discovered, and that is a modification of the carbon nanotubes [56]. Furthermore, there are numerous approaches for covalent functionalization of carbon nanotubes among other oxidation reactions, oxidized carbon nanotubes with esterification and amidation reactions, halogenation, ozonolysis, plasma activation, electrophilic/nucleophilic additions, mechanic/electrochemical functionalization, polymer grafting, nucleophilic/radical additions, treatment with ionic liquid and oxidized carbon nanotubes with complexation reactions [57]. However, for biomedical applications, strategies developed particularly for covalent functionalization include surface oxidation of carbon nanotubes, radical additions, cycloaddition reactions and its followed functionalization with biologically related molecules [58].

6.1.1 Oxidation Treatment

Covalent modification approach with various oxidizing acids/oxidants is most common because of their capability of attaching the preferred function groups, hydroxyl or carboxyl, on the surface of carbon nanotubes as well as tailoring the surface composition. Carbon nanotubes with the functional groups attached not only provide a positive effect for their dispersibility in different solvents but also the contact with varies compounds, for example, polymers. Some of the well-known oxidizing agents for the oxidation of carbon nanotubes are sulphuric acid (H_2SO_4), nitric acid (HNO_3), tri-oxygen (O_3), hydrogen peroxide (H_2O_2) and potassium permanganate ($KMnO_4$). Oxidizing agents like H_2SO_4, HNO_3 and $KMnO_4$ generate carboxyl groups while H_2O_2 and O_3 produce hydroxyl groups into the nanotubes [59]. However, hydrogen peroxide and tri-oxygen are categorized under mild oxidant. As an example covalent amide connection established when a mixture of single-walled carbon nanotubes (oxidized) and PEG-NH_2 sonicated for the duration of 30 min plus centrifugation to generate NH_2-PEG-modified carbon nanotubes. Moreover, oxidation of single-walled carbon nanotubes with addition covalent conjugation with amino acids shows sp^3 carbon atoms [14]. However, the outcome of two oxidizing agents generates two functional groups connect con-currently on the nanotubes. Further functionalization makes these groups to be used as anchor sites either for esterification or amidation reactions which can be applicable widely for the conjugation of water-soluble organic molecules, hydrophilic polymers such as nucleic acids, peptides or polyethylene glycol, as a result, multifunctional carbon nanotubes manufactured [60]. Multifunctional carbon nanotubes possess greater chemical reactivity as well as selectivity, therefore considered better and preferable compared to other functionalized carbon nanotubes [61]. Another method for the oxidation of graphene also allowed as an alternative for the carbon nanotubes oxidation. As graphene and MWNTs both have sp^2-hybridized carbons, therefore it makes them easier to oxidize. The most well-known approach for the oxidation of graphene is Hummers. As a matter of fact, various improved Hummers method has been used for unzipping carbon nanotubes with well-developed oxidation mechanisms. Consequence of mechanical mixing surely enhances the dispersion but also cause damage and length shorten of carbon nanotubes. Since, scanning electron microscope (SEM) is used for the morphology of MWNTs before and after oxidation treatment. With an outer diameter range from 10 to 20, 30 to 50, and >50 nm clearly showed varies in tube diameter as well as roughness along the walls of the tube after gone through acids treatment. In addition, MWNTs diameter slowly narrowed due the oxidation process (cited from [62]). However, treatment like ultraviolet [63], plasma [64] and microwave irradiation have been used either for incorporation of surface oxygen/for the attachment of various functional groups to the carbon nanotubes. In terms of mechanical mixing, for example, conventional reflux method is usually used at the time of oxidation. In comparison with reflux, functionalization of carbon nanotubes with ultra-sonication use may help to generate carboxyl, carbonyl and hydroxyl content greater on the MWNTs. It is because sonication method develops the greater surface area as well as defects sites for the functional group [65]. Experimental results showed that the ultra-sonication help to decrease the stacking of carbon nanotubes in bundles, and destroyed the tubes structure integrity [59].

Some of the drawbacks of oxidation treatment are that it grows defects on nanotubes surface, CNTs oxidization leading Hole doping, and introduce impurity states at the Fermi level. On the other hand, oxidation treatment helps to remove the raw materials impurities, cut/shorten the length and expand the CNTs. However, minimizing the CNTs length through oxidation treatment comprehensively based on the rate and extent of the reaction which let the rise of new length distribution. Carbon nanotubes with open and oxygenated ends are the consequences of cutting mechanism [66]. However, the oxidative stability particularly depends on diameter and production process of the carbon nanotubes.

6.1.2 Cycloaddition Reaction

Cycloaddition reaction is another well-known process that includes direct additional reaction, for example, 1, 3-Dipolar, carbene, nitrene etc. In cyclo-addition reactions, sidewalls are the primary objective where reaction occurred. The outcome of this covalent approach upgrades carbon nanotubes strength as it the helps to enhance the solubility of water, different organic solvents etc. [58]. However, analogues of carbenes are nitrenes, compounds of electrophile reagents which add-up C=C bond. Thermolysis of alkyl azidoformates required for the functionalization of carbon nanotubes on the side walls through nitrenes. Development of pyrrolidine rings on the surface of carbon nanotubes with the use of 1, 3-Dipolar cycloaddition of azomethine ylides [67]. For many biomedical applications, these pyrrolidine rings can be substituted with various functional groups, for example, peptides, therapeutic agents, fluorescent molecules, etc. [58]. Like nitrenes, functionalization of carbon nanotubes using carbenes followed the same path. At first, deprotonation of imidazolinium cation produced nucleophilic carbenes, and then each −ve charge/moiety is replaced to the tubes, and at last carbon nanotubes functionalized [67]. Moreover, this approach sub-categorized with varies direct additional reaction, like amidation, esterification, thiolation, halogenation, and hydrogenation [68]. Like in the process of thiolation, CNTs involved with the thiol group through subsequent carboxylation sonication, reduction with the supports of $NaBH_4$, and chlorination with $SOCl_2$, led by thiolation with Na_2S and NaOH mixture to the CNT open end. The aftermath of cycloaddition reaction, enhance the strength of nanotubes and improve the solubility in aqueous solution, many organic solvents etc. [58].

6.1.3 Radical-Additions

Radical addition reaction falls under the category of covalent functionalization approach. A chemical agent such as aryl diazonium, sodium nitrite etc. is involved in radical addition reaction. Among these chemical agents, aryl diazonium coupling is more preferable due to its easiness and higher yield. In aryl diazonium coupling, first prepared aryl diazonium salts [58]. In 1858, Peter Griess discovered aryl diazonium compound while he was manufacturing a product. The word 'diazo'

Table 6 Surface modification approaches for carbon nanotubes using aryl diazonium salts

Approaches	Materials	Mechanism	References
Chemical	SWNTs and MWNTs	Under iso-amyl nitrite, in-situ produced aryl-diazonium salts, reaction temperature 60 °C	[70]
	Carbon nanotubes	Aryl-diazonium reduction through H_3PO_2 (hypo phosphorus)	[70]
	Carbon black	In-situ produced aryl-diazonium salts in water under $NaNO_2$ and excess HCl	[70, 92]
Microwave	SWNTs and MWNTs	Arene radical reaction	[70]
Thermal	SWNTs	The mechanism was not notified in the manuscript	[70, 93]

originated from French and means di-nitrogen [69]. Furthermore, Jean Pinson, the modern surface chemistry father, could be accredited for aryl diazonium salts. Since its discovering, this salt has significantly considerable for organic synthesis of a number of vital compounds, for example long known azo dyes. This salt has become an ideal diazonium salt for the illustration of proof of new concepts, for example, modification of the surface, bio-sensor, clay related nano-fillers etc. At present, commercially available aryl diazonium include 4-nitrobenzenediazonium tetra-fluoroborate, 4-bromobenzenediazonium tetra-fluoroborate, 4-formyl benzene diazonium hexafluorophosphate and 4-aminodiphenylamine diazonium sulfate. However, aryl diazonium can be prepared by the introduction of $NaNO_2$ aqueous solution to aromatic amine solution under additional hydrochloric acid which later cooled down below 5 °C in a water bath [70]. In addition, CNTs can be envisaged through modification with isolated aryl diazonium and in situ prepared aryl diazonium from aromatic amines under $NaNO_2$ (medium acidic). On the other hand, in situ prepared aryl diazonium which guarantees significant grafting density of aryl groups on CNTs through pure iso-amyl-nitrite. This alternative route generates functionalized CNTs of higher solubility and processability in organic solvents (functionalization degree of one out of twenty atoms of carbon) and polymeric blends, respectively. This approach proved to be effective due to less time for the reaction as well as less consumption of solvent. Scalable is an option under this approach therefore paving the way for its use in more application, for example, growth of structural materials. Therefore, this approach seldom considered for application related to biological [58]. Table 6; listed the approaches and reaction mechanism for the modification of carbon nanotubes using aryl diazonium salts.

6.2 Non-covalent Approach for CNTs

Some other advantages of using non-covalent approach include biocompatibility, stability in very biological solutions and a functional group that required additional bio-conjugation. Lately, long-range single layers graphene has been prepared using non-covalent approach (π–π stacking) for a wide range of application. Since carbon nanotubes hold significant specific area which benefits them with high loading capacity along the following molecules. Through the route of non-covalent functionalization, carbon nanotubes have been positively covered/absorbed by amphiphilic molecules such as pyrene, derivatives of naphthalene, proteins, RNA, DNA, polymers, peptides, and surfactants [71]. Carbon nanotubes dispersion generally relies on chemical characteristics, amphiphilic molecules and solvent category and concentration as well as dispersing conditions [58]. Lightweight surfactant, few organic molecules as well as amphiphilic polymers are used as agents for non-covalent approach functionalization. Molecules of surfactants for the carbon nanotubes modification react with carbon nanotubes hydrophobic side and results in replacing to hydrophilic. This interaction of surfactants with carbon nanotubes makes them soluble not only in water but also in the wide range of solvents [72]. Synthetic peptides use also another route to disperse carbon nanotubes, as they capable to cover and solubilize the tubes. Some other surfactants capable to disperse carbon nanotubes include Odium dodecyl sulfate, cetyl-trimethyl ammonium bromide, non-ionic Tween-20 [73], Gum Arabic, salmon sperm DNA, chitosan [74]. Among all carbon nanotubes types, particularly for functionalization multi-walled carbon nanotubes are considered suitable for this approach as the damage is slightly low [75].

Moreover, functionalization of carbon nanotubes through non-covalent approach using meso-porphine and 5, 15-bis-porphyrin enhance the properties of tubes like biocompatibility, solubility in aqueous solution, luminescence etc. [73]. Among all poly-ethylene-mine (a hydrophilic character with polar head) use for carbon nanotubes functionalization and improve nanotubes water solubility significantly [76].

Furthermore, aromatic molecules are more strongly absorbed on the surface of graphite compared to aliphatic because of the π–π interaction among graphite surface and amphiphilic molecules of aromatic components as well as electrostatic and van der Waals interaction involvement [77]. This effect noticed in pyrene-containing molecules which is a non-covalent modification agent. In addition, to attaining successful dispersion of carbon nanotubes, surfactants (cationic, anionic/non-ionic) have been widely used. However, hydrophobic regions length and hydrophilic structure of the surfactants are the factors on which guaranteed carbon nanotubes dispersion based. Moreover, higher critical mi-cellar concentration, minor stability and partial interaction with cellular proteins are few limitations surfactants face in a biological environment. Several of these limitations can be avoided through the PEG-modified phospholipids [58] use which holds different functional groups that required additional functionalization with targeting and therapeutic molecules.

7 Conclusion and Perspectives

The discovery of carbon nanomaterial, particularly CNTs has aroused the attention of many researchers to study in-depth the capability of this remarkable material. In addition, functionalization approaches open completely new chapter for considering CNTs in many application where initially restricted to use. However, massive successful researches have carried out, and the outcome of these researches categorized CNTs as distinct biomaterials that hold potential to medical applications in bone engineering and orthopaedics techniques due to their outstanding capability of accelerating bone repair/restoration. Additionally, experimental results reflect that f-CNTs are capable to transmit cells without apparent cytotoxicity that eligible f-CNTs to use as a delivery vehicle for a range of biologically energetic molecules such as drugs, deoxyribonucleic/ribonucleic acid, and protein. Moreover, functionalization techniques assist in removing metallic containments and enhance the biocompatibility of CNTS, and allow them to be considered as a promising material for application related to bone tissue engineering.

In the past, CNTs were used with polymers but due to the issues like toxicity, aggregation restricted its use with polymers. Some of the experiments conducted in the past based on CNTs-polymers were MWCNT-Polycaprolactone, Polypropylene-MWCNT-nHA, Sodium hyaluronate-SWCNTs etc. However, functionalization of CNTs will certainly solve such issues and will allow its use as filler material with different polymers, and consider them appropriate for bone defects, replacement or loss.

References

1. Bawa R, Audette GF, Rubinstein I (2016) Handbook of clinical nanomedicine: nanoparticles, imaging, therapy, and clinical applications. CRC Press, Boca Raton
2. Adlakha-Hutcheon G, Khaydarov R, Korenstein R, Varma R, Vaseashta A, Stamm H et al (2009) Nanomaterials, nanotechnology. Nanomaterials: Risks and Benefits. Springer, Berlin, pp 195–207
3. Schaefer H-E (2010) Nanoscience: the science of the small in physics, engineering, chemistry, biology and medicine. Springer, Berlin Heidelberg, pp 615–735
4. Yang Y, Yang X, Yang Y, Yuan Q (2018) Aptamer-functionalized carbon nanomaterials electrochemical sensors for detecting cancer relevant biomolecules. Carbon 129:380–395
5. Liu Y, Dong X, Chen P (2012) Biological and chemical sensors based on graphene materials. Chem Soc Rev 41(6):2283–2307
6. Trung TQ, Lee NE (2016) Flexible and stretchable physical sensor integrated platforms for wearable human-activity monitoring and personal healthcare. Adv Mater 28(22):4338–4372
7. Yang W, Ratinac KR, Ringer SP, Thordarson P, Gooding JJ, Braet F (2010) Carbon nanomaterials in biosensors: should you use nanotubes or graphene? Angew Chem Int Ed 49 (12):2114–2138
8. Weiss NO, Zhou H, Liao L, Liu Y, Jiang S, Huang Y et al (2012) Graphene: an emerging electronic material. Adv Mater 24(43):5782–5825

9. Backes C (2012) Introduction: noncovalent functionalization of carbon nanotubes: fundamental aspects of dispersion and separation in water. Springer, Berlin Heidelberg, pp 1–37
10. Zamolo VA, Vazquez E, Prato M (2013) Carbon nanotubes: synthesis, structure, functionalization, and characterization. In: Siegel JS, Wu Y-T (eds) Polyarenes II. 350, pp 65–109, Springer, Cham
11. Yadav Y, Kunduru V, Prasad S (2008) Carbon nanotubes: synthesis and characterization. In: Morris JE (ed) Nanopackaging: nanotechnologies and electronics packaging. Springer US, Boston, MA, pp 325–344
12. Rezakazemi M, Amooghin AE, Montazer-Rahmati MM, Ismail AF, Matsuura T (2014) State-of-the-art membrane based CO_2 separation using mixed matrix membranes (MMMs): an overview on current status and future directions. Prog Polym Sci 39(5):817–861
13. Kong J, Zhou C, Morpurgo A, Soh HT, Quate CF, Marcus C et al (1999) Synthesis, integration, and electrical properties of individual single-walled carbon nanotubes. Appl Phys A 69(3):305–308
14. Sun H, She P, Lu G, Xu K, Zhang W, Liu Z (2014) Recent advances in the development of functionalized carbon nanotubes: a versatile vector for drug delivery. J Mater Sci 49(20):6845–6854
15. Liu Y, Zhao Y, Sun B, Chen C (2012) Understanding the toxicity of carbon nanotubes. Acc Chem Res 46(3):702–713
16. Schafer FQ, Qian SY, Buettner GR (2000) Iron and free radical oxidations in cell membranes. Cellular and molecular biology (Noisy-le-Grand, France) 46(3):657
17. Basiuk EV, Basiuk VA (2015) Solvent-free functionalization of carbon nanomaterials. In: Basiuk VA, Basiuk EV (eds) Green processes for nanotechnology: from inorganic to bioinspired nanomaterials. Springer, Cham, pp 163–205
18. Krishna V, Stevens N, Koopman B, Moudgil B (2010) Optical heating and rapid transformation of functionalized fullerenes. Nat Nanotechnol 5(5):330
19. Bai RG, Ninan N, Muthoosamy K, Manickam S (2017) Graphene: a versatile platform for nanotheranostics and tissue engineering. Progress in Materials Science
20. Egli RJ, Luginbuehl R (2012) Tissue engineering-nanomaterials in the musculoskeletal system. Swiss Med Wkly 142:w13647
21. Cowin SC (2001) Bone mechanics handbook. CRC Press, Boca Roton
22. Currey J (2002) Bones: structure and mechanics. Princeton University Press, Princeton, NJ
23. Behari J (1991) Solid state bone behaviour. Prog Biophys Mol Biol 56(1):1–41
24. Rouhi G (2006) Theoretical aspects of bone remodeling and resorption processes. Ph.D. Thesis, University of Calgary
25. Bartel D, Davy D, Keaveny T (2006) Orthopaedic biomechanics mechanics and design in musculoskeletal systems. Pearson Education Inc., Upper Saddle River
26. Lakes R, Saha S (1979) Cement line motion in bone. Science 204(4392):501–503
27. van der Meulen MC (2000) Mechanics in skeletal development, adaptation and disease. Philos Trans Royal Soc Lond A Math Phys Eng Sci 358(1766):565–578
28. Guldberg R, Caldwell N, Guo X, Goulet R, Hollister S, Goldstein S (1997) Mechanical stimulation of tissue repair in the hydraulic bone chamber. J Bone Miner Res 12(8):1295–1302
29. Burger EH, Klein-Nulend J (1999) Mechanotransduction in bone—role of the lacuno-canalicular network. FASEB J 13(9001):S101–S12
30. Parfitt A (1995) Problems in the application of in vitro systems to the study of human bone remodeling. Calcif Tissue Int 56(1):S5–S7
31. Standring S (2015) Gray's anatomy e-book: the anatomical basis of clinical practice. Elsevier, Amsterdam
32. Patka P, Haarman HJTM, van der Elst M, Bakker FC (2000) Artificial bone. In: Wise DL, Trantolo DJ, Lewandrowski K-U, Gresser JD, Cattaneo MV, Yaszemski MJ (eds) Biomaterials engineering and devices: human applications, vol 2, Orthopedic, Dental, and Bone Graft Applications, pp 95–109. 2 Totowa, Humana Press, NJ

33. Autograft (2001) In: Schwab M (ed) Encyclopedic reference of cancer, p 83. Springer, Berlin, Heidelberg
34. Isograft (2001) In: Schwab M (ed). Encyclopedic reference of cancer, p 468. Springer, Berlin, Heidelberg
35. Allograft (2001) In: Schwab M (ed) Encyclopedic reference of cancer, p 38. Springer, Berlin Heidelberg
36. Kabbashi N, Jamal Ibrahim D, Rosli NF (2011) Statistical analysis for removal of cadmium from aqueous solution at high pH. Aust J Basic Appl Sci 5(6):440–446
37. Syahrom A, Kadir MRA, Abdullah J, Öchsner A (2013) Permeability studies of artificial and natural cancellous bone structures. Med Eng Phys 35(6):792–799
38. Saijo H, Kanno Y, Mori Y, Suzuki S, Ohkubo K, Chikazu D et al (2011) A novel method for designing and fabricating custom-made artificial bones. Int J Oral Maxillofac Surg 40(9):955–960
39. Kamachimudali U, Sridhar T, Raj B (2003) Corrosion of bio implants. Sadhana 28(3–4):601–637
40. Trebše R (2012) Biomaterials in artificial joint replacements. Infected total joint arthroplasty. Springer, Berlin, pp 13–21
41. Virtanen S, Milošev I, Gomez-Barrena E, Trebše R, Salo J, Konttinen Y (2008) Special modes of corrosion under physiological and simulated physiological conditions. Acta Biomater 4(3):468–476
42. De Volder MF, Tawfick SH, Baughman RH, Hart AJ (2013) Carbon nanotubes: present and future commercial applications. Science 339(6119):535–539
43. Grace T (2003) An introduction to carbon nanotubes. Summer, Stanford University
44. Pénicaud A (2014) Solubilization of fullerenes, carbon nanotubes, and graphene. Making and exploiting fullerenes, graphene, and carbon nanotubes. Springer, Berlin, pp 1–35
45. Rao CK, Rao L (2017) Critical velocities in fluid-conveying single-walled carbon nanotubes embedded in an elastic foundation. J Appl Mech Tech Phys 58(4):743–752
46. Yu O, Daoyong L, Weiran C, Shaohua S, Li C (2009) A temperature window for the synthesis of single-walled carbon nanotubes by catalytic chemical vapor deposition of CH_4 over Mo_2-Fe 10/MgO catalyst. Nanoscale Res Lett 4(6):574
47. Qingwen L, Hao Y, Yan C, Jin Z, Zhongfan L (2002) A scalable CVD synthesis of high-purity single-walled carbon nanotubes with porous MgO as support material. J Mater Chem 12(4):1179–1183
48. Ahmed W, Jackson MJ (2016) Surgical tools and medical devices. Springer, Berlin
49. Radushkevich L, Lukyanovich V (1952) Carbon structure formed under thermal decomposition of carbon monoxide on iron. Zh Fiz Khim 26(1):88–95
50. Shin Y-H, Song J-W, Lee E-S, Han C-S (2007) Imaging characterization of carbon nanotube tips modified using a focused ion beam. Appl Surf Sci 253(16):6872–6877
51. Vajtai R (2013) Springer handbook of nanomaterials. Springer Science & Business Media, Berlin
52. Huczko A (2002) Synthesis of aligned carbon nanotubes. Appl Phys A 74(5):617–638
53. Chauhan SK, Shukla A, Dutta S, Gangopadhyay S, Bharadwaj LM (2012) Carbon nanotubes for environmental protection. Springer, Environmental Chemistry for a Sustainable World, pp 83–98
54. Syrgiannis Z, Melchionna M, Prato M (2015) Covalent carbon nanotube functionalization. In: Kobayashi S, Müllen K (eds) Encyclopedia of polymeric nanomaterials. Springer, Berlin Heidelberg, pp 480–487
55. Yang Y, Qiu S, Xie X, Wang X, Li RKY (2010) A facile, green, and tunable method to functionalize carbon nanotubes with water-soluble azo initiators by one-step free radical addition. Appl Surf Sci 256(10):3286–3292
56. Mananghaya MR, Santos GN, Yu DN (2017) Solubility of amide functionalized single wall carbon nanotubes: a quantum mechanical study. J Mol Liq 242:1208–1214
57. Giliopoulos DJ, Triantafyllidis KS, Gournis D (2013) Chemical functionalization of carbon nanotubes for dispersion in epoxy matrices. In: Paipetis A, Kostopoulos V (eds) Carbon

nanotube enhanced aerospace composite materials: a new generation of multifunctional hybrid structural composites, pp 155–183. Springer: Dordrecht, Netherlands
58. Erol O, Uyan I, Hatip M, Yilmaz C, Tekinay AB, Guler MO (2017) Recent advances in bioactive 1D and 2D carbon nanomaterials for biomedical applications. Nanomedicine: Nanotechnology, Biology and Medicine
59. Liang S, Li G, Tian R (2016) Multi-walled carbon nanotubes functionalized with a ultrahigh fraction of carboxyl and hydroxyl groups by ultrasound-assisted oxidation. J Mater Sci 51(7):3513–3524
60. Battigelli A, Ménard-Moyon C, Da Ros T, Prato M, Bianco A (2013) Endowing carbon nanotubes with biological and biomedical properties by chemical modifications. Adv Drug Deliv Rev 65(15):1899–1920
61. Zhao Z, Yang Z, Hu Y, Li J, Fan X (2013) Multiple functionalization of multi-walled carbon nanotubes with carboxyl and amino groups. Appl Surf Sci 276:476–481
62. Khani H, Moradi O (2013) Influence of surface oxidation on the morphological and crystallographic structure of multi-walled carbon nanotubes via different oxidants. J Nanostruct Chem 3(1):73
63. Martín O, Gutierrez HR, Maroto-Valiente A, Terrones M, Blanco T, Baselga J (2013) An efficient method for the carboxylation of few-wall carbon nanotubes with little damage to their sidewalls. Mater Chem Phys 140(2–3):499–507
64. Zschoerper NP, Katzenmaier V, Vohrer U, Haupt M, Oehr C, Hirth T (2009) Analytical investigation of the composition of plasma-induced functional groups on carbon nanotube sheets. Carbon 47(9):2174–2185
65. Saito T, Matsushige K, Tanaka K (2002) Chemical treatment and modification of multi-walled carbon nanotubes. Physica B 323(1–4):280–283
66. Dillon AC, Gennett T, Jones KM, Alleman JL, Parilla PA, Heben MJ (1999) A simple and complete purification of single-walled carbon nanotube materials. Adv Mater 11(16):1354–1358
67. Morelos-Gómez A, Tristán López F, Cruz-Silva R, Vega DÃ-az SM, Terrones M (2013) Modified carbon nanotubes. In: Vajtai R (ed) Springer Handbook of Nanomaterials, pp 189–232. Springer, Berlin Heidelberg
68. Hirsch A, Vostrowsky O (2005) Functionalization of carbon nanotubes. Functional molecular nanostructures. Springer, Berlin, pp 193–237
69. Trusova ME, Kutonova KV, Kurtukov VV, Filimonov VD, Postnikov PS (2016) Arenediazonium salts transformations in water media: coming round to origins. Resour Efficient Technol 2(1):36–42
70. Mohamed AA, Salmi Z, Dahoumane SA, Mekki A, Carbonnier B, Chehimi MM (2015) Functionalization of nanomaterials with aryldiazonium salts. Adv Coll Interface Sci 225:16–36
71. Backes C, Hirsch A (2010) Noncovalent functionalization of carbon nanotubes. Wiley, Chichester, UK, pp 1–48
72. Composites C. Functionalization of CNTs 2018 (cited 11 Mar 2018). Available from: https://sites.google.com/site/cntcomposites/functionalization-of-cnts
73. Ferreira FV, Cividanes LDS, Brito FS, de Menezes BRC, Franceschi W, Simonetti EAN et al (2016) Functionalization of carbon nanotube and applications. Functionalizing Graphene and carbon nanotubes: A review. Springer, Cham, pp 31–61
74. Bianco A, Sainz R, Li S, Dumortier H, Lacerda L, Kostarelos K et al (2008) Biomedical applications of functionalised carbon nanotubes. In: Cataldo F, Da Ros T (eds) Medicinal chemistry and pharmacological potential of fullerenes and carbon nanotubes. Springer, Dordrecht, Netherlands, pp 23–50
75. Kasperski A, Weibel A, Estournès C, Laurent C, Peigney A (2014) Multi-walled carbon nanotube–Al_2O_3 composites: covalent or non-covalent functionalization for mechanical reinforcement. Scripta Mater 75:46–49

76. Behnam B, Shier WT, Nia AH, Abnous K, Ramezani M (2013) Non-covalent functionalization of single-walled carbon nanotubes with modified polyethyleneimines for efficient gene delivery. Int J Pharm 454(1):204–215
77. Sanz V, Borowiak E, Lukanov P, Galibert AM, Flahaut E, Coley HM et al (2011) Optimising DNA binding to carbon nanotubes by non-covalent methods. Carbon 49(5):1775–1781
78. Jeon I-Y, Chang DW, Kumar NA, Baek J-B (2011) Functionalization of carbon nanotubes. Carbon nanotubes-Polymer nanocomposites: InTech
79. Eatemadi A, Daraee H, Karimkhanloo H, Kouhi M, Zarghami N, Akbarzadeh A et al (2014) Carbon nanotubes: properties, synthesis, purification, and medical applications. Nanoscale Res Lett 9(1):393
80. Pal S (2014) Biomaterials and its characterization. Design of artificial human joints & organs. Springer US, Boston, MA, pp 51–73
81. Tibbetts GG (2001) Vapor-grown carbon fiber research and applications: achievements and barriers. Carbon filaments and nanotubes: common origins, differing applications?. Springer, Berlin, pp 1–9
82. Ci L, Wei J, Wei B, Xu C, Liang J, Wu D (2000) Novel carbon filaments with carbon beads grown on their surface. J Mater Sci Lett 19(1):21–22
83. Ren Z, Lan Y, Wang Y (2012) Aligned carbon nanotubes: physics, concepts, fabrication and devices. Springer Science & Business Media, Berlin
84. Ko FK, Kuznetsov V, Flahaut E, Peigney A, Laurent C, Prinz VY, et al (2004) Formation of nanofibers and nanotubes production. Nanoeng Nanofibrous Mater. 169:1–129. Springer, Berlin
85. Oberlin A, Endo M, Koyama T (1976) Filamentous growth of carbon through benzene decomposition. J Cryst Growth 32(3):335–349
86. Demoncy N, Stephan O, Brun N, Colliex C, Loiseau A, Pascard H (1998) Filling carbon nanotubes with metals by the arc-discharge method: the key role of sulfur. Eur Phys J B-Condens Matter Complex Syst 4(2):147–157
87. Fonseca A, Nagy J (2001) Carbon nanotubes formation in the arc discharge process: carbon filaments and nanotubes: common origins, differing applications? p 75–84. Springer, Berlin
88. Hu J, Bando Y, Xu F, Li Y, Zhan J, Xu J et al (2004) Growth and field-emission properties of crystalline, thin-walled carbon microtubes. Adv Mater 16(2):153–156
89. Ren Z, Lan Y, Wang Y (2013) Carbon nanotubes: Aligned carbon nanotubes: physics, concepts, fabrication and devices. Springer, Berlin Heidelberg, pp 7–43
90. Iijima S, Ichihashi T (1993) Single-shell carbon nanotubes of 1-nm diameter. Nature 363 (6430):603
91. Joselevich E, Dai H, Liu J, Hata K, Windle AH (2008) Carbon nanotube synthesis and organization. Carbon nanotubes. Springer, Berlin, Heidelberg, pp 101–165
92. Lyskawa J, Grondein A, Bélanger D (2010) Chemical modifications of carbon powders with aminophenyl and cyanophenyl groups and a study of their reactivity. Carbon 48(4):1271–1278
93. Leinonen H, Lajunen M (2012) Direct functionalization of pristine single-walled carbon nanotubes by diazonium-based method with various five-membered S-or N-heteroaromatic amines. J Nanopart Res 14(9):1064

Inorganic Nanocomposite Hydrogels: Present Knowledge and Future Challenge

Nasrin Moini, Arash Jahandideh and Gary Anderson

Abbreviation

(1-ethyl-3-(3-dimethyl aminopropyl) carbodiimide	EDC
2-acrylamido-2-methylpropane sulfonic acid	AMPS
Acrylamide	AAm
Acrylic acid	AA
Carbon nanotube	CNT
Carboxy methyl cellulose	CMC
Cellulose nanocrystal	CNC
Cetyl trimethyl ammonium bromide	CTAB
Copolymer	co
Double network	DB
Ethylene glycol dimethacrylate	EGDMA
Graft	g
Graphene oxide	GO
Graphene	G
Hydroxyapatite	nHA
Hydroxyethoxyethyl metha-crylate	HEEMA
Hydroxyethyl methacrylate	HEMA
Interpenetrating polymer network	IPN
Lower critical solution temperature	LCST
Magnetic field	MF
Montmorillonite	MMT
N-(2-hydroxypropyl) methacrylamide	HPMA
N,N-dimethylacrylamide	DMA
Nanocomposite Hydrogel	NCH

N. Moini (✉) · A. Jahandideh
Adhesive and Resin Department, Iran Polymer and Petrochemical Institute (IPPI), P.O. Box. 14975-112, Tehran, Iran
e-mail: N.Moini@ippi.ac.ir

G. Anderson
Agricultural and Biosystems Engineering Department, South Dakota State University, Brookings, SD 57007, USA

© Springer Nature Switzerland AG 2019
Inamuddin et al. (eds.), *Sustainable Polymer Composites and Nanocomposites*, https://doi.org/10.1007/978-3-030-05399-4_28

Nanocomposite	NC
Nanoparticle	NP
N-isopropyl acrylamide	NIPAm
Poly(acrylic acid)	PAA
Poly(dimethylacrylamide)	PDMA
Poly(ethylene glycol) acrylate	PEGA
Poly(ethylene glycol) diacrylate	PEGDA
Poly(ethylene glycol) dimethacrylate	PEGDMA
Poly(ethylene glycol)	PEG
Poly(ethylene oxide)	PEO
Poly(fluorine)	PF
Poly(methacrylic acid)	PMAA
Poly(methyl methacrylate)	PMMA
Poly(N-isopropyl acrylamide)	PNIPAm
Poly(N-vinyl-2-pyrrolidone)	PVP
Polyaniline	PAN
Polycarbonate	PC
Polyvinyl alcohol	PVA
Sodium acrylate	SA
Sodium n-dodecyl sulfate	SDS
Tetraethyl orthosilicate	TEOS
Vinyl acetate	VAc

1 Introduction

Hydrogels are hydrophilic physically or chemically crosslinked polymer networks, capable of absorbing and retaining the various amount of aqueous fluids [1, 2]. This crosslinked nature affects the components rheometrical properties, and consequently, results in a non-Newtonian behavior (viscoelastic or even pure elastic) of the dissolved polymeric chains [3]. This behavior promotes swelling properties of hydrogels making it a candidate for versatile applications, including biomedicine, electronic, separation and water treatment, etc. [1]. The crosslink and charge densities of the network are of the essential factors for tailoring the hydrogels properties for a special application. Hydrogels can be articulated in a variety of physical forms, including slabs, microparticles, nanoparticles, coatings, and films [4].

Wichterlie and Lim [5] first reported the synthesis of hydrogels with controlled properties, such as swell and shrinkage, over several orders of magnitude. This initial discovery provided the foundations for a generation of the *stimuli-responsive* biocompatible systems based on glycolmethacrylate, which could be in a porous state or modified by acrylamide fibers. Great progress has been made in synthesizing and developing new types of hydrogels to meet the desired controllable properties and overcome material deficiencies. To be highly sensitive to stimuli, such as solvent composition, the ionic strength of solutes, pH, temperature, electric field, and light could provide fabrication of various types of stimuli-responsive hydrogels [6]. Despite the great properties, these soft materials suffer from poor mechanical properties. One of the most used and efficient approach to boost the mechanical properties of hydrogels is to use inorganic nanoparticles which have great strength in the hydrogel structure [7].

In this chapter, the classification and desired properties of general hydrogels will be introduced. Various methods of nanoparticle preparation have been summarized. Different methods of nanocomposite hydrogel preparation, considering the method of nanoparticle insertion into the hydrogel structure will be presented. In addition, the application of these nanocomposite hydrogels based on the employed nanoparticles will be explained. The nanocomposite hydrogels based on organic or polymeric nanoparticles are out of the scope of this chapter.

1.1 Classification of Hydrogels

Hydrogels are classified based on different features, including the source, the network electrical charge, crosslinking type, method of preparation, polymeric composition, and the configuration of polymer networks.

Considering the source, hydrogels can be classified into natural, semi-synthetic, and synthetic polymers. While natural polymers are highly biocompatible, they lack reliability and consistency, due to the inherent inconsistencies, which stems from their natural origin. On the other hand, synthetic polymers are highly reproducible materials with tunable chemical and physical properties. However, compared to natural biopolymers, they suffer from poor biocompatibility: cytotoxic or non-biocompatible monomers and organic solvents are often required during their processing. Emission of toxic by-products, i.e. unreacted monomers or products of hydrolysis, during their lifecycle is another important issue. The choice of using natural or synthetic materials for the production of hydrogels is dependent on the aspired properties and applications [8].

Natural polymer sources consist of several minerals and animal-based or plant-based materials, and the most common examples are starch, cellulose, collagen, alginate, elastin, gelatin, lignin, chitosan, and different gum silicates.

Synthetic hydrogels are prepared from various monomers, including hydroxyethyl methacrylate (HEMA), hydroxyethoxyethyl methacrylate (HEEMA), ethylene glycol dimethacrylate (EGDMA), N-vinyl-2-pyrrolidone (NVP), N-isopropyl acrylamide (NIPAAm), vinyl acetate (VAc), acrylic acid (AA), acrylamide (AAm), N-(2-hydroxypropyl) methacrylamide (HPMA), ethylene glycol (EG), PEG acrylate (PEGA), PEG methacrylate (PEGMA), PEG diacrylate (PEGDA), and PEG dimethacrylate (PEGDMA) [9].

Three different integrated parts of the hydrogel preparation include monomers, initiators, and crosslinkers [10]. Any technique, which causes crosslinking in polymers, can be employed to produce a hydrogel. The free-radical crosslinking polymerization/copolymerization is one of the most common techniques employed to produce hydrogels. In this technique, the network is created by the reaction of hydrophilic monomers and multifunctional crosslinkers in the presence of radical initiators [11]. Water-soluble linear polymers of both natural and synthetic origins are crosslinked to form hydrogels in several ways [10]: (a) via chemical reactions (bulk or surface crosslinking [12–16]), (b) by employing ionizing radiation to generate main-chain free radicals and recombination to form crosslink junctions, and (c) via physical interactions, such as entanglements, electrostatics, and crystallite formation. In addition, various polymerization methods, including bulk, solution, graft, or emulsion polymerization, as well as inverse suspension polymerization can be used for gel preparation [11].

1.2 Feature Characteristics of Hydrogels

The desired properties of hydrogels and their corresponding factors have been summarized in Fig. 1. The feature properties of hydrogels include aqueous solution absorption capacity, absorbency rate, the extent of soluble fraction and residual monomers, biodegradability, and biocompatibility, and mechanical strength of the swollen gel [11]. The thermodynamics of swelling can reflect the influence of

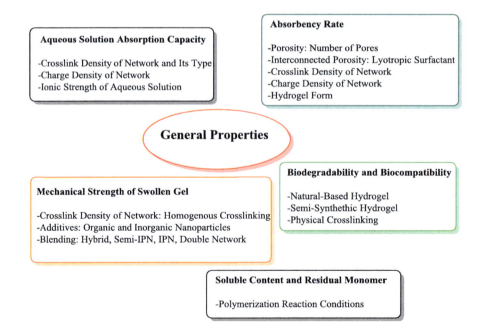

Fig. 1 Feature properties of hydrogels and their most effective parameters

crosslinking and the charge densities [17]. If the hydrogel is subjected to be used in biomedical and pharmaceutical applications, special attention must be paid to its swelling characteristics. In this case, detailed knowledge of swelling properties is essential, as the equilibrium degree of swelling affects several features of hydrogels, including its solute diffusion parameters, surface-dependent properties, optical properties as well as its mechanical properties [2].

Equilibrium swelling theory and the network characteristics of a single polymer network can be based on the contribution of mixing, network conformation (elastic), and ions. In terms of the free energy of the system, the total Gibbs free energy change (ΔG) upon swelling would be [17]:

$$\Delta G = \Delta G_{elastic} + \Delta G_{mix} + \Delta G_{ion} \qquad (1)$$

Here, $\Delta G_{elastic}$ is the Gibbs free energy contribution of the elastic retractive forces (representing network crosslink density effect); ΔG_{mix} represents the thermodynamic compatibility of the polymer and the swelling agent (water), and ΔG_{ion} is the ionic contribution of poly (electrolyte) hydrogels (representing the network charge density effect) [17].

Differentiation of Eq. 1 with respect to the number of water molecules in the system, yields the chemical potential change of water, in terms of the elastic, mixing, and ion contributions during swelling of the gel [2].

$$\mu_1 - \mu_{1,0} = \Delta\mu_{elastic} + \Delta\mu_{mix} + \Delta\mu_{ion} \qquad (2)$$

Here, μ_1 is the chemical potential of water within the gel and $\mu_{1,0}$ is the chemical potential of pure water. Equilibrium is reached when the chemical potentials of water inside and outside of the gel are equal. Therefore, the elastic, mixing and ion contributions to the chemical potential would balance at equilibrium. The chemical potential change upon mixing can be determined based on the Flory–Huggins theory [2, 18]. The elastic contribution of the crosslinked structure is usually described by the rubber elasticity theory and its variations [2]. The third contribution is taken as the ideal Donnan description for the ionic effect of polyelectrolytes [17]. This model has been illustrated in detail by Koetting et al. [19].

Even the *screening effect* of the ionic moieties, which is due to the high concentration of ions in an area next to the backbone that inhibits the ionization in polyelectrolytes, could influence the swelling ratio; often leads to the divergence between theory and practice [18, 20].

Swelling of polymer gels has adverse effects on its mechanical strength. Some hydrogels fracture under their own weight upon swelling. Inhomogeneity in hydrogels is often referred to as the inhomogeneous crosslinked density distribution of the gel. Inhomogeneity can decrease the hydrogels optical clarity, strength, ionization degree, electrostatic repulsion, and mobile counter ion number [21, 22].

On the other hand, several applications exist in which the hydrogels are required to sustain an external force in addition to their initial weight. *Tough hydrogels*, which reversibly deform to large extents without failure, are candidates for these applications. The synthetic hydrogels are often brittle, while the biological tissues, such as muscles and tendons, are tough. These biological hydrogel materials offer a high swelling degree and low modulus, along with high extensibility and high toughness [23]. Recently, this combination of properties has been achieved with synthetic hydrogels [22].

Energy dissipation and high stretchability are the fundamental factors in the design of tough hydrogels [24]. So far, synthesis of different tough hydrogels have been introduced, including interpenetrating polymer network (IPN) and double network (DN) hydrogels, ionically crosslinked hydrogels, covalently and physically crosslinked nanocomposite hydrogels, slide ring hydrogels, tough tetra-PEG hydrogels, and dendritic polymer hydrogel adhesives [25].

2 Nanocomposite Hydrogels

Nanocomposite hydrogels are generally designed to promote the mechanical strength of the swollen state [7, 11]. They may also offer advanced properties such as stimuli-responsiveness [26] and self-healing [27]. Several nanoparticles have been employed in the production of nanocomposite hydrogels, including nanoparticles of inorganic ceramics, metal and metal oxides, polymer-based materials, active glass, and carbon-based nanomaterials.

2.1 Nanoparticles Preparation

Inspired by nature, the nanoscale design of polymeric materials, mainly in biomedical fields, has made great progress [28]. In this regard, nanoparticle size of compounds (0.1–100 nm), their surface area, and quantum tunnelling effects are the most important factors which confer special properties to the product [21]. Nanoparticles have been classified based on their nano dimension, source (mineral or synthetic), morphology (crystalline or amorphous), and chemistry (clays, metals, metal oxide, carbon-based, etc.). The nano-sized compartments could be arranged in one (layered, e.g. clays), two (fibrous, e.g. carbon nanotubes) or three (particulates, e.g. silica) dimensions [21].

The size of nanoparticles can strongly affect the nanocomposites properties. Moreover, the agglomeration of the nanoparticles in the suspension and their dispersion are other challenges in this field [28].

Several preparation methods have been introduced for preparation of nanoparticles: physical and chemical methods (based on chemical reaction), gas-phase, liquid-phase, and solid-phase methods (based on the state of the reaction system). Gas-phase methods include: (a) gas phase evaporation method, using heating, plasma, electron beam, and laser, (b) chemical vapour reaction, induced by heating, laser, and plasma, (c) chemical vapour condensation, and (d) sputtering methods [21].

Liquid-phase methods are the important nanoparticle preparation methods. They are based on precipitation, emulsion, hydrolysis, spray, pyrolysis, sol-gel, radiation chemical synthesis, oxidation-reduction, etc. while solid state reaction methods include milling, stripping, spark discharge, and thermal decomposition [21].

Clays, which are characterized by their crystal structure and their charge, are the most commonly used nanoparticles. The crystallinity of the clays is proportional to the composite mechanical properties. Kaolin, Serpentines, Micas ('Mica' is a generic term applied to a group of complex aluminosilicate), Chlorites and Vermiculites, Glauconite, Sepiolite, Palygorskite and Attapulgite, Bentonite, and Montmorillonite (MMT) are crystalline clays. Compared to Kaolins with uniform chemical composition, the smectites such as Bentonite and Montmorillonite have a wide range of composition and cation exchange properties [22]. Clay particles are often plate-shaped, in which the layers are *intercalated* by an intercalant (organic or inorganic material with the opposite charge of layers, e.g. onium salt), in order to induce spacing between layers (ionically induced distance is at least 1.5 nm). *Exfoliation* is another step to disperse individual platelets (an *intercalated layered*) in a polymer matrix with the distance above 8.8 nm [21].

There are some advantages and disadvantages of utilization of mineral and synthetic clays. The mineral clays are available, and can be produced by well-known technologies; whereas, the inconsistencies in composition, difficulties in the removal of amorphous clays, poor reproducibility, and the crystallographic defects which prevent total exfoliation, are of their disadvantages. On the other side, the synthetic clays, such as Laponite, would have controlled compositions and shapes, high aspect ratio, and reasonable reproducibility. The production of synthetic clays is a developing technology, resulting in a higher price for these nanoparticles than the mineral nanoparticle clays [23].

The formation of stable dispersions in the polymer matrix is even more important than the size of the nanoparticles; agglomeration of nanoparticles leads to inferior properties [29]. In order to avoid and control agglomeration in hydrogels, the in situ formation of nanoparticles (i.e. silver nanoparticles) is employed. The nanoparticles formed in this method are surrounded by the polymer matrix, and subsequently, the possibility of agglomeration is declined [30]. The in situ methods for preparation of nanoparticles, such as reduction of metal oxides, suffer from poor control over the nanoparticle properties. The risk of contamination by unconverted precursor material and/or by-products also exists in these methods [31]. Two-step preparation method has its own advantages. It is possible to optimize the synthesis conditions for each individual component (e.g. size and shape of the nanocrystals). However, the dispersion of nanoparticles by this method is a challenge and adversely affect the nanocomposite properties [31].

Modification and functionalization of nanoparticles are often employed to overcome the incompatibility of hard and soft phases of the composite hydrogels. The phase-separation is an inevitable phenomenon which occurs when compounds are incompatible. It also results in mechanical deficiency at the interfacial surface, due to the presence of losing uniform stress distribution. Therefore, various methods have been suggested to modify nanoparticles to promote the mechanical properties. The nanoparticles could be modified by various functional groups [32, 33] such as amine [34], carboxylic acid [35], or silicone-based compounds [36].

2.2 Nanocomposite Hydrogel Preparation Methods

In general, the addition of filler compounds may be added to improve the mechanical strength of polymers. The high modulus of elasticity of the inorganic part improves the toughness of the polymer. Moreover, the nano-sized particles can significantly reform the properties, due to the scale of modification. Therefore, the combination of these two features is believed to improve mechanical properties [25] and also confer stimuli responsiveness to the hydrogels [37].

Based on the final application considered for the nanoparticle-hydrogel composite, the distribution of nanoparticles in the gels is achieved through several means, including (a) the formation of hydrogel in a nanoparticle suspension, (b) physical introduction of nanoparticles in the prepared gels, (c) in situ formation of reactive nanoparticles, (d) employing nanoparticles as the multifunctional crosslinking agents, and (e) employing nanoparticles and conductive additives along with polymeric binders [30]. Figure 2 illustrates these approaches schematically.

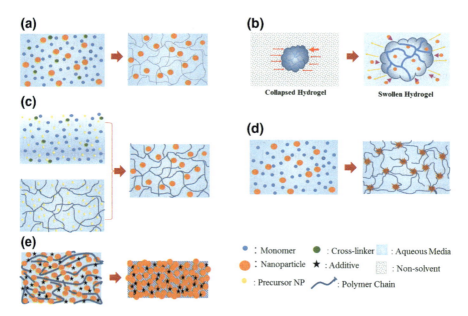

Fig. 2 Schematic representation of various methods of nanocomposite hydrogel preparation, **a** the formation of hydrogel in a nanoparticle suspension, **b** physical introduction of nanoparticles in the prepared gels, **c** in-situ formation of reactive nanoparticles, **d** employing nanoparticles as the multifunctional crosslinking agents, and **e** employing nanoparticles and conductive additives along with polymeric binders

2.2.1 Formation of a Hydrogel in a Nanoparticle Suspension

In this method, the polymerization is performed in the suspension solution of the monomers and the dispersed nanoparticles. Generally, the dispersion of nanoparticles is achieved by sonication of the suspension. In this part, the nanoparticles do not interfere with the polymerization. Ferromagnetic [38, 39] and gold nanoparticles [40] are examples of these kinds of particles. Gold nanoparticles were employed in the preparation of optically responsive hydrogel composites [41]. The weak interaction between particles and polymeric media leads to nanoparticles leaching during swelling, which is the main limitation of this method [30].

2.2.2 Physical Introduction of Nanoparticles into the Prepared Gels

Physical incorporation of nanoparticles into the gel during the polymerization, especially in non-conventional ones, is not always possible and sometimes there are obstacles to the addition of NPs during the gelation process. The electropolymerization and Au nanoparticles (Au-NP) are the best examples [30]. The Au nanoparticles cannot be used during the hydrogel preparations through electropolymerization, due to the agglomerations of Au in the electric field. To overcome this problem, the Au-NPs are doped into the hydrogel, via a "breathing in" mechanism. This mechanism consists of two steps; first, the highly aqueous swollen hydrogel is placed in an aprotic solvent (i.e. acetone) for 2 min, which is called the "breathing out" step. At the second step, the shrunk gel will be placed in an aqueous solution of citrate-stabilized Au-nanoparticles (13 nm) for 2 min (the nanoparticles have been stabilized by a chelating agent). This aqueous solution causes a swollen gel to form in the solution and the process is called "breathing in" of the suspended nanoparticles. The procedure could be repeated several times without any undesirable release of NPs during the "breathing" procedure. Since the nanoparticles have been stabilized with hydrogen bonding and physical entanglements that were induced by the chelating agent [42]. Guo et al. used the breathing mechanism for incorporation of Au-NPs into the porous anodic aluminium oxide film. For this purpose, the acrylamide gel was prepared by electropolymerization, and then, it was used as the media to absorb Au-NPs into the pores via breathing mechanism. Afterwards, the calcination is performed and the Au-NPs are trapped in the anodic aluminium oxide film [43].

2.2.3 In Situ Formation of Reactive Nanoparticles

In this approach, the nanoparticle precursors undergo a reaction to produce well-dispersed nanoparticles. It can also be called "in situ" nanoparticle formation, mostly via a reduction reaction. This method provides quite uniform distribution of metal or metal oxide NPs. In fact, the hydrogel media inhibits the aggregation of NPs during the reaction. Often, Ag-NPs are made for antibacterial purposes, or for the addition of optical and electrical properties. Studies have shown that the size of produced Ag nanoparticles has not been varied by changing the Ag+ ion concentration. In addition, aggregation of particles has not been reported. Therefore, the Ag-NPs are created through in situ techniques [44–46].

The in situ reduction reaction could be performed during the polymerization or even after that. In antibacterial cryogels based on Ag NPs, the cryogel is swollen to the equilibrium state in $AgNO_3$ solution (for 24 h) and then the gel was added to the $NaBH_4$ solution to prepare in situ Ag NPs after proper preparation of the cryogel [47].

The mangogels were most of the time undergone the in situ preparation of ferromagnetic NPs. To exemplify, first, the metal ions were bounded to the gel then reaction with NaOH results in the in situ formation of magnetic NPs [48, 49].

2.2.4 Nanoparticles as the Multifunctional Crosslinking Agents

The employment of nanoparticles as multifunctional crosslinking groups, present at the nanoparticle surface, is one of the most interesting aspects in the nanocomposite hydrogel field. The clays with a hydrophilic nature have a great potential to be used as multifunctional crosslinkers. Various clays have been *organomodified* with different intercalants, to be used for the preparation of superabsorbent nanocomposite hydrogels, such as Bentonite, Montmorillonite (MMT), and Sepiolite [11]. In this method, the effective intercalant should be of opposite charge. For instance, MMT can be intercalated with cationic, low molecular weight materials or polymeric materials. Alkyl ammonium salts, such as hexadecyl ammonium chloride and (3-acrylamidopropyl) trimethyl ammonium chloride, or polymeric materials, such as chitosan and poly (dimethyldiallylammonium chloride), have been currently employed in this method [50]. During radical polymerization, the radical transfer to the surface of the NPs is an ordinary phenomenon, which may trigger the grafting at the surface of the particle. For this reason, the clays are called "radical killers" because they retard the radical polymerization and gelation [11]. In addition, ultrahigh mechanical properties of these nanocomposites (NC) hydrogels can be attributed to the multiple non-covalent effects between clay nano-sheets (Clay-NS) and the polyacrylamide chains [51].

Zhang et al. prepared a four-component semiconductor nanoparticle-based hydrogel, via self-initiated polymerization under sunlight. The system consists of four components: water, n,n-dimethylacrylamide (DMA), water-soluble semiconductor nanoparticles (NP), such as ZnO, TiO_2, and clay-nano-sheet (Clay-NS). NPs initiated the polymerization by sunlight. Since the crosslinking on the NP surfaces was insufficient to change the viscous behaviour to viscoelastic, clay-NSs were employed to achieve a three-dimensional structure [52].

Au NPs are also of biocompatible materials which have been modified to play the role of multifunctional (vinyl or carboxylate functionalized) crosslinkers [32, 53]. In an approach other than in situ formation of Au NPs the nanoparticles were functionalized by vinyl groups to design multifunctional Au NP crosslinkers.

The PNIPAm nanocomposite hydrogels have the thermos-switchable electrical properties [32].

The Au NPs could be functionalized by carboxylic acid using mercapto compounds and dispersed in collagen media. The employment of "zero-length" linkers is a recent method that chemically crosslinks collagen and Au nanoparticles. In fact, the coupling agents are capable of forming peptide bonds between the collagen molecules. EDC (1-ethyl-3-(3-dimethyl aminopropyl) carbodiimide) is one example of these coupling agents. Multiple carboxyl groups are present at the surface of Au nanoparticles which enable the nanoparticles to form multiple links with the collagen developing novel properties [53].

Silica nanoparticles have been employed as multifunctional crosslinkers that adhere to different parts of the gel [54–56]. The silica NPs solution has been used to adhere the PDMA-based hydrogel surfaces. In that study, a droplet of silica solution binds the swollen polymer network. In this case, the diameter of the employed nanoparticle must be comparable with the gel network mesh size (crosslink density). The network chains can be adsorbed on the nanoparticles surfaces binding the particles to gel pieces. Particles act as *connectors* between gel surfaces. The adsorbed chains also form bridges between particles. Particle adsorption is considered to be irreversible as the binding to the gel networks occurs through numerous attachments. Monomers have the ability to detach from a particle surface, and other repeating units (from the same or a different network strand) could be replaced. Such exchange processes allow large deformations and energy dissipation under stress [54].

2.2.5 Nanoparticles and Conductive Additives Along with Polymeric Binders

NC hydrogels of graphite and silica have been used in the production of rechargeable lithium-ion batteries. More recently, electrodes of Si nanoparticle (SiNP) slurries have received attention. In these systems, the binders are used to fix the active material to the anode. The more interactions (either hydrogen bond or ionic) between binder and SiNPs, the more efficient the batteries are. Different binders, such as PVDF, PAA, CMC, alginate and phytic acid, are present during the polymerization and formation of conductive NC hydrogels [57–60].

NC hydrogels have been used in the production of rechargeable lithium-ion batteries with different nanoparticles like graphite and silica [60]. More recently, electrodes made from Si nanoparticle (SiNP) slurries instead of graphite (traditional anode), have received attention. In these systems, the binders are used to hold the active material together in the anode. The point is that the binder concentration in the slurry is much less than the nanoparticle concentration and other additive concentrations (15 wt% binder vs. 43% Si NP and 42% C as conductive martial [58]). The binder could be a previously prepared polymer [59] or synthesized in situ during anode preparation [57]. The more interactions (either hydrogen bond or ionic) between the binder and SiNPs, the more efficiency of the batteries are owing to less volume change (for high capacity batteries). The polymeric binders could act as dual-functional materials to improve binding and conductivity of the electrodes. The life cycle of the electrode can be extended by enhancing mechanical and electrical properties (integrity). Different binders, such as PVDF, PAA, CMC, alginate and phytic acid, are present during the polymerization and formation of conductive NC hydrogels [57–60].

2.3 How Nanoparticles Improve Mechanical Strength of Hydrogels?

Mechanisms of energy dissipations in tough hydrogels include reversible crosslinking or fracture of polymer chains, the transformation of domains in polymer chains or crosslinkers, as well as fracture and pullout of fibers or fillers. Mechanisms of maintaining elasticity in tough hydrogels can be attributed to the interpenetration of long-chain networks, hybrid physical and chemical crosslinkers, high-functionality of crosslinkers, and the presence of networks with long monodisperse polymer chains [24]. Many tough hydrogels have been produced based on these mechanisms, such as hydrogels of poly (vinyl alcohol) with crystalline domains, double-network systems, hydrogels with hybrid chemical and physical crosslinkers, hydrogels with high-functional crosslinkers, systems with transformable domains, topological hydrogels with sliding crosslinkers, and tetra-arm polymer hydrogels [25]. It is believed that employing a combination of these techniques would be a promising strategy to design next-generation tough hydrogels. For instance, a tough hydrogel may integrate fiber reinforcement at the macro-/meso-scale, high-functionality crosslinkers at the micro-scale, and hybrid crosslinkers at the nanoscale [24].

Examples of high-functional crosslinkers which are used for production of tough hydrogels include crystalline domains in polymer networks [i.e. poly(vinyl alcohol)], exfoliated nano-clays that can crosslink various polymers (i.e. polyacrylamide [61], poly(N-isopropylacrylamide [62], and poly(ethylene glycol) [63]), polyacrylamide crosslinked by chitosan nanofibers [64], and graphene oxide [65].

It is believed that nanoparticles act as multiple crosslinkers. The suggested mechanisms explaining the toughness these hydrogels exhibit will be explained in the following. Based on Flory's network theory, the number of polymer chains that can be crosslinked by a crosslinker is defined as the functionality of the crosslinker. Typically, common physical and chemical crosslinkers have functionalities less than 10; in addition, generally, a single polymer chain connects two adjacent common crosslinkers. When polymer chains are ruptured under deformation, the connections between crosslinkers are eliminated, and consequently, fracture of the network is commenced. In order to achieve high elasticity in hydrogels, large crosslinkers with very high functionality (e.g. over 100) must be incorporated into the polymer networks. In these networks, multiple polymer chains may connect two adjacent crosslinkers; and these chains usually do not have uniform lengths. In other words, parallel to the polymerization, *grafting* happens on the surface of the nanoparticles and multiple crosslinking occurs. Therefore, as the polymer networks are deformed, relatively short chains may be ruptured or detached from the high-functionality crosslinkers, but the long chains can still maintain the elasticity of the hydrogels [24].

Depending on the employed nanoparticles, the physical adsorption and desorption of building blocks of polymer chains may increase the non-covalent and reversible interactions (which act like crosslinks). Poly(acrylamide)s and silicates show a great potential for these interactions. The shear thinning characteristics (induced by reversible interactions) of this type of hydrogel composite can expand the hydrogel composite application, especially in the biomedical field [54, 55, 66, 67].

2.4 Characterization Methods of Nanocomposites Hydrogels

Various techniques have been employed to characterize nanocomposite hydrogels structural, mechanical, thermal, electrical, and optical properties. The physicochemical structure of these hydrogels is often studied by X-ray diffraction (XRD), transmission electron microscopy (TEM), small angle neutron scattering (SANS), and Fourier-transform infrared spectroscopy (FTIR) [11].

Based on the XRD data, the lattice related data (crystal planes, shape, and constants) can be obtained [61, 68–70]. XRD has also been used to study the extent of intercalation and exfoliation. The interlayer spacing (d) is commonly determined from the XRD patterns as the arbitrary intensity versus 2θ, based on Bragg's law [71]. To study the diffracting angles less than $0.05°$, small-angle X-ray scattering (SAXS) and SANS have been employed, mainly because of their sensitivity and ability to make long-range measurements. SANS can be adopted on various types of specimens and may be assisted to probe the polymer-clay interaction, interfacial polymer conformations, phase transition, and gelation mechanism [71–73].

While the XRD/WAXS does not reveal useful information for $2\theta \cong 2$, TEM provides visual evidence of the nanoparticle distribution. TEM can also provide useful information on crystals development, morphology, and size [48]. The interactions between the clay platelets, the intercalating agent, and the polymer can be studied by FTIR techniques [74].

The thermal behaviour of the nanocomposite hydrogels can be studied by Differential Scanning Calorimetry (DCS) and Thermogravimetric analysis (TGA). Often, the mechanical properties of the nanocomposite hydrogels are investigated by dynamic mechanical analysis (DMA), oscillatory rheometry, and compression and tensile tests [11].

2.5 Types of Nanocomposite Hydrogels and Their Applications

2.5.1 Inorganic Ceramics and Non-metal Nanoparticles

In the last decades, inorganic ceramics mostly used as counterparts have attracted attention in hydrogel field on account of their versatility, great mechanical strength, biocompatibility and reasonable price. A range of bioactive nanoparticles has been reported to be used in various biomedical applications. These bioactive nanoparticles include hydroxyapatite (nHA), synthetic silicate nanoparticles (i.e. Laponite), bioactive glasses (SiO_2, Na_2O, CaO, MgO, P_2O_5), silica, calcium phosphate, glass ceramic, and b-wollastonite [51].

Table 1 presents typical clay-based nanocomposite hydrogel preparations and their application. These clay-based NPs are used most in inorganic ceramics in the nanocomposite hydrogel field. They act as multifunctional crosslinkers giving great potential for functionalization and grafting. Various polymeric networks like PAA, PAAm, PEG, and chitosan have been incorporated. Based on Table 1 data, the majority of hydrogels were synthesized by aqueous solution polymerization via free radical copolymerization, which has been induced by redox initiators. Meanwhile, the inverse suspension polymerization has rarely been used to produce granular hydrogels.

The employment of the synthetic silicate nanoparticles (nanoclays) has great influence on the physical and mechanical properties of the hydrogels, which can be attributed to their anisotropic, plate-like nature, and their high aspect-ratio [56, 66]. Studies showed that the addition of silicates will improve the elongation of the polymeric hydrogels, mainly due to the formation of physically crosslinked networks [56]. The physical adsorption and desorption of building blocks of polymer chains can enhance the non-covalent and reversible interactions, which act as crosslinks. The unique bioactive properties of the synthetic silicates make it capable of being employed as an injectable tissue repair matrices, bioactive fillers, or therapeutic agents [51]. The nanocaly incorporation even renders greater mechanical properties, stretchability, and self-healing properties of the hydrogel [27].

The composite hydrogels, as drug release vehicles, may increase the biocompatibility by "hiding" the nanoparticles within the hydrogel, and also by preventing nanoparticle movement from their targeted site in vivo. This morphology of hydrogel phase (e.g. porosity) can also control the kinetics release profile and balance the burst release [4].

A class of SAPs (i.e. superporous hydrogels and SPHs) has been developed by Chen et al. [75] to be used in pharmaceutical applications. A variety of techniques have been employed to synthesis porous hydrogels, including foaming, microemulsion polymerization, porogen incorporations, freeze-drying, and phase-separation [76–78]. Typical SEM images of thermally dried particles of porous and non-porous superabsorbent hydrogels at different magnifications are displayed in Fig. 3.

Table 1 Various methods of clay preparation before nanocomposition and its incorporation in hydrogel networks

NP	NP preparation	Polymeric matrix	Polymerization method	NCH preparation	Application and feature	References
Kaolin	–	P(AAm-co-AMPS)	Thermal-induced (persulfate) aqueous solution polymerization	Method d	Super absorbency	[132]
	Suspended in SDS	PAA	Ultrasound induced (persulfate) solution polymerization	Method d	Removal of brilliant green dye from water	[83]
	–	Collagen-co-PAA-PSA	Thermal-induced (persulfate) solution graft copolymerization	Method d	Enhancement of gel strength and super absorbency	[133]
	–	Urea formaldehyde and polyphosphate potassium in core: P(AA-co-AAm) in shell	Inverse suspension polymerization	Method d	Controlled release and super absorbency by core-shell structure	[134]
	& MMT, sedimentation with water, filtered and then dried at 80 °C	P(AAm-co-AMPSNa)	Redox initiation (persulfate) and aqueous solution polymerization	Method d	Removal of Cu(II), Cd(II), and Pb(II) ions	[68]
	–	PAA	Redox initiation (persulfate) and aqueous solution polymerization	Method d	Porous, highly swelling rate and capacity	[77]
	& MMT, Mica modified and intercalated by quaternary alkylammonium-exchanged (3-acrylamidopropyl) trimethylammonium chloride)	PAA	Inverse suspension polymerization	Method d	Reactive clays, super absorbency	[135]

(continued)

Table 1 (continued)

NP	NP preparation	Polymeric matrix	Polymerization method	NCH preparation	Application and feature	References
Mica	Modified and intercalated by quaternary alkylammonium-exchanged	PAA	Inverse suspension polymerization	Method d	Reactive clays, Superabsorbency	[136]
Attapulgite	– Organification with hexadecyltrimethyl ammonium bromide (HDTMABr) – Acidified	PAAm	Thermal-induced (persulfate) aqueous solution polymerization	Method d	Salt-resistant Super absorbency	[137]
	–	Chitosan PAA-itaconic acid (IA)	Redox initiation (Vc/Peroxide) aqueous solution polymerization	Method d	Adsorption of malachite green	[81]
	–	N-Succinylchitosan-PAAm	Inverse suspension polymerization (persulfate)	Method d	Thermal stability and superabsorbency	[138]
	–	PAA	Thermal-induced (persulfate) aqueous solution polymerization	Method d	Superabsorbency	[139]
	–	Chitosan-PAA	Aqueous solution polymerization (persulfate)	Method d	Adsorption of Hg(II) ions from aqueous solution	[91]
	–	Chitosan PAA	Redox initiation (Vc/Peroxide) aqueous solution polymerization	Method d	Adsorption of La(III) and Ce(III)	[92]
	–	PAA hydroxypropyl cellulose	Redox initiaton (Fenton reagent) graft polymerization	Method d	Adsorption-desorption of the rare earth elements, La (III) and Ce(III)	[94]

(continued)

Table 1 (continued)

NP	NP preparation	Polymeric matrix	Polymerization method	NCH preparation	Application and feature	References
MMT	Dispersed in water under ultrasonication for 1 h	PAAm	Redox initiation (persulfate) and aqueous solution polymerization	Method d	Stretchability, toughness, and self-healing	[61]
		PAA	Redox initiation (persulfate) and aqueous solution polymerization	Method d	Negative impact of nanocomposite on residual monomer	[80]
	Homogenized with sodium cations	Chitosan-g-PAAm	Thermal-induced (persulfate) aqueous solution polymerization	Method d	Antibacterial and superabsorbency	[140]
	Intercalated by chitosan	PAMPS	Thermal-induced (persulfate) aqueous solution polymerization	Method d	Nontoxicity, high gel strength, superabsorbency	[141]
	Suspension modified with a sodium carbonate powder	PAA	Redox initiation (persulfate) and aqueous solution polymerization	Method d	Superabsorbency	[142]
	Ultrasonicated in water	N-isopropyl acrylamide (NIPAm) and acrylic acid (AA) onto CMC was	Thermal-induced (persulfate) graft copolymerization	Method d	Removal of Cu(II) and Pb (II) ions thermoresponsive	[62]
		Polysaccharide pullulan (PULL) with polyvinyl alcohol (PVA)	– No polymerization – Electrospinning	Method d	Supecansorbency	[143]

(continued)

Table 1 (continued)

NP	NP preparation	Polymeric matrix	Polymerization method	NCH preparation	Application and feature	References
Zeolite	& Nano Ag Ultrasound for 1 h in acetone	PAA PAAm	Redox initiation (persulfate) and aqueous solution polymerization	Method d and magnetron sputtering method	Antibacterial Removal of chemical oxygen demand (COD) wastewater	[126]
	–	CMC PAA	Thermal-induced (persulfate) graft polymerization	Method d	Controlled delivery of zinc micronutrient	[144]
	Prepared in 2,2,2-Trifluoroethanol	homo-poly (butylene succinate)	Electrospun fibers	Method d	Antimicrobial drug-delivery	[145]
Bentonite	Intercalated by the hydrochloride solution of AAm	PAA crosslinker: sugar	Thermal-induced (persulfate) solution polymerization	Method d	Superabsorbency	[146]
	–	PMAA-grafted-cellulose	Thermal-induced (persulfate) graft polymerization	Method d	Removal and recovery of thorium(IV)	[93]
Halloysite	–	Chitosan PAA	Thermal-induced (persulfate) graft polymerization	Method d	Adsorbent to remove ammonium from synthetic wastewater	[87]
Rectorite	–	Chitosan PAA	Thermal-induced (persulfate) graft polymerization	Method d	Adsorbent to remove ammonium from synthetic wastewater	
Illite Smectite	Dispersed in the aqueous solution of CTAB	sodium alginate-g-P (SA-co-styrene)	Micellar solution polymerization (with SDS)	Method d	Adsorbing methylene blue	[88]

(continued)

Table 1 (continued)

NP	NP preparation	Polymeric matrix	Polymerization method	NCH preparation	Application and feature	References
Sepiolite	Dispersed in water and sonicated	kappa-carrageenan-g-PAAm	Thermal-induced (persulfate) solution polymerization	Method d	Adsorption of cationic dye	[84]
Tourmaline	–	PVA PAA	Thermal-induced (persulfate) solution polymerization	Method d	Adsorption capacity for Pb^{2+} and Cu^{2+}	[89]
Hydrotalcite	Synthesized by urea method and intercalated by sodium methyl allyl sulfonate (SMAS)	PAA PAAm	Inverse suspension polymerization	Method d	Salt-resistance	[147]
Hydroxyapatite	Vortexing and sonication to	PEG diacrylate	Cross-linked via photopolymerization. (IRGACURE initiator)	Method d	Highly extensible, tough, and elastomeric nanocomposite	[63]
	Prepared by precipitation of disodium hydrogen phosphate and calcium chloride	Cellulose-g- polyacrylamide	Suspension polymerization (persulfate)	Method c	Removal of Cu (II)	[90]
Laponite	–	PDMA PNIPAm	Redox initiation (persulfate) and aqueous solution polymerization	Method d	Self-healable	[27]
		Telechelic dendritic macromolecule (binder) PSA	– No polymerization – Solution mixing	Method d	Self-healing and moldable for biomedical application	[148]
	Mixed with Semiconductor NP (TiO_2, ZnO, CdTe),	PDMA	Self-initiated polymerization under sunlight	Method d	Semiconductor NP-based Hydrogels	[52]

(continued)

Table 1 (continued)

NP	NP preparation	Polymeric matrix	Polymerization method	NCH preparation	Application and feature	References
	Exfoliated	Sulfobetain polymers (poly dimethyl (acrylamidopropyl) ammonium propane & butane sulfonate)	Redox initiation (persulfate) and aqueous solution polymerization	Method d	Swelling and deswelling	[149]
	–	PEG alginate	Ionically crosslinked by Ca ions	Method d	Fabrication by 3D printing (clay provide viscosity and shear thinning ability)	[150]
	Gel/solution exfoliation method	PEO	–	Method d	Cell cultivation – Biotechnological applications such as injectable matrices, biomedical coatings, drug delivery, and regenerative medicine	[67]

These nanocomposite hydrogels prepared by in-situ formation of reactive nanoparticles (method c) and by nanoparticles with the ability of multifunctional crosslinking methods (method d)

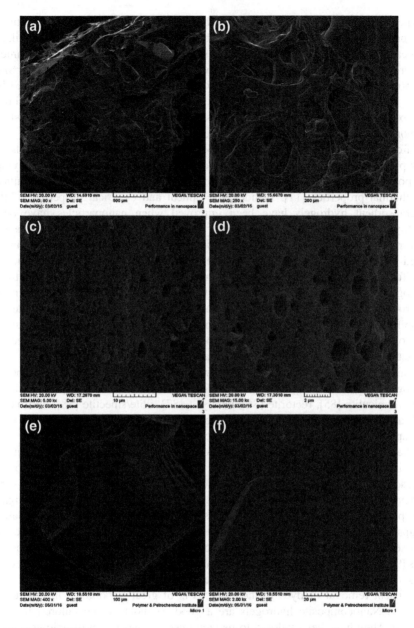

Fig. 3 Typical SEM images of porous superabsorbent hydrogel particles in **a** 500 μm magnification, **b** 200 μm magnification, **c** 10 μm magnification, **d** 2 μm magnification, and non-porous superabsorbent hydrogel particle in **e** 100 μm magnification, **f** 10 μm magnification

The SPHs were originally intended to be used in gastric retention applications. The prompt swelling of the SPHs can be attributed to the increased capillary permeability of the interconnected pore structure. The weakness in mechanical properties was mainly overcome by the development of the second-generation SPH composite (SPHCs) and the third-generation SPH hybrids [1]. Demirtas et al. [79] have synthesized and characterized polyacrylamide-based SPHCs-containing hydroxyapatite. The compressive modulus of this SPHC was 6.59 N/mm^2, where the non-composite SPH has a compressive modulus of ~ 0.63 N/mm^2.

Effects of composition and nanocomposition of superabsorbent hydrogels on their properties have also been investigated. In accordance with the other polymer composites, the presence of nanoparticles can significantly enhance the mechanical and thermal properties [11]. Besides, the optical and electrical properties of the superabsorbent hydrogels have been improved when nanoparticles have been employed [1]. However, the absorbency properties, i.e. free swelling capacity, swelling rate, reswellability, and saline sensitivity, have been pervasively influenced by the type and the content of the employed clay. Since the clays act as multifunctional crosslinkers, their hydrogels can be brittle. Moreover, in these hydrogels, employing a higher content of the clay may result in a reduction in the absorbency [11]. Composite hydrogels generally possess a slower swelling rate but exhibit a higher saline absorbency, due to the clay nature or organomodification. Moreover, the employment of the nanocomposite would have other negative impacts, such as an increase in the residual monomer content [80].

Some common inorganic components which have been used in nanocomposite hydrogels include attapulgite [81], montmorillonite [68, 82], kaolin [83], sepiolite [84], vermiculite [85], rectorite [86], halloysite[87], illite/smectite [88], tourmaline [89], and hydroxyapatite [90]. The nanocomposite superabsorbent hydrogels were also applied in water treatment and purifications processes.

These nanocomposite materials have been used for the adsorption of heavy metal ions; i.e. Pb^{2+}, Cu^{2+} [62, 68], and Hg^{2+} [91], as well as for the elimination of radioactive and rare earth elements; i.e. thorium (IV) and lanthanide (III) [92–95]. Compared to other low-cost adsorbents, fast adsorption kinetics and higher adsorption capacities are provided by the super-hydrophilic network and chelating groups of the hydrogels [49]. Other advantages of using adsorptive hydrogels include easy separation and effective desorption for the recovery and enrichment of pollutants [96]. The incorporation of the proper amount of inorganic components, not only boosts the adsorption rate but also improves the adsorption capacity [62, 68].

Efficient removal of dyes forms effluents is of great importance in textile the industry. Nanocomposite hydrogels are a good candidate to be tailored for this purpose. They have been tailored to be employed for separation of anionic (Silica sol) [97] and cationic dyes (Titania) [98]. Successful removal of dyes, such as malachite green (Attapulgite) [94], methylene blue (MMT) [82], methylene orange (MMT) [82], and Safranine-T (Titania) [98], has been reported using composite hydrogels. Figure 4 displays the swelling, and the methylene blue absorption of superabsorbent polymer films, from different perspectives.

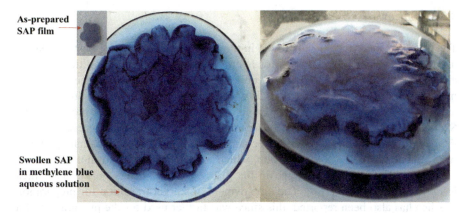

Fig. 4 Typical illustration of dye absorbency of superabsorbent hydrogels: absorption of methylene blue solution by superabsorbent polymer at different perspective

Tracking the migration of the hydrogel, which has been used as a cartilage repairing material, is a challenging task because the sensitive monitoring must be non-invasive [99]. The employment of hydrogels with desirable fluorescence properties would solve this problem. In this regard, different concepts have been employed, including polymerization of fluorescent monomers, functionalization of polymers with organic fluorophores, using fluorescent carbon dots (CDs), and semiconductor nanocrystallites [quantum dots (QDs)] [35, 99–101]. Polysaccharides are suitable candidates to be used in the synthesis of fluorescent polymeric materials. Alginate, chitosan, hyaluronic acid, dextran, and cellulose are polysaccharides commonly utilized for this purpose [99]. It has been shown that the fluorescent monomers and functionalized polymers appear to lose their luminescent properties. Therefore, the CDs are often preferred as they present high fluorescence, chemical stability, biocompatibility, and low toxicity. Recently, a novel CDs/PAM composite hydrogel with both excellent mechanical and fluorescence properties has been prepared [102].

Detection of enzymes by nanotools is of great importance; stable sensing that is very sensitive could be provided by self-assembled NPs in hydrogel media. Ruiz-Palomero et al. have reported laccase enzyme detection by immersing graphene quantum dots (GQDs) into nanocellulosic hydrogels. Noncovalent interactions between the sensor (GQDs/NC) and the analyte (laccase) have led to the stable and sensitive detection of the analyte [103].

The tracking problem can be addressed with semiconductor nanocrystals, with size-dependent emission property, can be produced in a simple synthetic process. For example, the size-dependent emission property of the CdSe nanocrystals results in blue to red emissions, with very pure colour. Chang et al. used CdSe/ZnS nanoparticles [quantum dots (QDs)] embedded in the cellulose matrices. The cellulose–QDs hydrogels displayed strong photoluminescence emission besides good compression strength [35].

2.5.2 Silicon-Based Nanoparticles

The presence and necessity of silicon in the human body are of the best reasons to use these inorganic nanoparticles for hydrogel preparation [51]. Addition of nHA and silica to the poly(ethylene glycol) (PEG) media enhanced the elasticity, mechanical strength, biological stability, and cell adhesion [63, 66]. In this case, ionic interactions could be the reason for elasticity enhancement in the gels [51]. Synthesized and modified silica nanoparticles were also used to prepare highly flexible Poly(acrylic acid)-based nanocomposite hydrogels. The silica nanoparticles were functionalized by a vinyl group, which may act as filler and multifunctional crosslinker. Entanglement trapped in the glassy polymer layers on the nanosilica leads to this flexibility [104]. The employment of sol-gel transition for preparation of silica has also been reported; this silica was further used for the preparation of a keratin-silica hydrogel, with the potential to be used for wound dressing [105].

The mesoporous silica nanoparticles (MSN), as drug vectors, have been incorporated into 3D scaffolds. The functionalized MSN has been synthesized and evaluated by different techniques: BET model for measuring the surface area, dynamic light scattering for measuring particle size, and TEM for evaluating the particle shapes. The matrix-forming self-assembling peptide, different MSNs, and precursor cells were combined to prepare injectable cell- and MSN-containing scaffolds. For this purpose, the self-assembly and coordination interactions between cells, matrices, and nanoparticles are required. Surface functionalization of MSN as well as its size, have a great impact on its nanocarrier internalization characteristics. The COOH-functionalized MSN exhibits less sensitivity to the hydrogel matrix, and in comparison with the monolayer cell culture, its internalization has been strongly enhanced in the hydrogel matrix [34].

Table 2 represents the typical silicon-based NCH preparation methods and their application. These NCHs have introduced novel properties to the hydrogel networks including toughness stretchability and self-healing which is owed to their multifunctional crosslinking role. These NPs could inherently enhance the mechanical properties of soft hydrogels. Various concepts have been explored for preparation of stretchable tough hydrogels, i.e. nanocomposition [56], micellar copolymerization [106], and hydrophobic association [107]. An example of a highly stretchable tough hydrogel under tension has been displayed in Fig. 5. The hydrogel film can be extended to desirable elongations (the displayed elongation ratio is more than 15 mm/mm).

2.5.3 Carbon-Based Nanoparticles

Carbon nanotubes (CNTs) and graphene, as the most used carbon-based nanoparticles, have attracted much attention to be used in various biomedical applications, such as actuators, conductive tapes, biosensors, tissue engineering scaffolds, etc. Table 3 summarizes typical carbon-based NCH preparation methods and their applications. In contrast to clays, the CNTs exhibit hydrophobic nature; therefore,

Table 2 Various preparation methods of silicon-based NP before nanocomposition and its incorporation into the hydrogel networks

NP	NP preparation	Polymeric matrix	Polymerization method	NCH preparation	Application and feature	References
Silica sol	–	PAAm Poly(diallyl dimethyl ammonium chloride) (DADMAC)	Thermal-induced (persulfate) aqueous solution polymerization	Method a	Removal of methyl orange from aqueous solutions	[97]
Silica CNT CNC	1-Sol-gel 2-SiO_2/Na_2O	PDMA	Thermal-induced (persulfate) aqueous solution polymerization	Method d	Adhesives for gels and biological tissues	[54]
Silica	Vinyl hybrid silica prepared by sol-gel reaction	PAAm	Thermal-induced (persulfate) aqueous solution polymerization	Method d	Highly stretchable and super tough nanocomposite	[151]
Silica	Vinyl silica prepared by methacryloxypropyl trimethoxy silane	PAA	Thermal-induced (persulfate) aqueous solution polymerization	Method d	Highly stretchable and super tough nanocomposite	[104]
Silica	Prepared by TEOS sol-gel reaction	Keratin	Siloxane network	Method d	Wound dressing	[105]
Mesoporous Silica	– Sol-gel reaction in alkaline solution (methanol and water) in the presence of CTAB – NH_2 and COOH functionalized NPs	Peptide	–	Method d	Drug vectors into injectable 3D scaffold	[34]

(continued)

Table 2 (continued)

NP	NP preparation	Polymeric matrix	Polymerization method	NCH preparation	Application and feature	References
Sodium Silicates	–	Colloidal silica	Siloxane network	Method d	Cell encapsulation	[152]
Silica	–	PDMA	Redox initiation (persulfate) and aqueous solution polymerization	Method d	Resilient and stretchable hydrogel	[55]
Nanocrystalline silicon	Ultrasonication in presence of monomer under dry nitrogen and then degassing by freeze-pump-thaw cycles	– Ethylene glycol dimethylacrylate – 2-hydroxyethyl acrylate	Laser-induced thermal crosslinking Polymerization	Method d	Semiconductor by green process	[153]

These nanocomposite hydrogels prepared by the formation of NCH in the NPs solution (method a) and by nanoparticles with the ability of multifunctional crosslinking methods (method d)

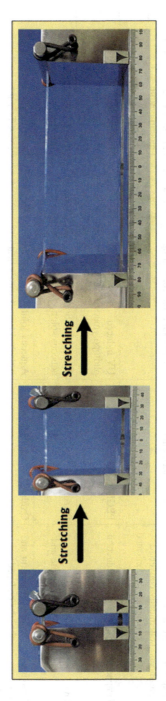

Fig. 5 Typical highly stretchable tough hydrogels under tension; the as-prepared hydrogel film (left) has been elongated to elongation ratio of ~15 mm/mm

Table 3 Typical carbon-based NCH preparation methods and their applications

NP	NP preparation	Polymeric matrix	Polymerization method	NCH preparation	Application and feature	References
GO	– Synthesized using Hummers' method – Sonicated	PAA	Dual ionic crosslinking Solution polymerization	Method d	Self-healable, super tough	[109]
GO	Synthesized using Hummers' method Sonicated	PNIPAm	Chemically and ophysically crosslinked Redox initiation (persulfate) and aqueous solution polymerization	Method d	Near-infrared light-responsive ultrahigh tensibility	[111]
Graphene nanosheets	Reduced GO to chemically converted grapheme and sonicated	Methacrylated chitosan	UV-crosslinkable conducting	Method d	3D structured biocompatible scaffold by 3D printing	[154]
Graphene	Simultaneous reduction of GO	PEGDA	UV-induced photopolymerization	Method d	Electrically conductive hydogel	[110]
GO	Synthesized by oxidizing graphite and ultasonication	PAAm	Redox initiation (persulfate) and aqueous solution polymerization	Method d	Tough and highly stretchable	[65]
Graphene quantum dots	Synthesis of sulfur and nitrogen codoped graphene quantum dot	Oxidize Nanocellulose (NC)	Aqueous solution in ultrasonic bath then heated with a heat gun	Method d	Eco-friendly and cost-efficient Nanotools detecting the laccase enzyme	[103]
GO	–	Poly (acryloyl-6-aminocaproic acid)	$-Ca^{2+}$ induces the formation of the 3D cross-linked Thermal-induced (persulfate) aqueous solution polymerization	Method d	Superior mechanical properties and self-healing Drug delivery	[155]

(continued)

Table 3 (continued)

NP	NP preparation	Polymeric matrix	Polymerization method	NCH preparation	Application and feature	References
CNT	Coating by gelatin methacrylate for better dispersion	Gelatin methacrylate	Hybrid hydrogel Photo-induced polymerization (Irgacure)	Method d, e	Cell encapsulation	[108]
CNT	CNT dispersed in PEG compounds, various surfactant and then sonicated	4-arm PEGAA PEG dithiol (linear crosslinker)	PEG remains in triethanolamine (TEA) in phosphate buffered saline for a week and then the same molar ratio of PEGAA and PEGdithiol reacted	Method a	Neural tissue engineering, sensor technology as electrode coatings, and drug delivery	[156]
CNT	Pristine CNT No surfactant	PMAA	In situ radical polymerization (aqueous dispersion)	Method a	Pulsatile drug delivery	[157]
CNT TiO$_2$	First TiO$_2$ dispersed and sonicated in deionized water then polymer solution and CNT solution were added	Poly (3,4-ethylenedioxythiophene (PEDOT) poly (4-styrenesulfonate) (PSS)	Radical polymerization	Method a	Flexible lithium ion battery electrodes	[158]

The NCH has been prepared by formation of hydrogel in NPs suspension (method a), in situ formation of NPs (method c), acting as multifunctional crosslinking (method d) and the polymeric network binding NPs (method e) dispersion of CNTs in hydrogels is a potential challenge. In order to induce hydrophilicity in these systems, different approaches have been employed: modification of CNT surfaces using various polar groups, i.e. amines (NH_2), hydroxyls (OH), and carboxyls (COOH), use of single-stranded DNA (ssDNA), proteins, and surfactants, as well as by grafting hydrophilic polymer chains to the CNT surfaces. The high electrical conductivity of CNT-reinforced hydrogels makes these components ideal candidates to be engineered for various electorally conductive tissues, including cardiac tissues, nerves or muscles [51]. In hybrid nanocomposites, the slightest amount of COOH-functionalized CNTs may significantly increase the tensile strength of methacrylated gelatin hydrogels in the interconnected porous structures. The addition of CNT does not interfere in the porosity and cell-growth ability of the hydrogel. The highly aligned structure with tight intercellular junctions, along with the electrical conductivity of the CNTs results in the formation of this conductive network [108].

Graphene sheets can also provide high mechanical strength (by acting as crosslinking agents), and excellent conductivity of heat and electricity in hydrogels. In order to ensure the miscibility of the hydrophilic polymer and graphene sheets, the sheets are often treated by strong oxidizers to form graphene oxide (GO). Thus, GO can be crosslinked to the media, both physically and covalently. The GO can be employed to produce stimuli-responsive hydrogels [51]. Generally, GO is introduced into the nanocomposite hydrogels by physical mixing, in solutions, or by in situ polymerization of water-soluble monomers. The latter strategy is believed to be more efficient in terms of obtaining tough and stretchable nanocomposite hydrogels [109].

Electrically conductive hydrogels of polyethylene diacrylate, containing GO, has also been prepared by photopolymerization. In these conductive gels, the GO crosslinker has introduced electrical properties to the network [110]. The ability of GO to form crosslinks leads to the design of self-healable and tough poly (acrylic acid)-based nanocomposite hydrogels using GO and Fe^{3+} ions. Amazing dual cross-linking effects through dynamic ionic interactions has been developed: Fe^{3+} ionically crosslinked to carboxylic acid groups of the hydrogel backbone and then linking GO nanosheets to the backbone through Fe^{3+} coordination. The proposed mechanism explains the tough and self-healable behaviour of the gels based on the energy dissipation through dynamic *breakage* and *recombination* of ionic interactions. Furthermore, the GO nanosheets, coordinated on to the backbones, act as stress-transfer centres, which transfer the stress to the polymer matrix, and meanwhile, they maintain the configuration of the hydrogels [109]. These nanocomposite hydrogels can facilitate the development of asoft materials to be used in various biomedical applications.

In another research, a combination of GO nanosheets and thermoresponsive poly(N-isopropylacrylamide) (PNIPAM) polymeric networks, resulted in light-responsive nanocomposite hydrogels with ultrahigh tensile strength. These novel properties which are essential for designing smart actuators, remote light-controlled devices, and artificial muscle, can be attributed to the GO nanosheets, which are physically crosslinked to the amide groups of the PNIPAM chains via hydrogen bonds in the presence of the chemical crosslinker of N, N-methylene bis acrylamide (MBA) [111].

2.5.4 Metal and Metal Oxide Nanoparticles

Various types of nanocomposite hydrogels have been fabricated, using different metallic and metal-oxide nanoparticles, including gold, silver, copper, iron oxide (Fe_3O_4–Fe_2O_3), ZnO, titania (TiO_2), alumina, etc. [51, 112]. In fact, the physical interactions between polymer and nanoparticles are not sufficient to provide enough mechanical strength. However, functionalization of nanoparticles will improve the mechanical strength by providing multiple crosslinking nodes in the network [51]. According to their physical, electrical, magnetic, and antimicrobial properties, various applications for imaging agents, conductive scaffolds, drug delivery systems, switchable electronics, actuators, and water treatments have the uses been developed for these hydrogels [112]. Mechanical properties also can be manipulated via magnetic field induction [26].

Thermo-responsive magnogel based on PNIPAm has been synthesized for drug delivery application; anti-cancer therapeutic drugs can be released by a magnetic field and temperature variation. In other to stabilize the Fe_2O_3 NPs, various modifications have been employed including oleic-acid, polyhedral oligomeric silsesquioxane (POSS) and nitro-dopamine PEG-dicarboxylic acid functionalization. This functionalization results in the better responsive performance of the nanocomposite hydrogel [113]. Table 4 represents the metal oxid-based NCH preparation methods especially for magnogels (ferrogels) and their applications.

Recently, ferrogels (magnetic hydrogels), which contain immobilized nano-magnetic particles (e.g. γ-Fe_2O_3, Fe_3O_4, Co Fe_2O_4), have attracted considerable attention. The magnetic hydrogels can quickly respond to an external magnetic field (MF), which acts as a distance-force (noncontact-or-remote force) device/system. This characteristic facilitates the incorporation of these hydrogels in various biomedical applications [38], i.e. in tissue engineering and cell/drug delivery. During the cell-growth process, limitation of available cells in the porous hydrogel is a great challenge which can be overcome by implementation of magnetic hydrogels. In fact, the required biological agents can be bound to the magnetic nanoparticles (MNPs) by applying external MF. In addition, the magnetic scaffolds can be stimulated by physical cues via interaction between MNPs and an alternating magnetic field (AMF) [37].

In order to microfabricate and 3D print hydrogels, two main strategies have been employed using microengineered hydrogels for tissue engineering including: "top-down" (hydrogel formed and cell cultivated in the fill media; for larger hydrogels) and "bottom-up" (every part of the microgels contains cells and then they are assembled into the desired shape; for microhydrogels) [114]. Magnetic microgels can be assembled in a way to form complex tissue structures in a controlled manner via MF. Several assembling techniques have been explored, including those based on microfluidics [114], nanotextured surfaces

Table 4 Typical most-used metal oxide NPs, their NCH preparation methods and their applications

NP	NP preparation	Polymeric matrix	Polymerization method	NCH preparation	Application and feature	References
ZnO	Calcination (at 850 °C) of entrapped "zinc acetate di-hydrate" in PAA gel	PAA	Thermal-induced (persulfate) aqueous solution polymerization	Method e	Biotechnology	[70]
ZnO	$ZnNO_3$ hydrolyzed and then turn to ZnO crystals by heating	CMC	Crosslinked by maleic, succinic, and citric acid used as crosslinker	Method c	Antibacterial	[122]
ZnO	Synthesized by zinc acetate dihydrate and sodium hydroxide	Chitosan	NC hydrogels were lyphophilzed	Method b	Wound dressing	[118]
ZnO	Synthesized by zinc acetate dihydrate and sodium hydroxide	β-chitin	Freeze-dried	Method a	Wound healing	[121]
ZnO	Synthesized using zinc acetylacetonate monohydrate, oleylamine and oleic acid at 240 °C under argon flow	PNIPAm	– Linear polymer by radical polymerization – UV-induced crosslinking reaction	Method d	Hydrogel as antibacterial coating	[31]
CdSe/ZnS	COOH-functionalized Core-shell QD by octyl-amine modified PAA	Cellulose	Crosslinked by sodium hydroxid and urea ligand formation	Method d	Fluorescent semiconductors and optoelectronics	[35]
Fe_3O_4 magnetic particles	Sonicated in PVA/DMSO solution	PVA	Freezing-thawing cycles	Method a	Magnetic properties for Controlled Release of Drug	[38]

(continued)

Table 4 (continued)

NP	NP preparation	Polymeric matrix	Polymerization method	NCH preparation	Application and feature	References
magnetite (Fe_3O_4) and maghemite (Fe_2O_3)	Synthesized in hydrogel network by NaOH	PAMPS	Photo-induced radical polymerization	Method c	Magnetic response hydrogels for water treatment	[49]
Ferrimagnet	Ferrofluid mixed with hydrophobic phase	Hydrophobic phase: styrene and divinyle benzene Hydrophilic phase: AAm and MBA	Anisotropic microfluidic device. – The hydrophobic monomer core with magnetic material is encapsulated by a hydrophilic monomer droplet suspended in fluorocarbon oil	Method a	Rotational control by applying an external field in biomedical application	[39]
magnetite (Fe_3O_4) and maghemite (Fe_2O_3)	In situ mineralization and coprecipitation of $FeCl_2$ and $FeCl_3$	Gelatin	Solution and physical crosslinking	Method c	Actuators	[48]
Fe_3O_4 magnetic	$FeCl_3·6H_2O$ and sodium oleate Iron-oleate complex then it was treated by oleic acid up to 318 °C to form hydrophobic Fe_3O_4. Their surface were treated by PEG and polyhedral oligomeric silsesquioxane	PNIPAm poly-ethylene glycol (PEG)	Redux initiation aqueous solution polymerization	Method d (functionalized NP)	Theranostic application e.g. long distance control of drug delivery by MF	[113]

(continued)

Table 4 (continued)

NP	NP preparation	Polymeric matrix	Polymerization method	NCH preparation	Application and feature	References
$\gamma\text{-}Fe_2O_3$	–	Chitosan modified by catechol	Chemically and physically crosslinked by pH variation	Method d (physical crosslinking: complex)	pH and magnetic responsive Drug delivery	[159]
Fe_3O_4	Coprecipitated in basic media alginate-coated in a suspension	Alginate and PAAm	Redox initiation and aqueous solution polymerization then ionically crosslinked by Fe$(NO_3)_3$	Method a	Soft robotics, clinical operations, tough, and stretchable magnogel	[160]

The NCH has been prepared by formation of hydrogel in NPs suspension (method a), in situ formation of NPs (method c), acting as multifunctional crosslinking (method d) and the polymeric network binding NPs (method e) (micromolding) [114], acoustic and magnetic fields (photolithography) [115], and surface tension (emulsification) [39].

The magnetic gels have shown the ability of pulsatile release of drugs, through low-frequency oscillatory MFs. Recently, an intelligent Fe_3O_4 MNP-PVA hydrogel has been designed to control the drug release by "on" and "off" modes [37, 116]. When the MF was intensified, the on-off magnetization can change the volume of the ferrogels, and subsequently, affect the swelling ratio. Upon applying an MF, the nanoparticles tend to get agglomerated together [37]. The reduced porosity and volume result in "close" configuration, and subsequently, the rate of drug delivery will be minimum. On the "off" mode, the volume and swelling ratio are increased, and the drug would be delivered at it's highest rate. Furthermore, the magnetic-sensitive hydrogel has shown outstanding flexibility and elasticity [38]. The magnetic-thermosensitive hydrogels have also been used for controlling the drug release rate. The AMF can adjust the hydrogels temperature. By increasing the temperature (above lower critical solution temperature (LCST)), the network will collapse and the drug diffusion would be "off"; and then, by reducing the temperature, the drug diffusion status would be "on" and the drug can be released [37]. This ability of magnetic gels to raise and control the temperature remotely has been used in cancer therapy. The concentration of Iron Oxide and the amplitude of MF are influential factors for controlling the generated heat. The local hyperthermia feature of the magnetic hydrogels makes them a promising injectable hydrogel system, especially for cancer therapy. The functionalized magnetic NPs provide better performance during swelling and deswelling phenomenon (considering no release of NPs, and subsequently less toxicity for biological application) [26].

Mechanical properties also can be manipulated via magnetic field induction [26]. Thermo-responsive magnogels based on PNIPAm have been synthesized for drug delivery application; the anti-cancer therapeutic drugs can be released by the magnetic field and temperature variation. In other to stabilize the Fe_2O_3 NPs, various modifications have been employed including oleic-acid, polyhedral oligomeric silsesquioxane (POSS) and nitro-dopamine PEG-dicarboxylic acid functionalization. This functionalization results in better responsive performance of nanocomposite hydrogel [113].

The poly(2-acrylamido-2-methyl-1-propane sulfonic acid) P(AMPS) magnetic composites have been prepared and employed for removal of toxic metals. The iron oxide can confer ferromagnetic property into the gel. These gels can be employed for absorbing toxic ions, i.e. Pb^{2+}, Cd^{2+}, Co^{2+}, Ni^{2+}, Cu^{2+} and Cr^{3+} [49].

Colonization of microorganism on the surface of the medical devices, such as implants, may result in severe infections. Therefore, the antibacterial media with the lowest cytotoxicity would be of great importance. Various metal and metal oxides nanoparticles such as silver, gold, copper, TiO_2, and ZnO have been applied in the hydrogels to improve their antimicrobial properties [117].

Zinc oxide is the metal oxide which has been used in hygienic applications such as cosmetics. Recently, the ZnO NPs have also been used as antibacterial agents against both gram-positive [118] and gram-negative bacteria [119] and showed no cytotoxicity at concentrations of up to 10 wt% NP. Various antibacterial mechanisms have been proposed to explain their antibacterial role, such as the formation of reactive oxygen species (ROS), and the release of Zn^{2+} ions [119, 120]. The ZnO NPs with uniform crystal structures have been successfully incorporated into poly(N-isopropylacrylamide) (PNIPAAm) hydrogel layers. In order to better disperse the NPs, the surfactants mediums of oleylamine and oleic acid have been employed. The antibacterial properties of these hydrogels can be altered by changing the thickness of the NP film [31]. Recently, β-chitin hydrogel/nZnO composite bandage has been fabricated and used for wound dressing applications. It has shown antibacterial effects against *E. coli* and *S. aureus*; however, the cytotoxicity of the hydrogels has been increased at elevated concentrations of ZnO NPs [121]. Furthermore, a flexible and microporous chitosan-ZnO nanocomposite hydrogel has been developed and tested for wound dressing purposes. In vivo studies revealed that this nanocomposite hydrogel has great potential to be used as a bandage for burn wounds, chronic wounds, and diabetic foot ulcers [118]. In another approach, the carboxymethyl cellulose (CMC)-based nanocomposite hydrogels were prepared by oligomeric acrylic acids, such as maleic, succinic or citric acid. The ZnO nanoparticles were synthesized in the presence of CMC to avoid agglomeration, which is known to be the main problem in zinc oxide nanocomposite hydrogel production. In fact, the polysaccharide structure, which has numerous hydrogen bonds, can effectively act as a template for nanoparticle growth. This nanocomposite hydrogel has shown a promising swelling ratio, and also great antibacterial activity against both gram positive and gram negative bacteria [122].

Metals are also used in the hydrogel network to add more functionality such as biocidal and electrical activity to the soft material. Table 5 summarizes preparation methods of the most-used metals in NCHs and their applications. Silver is the most studied antiseptic agent and has a long history in activity against gram-positive and gram-negative bacteria, fungi, protozoa, and certain viruses. Since the silver ions have shown concentration-dependent toxicity, care must be taken in the incorporation of silver in medical devices [29]. However, silver does not have the hazards associated with the accumulation of other heavy metals [29]. Factors, affecting the biocidal activity of silver nanoparticles (Ag NP) include particle size, the shape of the particle, and the dispersion status. The smaller particle size offers larger surface-to-volume ratio, which consequently enhances the antibacterial activity. In addition, the Ag NPs with triangular architecture has shown a better antibacterial effect against *E. coli* compared to the rod or spherical-shaped Ag NPs. Furthermore, for biocidal activity, the binding of Ag NPs to polymer networks is more important for effectiveness than the size of nanoparticles. The utilization of polymer supports for Ag NPs results in an increase in the stability of the particles, preserve them against aggregation, and also, increases their biocompatibility [29]. The antimicrobial activity of silver can be attributed to the strong bonding between the silver ion and the biological molecules containing sulfur, oxygen, or nitrogen. It is believed that the complex, which is formed between proteins of bacteria and the silver ion, can interfere with the metabolism and eventually, it disturbs the power functions of the bacteria. It can also prevent the cellular reproduction by interaction

Table 5 Typical most-used metals and their NCH preparation methods and applications

NP	NP preparation	Polymeric matrix	Polymerization method	NCH preparation	Application and feature	References
Ag	In situ reduction (AgNO$_3$, Acetic acid, NaBH$_4$ as the Ag NP-forming agent)	Chitosan/sodium tripolyphosphate	No polymerization Ionically crosslinked by polyanions	Method c	Drug delivery Antibacterial	[69]
Ag	Reduction of Ag ions in swollen cryogel	Methacrylated carboxymethyl chitosan PEGDA	Photopolymerization on ice, and lyophilizing (cryogel)	Method c	Antibacterial tissue eng.	[47]
Ag	Reduction of Ag ions in swollen hydrogel	PVA Sodium alginate PAAm	Thermal-induced (persulfate) aqueous polymerization (IPN)	Method c	Antibacterial	[44]
Ag	Reduction of Ag ions in swollen hydrogel	PNIPAM PSA	Redox initiation (persulfate) Aqueous polymerization	Method c	Antibacterial	[45]
Ag	Maleimide-coated silver nanoparticles	Furfuryl-gelatin	Diels-Alder Cycloaddition-based cross-linking	Method d	Drug delivery systems, improved mechanical properties	[124]
Ag	Reduction of Ag ions in swollen hydrogel	poly (vinyl pyrrolidone) (PVP) PAAm	IPN redox initiation and aqueous polymerization	Method c	Antibacterial	[46]
Ag	Reduction of Ag ion	PAA	In situ polymerization and reduction of Ag ion	Method c	Conductive hydrogels	[127]
Ag	(PVP)-coated silver nanoparticles	PAAm	Redox initiation (persulfate) and aqueous polymerization	Method e	optical biosensor	[128]
Ag	Reduction by one-pot green chemistry technique using starch and glucose	PVA	Freeze-thaw	Method a	Antimicrobial dressing scaffold	[123]

(continued)

Table 5 (continued)

NP	NP preparation	Polymeric matrix	Polymerization method	NCH preparation	Application and feature	References
Ag	Reduction of Ag ions in cross-linked hydrogel	Dialdehyde hemicelluloses Chitosan	–	Method a	Antibacterial	[161]
Ag	Reduction of Ag ions in hydrogel	Chitosan (CTS) and acrylic acid (AA)	Redox initiation (fentone) and solution polymerization	Method c	Catalytic reduction of organic dyes	[129]
Ag	Using biodegradable gelatin as a stabilizing agent	Gelatin PNIPAm	Redox initiation (persulfate) and aqueous polymerization	Method c	Antibacterial thermosensitive	[125]
Au Pt	Citrate-reduced gold prepared by Frens method PVP-protected Pt NPs prepared	PAAm	No polymerization	Method b Breathing in	Dispersed metal nanoparticles in porous anodic aluminum oxide	[43]
Au	Solution	PAAm	Electropolymerization in the presence of $ZnCl_2$	Method b Breathing in	Biosensors and solvent-switchable electrical properties	[42]
Au	Tiopronin (N-(2-Mercaptopropionyl) glycine) protected gold nanoparticles	Collagen	Nanoparticles crosslinked to collagen via EDC (1-ethyl-3-(3-dimethyl aminopropyl) carbodiimide) coupling agent	Method d	Photothermal therapies, imaging, and cell targeting	[53]
Au	Reduced (by $NaBH_4$) and then functionalized (by allyl mercaptan)	PNIPAm	Thermal-induced solution polymerization (AIBN and THF)	Method d	Tunable thermo-switchable electrical properties	[32, 33]
Cu	Synthesized by reduction reaction (starch capping) and modified a silica coating by inverse emulsion	Starch	Using urea	Method c	Antibacterial	[36]

The NCH has been prepared by formation of hydrogel in NPs suspension (method a), physical incorporation of NPs into the hydrogel networks (method b), in situ formation of NPs (method c), acting as multifunctional crosslinking (method d), and the polymeric network binding NPs (method e) with DNA [29]. The immobilization of Ag NPs has also been reported in synthetic hydrogels, such as poly(acrylic acid) and poly(methacrylate) [30]. Recently, the introduction of Ag NPs into different bio-based and biocompatible systems, such as (PVA) [123], gelatin/chondroitin sulfate [124], PVA/sodium alginate/poly(acrylamide) [44], and gelatin/N-isopropylacrylamide [125], has gained more attention.

Various studies deal with a physical dispersion of Ag NPs via in situ synthesis, in which the Ag NPs were often dispersed into the zeolite–poly (acrylamide-co-acrylic acid) hydrogels, using radical graft copolymerization and magnetron sputtering methods [126]. These antibacterial hydrogels were used for water treatment applications. Generally, the silver nanocomposite hydrogels, which have been prepared via radical graft copolymerization, display better biocidal activity. This feature is in accordance with their better dispersion which has also been confirmed by XRD result [126]. In another approach, the silver nanoparticles were synthesized in situ in the swollen hydrogel media. The superabsorbent hydrogels were based on poly(vinyl alcohol) (PVA), sodium alginate (Na–Alg), and poly (acrylamide). The Ag ion loading is proportional to the antimicrobial activity, and can be affected by the concentration of a silver ion, crosslinking density of hydrogel network, as well as the Na-Alg/PVA ratio [44]. Chitosan nanocomposite hydrogel beads have also been prepared by synthesizing Ag NPs in situ. The chitosan in these systems has been ionically crosslinked to the sodium tripolyphosphate. The amine and hydroxyl groups confer on to the chitosan the potential to interact with various metal cations, i.e. Ag^+, Zn^{2+}; therefore, the Ag NPs are expected to be distributed uniformly in the hydrogel beads, which is also consistent with the XRD results. Moreover, the effect of Ag NPs on the drug loading has been evaluated. It has been shown that an increase in the Ag NPs content would result in a decrease in the drug loading. It is attributed to the variation of the less porous structure of the chitosan induced by interactions between Ag ions and chitosan [69]. As it has been previously stated, the Ag NPs are incorporated into the polymer matrix physically; as a result, the continuous release of NPs to the surrounding environment is plausible. To overcome this deficiency, the Ag NPs have been covalently bonded to the furan-modified gelatin. The benzotriazole maleimide has been employed to cap the Ag NPs during in situ formation, and then, the click chemistry, Diels–Alder (DA) cycloadditions, were employed to crosslink the furan-modified gelatin to provide a mild reaction condition [124]. Multiple crosslinking effects on Ag NPs has increased the elastic modulus almost three folds. In addition to the antimicrobial applications, the Ag NPs embedded in the hybrid hydrogels, have also been employed as optoelectronic [127, 128] and catalytic materials [129]. For example, in the glucose-responsive Ag NP hydrogels, the absorbance strength of the localized surface plasmon resonance (LSPR) is decreased by an increase in the concentration of glucose. This property has been employed for the production of optical enzyme biosensors [128].

Gold is another metal element which has been used in biomedical application [30, 51, 53]. The Au NP hydrogels have been used as stimuli-responsive and switchable conductive materials. The distance variation of Au NPs during external stimulation (e.g. temperature and pH) is the main reason for the changes in its conductivity. The Au NPs have also been used for antibacterial applications, remote-controlled microfluidic valves, and surface plasmon resonance (SPR)-based sensors. However, the high-cost Au has limited incorporation of Au NPs in large-scale applications [30]. In general, the Au NPs cannot enhance the mechanical properties of the hydrogel. However, Au NPs in thiol-modified biomacromonomers can improve the gel strength. In order to hydrogel form and reform during and after printing, this nanocomposite hydrogel has been designed to represent *dynamic* crosslinking by functionalization of the Au NPs to act as multiple crosslinking agents [130].

It is believed that Au NPs are biocompatible and capable of being easily functionalized by biomolecules. The size and shape of these biocompatible NPs can be engineered; the combination of these unique features make Au NPs excellent candidates use in various applications like new contrast agents for imaging and novel photothermal therapies, biomedical applications, and drug delivery [53].

3 Summary and Outlooks

Hydrogels have tremendous potential to be tuned to obtain the desired physicochemical properties in contact with an aqueous media. The need for engineered hydrogels with specific properties results in outstanding developments in the conventional hydrogels which often offer poorer properties. The softness of the hydrogels provides necessary and sufficient resemblance to the biological and natural systems, while this softness would defiantly affect the mechanical properties which in turn restrict the application of the gels. Inspired by nature, nano-modification can promote the properties and performance extensively. The nano-scale incorporation of minerals, as a hard component, not only confer better mechanical properties but also introduce other functionalities, such as stretchability and stimuli-responsiveness into the hydrogels. The facile techniques of nanomodification can be simply employed to shift from conventional hydrogels into smart ones.

The stimuli-responsive hydrogels trigger the idea of smart hydrogels as multifunctional materials. The nanocomposite properties can be manipulated by altering the pH and the temperature of the surrounding media. In addition, the mesh size of the nanocomposite hydrogels network can be regulated by electrical and magnetic fields. This on-off behaviour enabled the gel to change its shape, to move, and also to releases certain drugs. Quantum dot nanoparticles offer photoluminescent properties to the nanocomposite hydrogels. The supramolecular-like behaviour and reversible crosslinking, introduced by the hydrogen bonding or ionic interactions, have resulted in self-healable hydrogels.

The engineered hydrogels have received much attention and are used in various applications, including actuators, biosensors, controllable drug delivery, artificial muscles, etc. The clinical applications of hydrogels can limit the type of hydrogels used. Various pristine and modified bio-based macromolecules have been introduced into the networks to design engineered hydrogels, which are tough, stretchable, resilient, or self-healable. Recently, click-chemistry has assisted in producing biobased hydrogels [109, 110] with various functionalities, including photopatternable, antibacterial, antifungal, and anticancer. This chemistry provides an efficient, selective, and mild situation to prepare gels with improved properties; the Au nanocomposite hydrogels with superior mechanical and electrical properties, are of examples of the employment of the click-chemistry (Diels-Alder reaction). The click-chemistry, "thiol-ene" rection has been employed to prepare 3D structured cell encapsules, using acylated-modified, sulfobetaine-derived starch, and dithiol functionalized PEG [131]. The mechanical and swelling properties, as well as gelation time, can be easily tuned in physiological condition.

Another aspect of the formation of hydrogels is the engagement of relatively simple radical polymerization. This advantage has extended hydrogel application extensively, owing to the ability to construct complicated structures precisely. Also, 3D printing is a newfound way in hydrogel fields to produce well-defined volumetric objects. The important factors in 3D printing include viscosity, gelation mechanism, and speed. The incorporation of nanoclays is an efficient way to modify the viscosity and shear-thinning effect of the hydrogel.

References

1. Zohuriaan-Mehr MJ, Omidian H, Doroudiani S, Kabiri K (2010) Advances in non-hygienic applications of superabsorbent hydrogel materials. J Mater Sci 45(21):5711–5735
2. Peppas NA (2012) Hydrogel. Introd Mater Med, pp 35–42
3. Akhtar MF, Hanif M, Ranjha NM (2016) Methods of synthesis of hydrogels... A review. Saudi Pharm J 24(5):554–559
4. Hoare TR, Kohane DS (2008) Hydrogels in drug delivery: progress and challenges. Polym (Guildf) 49(8):1993–2007
5. Wichterle O, Lim D (1960) Hydrophilic gels for biological use. Nature 185(4706):117–118
6. Chu L, Ju X, Xie R, Wang W (2013) Smart hydrogel functional materials
7. Schexnailder P, Schmidt G (2009) Nanocomposite polymer hydrogels. Colloid Polym Sci 287(1):1–11
8. Utech S, Boccaccini AR (2016) A review of hydrogel-based composites for biomedical applications: enhancement of hydrogel properties by addition of rigid inorganic fillers, vol 51, no 1. Springer US
9. Laftah WA, Hashim S, Ibrahim AN (2011) Polymer hydrogels: a review. Polym Plast Technol Eng 50(14):1475–1486

10. Ahmed EM (2015) Hydrogel: preparation, characterization, and applications: a review. J Adv Res 6(2):105–121
11. Kabiri K, Omidian H, Zohuriaan-Mehr MJ, Doroudiani S (2011) Superabsorbent hydrogel composites and nanocomposites: a review. Polym Compos 32(2):277–289
12. Moini N, Kabiri K (2015) Effective parameters in surface cross-linking of acrylic-based water absorbent polymer particles using bisphenol A diethylene glycidyl ether and cycloaliphatic diepoxide. Iran Polym J 24(11):977–987
13. Moini N, Kabiri K, Zohuriaan-Mehr MJ (2018) Surface treatment of superabsorbent. US Pat. 2018/0008960
14. Moini N, Kabiri K, Zohuriaan-Mehr MJ (2015) Practical improvement of SAP hydrogel properties via facile tunable cross-linking of the particles surface. Polym Plast Technol Eng 55(3):278–290
15. Moini N, Kabiri K, Zohuriaan-mehr MJ, Esmaeili N (2015) Simple and efficient approach for recycling of fine acrylic-based superabsorbent waste. Polym Bull 73(4):1119–1133
16. Moini N, Kabiri K, Zohuriaan-Mehr MJ, Omidian H, Esmaeili N (2017) Fine tuning of SAP properties via epoxy-silane surface modification. Polym Adv Technol 28(9):1132–1147
17. Buchholz FL, Graham AT (1998) Modern superabsorbent polymer technology. Wiley, Hoboken
18. Buchholz FL, Peppas NA (1994) Superabsorbent polymers. ACS, Washington DC
19. Koetting MC, Peters JT, Steichen SD, Peppas NA (2015) Stimulus-responsive hydrogels: theory, modern advances, and applications. Mater Sci Eng R Reports 93:1–49
20. Zeldovich KB, Khokhlov AR (1999) Osmotically active and passive counterions in inhomogeneous polymer gels. Macromolecules 32(10):3488–3494
21. Rutz AL, Shah RN (2016) Polymeric hydrogels as smart biomaterials
22. Naficy S, Brown HR, Razal JM, Spinks GM, Whitten PG (2011) Progress toward robust polymer hydrogels. Aust J Chem 64(8):1007–1025
23. Zhang YS, Khademhosseini A (2017) Advances in engineering hydrogels. Science, 356 (6337)
24. Zhao X (2014) Multi-scale multi-mechanism design of tough hydrogels: building dissipation into stretchy networks. Soft Matter 10(5):672–687
25. Peak CW, Wilker JJ, Schmidt G (2013) A review on tough and sticky hydrogels. Colloid Polym Sci 291(9):2031–2047
26. Ilg P (2013) Stimuli-responsive hydrogels cross-linked by magnetic nanoparticles. Soft Matter 9(13):3465
27. Haraguchi K, Uyama K, Tanimoto H (2011) Self-healing in nanocomposite hydrogels. Macromol Rapid Commun 32(16):1253–1258
28. Gupta RB, Kompella UB (2006) Nanoparticle technology for drug delivery, vol 159. New York
29. Monteiro DR, Gorup LF, Takamiya AS, Ruvollo-Filho AC, de Camargo ER, Barbosa DB (2009) The growing importance of materials that prevent microbial adhesion: antimicrobial effect of medical devices containing silver. Int J Antimicrob Agents 34(2):103–110
30. Thoniyot P, Tan MJ, Karim AA, Young DJ, Loh XJ (2015) Nanoparticle–hydrogel composites: concept, design, and applications of these promising, multi-functional materials. Adv Sci 2(1–2):1–13
31. Schwartz VB et al (2012) Antibacterial surface coatings from zinc oxide nanoparticles embedded in poly (N-isopropylacrylamide) hydrogel surface layers. Adv Funct Mater 22 (11):2376–2386
32. Zhao X et al (2006) A kind of smart gold nanoparticle–hydrogel composite with tunable thermo-switchable electrical properties. New J Chem 30(6):915–920
33. Zhao X, Ding X, Deng Z, Zheng Z, Peng Y, Long X (2005) Thermoswitchable electronic properties of a gold nanoparticle/hydrogel composite. Macromol Rapid Commun 26 (22):1784–1787

34. Baumann B, Wittig R, Lindén M (2017) Mesoporous silica nanoparticles in injectable hydrogels: factors influencing cellular uptake and viability. Nanoscale 9:12379–12390
35. Chang C, Peng J, Zhang L, Pang D-W (2009) Strongly fluorescent hydrogels with quantum dots embedded in cellulose matrices. J Mater Chem 19(41):7771
36. Villanueva ME et al (2016) Antimicrobial activity of starch hydrogel incorporated with copper nanoparticles. ACS Appl Mater Interfaces 8(25):16280–16288
37. Li Y et al (2013) Magnetic hydrogels and their potential biomedical applications. Adv Funct Mater 23(6):660–672
38. Liu TY, Hu SH, Liu TY, Liu DM, Chen SY (2006) Magnetic-sensitive behavior of intelligent ferrogels for controlled release of drug. Langmuir 22(14):5974–5978
39. Chen CH, Abate AR, Lee D, Terentjev EM, Weitz DA (2009) Microfluidic assembly of magnetic hydrogel particles with uniformly anisotropic structure. Adv Mater 21(31):3201–3204
40. Jayaramudu T, Raghavendra GM, Varaprasad K, Sadiku R, Raju KM (2013) Development of novel biodegradable Au nanocomposite hydrogels based on wheat: for inactivation of bacteria. Carbohydr Polym 92(2):2193–2200
41. Kim J-H, Lee TR (2006) Discrete thermally responsive hydrogel-coated gold nanoparticles for use as drug-delivery vehicles. Drug Dev Res 67:61–69
42. Pardo-Yissar V, Gabai R, Shipway AN, Bourenko T, Willner I (2001) Gold nanoparticle/hydrogel composites with solvent-switchable electronic properties. Adv Mater 13(17):1320–1323
43. Guo YG, Hu JS, Liang HP, Wan LJ, Bai CL (2003) Highly dispersed metal nanoparticles in porous anodic alumina films prepared by a breathing process of polyacrylamide hydrogel. Chem Mater 15(22):4332–4336
44. Ghasemzadeh H, Ghanaat F (2014) Antimicrobial alginate/PVA silver nanocomposite hydrogel, synthesis and characterization. J Polym Res 21(3):355–369
45. Mohan YM, Lee K, Premkumar T, Geckeler KE (2007) Hydrogel networks as nanoreactors: a novel approach to silver nanoparticles for antibacterial applications. Polymer (Guildf) 48 (1):158–164
46. Murthy PSK, Mohan YM, Varaprasad K, Sreedhar B, Raju KM (2008) First successful design of semi-IPN hydrogel-silver nanocomposites: a facile approach for antibacterial application. J Colloid Interface Sci 318(2):217–224
47. Zou X et al (2017) Preparation of a novel antibacterial chitosan-poly(ethylene glycol) cryogel/silver nanoparticles composites. J Biomater Sci Polym Ed 28(13):1324–1337
48. Helminger M et al (2014) Synthesis and characterization of gelatin-based magnetic hydrogels. Adv Funct Mater 24(21):3187–3196
49. Ozay O, Ekici S, Baran Y, Aktas N, Sahiner N (2009) Removal of toxic metal ions with magnetic hydrogels. Water Res 43(17):4403–4411
50. Ullah F, Othman MBH, Javed F, Ahmad Z, Akil HM (2015) Classification, processing and application of hydrogels: a review. Mater Sci Eng, C 57:414–433
51. Gaharwar AK, Peppas NA, Khademhosseini A (2014) Nanocomposite hydrogels for biomedical applications. Biotechnol Bioeng 111(3):441–453
52. Zhang D, Yang J, Bao S, Wu Q, Wang Q (2013) Semiconductor nanoparticle-based hydrogels prepared via self-initiated polymerization under sunlight, even visible light. Sci Rep 3(1):1399
53. Castaneda L, Valle J, Yang N, Pluskat S, Slowinska K (2009) Collagen crosslinking with Au nanoparticles. Biomacromolecules 9(12):3383–3388
54. Rose S, Prevoteau A, Elzière P, Hourdet D, Marcellan A, Leibler L (2013) Nanoparticle solutions as adhesives for gels and biological tissues. Nature 505(7483):382–385
55. Carlsson L, Rose S, Hourdet D, Marcellan A (2010) Nano-hybrid self-crosslinked PDMA/silica hydrogels. Soft Matter 6(15):3619

56. Gaharwar AK, Rivera CP, Wu CJ, Schmidt G (2011) Transparent, elastomeric and tough hydrogels from poly(ethylene glycol) and silicate nanoparticles. Acta Biomater 7(12):4139–4148
57. Wu H et al (2013) Stable Li-ion battery anodes by in-situ polymerization of conducting hydrogel to conformally coat silicon nanoparticles. Nat Commun 4:1943–1946
58. Magasinski A et al (2010) Toward efficient binders for Li-ion battery Si-based anodes: polyacrylic acid. ACS Appl Mater Interfaces 2(11):3004–3010
59. Liu G et al (2011) Polymers with tailored electronic structure for high capacity lithium battery electrodes. Adv Mater 23(40):4679–4683
60. Bridel JS, Azaïs T, Morcrette M, Tarascon JM, Larcher D (2010) Key parameters governing the reversibility of Si/carbon/CMC electrodes for Li-ion batteries. Chem Mater 22(3):1229–1241
61. Gao G, Du G, Sun Y, Fu J (2015) Self-healable, tough, and ultrastretchable nanocomposite hydrogels based on reversible polyacrylamide/montmorillonite adsorption. ACS Appl Mater Interfaces 7(8):5029–5037
62. Özkahraman B, Acar I, Emik S (2011) Removal of Cu 2+ and Pb 2+ ions using CMC based thermoresponsive nanocomposite hydrogel. Clean - Soil, Air, Water 39(7):658–664
63. Gaharwar AK, Dammu SA, Canter JM, Wu C-J, Schmidt G (2011) Highly extensible, tough, and elastomeric nanocomposite hydrogels from poly(ethylene glycol) and hydroxyapatite. Biomacromolecules 12:1641–1650
64. Jayakumar R, Menon D, Manzoor K, Nair SV, Tamura H (2010) Biomedical applications of chitin and chitosan based nanomaterials - a short review. Carbohydr Polym 82(2):227–232
65. Liu R, Liang S, Tang X-Z, Yan D, Li X, Yu Z-Z (2012) Tough and highly stretchable graphene oxide/polyacrylamide nanocomposite hydrogels. J Mater Chem 22(28):14160
66. Gaharwar AK et al (2013) Bioactive silicate nanoplatelets for osteogenic differentiation of human mesenchymal stem cells. Adv Mater 25(24):3329–3336
67. Gaharwar AK et al (2012) Physically crosslinked nanocomposites from silicate-crosslinked PEO: mechanical properties and osteogenic differentiation of human mesenchymal stem cells. Macromol Biosci 12(6):779–793
68. Kaşgöz H, Durmuş A, Ka A (2008) Enhanced swelling and adsorption properties of AAm-AMPSNa/clay hydrogel nanocomposites for heavy metal ion removal. Polym Adv Technol 19:213–220
69. Yadollahi M, Farhoudian S, Namazi H (2015) One-pot synthesis of antibacterial chitosan/silver bio-nanocomposite hydrogel beads as drug delivery systems. Int J Biol Macromol 79:37–43
70. Mekewi M, Shebl A, Imam AI, Amin MS, Albert T (2012) Screening the insecticidal efficacy of nano ZnO synthesized via in-situ polymerization of crosslinked polyacrylic acid as a template. J Mater Sci Technol 28(11):961–968
71. Utracki LA (2004) Clay-containing polymeric nanocomposites, vol 1. Shropshire, United Kingdom
72. Nelson A, Cosgrove T (2004) A small-angle neutron scattering study of adsorbed poly (ethylene oxide) on laponite. Langmuir 20(6):2298–2304
73. Loizou E et al (2005) Large scale structures in nanocomposite hydrogels. Macromolecules 38(6):2047–2049
74. Kevadiya BD, Joshi GV, Patel HA, Ingole PG, Mody HM, Bajaj HC (2010) Montmorillonite-Alginate nanocomposites as a drug delivery system: Intercalation and in vitro release of vitamin B1 and vitamin B6. J Biomater Appl 25(2):161–177
75. Chen J, Park H, Park K (1999) Synthesis of superporous hydrogels: hydrogels with fast swelling and superabsorbent properties. J Biomed Mater Res - Part A 44:53–62
76. Bao J, Chen S, Wu B, Ma M, Shi Y, Wang X (2015) A novel foaming approach to prepare porous superabsorbent poly(sodium acrylic acid) resins. J Appl Polym Sci, 132(3):n/a–n/a

77. Kabiri K, Zohuriaan-Mehr MJ (2004) Porous superabsorbent hydrogel composites: synthesis, morphology and swelling rate. Macromol Mater Eng 289(7):653–661
78. Omidian H, Rocca JG, Park K (2005) Advances in superporous hydrogels. J Control Release 102(1):3–12
79. Demirtaş TT, Karakeçili AG, Gümüşderelioğlu M (2008) Hydroxyapatite containing superporous hydrogel composites: synthesis and in-vitro characterization. J Mater Sci 19:729–735
80. Kabiri K et al (2011) Superabsorbent polymer composites: does clay always improve properties? J Mater Sci 46(20):6718–6725
81. Zheng Y, Zhu Y, Wang A (2014) Highly efficient and selective adsorption of malachite green onto granular composite hydrogel. Chem Eng J 257:66–73
82. Wang L, Zhang J, Wang A (2008) Removal of methylene blue from aqueous solution using chitosan-g-poly(acrylic acid)/montmorillonite superadsorbent nanocomposite. Colloids Surf A Physicochem Eng Asp 322(1–3):47–53
83. Shirsath SR, Patil AP, Patil R, Naik JB, Gogate PR, Sonawane SH (2013) Removal of brilliant green from wastewater using conventional and ultrasonically prepared poly(acrylic acid) hydrogel loaded with kaolin clay: a comparative study, vol 20, no 3. Elsevier B.V
84. Mahdavinia GR, Asgari A (2013) Synthesis of kappa-carrageenan-g-poly(acrylamide)/sepiolite nanocomposite hydrogels and adsorption of cationic dye. Polym Bull 70(8):2451–2470
85. Liu Y, Zheng Y, Wang A (2010) Enhanced adsorption of methylene blue from aqueous solution by chitosan-g-poly (acrylic acid)/vermiculite hydrogel composites. J Environ Sci 22 (4):486–493
86. Zheng Y, Wang A (2009) Evaluation of ammonium removal using a chitosan-g-poly (acrylic acid)/rectorite hydrogel composite. J Hazard Mater 171(1–3):671–677
87. Zheng Y, Wang A (2010) Enhanced adsorption of ammonium using hydrogel composites based on chitosan and halloysite. J Macromol Sci Part A Pure Appl Chem 47(1):33–38
88. Wang Y, Wang W, Wang A (2013) Efficient adsorption of methylene blue on an alginate-based nanocomposite hydrogel enhanced by organo-illite/smectite clay. Chem Eng J 228:132–139
89. Zheng Y, Wang A (2010) Removal of heavy metals using polyvinyl alcohol semi-IPN poly (acrylic acid)/tourmaline composite optimized with response surface methodology. Chem Eng J 162(1):186–193
90. Saber-Samandari S, Saber-Samandari S, Gazi M (2013) Cellulose-graft-polyacrylamide/hydroxyapatite composite hydrogel with possible application in removal of Cu(II) ions, vol 73, no 11. Elsevier Ltd
91. Wang X, Wang A (2010) Adsorption characteristics of chitosan-g-poly(acrylic acid)/attapulgite hydrogel composite for Hg(II) ions from aqueous solution. Sep Sci Technol 45 (14):2086–2094
92. Zhu Y, Zheng Y, Wang A (2015) Preparation of granular hydrogel composite by the redox couple for efficient and fast adsorption of La(III) and Ce(III). J Environ Chem Eng 3 (2):1416–1425
93. Anirudhan TS, Suchithra PS, Arauf T (2012) Kinetic and equilibrium profiles of adsorptive recovery of thorium (IV) from aqueous solutions using poly (methacrylic acid) grafted cellulose/bentonite superabsorbent composite kinetic and equilibrium profiles of adsorptive recovery of thorium (IV), no IV
94. Zhu Y, Zheng Y, Wang A (2015) A simple approach to fabricate granular adsorbent for adsorption of rare elements. Int J Biol Macromol 72:410–420
95. Wang M, Li X, Hua W, Shen L, Yu X, Wang X (2016) Electrospun poly(acrylic acid)/silica hydrogel nanofibers scaffold for highly efficient adsorption of lanthanide ions and its photoluminescence performance. ACS Appl Mater Interfaces 8(36):23995–24007
96. Zheng Y, Wang A (2015) Superadsorbent with three-dimensional networks: from bulk hydrogel to granular hydrogel. Eur Polym J

97. Yang X, Ni L (2012) Synthesis of hybrid hydrogel of poly(AM co DADMAC)/silica sol and removal of methyl orange from aqueous solutions. Chem Eng J 209:194–200
98. Dhodapkar R, Rao NN, Pande SP, Nandy T, Devotta S (2007) Adsorption of cationic dyes on Jalshakti®, super absorbent polymer and photocatalytic regeneration of the adsorbent. React Funct Polym 67(6):540–548
99. Ma X, Sun X, Chen J, Lei Y (2017) Natural or natural-synthetic hybrid polymer-based fluorescent polymeric materials for bio-imaging-related applications. Appl Biochem Biotechnol 183(2):461–487
100. Zheng Y et al (2003) Immobilization of quantum dots in the photo-cross-linked poly (ethylene glycol)-based hydrogel, J Phys Chem B, pp 10464–10469
101. Cheng Z, Liu S, Beines PW, Ding N, Jakubowicz P, Knoll W (2008) Rapid and highly efficient preparation of water-soluble luminescent quantum dots via encapsulation by thermo-and redox-responsive hydrogels. Chem Mater 20:7215–7219
102. Wang YQ et al (2017) Nanocomposite carbon dots/PAM fluorescent hydrogels and their mechanical properties. J Polym Res, 24(12)
103. Ruiz-Palomero C, Benítez-Martínez S, Soriano ML, Valcárcel M (2017) Fluorescent nanocellulosic hydrogels based on graphene quantum dots for sensing laccase. Anal Chim Acta 974:93–99
104. Yang J, Han C-R, Duan J-F, Xu F, Sun R-C (2013) Insitu grafting silica nanoparticles reinforced nanocomposite hydrogels. Nanoscale 5(22):10858
105. Kakkar P, Madhan B (2016) Fabrication of keratin-silica hydrogel for biomedical applications. Mater Sci Eng, C 66:178–184
106. Algi MP, Okay O (2014) Highly stretchable self-healing poly(N, N-dimethylacrylamide) hydrogels. Eur Polym. J 59:113–121
107. Haque MA, Kurokawa T, Kamita G, Gong JP (2011) Lamellar bilayers as reversible sacrificial bonds to toughen hydrogel: Hysteresis, self-recovery, fatigue resistance, and crack blunting. Macromolecules 44(22):8916–8924
108. Shin SR, Bae H, Cha M, Mun Y, Chen Y, Tekin H (2012) Carbon Nanotube Reinforced Hybrid Microgels as Sca ff old Materials for Cell. ACS Nano 6(1):362–372
109. Zhong M, Liu Y-T, Xie X-M (2015) Self-healable, super tough graphene oxide–poly(acrylic acid) nanocomposite hydrogels facilitated by dual cross-linking effects through dynamic ionic interactions. J Mater Chem B 3(19):4001–4008
110. Fabbri P et al (2012) In-situ graphene oxide reduction during UV-photopolymerization of graphene oxide/acrylic resins mixtures. Polym (United Kingdom) 53(26):6039–6044
111. Shi K et al (2015) Near-infrared light-responsive poly(N-isopropylacrylamide)/graphene oxide nanocomposite hydrogels with ultrahigh tensibility. ACS Appl Mater Interfaces 7(49): 27289–27298
112. Schexnailder P, Schmidt G (2009) Nanocomposite polymer hydrogels. Colloid Polym Sci 287:1–11
113. Jaiswal MK et al (2014) Thermoresponsive magnetic hydrogels as theranostic nanoconstructs. ACS Appl Mater Interfaces 6(9):6237–6247
114. Khademhosseini A, Langer R (2007) Microengineered hydrogels for tissue engineering. Biomaterials 28(34):5087–5092
115. Gurkan UA, Tasoglu S, Kavaz D, Demirel MC, Demirci U (2012) Emerging technologies for assembly of microscale hydrogels. Adv Healthc Mater 1(2):149–158
116. Satarkar NS, Biswal D, Hilt JZ (2010) Hydrogel nanocomposites: a review of applications as remote controlled biomaterials. Soft Matter 6(11):2364
117. Jain A, Duvvuri LS, Farah S, Beyth N, Domb AJ, Khan W (2014) Antimicrobial polymers. Adv Healthc Mater 3(12):1969–1985
118. Sudheesh Kumar PT et al (2012) Flexible and microporous chitosan hydrogel/nano ZnO composite bandages for wound dressing: In vitro and in vivo evaluation. ACS Appl Mater Interfaces 4(5):2618–2629
119. Li M, Zhu L, Lin D (2011) Toxicity of ZnO nanoparticles to *Escherichia coli*: mechanism and the influence of medium components. Environ Sci Technol 45:1977–1983

120. Zhang L et al (2010) Mechanistic investigation into antibacterial behaviour of suspensions of ZnO nanoparticles against E. *coli*. J Nanopart Res 12:1625–1636
121. Sudheesh Kumar PT et al (2013) Evaluation of wound healing potential of β-chitin hydrogel/nano zinc oxide composite bandage. Pharm Res 30(2):523–537
122. Hashem M, Sharaf S, Abd El-Hady MM, Hebeish A (2013) Synthesis and characterization of novel carboxymethylcellulose hydrogels and carboxymethylcellulolse-hydrogel-ZnO-nanocomposites. Carbohydr Polym 95(1):421–427
123. Bhowmick S, Koul V (2016) Assessment of PVA/silver nanocomposite hydrogel patch as antimicrobial dressing scaffold: synthesis, characterization and biological evaluation. Mater Sci Eng, C 59:109–119
124. García-Astrain C et al (2015) Biocompatible hydrogel nanocomposite with covalently embedded silver nanoparticles. Biomacromolecules 16(4):1301–1310
125. Manjula B, Varaprasad K, Sadiku R, Ramam K, Reddy GVS, Raju KM (2014) Development of microbial resistant thermosensitive Ag nanocomposite (gelatin) hydrogels via green process. J Biomed Mater Res - Part A 102(4):928–934
126. Zendehdel M, Zendehnam A, Hoseini F, Azarkish M (2015) Investigation of removal of chemical oxygen demand (COD) wastewater and antibacterial activity of nanosilver incorporated in poly (acrylamide-co-acrylic acid)/NaY zeolite nanocomposite. Polym Bull 72(6):1281–1300
127. Devaki SJ, Narayanan RK, Sarojam S (2014) Electrically conducting silver nanoparticle-polyacrylic acid hydrogel by in situ reduction and polymerization approach. Mater Lett 116:135–138
128. Endo T, Ikeda R, Yanagida Y, Hatsuzawa T (2008) Stimuli-responsive hydrogel-silver nanoparticles composite for development of localized surface plasmon resonance-based optical biosensor. Anal Chim Acta 611(2):205–211
129. Zheng Y, Wang A (2012) Ag nanoparticle-entrapped hydrogel as promising material for catalytic reduction of organic dyes. J Mater Chem 22(32):16552
130. Skardal A, Zhang J, McCoard L, Oottamasathien S, Prestwich GD (2010) Dynamically crosslinked gold nanoparticle-hyaluronan hydrogels. Adv Mater 22(42):4736–4740
131. Dong D et al (2016) In Situ 'clickable' Zwitterionic Starch-Based Hydrogel for 3D cell encapsulation. ACS Appl Mater Interfaces 8(7):4442–4455
132. Zhu H, Yao X (2013) Synthesis and characterization of poly(Acrylamide-co-2-Acrylamido-2-Methylpropane Sulfonic Acid)/kaolin superabsorbent composite. J Macromol Sci Part A 50(2):175–184
133. Pourjavadi A, Ayyari M, Amini-Fazl MS (2008) Taguchi optimized synthesis of collagen-g-poly(acrylic acid)/kaolin composite superabsorbent hydrogel. Eur Polym J 44 (4):1209–1216
134. Liang R, Liu M, Wu L (2007) Controlled release NPK compound fertilizer with the function of water retention. React Funct Polym 67(9):769–779
135. Lee WF, Chen YC (2005) Preparation of reactive mineral powders used for poly(sodium acrylate) composite superabsorbents. J Appl Polym Sci 97(3):855–861
136. Lee WF, Chen YC (2005) Effect of intercalated reactive mica on water absorbency for poly (sodium acrylate) composite superabsorbents. Eur Polym J 41(7):1605–1612
137. Zhang J, Chen H, Wang A (2005) Study on superabsorbent composite. III. Swelling behaviors of polyacrylamide/attapulgite composite based on acidified attapulgite and organo-attapulgite. Eur Polym J 41:2434–2442
138. Li P, Zhang J, Wang A (2007) A NovelN-Succinylchitosan-graft-Polyacrylamide/Attapulgite composite hydrogel prepared through inverse suspension polymerization. Macromol Mater Eng 292(8):962–969
139. Li A, Wang A, Chen J (2004) Studies on Poly(acrylic acid)/Attapulgite superabsorbent composite. I. Synthesis and characterization. J Appl Polym 92:1596–1603

140. Ferfera-Harrar H, Aiouaz N, Dairi N, Hadj-Hamou AS (2014) Preparation of chitosan-g-poly (acrylamide)/montmorillonite superabsorbent polymer composites: studies on swelling, thermal, and antibacterial properties. J Appl Polym Sci 131(1):1–14
141. Kabir K, Mirzadeh H, Zohuriaan-Mehr MJ, Daliri M (2009) Chitosan-modified nanoclay-poly(AMPS) nanocomposite hydrogels with improved gel strength. Polym Int 58:1252–1259
142. Li L, Liu PS, Zhou NL, Zhang J, Wei SH, Shen J (2006) Synthesis and properties of a poly (acrylic acid)/montmorillonite superabsorbent nanocomposite. J Appl Polym Sci 102 (6):5725–5730
143. Islam MS, Rahaman MS, Yeum JH (2015) Electrospun novel super-absorbent based on polysaccharide-polyvinyl alcohol-montmorillonite clay nanocomposites. Carbohydr Polym 115:69–77
144. Sarkar DJ, Singh A, Mandal P, Kumar A, Parmar BS (2015) Synthesis and characterization of poly (CMC-g-cl-PAam/Zeolite) superabsorbent composites for controlled delivery of zinc micronutrient: swelling and release behavior. Polym Plast Technol Eng 54(4):357–367
145. Hwang SY, Yoon WJ, Yun SH, Yoo ES, Kim TH, Im SS (2013) Fabrication of superabsorbent ultrathin nanofibers using mesoporous materials for antimicrobial drug-delivery applications. Macromol Res 21(11):1281–1288
146. Zhang J, Yuan K, Wang YP, Gu SJ, Zhang ST (2007) Preparation and properties of polyacrylate/bentonite superabsorbent hybrid via intercalated polymerization. Mater Lett 61(2):316–320
147. Zhang Y, Fan L, Zhao P, Zhang L, Chen H (2008) Preparation of nanocomposite superabsorbents based on hydrotalcite and poly(acrylic-co-acrylamide) by inverse suspension polymerization. Compos Interfaces 15(7–9):747–757
148. Wang Q et al (2010) High-water-content mouldable hydrogels by mixing clay and a dendritic molecular binder. Nature 463(7279):339–343
149. Haraguchi K, Ning J, Li G (2015) Swelling/deswelling behavior of zwitterionic nanocomposite gels consisting of sulfobetaine polymer–clay networks. Eur Polym J 68:630–640
150. Hong S et al (2015) 3D Printing: 3D Printing of Highly Stretchable and Tough Hydrogels into Complex, Cellularized Structures. Adv Mater 27(27):4034–4040
151. Shi F-K, Wang X-P, Guo R-H, Zhong M, Xie X-M (2015) Highly stretchable and super tough nanocomposite physical hydrogels facilitated by the coupling of intermolecular hydrogen bonds and analogous chemical crosslinking of nanoparticles. J Mater Chem B 3 (7):1187–1192
152. Ahmed NB et al (2017) The physics and chemistry of silica-in-silicates nanocomposite hydrogels and their phycocompatibility. J Mater Chem B 5(16):2931–2940
153. Deubel F, Steenackers M, Garrido JA, Stutzmann M, Jordan R (2013) Semiconductor/polymer nanocomposites of acrylates and nanocrystalline silicon by laser-induced thermal polymerization. Macromol Mater Eng 298(11):1160–1165
154. Sayyar S, Gambhir S, Chung J, Officer DL, Wallace GG (2017) 3D printable conducting hydrogels containing chemically converted graphene. Nanoscale 9(5):2038–2050
155. Cong H, Wang P, Yu S (2013) Stretchable and self-healing graphene oxide-polymer composite hydrogels: a dual-network design. Chem Mater 25:3357–3362
156. Shah K, Vasileva D, Karadaghy A, Zustiak SP (2015) Development and characterization of polyethylene glycol—carbon nanotube hydrogel. J Mater Chem B 3:7950–7962
157. Servant A, Methven L, Williams RP, Kostarelos K (2013) Electroresponsive polymer-carbon nanotube hydrogel hybrids for pulsatile drug delivery in vivo. Adv Healthc Mater 2(6):806–811
158. Chen Z et al (2014) A three-dimensionally interconnected carbon nanotube-conducting polymer hydrogel network for high-performance flexible battery electrodes. Adv Energy Mater 4(12):1–10

159. Ghadban A et al (2016) Bioinspired pH and magnetic responsive catechol-functionalized chitosan hydrogels with tunable elastic properties. Chem Commun 52(4):697–700
160. Haider H et al (2015) Exceptionally tough and notch-insensitive magnetic hydrogels. Soft Matter 11(42):8253–8261
161. Guan Y, Chen J, Qi X, Chen G, Peng F, Sun R (2015) Fabrication of Biopolymer Hydrogel Containing Ag Nanoparticles for Antibacterial Property. Ind Eng Chem Res 54(30): 7393–7400

Processing, Characterization and Application of Natural Rubber Based Environmentally Friendly Polymer Composites

Nayan Ranjan Singha, Manas Mahapatra, Mrinmoy Karmakar and Pijush Kanti Chattopadhyay

Abbreviations

BF	Bamboo fiber
CB	Carbon black
CF	Coir/coconut fiber
CV	Conventional vulcanization
DMA	Dynamic mechanical analysis
EAB	Elongation at break
EV	Efficient vulcanization
IF	Isora fiber
IPN	Interpenetrating polymer network
LW	Leather waste
MBTS	2-mercaptobenzothiazole disulfide
NC	Nanocellulose
NCP	Nanocomposite
NF	Nanofiller
NRC	Natural rubber composite
NR	Natural rubber
NW	Nanowhisker
OPF	Oil palm fiber
PC	Polymer composite
PLA	Poly(lactic acid)
PLF	Pineapple leaf fiber

N. R. Singha (✉) · M. Mahapatra · M. Karmakar
Advanced Polymer Laboratory, Department of Polymer Science and Technology, Government College of Engineering and Leather Technology (Post-Graduate), Maulana Abul Kalam Azad University of Technology, Salt Lake, Kolkata 700106, West Bengal, India
e-mail: drs.nrs@gmail.com

P. K. Chattopadhyay
Department of Leather Technology, Government College of Engineering and Leather Technology (Post-Graduate), Maulana Abul Kalam Azad University of Technology, Salt Lake, Kolkata 700106, West Bengal, India

© Springer Nature Switzerland AG 2019
Inamuddin et al. (eds.), *Sustainable Polymer Composites and Nanocomposites*, https://doi.org/10.1007/978-3-030-05399-4_29

PP Polypropylene
RG Rubber granulate
RRG Recycled rubber granulate
SBR Styrene butadiene rubber
SEV Semi-efficient vulcanization
SF Sisal fiber
TMTD Tetramethyl thiuram disulfide
TSH Toluenosulfohydrazina
TS Tensile strength
ZMB Zinc-2-mercaptobenzothiazole

1 Introduction

Materials, together with energy and information, are considered to be the skeleton of the world economy of the twenty-first century. Among these materials, PCs, usually constituting of a polymer and one or more solid fillers, have been widely used for several years (Fig. 1) [1–5]. Instead of having several advantages, including a combination of the main properties of the two or more solid phases, PCs suffer from several limitations, such as difficulty in reuse and recycling. In fact, once the PCs become useless, these are commonly disposed directly or incinerated. However, both of these techniques are costly as well as difficult and possess high environmental impact. Indeed, such problems have begun to be evident for the last 15 years, and hence the recent scientific research has been inclined to look for the new alternatives, such as replacement of the traditional PCs with environment friendly PCs having lower environmental impact, often referred to as 'eco-composites' or 'green composites' (Fig. 1). In fact, the ecological damage, such as global warming and plastic pollution, caused by the conventional petroleum-based polymer products has encouraged the use of renewable and biodegradable materials by both scientific and industrial communities. Moreover, replacement of orthodox microcomposites by NCP has gained high insight in the last two decades to overcome the limitations of micrometre-scale, via designing novel materials and structures with unprecedented flexibility, elevated physical properties, and considerable industrial impact [6]. Indeed, the term NCP describes a class of two-phase materials, in which one of the phases has at least one dimension lower than 100 nm (Fig. 1). Thus, green chemistry coupled with nanotechnology to produce 'green polymeric NCP', based on derived raw materials of natural sources of plant or animal origin and rigid nano particles are of great interest in scientific, academic, and industrial fields because of the environmental and technological concerns. Moreover, because of the high surface to volume ratio, green NCPs exhibit unique mechanical, electrical, and thermal properties along with the environmental safety. Furthermore, in recent years, extensive

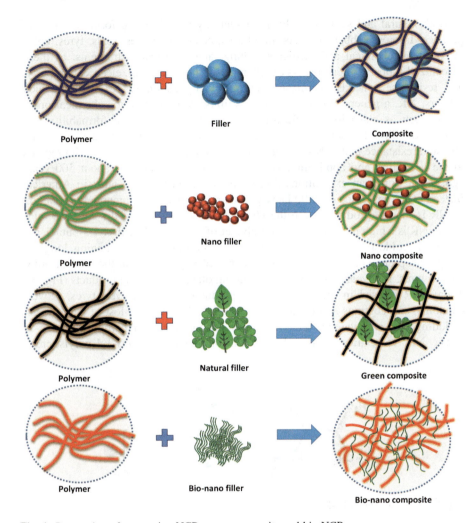

Fig. 1 Preparation of composite, NCP, green composite and bio-NCP

efforts have been devoted to develop, characterize, and utilize bio-based materials, and bio-based NCPs belonging to the new era of bio-based materials. These have attracted both industrial and academic attention because of the increasing interest in developing new sustainable, biodegradable, biocompatible, and environment-friendly nanomaterials. In fact, bio-NCPs can be considered as a subset of polymer NCPs where the NFs, the matrix or both come from bio-based renewable resources (Fig. 1). Moreover, incorporation of NFs into biopolymers provides the practical way to improve the properties of such bioplastics, to make them competitive with petroleum-derived materials.

Both natural and synthetic rubbers, essentially in vulcanized form, are used to produce different rubber products, like hoses, cushions, gloves, seals, tyres, belts, diving gear, chemical and medicinal tubing, and electrical instruments, as these vulcanized rubbers, often possess excellent flexibility, elasticity, electrical property, and resistance towards chemicals and crack propagation [7–10]. Among such rubbers, NR, a general purpose rubber, exhibits excellent physico-chemical properties, such as elasticity and flexibility, as well as magnificent formability and biodegradability. In fact, superior strength, elasticity, flexibility, resilience, and abrasion resistance makes NR as one of the most important elastomers with regards to the versatility and application volume. NR was first reported about 500 years ago, when European expeditions first experienced rubbers and latex in America. Instead of the existence of 2500 latex producing plants, the commercial production of NR is mainly produced from the *Hevea brasiliensis* tree of the Amazon rainforest. NR is a high molecular weight polymer of isoprene (2-methyl-1,3-butadiene) and is the oldest known rubber (Fig. 2). Accordingly, NR, containing long *cis* 1,4-polyisoprene chains, finds a large number of applications in the field of automotive tyres, footwears, and for manufacturing other engineering products (Fig. 3). Today, NR is used in producing 50,000 products, like adhesives, tyres, gloves, condoms, and coatings and the applications are still growing. Most of the rubbers, including NR, are available as aqueous dispersions of rubber particles, known as latex. NR is extracted as white emulsion containing *cis*-1,4-polyisoprene

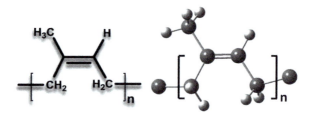

Fig. 2 2D/3D structure of NR

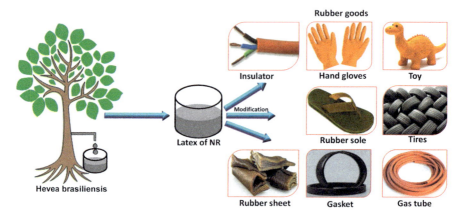

Fig. 3 Modifications and uses of NR

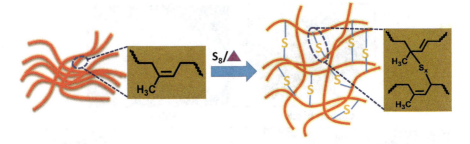

Fig. 4 Vulcanization of NR

nanoparticles, usually possessing approximately 100 nm diameter. However, sticky and inelastic uncured NR is useless. Therefore, uncured NR is vulcanized or cured to make it a more durable crosslinked material. Theoretically, vulcanization is a chemical process by which long chains of rubber molecules are crosslinked, leading to the transformation of the soft, weak plastic-like material into a strong elastic product of high and reversible deformability along with good mechanical properties because of strain-induced crystallization, low hysteresis, extraordinary dynamic properties, and fatigue resistance. During vulcanization, rubber is heated with sulfur or other equivalent curatives with/without accelerators. Such additives reinforce the polymer chains via generating crosslinks between individual polymer chains to attain improved elasticity, resilience, and enhanced mechanical properties (Fig. 4). Indeed, in the traditional vulcanization process of NR, CB is commonly employed as the reinforcing filler for achieving improved strength, weather resistance, and stiffness, leading to the production of traditional NR composites (NRC). Despite possessing several advantages, including the capability to produce reinforced NRCs, CBs are non-biodegradable petrochemical-based products, which consume a substantial amount of energy during their production. Therefore, rubber manufacturers are in search of new reinforcing fillers, which should be renewable, readily available, cheap, light-weight, and biodegradable to achieve environmentally friendly NR bases PCs (Fig. 5).

Till date, rubber-based NCPs have been studied lesser than the plastic-NCPs, in which most of the research are focusing on the use of either nanoclays or carbon nanotubes as reinforcements. Over the past decade, in order to replace CB to synthesize NR composites, the use of wood, cellulosic fibers, and their derivatives as organic fillers has attracted much attention. In this context, cellulose, the most abundant natural polymer, is used to prepare high strength nanoparticles because of the outstanding properties, such as biocompatibility, required chemical stability, superior thermal stability, and environmental benignancy. In fact, cellulose nanofibers can also be used as a matrix to form green NCPs because of the natural

Fig. 5 Different types of fillers

abundance, renewability, and biodegradability. In this context, different workers investigated the potentials of biodegradable organic fillers, such as sisal, [11] coir, oil palm, [12] isora, BFs, starch, carrageenan, rattan, pistachio, peanut shell powder, and coconut shell powder, as the reinforcing additives in NRCs. In the recent past, the potential of NC, originated from microcrystalline cellulose, bamboo residue of newspaper production unit, [13] rachis of the date palm tree, sugar cane bagasse kraft pulp, and jute fiber, was explored as reinforcing additive in fabricating eco-friendly NR latex based NCPs. In fact, very few NR-based NCPs with bio-based nano-reinforcements, such as chitin whiskers, cellulose whiskers of Syngonanthus nitens (*Capim dourado*), starch nanocrystals, rachis of a palm tree, sisal, and bagasse are found in the literature. As expected, the NC not only provided the superior mechanical strength in rubber NCPs but also it increased the rate of degradation of rubber in the soil when disposed at the end of life. Thus, being a bio-based polymer, the use of bio-nano reinforcements in NR is beneficial in the development of bio-based and green NCPs.

Theoretically, in order to attain high specific strength, modulus, and dimensional stability in NRCs, the compatibility between NR and filler, and their interfacial adhesion should be sufficient enough to ensure strong interaction between filler and NR matrix. Since the surface of the added filler is often polar and hydrophilic and NR is non-polar and relatively hydrophobic, the interfacial adhesion between filler and NR can be improved by modifying the surface characteristics of both the components by physical or chemical treatment (Fig. 6). For instance, the interfacial bonding between NR and filler(s), such as rice husk, jute, sisal, and silk fiber, was improved by physical treatment, like electron beam and gamma irradiation. For instance, the hydrophobicity of NR matrix and hydrophilicity of cellulose NCPs are inherently incompatible and insufficient molecular-scale interactions can resist the

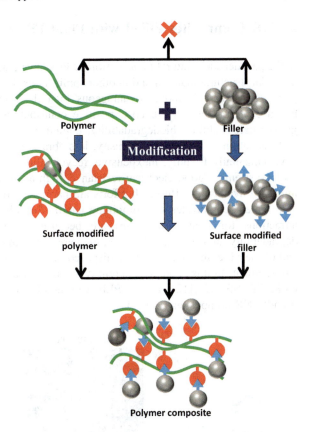

Fig. 6 Surface modification of NR/filler or both improving homogeneous dispersion and interfacial adhesion

entire activity of the material. Moreover, cellulose NCP aggregates act as stress concentrator and produce poor dispersion within the matrix to deteriorate the properties of composites. Conventionally, the chemical grafting of specific moieties on the cellulose NCP surface was used to control the interfacial adhesion and interactions. Alternatively, modification of NR chains in contact with the cellulose NCP filler could also be carried out.

Furthermore, attempts have been made to fabricate polymer-rubber composites wherein ground particles of waste tyres have effectively been utilized as potent fillers. Though waste tyres are pollutants, the ground used tyres should be treated as a source of sustainable materials, instead of a pollutant. In this regard, eight different types of NR are presently used as basic components in tyre manufacture. Therefore, a significant portion of the waste tyre should contain NR as the important constituent. Accordingly, NR based ground waste tyre can be used as an important ingredient in preparing NR based environmental friendly PCs.

2 NR Composites Filled with Plant Fibers

In the past decades, natural fiber reinforced PCs have gained substantial interest as a potential structural material and in other applications. Use of natural fiber as filler in polymeric matrix is more advantageous than the conventional inorganic fillers because of low energy the cost, positive contribution to global carbon budget, greater deformability, biodegradability, combustibility, recyclability, fair thermal and insulation properties, low density, less abrasiveness to processing equipment, environmentally friendly, inexpensive, recyclability, non-toxicity, flexibility, high specific strength, good electrical resistance, high acoustic insulation property, and universal availability. Research efforts are harnessed to develope fully biodegradable 'green' composites via combining natural-/bio-fibers with rubber. The major attractions for green composites are because of environment-friendly nature, degradability, and sustainability, i.e., they are truly 'green' in every respect. At the end of their life, they can easily be disposed of or composted without damaging the environment. In this context, different types of plant-derived natural fibers, such as OPF, CF, SF, BF, [12] IF, and PLF were employed to prepare environmentally friendly NR composites (Fig. 7).

Fig. 7 Cellulose nanofibers from different plants

Natural fibers are superior to synthetic ones with regards to properties, such as biodegradability, lightweight, low toxicity, cost, and availability. In order to achieve adequate adhesion between fibers and matrix, studies were focused on the treatment of fibers to improve bonding between fiber and matrix. Various methods, such as mercerization and corona, plasma, heat, and silane treatment were reported to enhance the bonding in natural fiber composites. Moreover, various studies were conducted on hybrid NR composites filled with multiple type natural fibers derived from plant resources [11]. The main objective of fabricating a hybrid composite was to extract the combined advantages of two or more different types of fibers so that the advantages of one type of fiber could compensate the limitations of the other. As a consequence, the proper balance between performance and cost could be achieved through proper material designing. Meanwhile, the individual characteristics of plant-derived natural fibers were described one after another. In this context, physical properties and chemical constituents of different types of natural fibers are summarized in Table 1.

Table 1 Composition and properties of different natural fibers extracted from plants

Fibers	IF	CF	OPF	SF	BF	PLF
Diameter (μm)	10	100–400	150–250	103	–	55–95
Density (g cm^{-3})	1.20–1.30	1.20	0.70–1.55	1.50	0.60–1.10	0.80–1.60
TS (MPa)	500–600	175	50–400	511–635	140–230	400–627
Young modulus (GPa)	–	4.00–6.00	3.20	9.40–22.00	11.00–17.00	1.44
EAB (%)	5.0–6.0	15.0–40.0	25.0	2.0–2.5	–	14.5
Moisture content (%)	6.0–7.0	10.0–12.0	16.0	–	11.7	–
Micro fibrillar angle (°)	20.00–26.00	3.49	42.00–46.00	20.00–25.00	–	–
Holocellulose (%)	–	–	68.3–86.3	–	73.3	–
Cellulose (%)	74	32–43	65	65	26–43	81
Hemicellulose (%)	–	0.15–0.25	–	12.00	30.00	–
Alpha-cellulose (%)	–	–	41.9–60.6	–	48.2	–
Lignin (%)	23.0	40.0–45.0	29.0	9.9	21.0–31.0	12.7
Fat (%)	1.09	0.30	–	–	–	–
Pentosan (%)	–	–	17.8–20.3	–	20.3	–
Waxes	–	–	–	2	–	–

3 Coir/Coconut Fiber (CF)

The lignocellulosic fiber, originated from the fibrous mesocarp of the fruit of the tropical coconut trees (*Cocos nucifera*), is called the coir. CF is more profitable than other natural fibers, including high weather resistance due to the presence of higher amount of lignin and poor water absorptivity because of the lesser cellulose content. Moreover, this fiber can be stretched beyond the elastic limit without rupture, because of the helical arrangement of micro-fibrils at 45°.

4 Oil Palm Fiber (OPF)

OPF, a waste material of oil extraction, is a lignocellulosic fiber obtained from the empty fruit and bunch fibrous mesocarp of oil palm tree (*Elaeis guineensis*). Of the different fiber sources in oil palm tree, empty fruit bunch can yield up to 73% fibers and hence, it is preferable in terms of availability and cost. Palm oil industry has to dispose of about 1.1 ton of empty fruit bunch per each ton of oil produced. In fact, retting process is utilized for extracting OPF from empty fruit bunch. Of the different commonly used retting processes, such as mechanical retting (hammering), chemical retting (boiling with chemicals), steam/vapor/dew retting and water/microbial retting, water retting is the most popular. OPFs are hard, tough, and show similarity to CFs in cellular structure. The central part of the transverse section of OPF contained a lacuna like portion surrounded by porous tubular structures. The morphology of fibrous surface, containing pores having average diameter of 0.07 μm, is essential for stronger mechanical bonding with matrix resin in composite fabrication.

5 Sisal Fiber (SF)

Sisal is an agave (*Agave sisalana*) and commercially produced in Brazil and East Africa. SF is one of the strongest fibers and can be used for several applications. SF possesses excellent ageing resistance.

6 Bamboo Fiber (BF)

Bamboo (*Bambusa Shreb.*), a perennial plant, grows up to 40 m in monsoon climates. Bamboo is most commonly used in construction, carpentry, weaving, and plaiting. Fabrication and studies on BF filled NR composites have been conducted. In fact, overall mechanical properties of BF, abundantly available in Asia, are comparable to wood.

7 Isora Fiber (IF)

For the first time, Mathew et al. investigated the applicability of the IF, extracted from the bark of the *Helicteres isora* plant by retting, as a reinforcing additive in preparing NR composites [12]. In fact, Isora shrubs grow in many parts of South India, especially in Kerala. Isora resembles jute in appearance but possesses superior strength, durability, and lustre. It has the better TS than some other natural fibers.

8 Pineapple Leaf Fiber (PLF)

Pineapple (*Ananas comosus*), a tropical plant of Brazil, is cellulose-rich, relatively inexpensive and highly abundant. PLF, the waste product of pineapple cultivation is relatively inexpensive and readily available for industrial purposes. PLF exhibits high specific strength and stiffness because of high cellulose content of 70–80% and comparatively low microfibrillar angle. However, because of the hydrophilic character of its cellulose structure and high susceptibility to water absorption, mostly at elevated temperatures, inadequate bonding between PLF and the hydrophobic matrix is the major problem associated with the application of PLF as filler in NR.

9 Processing

In general, natural fiber-filled composites are processed in an open two-roll mill. For instance, OPF filled composites were prepared in a laboratory two roll mill (150–300 mm) at a nip gap of 1.3 mm. Followed by initial mastication of NR, untreated chopped OPF is added along with other main ingredients like accelerator, activator, and vulcanizing agent. In this context, the fibers were added towards the end of the mixing process, so that to ensure minimum breakage of the fiber during mixing.

Prior to the addition of fibrous fillers, the usual practice is to execute optimal cleaning, washing, and drying of the plant fibers, followed by size reduction via chopping or any other suitable mechanical operation. However, to enhance the bonding in natural fiber composites, fibers are pre-treated before incorporating in the NR matrix. For instance, chemical treatment, such as mercerization of IF requires 3–4 h of continuous heating at 80 °C in 10% aqueous NaOH solution, followed by washing with water and drying in an air oven at 70 °C. In case of acetylation, the same fiber is initially treated by alkali and thereafter the alkali-treated fibers are soaked in glacial CH_3COOH for 1 h. Later, the material is decanted and then soaked in acetic anhydride containing two drops of concentrated

H_2SO_4 for 5 min. Finally, the acetylated fibers are filtered, washed, and dried in an air oven at 70 °C. As a consequence of chemical treatment, certain physical and microstructural changes occur on the fiber surface, such as dissolution and leaching out of fatty acids and lignin components of the fiber. For example, as a result of mercerization of IF, a considerable quantity of uranic acid, a constituent of hemi-cellulose (xylan), can be removed from the fiber. As a consequence of acetylation, substantial esterification of O–H of IF is actuated. Similarly, prior to the compounding process, raw CFs are also undergone various chemical treatments to remove coir pith and other undesirable materials, and thereby improving the binding of CF with NR.

10 Characterization

10.1 Mechanical Properties

Effect of OPF length on the mechanical properties of the NR compounds are analyzed by stress-strain measurement. Both TS and EAB are maximized when the length of OPF becomes 6 mm (Table 2). At higher fiber lengths, significant deterioration in the mechanical properties is observed because of the entangling tendency of the longer fibers. Altogether, 6 mm was found to be the optimum fiber length for OPF reinforced NR matrix. Moreover, the mechanical properties of the OPF filled NR composites in the longitudinal direction are superior to the transverse direction. However, incorporation of OPF in NR matrix decreases TS and EAB. The intrinsic high strength of NR, related to the strain-induced crystallization, is disrupted when fibers are incorporated into NR, thereby destroying the regular arrangement of rubber molecules, resulting in the deterioration of crystallization. The mechanical properties of the CF-reinforced NR composites in the longitudinal direction are superior to those in the transverse direction. However, in this case, the optimum length for CF is maintained at 10 mm for achieving good reinforcement in NR composites. Moreover, to maximize the fiber orientation and tensile properties of the CF-reinforced NR composites, CFs are immersed in 5% NaOH solution for 48 h. Similarly, NaOH and benzoyl peroxide is used to treat the surfaces of PLF. It is found that all surface modifications enhance adhesion and tensile properties of PLF-NR composites. In fact, treatment with 5% NaOH and 1% benzoyl peroxide provides the best improvement of composite strength by 28 and 57% respectively, when compared with that of untreated fiber. Similarly, the adhesion between the BF and NR can be enhanced by the use of a bonding agent, such as silane coupling agents, phenol formaldehyde, and hexamethylenetetramine, leading to improved tensile modulus and overall mechanical properties of BF-NR composites.

The influence of the ratio of two lignocelluloses fibers, i.e., sisal and oil palm on the tensile properties of NR composites have already been reported. The mechanical properties are found to be more dependent on SF than oil palm, because of the superior tensile properties of SF than OPF (Table 1). Moreover, since the microfibrillar angle of SF is lesser (20°) than OPF (44°), the reinforcing ability of sisal is more than that of oil palm in any polymeric matrix. Furthermore, the surface area of the fiber in a unit area of the composite is higher in SF filled composite than OPF filled composite because the diameter of SF is lesser than that of OPF (Table 1). Hence, physical interaction, as well as stress transfer in the unit area, is higher for SF filled composites. Altogether, it has been noted that TS of SF-OPF-NR hybrid composite is lesser than pure gum. In this context, the mechanical properties of OPF-sisal-NR composite, studied by Jacob et al. are tabulated in Table 2.

10.2 Dynamic Mechanical Properties

Most rubber articles, such as automobile tyres, springs, and dampers, undergo cyclic loading or cyclic deformation, and hence, dynamic properties are crucial for evaluating the real-time service performance of those articles. In this context, a complete description of the viscoelastic properties is derived via dynamic experiments conducted over a range of time, temperature or frequency. It is observed that the stress relaxation rate of OPF-SF-NR hybrid composites decreases with increase in the fiber content. In fact, the relaxation of rubber molecules in the gum compound is hindered because of the influence of fiber-rubber interface formed via addition of fibers. Notably, at all temperatures, storage modulus of OPF-SF-NR composites enhances with the rise in fiber content [11]. Invariably, the unfilled NR compound, containing only rubber phase, makes the material better flexible to impart low stiffness, and thus, low storage modulus. Once the fiber was added, the stiffness of the composite increases as fibers allow greater stress transfer at the fiber-rubber interface, resulting in higher storage modulus. The loss modulus also increases with fiber loading, to reach up to 756 MPa at 50 phr fiber loading, whereas gum has loss modulus of 415 MPa. In this context, the damping factor decreases with fiber loading because of lower flexibility and lower degrees of molecular motion caused by incorporation of fibers in a rubber matrix.

Short CF reinforced NR composites with poor interfacial bonding tend to dissipate higher energy than those with fair interfacial bonding. The composite, containing fibers subjected to bleaching, exhibits very high tan δ values in the low-temperature range but the low values at high-temperature region. This proves that such composites are good elastomeric compounds at higher temperatures. However, composites of resorcinol-formaldehyde-latex treated CFs exhibit low tan δ values at both low and high temperatures, suggesting low damping and hence, good interfacial bonding. Moreover, with the increased fiber loading, glass transition temperature (T_g) of the composites continuously shifts towards higher temperature, because of increased immobilization of the polymer chains adhered to the treated CFs.

Table 2 Tensile properties of various polyamide-filled composites and NCPs

Composites/nano-composites	TS (MPa)	EAB (%)	Tensile modulus or Young's modulus (Mpa)	Bending strength (Mpa)	Bending Modulus (Mpa)	Impact strength (kJ m^{-2})	Hardness (shore A)	Abrasion resistance (mm^3/40 m)	Hysteresis (MPa)
Non-purified NR/silk	11.81 ± 0.31	21 ± 5	–	–	–	–	–	–	–
Purified NR/silk	8.49 ± 0.09	24 ± 5	–	–	–	–	–	–	–
Non-purified NR/nylon	6.54 ± 0.17	76 ± 12	–	–	–	–	–	–	–
Purified NR/nylon	5.48 ± 0.16	77 ± 10	–	–	–	–	–	–	–
Silk fiber/PP:NR (50:50)	41.10	25	901.5	37.1	1413.3	23.0	93.0	–	–
Silk fiber/PP:NR (75:25)	42.40	21	955.3	39.2	1564.8	26.1	93.0	–	–
Silk fiber/PP:NR (90:10)	45.10	18	1278.8	42.9	2132.4	21.9	94.0	–	–
γ irradiated (250 krad) silk fiber/PP:NR (50:50)	46.60	–	1313.8	44.3	1620.1	–	–	–	–
γ irradiated (250 krad) silk fiber/PP:NR (75:25)	47.10	–	1410.1	46.1	2200.6	–	–	–	–
γ irradiated (250 krad) silk fiber/PP:NR (90:10)	48.30	–	1553.5	48.9	2620.8	–	–	–	–
Soy particle basic/NR (10 phr)	21.00	550	1.9	–	–	–	–	–	–
Soy particle basic/NR (20 phr)	21.10	510	2.4	–	–	–	–	–	–

(continued)

Table 2 (continued)

Composites/nano-composites	TS (MPa)	EAB (%)	Tensile modulus or Young's modulus (Mpa)	Bending strength (Mpa)	Bending Modulus (Mpa)	Impact strength (kJ m^{-2})	Hardness (shore A)	Abrasion resistance (mm^3/40 m)	Hysteresis (MPa)
Soy particle basic/NR (30 phr)	19.40	400	3.8	–	–	–	–	–	–
Soy particle basic/NR (40 phr)	14.60	280	6.4	–	–	–	–	–	–
Soy particle acidic/NR (10 phr)	20.80	550	1.9	–	–	–	–	–	–
Soy particle acidic/NR (20 phr)	21.90	550	2.4	–	–	–	–	–	–
Soy particle acidic/NR (30 phr)	19.80	440	4.2	–	–	–	–	–	–
Soy particle acidic/NR (40 phr)	14.40	340	6.0	–	–	–	–	–	–
NR/CB	18.60 ± 1.10	759 ± 19	–	–	–	–	42.0 ± 1.6	77.10 ± 4.10	–
NR/CB/leather-60 phr	12.20 ± 1.10	84 ± 2	–	–	–	–	72.7 ± 0.9	213.00 ± 11.90	–
NR/CB/leather-80 phr	8.60 ± 0.80	45 ± 1	–	–	–	–	76.0 ± 1.7	178.90 ± 3.30	–
NR foam	1.38	409	0.0026	–	–	–	–	–	0.126
NR foam/leather waste–20 phr	–	395	–	–	–	–	–	–	0.206
NR foam/leather waste–40 phr	–	74	–	–	–	–	–	–	3.459
NR foam/leather waste–60 phr	2.18	58	0.1206	–	–	–	–	–	7.552
NR	4.70 ± 0.90	700 ± 65	–	–	–	–	56.6 ± 1.1	207.21 ± 33.74	–

(continued)

Table 2 (continued)

Composites/nano-composites	TS (MPa)	EAB (%)	Tensile modulus or Young's modulus (Mpa)	Bending strength (Mpa)	Bending Modulus (Mpa)	Impact strength (kJ m^{-2})	Hardness (shore A)	Abrasion resistance (mm^3/40 m)	Hysteresis (MPa)
NR/leather waste–20 phr	5.50 ± 0.20	679 ± 15	–	–	–	–	61.3 ± 1.2	112.73 ± 24.01	–
NR/leather waste–40 phr	6.90 ± 0.30	543 ± 17	–	–	–	–	63.6 ± 1.2	135.66 ± 52.66	–
NR/leather waste–60 phr	8.90 ± 0.60	435 ± 29	–	–	–	–	64.5 ± 0.5	197.21 ± 46.82	–
NR/leather waste–80 phr	9.20 ± 0.70	366 ± 24	–	–	–	–	67.3 ± 0.5	215.21 ± 14.07	–
OPF-sisal-NR–5 phr	–	–	–	–	–	–	–	–	–
OPF-sisal-NR–10 phr	1.75	650	–	–	–	–	–	–	–
OPF-sisal-NR–30 phr	7.50	800	–	–	–	–	–	–	–
OPF-sisal-NR–50 phr	3.25	–	–	–	–	–	–	–	–
OPF-NR–5 phr	19.20	1082	–	–	–	–	–	–	–
OPF-NR–10 phr	–	–	–	–	–	–	–	–	–
OPF-NR–30 phr	–	496	–	–	–	–	–	–	–
OPF-NR–50 phr	7.28	–	–	–	–	–	–	–	–

11 Application

Significant research is currently underway around the world to address and overcome the obstacles to developing biocomposite materials with improved performance for global applications. Interfacial adhesion between natural fibers and matrix plays the pivotal role for the overall performance, since the final properties of the composites totally depend on it. Recently, CFs bonded with NR latex are being used in seats of the Mercedes Benz A-class model.

12 NC Reinforced NR NCPs

Recently, NCbased reinforcement in NCPs is gaining high insight. Besides low cost, density, and energy consumption, renewability, high specific properties, biodegradability, and relatively good surface reactivity, it shows better properties as a reinforcing phase in NCPs than micro-/macro-cellulose composites. The concept of cellulosic NF reinforced polymer materials has shown rapid advances and considerable interest in the last decade, because of their renewable character, high mechanical properties, low density, availability, and diversity of sources. Because of the perfect balance between flexibility and stiffness, NR matrix is used as a model system to study the effect of cellulose NFs reinforcement. Currently, NC reinforced NR-NCPs is one of the important categories of lignocellulosic fibers mediated rubber composite materials. In fact, the characteristics of NC depend on the origin of fibers and the isolation methods. Cellulosic nanoparticles consist of either cellulose whiskers or microfibrillated cellulose. The nanoscale dimensions of cellulose crystals enable cellulose NCPs to impart unique characteristics. The extensive research work is devoted to cellulose nanoparticles obtained by either (i) disintegration shearing for microfibrillated cellulose or (ii) chemical acid hydrolysis treatment for cellulose nanocrystals or whiskers. Cellulose of various sources has been utilized to produce such cellulosic nanoparticles. Generally, the elongated rod-like high-purity single cellulose nanocrystals are produced from different sources, whose dimensions depend on the nature of cellulose source and hydrolysis conditions. In fact, the diameter and length typically lie within 5–10 and 100–500 nm, respectively. Cellulosic nanoparticles can be classified into two main groups: (i) cellulose nanocrystals, as obtained by acid treatment and (ii) cellulose nanofibers, synthesized through mechanical disintegration (Fig. 8). Both cellulose nanocrystals and cellulose nanofibers are used for different applications depending on their properties. Indeed, nanofibrillated cellulose has been used in these NCPs, including NR based NCPs, because of typically ultra-high strength and environmental friendliness of NC.

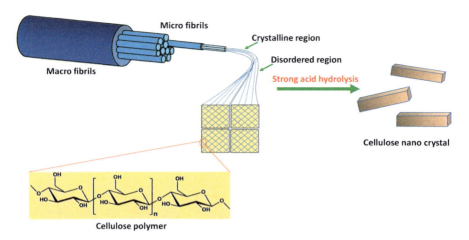

Fig. 8 Different forms of cellulose: process modification steps

13 Processing

Bio-based rubber composites are fabricated through the standard rubber processing operations, such as compression moulding, injection moulding, and extrusion. NC whisker reinforced NR-NCPs are prepared by applying the conventional rubber compounding method, after extracting cellulose NWs from bamboo pulp residue collected from newspaper production unit [13]. Initially, cellulose NWs are prepared via acid hydrolysis of the bamboo pulp and thereafter, the cellulose NW based NR NCPs have been produced via a two-step process. In the first step, a master-batch is prepared via dispersing the cellulose NWs in NR latex, followed by coagulation of the dispersion. In the next step, this coagulated master-batch is compounded with solid NR and vulcanizing agents in a two-roll mill and thereafter the NR compound is cured via compression moulding at 150 °C. Earlier, employing the similar acid hydrolysis protocol, cellulose whiskers were extracted from cellulose fibers present in cassava bagasse. Thereafter, NW-NR composites have been prepared by initial mixing of NWs into NR latex, followed by casting and evaporation. Applying the same methodology, NC-NR-NCPs are fabricated, wherein the added cellulose nanoparticles were initially extracted from soy hulls by acid hydrolysis.

NR-NCPs filled with cellulose nanoparticles, i.e. whiskers and microfibrils, are fabricated, where cellulose whiskers are extracted via bleaching the purified cell wall of the rachis of date palm tree, whereas the cellulose microfibrils are obtained through the disintegration of the bleached and purified cell wall by microfluidizer. Later, NR-NCP films have been prepared by casting/evaporation method, comprising of cellulose whiskers of 84–102 and 4–12 nm length and width, respectively. Indeed, these whiskers are initially isolated from bleached sugar cane bagasse kraft pulp. Thereafter, these purified whiskers are admixed to NR latex, followed by casting/evaporation method for synthesizing NCPs. In an almost similar fashion, the NR-NC type composites are prepared after isolating the NC from raw jute fiber by steam explosion.

The direct extrusion method can be adopted in preparing cellulose-rubber NCPs. The lyophilized NC, produced from microcrystalline cellulose, is initially pulverized in a grinder to obtain NC powder, and then the pulverized NC powder is mixed with NR in an extruder to generate the bio-composite. While preparing bio-composite, the temperature for direct extrusion is restricted to 160 °C to avoid thermal degradation of NC whiskers. Nevertheless, such NC based NCPs are generally prepared via casting method comprising of two processing steps, in which the first step is associated with the mixing of NC suspension with rubber latex for a stipulated period (i.e. 0.5–12 h) to produce uniform NC dispersion. However, in the second step, this uniform aqueous NC dispersion is cast in a formulation mould and dried to produce NC films. In between these two steps, degassing and water evaporation should be carried out, based on the concentration and bubbles in the suspension. In addition, a combination of casting and extrusion can also be attempted, wherein the NC dispersion is cast and dried to produce NC based film and thereafter, the film is cut and extruded with NR.

Recently, attempts have been made to prepare NC-oxidized NR-NCPs via dispersing NC into oxidized NR latex, followed by usual casting and evaporation. In fact, NR latex suspensions are oxidized with a $KMnO_4$ solution to introduce O–H groups at the NR chains and thereby to increase the possibility of H-bonding between hydroxylated polyisoprene chains and NC. In the recent past, few publications have reported the properties of NR as a matrix to synthesize green, conductive, and flexible NCPs. Graphene sheets are introduced in water suspensions of NC prior to the reduction of the particles by adding hydrazine hydrate. The resultant hybrid suspension is mixed with NR latex and dried to form a structured conductive film. The similar strategy of coating nanoparticles may be employed to obtain polyaniline modified cellulose nanofibers, in which polyaniline is attached to the cellulose nanofiber surface by in situ polymerization, and NR-NCP is obtained by applying the casting and evaporation technique.

Recently, attempts have been made to prepare NR/Regenerated Cellulose hybrids comprising of a cellulose-rich phase and NR latex particles. Such hybrid was obtained by simply co-precipitating the mixture of NR latex and cellulose alkaline—urea—aqueous solution. As a result, honeycomb-like structural moieties of Regenerated Cellulose were noted to become homogeneously distributed in the hybrid matrix wherein Regenerated Cellulose and NR phases interlaced/interpenetrated each other to form a semi-IPN/fullIPN structure.

14 Characterization

14.1 Biodegradability

Biodegradability of the cellulose whisker filled NR-NCPs in soil was noted to become significantly enhanced as compared to that of the unfilled NR. It is well known that the NR degrades in nature by specific microorganisms, such as *Streptomyces coelicolor* 1A and *Nocardia farcinica* strain E1, in a slow process and accordingly the growth of these rubber utilizing bacteria is also slow. However, as the biodegradation of cellulose is faster than rubber, the cellulose component in the NR NCPs films is rapidly consumed by the microorganisms, producing porosity, void formation, and the loss of the integrity of the rubber matrix. Thus, the rubber matrix would be broken down into smaller particles and accordingly smaller and less organized rubber particles become more susceptible to the bacterial degradation. Similarly, jute fiber originated NC filled NR demonstrated a significantly higher level of biodegradability over the unfilled NR. As expected, the non-crosslinked NR composite showed a higher degree of biodegradation when compared with the crosslinked NR composites. In this context, a compost system of increased degradation potential, constituting of the complex biological environment having relatively higher microbial diversity, was utilized to enhance biodegradation of these composite materials. As a result of composting, quicker deterioration of the whole composite material, including the interior part of the composite, was achieved by means of rapid biodegradation of NC. Accordingly, NR-NC biocomposite envisaged lower TS retention when compared with the neat NR, as unreinforced NR showed the higher resistance to the microorganism attacks in comparison to that of the NC filled composite.

14.2 Mechanical Properties

As compared to the mere PLA materials, a strong increase in EAB is observed when 10 wt% of NR is added in PLA matrix. Such ductile behaviour and EAB of PLA-NR blend is effectively conserved in the PLA grafted NC filled PLA-NR-NC bio-NCPs, as grafted short chains of PLA on NC act as the effective compatibilizer between NC and PLA phases, even though the PLA grafted NC are preferentially located in the PLA phase (Table 3).

The stress-strain behaviour of cellulose whisker filled NR NCPs was considerably different from neat NR. In fact, a non-linear mechanical behaviour of NR-cellulose whisker NCPs is observed in the tensile test performed at room temperature. Stress-strain curves clearly demonstrate the stiffening effect of the cellulose whiskers in the NR NCPs. Both Young's modulus and TS significantly increase upon whisker addition, while the EAB decreases. Such high reinforcing effect of cellulose whisker can be assigned to the mechanical percolation phenomenon of cellulose whiskers, which forms a stiff continuous network of cellulosic nanoparticles linked through hydrogen bonding. Such improvement in Young's modulus and TS is also explained by a mechanism based on the formation of the Zn-cellulose complex (Fig. 9). The three-dimensional network of cellulose nanofibers (cellulose-cellulose-/Zn-cellulose-network) in the NR matrix can play the pivotal role to enhance the properties of the crosslinked NCPs. In this context, relative improvements in mechanical properties, demonstrated by cellulose whiskers isolated from various sources, have been demonstrated in Table 2. Herein, the aspect ratio of different cellulose whiskers is an important factor that guides the variegated mechanical properties of NR NCPs filled with cellulose whiskers isolated from different resources. Thus, as compared to cellulose whiskers originated from starch, rachis of date palm tree and *Capim dourado*, relatively lower aspect ratio of cellulose whiskers isolated from bagasse could be the reason behind the lower TS properties of NR NCPs filled with cellulose whiskers isolated from bagasse. Similarly, mild hydrolysis is preferable in enhancing the extraction yield of NC crystals from soy hulls as well as to maintain the crystallinity of native cellulose and obtain high aspect ratio NC crystals.

Accordingly, a high reinforcing effect is observed even at low filler contents when high aspect ratio NC crystals are used to prepare NCPs with a NR matrix by casting/evaporation. For instance, by adding only 2.5 wt% NC crystals, the storage tensile modulus at 25 °C of the NCP was about 21 times higher than that of the neat NR matrix. However, it has to be kept in mind that the ultimate strength is not only dependent on the chemical interactions between the matrix and the NF but also contributed by the physical entanglements of the NC having a high aspect ratio. For this reason, the modulus of the 5% composite gives a comparative increase of fourfold with its pristine matrix. But the 10% composite showed a modulus of 9.6 MPa and the comparative increase from 5 to 10% NCP is only 2.5 fold (Table 2).

Besides, mechanical properties of NC based polymer NCPs are function of the filler dispersibility and compatibility of NC with the matrix. Moreover, because of the presence of numerous hydroxyl groups on the surface of NC, NC possesses strong tendency to form an aggregate, and hence dispersion of NC is really difficult in the nonpolar or hydrophobic polymer matrices. Improper distribution of NC may lead to the formation of NC aggregates which can act as stress concentrator, resulting in poor performance of the NCP. Thus, the formation of NC aggregates should be avoided to achieve effective reinforcement. In this regard, considerable enhancement in both TS and modulus values was observed with the increased addition of cellulose NWs, accompanied by a moderate decrease in EAB, as NCPs were devoid of any micro-scaled aggregates of cellulose NWs. Furthermore, mechanical properties of NC filled NR NCPs can be improved by enhancing the interfacial interactions between NR and NC, via introducing a limited extent of –OH groups in the NR chains via oxidization of the NR. However, uncontrolled oxidation-mediated generation of a huge number of –OH, led to severe deterioration of the mechanical properties. As discussed earlier, the percolation phenomenon of NC in NR was effectively modified by the introduction of graphene, leading to the formation of an assembled conductive structure that played a key role to improve electrical conductivity and mechanical properties of the cellulose mediated NR/graphene composites. Such unique 'fragile' but effective conductive network with low percolation threshold facilitated the disconnection and reestablishment of conductive paths in presence of organic solvents. Thus, the composite, having such sensitive conductive network, can function as high-performance sensing materials with superior resistivity responses for organic liquids.

To improve the interfacial interactions and compatibility between NR and NC, attempts were made to introduce the cross-linkable mercapto-groups onto the surface of cotton originated cellulose nanocrystals by esterification [14]. In comparison to biocomposites based on NR filled with unmodified cellulose nanocrystals, the NR NCPs having modified cellulose nanocrystals showed a 2.4-fold increase in TS and 1.6-fold increase in EAB. Indeed, in the modified cellulose nanocrystals, mercaptoundecanoyl groups were introduced at the surface, leading to attachment of long hydrocarbon chain on the surface of modified cellulose nanocrystals, which reduced the hydrophilic nature of the cellulose nanocrystals and consequently improved the compatibility of the modified cellulose nanocrystals with the hydrophobic NR matrix. The cross-linking of NR with modified NC surface, through the thiol functionalities on the nanocrystal surface, increased strength and toughness of the NR/modified cellulose nanocrystal composites as summarized in Table 3. Synergistic effect of cross-linking at the filler–matrix interface together with reinforcement in NR/modified cellulose nanocrystal.

NCPs offered by the thiol-modified cellulose nanocrystals is expressed in these results. Indeed, formation of covalent thioether (C–S) bonds at the NR/modified cellulose nanocrystal composite interface was identified from the FTIR results which suggested the reaction of –SH groups of mercaptoundecanoyl group in modified cellulose nanocrystals with the double bonds of NR (Fig. 10) [14]. Likewise, both cellulose nanofibrils and polyaniline treated cellulose nanofibrils were highly effective in improving overall mechanical properties of NR NCPs (Table 3). However, both Young's modulus and TS are lower for cellulose nanofibril/polyaniline-reinforced NCPs as compared to cellulose nanofibril reinforced NR. This result can be explained by the fact that cellulose nanofibril is more hydrophobic than cellulose nanofibril/polyaniline resulting in a higher level of adhesion with the NR matrix. Moreover, comparative reinforcing abilities of both cellulose whiskers and microfibrillated cellulose in NR composites were determined, and it was observed that the reinforcing effect was higher for NCPs filled with microfibrillated cellulose over the whisker filled NR. Again, relatively higher aspect ratio and the possible presence of entanglements in microfibrillated cellulose were the major factors behind the greater reinforcing ability of microfibrillated cellulose over the whiskers. Moreover, the presence of residual lignin, extractive substances and fatty acids at the surface of microfibrillated cellulose was also suggested to promote higher adhesion level with the NR matrix. Later, in order to investigate the role of fatty acids in enhancing the reinforcing capability, attempts were also made to achieve highly efficient reinforcement of NR with cellulose

Table 3 Tensile properties of modified and unmodified composites

Composites	TS (MPa)	Young modulus (MPa)	Work-of fracture (MJ m^{-3})	EAB (%)	Stress at 100% (MPa)	Stress at 200% (MPa)	Crosslink density (mol L^{-1})
NR[a]	2.40 ± 0.40	1.01 ± 0.08	1.45 ± 0.41	910 ± 174	–	–	–
NR/CNCs[b]-2	3.30 ± 0.90	1.05 ± 0.03	1.89 ± 0.67	975 ± 120	–	–	–
NR/CNCs[b]-5	3.60 ± 0.40	1.10 ± 0.08	1.73 ± 0.48	960 ± 200	–	–	–
NR/CNCs[b]-10	4.20 ± 0.80	1.75 ± 0.38	1.56 ± 0.32	750 ± 125	–	–	–
NR/m-CNCs[c]-2	6.80 ± 1.50	1.49 ± 0.31	2.97 ± 0.39	1220 ± 30	–	–	–
NR/m-CNCs[c]-5	9.60 ± 2.00	1.53 ± 0.26	4.18 ± 0.75	1270 ± 157	–	–	–
NR/m-CNCs[c]-10	10.20 ± 1.30	1.86 ± 0.12	4.60 ± 0.57	1210 ± 110	–	–	–
NR[a]	3.52	–	–	860	1.90	2.05	1.39 × 10^{-3}
1% NR[a] nano-composite	3.81	–	–	750	2.20	2.30	1.61 × 10^{-3}
2% NR[a] nano-composite	4.15	–	–	620	2.60	2.60	1.91 × 10^{-3}
3% NR[a] nano-composite	4.25	–	–	410	3.05	3.40	2.20 × 10^{-3}
NR[a]	1.08 ± 0.18	0.65 ± 0.02	–	698 ± 45	–	–	–
NR/CNF[d] (95/05)	2.72 ± 0.21	3.12 ± 0.34	–	527 ± 28	–	–	–
NR/CNF[d] (90/10)	4.03 ± 0.30	8.15 ± 0.90	–	462 ± 14	–	–	–
NR[a]	1.72 ± 0.39	1.33 ± 0.39	–	878 ± 57	–	–	–
NRC[e]	2.08 ± 0.45	7.47 ± 1.67	–	684 ± 69	–	–	–
ONR2C[f]	2.37 ± 0.42	7.92 ± 1.02	–	703 ± 43	–	–	–
ONR3C[f]	2.18 ± 0.93	8.36 ± 0.85	–	697 ± 40	–	–	–
ONR4C[f]	0.36 ± 0.05	5.02 ± 1.13	–	570 ± 50	–	–	–
ONR5C[f]	0.11 ± 0.02	0.72 ± 0.03	–	202 ± 58	–	–	–

(continued)

Table 3 (continued)

Composites	TS (MPa)	Young modulus (MPa)	Work-of fracture (MJ m^{-3})	EAB (%)	Stress at 100% (MPa)	Stress at 200% (MPa)	Crosslink density (mol L^{-1})
NR[a]	–	0.50 ± 0.15	–	575 ± 35	0.56 ± 0.12	–	–
NR-CNW1[g]	–	1.70 ± 0.50	–	408 ± 49	0.86 ± 0.06	–	–
NR-CNW2.5[g]	–	2.80 ± 0.40	–	358 ± 22	1.17 ± 0.24	–	–
NR-CNW5[g]	–	8.40 ± 1.10	–	231 ± 53	2.71 ± 0.10	–	–
NR-CNW10[g]	–	118.00 ± 6.00	–	16 ± 3	8.93 ± 1.23	–	–
NR-CNW15[g]	–	187.00 ± 0.50	–	14 ± 1	12.15 ± 1.48	–	–
NR-MF-1[h]	–	1.27 ± 0.00	–	209 ± 29	0.70 ± 0.13	–	–
NR-MF-2.5[h]	–	10.52 ± 0.66	–	15 ± 5	0.80 ± 0.26	–	–
NR-MF-5[h]	–	35.46 ± 5.79	–	14 ± 2	2.17 ± 0.38	–	–
NR-MF-7.5[h]	–	121.20 ± 8.80	–	8 ± 2	4.15 ± 0.71	–	–
NR-NR-MF-10[h]	–	172.00 ± 62.00	–	7 ± 2	5.99 ± 2.56	–	–
NR-MF-15[h]	–	233.00 ± 57.00	–	4 ± 1	6.26 ± 2.70	–	–
NR[a]	1.60 ± 0.20	1.30 ± 0.15	–	912 ± 19	–	–	–
2.5% NR[a] nano-composite	5.20 ± 0.15	4.20 ± 0.25	–	576 ± 23	–	–	–
5% NR[a] nano-composite	6.80 ± 0.18	6.30 ± 0.22	–	413 ± 22	–	–	–
7.5% NR[a] nano-composite	9.80 ± 0.24	8.10 ± 0.35	–	275 ± 12	–	–	–
10% NR[a] nano-composite	12.20 ± 0.36	9.60 ± 0.31	–	144 ± 5	–	–	–
NR[a]	0.59 ± 0.08	0.60 ± 0.10	–	611 ± 71	0.19 ± 0.04	–	–
NR1%CNC[b]	0.89 ± 0.04	1.90 ± 0.30	–	396 ± 21	0.27 ± 0.01	–	–

(continued)

Table 3 (continued)

Composites	TS (MPa)	Young modulus (MPa)	Work-of fracture (MJ m^{-3})	EAB (%)	Stress at 100% (MPa)	Stress at 200% (MPa)	Crosslink density (mol L^{-1})
NR2.5%CNC[b]	1.11 ± 0.13	3.70 ± 0.10	–	485 ± 35	0.34 ± 0.02	–	–
NR5%CNC[b]	3.03 ± 0.11	18.10 ± 2.80	–	552 ± 9	0.54 ± 0.05	–	–
NR[a]	0.63 ± 0.14	0.69 ± 0.19	–	747 ± 38	–	–	–
NR/5RC[i]	1.36 ± 0.17	1.77 ± 0.21	–	612 ± 45	–	–	–
NR/10RC[i]	2.38 ± 0.19	2.63 ± 0.31	–	570 ± 32	–	–	–
NR/15RC[i]	4.81 ± 0.28	4.59 ± 0.45	–	505 ± 29	–	–	–
NR/20RC[i]	5.44 ± 0.37	11.70 ± 1.80	–	484 ± 24	–	–	–
NR/25RC[i]	5.36 ± 0.42	15.70 ± 1.70	–	370 ± 18	–	–	–
NR/30RC[i]	6.03 ± 0.61	20.80 ± 2.90	–	225 ± 11	–	–	–
NR[a]	1.60 ± 0.20	1.30 ± 0.15	–	912 ± 19	–	–	–
NR/2.5%NC[j]	5.20 ± 0.15	4.20 ± 0.25	–	576 ± 23	–	–	–
NR/5%NC[j]	6.80 ± 0.18	6.30 ± 0.22	–	413 ± 22	–	–	–
NR/7.5%NC[j]	9.80 ± 0.24	8.10 ± 0.35	–	275 ± 12	–	–	–
NR/10%NC[j]	12.20 ± 0.36	9.60 ± 0.31	–	144 ± 5	–	–	–
NR[a]	16.10 ± 1.40	1.70 ± 0.00	–	623 ± 14	–	–	–
NR[a] + CNF[d]1%	20.80 ± 3.10	2.20 ± 0.10	–	658 ± 41	–	–	–
NR[a] + stCNF[k]1%	15.20 ± 2.10	5.0 ± 0.40	–	513 ± 40	–	–	–
NR[a] + oleCNF[l]1%	18.80 ± 1.70	5.4 ± 0.80	–	531 ± 25	–	–	–
NR[a] + CNF[d]3%	28.40 ± 2.80	3.60 ± 0.30	–	713 ± 44	–	–	–
NR[a] + stCNF[k]3%	22.40 ± 2.50	9.60 ± 0.80	–	537 ± 37	–	–	–
NR[a] + oleCNR[l]3%	25.60 ± 1.00	12.70 ± 1.90	–	492 ± 12	–	–	–
NR[a] + CNF[d]5%	30.30 ± 0.40	4.40 ± 0.10	–	718 ± 6	–	–	–

(continued)

Table 3 (continued)

Composites	TS (MPa)	Young modulus (MPa)	Work-of fracture (MJ m^{-3})	EAB (%)	Stress at 100% (MPa)	Stress at 200% (MPa)	Crosslink density (mol L^{-1})
NR^a + stCNF^{k}5%	28.90 ± 1.40	18.30 ± 1.00	–	530 ± 30	–	–	–
NR^a + oleCNF^{l}5%	16.70 ± 2.40	27.70 ± 4.40	–	251 ± 69	–	–	–
NR^a	9.20 ± 1.30	1.70 ± 0.20	–	554 ± 9	–	–	–
NR^a-CNW^{g}2.5	14.00 ± 2.10	2.60 ± 0.10	–	539 ± 14	–	–	–
NR^a-CNW^{g}5	14.50 ± 2.60	3.00 ± 0.30	–	477 ± 13	–	–	–
NR^a-CNW^{g}10	17.30 ± 1.40	3.80 ± 0.20	–	455 ± 11	–	–	–

^{a}Natural rubber, ^{b}natural rubber/cellulose nanocrystals, ^{c}natural rubber/modified cellulose nanocrystals, ^{d}natural rubber/cellulose nanofibrils, ^{e}natural rubber cellulose whiskers, ^{f}natural rubber NCPs, ^{g}natural rubber-cellulose nanowhiskers, ^{h}natural rubber/regenerated cellulose, ^{i}natural rubber nano-cellulose

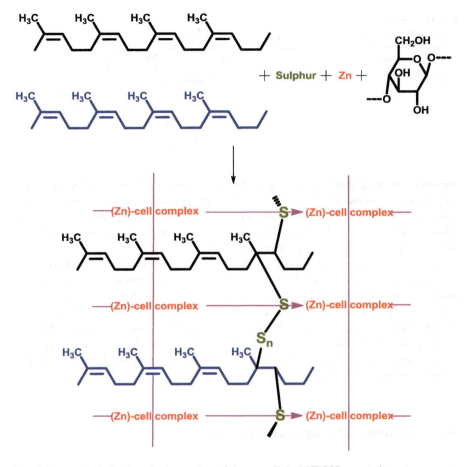

Fig. 9 Proposed mechanism for interaction of the cross linked NR/NC composite

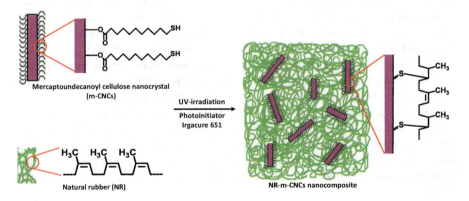

Fig. 10 Illustration of the structure for the NR-m-CNCs NCPs

Fig. 11 Diagram of sulfur vulcanization reaction between polyisoprene and CNFs incorporating unsaturated fatty acids (oleic acid)

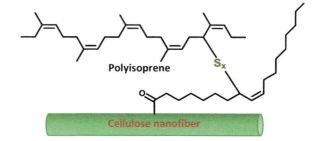

nanofibers bearing unsaturated fatty acids, which crosslinked with sulfur using the polyisoprene double bonds (Fig. 11) [15]. In particular, the incorporation of unsaturated fatty acid groups such as oleoyl on the cellulose nanofiber surfaces was effective after sulfur vulcanization because of the creation of crosslinks with the sulfur via the polyisoprene double bonds (Table 3).

Recently, significant improvements in both Young's modulus and TS have been reported in case of semi-IPN/IPN type NR/Regenerated Cellulose hybrids having microstructures comprising of unique honeycomb-like structure that encouraged the formation of intense physical entanglements/interlocks within the matrix and thereby promoted the polymer–filler interaction (Table 3). Indeed, the stretching of NR/Regenerated Cellulose hybrids was effectively hindered because of interlocking effect imposed by the honeycomb-like structure of Regenerated Cellulose on the slippage of NR domains (Fig. 12).

14.3 Dynamic Mechanical Properties

Though the mechanical properties of cellulose whisker filled NR NCPs markedly differ from neat NR, DMA results did not exhibit any significant change in the T_g of the rubber matrix. However, above T_g, a higher increase of the storage tensile modulus is observed in NR NCPs filled with an increasingly higher amount of cellulose whiskers, which could be related to the increased whiskers/whiskers interaction probability and density of the cellulosic network. In fact, good interaction between cellulose NWs and NR chains was generally reflected in the increased storage modulus, along with the slight positive shift in tan δ peak position of the NCP, if NCPs were devoid of any micro-scaled aggregates.

Similarly, formation of covalent thioether (C–S) bonds at the NR/modified cellulose nanocrystal composite interface in NR/modified cellulose nanocrystal composite is also responsible for the improved modulus compared to NR/cellulose nanocrystal composite in the transition region [14]. A significant reinforcing effect is observed and the rubbery modulus increased upon cellulose whiskers addition in NR.

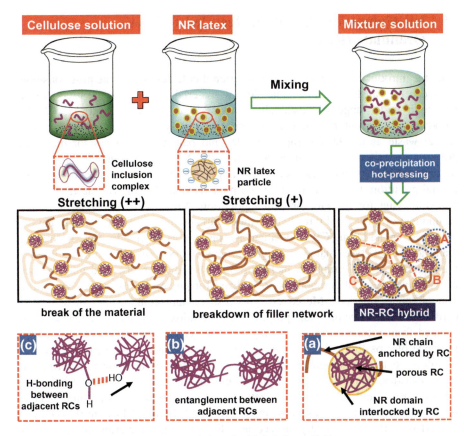

Fig. 12 Schematic illustration of NR reinforced with RC from alkaline-urea-aqueous system

15 Application

NR composites, fabricated with 3D interconnected graphene-based conductive networks, could be utilized as an eco-friendly strategy for fabrication of liquid sensors capable of sensing, discriminating, and monitoring various solvent leakage from the chemical industries.

15.1 Packaging

Because of the 100% disintegrating ability, PLA-NR-NC bio-NCPs have the potential to be utilized as biodegradable packaging materials.

16 NR Composites Based on Recycled Rubber Granulate (RRG)

With the introduction of environmentally friendlier technologies, the post-consumer tyres can be transformed into valuable raw materials, which can be used to synthesize a wide range of sustainable polymeric composites. In 1853, Charles Goodyear, the inventor of rubber vulcanization, firstly reported the use of ground rubber waste as a rubber compound filler and patented a process for moulding polymer materials obtained from RG and NR. Nowadays, dynamic increase in rubber wastes, especially as used tyres, is one of the major source for environmental pollution. Complex structure and poor recyclability of tyre materials is the potential issue for the environmental pollution. Vulcanized rubber is the major component of these used tyres, which makes up about 70–80% of the total mass of the tyres. In addition, CB and silica are present as fillers. In fact, during the manufacture of tyre materials, eight different types of NRs, thirty different high-quality synthetic rubbers, including SBR, butadiene rubber and butyl rubber, and various chemical compounds are utilized for processing as well as vulcanization of rubber matrices. Moreover, tyres also contain components, such as steel cord and fibers, made of nylon, polyester, and cellulose, which are mandatory to be isolated from rubber during recycling of waste tyre. Furthermore, because of the presence of crosslinks, the tyres are insoluble and infusible and hence, cannot be re-processed by the simple process that is generally used in manufacturing the thermoplastics. Thus, for the sustainable management of used tyres, grinding of rubber wastes, and subsequent utilization of these granulates as a component or filler can be opted to process new 'environmentally friendly' PCs. In this context, ground tyre rubber is usually utilized as a filler for manufacturing composites of thermosets, thermoplastics, and virgin rubber.

Continuous attempts have been made to improve the properties of RG filled NR composites. In this regard, materials containing only ground rubber wastes are developed by sintering the RGs at elevated temperature and pressure. By the application of sintering process, not only significantly higher quantity of rubber wastes can be recycled, but also no fresh rubber material is required to produce such materials. Indeed, this group of rubber materials becomes very attractive as these composites are environmentally friendlier and more economical than the composites comprising of ground rubber wastes and a newly incorporated rubber components.

17 Processing

NR-SBR based compounds have been synthesized using RRG as filler, replacing the conventionally used CB (N-220). Initially, RRGs were generated via grinding the waste tyres. Thereafter, NR based compounds are prepared in a mixer with the rotor speed of 60 rpm at 60 °C for 6 min, followed by the addition of other ingredients in a two-roll mill. Thereafter, the compound is moulded into sheets, maintaining the vulcanizing temperature and time of 150 °C and 10 min, respectively. In an almost similar fashion, NR, RG, and various ingredients have mixed in a laboratory two-roll mill, followed by moulding at 140 °C and optimum cure time evaluated earlier by disc rheometer. Thereafter, the ageing studies of those moulded samples have also been conducted at 100 °C for 36 h in a hot air oven. In order to analyze the influence of vulcanization, three different types of NR vulcanizates have been prepared via different vulcanizing systems, viz. conventional (CV), semi-efficient (SEV) and efficient (EV) curing system. In this context, ground rubber particles are produced in the laboratory from fully cured NR vulcanizates by a mechanical crusher having rotary type cutters. Thereafter, by means of a two-roll mill, different RG particles are admixed into the NR vulcanizates produced by three different vulcanizing systems, followed by the usual molding operation in a hot press at 150 °C for a predetermined optimum curing time.

On the other hand, sintered RGs are obtained at an elevated temperature (80–240 °C) and high pressure (0.5–26.0 MPa). In fact, the RGs are press moulded at high pressure (0.5–26.0 MPa), leading to consolidated adhesion of grains and improvement in their mutual attachment. Side by side, at the elevated temperature (80–240 °C), the crosslinks in RGs are broken up and the main chains are also partially disrupted to generate radicals. Consequently, in the next stage, new bonds connecting the individual grains of granulate can be generated via rearrangement of the generated radicals, which ultimately produces a homogenous rubbery material.

18 Characterization

According to the earlier studies, properties of RGs are mostly dependent on the size-reduction methods (i.e., grinding at cryogenic or ambient-temperature grinding), grain dimension, extent of crosslinking, quantity of filler, and the type of NR that are used to produce such RGs.

18.1 Mechanical Properties

Incorporation of ground rubber tyres into NR reduces TS, EAB and tear resistance of the vulcanizates. The effect is more pronounced for composites filled with larger particles (Table 4). This is reasonable since as the particle size decreases, surface area increases, and flaw size in the matrix also decreases. However, it is found that smaller the particle, poorer is the ageing property. For example, NR mixed with ground rubber tyre of <52 mesh particles (650–450 µm) registers the retention of TS and tear strength of 71 and 78%, respectively, while the NR mixed with ground rubber tyre of 100–150 mesh particles (150–100 μm) exhibits only 31 and 46% retention of TS and retention of tear strength, respectively. Furthermore, metals present in ground rubber tyres possess detrimental effect on the physical properties of aged vulcanizates. In fact, the occurrence of huge amounts of metals in fine-grained rubber dust is attributed to the metallic impurities generated via grinding of the residual steel wire beads present in the ground rubber tyre.

Mechanical properties and the vulcanization characteristics of rubber mixtures, obtained from NR and RGs, are dependent on the type of crosslinking agent applied at the time of their vulcanization. However, except EAB, all the properties of ground RG filled vulcanizates are adversely affected (Table 4). Notably, for NR vulcanizates prepared via SEV system, the relative decrease in TS, modulus, and tear strengths are smaller than those of NR vulcanizates fabricated by CV and EV systems. The composites containing crosslinked RGs by sulfur possess greater TS, EAB, and tear strength over the composites obtained from RGs crosslinked with a

Table 4 Tensile properties of various RRP filled NR composites and NCPs

Composites/NCPs	TS (MPa)	EAB (%)	Tensile modulus or Young's modulus (Mpa) at 100% elongation	Tear strength (kN m^{-1})
NR (unfilled)	14.00	1175.00	–	28.20
NR (unfilled and aged)	8.00	770.00	–	20.30
NR filled with 30 phr 150–100 μm RRP particles	8.00	860.00	–	21.20
NR filled with 30 phr 150–100 μm RRP particles (aged)	2.50	400.00	–	9.70
NR filled with 30 phr 650–450 μm RRP particles	2.20	430.00	–	12.40
NR filled with 30 phr 650–450 μm RRP particles (aged)	1.60	230.00	–	9.70
NR (conventional vulcanizate)	27.70	387.00	5.20	117.00
NR (conventional vulcanizate) with 50 phr RRP	17.60	368.00	2.55	73.00
NR (semi-efficient vulcanizate)	25.70	396.00	3.63	111.00
NR (semi-efficient vulcanizate) with 50 phr RRP	21.60	417.00	2.35	86.00
NR (efficient vulcanizate)	23.90	451.00	2.45	92.00
NR (efficient vulcanizate) with 50 phr RRP	15.90	461.00	1.37	61.00

peroxide. In general, it is noted that the mechanical properties significantly deteriorates with the elevated RG content of the composites. However, with the increase in the RG content, from 10 to 50 phr, EAB increases from ca 320–360 to 360–400%. Indeed, overall deterioration of mechanical properties for increasingly filled vulcanizates is closely associated with the continuous reduction of crosslink densities for vulcanizates filled with the higher amount of RGs. Moreover, significantly deteriorated mechanical properties of composites, bearing peroxide crosslinked granulate, are related to the prevalence of weaker adhesion force among fillers and the matrices. In this context, the reasonable deterioration of mechanical properties has also been noticed with the increasing particle size and extent of RGs in NR composites. However, partial replacement of CB by RRP in NR composites, amounting up to 15 wt%, cannot affect TS, EAB, and hardness of the composites. Nevertheless, the increasingly higher extent of replacement of CB by RRG adversely affects the overall mechanical properties. The EAB of RG filled NR composites increases with the increasing amount of RG.

The mechanical properties of sintered rubbers are predominantly dependent on the temperature, processing time, type of RGs and their grain size. Again, materials obtained from NR based RGs can attain TS within 3.5–6.5 MPa and EAB of 330–530%. Further improvement of these sintered materials can be possible if NR granulates are moulded for 20–30 min at 200 °C and 8.6 MPa. Moreover, the mechanical properties of sinters, based on NR granulate, can be elevated by adding organic acids of low molecular weight, such as benzoic acid, salicylic acid, maleic acid/anhydride, phthalimide, and phthalic anhydride [16]. In fact, investigation of the mechanisms of sintering and the underlying factors behind the enhancement of the properties during incorporation of additives, such as benzoic acid, salicylic acid, maleic acid/anhydride, phthalimide and phthalic anhydride, have already been reported that contains the possible way of breaking and reconstitution of crosslinks during sintering of RGs based on NR (Fig. 13). It is well-known that the mechanical properties of the rubber decrease with the increase in the formation of conjugated double (Fig. 14) bonds because of reversion. Thus, mechanical properties of sintered NR are significantly inferior to that of composites made from fresh NR. In fact, the

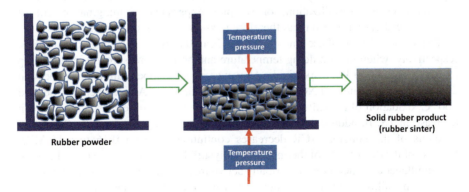

Fig. 13 Process for sintering of rubber

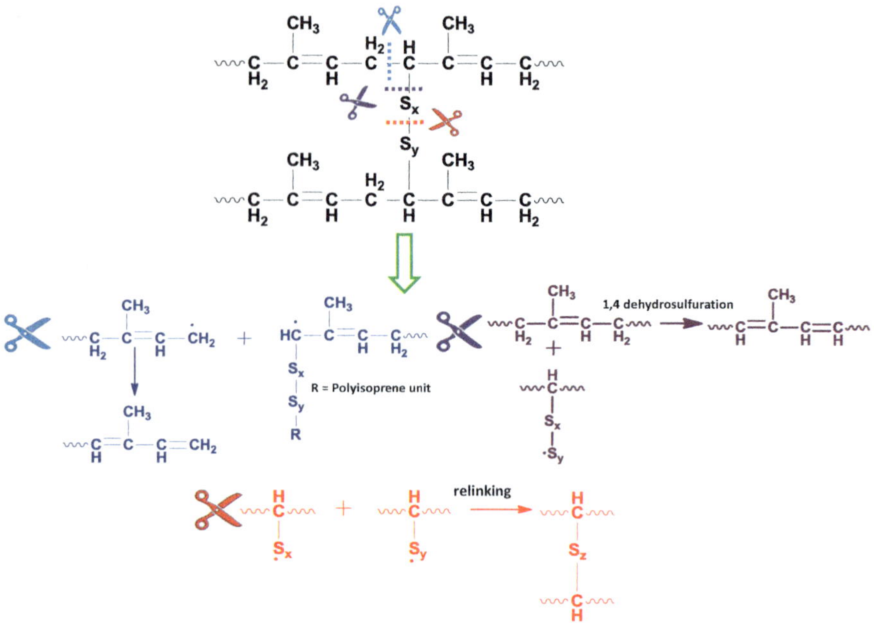

Fig. 14 Breaking of cross-link on heating

sintered rubbers have inferior mechanical properties because of the energy involved in void propagation and strain-induced crystallization. At the time of stretching, the network rubber chains demonstrate their inherent tendency to orient themselves in the stretching direction, which encourages formation of crystallites. These crystallites bind simultaneously with a multiple number of neighbouring network chains, resulting in the crosslinking network for high TS and EAB. Because of the lack of any chain entanglement, sintered rubber absorbs lesser energy to propagate a defect/void than the energy required to stretch the backbone chains to achieve strain-induced crystallization. Accordingly, the samples rupture before the commencement of strain-induced crystallization, or samples experience a marginal extent of strain-induced crystallization and therefore possess a lower strength and EAB.

The TS and EAB of these sinters increase from 0.8 to 1.3 MPa and 100 to 270%, respectively, when the moulding temperature and applied pressure are increased up to 200 °C and 6000 kg, respectively. However, as the temperature is raised beyond 200 °C, TS of the sintered NCR deteriorates considerably. At a temperature beyond 200 °C, oxidation of the sulfur from rubber may be the possible reason behind such decrease in TS. In addition, with the increase in moulding temperature, the elastic modulus of the sintered NCR decreases continuously, from 0.26 to 0.15 MPa, because of the destruction of thermolabile polysulfide linkages ($-S_x-$) in rubber with the simultaneous generation of thermally active free sulfur. The optimum parameters of sintering allowed to achieve the best mechanical properties, are 200 °C and 6000 kg.

19 Application

According to the analysis of tyre recycling market, RGs are no longer considered as a cheap filler but as a valuable component for manufacturing sustainable rubber composites. RGs are nowadays effectively utilized in widespread applications, such as molded/extruded products (wheels, gasket, sole), mulch, animal bedding, playgrounds, artificial sports surfacing, and automotive industries. However, in some applications, especially in products of higher quality and strength, e.g. in new tyres, the use of RGs are limited. Therefore, these rubber composites, obtained using RGs, are practically used to manufacture cheap articles, where strength is not important, such as floor tiles and other flooring materials, washers, windscreen wipers, tapes, cable housings, moulds, and footwear soles. Besides, sintered RGs can also be utilized in manufacturing washers, roofing materials, insulation boards, shoe soles, and solid tyres.

20 NR Composites Containing Proteins

Proteinous substances, such as silk, soy, and leather waste (LW), can be utilized as potential fillers to produce environment-friendly NR composites. The significantly huge quantity of a variety of leather solid wastes, such as buffing dust, shaving dust, and fleshings, are generated during various stages of the leather tanning process. In fact, a massive portion of these solid wastes comprise of collagenous matter and these collagenous LWs can be used as an inert filler for manufacturing environment-friendly polymeric composites. On the other hand, silk, like *Bombyx mori*, is basically a protein comprising of natural polymer fiber, used in textile production. Silk, in its natural form, is composed of a filament core protein, silk fibroin, and a glue-like coating consisting of a family of sericin proteins. Again, soy protein is a low-cost raw material, which is derived from natural resources, such as soybean.

21 Processing

Fabrication of green elastomeric composites, based on NR and silk textiles, is carried out by sandwiching a single layer of textile between layers of NR. Initially, NR samples are compressed at 70 °C for 10 min in order to obtain 1 mm thick sheets. Thereafter, silk fabric is sandwiched between two rubber sheets and the sandwich sample is compressed at 70 °C for 10 min, allowing the rubber to get impregnated with the silk fabric (Fig. 15). Earlier, silk fiber reinforced PP and NR blend composites have been prepared and characterized to monitor the environmental and gamma radiation effect on the mechanical properties of the composites comprising of NR, polypropylene, and silk. Initially, a varying amount of NR lumps are cut into small pieces and blended with PP in an extruder at 180–200 °C

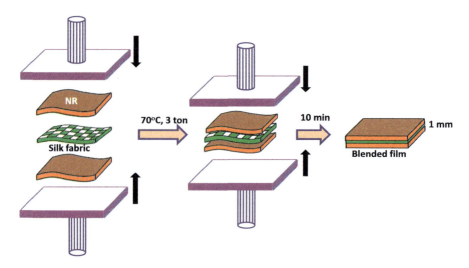

Fig. 15 Preparation of silk fabric blended NR film

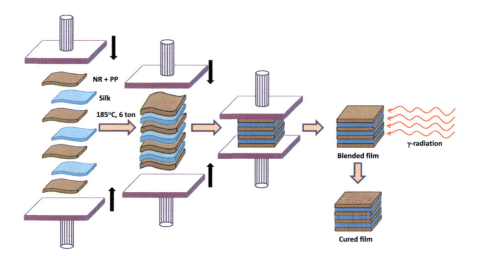

Fig. 16 Preparation of blended and cured film

to produce different compositions. Thereafter, these blends of varying compositions are cold pressed at 12-ton pressure to prepare films of desired thickness. To prepare composites, three layers of silk fibers are sandwiched among four layers of the blended films using hot pressing at 185 °C and 6 ton pressure, followed by cold pressing at 6 ton pressure (Fig. 16). Finally, in order to improve the mechanical properties, these composites are exposed to varying doses of gamma radiation.

To fabricate soy protein reinforced NR composites, initial hydrolysis of soy protein under different conditions is done, followed by microfluidization and

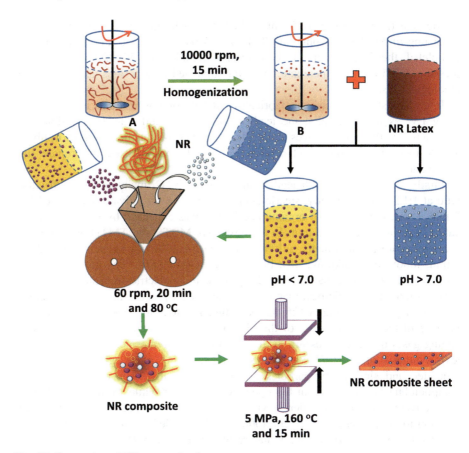

Fig. 17 Preparation of NR composite sheet

associated size reduction. Thereafter, these particles of near uniform sizes are incorporated as reinforcing additives during the preparation of NR composites in an internal mixer. In this context, initially, an alkaline dispersion of soy protein in distilled water is prepared by continuous stirring for 1 h at 60 °C. Thereafter, the dispersion is homogenized at 10,000 rpm for 15 min, followed by passing through a microfluidizer for several times (10 cycles). Later, the dispersion is admixed with NR latexes in both alkaline and acidic conditions to produce NR composite particles of varied pH (Fig. 17). In the next step, these dried composite particles, prepared under both alkaline and acidic conditions, are used as reinforcing ingredients in preparing NR composites by Brabender mixer, operated at 60 rpm for 20 min at 80 °C. Finally, the NR compounds are compression moulded in a window-type mould at 5 MPa and 160 °C for 15 min.

Processing of LW filled NR composites is performed in an open mixing mill or a rubber mixer for 20 min at 40 °C according to the ASTM D 3182 method. The composites are made of CB (60 phr), LW (60 or 80 phr), zinc oxide (5 phr), stearic acid (3 phr), sulfur (2.5 phr), ZMB-2 (1phr), and an accelerating system consisting of MBTS (1.2 phr) and TMTD (0.4 phr). Once the mixing process is completed, the formulations are compression-moulded at 150 °C with a closing pressure of 7.5 ton in a pneumatic press for stipulated time periods determined via rheological assays.

In an almost similar fashion, fabrication of NR composite foams is carried out using LW as filler. Initially, the leather fibers are shredded to 16 mm diameter using a mill with rotating knives and a 30 mesh sliver for obtaining both short fibers and leather fiber granules. Thereafter, a two-roll mill is used to prepare the composites.

Firstly, NR is milled with leather shavings at 65 °C, maintaining the friction ratio at 1:1.25. Subsequently, stearic acid is added as a co-activator and mixed for 5 min. Later, the activity of organic accelerator was enhanced by mixing an activator, i.e., zinc oxide for 7 min. In the next step, MBTS and TMTD are added as accelerators for 3 min. Finally, sulfur is added as a vulcanizing agent and mixed for 5 min, followed by addition of a blowing agent (i.e. TSH). The compounds are vulcanized and foamed via heat transfer process in an electrically heated hydraulic press to mould into microcellular rubber foam. This process involves a simultaneous curing and foaming at 125 °C for 7 min. The foaming process occurs in NR because of the decomposition of the blowing agent (toluenosulfohydrazine) generating gases that exert pressure in the surrounding polymer network. In this context, synthesis of NR-LW composites, via employing a variegated quantity of grounded and sieved LW particles as fillers, have also been carried out. The compounding is carried out in the open two-roll mixer while various ingredients, like activator, accelerator, filler, a vulcanizing agent, are added sequentially, in addition to the polyethylene glycol 4000, which is added as an acid neutralizer or antioxidant to neutralize the acidic nature of the added LW (Fig. 18).

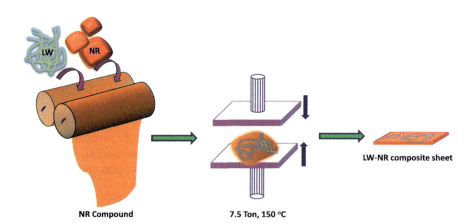

Fig. 18 Preparation of LW-NR composite

22 Characterization

22.1 Mechanical Properties

The reinforcing ability of the prepared green elastomeric composites is examined and compared with nylon textile reinforced NR composites. NR-silk composites exhibit superior mechanical properties than NR-nylon composites (Table 2). In fact, TS of nylon reinforced NRs are only 35–45% of the TS of the corresponding silk reinforced rubbers. Moreover, both nylon and silk-based non-purified NR composites exhibit significantly higher TS than the corresponding purified NR composites. Two important aspects are mainly responsible for such improvement in mechanical properties for non-purified NR based composites. First of all, better impregnation of rubber with silk results when non-purified NR is used as the matrix. In addition, protein impurities in non-purified NR increase the possibility of protein-protein interactions in NR-silk composites.

On contrary, mechanical properties of NR-PP-silk composites decrease significantly as the NR content in the matrix is increased (Table 2). However, once the composites are exposed to a certain limiting dose of gamma irradiation up to the maximum of 250 krad, both tensile and bending properties improve significantly. However, higher radiation dose, i.e. >250 krad, deteriorates the overall mechanical properties of all the composites. In fact, at higher radiation dose, bond scission occurs, which is responsible for the decreased mechanical strength of the composites. On the other hand, if the applied radiation dose is not allowed to go beyond 250 krad, the radiation-induced free radicals of silk, NR, PP might produce new bonds/crosslinks, which are responsible for the increased mechanical strength of composites.

Among the soy protein filled particulate NR composites, the less hydrolyzed soy protein particles yield NR composites with greater TS, Young's modulus, and toughness, while the highly hydrolyzed soy protein particles yield composites with greater elongation (Table 2). It is also observed that the NR composites, filled with soy-NR latex particles prepared under alkaline condition, provide greater TS, Young's modulus, and toughness, whereas NR composites, filled with soy-NR latex particles prepared under acidic condition, are of higher EAB. In this regard, when soy-NR latex particles are prepared via coagulation under acidic condition, the particle size of the filler is increased because of the enhanced aggregating tendency of soy protein particles, as soy protein particles approach to the characteristic isoelectric point at pH 4.5. Thus, the increased particle size of filler, resulting from enhanced aggregation, impairs TS, modulus of the composites filled by soy-NR latex particles coagulated under acidic condition.

Strain-strain test is also conducted for LW-NR composites. TS and EAB values of all the LW treated samples are severely reduced (Table 2), as compared to the control sample devoid of LW. Indeed, these parameters are noted to be further deteriorated with the increased addition of LW component in LW-NR composites. However, both shore A hardness and abrasion resistance of all the LW-NR composites samples is improved substantially (Table 2), as the composite material becomes increasingly more compact with the addition of fibrous protein waste. In this context, Santosh et al. have also reported the increased level of hardness and abrasion resistance in the LW filled NR samples, along with the expected rise in TS and associated loss in EAB (Table 2).

The cyclic stress-strain compression analyses for NR composite foams, filled with LW, are also carried out in which the samples are submitted to five compression-decompression cycles. It is observed that increased proportion of LW in the polymeric matrix leads to the increased hysteresis values (Table 2), suggesting reluctance of the system to return back to the original shape. The energy dissipation during compression-decompression deteriorates the matrix. The hysteresis values are measured via estimating the area related to the cyclic compression curves, attributed to the interfacial interaction of waste with NR. This interaction decreases the effectivity of stress transfer from rubber and consequently increases the work essential to effect the deformation of the composites, resulting in the enhancement of Young's modulus (Table 2). In this context, increasing amount of LW improves the strength at rupture or TS from 1.38 for NR to 2.18 MPa for NR composite foams filled with 60 phr LW, which is attributed to the interfacial adhesion between the LW and NR, i.e. the strength is transferred from NR matrix to LW and therefore required more stress to attain the rupture. Nevertheless, the deformation at rupture decreases from NR (408.7%) to NR composite foams filled with 20 phr LW (394.9%) because of due to the reduced mobility of rubber chain. Such phenomenon is attributed to the presence of filler which enhances the rigidity of the polymer, as evidenced from the increase in respective Young's modulus from 0.0026 for NR to 0.1206 MPa for NR composite foams filled with 60 phr LW (Table 2). Accordingly, the EAB values for 40 and 60 phr LW filled NR composite foams deteriorate to 73.8 and 57.6%, respectively (Table 2).

22.2 Dynamic Mechanical Properties

NR-silk composites exhibit improved dynamic mechanical properties than NR-nylon composites. For the soy-NR latex particle filled composites, the experimental reinforcement factors are compared with calculated reinforcing factors, evaluated by the Einstein-Smallwood (Eq. 1) and Guth-Gold equations (Eq. 2). Almost similar reinforcing factors are obtained for the composites prepared under alkaline or acidic conditions, and are greater than the calculated reinforcing factors, as evaluated using Guth-Gold equation. In fact, composites, having 30 and 40% fillers, exhibit greater reinforcement factors than that predicted by the Guth-Gold equation, indicating greater interparticle interactions to produce stiffer filler network.

$$G' = G'_0(1 + 2.5\phi) \tag{1}$$

$$G' = G'_0\left(1 + 2.5\phi + 14.1\phi^2\right) \tag{2}$$

22.3 Biodegradability

NR-PP-silk composites become more biodegradable with the increased incorporation of NR in PP. For instance, after 24 weeks of soil burial, silk fiber reinforced PP composite loses 10.20% TS whereas silk reinforced PP and NR blend (50:50) composite loses 24.3% TS. Similarly, under the same condition, silk fiber reinforced PP composite losses 13% BS, whereas silk reinforced PP and NR blend (50:50) composite loses 29.2% BS.

23 Application

The developed green materials, such as NR-silk composites, may be suitable for applications where damping, waterproofing, or high-pressure capacities in elastomeric tubing (such as in high-end bicycle tyres), alongside high mechanical properties is desired. Because of the improved values of conductivity, LW-NR composites exhibit the potential to be used as an antistatic flooring. In this way, the development of these innovative composites based on NR, CB, and LW can be an excellent option for allocation to hazardous LW within the composites and thereby minimization of environmental impact. Again, the NR composite foams filled with LWs can be directed to diversified foam related applications and thereby the mechanical/thermal properties can be modulated by controlling the foam density and other characteristic features.

24 Conclusions

Environmentally friendly NR composites filled with natural-organic fillers induce improved environmental impact and thereby accentuate the biodegradable character of these 'green' composites. By developing such 'green' composites, the use of mineral-inorganic fillers obtained from petroleum-based non-renewable resources can be minimized or obviated. At present, many research works are concentrated upon the mechanical property enhancement of these 'green' NR composites through chemical modification of the filler, use of adhesion promoters, and additives. Improvement of interfacial adhesion between natural fibers and NR matrix will remain the key issue in terms of overall performance since it determines the final properties of the composites. Further research is in progress to overcome the obstacles, which includes moisture absorption, inadequate toughness, and reduced long-term stability for outdoor applications. In particular, the major attention in near future would be to ensure that the different weathering conditions, such as temperature, humidity, and UV radiation, should not be able to deteriorate the service life of these environmentally friendly NRCs.

Acknowledgements The corresponding author gratefully acknowledges the Department of Science and Technology (DST), Government of India (YSS/2015/000886).

References

1. Chattopadhyay PK, Das NC, Chattopadhyay S (2011) Influence of interfacial roughness and the hybrid filler microstructures on the properties of ternary elastomeric composites. Compos Part A-Appl Sci 42:1049–1059
2. Chattopadhyay PK, Chattopadhyay S, Das NC et al (2011) Impact of carbon black substitution with nanoclay on microstructure and tribological properties of ternary elastomeric composites. Mater Des 32:4696–4704
3. Mondal M, Chattopadhyay PK, Chattopadhyay S et al (2010) Thermal and morphological analysis of thermoplastic polyurethane-clay nanocomposites: comparison of efficacy of dual modified laponite vs. commercial montmorillonites. Thermochim Acta 510:185–194
4. Praveen S, Chattopadhyay PK, Albert P et al (2009) Synergistic effect of carbon black and nanoclay fillers in styrene butadiene rubber matrix: development of dual structure. Compos Part A-Appl Sci 40:309–316
5. Singha NR, Das P, Ray SK (2013) Recovery of pyridine from water by pervaporation using filled and crosslinked EPDM membranes. J Ind Eng Chem 6:2034–2045
6. Yu P, He H, Luo Y et al (2017) Reinforcement of natural rubber: the use of in situ regenerated cellulose from alkaline-urea-aqueous system. Macromolecules 50:7211–7221
7. Karmakar M, Mahapatra M, Singha NR (2017) Separation of tetrahydrofuran using RSM optimized accelerator-sulfur-filler of rubber membranes: systematic optimization and comprehensive mechanistic study. Korean J Chem Eng 34:1416–1434
8. Mahapatra M, Karmakar M, Mondal B et al (2016) Role of ZDC/S ratio for pervaporative separation of organic liquids through modified EPDM membranes: rational mechanistic study of vulcanization. RSC Adv 6:69387–69403
9. Singha NR, Ray S, Ray SK et al (2011) Removal of pyridine from water by pervaporation using filled SBR membranes. J Appl Polym Sci 121:1330–1334
10. Singha NR, Ray SK (2012) Removal of pyridine from water by pervaporation using crosslinked and filled natural rubber membranes. J Appl Polym Sci 124:E99–E107
11. Jacob M, Francis B, Thomas S et al (2006) Dynamical mechanical analysis of sisal/oil palm hybrid fiber-reinforced natural rubber composites. Polym Compos 27:671–680
12. Joseph S, Joseph K, Thomas S (2006) Green composites from natural rubber and oil palm fiber: physical and mechanical properties. Int J Polym Mater 55:925–945
13. Visakh PM, Thomas S, Oksman K et al (2012) Crosslinked natural rubber nanocomposites reinforced with cellulose whiskers isolated from bamboo waste: processing and mechanical/thermal properties. Compos Part A-Appl Sci 43:735–741

14. Kanoth BP, Claudino M, Johansson M et al (2015) Biocomposites from natural rubber: synergistic effects of functionalized cellulose nanocrystals as both reinforcing and cross-linking agents via free-radical thiol-ene chemistry. ACS Appl Mater Interfaces 7:16303–16310
15. Kato H, Nakatsubo F, Abe K et al (2015) Crosslinking via sulfur vulcanization of natural rubber and cellulose nanofibers incorporating unsaturated fatty acids. RSC Adv 5:29814–29819
16. Tripathy AR, Morin JE, Williams DE et al (2002) A novel approach to improving the mechanical properties in recycled vulcanized natural rubber and its mechanism. Macromolecules 35:4616–4627

Electrical Properties of Sustainable Nano-Composites Containing Nano-Fillers: Dielectric Properties and Electrical Conductivity

Sabzoi Nizamuddin, Sabzoi Maryam, Humair Ahmed Baloch, M. T. H. Siddiqui, Pooja Takkalkar, N. M. Mubarak, Abdul Sattar Jatoi, Sadaf Aftab Abbasi, G. J. Griffin, Khadija Qureshi and Nhol Kao

1 Introduction

Nanocomposite materials have received much recognition in the last two decades due to their extraordinary characteristics by combining properties of different types of materials. Nanocomposites are made by mixing two or more phases such as fibers, layers, or particles, where a minimum of one phase is in the nanometer size range. The characteristics of nanocomposites are dependent on the characteristics of the nanofiller and parent material as well as upon morphological, mechanical and interfacial characteristics between the nanofillers and the parent materials [1, 2]. Nanocomposites can be utilized in a wide range of applications including aerospace, automotive, defence, energy, infrastructure, sporting goods and trans-

---

S. Nizamuddin · H. A. Baloch · M. T. H. Siddiqui · P. Takkalkar · S. A. Abbasi
G. J. Griffin (✉) · N. Kao
School of Engineering, RMIT University, Melbourne, VIC 3000, Australia
e-mail: gregory.griffin@rmit.edu.au

S. Maryam
Department of Electrical Engineering, Quaid-e-Awam University of Engineering, Science, and Technology, Nawabshah, Sindh, Pakistan

N. M. Mubarak
Department of Chemical Engineering, Faculty of Engineering and Science, Curtin University, 98009 Sarawak, Malaysia

A. S. Jatoi
Department of Chemical Engineering, Dawood University of Engineering and Technology, Karachi, Pakistan

K. Qureshi
Department of Chemical Engineering, Mehran University of Engineering and Technology, Jamshoro, Sindh, Pakistan

© Springer Nature Switzerland AG 2019
Inamuddin et al. (eds.), *Sustainable Polymer Composites and Nanocomposites*, https://doi.org/10.1007/978-3-030-05399-4_30

portation sectors due to their light weight, high strength, high durability, and design and process flexibility [3].

The discovery of a variety of nano-scale materials may offer a number of new composites with specific properties. Nanoparticles possess great potential to be utilized as filler materials to improve the electrical, mechanical and physical properties of nanocomposites. It is important that at least one dimension of the filler material be of nanometer order for the synthesis of nanocomposite materials but the final product can be of any size either nano, micro or macroscopic in size [2]. Carbon black, carbon fibre and carbon nanotubes have been effectively investigated as fillers to fabricate nanocomposites [4]. For example, aligned carbon nanotubes in polymeric composites result in enhanced thermal transport [5] and strong luminescence [6] favouring the direction along the nanotube axis. A number of polymer/carbon nanotubes have successfully been fabricated by incorporating carbon nanotubes in polymer matrices including polyamide, polystyrene, polycarbonate, polyethylene, polypropylene and polylactide.

Extraordinary electrical and dielectric (high dielectric constant and low dielectric loss) properties of nanosize particles with high conductivity are considered as a gateway for substituting the monolithic metals in various emerging applications such as e-paper, microelectronics, organic light emitting diodes, antistatic coatings, sensors and touch screens. The above-mentioned applications can be achieved by obtaining the maximum electrical conductivity of nanocomposites with the lowest possible filler concentration [7]. It is reported in the literature that the electric/dielectric properties of nanocomposites were greatly enhanced by incorporating nanofillers including carbon nanotubes, carbon nanofibers, metal particles and carbon black [8]. Therefore, this study mainly overviews the electrical conductivity of nanocomposites containing nanofillers. Further, dielectric properties (dielectric constant, dielectric loss and tangent loss) of nanocomposites are discussed in detail. In addition, a critical discussion on evaluation of electrical conductivity and dielectric properties of nanocomposites with nanofillers is provided. Conclusions are drawn in the light of literature; with recommendations being henceforth provided.

2 Nanocomposites with Nanofillers

Polymer nanocomposites are a branch of nanotechnology and composite science which involves the development of a new category of materials. These materials possess superior properties which improve the performance of these materials with multi-functionalities. The polymer nanocomposite is fabricated by incorporating nanofillers in a polymer matrix at different concentrations [9–13]. The addition of nanofillers helps to improve the physical, mechanical, chemical and thermal, barrier, electrical and dielectric properties of the composite [14–22]. The inherent properties of polymers like light weight, transparency and flexibility are combined in a nanocomposite with other properties.

For decades, the development of polymer composites has been one of the most important areas in materials science. When fillers are used at the nanometer scale (for the production of entities that enhance or increase polymer properties), i.e. at least one dimension is between 1 and 100 nm, it is called a nanofiller, and the composite becomes a nanocomposite [23]. In this regard, various researchers work on different nanofillers for nanocomposites fabrication [24]. Among others, the carbon nanofillers including carbon nanotubes, carbon nanofibers, and metallic nanowires get appreciable interest in the field of polymers [25]. Carbon nanotubes have a high tensile strength and are considered 100 times stronger than steel with only one-sixth of the weight, so they can be the strongest fibers and the smallest known. They also have high conductivity, high surface area, unique electronic properties and polymer adsorption potential [26]. The applications of the carbon nanotubes that are currently studied include polymeric compounds (conductive and structural fillers), electron field emitters (flat screens), electromagnetic shielding, batteries, supercapacitors, hydrogen storage and structural compounds [27].

Carbon nanotubes can be produced from different material which includes silicon and germanium, but the development and application of carbon nanotubes remains the main focus of activity [27]. Carbon nanotubes hold outstanding electronic, optical, thermal and mechanical properties [28]. Its tubular structure prevents the diffusion of cracks and dislocations, providing the carbon nanotubes with a great potential for producing materials with high Young's modulus [29]. The partial sp2-sp3 hybridization of the C-C bonds of carbon nanotubes can induce high flexibility [30]. Therefore, the potential application of carbon nanotubes has always attracted great interest. Carbon nanotubes are also excellent nanofillers for polymeric nanocomposites [31, 32]. Even if a small number of carbon nanotubes are added to the polymer, the mechanical properties can be significantly improved, and even an electrically insulating polymer can be converted into a conductive composite material by addition of carbon nanotubes as filler [33].

Nanowires can be produced from conductive (e.g. metal) or semi-conductive (e.g. carbon) materials using a variety of production techniques. They have a unique crystalline structure and typical diameter of tens of nanometers with high aspect ratio. They are used as interconnectors for the transport of electrons in nano-electronic devices. Several metals have been used to make nanowires including cobalt, gold and copper. Polymer and filler interaction can be increased through polymer grafting onto the nanofiller. For grafting, two methods are widely used for matrix preparation; the first method commonly takes place by the reaction of polymer chains with the surface of nanofiller. For example, polystyrene grafting to single-walled, oxidized carbon nanotubes and PVA grafting to multiwalled carbon nanotubes activated by carbodiimide. Other techniques include reactions with nanofillers that are oxidized by amidation or esterification, cycloaddition, nucleophilic addition, condensation and other chemical reactions. A major shortcoming of the first step is lower enthalpy density due to macromolecular steric hindrance caused by polymer chains attached to the surface of nanotubes. The second technique of grafting encompasses polymerization of the surface of the nanofillers. In this process, three-dimensional barriers are not problematic and polymers with higher molecular weight can be prepared. However, this method requires precise control of the polymerization reaction. "Grafting" methods include radical polymerization transfer, additional separation chain transfer, free radical polymerization, ring opening polymerization, polycondensation, cationic/anionic polymerization, redox/oxidation, and metallocene and radical polymerization of nitrogen oxide. These methods of grafting are described and well explained in various outstanding reviews reported in the literature [34].

3 Dielectric Properties of Nanocomposites

Dielectric material properties define energy transmission, absorption and reflection to the electric field of electromagnetic waves. These characteristics measure the capability of materials for dissipation of electromagnetic energy as heat [35]. Dielectric characteristics are highly dependent on the nature of the materials such as structure and composition of materials, on the other hand, it is also dependent on process conditions like temperature and frequency [36]. Dielectric characteristics of different materials also depend on bulk density, moisture and association of permanent dipole moment with water and other constituent molecules [37]. An understanding of dielectric characteristics might support the design of microwave systems and predict the yield of microwave processes [38]. Knowledge of dielectric characteristics leads to applications for materials including electronic packaging and sensors [38].

The dielectric properties are dielectric loss factor, dielectric constant, tangent loss (tan δ) and penetration depth. The dielectric constant measures the dielectric's capacity for storing electrical energy. For example, the Polymer-ceramic nanocomposites hold strong potential for improving the performance in energy storage capacitors, hybrid electric vehicles and kinetic energy weapons, since they contain a high-breakdown-strength polymer matrix and high-dielectric-permittivity ceramic nanofillers and thus can reach a high level of energy-storage density as shown in Fig. 1. The loss factor estimates loss of electrical energy in dielectrics. Penetration depth is a measurement of the depth microwaves can penetrate materials at a specific frequency. The tangent loss is the determination of the capability of the materials for transforming the electromagnetic energy into internal energy at a particular temperature and frequency [39]. The tangent loss is calculated as the ratio of dielectric loss factor to dielectric constant. Penetration depth is calculated by an equation: DP = $\lambda_0/2\pi$ (ε') 0.5 [1 + (Tan δ) 20.5 − 1] −0.5. The relaxation time and static permittivity are calculated via the equation: ε'r = −(ωε"r)τ + εs.

The loss factor and dielectric constant depends highly on the frequency, magnitude and material's efficiency for interacting with microwaves. Therefore, the frequency is a fundamental factor for determining dielectric heating of materials [40]. The amount of moisture existing in a material strongly affects the dielectric characteristics of materials. It has been observed that the higher the moisture content the lower will be the dielectric properties [41]. The physicochemical composition, density and temperature also have a significant influence on dielectric characteristics of materials [39].

3.1 Dielectric Constant

The dielectric constant is an important property of materials which measures their ability to store electrical energy [42]. The dielectric constant for pure polymers usually falls between 2 and 10. Generally, a high dielectric constant is required for all types of polymers except for fast static dissipation and high-speed integrated circuit applications, which requires a low dielectric constant [43]. A high dielectric constant can be achieved by reinforcing the conductive filler in the polymer matrix. For instance, graphite nanoplatelet (GNP), as shown in Fig. 2, is one promising nanofiller to fabricate polymer/GNP nanocomposites with a high dielectric constant due to strong interfacial polarization.

A number of studies have been carried out for improving the dielectric constant values of nanocomposites using suitable nanofiller. Huang et al. [44] reported the preparation of three-phase nanocomposites containing poly(vinylidene fluoride)

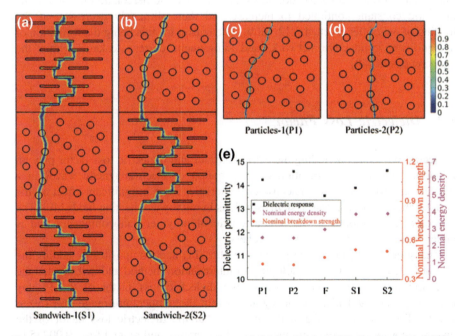

Fig. 1 The breakdown paths of nanocomposites with (a) Sandwich-1 (S1), (b) Sandwich-2 (S2), (c) Particles-1 (P1), (d) Particles-2 (P2) microstructures. (e) Dielectric permittivity, nominal breakdown strength and nominal energy density of nanocomposites with different microstructures of nanofillers [63]

(PVDF), nanoparticles of barium titanate (BT), and β-silicon carbide (β-SiC). The results showed that the three-phase nanocomposites demonstrated considerably higher dielectric constant than the two-phase PVDF/BT nanocomposites. The addition of the β-SiC whiskers caused a drastic improvement of the dielectric constant of the nanocomposite at a certain loading level. The value of dielectric constant of the three-phase system was highest (325) at a β-SiC loading of 17.5 vol. %. However, further addition of β-SiC (20 vol.%) lead to a reduction in the dielectric constant of the PVDF/BT/β-SiC composite to 253. This could possibly be due to the generation of voids and porosity in the composite [45].

He et al. [46] fabricated a novel nanocomposite with poly(vinylidene fluoride) (PVDF) as the matrix and exfoliated graphite nanoplates (xGnPs) as the conductive nanofiller. The results (refer Fig. 2.) clearly showed an increase in the dielectric constant of the PVDF/xGnP nanocomposites and this can be credited to the uniform dispersion of the xGNPs within the PVDF matrix with consequent development of micro-capacitors with the increase in xGNP concentration [47]. This network advancement was explained by three stages (I, II, and III). There is gradual increase initially in the dielectric constant of the nanocomposites when compared to the unfilled system; this is due to the initiation of the development of the micro-capacitance assemblies.

The dielectric constant continued to rise with the increasing nanofiller concentration until it reached percolation threshold. Beyond the percolation threshold, the dielectric constant continued to increase and then ultimately decreased as the loading of nanofiller reached 3.12 vol.%. The reason for this being the formation of a significant conductive network, and subsequent current leakage within the nanocomposite [48].

3.2 Dielectric Loss Factor

Dielectric properties including dielectric constant and dielectric loss factor play an important role in the simulation of propagation of electromagnetic waves in a wireless environment. The dielectric characteristics have a significant effect on the performance of the wireless system as well as in microwave heating fields [49]. The dielectric loss factor is a vital characteristic of polymer nanocomposites and is under investigation by various researchers.

Zhang et al. [50] fabricated polyamide-clay (PI) nanocomposites by adding the clay at 1, 3, 10 and 20 wt% employing an intercalation method and investigated the dielectric loss factor. Figure 3 shows the dielectric loss factor of the PI-clay nanocomposites at 1 kHz frequency and within the temperature range of −150 to 150°C. The findings of the study revealed that the dielectric loss factor of the nanocomposites increased with increase in temperature and ranged from 0.00235 to 0.0335. The decrease of dielectric loss as the temperature decreases was due to the freezing of polyamide molecules, which results in complicated segmental movements.

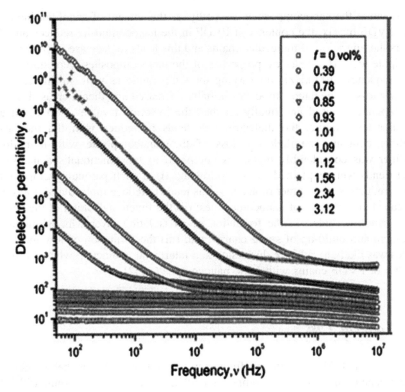

Fig. 2 The dependence of dielectric constant on frequency for PVDF/xGnP nanocomposite at room temperature [46]

The extent of adhesion and interactions at the interface between the filler and polymer play an important role in improving the dielectric properties of their nanocomposites, which include dielectric permittivity and dielectric loss. Zhang et al. [51] reported a simple method to prepare nanocomposites with $B_aT_iO_3$ (BT) nanofiber and ferroelectric polyvinylidene fluoride (PVDF) matrix. The $B_aT_iO_3$ (BT) nanofiber has the advantage of high dielectric constant. Unmodified and fluorosilane modified BT nanofibers were incorporated in PVDF matrix at different concentrations of the nanofibers. The dielectric properties of the nanocomposites were studied with respect to frequency and temperature. The dielectric loss of the nanocomposites reduced with an increase in the concentration of functionalised BT nanofibers, particularly in the frequency ranging from 102 to 107 Hz. The nanocomposite with 20 v% of modified BT exhibited the lowest dielectric loss tangent. This can be possible due to two reasons. Firstly, the surface functionalization of BT nanofiber enhances the interfacial interaction between the nanofiber and the PVDF matrix; this limits the mobility and build-up of space charge within the nanocomposites. Secondly, as the content of nanofiber increases in the PVDF matrix, the contents of PVDF in the nanocomposite reduces and this hinders the migration of molecular chains and this leads to decrease of β-relaxation and dipole loss. The dielectric properties of the nanocomposites were studied at a fixed frequency (100 kHz) and varying the temperature as well. In this case, the dielectric loss of the nanocomposites initially decreased and then increased later as the temperature raised and finally touched the lowermost value at 60°C. At any particular temperature, the dielectric loss tends to reduce with the increasing nanofiber content. The dielectric loss of the nanocomposite with unmodified nanofiber was considerably higher as compared to the functionalised nanofiber. These trends were explained by the following: (i) at each particular temperature, there is reduction of polymer molecules, this tends to reduce molecular dipoles, and this leads to dipole loss of nanocomposites; (ii) the functionalization of the surface of the nanofiber aided in the formation of an isolating layer which limited the movement and build-up of space charge; and (iii) the formation of low molecular dipoles was likely to enhance dispersion and interfacial interaction which limits the mobility of PVDF chains within the nanocomposite [52].

3.3 Tangent Loss

Pradhan et al. [53] reported the frequency dependent dielectric behaviour of a polyethylene oxide based nanocomposite (PPNCE) with montmorillonite (DMMT) modified with dodecyl amine and polyethylene glycol (PEG) as the plasticizer. The PPNCE films were cast by using a tape casting technique, where the DMMT concentration was fixed at 5wt% and PEG at six different concentrations was added (0, 5, 10, 20, 30 and 50 wt%). Figure 4 shows the frequency dependency of tangent loss of PPNCE films at ambient temperature with the varying concentration of PEG.

As the PEG content increased the tangent loss peak shifts in the direction of higher frequency. The loss spectra can be related to the peak appearing at typical frequencies for PPNCE samples with and without PEG content. These peaks indicate that there are relaxing dipoles within all the nanocomposites and the strength and particular frequency of relaxation are influenced by the specific dipolar relaxation. The addition of plasticizer generally causes an increase in the amorphous content within the nanocomposite. The small and mobile PEG molecules tend to rapidly increase the segmental movement by increasing the available free volume. The reasonably fast segmental movement together with mobile ions speed up the transport properties with the addition of plasticizer. The peaks shifting towards higher frequency with an increase in plasticizer content eventually reveals that the relaxation time is decreasing.

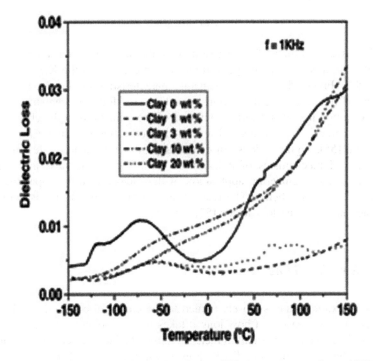

Fig. 3 Effect of temperature on dielectric loss factor of PI-clay nanocomposite at 1 kHz [50]

3.4 Static Permittivity

Static permittivity can be defined as polarization effect under dc conditions. When the sinusoidal electric field is functional, polarization and static differ under AC environments. When the sinusoidal electric field was applied, medium polarization and static are different under AC conditions. The polarization of the dielectric does not always respond in time to the changes in the applied electric field due to the thermal alterations that randomize the orientation of the dipole and the rotation of the molecule in the viscous mediums due to their interaction to neighbours. Therefore, the response of the material to the outer field is causal every time (after application of the field) and it depends on the frequency of field that can be shown through the phase difference. Therefore, the dielectric constant is commonly considered as a complex function of the frequency of the applied field.

$$D_0 e^{-j\omega t} = \varepsilon(\omega) E_0 e^{-j\omega t}$$

where j is the imaginary unit, ω is electromagnetic field frequency, and 't' is time. The above equation can be rewritten as with conversion

$$\varepsilon_\tau = \varepsilon'_\tau - \varepsilon''_\tau$$

For the causal and linear dielectric responses, the relationship between imaginary parts and real parts of complex permittivity is represented by the Kramers-Kronig relation [54]. Figure 5 shows the general characteristics of the dependence on frequency of real and virtual dielectric constants for four mechanisms of polarization. Although it shows the characteristics of the transitions in the characteristic peaks in εr 'and εr', these peaks and several characteristics are usually wider in the real material [55]. Figure 5 does not denote any particular material, few materials demonstrate all mechanisms of polarization [56]. The energy loss is calculated through εr. In the engineering application of the dielectric in the capacitor, for a given εr', εr' is always better. The relative size of εr 'relative to εr' is defined as tanδ, which is called the loss factor [55] (Fig. 5).

3.5 Relaxation Time

Dielectric relaxation time is closely related to the electrical conductivity. In semiconductors, it is a measure of how long it takes for a conductive process to occur. This relaxation time is very small in metals and may be large in semiconductors and insulators. The relaxation time distribution was used as a rough indication of the heterogeneity of the polymer/filler interface. Maxwell-Wagner-Sillars interfacial

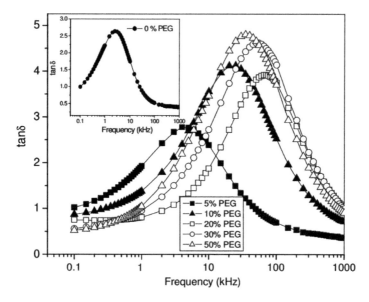

Fig. 4 Variation in Tangent loss of PPNCE thin films with respect to frequency for various concentration of PEG at the room temperature [53]

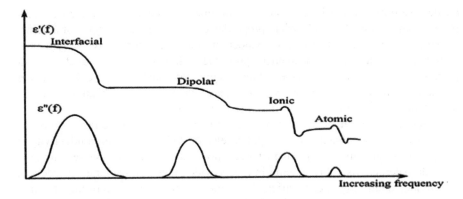

Fig. 5 Demonstration of permittivity real and imaginary dependence with the presence of orientational, ionic, electronic polarization and interfacial mechanisms [55]

polarization (MWS) can be compared with its larger dielectric strength. The presence of the filler/polymer interface increases the MWS relaxation time distribution because each interface has different interface geometry and is polarized at different time scales. However, the relaxation of the polymer melts originates from the split amorphous dynamics. The temperature shows a fairly uniform chain, much higher than the Tg value, whose nylon 12 Tg is approximately 508 °C [57].

When current flows through the biomaterial interface, the charge can accumulate at the interface between the two dielectric materials with different relaxation times ($\tau = \varepsilon/\sigma$, where ε is the permittivity and σ is the conductivity). The nanocomposite has a large interfacial area and provides many sites for the MWS enhancement effect compared to microcomposites [58].

4 The Electrical Conductivity of Nanocomposite

Nanocomposites are materials in which at least one of the constituents has dimensions in the nanometre range (1 nm = 10–9 m) [59]. In the last few years, polymer nanocomposites are getting remarkable industrial and academic recognition due to their light-weight, higher mechanical and thermo-mechanical characteristics [60]. Light-weight nanocomposites can exhibit few functional characteristics including magnetic and electrical properties. Nanocomposites are electrically conductive polymers and they can offer an enormous range of electrical conductivities parallel to traditional inorganic conductors and semiconductors. These polymers can be used in a wide range of industrial applications. Nanocomposites may be distributed into two classes: one of the composites which are comprised of intrinsically conductive polymers whereas other which can be manufactured by the addition of conducting nanofillers. Composites containing conductive polymer matrix may offer several distinct functional characteristics, for instance, their optical and electrical properties can vary to the response of external stimuli. The main disadvantage of nanocomposites with conductive polymers is their high cost and poor mechanical properties. However, the conductive composites with insulating polymer matrix are more stable and exhibit better mechanical properties. Currently, there are widespread applications of the nanocomposites with conducting polymer matrices [61].

Generally, the term electrical properties of materials refer to the response of the material under the influence of an applied electric field. Electrical conductivity is the physical quantity which characterizes how easily electric charge can flow through materials. Its values extend in a very wide range approaching 30 orders of magnitude, qualifying the electrical behaviour of materials. According to their conductivity, materials can be classified as conductors, semiconductors, and insulators (dielectrics). In nanocomposites, the matrix is considered as a continuous phase of the system and in most cases is the phase with a high volume fraction of the material. With this point of view, the nature of the matrix defines, at least at first approximation, the electrical behaviour of the composite material. Thus, metal matrix composites are obviously conductors, while polymer composites and in some cases ceramic composites are classified as insulators.

4.1 Characteristics of Electrically Respond Polymer Nanocomposites

The electrical response of any polymer matrix refers to its conductivity and dielectric behaviour. Nanocomposites have attracted the attention of researchers as an engineering material due to its thermo-mechanical performance. Work started with the manufacturing of the nanocomposites and later it expanded to its properties such as mechanical, electrical and thermal characteristics. Two were starting points of this investigation, examining the effect of nanofiller on the mechanism of polarization, molecular mobility, interfacial effect and improvement of conductivity in nanocomposites incorporating conductive inclusions. Considering the attractiveness of the nanofiller or nano-inclusions, polymer nanocomposites can be classified into two major categories: (i) the insulating matrix-dielectric reinforcing phase and (ii) the insulating matrix-conductive reinforcing phase. Conversely, the nanofiller can be ceramic, metallic as well as an allotropic type of carbon. A stimulating collection of ceramic inclusion is so-called active and functional dielectrics, for instance, pyroelectric, ferroelectric and piezoelectric elements. The existence of such type of nanofillers may provide functionality to the performance of the system. Although it is not easy to say which type of polymer nanocomposite drawn the attention of researchers from all over the world, it is recognized that several studies mentioned electrical characteristics of carbon/polymer matrix nanocomposites.

4.2 Parameters Influencing Electrical Conductivity of Nanocomposites

Mostly, the addition of conducting nanofiller should be as lower as possible because more filler can cause processing difficulty and also mechanical properties can be affected. But, at the same time, it should be kept in mind that the amount of filler is desired to be in acceptable quantity for getting the continuous conducting network. Therefore, it is an ideal to attain conducting composites with lower percolation threshold [62]. The electrical conductivity and percolation threshold of any conductive polymeric nanocomposite are dependent of various factors including physical properties of filler (surface properties, electrical conductivity, and aspect ratio), distribution and dispersion fillers, physical characteristics of the polymer matrix, including polarity, viscosity and crystalline and alignment and orientation of fillers. It is important to note that many of above-mentioned parameters are considerably affected by the fabrication methods of the nanocomposites.

5 Conclusion

Various nanomaterials hold tunable physical, chemical, thermal, mechanical, dielectric and electrical properties. These properties, especially dielectric and electrical properties, are important characteristics of such nanomaterials for their utilization as nanofiller for the synthesis of nanocomposite with numerous applications. This chapter deals in detail with an overview of the dielectric properties and electrical conductivity of nanocomposite synthesized by addition of different nanofillers.

References

1. Hussain F, Hojjati M, Okamoto M, Gorga RE (2006) Polymer-matrix nanocomposites, processing, manufacturing, and application: an overview. J Compos Mater 40:1511–1575
2. Venkatesh DN, Priya VK, Bhavitha K (2016) Polymer-matrix nanocomposites, processing, manufacturing and application: an overview
3. Choudhary V, Gupta A (2011) Polymer/carbon nanotube nanocomposites. In: Carbon nanotubes-polymer nanocomposites, Intech
4. Rezakazemi M, Zhang Z (2018) 2.29 Desulfurization materials. In: a2-Dincer I (ed) Comprehensive energy systems, Elsevier, Oxford, pp. 944–979
5. Song W-L, Wang W, Veca LM, Kong CY, Cao M-S, Wang P, Meziani MJ, Qian H, LeCroy GE, Cao L (2012) Polymer/carbon nanocomposites for enhanced thermal transport properties–carbon nanotubes versus graphene sheets as nanoscale fillers. J Mater Chem 22:17133–17139

6. Zhou B, Lin Y, Veca LM, Fernando KS, Harruff BA, Sun Y-P (2006) Luminescence polarization spectroscopy study of functionalized carbon nanotubes in a polymeric matrix. J Phys Chem B 110:3001–3006
7. Kumar V, Rawal A (2016) Tuning the electrical percolation threshold of polymer nanocomposites with rod-like nanofillers. Polymer 97:295–299
8. Sun L-L, Li B, Zhao Y, Zhong W-H (2010) Suppression of AC conductivity by crystalline transformation in poly (vinylidene fluoride)/carbon nanofiber composites. Polymer 51:3230–3237
9. Dashti A, Harami HR, Rezakazemi M (2018) Accurate prediction of solubility of gases within H2-selective nanocomposite membranes using committee machine intelligent system. Int J Hydrogen Energy 43:6614–6624
10. Rezakazemi M, Sadrzadeh M, Mohammadi T, Matsuura T (2017) Methods for the preparation of organic-inorganic nanocomposite polymer electrolyte membranes for fuel cells. In: Inamuddin D, Mohammad A, Asiri AM (eds) Organic-Inorganic composite polymer electrolyte membranes. Springer International Publishing, Cham, pp 311–325
11. Rezakazemi M, Vatani A, Mohammadi T (2016) Synthesis and gas transport properties of crosslinked poly(dimethylsiloxane) nanocomposite membranes using octatrimethylsiloxy POSS nanoparticles. J Nat Gas Sci Eng 30:10–18
12. Rezakazemi M, Vatani A, Mohammadi T (2015) Synergistic interactions between POSS and fumed silica and their effect on the properties of crosslinked PDMS nanocomposite membranes. RSC Adv 5:82460–82470
13. Rezakazemi M, Razavi S, Mohammadi T, Nazari AG (2011) Simulation and determination of optimum conditions of pervaporative dehydration of isopropanol process using synthesized PVA–APTEOS/TEOS nanocomposite membranes by means of expert systems. J Membr Sci 379:224–232
14. Sodeifian G, Raji M, Asghari M, Rezakazemi M, Dashti A (2018) Polyurethane-SAPO-34 mixed matrix membrane for CO_2/CH_4 and CO_2/N_2 separation. Chin. J. Chem. Eng
15. Rezakazemi M, Dashti A, Asghari M, Shirazian S (2017) H 2 -selective mixed matrix membranes modeling using ANFIS, PSO-ANFIS, GA-ANFIS. Int J Hydrogen Energy 42:15211–15225
16. Rezakazemi M, Amooghin AE, Montazer-Rahmati MM, Ismail AF, Matsuura T (2014) State-of-the-art membrane based CO_2 separation using mixed matrix membranes (MMMs): an overview on current status and future directions. Prog. Polym. Sci 39(5) 817–861
17. Baheri B, Shahverdi M, Rezakazemi M, Motaee E, Mohammadi T (2014) Performance of PVA/NaA mixed matrix membrane for removal of water from ethylene glycol solutions by pervaporation. Chem Eng Commun 202:316–321
18. Shahverdi M, Baheri B, Rezakazemi M, Motaee E, Mohammadi T (2013) Pervaporation study of ethylene glycol dehydration through synthesized (PVA-4A)/polypropylene mixed matrix composite membranes. Polym Eng Sci 53:1487–1493
19. Rostamizadeh M, Rezakazemi M, Shahidi K, Mohammadi T (2013) Gas permeation through H_2-selective mixed matrix membranes: experimental and neural network modeling. Int J Hydrogen Energy 38:1128–1135
20. Rezakazemi M, Mohammadi T (2013) Gas sorption in H_2-selective mixed matrix membranes: experimental and neural network modeling. Int J Hydrogen Energy 38:14035–14041
21. Rezakazemi M, Shahidi K, Mohammadi T (2012) Sorption properties of hydrogen-selective PDMS/zeolite 4A mixed matrix membrane. Int J Hydrogen Energy 37:17275–17284
22. Rezakazemi M, Shahidi K, Mohammadi T (2012) Hydrogen separation and purification using crosslinkable PDMS/zeolite a nanoparticles mixed matrix membranes. Int J Hydrogen Energy 37:14576–14589
23. Thomas S, Rouxel D, Ponnamma D (2016) Spectroscopy of polymer nanocomposites, William Andrew
24. Rezakazemi M, Sadrzadeh M, Matsuura T (2018) Thermally stable polymers for advanced high-performance gas separation membranes. Progr. Energy Combust. Sci. 66:1–41

25. Mutiso RM, Winey KI (2015) Electrical properties of polymer nanocomposites containing rod-like nanofillers. Prog Polym Sci 40:63–84
26. Maynard AD, Baron PA, Foley M, Shvedova AA, Kisin ER, Castranova V (2004) Exposure to carbon nanotube material: aerosol release during the handling of unrefined single-walled carbon nanotube material. J Toxicol Environ Health Part A 67:87–107
27. Njuguna J, Ansari F, Sachse S, Zhu H, Rodriguez V (2014) Nanomaterials, nanofillers, and nanocomposites: types and properties. In: Health and environmental safety of nanomaterials, Elsevier, pp. 3–27
28. Koster LJA, Mihailetchi VD, Ramaker R, Blom PW (2005) Light intensity dependence of open-circuit voltage of polymer: fullerene solar cells. Appl Phys Lett 86:123509
29. Koster L, Mihailetchi V, Blom P (2006) Ultimate efficiency of polymer/fullerene bulk heterojunction solar cells. Appl Phys Lett 88:093511
30. Wu P-T, Ren G, Jenekhe SA (2010) Crystalline random conjugated copolymers with multiple side chains: tunable intermolecular interactions and enhanced charge transport and photovoltaic properties. Macromolecules 43:3306–3313
31. Lenes M, Wetzelaer GJA, Kooistra FB, Veenstra SC, Hummelen JC, Blom PW (2008) Fullerene bisadducts for enhanced open-circuit voltages and efficiencies in polymer solar cells. Adv Mater 20:2116–2119
32. Li G, Shrotriya V, Huang J, Yao Y, Moriarty T, Emery K, Yang Y (2005) High-efficiency solution processable polymer photovoltaic cells by self-organization of polymer blends. Nat Mater 4:864
33. Huang J-H, Li K-C, Chien F-C, Hsiao Y-S, Kekuda D, Chen P, Lin H-C, Ho K-C, Chu C-W (2010) Correlation between exciton lifetime distribution and morphology of bulk heterojunction films after solvent annealing. J Phys Chem C 114:9062–9069
34. Spitalsky Z, Tasis D, Papagelis K, Galiotis C (2010) Carbon nanotube–polymer composites: chemistry, processing, mechanical and electrical properties. Prog Polym Sci 35:357–401
35. Boldor D, Sanders T, Simunovic J (2004) Dielectric properties of in-shell and shelled peanuts at microwave frequencies. Trans-Am Soc Agric Eng 47:1159–1170
36. Sosa-Morales M, Valerio-Junco L, López-Malo A, García H (2010) Dielectric properties of foods: reported data in the 21st century and their potential applications. LWT-Food Sci Technol 43:1169–1179
37. Stone ML, Maness NO (2006) Plant biomass estimation using dielectric properties, ASABE paper, 15
38. Tripathi M, Sahu J, Ganesan P, Dey T (2015) Effect of temperature on dielectric properties and penetration depth of oil palm shell (OPS) and OPS char synthesized by microwave pyrolysis of OPS. Fuel 153:257–266
39. Motasemi F, Afzal MT, Salema AA, Mouris J, Hutcheon R (2014) Microwave dielectric characterization of switchgrass for bioenergy and biofuel. Fuel 124:151–157
40. Salema AA, Yeow YK, Ishaque K, Ani FN, Afzal MT, Hassan A (2013) Dielectric properties and microwave heating of oil palm biomass and biochar. Ind Crops Prod 50:366–374
41. Sweeney JJ, Roberts JJ, Harben PE (2007) Study of dielectric properties of dry and saturated green river oil shale. Energy Fuels 21:2769–2777
42. Nizamuddin S, Mubarak N, Tiripathi M, Jayakumar N, Sahu J, Ganesan P (2016) Chemical, dielectric and structural characterization of optimized hydrochar produced from hydrothermal carbonization of palm shell. Fuel 163:88–97
43. Li B, Zhong W-H (2011) Review on polymer/graphite nanoplatelet nanocomposites. J Mater sci 46:5595–5614
44. Li Y, Huang X, Hu Z, Jiang P, Li S, Tanaka T (2011) Large dielectric constant and high thermal conductivity in poly (vinylidene fluoride)/barium titanate/silicon carbide three-phase nanocomposites. ACS Appl Mater Interfaces 3:4396–4403
45. Calame J (2006) Finite difference simulations of permittivity and electric field statistics in ceramic-polymer composites for capacitor applications. J Appl Phys 99:084101
46. Herzberg M, Kang S, Elimelech M (2009) Role of extracellular polymeric substances (EPS) in biofouling of reverse osmosis membranes. Environ Sci Technol 43:4393–4398

47. Dang Z-M, Wu J-P, Xu H-P, Yao S-H, Jiang M-J, Bai J (2007) Dielectric properties of upright carbon fiber filled poly (vinylidene fluoride) composite with low percolation threshold and weak temperature dependence. Appl Phys Lett 91:072912
48. Li Y-J, Xu M, Feng J-Q, Dang Z-M (2006) Dielectric behavior of a metal-polymer composite with low percolation threshold. Appl Phys Lett 89:072902
49. Lee AY, Tran VN (2006) Dielectric characterisation of high loss and low loss materials at 2450 MHz. In: Advances in microwave and radio frequency processing, Springer, pp. 77–84
50. Zhang Y-H, Dang Z-M, Fu S-Y, Xin JH, Deng J-G, Wu J, Yang S, Li L-F, Yan Q (2005) Dielectric and dynamic mechanical properties of polyimide–clay nanocomposite films. Chem Phys Lett 401:553–557
51. Zhang X, Ma Y, Zhao C, Yang W (2014) High dielectric constant and low dielectric loss hybrid nanocomposites fabricated with ferroelectric polymer matrix and $BaTiO_3$ nanofibers modified with perfluoroalkylsilane. Appl Surf Sci 305:531–538
52. Yang K, Huang X, Huang Y, Xie L, Jiang P (2013) Fluoro-polymer@ $BaTiO_3$ hybrid nanoparticles prepared via RAFT polymerization: toward ferroelectric polymer nanocomposites with high dielectric constant and low dielectric loss for energy storage application. Chem Mater 25:2327–2338
53. Pradhan DK, Choudhary R, Samantaray B (2008) Studies of dielectric relaxation and AC conductivity behavior of plasticized polymer nanocomposite electrolytes. Int J Electrochem Sci 3:597–608
54. Balanis C (1989) In: Advanced enginerring electromechanics, Wiley, New York
55. Kasap S (1997) Principles of Electrical Engineering Materials, p 184
56. Sadiku MN (2014) Elements of electromagnetics, Oxford university press
57. Lee YH, Bur AJ, Roth SC, Start PR, Harris RH (2005) Monitoring the relaxation behavior of nylon/clay nanocomposites in the melt with an online dielectric sensor. Polym Adv Technol 16:249–256
58. Yuan J-K, Yao S-H, Dang Z-M, Sylvestre A, Genestoux M, Bai1 J (2011) Giant dielectric permittivity nanocomposites: realizing true potential of pristine carbon nanotubes in polyvinylidene fluoride matrix through an enhanced interfacial interaction. J Phys Chem C, 115(13) 5515–5521
59. Paul D, Robeson LM (2008) Polymer nanotechnology: nanocomposites. Polymer 49:3187–3204
60. Coleman JN, Khan U, Blau WJ, Gun'ko YK (2006) Small but strong: a review of the mechanical properties of carbon nanotube–polymer composites. Carbon 44(9) 1624–1652
61. Lu X, Zhang W, Wang C, Wen T-C, Wei Y (2011) One-dimensional conducting polymer nanocomposites: synthesis, properties and applications. Prog Polym Sci 36:671–712
62. Al-Saleh MH, Sundararaj U (2009) A review of vapor grown carbon nanofiber/polymer conductive composites. Carbon 47:2–22
63. Cai Z, Wang X, Luo B, Hong W, Wu L, Li L (2017) Nanocomposites with enhanced dielectric permittivity and breakdown strength by microstructure design of nanofillers. Compos Sci Technol 151:109–114

Thermal Properties of Sustainable Thermoplastics Nanocomposites Containing Nanofillers and Its Recycling Perspective

Pooja Takkalkar, Sabzoi Nizamuddin, Gregory Griffin and Nhol Kao

1 Introduction

Commercialized fuel-based polymers have created immense adverse effects on the environment due to their non-renewable nature and emission of greenhouse gases, particularly carbon dioxide (CO_2) [1]. Pollution created by conventional polymers has risen to dangerous extents, predominantly in developing nations. These polymers, being non-biodegradable, are resistant to microbial degradation and hence they accrue in the surroundings. Furthermore, increases in oil prices have aided in stimulating interest in biodegradable polymers. Eco-friendly polymers were first introduced in the 1980s [2]. Over the past two decades, bio-based polymers have attracted considerable attention, mainly because of two major limitations with the use of conventional polymers. Firstly, the environmental pollution created by the increased reliance on fossil fuels and secondly the fact that the source of these petroleum-based polymers is limited and exhaustible [3].

Polymer nanocomposites are new engineering materials in which nanofillers [4], with at least one dimension less than 100 nm, are dispersed in a polymer to improve its properties [5–7]. Based on the appropriate application and final properties desired of the nanocomposite, the type and concentration of nanofiller can be studied. Nanocomposites possess more advanced properties than that of microcomposites as the incorporated nanofiller has a better aspect ratio and surface area. Biodegradable nanocomposites are considered more advantageous because they are light in weight, transparent, and have better mechanical, thermal and barrier properties than that of conventional composites, even at a very low concentration of nanofiller [8].

Nanocomposites have improved properties than the constituent polymers itself [9, 10]. The efficient dispersion of nanofiller in the polymer indicates that there is an

---

P. Takkalkar · S. Nizamuddin · G. Griffin · N. Kao (✉)
School of Engineering, RMIT University, Melbourne, Vict 3001, Australia
e-mail: nhol.kao@rmit.edu.au

increased interfacial interaction among the constituents [11]. Although the biodegradable polymers are environmentally friendly, their properties are inferior to conventional polymers; hence, nanocomposites with biodegradable polymers have been investigated to seek improved properties. The area of nanocomposites with biodegradable polymers has gained attention in the last two decades. A number of bio-based polymer matrices have been studied after incorporating organic and inorganic fillers to enhance their properties. The biodegradable polymers used as a matrix in preparation of nanocomposites are polybutyl succinate, polylactic acid, cellulose, starch, alginate, soy protein isolate, plant oil based polymers, polyhydroxy alkanoate and epoxies [12]. The properties which are aimed to be improved in these biodegradable nanocomposites include their thermal, mechanical, barrier, rheological and crystalline properties while maintaining their biodegradability. Depending on the particular application of these nanocomposites the appropriate nanofiller can be selected. The nanofillers are classified based on their source, shape, aspect ratio and crystallinity. Although the research in this area and the applications of biodegradable nanocomposites is in its infancy, it is expected to be enormous in future years.

The overall properties of the polymer nanocomposites are largely affected by-

i. Compatibility and interactions between the polymer matrix and nanofiller
ii. Nanofiller shape and aspect ratio
iii. Dispersion of nanofiller within the polymer matrix
iv. Modifications on the surface of nanofiller (if any).

This chapter provides a description of the thermal properties of sustainable thermoplastic nanocomposites; its recycling perspective is also considered.

2 Sustainable Thermoplastic Nanocomposite

Polymer scientists have shown increased interest towards developing various environmental friendly polymer nanocomposites which can potentially reduce the dependency on conventional polymers. These sustainable thermoplastic nanocomposites deserve attention as they aid in resolving concerns on issues such as the emission of greenhouse gases, depletion of fossil fuels and pollution [12]. Thermoplastic polymers are preferred over thermoset polymers due to various advantages such as low processing cost, design flexibility and easier moulding of complex parts [13]. Simple moulding techniques such as extrusion or injection moulding are widely used for fabrication of such types of composites.

The polymer matrix plays an important part for the performance of a polymer nanocomposite. In addition, the dispersion of fibers in the composite is also one of the crucial parameters for achieving the consistency of the product. Further, the properties of fibers, fibre-matrix interface and aspect ratio of fibres also govern the properties of nanocomposites [13]. The performance and properties of thermoplastic composites are also influenced by process parameters [14]. It is reported by Takase and Shiraishi [15] that tensile strength of polypropylene/wood composite changed non-linearly with mixing temperature, mixing rate and time. The dispersion of fibers and fiber length should be optimized in order to enhance the properties of the composite. It is important that a uniform dispersion of nanomaterial within the polymer should be carried out [16]. Several dispersion methods for thermoplastic polymers have been reported in the literature, but the easiest method is blending in a minimax injection moulder's crucible [17], where the rotary cylinder is used for mixing the polymer melt with fiber addition by hand.

The concept of combining nanomaterials as filler and polymer as a matrix in order to form different forms of nanocomposites has been gaining much recognition by researchers [18–28]. Nanomaterials such as nanotubes and nanoclays provide great potential for fabrication of a variety of different forms of composites, coatings, adhesives and sealant materials with specific properties. These nanoparticles can be successfully utilized as a filler in thermoplastic polymers in order to improve the mechanical, physical and thermal properties of the polymer.

In this context, polylactide (PLA) has been leading among other thermoplastic polymers due to its inherent advantages like good mechanical strength, renewability, biocompatibility, and biodegradability. It is a versatile polymer made from agricultural raw materials, which are fermented into lactic acid. Then, this lactide acid is ring-opening polymerized through cyclic dilactone (lactide) to the desired polylactic acid. The polymer is altered through different means, which improves the thermal stability of the polymer and decreases the residual amount of monomers [29]. The thermal stability of the polylactic acid also can be improved by reinforcing it with fibers. The thermal properties are essential characteristics to understand the behaviour of the raw material and the final product [30]. Therefore, the thermal properties of various thermoplastic nanocomposites have been elaborated in detail in the following sections.

3 Thermal Properties of Sustainable Nanocomposites Based on Types of Sustainable Polymers

3.1 Polylactic Acid (PLA) Based Nanocomposites

PLA is a thermoplastic biodegradable polymer derived from corn. Although it is a biodegradable polymer, its properties need to be optimized to make it suitable for commercial packaging applications. Polylactic acid nanocomposites with improved properties by incorporation of cellulose nanofibers (CNFs) were prepared by Frone et al. [31]. CNFs were obtained using the most commonly used acid hydrolysis method. Microcrystalline cellulose (MCC) with particle size 20 μm was hydrolysed to form CNF with diameter 11–44 nm. These CNF were further surface treated to form silanized (CNFS) after that, both the CNF and the CNFS were added in PLA at 2.5 wt% of each nanofiller were melt blended to form the nanocomposites. The effect of treated and untreated CNFs on nucleation characteristics of PLA was determined through differential scanning calorimetry (DSC) analysis. There was no substantial change in T_g, owing to the presence of nanofiller at very low concentrations. The PLA nanocomposite with CNF as nanofiller showed higher crystallinity due to the better nucleating effect created by CNF. In comparison, due to the better dispersion and adhesion of CNFS with PLA, the crystallinity was reduced as compared to CNF. Overall the degree of crystallinity was improved for PLA with both nanofillers, however, the effect was more pronounced with CNF nanofiller [31].

A similar study on cellulose nanocrystals (CNC) and silylated CNC (SCNC) in PLA was reported in the literature [32]. The authors prepared the nanocomposite with 1 and 2 wt% of each type of nanofiller through solvent casting with chloroform as a solvent. The samples with 1 and 2 wt% of CNC were labelled as PLLA-CNC-1 and PLLA-CNC-2, respectively and the samples with a similar concentration of SCNC were coded as PLLA-SCNC-1 and PLLA-SCNC-2, respectively. As expected, the bulk of the measurements showed there was no change in the melting temperature of all the nanocomposites, at around 171 °C. The degree of crystallinity (X_c) was determined using the enthalpy data from DSC. The X_c values were higher than pure PLA (14.3%) for all the nanocomposites. The X_c values were improved slightly for PLLA-CNC nanocomposites and they increased with filler loadings. Conversely, in the case of PLLA-SCNC nanocomposites, the X_c values were almost doubled, but decreased slightly with the increase in SCNC loading. The highest crystallinity was achieved for PLLA-SCNC-1 (30.4%). This was attributed to the improvement in nucleation effect triggered by homogeneous dispersion of the nanofiller within the PLLA matrix [32].

A recent study described the improvement in properties of PLA based nanocomposites with starch nanocrystals (SNC) as nanofillers at three different loadings of SNC- 1, 3 and 5 wt% [33]. The nanocomposites were prepared by solvent casting and evaporation techniques. The square-shaped SNC nanofiller used in this study was derived from acid hydrolysis of waxy maize starch, which consists of 99% amylopectin. The thermal properties of these nanocomposites were studied by means of TGA and MDSC. The TGA study revealed that all the nanocomposites were stable to process in the commercial processing range of polymers (25–240 °C). The MDSC results indicated a slightly higher degree of crystallinity for PLA-SNC-3 wt% but with further increase in the concentration of SNC, it declined. Also, a decline in the cold crystallization temperature was detected for all the nanocomposites. These changes were attributed to the enhanced nucleation effect created by crystalline SNC. Beyond the 3 wt% of SNC, the nanofillers had a tendency to aggregate and declines in thermal and rheological properties were observed. Similar results were reported for PLA and acetylated microcrystalline cellulose (Ac-MCC) based nanocomposites [34]. The optimum concentration obtained through rheological percolation threshold was at 2.5 wt% of Ac-MCC.

PLA nanocomposite with nanographite platelets (NGP) was prepared by melt blending at 180 °C [35, 36]. The concentration of NGP in PLA was varied from 1 to 10 wt%. The influence of NGP loading on thermal and crystallization properties was studied via MDSC. The NGP was unable to exhibit any nucleating effect or improve the overall crystallinity and thermal properties of the nanocomposites. However, a comparatively higher crystallinity and fusion enthalpy were noticed for nanocomposite with 5 wt% NGP. The researchers concluded that the melt compounding method alone is not sufficient to obtain a well-dispersed nanocomposite. A similar study was reported by the same group of authors [36], however, a melt intercalation and mixing technique were used to fabricate the PLA-NGP nanocomposites. In this study, the MDSC results showed that the melting temperature and glass transition temperatures were unaffected by the presence of NGPs, though the crystallisation temperature and degree of crystallinity were increased up to 5 wt% loading of NGP.

3.2 Thermoplastic Starch (TPS)

There is an environmental demand to create biodegradable plastics which have adequate properties and can replace commercial plastics. Nanocomposites prepared from starch as a matrix have great potential as they are available in abundance and are renewable. To overcome some of the inherent limitations of starch such as limited gas barrier properties, low heat distortion temperature and brittleness; polysaccharide nanofiller such as cellulose or starch nanocrystals can be incorporated in thermoplastic starch. The addition of biodegradable nanofiller not only improves the properties of the biodegradable polymer but maintains their biodegradability as well. Kaushik et al. [37] prepared biodegradable nanocomposites with thermoplastic starch as nanocomposite and cellulose nanofibers (CNFs) as nanofiller, at three different loadings of CNF, 5, 10 and 15 wt% by solution casting method. The CNFs reported in their study were extracted from wheat straw through a combination of steam explosion, chemical treatments (involving alkali and acid hydrolysis and bleaching) and mechanical treatments (homogenization). Morphological, structural, mechanical, thermal and moisture retention properties of the prepared nanocomposites were studied. The nano-dimensions of CNFs were measured through TEM, which showed fiber diameter in the range of 10–60 nm with a tendency to agglomerate. There was a significant improvement in the mechanical property of the nanocomposite with increase in CNF concentration; this was noted as a linear increase in tensile strength and modulus of the nanocomposite. The barrier property improved until a loading of 10 wt% and declined after a further increase in CNF concentration. This was credited to agglomeration of CNF fibers within the nanocomposite. The thermal properties of nanocomposites were studied through TGA and DSC. The thermal property study indicates the interaction between glycerol and CNF. The TGA results showed no change in onset degradation temperature until a concentration of CNF-10 wt% was reached, beyond that

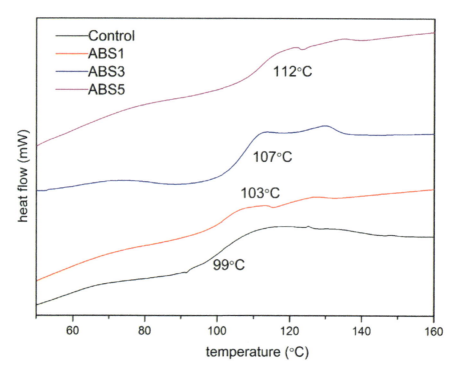

Fig. 1 DSC curves for Acrylonitrile butadiene styrene and Acrylonitrile butadiene styrene/organ-montmorillonite nanocomposite at 1, 3 and 5 wt% loading of nanofiller. Reprinted from Ref. Weng, Wang [54] with permission from Elsevier

the onset degradation temperature reduced when compared to pure starch. They presented two possible causes for this reduction- (i) the reduced flexibility of branched amylopectin hindered by nanocrystalline cellulose and (ii) accumulation of the plasticizer (glycerol) on the CNF surface [37].

Cao et al. [38] reported similar properties of solvent cast plasticized starch (PS) based nanocomposite with different concentrations of cellulose nanocrystals derived from flax fiber (FCNs). PS was reinforced with FCN at six different loadings 5, 10, 15, 20, 25, 30 wt% of FCN. The thermal properties of the prepared nanocomposites were recorded through DSC analysis. They focussed the thermal analysis on the changes in glass transition temperature (T_g). T_g was associated with two phases; T_{g1}—glycerol rich phase (−80 to −50 °C) and T_{g2}—starch-rich phase (30–60 °C). The T_{g1} remained unaffected by the presence of FCNs, however, the T_{g2} kept increasing as the FCN loadings were increased. This was ascribed to the high interaction of FCNs with PS at the interface which hinders the molecular mobility of PS chains.

3.3 Polycaprolactone (PCL)

Performance improvement of biocompatible polymers like polycaprolactone (PCL) by adding nanofillers like clay or nanocellulose is studied widely. PCL based nanocomposites were fabricated with unmodified cellulose nanowhiskers (CNWr) and poly-ester grafted cellulose nanowhiskers (CNWr-g-PCL) at three different concentrations of each nanofiller- 2, 4 and 8 wt% in PCL matrix [39]. The CNWr were obtained through alkaline and acid hydrolysis of ramie fibers. Finally, the nanocomposites were prepared by melt blending PCL and the nanofillers at 120 °C. There was no significant change in the melting and crystallisation characteristics of the nanocomposites with CNWr as nanofiller. The crystallization temperature increased slightly with the increase in the concentration of CNWr-g-PCL. The grafted nanofiller largely improved the thermo-mechanical properties of the prepared nanocomposite. A similar study of melt blended PCL nanocomposites was reported previously [40].

3.4 Polyamide/Clay Nanocomposites

In the last decade, the number of investigations about the thermal degradation of clay-based nanocomposites has increased markedly due to the importance of this property for polymers production. In particular, clay-based nanocomposites polymers have drawn more interest recently [41]. The nanofillers are added into polymers as they have a potential to improve the thermal properties, particularly increasing the flame or chemical resistance characteristics, enhance the bulk clarity and ionic conductivity as well as decrease the moisture, hydrocarbons or gases permeability [42]. However, the problem of polymers thermal degradation in nanocomposites is still an obstacle to produce anti-thermal degradation polymers. This is due to the high temperature used in the preparation process of the clay-nanocomposites. Normally, degradation of the clay-based-polymers occurs if the temperature used in the preparation is higher than the stable temperature of the organic material used to reinforce the polymer [42]. Therefore, different materials have been used to advance the thermal and the mechanical properties of polymer products. For example, clay modified with cationic surfactants, specifically quaternary ammonium-organic surfactants, dispersed at the nanoscale into the polymer to improve the mechanical compatibility and increase the interlayer spacing of the clay and, hence, decrease the forces between the single platelets and thus aid exfoliation [41].

Morgan et al. (2002) have successfully formed a clay-catalysed-carbonaceous-char and reinforced char by clay polystyrene nanocomposites. They stated that the enforcement of the char by clay has decreased the flammability of the nanocomposites of polystyrene [43]. In this case, the mass fraction of clay loading, 5%, was the optimal percentage amount to be loaded for improving the heat release rates among the 0, 2, 5, 10% mass fractions used in this study [43].

3.5 Polypropylene/Layered Silicates Nanocomposites

Polymer nanocomposites, compared to traditional composites, show a dramatic transformation in various properties at low loadings of nanofillers such as graphite nanoplatelets, carbon nanotubes, and nanosilicate layers [44]. The performance of these nanofillers is highly dependent on the uniform dispersion of the nanofillers and strong interactions at the interface between the polymer matrix and nanofillers. The graphene nanosheets, which is structurally similar to silicate layers and chemically analogous to carbon nanotubes [45], are considered as the most promising nanofiller in order to improve the barrier, mechanical as well as thermal properties. In spite of great potential of graphene nanofiller, the good dispersion is still a challenge for the effective reinforcement of polymers, specifically in nonpolar polymers such as polypropylene. Trokeson et al. [45, 46] fabricated fully-exfoliated polypropylene/graphite nanocomposites through a solid-state shear pulverization method. It was noticed that 2.5 wt% loading of graphite exhibited a 100% increase in Young's modulus and around 60% improvement in yield strength compared to pure polypropylene. Miltner et al. [47] successfully incorporated carbon nanotubes in polypropylene latex by sonication for fabrication of very well-dispersed polypropylene/carbon nanotubes nanocomposites.

Graphene-based polypropylene nanocomposite with improved mechanical and thermal properties were successfully fabricated by previous researchers [44]. An eco-friendly technique was utilized for preparation of polymer nanocomposite with well-dispersed graphene sheets within the polymer matrix, through coating graphene with polypropylene latex, and then melt-blending coated graphene with polypropylene matrix. It was observed that yield strength and Young's modulus of polypropylene were improved up to 75% and 74% respectively at 0.042% loading of graphene. It was reported that the thermal properties were improved after addition of graphene. For instance, the glass transition temperature was increased by 2.5 °C at 0.041% loading of graphene. Similarly, the thermal oxidative stability was also enhanced significantly up to 26 °C at 0.42% loading of graphene. Mounir et al. [48] synthesized graphene/polypropylene nanocomposite through melt mixing and investigated the effect of graphene loading on propylene characteristics (thermal properties). It was observed that an increase in graphene loading resulted in a significant improvement of mechanical and thermal properties of polypropylene. Further, graphene showed a significant effect on thermal stability of neat polypropylene and its composite. It was reported that the thermal degradation temperature of both the neat polypropylene and polypropylene/graphene nanocomposite occurs as a single step process with maximum degradation temperature of 460 °C. An increase in thermal stability of polypropylene after graphene loading, at initial degradation stage, was observed which was attributed to the hindered diffusion of volatile decomposition products within the nanocomposites and is highly dependent upon the interaction between chains of nanosheets and the polymer.

4 Thermal Properties of Sustainable Nanocomposites Based on Various Thermal Properties

4.1 Thermogravimetric/Differential Thermogravimetric Analysis (TGA/DTG)

TGA is an important method used to examine thermal decomposition behaviour of polymeric materials for high-temperature applications [49]. In TGA, the weight loss of a substance is monitored as a function with time or temperature. The weight loss is due to the generation of the volatile product post-degradation. The thermal properties are essential characteristics to understand the behaviour of the raw material and the final product [30]. The thermal stability of any polymer matrix can be improved by adding a small number of nano additives/nanofillers [40]. Wang et al. [50] noticed that thermal stability of Acrylonitrile butadiene styrene was improved by 5% of organ-montmorillonite.

Alemdar and Sain [51] synthesized wheat straw nanofiber/thermoplastic starch polymer nanocomposite and studied the thermal behaviour of neat polymer and nanocomposite at 5% loading of the nanofiber. It was observed that the degradation temperature for both the neat polymer matrix and nanocomposite is similar but less than that of each individual component. Similar behaviour for degradation temperature of thermoplastic starch and thermoplastic starch/lignocellulosic fiber composite was observed in the study conducted by Averous and Boquillon [52]. Yuan et al. [53] reported the effect of multiwall carbon nanotubes on thermal stability of polyamide 12. It was found that there were two stages of decomposition for neat polyamide 12 at 443 and 462 °C respectively. The addition of carbon nanotube at 1% loading stabilized the polymer matrix against first degradation stage; this is due to the carbon nanotube network capturing thermal radicals and transferring the thermal energy effectively between the polymer chains and carbon nanotube thus avoiding chemical decomposition of the polyamide 12. Similar enhancement in thermal stability of nanocomposite was detected by Weng et al. [54]. They concluded that the addition of nanofiller in the polymer increased the thermal stability by acting as a superior insulator or barrier to the volatiles produced during degradation. In addition, the onset temperature moved slightly to a higher temperature than that of pure matrix confirming that the thermal stability was improved by the reinforcement using nanofiller. Bera and Maji [55] prepared graphene oxide/polyurethane and reduced graphene oxide/polyurethane nanocomposites and measured their thermal properties with the help of TGA and DSC to understand the effect of nanofillers on the thermal properties of end product-nanocomposite. The results suggest that both the neat polyurethane and nanocomposites represented a two-stage thermal degradation pattern which corresponds to the degradation of hard and soft segments. Further, it was reported that the neat polyurethane was thermally stable up to 300 °C and the degradation of the soft segment for raw polyurethane occurred in the range of 300–405 °C, while degradation of the hard segment took place after 405 °C. The initial degradation temperature and T_{max} for graphene oxide/polyurethane nanocomposite at 0.10% loading of graphene oxide were increased by 6 and 8 °C; for reduced graphene oxide/polyurethane at 0.10% loading of reduced graphene oxide initial degradation temperature and T_{max} increased by 5 and 6 °C, respectively. Based on these findings, it was concluded that the graphene oxide provides more thermal stability than reduced graphene oxide at the same loading. An increase in thermal stability by addition of graphene oxide or reduced graphene oxide nanofillers is attributed to the physicochemical interaction between nanofiller and polyurethane, which leads to restricted motion of the polyurethane chains [56]. Furthermore, the volatiles produced during decomposition also remain in the material due to the high barrier properties of nanocomposite compared to raw polyurethane. Their findings were in agreement to literature in which a similar effect of nanofiller on thermal stability of nanocomposite is reported [57].

Arrieta, Fortunati [58] studied the thermal properties of binary and ternary nanocomposite films. The binary system involved neat PLA incorporated with unmodified and surface modified CNC, coded as PLA-CNC and PLA-CNCs, respectively. The ternary system consisted of PLA-PHB-CNC (PHB refers to poly (hydroxybutyrate)) and PLA-PHB-CNCs. The key observation was that the neat PLA and PLA-CNC had a single degradation step while the PLA-CNCs had dual degradation step. The first degradation step of PLA-CNC takes place at temperatures lower than the main degradation step, which might be associated with surfactant loss.

4.2 Differential Scanning Calorimetry (DSC)

DSC provides important data such as the heat capacity (ΔC_p) assigned to the polymer matrix, heat enthalpy (ΔH_m) assigned to the nanofiller, glass transition temperature (T_g), and the melting temperature (T_m). The addition of nanofiller in polymer matrix helps to improve the DSC results. It is reported in the literature [54] that the glass transition temperature (T_g) was improved with the loading of the nanofiller. It is reported that the T_g of a neat ABS polymer was 99 °C, which increased to 103, 107 and 112 °C at 1, 3 and 5% loading of nanofiller as shown in Fig. 1. According to Yu et al. [59], the DSC results showed that the PLA used in their study had a T_g around 57.7 °C and T_m of 149.1 °C. The T_g for neat PLA was observed at 57.7 °C, with an increase in loading of starch nanocrystals (StN) to grafted PCL (StN-g-PCL) it reduced to 47.8 °C for 5 wt% filler and eventually, it disappeared for higher loadings. This was ascribed to the increased flexibility of the PLA chains during the melting process of the PCL. The fabrication of nanocomposites, with PLA as a matrix, and 1 and 3%wt of unmodified (CNC) and surfactant-modified (s-CNC) cellulose nanocrystals as fillers, using solvent casting technique was reported by Fortunati et al. [60]. There is no significant change in the T_g as reported in the previous literature on PLA based nanocomposites. Conversely, a reduction in T_g values was detected in the PLA 1 s-CNC system (about 15 °C), though a less intense signal, which was hard to perceive, was recorded for the PLA 3 s-CNC nanocomposite. This effect may be credited to the presence of the surfactant that acts as a plasticizer of the polymer, decreasing the glass transition temperature.

The raw polyurethane did not have a sharp melting point (T_m) for the hard segment [61] due to its lower percentage i.e. 23%. Therefore, the T_m of second heating was considered as true T_m because the impurities in polymer get melted in first heating, therefore, the T_m of neat polymer or nanocomposites takes place in the second heating [55]. On the other hand, Gabr et al. [62] found that the addition of nanoclay as a nanofiller does not have any effect on T_m. They found that the T_m of both the neat polymer and nanocomposite was 165 °C as confirmed by the DSC heating curve of neat polypropylene and polypropylene/nanoclay composite as shown in Fig. 2b. Yu et al. [59] noticed that the neat PLA has a single sharp melting peak at 149.1 °C, however, dual peaks were observed for nanocomposites which were assigned to the presence of PCL. The dual peaks are indicative of the presence of interfacial layers based on the interactions between PLA, PCL, and StN. Gabr et al. [62] synthesized a polypropylene/organoclay nanocomposite up to 5% and studied the DSC analysis of both the raw polypropylene and nanocomposite as shown in Fig. 2a. The results revealed that crystallization peak temperature of neat polypropylene was 121.1 °C, which increased to 123.4 °C at 5% loading of nanoclay filler. An increase in crystallization peak temperature by the addition of organoclay is assumed to be due to the layers of organoclay acting as effective nucleating agents for crystallization of polypropylene. Due to the interaction between layers of organoclay and polypropylene, organoclay layers absorb molecule segments of polypropylene and some of the polypropylene is immobilized and

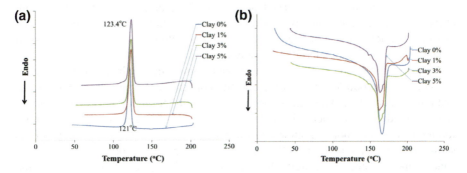

Fig. 2 **a** and **b** Cooling and heating curves for DSC of neat polypropylene and polypropylene/organoclay nanocomposite. Reprinted from Ref. Gabr, Okumura [62] with permission from Elsevier

this immobilization of polypropylene molecules helps in the crystallization of polypropylene. An increase in crystallization temperature by the addition of nanofillers in the polymer matrix was reported by Mingliang and Demin [63].

Robles, Urruzola [64] reported six different nanocomposites based on PLA and surface modified cellulose nanofibres and cellulose nanocrystals, obtained from blue agave bagasse. The nanocomposites were developed using a Haake Minilab twin-screw extruder with an L/D ratio of 24:1 at 170 °C. Two different surface modifications were done in order to modify the hydrophilic character of the nanofillers. The surface treatment was done using 3aminopropyl triethoxysilane and dodecanoyl chloride for cellulose nanofibers and cellulose nanocrystals, respectively. An improvement in crystallization of the PLA chains was noted with an increase in the loading of nanoreinforcement, without altering the overall crystallinity of the PLA matrix. The slight deviation in the T_c and the rise in the enthalpy of melting are ascribed to the good nucleation of the nanocellulose filler (fiber and crystals). This enables the PLA chains to flow freely around the crystals whilst maintaining a stable melting temperature.

4.3 Thermal Conductivity

Thermoplastic polymers have a wide range of applications to act as electrical and thermal insulators. To enhance the thermal and electrical conductivity of these polymers, it is suggested to add any nanofiller to obtain a nanocomposite, which has potential to overcome the limitations of these polymers in electrical and thermal applications [53]. Polymeric nanocomposites have the potential for utilization in a variety of thermal applications such as heat exchange and electronics thermal management. It is reported that the thermal conductivity of polymers can be enhanced significantly by reinforcement with small amounts of nanomaterials [65]. For example, single-wall carbon nanotube have thermal conductivity over 3000 W/m K; it is estimated by simulation that this value can be as high as around 6000 W/m K [66]. Another study reported that a factor of 250 enhancements in thermal conductivity was observed at 10% loading of multiwalled-carbon nanotubes [67, 68]. A number of studies have been conducted for improving the thermal conductivity of polymers by incorporating different amount of loadings of different types of nanomaterials in a polymer matrix.

Yuan et al. [53] prepared two different nanocomposites by incorporating carbon nanotube as a nanofiller in two different thermoplastic polymers i.e. polyamide 12 and polyurethane and investigated the effect of nanofiller on thermal conductivity. It was reported that the thermal conductivity was improved significantly by reinforcing 0–1% of carbon nanotube into the polymer matrix for both the polymers as shown in Fig. 3. Patton et al. [69] prepared vapour grown carbon nanofiber/epoxy nanocomposite and measured the thermal conductivity of pure resin and nanocomposite. It was found that the thermal conductivity of pure resin was 0.26 W/m K, which improved to 0.8 W/m K at 40 vol.% of carbon nanofiber. The

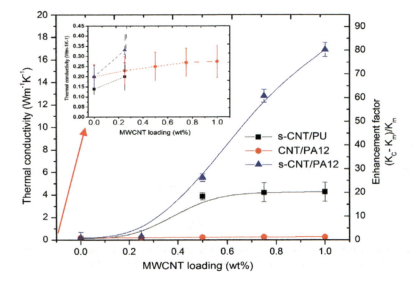

Fig. 3 Thermal conductivities of carbon nanotube/polyamide and carbon nanotube/polyurethane with 0–1% loading of carbon nanotube as a nanofiller. Reprinted from Ref. Yuan, Bai [53] with permission from Elsevier

small increment in thermal conductivity of nanocomposite was ascribed to the difficulty in transmitting thermal energy from fiber to fiber. Lafdi and Matzek produced 20% carbon nanofiber/epoxy resin and measured the thermal conductivity and observed a rise in thermal conductivity from 0.2 W/m K of neat resin to 2.8 W/m K of nanocomposite [70].

It is reported in the literature that the thermal conductivity depends on the temperature [71]. They fabricated polypropylene/single wall carbon nanotube and studied the dependence of thermal properties of both the neat polymer and nanocomposite on temperature. The research found that both the neat polypropylene and polypropylene/carbon nanotube nanocomposite showed a decreasing thermal conductivity with increasing temperature at lower temperatures but this trend reversed at higher temperatures—refer to Fig. 4. Further, the temperature dependence of thermal conductivity of both the raw polymer and nanocomposite can be modelled by bicubic regression polynomials. Also, the thermal conductivity increases with an increase in the amount of loading of nanofiller.

5 Recycling Perspective

In the past few years, considerable effort has been devoted globally to extend the applications of sustainable thermoplastic materials by conferring on them advanced properties through mixing and blending them with various nanofillers.

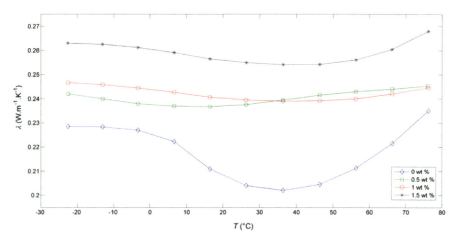

Fig. 4 Temperature dependence of thermal conductivity of pure polypropylene and nanocomposite. Reprinted from Ref. Ivan, Pavol [71] with permission from Elsevier

Thermoplastic polymers like polyolefins, polyamide, polyesters, and styrene polymers are the most representative commodity polymers for cost-efficient manufacturing techniques, outstanding thermo-mechanical properties and good compatibility with the environment, together with easy recycling. The thermal features of thermoplastics nanocomposites plays an important role in deciding its recyclability properties. Sustainable nanocomposites are considered the polymeric materials of this era and are used with organic or inorganic nanofillers with size usually of 1–100 nm. In particular, the higher surface area of the nanofillers than their counterparts allows efficient interfacial interaction with the polymer. The unique properties of the biodegradable nanofillers enhance the overall performance of the polymer matrix. These thermoplastics nanocomposites have a wide variety of applications due to their good thermal and mechanical properties as well as durability in the area of optoelectronic and chemiresistor devices, temperature sensors, linear polarizers, polymeric matrices, catalytic and other chemical sensors, functional materials, polyolefins, polyesters and polyamides. The current total production of plastics is very high and will continue to increase. Hence, due to the creation of huge quantities of plastic waste by industrial manufacturers and householders, the world is confronting a crisis in concerns to environmental, economic and petroleum affects of such production.

Old-style methods of disposal of plastics have negative influences on the environment, for instance, the combustion of unwanted polymers in the form of fumes and toxic gases as well as waste to underground water. The recycling practice is the best method to treat waste polymers and recycling commonly involves processing plastics to produce energy [72]. The overall recycling rate is 2.4 million tonnes for different polymers. Different technological processes have been reported for recycling of polymers such as chemical or feedstock recycling with energy recovery and mechanical recycling [73]. In plastic industries, mechanical recycling is a relatively simple and common method and is also preferable for adequate quantities of homogenous and separated wastes [74]. The conventional mechanical recycling process includes the separation, grinding and producing another material without changing its chemical nature; therefore, mechanical recycling is limited. Low molecular weight materials can be produced by chemical or thermal treatment during feedstock recycling [75], which is an attractive method to substitute for mechanical recycling. The energy recovery techniques mainly applies to plastic disposal wastes via incineration, which also contains a large number of combustible solids. After melting, a stray polyvinyl chloride (PVC) bottle within 10,000 polyethylene terephthalate (PET) bottles can cause the deterioration of the whole batch of PET bottles due to separation difficulties. Therefore, the Society of plastics industry passed the numerical coding system act in the 1980s for ease of separation to determine plastic types. Therefore, these material and chemical recycling of plastic is an appropriate solution to the problems of environmental pollution from various wastes on a worldwide industry level. This can be achieved by using a wide range of different processing techniques with low-cost parameters and has gained increasing importance in the scientific and industrial communities for thermoplastics nanocomposites. Among European countries, Germany has the highest number of recyclers and is globally regarded as the most advanced country for PVC/plastic recycling. The recycling of these polymers usually requires a suitable separation method in which polymer/plastic materials in the mixed solid wastes are separated into a homogeneous stream, which allows a wide range of applications from the recycling perspective.

A simple method is a separation technique of general polymers from plastic wastes is via hand sorting. The application of low content nanofillers is one of the widest and well-known methods to add value to recycled waste plastics to produce thermoplastic nanocomposites, which enhance the thermal stability as well as mechanical properties [76]. The surface area, strength, and viscosity of polymers increase with the addition of nanofiller, which affects interfacial interactions between the polymer and filler [77] and ultimately it increases the performance. The interfacial interactions improve the tensile properties of composites. Titanium dioxide was also incorporated via solution method to improve polymer stability of recycled plastic as well as discolouration resistance [78]. The addition of small organoclays also allows a change in chemical nature of polymer to enhance the recyclability of the polymer [79]. In this way, degradation is minimized with an improvement of some properties. Thus the recycling process involves different new developments and separation techniques for waste polymers with novel energy-recovery procedures for effective cost management.

6 Conclusion

The concept of combining nanomaterials as filler with polymer as a matrix in order to form different forms of nanocomposites has been gaining much recognition these days by researchers. The nanomaterials such as nanotubes and nanoclays provide great potential for fabrication of a variety of different forms of composites, coatings, adhesives and sealant materials with specific properties. Incorporating small quantity of nano additives/nanofillers tends to affect the thermal characteristics of the polymer matrix. Both the initial degradation temperature and T_{max} values for nanocomposites are greater than the neat polymer due to the presence of nanofiller in the nanocomposite. The thermal conductivity highly depends upon the temperature and is well-modelled by bicubic regression polynomials.

Recycling is the best way to treat and minimize waste polymer products and is achieved using various processes on plastics. Different technological processes have been reported for recycling of polymers such as chemical or feedstock recycling with energy recovery, and mechanical recycling. The recycling process involves new development and separation techniques for waste polymers with novel energy-recovery procedures for effective cost management from an economical standpoint.

References

1. Hajilary N, Shahi A, Rezakazemi M (2018) Evaluation of socio-economic factors on CO_2 emissions in Iran: factorial design and multivariable methods. J Clean Prod 189:108–115
2. Ghanbarzadeh B, Almasi H (2013) Biodegradable Polymers. In: Biodegradation—Life of Science
3. Madhavan KN, Nair NR, John RP (2010) An overview of the recent developments in polylactide (PLA) research. Bioresour Technol 101(22): p. 8493–501
4. Rezakazemi M, Zhang, Z (2018) 2.29 Desulfurization materials. In: a2–Dincer I (ed) Comprehensive energy systems. Elsevier, Oxford. p. 944–979
5. Pawelec Z, Bakar M (2013) Shaping mechanical and thermal properties of polymer nanocomposites. Problemy Eksploatacji
6. Rezakazemi M et al (2011) Simulation and determination of optimum conditions of pervaporative dehydration of isopropanol process using synthesized PVA–APTEOS/TEOS nanocomposite membranes by means of expert systems. J Membr Sci 379(1–2):224–232
7. Rezakazemi M et al (2011) CFD simulation of water removal from water/ethylene glycol mixtures by pervaporation. Chem Eng J 168(1):60–67
8. Bari SS, Chatterjee A, Mishra S (2016) Biodegradable polymer nanocomposites: an overview. Polym Rev 56(2):287–328
9. Rezakazemi M, Vatani A, Mohammadi T (2015) Synergistic interactions between POSS and fumed silica and their effect on the properties of crosslinked PDMS nanocomposite membranes. RSC Advances 5(100):82460–82470
10. Rezakazemi M, Vatani A, Mohammadi T (2016) Synthesis and gas transport properties of crosslinked poly(dimethylsiloxane) nanocomposite membranes using octatrimethylsiloxy POSS nanoparticles. J Nat Gas Sci Eng 30:10–18

11. Mittal V, Mittal V (2011) Nanocomposites with biodegradable polymers: synthesis, properties, and future perspectives. In: Mittal V Mittal V.e. (eds) Oxford, New York Oxford, Oxford University Press
12. Raquez J-M et al (2013) Polylactide (PLA)-based nanocomposites. Prog Polym Sci 38(10–11):1504–1542
13. Saheb DN, Jog JP (1999) Natural fiber polymer composites: a review. Adv Polym Technol 18(4):351–363
14. Rezakazemi M, Shahidi K, Mohammadi T (2014) Synthetic PDMS composite membranes for pervaporation dehydration of ethanol. Desalin Water Treat 54(6):1–8
15. Takase S, Shiraishi N (1989) Studies on composites from wood and polypropylenes II. J Appl Polym Sci 37(3):645–659
16. Tibbetts GG et al (2007) A review of the fabrication and properties of vapor-grown carbon nanofiber/polymer composites. Compos Sci Technol 67(7–8):1709–1718
17. Tibbetts GG, McHugh JJ (1999) Mechanical properties of vapor-grown carbon fiber composites with thermoplastic matrices. J Mater Res 14(7):2871–2880
18. Sodeifian G et al (2018) Polyurethane-SAPO-34 mixed matrix membrane for CO_2/CH_4 and CO_2/N_2 separation. Chin J Chem Eng
19. Rezakazemi M et al (2017) Methods for the preparation of organic–inorganic nanocomposite polymer electrolyte membranes for fuel cells. In: Inamuddin D, Mohammad A, Asiri AM (eds) Organic-inorganic composite polymer electrolyte membranes. Springer, Cham. p. 311–325
20. Rezakazemi M et al (2014) State-of-the-art membrane based CO_2 separation using mixed matrix membranes (MMMs): an overview on current status and future directions. Prog Polym Sci 39(5):817–861
21. Baheri B et al (2014) Performance of PVA/NaA mixed matrix membrane for removal of water from ethylene glycol solutions by pervaporation. Chem Eng Commun 202(3):316–321
22. Shahverdi M et al (2013) Pervaporation study of ethylene glycol dehydration through synthesized (PVA-4A)/polypropylene mixed matrix composite membranes. Polym Eng Sci 53(7):1487–1493
23. Rostamizadeh M et al (2013) Gas permeation through H_2-selective mixed matrix membranes: experimental and neural network modeling. Int J Hydrogen Energy 38(2):1128–1135
24. Rezakazemi M, Mohammadi T (2013) Gas sorption in H_2-selective mixed matrix membranes: experimental and neural network modeling. Int J Hydrogen Energy 38(32):14035–14041
25. Rezakazemi M, Shahidi K, Mohammadi T (2012) Sorption properties of hydrogen-selective PDMS/zeolite 4A mixed matrix membrane. Int J Hydrogen Energy 37(22):17275–17284
26. Rezakazemi M, Shahidi K, Mohammadi T (2012) Hydrogen separation and purification using crosslinkable PDMS/zeolite a nanoparticles mixed matrix membranes. Int J Hydrogen Energy 37(19):14576–14589
27. Dashti A, Harami HR, Rezakazemi M (2018) Accurate prediction of solubility of gases within H2-selective nanocomposite membranes using committee machine intelligent system. Int J Hydrogen Energy 43(13):6614–6624
28. Rezakazemi M et al (2017) H 2 -selective mixed matrix membranes modeling using ANFIS, PSO-ANFIS, GA-ANFIS. Int J Hydrogen Energy 42(22):15211–15225
29. Oksman K, Skrifvars M, Selin J-F (2003) Natural fibres as reinforcement in polylactic acid (PLA) composites. Compos Sci Technol 63(9):1317–1324
30. Aji I et al (2011) Thermal property determination of hybridized kenaf/PALF reinforced HDPE composite by thermogravimetric analysis. J Therm Anal Calorim 109(2):893–900
31. Frone AN et al (2013) Morphology and thermal properties of PLA–cellulose nanofibers composites. Carbohyd Polym 91(1):377–384
32. Pei A, Zhou Q, Berglund LA (2010) Functionalized cellulose nanocrystals as biobased nucleation agents in poly(l-lactide) (PLLA)—crystallization and mechanical property effects. Compos Sci Technol 70(5):815–821

33. Takkalkar P et al (2018) Preparation of square-shaped starch nanocrystals/polylactic acid based bio-nanocomposites: morphological, structural, thermal and rheological properties. In: Waste and biomass valorization
34. Mukherjee T et al (2013) Improved dispersion of cellulose microcrystals in polylactic acid (PLA) based composites applying surface acetylation. Chem Eng Sci 101:655–662
35. Narimissa E et al (2012) Morphological, mechanical, and thermal characterization of biopolymer composites based on polylactide and nanographite platelets. Polym Compos 33 (9):1505–1515
36. Narimissa E et al (2012) Influence of nano-graphite platelet concentration on onset of crystalline degradation in polylactide composites. Polym Degrad Stab 97(5):829–832
37. Kaushik A, Singh M, Verma G (2010) Green nanocomposites based on thermoplastic starch and steam exploded cellulose nanofibrils from wheat straw. Carbohyd Polym 82(2):337–345
38. Cao X et al (2008) Starch-based nanocomposites reinforced with flax cellulose nanocrystals. Express Polym Lett 2(7):502–510
39. Goffin A-L et al (2011) Poly (ε-caprolactone) based nanocomposites reinforced by surface-grafted cellulose nanowhiskers via extrusion processing: morphology, rheology, and thermo-mechanical properties. Polymer 52(7):1532–1538
40. Lepoittevin B et al (2002) Poly (ε-caprolactone)/clay nanocomposites prepared by melt intercalation: mechanical, thermal and rheological properties. Polymer 43(14):4017–4023
41. Stoeffler K et al (2013) Polyamide 12 (PA12)/clay nanocomposites fabricated by conventional extrusion and water-assisted extrusion processes. J Appl Polym Sci 130(3):1959–1974
42. Gupta B, Lacrampe M-F, Krawczak P (2006) Polyamide-6/clay nanocomposites: a critical review. Polym Polym Compos 14(1):13–38
43. Morgan AB et al (2002) Flammability of polystyrene layered silicate (clay) nanocomposites: carbonaceous char formation. Fire Mater 26(6):247–253
44. Song P et al (2011) Fabrication of exfoliated graphene-based polypropylene nanocomposites with enhanced mechanical and thermal properties. Polymer 52(18):4001–4010
45. Wakabayashi K et al (2008) Polymer - Graphite Nanocomposites: Effective Dispersion and Major Property Enhancement via Solid-State Shear Pulverization. Macromolecules 41 (6):1905–1908
46. Wakabayashi K et al (2010) Polypropylene-graphite nanocomposites made by solid-state shear pulverization: effects of significantly exfoliated, unmodified graphite content on physical, mechanical and electrical properties. Polymer 51(23):5525–5531
47. Miltner HE et al (2008) Isotactic polypropylene/carbon nanotube composites prepared by latex technology. Thermal analysis of carbon nanotube-induced nucleation. Macromolecules. 41(15): p. 5753–5762
48. El Achaby M et al (2012) Mechanical, thermal, and rheological properties of graphene-based polypropylene nanocomposites prepared by melt mixing. Polym Compos 33(5):733–744
49. Agustin MB et al (2014) Bioplastic based on starch and cellulose nanocrystals from rice straw. J Reinf Plast Compos 33(24):2205–2213
50. Wang S et al (2002) Preparation and thermal properties of ABS/montmorillonite nanocomposite. Polym Degrad Stab 77(3):423–426
51. Alemdar A, Sain M (2008) Biocomposites from wheat straw nanofibers: morphology, thermal and mechanical properties. Compos Sci Technol 68(2):557–565
52. Averous L, Boquillon N (2004) Biocomposites based on plasticized starch: thermal and mechanical behaviours. Carbohyd Polym 56(2):111–122
53. Yuan S et al (2016) Highly enhanced thermal conductivity of thermoplastic nanocomposites with a low mass fraction of MWCNTs by a facilitated latex approach. Compos A Appl Sci Manuf 90:699–710
54. Weng Z et al (2016) Mechanical and thermal properties of ABS/montmorillonite nanocomposites for fused deposition modeling 3D printing. Mater Des 102:276–283
55. Bera M, Maji PK (2017) Effect of structural disparity of graphene-based materials on thermo-mechanical and surface properties of thermoplastic polyurethane nanocomposites. Polymer 119:118–133

56. Thakur S, Karak N (2015) A tough, smart elastomeric bio-based hyperbranched polyurethane nanocomposite. New J Chem 39(3):2146–2154
57. Zhang J, Zhang C, Madbouly, SA (2015) In situ polymerization of bio-based thermosetting polyurethane/graphene oxide nanocomposites. J Appl Polym Sci 132(13)
58. Arrieta MP et al (2014) Multifunctional PLA–PHB/cellulose nanocrystal films: processing, structural and thermal properties. Carbohyd Polym 107:16–24
59. Yu J et al (2008) Structure and mechanical properties of poly(lactic acid) filled with (starch nanocrystal)- graft -poly(ε -caprolactone). Macromol Mater Eng 293(9):763–770
60. Fortunati E et al (2015) Processing of PLA nanocomposites with cellulose nanocrystals extracted from Posidonia oceanica waste: innovative reuse of coastal plant. Ind Crops Prod 67:439–447
61. Sadasivuni KK et al (2014) Dielectric properties of modified graphene oxide filled polyurethane nanocomposites and its correlation with rheology. Compos Sci Technol 104:18–25
62. Gabr MH et al (2015) Mechanical and thermal properties of carbon fiber/polypropylene composite filled with nano-clay. Compos B Eng 69:94–100
63. Mingliang G, Demin J (2009) Preparation and properties of polypropylene/clay nanocomposites using an organoclay modified through solid state method. J Reinf Plast Compos 28 (1):5–16
64. Robles E et al (2015) Surface-modified nano-cellulose as reinforcement in poly(lactic acid) to conform new composites. Ind Crops Prod 71:44–53
65. Gulotty R et al (2013) Effects of functionalization on thermal properties of single-wall and multi-wall carbon nanotube–polymer nanocomposites. ACS Nano 7(6):5114–5121
66. Kim P et al (2001) Thermal transport measurements of individual multiwalled nanotubes. Phys Rev Lett 87(21):215502
67. Nan C-W, Shi Z, Lin Y (2003) A simple model for thermal conductivity of carbon nanotube-based composites. Chem Phys Lett 375(5–6):666–669
68. Prasher RS et al (2009) Turning carbon nanotubes from exceptional heat conductors into insulators. Phys Rev Lett 102(10):105901
69. Patton R et al (1999) Vapor grown carbon fiber composites with epoxy and poly (phenylene sulfide) matrices. Compos A Appl Sci Manuf 30(9):1081–1091
70. Lafdi K, Matzek M (2003) Carbon nanofibers as a nano-reinforcement for polymeric nanocomposites. In: SAMPE Conference Preceding Materials and Processing
71. Ivan K et al (2016) Temperature dependence of thermal properties of thermoplastic polyurethane-based carbon nanocomposites. In: AIP Conference Proceedings. AIP Publishing
72. Raiisi-Nia MR, Aref-Azar A, Fasihi M (2013) Acrylonitrile–butadiene rubber functionalization for the toughening modification of recycled poly (ethylene terephthalate). J Appl Polym Sci 131(13)
73. Zhang Y, Broekhuis AA, Picchioni F (2009) Thermally self-healing polymeric materials: the next step to recycling thermoset polymers? Macromolecules 42(6):1906–1912
74. Hu X, Calo J (2006) Plastic particle separation via liquid-fluidized bed classification. AIChE J 52(4):1333–1342
75. Kameda T et al (2010) Chemical modification of rigid poly (vinyl chloride) by the substitution with nucleophiles. J Appl Polym Sci 116(1):36–44
76. Zare Y, Garmabi H (2012) Nonisothermal crystallization and melting behavior of PP/nanoclay/$CaCO_3$ ternary nanocomposite. J Appl Polym Sci 124(2):1225–1233
77. Zare Y et al (2014) An analysis of interfacial adhesion in nanocomposites from recycled polymers. Comput Mater Sci 81:612–616
78. Herrera-Sandoval G et al (2013) Novel EPS/TiO_2 nanocomposite prepared from recycled polystyrene. Mater Sci Appl 4(03):179
79. Orden MU et al (2014) Clay-induced degradation during the melt reprocessing of waste polycarbonate. J Appl Polym Sci 131(5)

Application of Sustainable Nanocomposites in Membrane Technology

Pravin G. Ingole

1 Introduction

Nowadays nanocomposite membrane technology is widely used in industrial application. The developments of polymer membrane using new generation materials that broaden the industrial applications of membrane processes entail an elevated level of control over a polymer base and nanoparticles addition in the support layer. Polymeric membrane-based separation processes provide a sustainable separation technique for solid/liquid/gas permeance and selectivity [32–35]. Membrane-based separation is economical and conventional base separation process. Especially the nanocomposite membrane-based separation technology is environment-friendly and economically viable. Development of nanocomposite membrane technology for diverse application is one of the best ways to resolve the current inescapable problems.

Currently, nanocomposite membrane technologies are used in gas separation, for example, functionalized TiO_2, SiO_2 NPs incorporated thin-film nanocomposite (TFN) membranes are widely used for mixture gas separation to enhance the gas permeance and selectivity [13, 33, 34]. Also same type of nanocomposite membrane materials are used for water purification [21], wastewater treatment [61], dye separation [88] water vapour separation [5, 31, 33, 34] drug separation [91] etc. NPs incorporated membranes are mechanically strong and thermally more stable compared to without NPs incorporated membranes. Generally, two kinds of nanocomposite membranes varieties are available either it is a flat sheet or hollow fiber shape. Addition of inorganic moieties in the polymer matrix is increases the flexibility and ductility of organic polymers. Recent decade researchers found that the commercialization of nanocomposite membrane is easiest way compared with

---

P. G. Ingole (✉)
Chemical Engineering Group, Engineering Sciences and Technology Division,
CSIR-North East Institute of Science and Technology, Jorhat 785006, Assam, India
e-mail: ingolepravin@gmail.com

another kind of membrane over enhanced flux and selectivity. Khalid et al. suggested that PEG-CNTs nanocomposite PSU membranes are more advanced for wastewater treatment [43]. Nanocomposite membranes also provide motivation to unite the qualities of inorganic nanomaterials and polymeric matrices for exceptional nanofiltration performance [55]. Recently researchers develop the low fouling ultrathin nanocomposite membranes for efficient removal of manganese and lithium [76, 82].

To develop the nanocomposite membrane technology in large scale, early it was the main task of the researcher and now the commercial technology is available in the market. The nanocomposite membranes having mechanical, thermal and swelling properties have developed by using cellulose nanocrystals and PVA [36]. Nanocomposite membrane technology is also applicable for fuel cell applications [7, 84, 86]. Antibacterial mixed matrix nanocomposite membranes fabricated using hybrid nanostructure of silver-coated multi-walled carbon nanotubes by Aani et al. [1]. Using nanocomposite anion exchange membranes, Fernandez-Gonzalez et al. studied the valorization of desalination brines by electrodialysis with bipolar membranes [26]. Within the broad array of commercially existing nanoscale materials, TiO_2 NPs are gained special interest for water desalination [25, 73]. In membrane distillation perfluorododecyl trichlorosilane (FTCS) was employed to modify the virgin polyvinylidene fluoride electrospun nanofiber membrane (PVDF ENM) [72]. TiO_2 (P25 and ST01) deposited on porous ceramic materials for photocatalytic degradation of organic substances in water, a three-phase catalytic membrane contactor (CMC) was implemented [47, 49].

Structural modification of the polymer membrane materials improves the membrane permeability, permselectivity along with mechanical straight and thermal stability. These properties would play a very significant role in membrane science and technology. Thin film nanocomposite membranes are one of the best examples in membrane science to resolve the various issues related to water [66, 70, 78], energy [80], pharmaceuticals [39], environment [10, 33, 34] etc. Further, incorporation of nanoparticles on the thin layer while polymerization especially graphene oxide (GO) membranes offer a wide range of opportunities. Such materials can be engineered to exhibit the desired for the separation characteristics because of ultimate thinness, flexibility, chemical stability and mechanical strength. Different to glassy polymers with a rigid backbone and a high portion of free volume (PTMSP) or with highly interconnected free volume polymers of intrinsic microporosity, GO materials can achieve high-flux and high-selectivity at the same time.

Nanoparticles incorporation to a polymer matrix control the permeability [60] throughout the subsequent sound effects: (a) they work as molecular sieves and amended the permeability [95], (b) also they interrupt the polymeric structure and increase the permeability [15]. One of the examples for Global warming is the result of increasing atmospheric concentration of greenhouse gases such as carbon dioxide (CO_2), methane (CH_4), nitrous oxides (NO_2), hydrofluorocarbons (HFCs), perfluorocarbon (PFC) and sulfurhexafluoride (SF6). These gases trap an increasing portion of terrestrial infrared radiation so, it is expected that global temperature will increase between 1.4 and 5.8 °C in 2100 if no policies on climate change are initiated. The temperature variation causes devastating effects in large and diverse areas of the globe such as possible variations in sea levels, changes in ecosystems, biodiversity loss, reduced crop yields and changes in global precipitation patterns, among other. Different types of membrane gas absorption processes will be tested for the removal of GHGs, both solid and in liquid suspension [18, 41]. Target GHGs will be carbon dioxide, nitrous oxide and methane [46]. This elimination is performed by adsorption and/or absorption processes. The experiment will involve separation and kinetic experiments (isotherms) with nanoparticles and membranes under different conditions of concentration, both in dry or in liquid media, bottled in a small volume and operating in discontinuous (microcosm systems) and perfectly airtight. In order to think in future industrial scale implementation, special attention will be focused on the immobilization of the nanoparticles in porous supports or membranes. In this chapter, the weight has been given to nanocomposite membranes preparation and their implementation in diverse applications including gas separation, water desalination, wastewater treatment, water vapour removal, and energy generation.

Nanocomposite materials especially nanocomposite membranes are facilitating speedy improvements in structural and functional materials diagonally all industries and most of the applications. Recently developed a new method of incorporating functional nanoparticles (10–15 nm) in polymer films, which has guided to the manufacture of a new method of thin film nanocomposite (TFN) membranes technology [33]. Super-hydrophilic nanoparticles synthesis and implementation is the first invention of TFN-based membranes. Introduction studies verify that TFN membranes separate the water vapour, with significant energy savings; and it has super-hydrophilic nature. The commercialization or large-scale productions of TFN membranes using prepared nanoparticles are possible without a major change in TFC membrane process and even cost is also not much higher. It will affect only 5–7% higher cost compare with TFC but the results showed the significant effect. Figure 1 represents the types of nanocomposite used in polymer and non-polymer base materials.

2 Types of Nanoparticles

2.1 Inorganic Metal Oxide and Hydroxide

Along with the numerous groups of nanoparticles, inorganic metal oxide and hydroxide have been of extensive attention from both technological and scientific point of view. Compared to the untainted materials the nano size synthesized metal oxide and hydroxides show the superior properties. Nowadays metal oxides and hydroxides are incorporated into other supports, such as polymeric materials for the applications like supercapacitor electrodes [42], polymer composites for aerospace applications [65], etc. Our group have the experience to synthesize SiO_2 NPs having

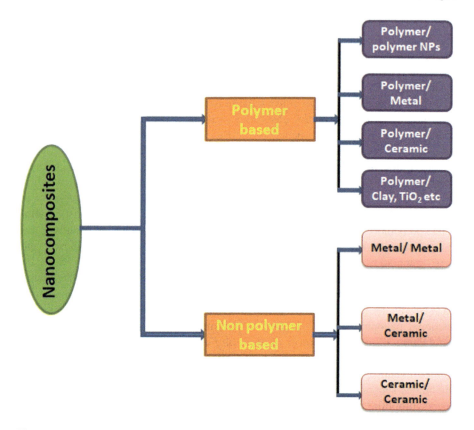

Fig. 1 Polymer and non-polymer based nanocomposite

particle size 10 to 15 nm and implemented it successfully in the nanocomposite membrane materials for diverse applications. Also, amorphous hydroxylated Silicon nanoparticles were synthesized in alcohol-based solvents to fabricate nanocomposite membranes have excellent surface hydrophilicity and roughness. After functionalization, nanoparticles showed higher cross-linking density, higher loading capacity, and high membrane performance.

Even though ceramic membranes play very imperative role in water treatment the polymer membrane technology has achieved noteworthy attention for water treatment applications because of advanced characteristics like its high flexibility, broad range of pore sizes and structures, easy developed process, low costs and easy to scale up [43, 61, 97].

2.2 Inorganic Nanoparticles to Prepare Polymeric Nanocomposite Membranes

In the polymeric composite membranes mechanical performance, lucidity, and thermal stability still remain the controlling limit for the various applications. Thus, the researchers need to develop strong, transparent and heat-resistant nanocomposite membranes for encouraging realistic outcomes. To develop the nanocomposite membrane, the main innovational target consist of concurrently obtaining high permeability and high selectivity at minimum costs, uniting reactions within the pore structures to avoid membrane fouling with avoiding further downstream unit operations, and rising membrane physical strength [50, 69, 84, 86].

Functional polymer membranes are usually premeditated and optimized with precise applications in researchers mind. Figure 2 represents the variety of nanocomposite membrane materials for diverse applications. The presence of functional groups on the surface of nanoparticles not only provides the hydrophilic nature but also reduced the Van der Waals interactive forces between nanoparticles [8]. Nanocomposite membranes, an innovative class of membranes prepared by coalescing polymeric materials with nanomaterials, are rising as a capable elucidation to resolve the various challenges. Especially several inorganic nanoparticles like zirconium phosphates, heteropolyacids, clays, ionic liquids, metal or metal oxides [4, 37, 67, 74], are of extraordinary attention for developing the composite materials. In the fuel cell application, the inorganic nanoparticles like zirconium phosphates, boron phosphate in the nanocomposite membranes do not only provide the water uptake but also provide an extra proton transport pathway [45].

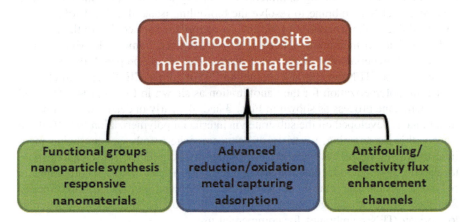

Fig. 2 Diverse nanocomposite membrane materials with various characteristic

3 Thin-Film Nanocomposite (TFN)

3.1 Thin-Film Nanocomposite (TFN) Membranes for Water Desalination

Many efforts have been dedicated to developing the advanced membrane technology to improve the performance of membrane in the form of flux, solute rejection and antifouling properties in the last 3 decades. Year after year research is going to progress and currently, researchers are focusing on nanocomposite membrane to improve the membrane properties. One of the attention was selected by researchers is to prepare the advance thin-film nanocomposite membrane along with support for the high flux, high rejection and antifouling property. Various conditions have been changed while interfacial polymerization (IP) to prepare TFN membranes, changing monomers, monomer concentrations, nanoparticles concentration, nanoparticles size, reaction times, applying chemical modification etc.

Nanomaterials are at present controlling the existing wave of original membrane material development because of the intrinsic explicit physicochemical features that make them apt for water treatment [97]. A number of nanoparticles like silica, graphene, zeolites, carbon nanotubes (both single and multiwalled), silver, metal-organic frameworks (MOFs), silicon and titanium dioxide are the mainly tested nanoparticles in existing and current research. The membranes prepared by using above-mentioned nanoparticles have been shows improved results in the form of permeability, rejection and antifouling properties [2, 24, 102]. Nanoporous silica incorporated membrane shown to reveal a high affinity for water and advanced hydrophilicity of TFN membrane [81]. A TiO_2 and silver nanoparticles have the main characteristic, is strong antimicrobial property so it is important material to develop the TFN membrane to resolve the biofouling issue [44, 96]. Biofouling is happened due to the formation of biofilm on membrane surface due to the intrinsic hydrophobicity of membrane materials. A metal-organic framework (MOF) is one of the best materials for water purification. Zhe et al. prepared the thin-film nanocomposite (TFN) membrane containing PSS-modified ZIF-8 nanoparticles via interfacial polymerization for the nanofiltration as shown in Fig. 3 [104]. The well TFN membrane process as shown in Figs. 3 and. 4 clearly understood that how the membrane is developed on the substrate via interfacial polymerization [94, 104]. To trounce this shortcoming, a variety of nanocomposite membranes are being modified to convey properties such as anti-fouling, hydrophilicity, self-cleaning, photocatalytic, and photodegradation using the nanoparticles (NPs) incorporated in polymeric membrane matrix or use in the interfacial polymerization process. Somehow still, it challenges the researchers to develop the cheapest nanomaterials to fabricate TFN membranes for commercial use.

Application of Sustainable Nanocomposites in Membrane Technology 941

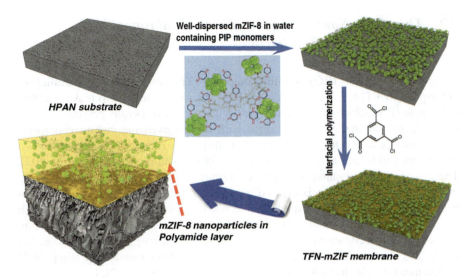

Fig. 3 Preparation process of thin-film nanocomposite (TFN) containing PSS-modified ZIF-8 nanoparticles via interfacial polymerization. Reprinted from Ref. [104], Copyright © 2017 American Chemical Society

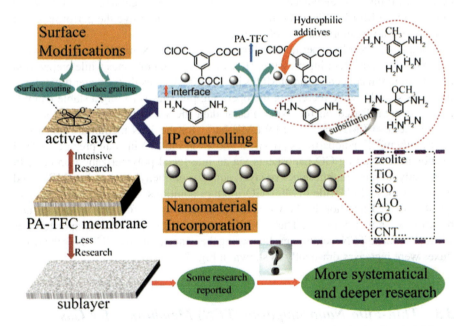

Fig. 4 Surface modification using different NPs to make nanocomposite layer. Reprinted from Ref. [94] Copyright © 2017 with permission from Elsevier

3.2 Thin-Film Nanocomposite (TFN) Membranes for Wastewater Treatment

In the reverse osmosis (RO) thin film composite (TFC) membranes are familiar for FO applications [14, 27, 28, 85, 87, 93, 99]. The technique used to make ultrathin polyamide (PA) selective rejection layer on the surface of porous polymer support is interfacial polymerization. Compare with market available commercial membranes (ex. Cellulose membrane) the thin film composite membranes are shown high permeability and also good resistance aligned with biodegradation [20, 103]. There are some disadvantages of TFC membranes while an operation like intrinsic internal concentration polarization (ICP), solute reverse diffusion and fouling has been found. Then researchers think there is need to develop such membranes which will resolve the above issues. So, Jeong et al. studied the concept of fabricating nanocomposite membranes and use it for RO application [38]. Furthermore, Ma et al. also develop the thin-film nanocomposite (TFN) membranes using NaY zeolite nanoparticles incorporation in the active layer while IP [57]. Later on, many research has been done by the researchers using TiO_2, silica, SiO_2, clay, carbon nanotubes, activated carbon, incorporated TFN membranes for wastewater treatment. The incorporation of NPs is useful for to improved surface hydrophilicity and because of it flux also enhanced drastically. In wastewater treatment researchers use functionalized MOF, CNT and other NPs like TiO_2 to improve the performance of nanocomposite membranes [56, 68, 104].

Comparative results of permeation through TFC and TFN membranes are shown in Fig. 5 on different operating pressure. Here in the TFN membranes while preparation added a different concentration of MOFs i.e. mZIFs. As a result, the water fluxes were increases sharply while increasing the operating pressure verifying a stable nanofiltration system. Based on the experimental data the water flux increases for TFC membrane 6.94 LMH bar^{-1} to 14.9 LMH bar^{-1} for the TFN membrane containing 0.10% w/v mZIF nanoparticles [104]. As shown in Fig. 6 Yin et al. prepared a TFN membrane containing GO nanosheets via an interfacial polymerization process. In their study, to prepare thin selective layer, aqueous *m*-phenylenediamine (MPD) and organic trimesoyl chloride (TMC)–GO mixture solutions were used [98]. A small quantity of GO addition is shown excellent results in the form of water flux and rejection as shown in Fig. 7. The GO NPs were added in thin film composite layer while IP to make TFN membranes. Increasing the concentrations of GO NPs the water fluxes were increases drastically as shown in Fig. 7.

3.3 Thin-Film Nanocomposite (TFN) Membranes for Gas Separation

In the gas separation, nanocomposite proposed an innovative direction to develop polymeric membrane with high performance. In gas separation, nanoporous

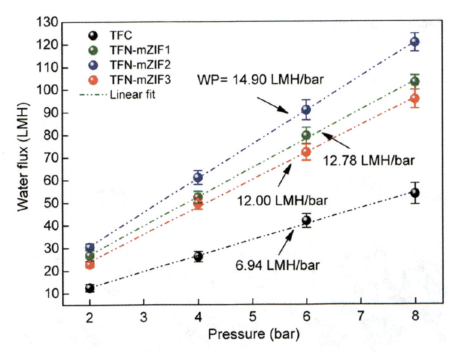

Fig. 5 Water flux of TFC, TFN-mZIF1, TFN-mZIF2, and TFNmZIF3 membranes. WP: water permeability. (Reprinted from Ref. [104] Copyright © 2017 American Chemical Society)

inorganic materials demonstrate high permeability and high selectivity because of their consistent nanopores. There are several ways to prepare the nano size, dense layer for separation after incorporating diverse NPs into polymeric materials while interfacial polymerization to improve gas permeation performance by troublesome the polymer chain packing [3, 19]. Under optimized conditions, TFN membranes performance is very high compared with TFC membranes in the gas separation due to their hydrophilic, smoother and more negatively charged nature. TFN membranes have the advantage to reduce energy consumption and make simpler operations in gas separation applications [53]. The selective TFN layer necessitates elevated selectivity and high gas permeability to reach proficient separation. As shown in Fig. 8, in mixture gas separation especially CO_2/N_2 the porous graphene (PG) nanosheets functionalized TFN shows enhanced CO_2 permeance and the CO_2/N_2 selectively compared to that of the membrane without PG separately. There are lots of literature is available on TFN membrane use in the field of water treatment but from last decades researchers started the application of gas separation using same kinds of membranes.

Figure 9 presents the permeability of O_2 on a logarithmic scale and the O_2/N_2 selectivity after adding inorganic moieties in the polymeric membranes. Wonderful enhancement in the permeability and selectivity had been achieved using diverse polymer materials. Similarly, our previous result also shows using

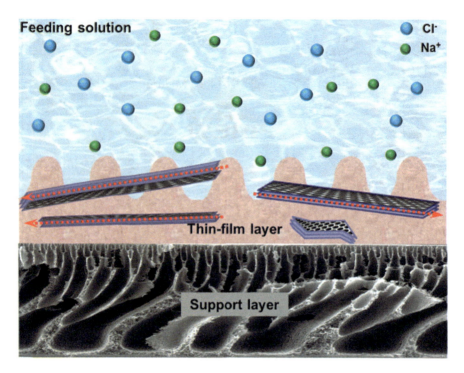

Fig. 6 Schematic illustration of the hypothesized mechanism of GO TFN membrane. Reprinted from Ref. [98] Copyright © 2016 with permission from Elsevier

PDMS CoSalen mixed matrix membrane achieved the 7.7 ideal gas selectivity and good permeance with defect-free membranes [17]. As seen in literature Ismail et al. reported an MWNTs/polymer thin film nanocomposite membranes are greatly improving the carbon capture capacity from N_2 and CH_4 [89, 90]. Xingwei et al. studies on TFN membranes have focused on using silica NPs for enhanced CO_2 separation from mixture gas separation [92]. The challenge is to develop TFN membranes with high-flux and high-selectivity is an urgent basis for cost-efficient CO_2 capture.

Thus, functional graphene oxide (GO) and/or graphene sheets contain a variety of functional groups, having excellent mechanical strength [51]. GO is a brilliant starting nanomaterial for developing size-selective, uniform and stable TFN membranes [16, 23, 40, 52, 59, 75, 77, 83]. In the TFN membrane, the GO nanoparticles are responsible for enhancing selectivity because selective pores in graphenes are allowed the separation of gas molecules.

Any TFN membranes, the NPs plays a key role in enhancing the separation performance. Most of the cases the functionalized NPs takes part in the interfacial polymerization process, also it is found that as a results chemical functionalization of the NPs pore frame could drastically improve the selectivity of mixture gases especially CO_2 over N_2 [77]. The O_2/CO_2 separation was done by using facilitated

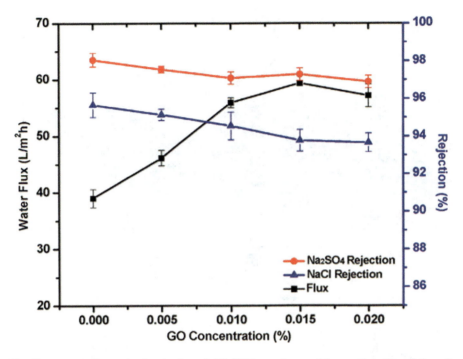

Fig. 7 Permeate flux and salt rejection of GO TFN membranes. The concentration of the salt solution is 2000 mg/L and the TMP is 300 psi. Reprinted from Ref. [98] Copyright © 2016 with permission from Elsevier

transport hollow fiber membranes. The hollow fiber membrane was coated by using poly(n-butyl methacrylate) and cobalt tetraphenylporphyrin complex. The prepared membrane shows 1.5 selectivity of O_2/CO_2 with a high O_2 permeance of 17 GPU at a pressure of 0.098 bar [16, 52].

3.4 Thin-Film Nanocomposite (TFN) Membranes for Fuel Cell Applications

It is well known the fuel cells are a chief technology for the nation's energy portfolio. Fuel cell contribution is a cleaner, more proficient substitute for combustion engines that exploited fossil fuels. Nanocomposite polymer electrolyte membrane (PEM) made up of nanosized inorganic building blocks in the organic polymer by the molecular level of hybridization is pertinent for fuel cell application. The researchers have selected the combined inorganic and organic solid including advance properties like mechanical and thermal stability containing inorganic backbone and specific chemical reactivity, ductility, dielectric, flexibility and processability of the organic polymer to make nanocomposites [83]. During the last ten

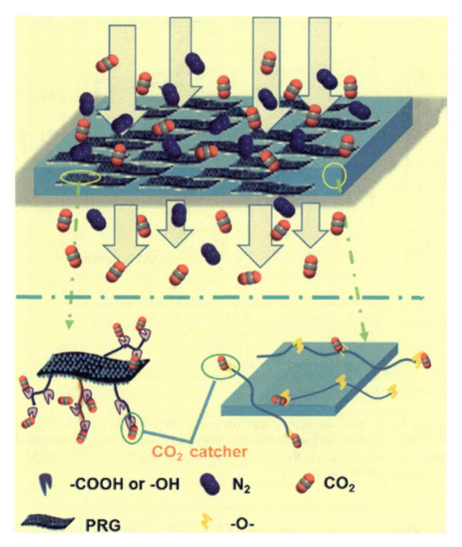

Fig. 8 Mechanism of gas molecules through PG-TFN membranes. Reprinted from Ref. [53] Copyright © 2017 with permission from Elsevier

years, zeolites have attracted a lot of attention and are more and more used in fuel cell applications [27, 28]. There are the criteria for selection of inorganic nanomaterials for fuel cell considering the hygroscopic characteristics, porosity, and pore connectivity, surface area these type of characteristics.

The important thing in the preparation of effective proton conducting nanocomposite membrane for fuel cell application is a covalent bond in between organic moieties and inorganic fillers. One more thing is required to make nanocomposite membrane for a fuel cell is the hydrolytically stable covalent bond between inorganic

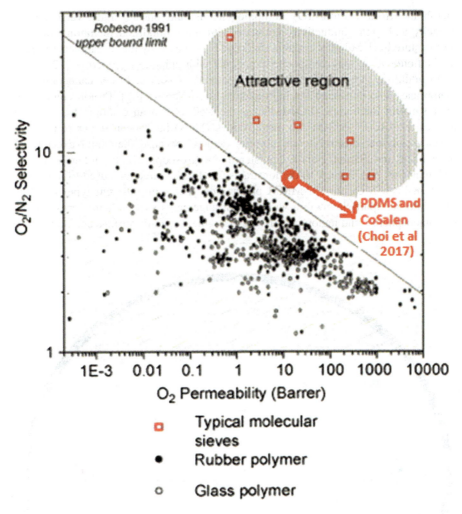

Fig. 9 Relationship between the O_2/N_2 selectivity and O_2 permeability for polymeric membranes and inorganic membranes (the dots indicate the performance of polymeric materials). Reprinted from Ref. [19] Copyright © 2007 with permission from Elsevier

and organic moieties [62]. There are the several ways to modify the organic components for the formation of a stable chemical bond with inorganic components for e.g. silylation (substituted silyl group (R_3Si) to a molecule). Reinholdt et al. studied the composite membranes prepared by using synthesized silica nanoparticles and two SPEEK polymers with sulfonation degrees of 69.4 and 85.0% are characterized for their proton conductivity and water uptake properties [71].

Nafion is one of the key materials for the fuel cell application. Modification of Nafion membrane, the inorganic nanoparticles such as zirconium oxide (ZrO_2), silica, and titanium dioxide (TiO_2) have been used successfully. Modified membrane from Nafion/ZrO$_2$ is homogeneous and shows high water uptake capacity and high conductivity compare with the unmodified membrane at high temperature [64]. Sulfated zirconia (S-ZrO$_2$) is also used by the researchers to make the Nafion/S-ZrO$_2$ nanocomposite membrane with enhanced properties [22].

In addition, the use of S-ZrO$_2$ nanomaterial in Nafion based nanocomposite membranes also enhanced the high-temperature response [63]. Proton conducting mixed matrix membrane (PC-MMM) is the well known an example for fuel cell applications. In PC-MMM the metal oxides (MOs) have been under scrutiny to develop polymer electrolyte membranes (PEMs) because they hold exceptional mechanical and thermal stability, outstanding hygroscopic ability and are in nature abundant [48, 54]. Figure 10 demonstrates the diverse directions used to modified/functionalized MOs for PC-MMM preparation [12]. The different types of MOs form into nanoparticles with a variety of arrangements such as nanohorns, nanorods, nanospheres, and nanotubes, in sort to augment specific surface area to volume

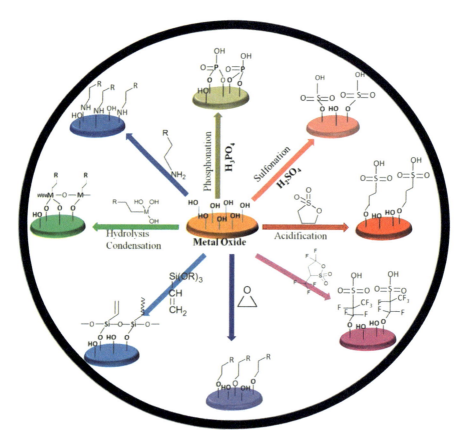

Fig. 10 Illustration of functionalization strategies used to modify metal oxides (MOs) for PC-MMM. Reprinted from Ref. [12] Copyright © 2016 adapted with permission from Elsevier Ltd

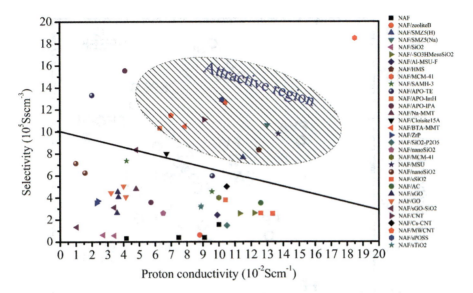

Fig. 11 Proton conductivity versus methanol crossover of PC-MMM composed with Nafion® matrix and inorganic particles at 30 °C and 100% relative humidity (RH). Reprinted from Ref. [12] Copyright © 2016 adapted with permission from Elsevier Ltd

ratio. Figure 11 summarizes the PC-MMM based Nafion® matrix and different inorganic fillers proton conductivity and methanol permeability. These types of inorganic fillers added membranes revealed advanced selectivity evaluated to pristine Nafion® membranes. The mesoporous fillers i.e. zeolites, aluminosilicate, MesoSiO$_2$, CNT that unites the benefit of porous and layered structure, was more successful in dropping the methanol permeability and rising the proton conductivity of the PC-MMM-based Nafion® matrix.

3.5 Thin-Film Nanocomposite (TFN) Membranes for Flue Gas Dehydration

Removal of the water vapour from the flue gas is a hard task for the researcher. Solid adsorbent materials are well known for water vapour adsorption but yet no low-cost technology is available in the market for high scale utilization. To develop the thin-film nanocomposite membranes, Ingole et al. used different types of NPs with various NPs sizes in a range of 10–100 nm in a polyamide (PA) thin film selective layer via in situ interfacial polymerization on the top of various polymer porous supports like polysulfone, polyethersulfone, polyethylene, polyetherimide etc. [29–31]. Various polymeric membrane studies for the flue gas dehydration had also been done by Metz et al. in details [58] (Fig. 12). Their studies teaches about the

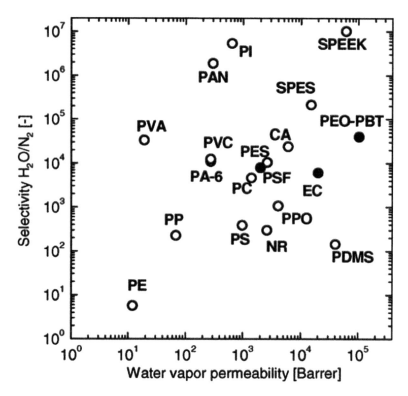

Fig. 12 Water vapour permeability and water vapour/N2 selectivity for various polymers at 30 °C. Reprinted from Ref. [58] Copyright © 2005 Adapted with permission from Elsevier Ltd

measurement of the permeation properties of highly permeable and highly selective polymers for water vapour/nitrogen gas mixtures, and also they reported the analysis of the mass transport of a highly permeable polymer is complicated by the presence of stagnant boundary layers at feed and permeate side. Sijbesma et al. reported that polymer membrane prepared by PEBAX® 1074, a block copolymer, and sulfonated poly(ether ether ketone) (SPEEK) polymers give extremely high separation factors and fluxes for the removal of water vapour from flue gasses [79]. Yun et al. also reported that hydrophilic thin film composite membranes are shown superior performances for flue gas dehydration by water vapour permeation [100, 101].

Furthermore, the flue gas dehydration using polymeric nanocomposite membranes was started by our group in detail. Thin film composite and thin film nanocomposite both types of membranes was targeted to achieve the best result. TFN membranes shows significant performance in the form of permeance and selectivity for flue gas dehydration. For the preparation of TFN membrane, Fig. 13 represented a general procedure for the interfacial polymerization to synthesize the TFN selective barrier layer. TFN membrane is more hydrophilic than TFC membrane so more water vapour has been collected on TFN layer as shown

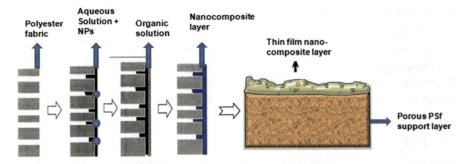

Fig. 13 Schematic illustration of the interfacial polymerization to synthesize the TFN selective barrier layer

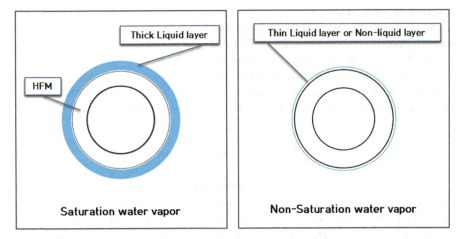

Fig. 14 The comparison, surface of the membrane in a saturated water vapour (TFN) with non-saturated water vapour (TFC)

schematically in Fig. 14. Hydrophilicity of both types of membranes was confirmed by contact angle measurement. After adding Si nanoparticles, TFN prepared from *m*-phenylenediamine and trimesoyl chloride (with 0.05% Si NPs) is more hydrophilic than TFC prepared from same monomers without Si NPs. The contact angle of TFC and TFN membranes were found 55.0° and 37.0°, respectively [9, 11].

The water vapour permeation test was conducted at 2 bar of pressure and 30 °C temperature with N_2 as a carrier gas. The feed gas was fed from the shell side while the permeate side was kept under vacuum. Relative and absolute humidity was measured using the Dew Point meter (HMT 334). At first, the dry gas was passed through the fibers till the steady state of humidity was attained in the membrane. The total flow rate was kept constant at 1000 cc/min. To study the effect of water activity, the wet gas was introduced into the module by using MFC (mass flow controller). The flow rate of wet gas was increased gradually to increase the relative

Table 1 Operating conditions

Operating conditions	
Feed pressure	2 kgf/cm^2
Oven temperature	30 °C
Carrier, dilution gas	N$_2$
Feed gas flow rate	1000 cc/min

Table 2 Membrane specifications

Membrane	Fiber strains	I.D. (μm)	O.D. (μm)	Area (cm^2)
PS$_f$ TFN membrane	5	1000	1400	47.5

humidity in the feed side while keeping the total flow rate constant. Retentate and permeate flow rates were measured via bubble flow meters. The experimental operating conditions are summarized in Table 1.

The membrane specifications are mentioned in Table 2.

The calculations was done using the below equations.

Water vapour permeance was calculated by first calculating the water vapour flow rates at the feed, retentate and permeate streams by using Eq. (1).

$$Q_{vapour} = \frac{Q_{N_2}\, \gamma_{H_2O}\, V_m}{M_{W,H_2O}} \qquad (1)$$

where Q_{N_2} (cm^3/s) was precised by bubble flow meter following retentate and permeate streams conceded during the iced cold trap. γ_{H_2O} is the absolute humidity (g/m^3) and V_m is the volume of 1 mol penetrant at standard temperature and pressure (22.4 L/mol), M_{W,H_2O} is the molecular weight of water (18 g/mol) and Q_{vapour} (cm^3(STP)/s) is the water vapour flow rate at the desired stream.

The permeance of a component P_i in the mixed gas stream can be premeditated by using Eq. (2).

$$P_i = \frac{Q_P}{\Delta P_i \times A} \qquad (2)$$

As results are shown in Fig. 15, the water vapour permeance and selectivity both increases until certain Si NPs concentrations but further after specific concentration of Si NPs the permeance become increases but selectivity decreases. The water vapour permeances ascended due to increased surface roughness coupled with lower contact angles contribute to excellent hydrophilic properties of TFN membranes [9, 11]. Due to more hydrophilic nature, the TFN membranes shows good water vapour permeance and selectivity until connections of Si NPS was 0.5% but furthermore, the permeance was increases but selectivity was decreased. The reason for this type of results is the agglomeration of NPs. After 0.5% NPs concentration in

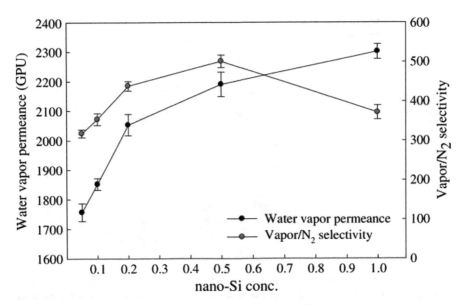

Fig. 15 Effect on the water vapour permeance and selectivity of TFN membranes at various Si nanoparticles concentration. Experimental conditions: temperature = 30 °C, operating pressure = 3 kgf/cm^2, feed water vapour activity = 0.7 ~ 0.8, total feed flow rate = 1000 cm^3/min. Reprinted from Ref. [9, 11] Copyright © 2017 adapted with permission from Elsevier Ltd

monomer solution the NPs agglomeration started and while TFN membrane preparation its shows the disadvantageous towards selectivity because of both N$_2$ and water vapour permeance increases so as a side effect the selectivity decreases [6, 9, 11].

TFN membranes prepared on the inner surface of the polymeric hollow fiber are extremely terrific materials for water vapour separation from flue gas because of their advanced selectivity. The TFN membranes prepared by using MPD and TMC as monomers along with the incorporation of functionalized MOF (NH$_2$–MIL–125 (Ti)) shows very interesting results [35]. The TFN selective layer was prepared the inner surface of the hollow fiber membrane. The schematic representations of the TFN membrane preparation on the inner surface of the PSf hollow fiber membranes are shown in Fig. 16. After incorporation of MOF (NH$_2$–MIL–125(Ti)) nanoparticles in the TFN layer, the performance of membranes was drastically enhanced. Results as shown in Fig. 17, the concentration of MOF (NH$_2$–MIL–125(Ti)) NPs increases from 0.01 to 0.1 w/w% in TFN membranes, the water vapour permeance was enhanced from TFC 785 GPU to TFN 2244 GPU, and the selectivity also jumped from 116 to 542 [35]. Furthermore, after addition of 0.1% NH$_2$–MIL–125 (Ti) NPs, the permeance is decreased because of agglomeration of nanoparticles in the monomer solution. Because of agglomeration of NH$_2$–MIL–125(Ti)) particles, the membrane structure become interrupted.

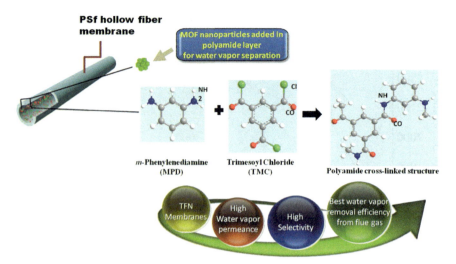

Fig. 16 Interfacial polymerization reaction between MPD (containing NH_2–MIL–125(Ti) MOF nanoparticles) and TMC to form a cross-linked structure on the inner side of PSf hollow fiber membrane. Reprinted from Ref. [35] Copyright © 2018 adapted with permission from Elsevier Ltd

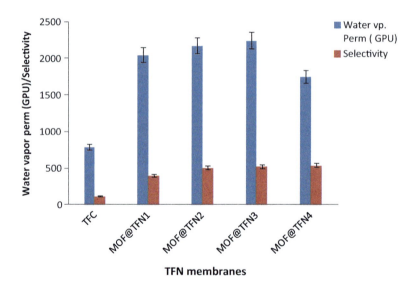

Fig. 17 Effect of NH_2–MIL–125(Ti) MOF nanoparticles concentration in TFN membranes on the performance as water vapour permeance and selectivity. Experimental conditions: temperature = 30 °C, operating pressure = 3 kg-f/cm², feed water vapour activity = 0.7−0.8, total feed flow rate = 1200 cm³/min. Reprinted from Ref. [35] Copyright © 2018 adapted with permission from Elsevier Ltd

4 Conclusions

Application of sustainable nanocomposites in membrane technology is the book chapter to bring a wide study of nanocomposite membrane technology. This pioneering book chapter text offers a fluent introduction to the field as well as an inclusive overview of fundamental facets and application area of nanocomposite membrane. Approaching the subject from the materials point of view, this book chapter:

- Discusses the history, synthesis, and characterization of nanocomposite membranes.
- Examines nanocomposite membranes for water desalination, wastewater treatment, gas separation, fuel cell applications, and flue gas dehydration applications.
- Judges processing challenges, including scalability issues and real implementations.

References

1. Aani SA, Gomez V, Wright CJ, Hilal N (2017) Fabrication of antibacterial mixed matrix nanocomposite membranes using hybrid nanostructure of silver coated multi-walled carbon nanotubes. Chem Eng J 326:721–736
2. Aghigh A, Alizadeh V, Wong HY, Islam MS, Amin N, Zaman M (2015) Recent advances in utilization of graphene for filtration and desalination of water: a review. Desalination 365:389–397
3. Ahn JY, Chung WJ, Pinnau I, Guiver MD (2008) Poly sulfone/silica nanoparticle mixed matrix membranes for gas separation. J Membr Sci 314:123–133
4. Al-bishri HM, Abdel-Fattah TM, Mahmoud ME (2012) Immobilization of [Bmim + Tf2 N] hydrophobic ionic liquid on nano-silica-amine sorbent for implementation in solid phase extraction and removal of lead. J Ind Eng Chem 18:1252–1257
5. An X, Ingole PG, Choi WK, Lee HK, Hong SU, Jeon JD (2017) Enhancement of water vapour separation using ETS-4 incorporated thin film nanocomposite membranes prepared by interfacial polymerization. J Membr Sci 531:77–85
6. An X, Ingole PG, Choi WK, Lee HK, Hong SU, Jeon JD (2018) Development of thin film nanocomposite membranes incorporated with sulfated β-cyclodextrin for water vapour/N_2 mixture gas separation. J Ind Eng Chem 59:259–265
7. Bae I, Oh KH, Yun M, Kang MK, Song HH, Kim H (2018) Nanostructured composite membrane with cross-linked sulfonated poly(arylene ether ketone)/silica for high-performance polymer electrolyte membrane fuel cells under low relative humidity. J Membr Sci 549:567–574
8. Bai L, Liang H, Crittenden J, Qu F, Ding A, Ma J, Du X, Guo S, Li G (2015) Surface modification of UF membranes with functionalized MWCNTs to control membrane fouling by NOM fractions. J Membr Sci 492:400–411
9. Baig MI, Ingole PG, Choi WK, Jeon JD, Jang B, Moon JH, Lee HK (2017) Synthesis and characterization of thin film nanocomposite membranes incorporated with surface functionalized Silicon nanoparticles for improved water vapour permeation performance. Chem Eng J 308:27–39

10. Baig MI, Ingole PG, Choi WK, Park SR, Kang EC, Lee HK (2016) Development of carboxylated TiO_2 incorporated thin film nanocomposite hollow fiber membranes for flue gas dehydration. J Membr Sci 514:622–635
11. Baig MI, Ingole PG, Jeon JD, Hong SU, Choi WK, Jang B, Lee HK (2019) Water vapour selective thin film nanocomposite membranes prepared by functionalized Silicon nanoparticles. Desalination 451:59–71
12. Bakangura E, Wu L, Ge L, Yang Z, Xu T (2016) Mixed matrix proton exchange membranes for fuel cells: state of the art and perspectives. Prog Polym Sci 57:103–152
13. Bhattacharya M, Mandal MK (2018) Synthesis of rice straw extracted nano-silica-composite membrane for CO2 separation. J Clean Prod 186:241–252
14. Bui NN, Lind ML, Hoek EMV, McCutcheon JR (2011) Electrospun nanofiber supported thin film composite membranes for engineered osmosis. J Membr Sci 385–386:10–19
15. Buonomenna MG, Yave W, Golemme G (2012) Some approaches for high performance polymer based membranes for gas separation: block copolymers, carbon molecular sieves and mixed matrix membranes. RSC Adv 2:10745–10773
16. Choi W, Ingole PG, Li H, Kim JH, Lee HK, Baek IH (2016) Preparation of facilitated transport hollow fiber membrane for gas separation using cobalt tetraphenylporphyrin complex as a coating material. J Cleaner Product 133:1008–1016
17. Choi W, Ingole PG, Li H, Park SY, Kim JH, Lee HK, Baek IH (2017) Facilitated transport hollow fiber membrane prepared by t-Bu CoSalen for O_2/N_2 separation. Microchemical J 132:36–42
18. Choi W, Ingole PG, Park JS, Lee DW, Kim JH, Lee HK (2015) H_2/CO mixture gas separation using composite hollow fiber membranes prepared by interfacial polymerization method. Chem Eng Res Des 102:297–306
19. Chung TS, Jiang LY, Li Y, Kulprathipanja S (2007) Mixed matrix membranes (MMMs) comprising organic polymers with dispersed inorganic fillers for gas separation. Prog Polym Sci 32:483–507
20. Chung TS, Zhang S, Wang KY, Su J, Ling MM (2012) Forward osmosis processes: yesterday, today and tomorrow. Desalination 287:78–81
21. Dalvi V, Tang YP, Staudt C, Chung TS (2017) Influential effects of nanoparticles, solvent and surfactant treatments on thin film nanocomposite (TFN) membranes for seawater desalination. Desalination 420:216–225
22. D'Epifanio A, Navarra MA, Weise FC, Mecheri B, Farrington J, Licoccia S, Greenbaum S (2010) Composite Nafion/sulfated zirconia membranes: effect of the filler surface properties on proton transport characteristics. Chem Mater 22:813–821
23. Du H, Li J, Zhang J, Su G, Li X, Zhao Y (2014) Separation of hydrogen and nitrogen gases with porous graphene membrane. J Phy Chem C 115:23261–23266
24. Duan J, Pan Y, Pacheco F, Litwiller E, Lai Z, Pinnau I (2015) High-performance polyamide thin-film-nanocomposite reverse osmosis membranes containing hydrophobic zeolitic imidazolate framework-8. J Memb Sci 476:303–310
25. Fan Y, Chen S, Zhao H, Liu Y (2017) Distillation membrane constructed by TiO_2 nanofiber followed by fluorination for excellent water desalination performance. Desalination 405: 51–58
26. Fernandez-Gonzalez C, Dominguez-Ramos A, Ibañez R, Chen Y, Irabien A (2017) Valorization of desalination brines by electrodialysis with bipolar membranes using nanocomposite anion exchange membranes. Desalination 406:16–24
27. Han G, Zhang S, Li X, Widjojo N, Chung TS (2012) Thin film composite forward osmosis membranes based on polydopamine modified polysulfone substrates with enhancements in both water flux and salt rejection. Chem Eng Sci 80:219–231
28. Han W, Kwan SM, Yeung KL (2012) Zeolite applications in fuel cells: water management and proton conductivity. Chem Eng J 187:367–371
29. Ingole PG, Baig MI, Choi W, An X, Choi WK, Lee HK (2017) Role of functional nanoparticles to enhance the polymeric membrane performance for mixture gas separation. J Ind Eng Chem 48:5–15

30. Ingole PG, Baig MI, Choi W, An X, Choi WK, Jeon JD, Lee HK (2017) Synthesis of superhydrophilic Nafion based nanocomposite hollow fiber membranes for water vapour separation. Chem Eng Res Des 127:45–51
31. Ingole PG, Baig MI, Choi WK, Lee HK (2016) Synthesis and characterization of polyamide/polyester thin-film nanocomposite membranes achieved by functionalized TiO_2 nanoparticles for water vapour separation. J Mat Chem A 4:5592–5604
32. Ingole PG, Bajaj HC, Singh K (2014) Membrane separation processes: optical resolution of lysine and asparagine amino acids. Desalination 343:75–81
33. Ingole PG, Choi WK, Lee GB, Lee HK (2017a) Thin-film-composite hollow-fiber membranes for water vapour separation. Desalination 403:12–23
34. Ingole PG, Pawar RR, Baig MI, Jeon JD, Lee HK (2017b) Thin film nanocomposite (TFN) hollow fiber membranes incorporated with functionalized acid-activated bentonite (ABn-NH) clay: towards enhancement of water vapour permeance and selectivity. J Mat Chem A 5:20947–20958
35. Ingole PG, Sohail M, Abou-Elanwar AM, Baig MI, Jeon JD, Choi WK, Kim H, Lee HK (2018) Water vapour separation from flue gas using MOF incorporated thin film nanocomposite hollow fiber membranes. Chem Eng J 334:2450–2458
36. Jahan Z, Niazi MBK, Gregersen ØW (2018) Mechanical, thermal and swelling properties of cellulose nanocrystals/PVA nanocomposites membranes. J Ind Eng Chem 57:113–124
37. Jeon SY, Yun JM, Lee YS, Kim HI (2012) Preparation of poly (vinyl alcohol)/poly (acrylic acid)/TiO_2/carbon nanotube composite nanofibers and their photobleaching properties. J Ind Eng Chem 18:487–491
38. Jeong BH, Hoek EMV, Yan Y, Subramani A, Huang X, Hurwitz G, Ghosh AK, Jawor A (2007) Interfacial polymerization of thin film nanocomposites: a new concept for reverse osmosis membranes. J Membr Sci 294:1–7
39. Ji Y, Ke J, Duan F, Chen J (2017) Preparation and application of novel multi-walled carbon nanotubes/polysulfone nanocomposite membrane for chiral separation. Des Wat Treat 87:179–187
40. Jiang D, Cooper VR, Dai S (2009) Porous graphene as the ultimate membrane for gas separation. Nano Let 9:4019–4024
41. Jo ES, An X, Ingole PG, Choi WK, Park YS, Lee HK (2017) CO_2/CH_4 separation using inside coated thin film composite hollow fiber membranes prepared by interfacial polymerization. Chinese J Chem Eng 25:278–287
42. Ke Q, Wang J (2016) Graphene-based materials for supercapacitor electrodes–A review. J Materiomics 2:37–54
43. Khalid A, Abdel-Karim A, Atieh MA, Javed S, McKay G (2018) PEG-CNTs nanocomposite PSU membranes for wastewater treatment by membrane bioreactor. Sep Purif Technol 190:165–176
44. Khorshidi B, Biswas I, Ghosh T, Thundat T, Sadrzadeh M (2018) Robust fabrication of thin film polyamide-TiO_2 nanocomposite membranes with enhanced thermal stability and anti-biofouling propensity. Sci Rep 8:784
45. Kim DJ, Jo MJ, Nam SY (2015) A review of polymer–nanocomposite electrolyte membranes for fuel cell application. J Ind Eng Chem 21:36–52
46. Kim KH, Ingole PG, Kim JH, Lee HK (2013) Separation performance of PEBAX/PEI hollow fiber composite membrane for SO_2/CO_2/N_2 mixed gas. Chem Eng J 233:242–250
47. Kochkodan VM, Rolya EA, Goncharuk VV (2009) Photocatalytic membrane reactors for water treatment from organic pollutants. J Wat Chem Technol 31:227–237
48. Kreuer K (2001) On the development of proton conducting polymer membranes for hydrogen and methanol fuel cells. J Membr Sci 185:29–39
49. Kumakiri I, Diplas S, Simon C, Nowak P (2011) Photocatalytic membrane contactors for water treatment. Ind Eng Chem Res 50:6000–6008
50. Lai CY, Groth A, Gray S, Duke M (2014) Nanocomposites for improved physical durability of porous PVDF membranes. Membranes (Basel) 4:55–78

51. Lee C, Wei X, Kysar JW, Hone J (2008) Measurement of the elastic properties and intrinsic strength of monolayer grapheme. Science 321:385–388
52. Li H, Choi W, Ingole PG, Lee HK, Baek IH (2016) Oxygen separation membrane based on facilitated transport using cobalt tetraphenylporphyrin-coated hollow fiber composites. Fuel 185:133–141
53. Li H, Ding X, Zhang Y, Liu J (2017) Porous graphene nanosheets functionalized thin film nanocomposite membrane prepared by interfacial polymerization for CO_2/N_2 separation. J Membr Sci 543:58–68
54. Li Q, He R, Jensen JO, Bjerrum NJ (2003) Approaches and recent development of polymer electrolyte membranes for fuel cells operating above 100 C. Chem Mater 15:4896–4915
55. Lv Y, Du Y, Chen ZX, Qiu WZ, Xu ZK (2018) Nanocomposite membranes of polydopamine/electropositive nanoparticles/polyethyleneimine for nanofiltration. J Membr Sci 545:99–106
56. Ma L, Dong X, Chen M, Zhu L, Wang C, Yang F, Dong Y (2017) Fabrication and water treatment application of carbon nanotubes (CNTs)-based composite membranes: a Review. Membranes (Basel) 7:16
57. Ma N, Wei J, Liao R, Tang CY (2012) Zeolite-polyamide thin film nanocomposite membranes: towards enhanced performance for forward osmosis. J Membr Sci 405–406:149–157
58. Metz SJ, van de Ven WJC, Potreck J, Mulder MHV, Wessling M (2005) Transport of water vapour and inert gas mixtures through highly selective and highly permeable polymer membranes. J Membr Sci 251:29–41
59. Meyer JC, Geim AK, Katsnelson MI, Novoselov KS, Booth TJ, Roth S (2007) The structure of suspended graphene sheets. Nature 446:60–63
60. Moore TT, Mahajan R, Vu DQ, Koros WJ (2004) Hybrid membrane materials comprising organic polymers with rigid dispersed phases. AIChE J 50:311–321
61. Morsi RE, Alsabagh AM, Nasr SA, Zaki MM (2017) Multifunctional nanocomposites of chitosan, silver nanoparticles, copper nanoparticles and carbon nanotubes for water treatment: antimicrobial characteristics. Int J Biol Macromol 97:264–269
62. Nagarale RK, Shin W, Singh PK (2010) Progress in ionic organic-inorganic composite membranes for fuel cell applications. Polym Chem 1:388–408
63. Navarra MA, Abbati C, Scrosati B (2008) Properties and fuel cell performance of a Nafion-based, sulfated zirconia-added, composite membrane. J Power Sources 183:109–113
64. Navarra MA, Croce F, Scrosati B (2007) New, high temperature superacid zirconia-doped Nafion composite membranes. J Mat Chem 17:3210–3215
65. Njuguna J, Pielichowski K (2003) Polymer Nanocomposites for Aerospace Applications: Properties. Adv Eng Mat 5:769–778
66. Pandey N, Shukla SK, Singh NB (2017) Water purification by polymer nanocomposites: an overview. Nanocomposites 3:47–66
67. Park JT, Roh DK, Chi WS, Patel R, Kim JH (2012) Fabrication of double layer photoelectrodes using hierarchical TiO_2 nanospheres for dye-sensitized solar cells. J Ind Eng Chem 18:449–455
68. Pekakis PA, Xekoukoulotakis NP, Mantzavinos D (2006) Treatment of textile dyehouse wastewater by TiO_2 photocatalysis. Water Res 40:1276–1286
69. Pendergast MM, Hoek EMV (2011) A review of water treatment membrane nanotechnologies. Energy Environ Sci 4:1946–1971
70. Qu X, Alvarez PJ, Li Q (2013) Applications of nanotechnology in water and wastewater treatment. Water Res 47:3931–3946
71. Reinholdt MX, Kaliaguine S (2010) Proton exchange membranes for application in fuel cells: grafted silica/SPEEK nanocomposite elaboration and characterization. Langmuir 26:11184–11195
72. Ren LF, Xia F, Chen V, Shao J, Chen R, He Y (2017) TiO_2-FTCS modified superhydrophobic PVDF electrospun nanofibrous membrane for desalination by direct contact membrane distillation. Desalination 423:1–11

73. Safarpour M, Khataee A, Vatanpour V (2015) Thin film nanocomposite reverse osmosis membrane modified by reduced graphene oxide/TiO$_2$ with improved desalination performance. J Membr Sci 489:43–54
74. Saliby IE, Okour Y, Shon HK, Kandasamy J, Lee WE, Kim JH (2012) TiO$_2$ nanoparticles and nanofibres from TiCl$_4$ flocculated sludge: characterisation and photocatalytic activity. J Ind Eng Chem 18:1033–1038
75. Schrier J, Mcclain J (2012) Thermally-driven isotope separation across nanoporous graphene. Chem Phy Let 521:118–124
76. Seyedpour SF, Rahimpour A, Mohsenian H, Taherzadeh MJ (2018) Low fouling ultrathin nanocomposite membranes for efficient removal of manganese. J Membr Sci 549:205–216
77. Shan M, Xue Q, Jing N, Ling C, Zhang T, Yan Z, Zheng J (2012) Influence of chemical functionalization on the CO2/N2 separation performance of porous graphene membranes. Nanoscale 4:5477–5482
78. Shannon MA, Bohn PW, Elimelech M, Georgiadis JG, Mariñas BJ, Mayes AM (2008) Science and technology for water purification in the coming decades. Nature 452:301–310
79. Sijbesma H, Nymeijer K, Marwijk RV, Heijboer R, Potreck J, Wessling M (2008) Flue gas dehydration using polymer membranes. J Membr Sci 313:263–276
80. Son M, Park H, Liu L, Choi H, Kim JH, Choi H (2016) Thin-film nanocomposite membrane with CNT positioning in support layer for energy harvesting from saline water. Chem Eng J 284:68–77
81. Su VMT, Clyne TW (2016) Hybrid filtration membranes incorporating Nanoporous silica within a nanoscale alumina fibre scaffold. Adv Eng Mat 18:96–104
82. Sun D, Meng M, Qiao Y, Zhao Y, Yan Y, Li C (2018) Synthesis of ion imprinted nanocomposite membranes for selective adsorption of lithium. Sep Purif Technol 194:64–72
83. Tripathi BP, Shahi VK (2011) Organic–inorganic nanocomposite polymer electrolyte membranes for fuel cell applications. Prog Poly Sci 36:945–979
84. Wang F, Wu Y, Huang Y, Liu L (2018a) Strong, transparent and flexible aramid nanofiber/POSS hybrid organic/inorganic nanocomposite membranes. Compos Sci Technol 156:269–275
85. Wang KY, Chung TS, Amy G (2012) Developing thin-film composite forward osmosis membranes on the PES/SPSf substrate through interfacial polymerization. AIChE J 58:770–781
86. Wang M, Liu G, Cui X, Feng Y, Zhang H, Wang G, Zhong S, Luo Y (2018b) Self-crosslinked organic-inorganic nanocomposite membranes with good methanol barrier for direct methanol fuel cell applications. Solid State Ionics 315:71–76
87. Wang R, Shi L, Tang CY, Chou S, Qiu C, Fane AG (2010) Characterization of novel forward osmosis hollow fiber membranes. J Membr Sci 355:158–167
88. Wang Y, Zhu J, Dong G, Zhang Y, Guo N, Liu J (2015) Sulfonated halloysite nanotubes/polyethersulfone nanocomposite membrane for efficient dye purification. Sep Purif Technol 150:243–251
89. Wong K, Goh P, Ismail A (2015) Gas separation performance of thin film nanocomposite membranes incorporated with polymethyl methacrylate grafted multi-walled carbon nanotubes. Int Biodeter Biodegr 102:339–345
90. Wong K, Goh P, Ng B, Ismail A (2015) Thin film nanocomposite embedded with polymethyl methacrylate modified multi-walled carbon nanotubes for CO$_2$ removal. RSC Advances 5:31683–31690
91. Wu X, Wu Y, Dong H, Zhao J, Wang C, Zhou S, Luc J, Yan Y, Li H (2018) Accelerating the design of molecularly imprinted nanocomposite membranes modified by Au@polyaniline for selective enrichment and separation of ibuprofen. App Sur Sci 428:555–565
92. Xingwei Y, Zhi W, Juan Z, Fang Y, Shichun L, Jixiao W, Shichang W (2011) An effective method to improve the performance of fixed carrier membrane via incorporation of CO$_2$-selective adsorptive silica nanoparticles. Chin J Chem Eng 19:821–832

93. Xiong S, Zuo J, Ma YG, Liu L, Wu H, Wang Y (2016) Novel thin film composite forward osmosis membrane of enhanced water flux and anti-fouling property with N-[3-(trimethoxysilyl) propyl] ethylenediamine incorporated. J Membr Sci 520:400–414
94. Xu GR, Xu JM, Feng HJ, Zhao HL, Wu SB (2017) Tailoring structures and performance of polyamide thin film composite (PA-TFC) desalination membranes via sublayers adjustment-a review. Desalination 417:19–35
95. Yang X, Fraser T, Myat D, Smart S, Zhang J, da Costa JCD, Liubinas A, Duke M (2014) A pervapouration study of ammonia solutions using molecular sieve silica membranes. Membranes (Basel) 4:40–54
96. Yang Z, Wu Y, Wang J, Cao B, Tang CY (2016) In Situ Reduction of Silver by Polydopamine: A novel antimicrobial modification of a thin-film composite polyamide membrane. Environ Sci Technol 6:9543–9550
97. Yin J, Deng B (2015) Polymer-matrix nanocomposite membranes for water treatment. J Membr Sci 479:256–275
98. Yin J, Zhu G, Deng B (2016) Graphene oxide (GO) enhanced polyamide (PA) thin-film nanocomposite (TFN) membrane for water purification. Desalination 379:93–101
99. Yip NY, Tiraferri A, Phillip WA, Schiffman JD, Elimelech M (2010) High performance thin-film composite forward osmosis membrane. Environ Sci Technol 44:3812–3818
100. Yun SH, Ingole PG, Kim KH, Choi WK, Kim JH, Lee HK (2014) Properties and performances of polymer composite membranes correlated with monomer and polydopamine for flue gas dehydration by water vapour permeation. Chem Eng J 258:348–356
101. Yun SH, Ingole PG, Kim KH, Choi WK, Lee HK (2015) Synthesis of cross-linked amides and esters as thin film composite membrane materials yields permeable and selective for water vapour/gas separation. J Mater Chem A 3:7888–7899
102. Zhao H, Qiu S, Wu L, Zhang L, Chen H, Gao C (2014) Improving the performance of polyamide reverse osmosis membrane by incorporation of modified multi-walled carbon nanotubes. J Membr Sci 450:249–256
103. Zhao S, Zou L, Tang CY, Mulcahy D (2012) Recent developments in forward osmosis: opportunities and challenges. J Membr Sci 396:1–21
104. Zhu J, Qin L, Uliana A, Hou J, Wang J, Zhang Y, Li X, Yuan S, Li J, Tian M, Lin J, Van der Bruggen B (2017) Elevated performance of thin film nanocomposite membranes enabled by modified hydrophilic MOFs for nanofiltration. ACS Appl Mater Interfaces 9:1975–1986

Reliable Natural-Fibre Augmented Biodegraded Polymer Composites

Ritu Payal

1 Introduction

Since the early 19th century investigations on natural fibre composites have been going on but get an acknowledgement in 1980s. Fibres are hair alike materials which are similar to thread pieces, continuous filaments or can exist as distinct extend pieces. They may be twisted into filaments, yarn or cord, as a component of composite materials, can also be matted into sheets to yield a wide variety of products. Biodegradable polymer composites obtained by incorporating natural resources have witnessed a tremendous research interest nowadays. The biocomposite polymers are widely emergent areas in polymer science which perceived huge attention for their use in various applications such as automobiles, fossil plastic materials, building industries, railway coach interiors, packaging, storage devices and aerospace [1, 2].

Compared with traditional fibres (glass, aramid and carbon fibres), natural fibres (e.g. banana, coir, flax, hemp, henequen, kenaf, jute, sisal, kapok, and many more—Fig. 1) offers many advantages, for instance, abundance and lower cost of materials, decrease in density, biodegradability, nominal health hazards, sustainability, flexibility and less machine wear during processing, comparatively high tensile and stretch modulus, extraordinary stiffness and strength [1]. Additionally, biodegradable, eco-friendly and renewable properties of natural fibres assist their disposal by incineration or composting and alleviate the use of non-biodegradable polymers. In addition, these fibres are environmentally benign as their dimensional structure encompass confiscated atmospheric carbon dioxide and emit lower energy comparative to industrially manufactured synthetic fibres [3].

R. Payal (✉)
Department of Chemistry, Rajdhani College,
University of Delhi, Delhi 110015, India
e-mail: ritupayal.10@gmail.com

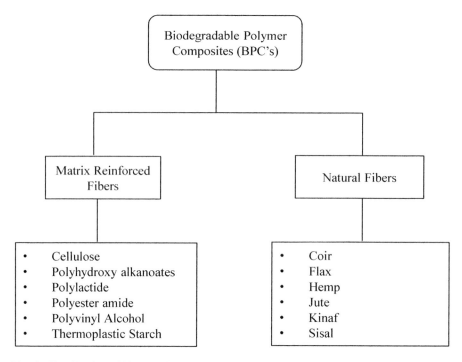

Fig. 1 Classification of biodegradable polymer composites

BPC's also showed fairly good mechanical properties, has high specific strength, and a good amount of tensile strength as an outcome of interfacial adhesion amid the matrix and fibres. Generally, the tensile strengths of BPC's increase on increasing the fibre content reached an optimum value and later a drop in value is witnessed. Also, the tensile properties of composites are prominently boosted by adding fibres to a polymer matrix as fibres have the much greater strength and stiffness relative to the matrices. Furthermore, it was established that the composites having fibres in the perpendicular direction deliver inferior tensile strength compared to the composites having fibres in the parallel direction [3–5]. Hence, it is required to alter the surface of fibre by means of chemical modifications to enhance the adhesion amid fibre and matrix by employing suitable processing methods and parameters to produce optimum composite products.

This chapter briefly deals with the reported works on the characterization of natural fibres, along with the comparative properties of natural and synthetic fibres (Table 1), advantages and disadvantages of BPC's in addition to manufacturing techniques, chemical and physical treatments and their applications in several areas.

Table 1 Properties of various natural-and synthetic fibres [1, 7–9]

Fibre	Density (g/cm^3)	Elongation	Elastic modulus (GPa)	Specific elastic modulus (GPa)	Tensile strength (MPa)
Aramid	1.4	3.3–3.7	63–67	33–36	3000–3150
Carbon	1.4	1.4–1.8	230–240	164–171	4000
Cotton	1.5–1.6	7.0–8.0	5.5–12.6	3.1–5.8	400
Coir	1.2	30	40	20.4	593
E-glass	2.5	0.5	70	28	2000–3500
S-glass	2.5	2.8	86	41.2	4570
Flax	1.5	2.7–3.2	27.6	26–46	500–1500
Hemp	1.47	2–40	70	47	690
Jute	1.3	1.5–1.8	26.5	7–21	393–773
Kenaf	1.45	1.6	53	36.5	930
Polyester	1.2–1.5	2.0–4.5	2	–	40–90
Polyhydroxy alkonates	1.1–1.4	1–6	3–6	–	35–100
Sisal	1.5	2.0–2.5	9.4–22	6.3–14.7	511–635

2 Classification of Fibres

Mostly, polymers may be categorized into two diverse classes: thermoplastics and thermosetting. Both types of materials were widely used as matrices for fabrication of bio-fibres. Polyethylene (PE), polyvinyl chloride (PVC), and polypropylene (PP) are frequently used thermoplastics; whereas epoxy, phenolic, and polyester resins are generally used thermosetting polymers for the manufacture of composites. Now a day, fibres obtained from natural sources like jute, flax, coir, hemp, kenaf, sisal etc. are extensively used in the fabrication of composites.

2.1 Drawbacks of Natural Fibres

The natural fibres suffer limitations, namely: (1) strength, (2) water absorption, (3) thermal stability

1. **Strength**: The tensile strength of BPC's formed using natural fibres is very low comparative to glass, aramid and carbon fibres. This is due to the incompatibility amid the fibre-resin matrix, which in turn reduces the wettability of the fibres and creates a challenge in their productions. Although, by comparing the specific strength of both natural and synthetic fibres, it is observed that there is barely any difference.

2. **Water absorption**: Another limiting factor for the assembly of BPC's is the absorption of water from the atmosphere or direct contact with the surroundings. The difference in the polarity of hydrophilic natural fibres and hydrophobic polymer matrix provides minimal interactions and results in the formation of aggregates. The absorption allows the distortion of surfaces of resulting composites via swelling and forming voids, thereby alleviating the interfacial adhesion and weakens the strength in addition to proliferation in the mass of composites. The wettability also permits the growth of fungi on/in the composites resulting in decay of their structure [6]. The high water absorption indicates inadequate resistant to moisture, which provides reduced tensile properties to reinforced natural fibre composites (Table 1).
3. **Thermal stability**: Natural fibres are of restricted thermal stability and lead to microcracking and, consequently, thermal degradation may possibly occur in the course of BPC's processing at an elevated temperature, specifically in the cases of hot compression and thermal extrusion processes.

2.2 Advantages of Natural Fibres

The biggest advantage of using natural fibres for the fabrication of BPC's is the low cost of materials, reduced density, and sustainability. Natural fibres are easy to cultivate and can be grown within few months and are cost effective. They also have the perspective as a cash crop for the agriculturalists. Natural fibres are eco-friendly, lightweight, consumes lesser energy, non-abrasive, non-carcinogens, strong, renewable, less health risk, non-irritation to the skin, recyclable and biodegradable, and have small processing time [2].

The fabrication of biocomposites uses various techniques which include: compression moulding, extrusion (extensively used for green biocomposite), filament winding, injection moulding, machine press, pultrusion, resin transfer moulding, sheet moulding compound. Thermoplastic bio-composites can be mainly processed through compounding, compression moulding, extrusion, injection, and vacuum consolidation. Alternatively, thermosetting biocomposites are manufactured by compression moulding, vacuum assisted resin transfer moulding, hand lay-up, pultrusion, resin transfer moulding, and vacuum bagging [10–12].

2.3 Strategies for Surface Modification in Natural Fibres

Surface modification is one of the crucial processes in the fabrication of biocomposites since natural fibres possess hydrophilic characteristic and with the aim of improving the compatibility of a hydrophobic polymer matrix with the natural fibres, this surface modification is needed. Surface modification is categorized into types, namely chemical and physical methods. The surface modification eases fibre dispersion within polymer matrix along with enhancement in their interaction. Various techniques described in the literature surveys for altering fibre-matrix adhesion consist of acetylation [13], acrylic acid treatment [14], bleaching [15], esterification [16], grafting of monomers using maleic anhydride [17], and using bi-functional molecules [18]. The usage of coupling agents such as isocyanates [19], chitosan [20], maleated polypropylene [21], silanes [22], titanates [23], zirconates [24], etc. have been testified in previous studies for the improvement of conventional polymer composites.

2.3.1 Chemical Techniques

A wide range of chemical techniques is reported in the literature but alkaline-, silane-, esterification-, and isocyanate treatments are some of the frequently used chemical techniques, which are described as follows:

1. **Alkaline treatment**: Alkaline treatment is a unique approach to be used as a chemical treatment of BPC's. This method is also known as mercerization and is usually used to reinforce thermoplastics and thermosets with natural fibers by removing lignin, wax and oil from the external surface of the fibers. This methodology disrupts the hydrogen bonding in the structural framework of BPC's, providing surplus positions for mechanical interlocking, henceforth support surface irregularities and boost fibre and matrix diffusion at the boundaries. In alkaline treatment, the fibres are dipped in sodium hydroxide (NaOH) solution for a fixed time period depending upon the interaction between the two. The resulting interaction facilitates the ionization of the hydroxyl group to the alkoxide group (Fig. 2), which is then stirred continuously in a binary solvent such as water-ethanol solution (80:20) for 1 h and the solution is kept undisturbed for $\sim$3 h. Consequently, the fibers are washed repeatedly with distilled water followed by drying in a hot air oven at 80 °C for $\sim$5 h [25, 26].

The literature surveys reported that the natural fibers such as coir, ramie, sisal, jute were utilized to fabricate BPC's using alkaline treatment and it was established that these composites possess high storage as well as significant flexural modulus [27].

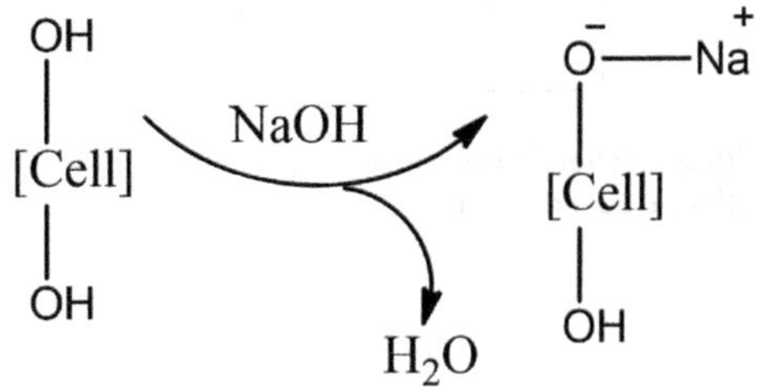

Fig. 2 Ionization of the hydroxide group used in alkali treatment of the fibers

2. **Silane treatment**: Earlier silanes (with chemical formula SiH_4) were reinforced with glass fibres for the augmentation of polymer composites. Silanes reduce the prevalence of hydrogen bonding in the matrix-fiber structural complex. Presence of moisture allows hydrophilic alkoxy groups to form silanols, which in turn responds to the hydroxyl groups of the fibre and form a stable, cross-linked network due to the covalently bonded structure with the cell wall (Fig. 3a, b).

As a consequence, hydrophobic fibre surface is generated; consecutively intensify the compatibility with the polymer matrix (Fig. 3c). A significant enhancement in tensile strength was discovered as a consequence of strong interfacial bond produced by the acid and water conditions during the course of fibre pretreatment [22]. The modification of kenaf and polyester fibre composites finds mention in the literature using silane treatment method [28, 29]. The fiber modification offers higher storage modulus and lowers tan δ values than those with untreated fiber indicating a greater interfacial interaction between the matrix resin and the fiber.

3. **Esterification**: Esterification involves the modification of the surface of fibres especially wood composites with organic acid anhydrides (Fig. 4). Acetylation

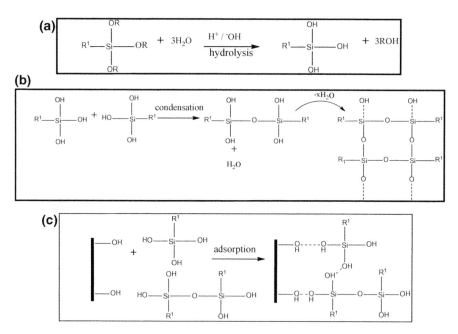

Fig. 3 Interaction of silanes with natural fibers via different processes: **a** hydrolysis, **b** condensation, **c** adsorption [30]

with acetic anhydride (non-cyclic anhydrides) and maleic anhydride (cyclic anhydrides) is broadly defined in the literature [17]. In Acetylation process, a hydroxyl group is converted into an ester group via association of the free hydroxyl groups present in wood composites with the carboxylic group of the anhydrides.

4. **Isocyanate treatment**: Isocyanates are organic compounds having isocyanate group–N=C=O in their structure. These are highly reactive with polar groups such as hydroxyl (–OH), amino (–NH_2) and others (Fig. 5). Isocyanates are not considered feasible treatment method for natural fibres but testified a meek improvement in strength of polymer matrix relative to the unaided matrix.

2.3.2 Physical Techniques

Physical approaches [29, 31] mentioned in the literature involves the corona or plasma treatments for amending conventional polymers. In recent years, a "greener" alternative is available for the expansion of polymer composites, which comprises plasma treatment for reinforcement of natural fibres. Some of the chemical techniques mentioned previously proved to be harmful, e.g. isocyanates are hazardous, and as a result, such agents are not viable for the augmentation of BPC's. Therefore, physical methods involving plasma treatment provides a better option for treating natural fibres. Plasma has the tendency to modify the properties of surface/interface of natural fibres via formation of free radicals (*for instance* electrons, ions, etc.) on their surfaces by the bombardment with high energy particles inside the stream of plasma (Fig. 6) [32].

Using physical techniques various surface properties, for example, chemistry, wettability, and surface irregularities of composites can be improved without using

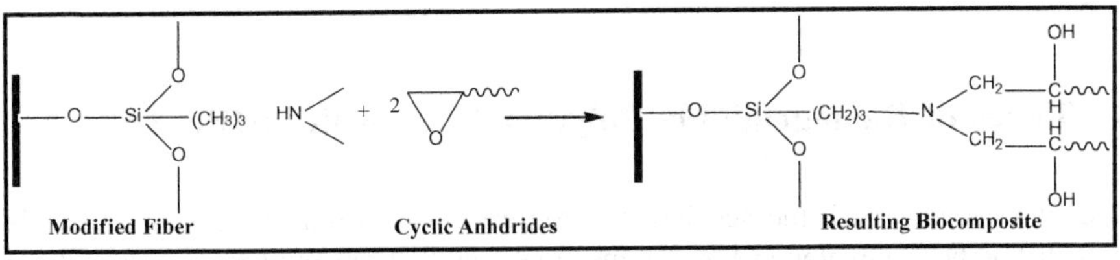

Fig. 4 Coupling reaction of modified natural fibers with anhydrides

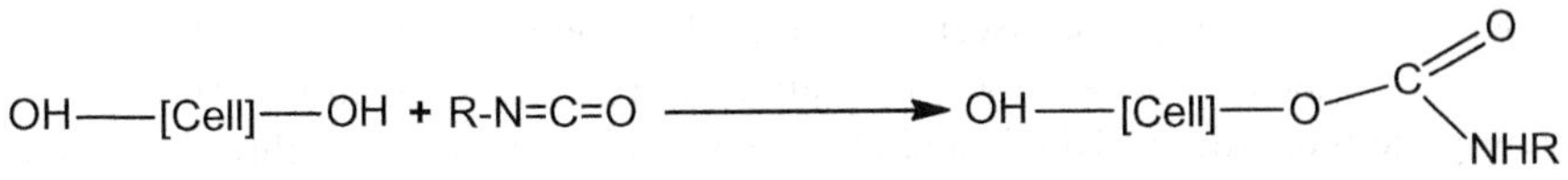

Fig. 5 Isocyanation of fibers

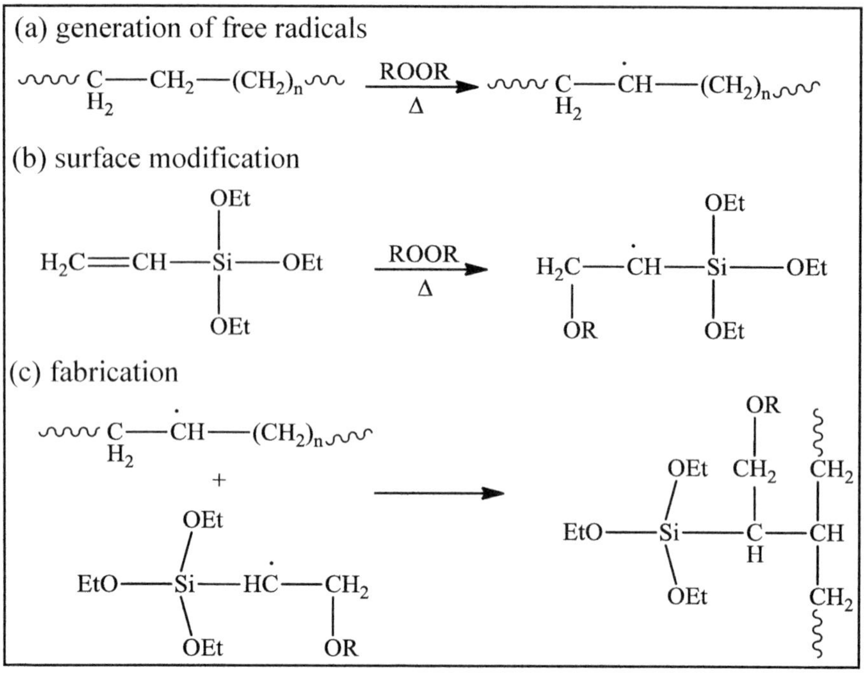

Fig. 6 Fabrication of silanes onto polymer surface by free radicals (where Et = ethyl group)

solvents or employing any other harmful materials. An alternate approach for surface modification by plasmas is to alter the carrier gas or by depositing free radicals or other reactive species on the shells of natural fibres [33]. Furthermore, this can be stimulated by embedding monomers or polymers on the surface of the reactive natural fibre, which accelerates its aptness for the polymer matrix.

3 Types of Biodegradable Polymer Composites (BPC'S)

Over the decades with the accelerated progress of biocomposites, a considerable interest has been developed for polymer matrices that are reinforced with natural fibres. Owing to difficulties in discarding the non-biodegradable polymers, researches are continuously carried out to fortified new biodegradable polymer composite materials from natural resources.

Conversely, natural fibre reinforced composites (BPC's) are manufactured from biodegradable polymers to overcome the shortcomings of non-biodegradable polymers composites. Biodegradable matrices were enclosed with natural fibres to amplify the properties of BPC's and they are scientifically sound, lightweight, high mechanical properties and cost-effective. Some of the BPC's along with their synthetic method and properties are explained as follows.

3.1 Coir Fibre Reinforced Composite

Coir is generated by the husk of coconut fruit fibre. The life expectancy of coir is more comparative to other natural fibres as a result of high lignin content. Coir fibre reinforced polyester matrix was tested for their interfacial adhesion characteristics against different ageing solutions and was found to display excellent interfacial adhesion under arid conditions [34]. Coir fibre reinforced polymer composites are developed for industrial and various household applications such as automotive interior, helmets and post boxes, packing material, paneling and roofing as building materials, mirror casing, projector cover, storage tank, paperweights, voltage stabilizer covers.

The efficiency of coir fibre fabricated epoxy composites is dependent on alkali treatment in addition length of the fibre. Coir fibres having lengths 10, 20 and 30 mm were cured with NaOH for 10 days. Fibre length was Alkali treated composite along with increased fibre had Coir fibre having length of 30 mm and 8% alkali concentrations showed better impact strength (27 kJ/m^2) [35, 36]. Pretreated coir based composite have far better mechanical and flexural properties than the untreated one. On increasing fibre content the flexural strength of composite decreases as the matrix is inadequate to shield the complete surface of the coir fibre (Fig. 7).

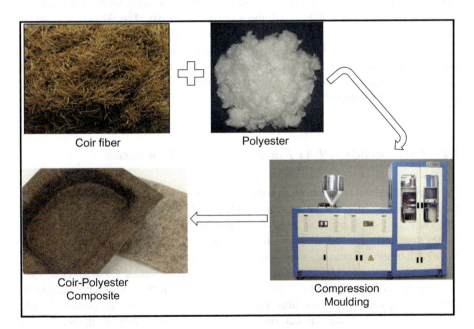

Fig. 7 Coir reinforced polyester composite formation via compression moulding

3.2 Cellulose Fibre Reinforced Composite

Cellulose and its derivatives (lignin, hemicellulose, pectin etc.) are semi-crystalline polysaccharides that impart hydrophilic nature as well as tendency to hold the fibre. These materials have found wide use in the potential matrices-composites as strong molecular interactions exist at the interface of cellulose fibres-polymer matrix composites, which in turn results in strong interfacial adhesion. The composites may be treated via blow and rotation moulding, extrusion, injection moulding, etc. to form fundamental components [18, 33].

3.3 Jute Fibre Reinforced Composite

Jute has wood like characteristics, has a high aspect ratio, good insulation properties, strength to weight ratio. In view of above-mentioned properties, jute fibre reinforced polymer composite has tested for grooved sheet, door, furniture, roofing, floor tiles, I-shaped beam, recovery of underground drain pipes, window, and water pipes [37].

The jute fibre reinforced PP composites were analyzed for their mechanical properties. The washing of fibres preceded by alkaline treatment and bleaching revealed intensification in tensile strength and tensile modulus with an increase in % weight fraction and NaOH % of fibres in the PP matrix (Fig. 8a).

Jute fibres stiffen epoxy composites were examined and results were also compared with bamboo fibre supported epoxy composites and formers were found to have a higher strength. Additionally, upon alkaline treatment jute fibre reinforced epoxy composites encompass enhanced mechanical properties of bamboo fibre reinforced epoxy composites (Fig. 8b, c).

3.4 Poly Lactide (PLA) Fibre Reinforced Composite

The blend of natural fibre and PLA bids an excellent response to preserve the viable economic and ecological development. Ochi [38] carried out a study on PLA composite and explored that PLA fabricated unidirectional biodegradable composite materials showed tensile strengths of 223 MPa. He also evaluated the biodegradability of same via composting and the conducting tests revealed a 38% decrease in composites weight after a time period of four weeks. Oksman et al. [39] reported the fabrication of PLA-Flax composites and matched them with frequently used polypropylene (PP) flax fibre composites (PP-Flax). The comparative study marked 50% higher mechanical properties of PLA-Flax fibre composites over PP-Flax fibre composites.

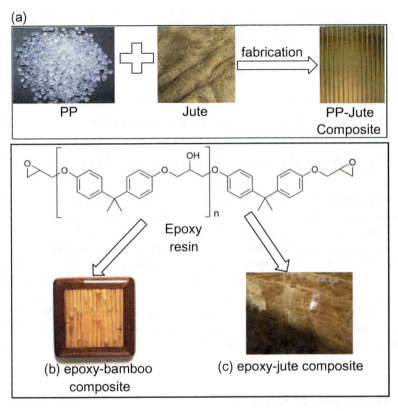

Fig. 8 **a** Jute enforced PP composites, **b** epoxy-bamboo composites, **c** epoxy-jute composite

PLA-Natural fibre composites containing >30% weight fibre showed an improved tensile modulus with lesser tensile strength on comparing with untreated PLA. This was ascribed as a result of poor interfacial interaction flanked by hydrophilic cellulose fibres and hydrophobic PLA matrix, as well as an insufficient fibre dispersion caused by the elevated amount of fibre agglomeration [40]. Hu and Lim [41] inspected that tensile properties of hemp fibre enforced PLA composite significantly improved upon alkali treatment than that of untreated composites. The composites produced using 40% treated fibre has approximately twice tensile modulus comparative to neat PLA (35 GPa). Fabrication of PLA with natural fibres is most likely done by conventional methods such as blow- and injection moulding, extrusion, as well as film-forming operations.

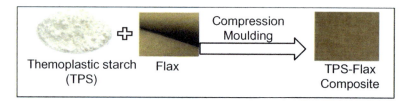

Fig. 9 Thermoplastic starch-flax composite using compression moulding

3.5 Polyhydroxyalkanoates Fibre Reinforced Composite

Polyhydroxyalkanoates (PHAs) are found to be renewable and biodegradable and represent a class of polyesters that are manufactured by bacterial action [42]. The fabrication of the green composites was carried out by Singh and Mohanty [43] by means of injection moulding succeeding the extrusion compounding of bacterial polyester (PHBV) i.e., poly(hydroxybutyrate-co-valerate) with 30–40 weight percentage of bamboo fibre. They also examined morphological, mechanical, and thermomechanical properties and corroborate that the tensile modulus and storage modulus of PHBV composites amplify progressively with increasing fibre loading. Moreover, the tensile strength of bacterial polyester was reduced by adding bamboo fibre due to insufficient interfacial interaction amid fibre and matrix.

3.6 Thermoplastic Starch (TPS)

One of the utmost popular eco-friendly biodegradable polymer-thermoplastic starch can be used as a matrix in fabricated of biocomposites [44, 45]. In the quest for improved performance of biodegradable and environmental acceptable TPS polymer, natural clays can be accumulated on to its surface to produce nanocomposites. Ecologically acceptable filler (such as clay) improve the properties of TPS so that it can be used in various applications. It has been established that the tensile strength of TPS showed an increase from 2.6 to 3.3 MPa on treatment with 5 wt% sodium montmorillonite (Fig. 9).

4 Conclusions

Natural fibre fabricated biodegradable polymer composites (BPC's) are becoming scientifically sound for a plethora of applications as they are lightweight, environmentally benign, and possess good mechanical properties. The modification of the surface of fibres is presently an area of research. The mechanical properties of biodegradable fibres such as sisal, kenaf, hemp, coir, jute, reinforced composites have been discussed. It was established from the various studies that the tensile modulus and strength raises with an increase in fibre content. It was also accepted that the specific properties of natural fibre composites were far superior relative to synthetic fibre reinforced composites (glass, carbon and aramid). This advocates that the natural fibre composites would prove to be a potential candidate to substitute glass and alike materials in various applications. However, hydrophilic characteristics of biodegradable polymers challenge them to be a good candidate for outdoor applications due to reduced adhesion between natural fibres and matrix resins. However, BPC's are in demand these days and open their possibilities as an excellent candidate in a wide range of applications, such as an automobile, constructional and household applications.

References

1. Wambua P, Ivens J, Verpoest I (2003) Natural fibres: can they replace glass in fibre reinforced plastics. Compos Sci Technol 63:1259–1264
2. Malkapuram R, Kumar V, Yuvraj SN (2008) Recent development in natural fibre reinforced polypropylene composites. J Reinf Plast Compos 28:1169–1189
3. Li X, Tabil LG, Panigrahi S, Crerar WJ (2009) The influence of fibre content on properties of injection molded flax fibre-HDPE biocomposites. Can Biosyst Eng 08–148:1–10
4. Holbery J, Houston D (2006) Natural-fibre-reinforced polymer composites in automotive applications. JOM (TMS) 58(11):80–86
5. Ahmad I, Baharum A, Abdullah I (2006) Effect of extrusion rate and fibre loading on mechanical properties of twaron fibre-thermoplastic natural rubber (TPNR) composites. J Reinf Plast Compos 25:957–965
6. Pickering K (2008) Properties and performance of natural-fibre composites, 1st ed. Woodhead Publishing
7. Hajnalka H, Racz I, Anandjiwala RD (2008) Development of HEMP fibre reinforced polypropylene composites. J Thermoplast Compos Mater 21:165–174
8. Ahmad I, Baharum A, Abdullah I (2006) Effect of extrusion rate and fibre loading on mechanical properties of twaron fibre-thermoplastic natural rubber (TPNR) composites. J Reinf Plast Compos 25:957–965
9. Nabi SD, Jog JP (1999) Natural fibre polymer composites: a review. Adv Polym Technol 18:351–363
10. Pickering KL (2008) Properties and performance of natural-fibre composites CRC Press, Florida
11. Odian G (2004) Principles of polymerization, 4th edn. Wiley, New Jersey
12. Cardon LK, Ragaert KJ, Koster RP (2010) Design and fabrication of biocomposites. Woodhead Publishing, Biomedical Composites, pp 25–43
13. Hill CAS (2006) Chemical modification of wood (I): acetic anhydride modification 3.1. In: Cas H (ed) Wood modification: chemical, thermal and other processes. Wiley, New Jersey, pp 45–76
14. Li X, Panigrahi S, Tabil LG (2009) A study on flax fibre-reinforced polyethylene biocomposites. Appl Eng Agr 25:525–531
15. Tripathy S, Mishra S, Nayak S (1999) Novel, low-cost jute-polyester composites. Part 1: processing, mechanical properties, and SEM analysis. Polym Compos 20(1):62–71
16. Li X, Tabil LG, Panigrahi S (2007) Chemical treatments of natural fibre for use in natural fibre-reinforced composites: a review. J Polym Env 15(1):25–33

17. Panigrahy BS, Rana A, Chang P, Panigrahi S (2006) Overview of flax fibre reinforced thermoplastic composites. Can Biosyst Eng J 06–165:1–12
18. Belgacem MN, Gandini A (2004) The surface modification of cellulose fibres for use as reinforcing elements in composite materials. Compos Interfaces 12(1–2):41–75
19. Wang L, Duan Y, Zhang Y, Huang R, Dong Y, Huang C, Zhou B (2016) Surface modification of poly-(p-phenylene terephthalamide) pulp with a silane containing isocyanate group for silicone composites reinforcement 21(6):505–511
20. Shah BL, Selke SE, Walters MB, Heiden PA (2008) Effects of wood flour and chitosan on mechanical, chemical, and thermal properties of polylactide. Polym Compos 29(6):655–663
21. Sun-M, Lai F-CY, Yeh Wang, Hsun-C, Chan, Hsiao-F, Shen (2003) Comparative study of maleated polyolefins as compatibilizers for polyethylene/wood flour composites. Appl Polym Sci 87:487–496
22. Abdelmouleh M, Boufi S, Salah AB, Belgacem MN, Gandini A (2002) Interaction of silane coupling agents with cellulose. Langmuir 18:3203–3208
23. Gomez JA (1989) Oligomeric titanates as coupling agents for fibre-reinforced composites. Doctoral Dissertations, University of Connecticut. http://opencommons.uconn.edu/dissertations/AAI9023893
24. Vishnyakov LP, Moroz VP, Pisarenko VA, Samelyuk AV (2007) Composites with zirconium matrix reinforced with boron and silicon carbide fibres. Powder Metall Metal Ceram 46(1–2):38–42
25. Oh JT, Hong JH, Ahn Y, Kim H (2012) Reliability improvement of hemp based bio-composite by surface modification. Fibres Polym 13(6):735–739
26. Natrajan S, Moses JJ (2012) Surface modification of polyester fabric using polyvinyl alcohol in alkaline. Ind J Fibre Tex Res 37:287–291
27. Faruk Omar, Bledzki Andrzej K, Fink Hans-Peter, Sain Mohini (2012) Biocomposites reinforced with natural fibers: 2000–2010. Prog Polym Sci 37:1552–1596
28. Lee BH, Kim HS, Lee S, Kim HJ, Dorgan JR (2009) Bio-composites of kenaf fibers in polylactide: role of improved interfacial adhesion in the carding process. Compos Sci Tech 69:2573–2579
29. Pothan LA, Thomas S (2003) Polarity parameters and dynamic mechanical behavior of chemically modified banana fiber reinforced polyester composites. Compos Sci Tech 63:1231–1240
30. Xie Y, Hill CAS, Xiao Z, Militz H, Mai C (2010) Silane coupling agents used for natural fiber/polymer composites: a review, compos: part A 41:806–819
31. Dong S, Saphieha S, Schreiber HP (1992) Rheological properties of corona modified cellulose/ polyethylene composites. Polym Eng Sci 32(22):6
32. Lee KY, Delille A, Bismarck A (2011) Greener surface treatments of natural fibres for the production of renewable composite materials cellulose fibres: Bio- and nano-polymer composites. In: Kalia S, Kaith BS, Kaur I (eds) Springer, Berlin Heidelberg, pp 155–178
33. Nguyen MH, Kim BS, Ha JR, Song JI (2011) Effect of plasma and NaOH treatment for rice husk/PP composites. Adv Compos Mater 20(5):435–442
34. Yousif BF, Ku H (2012) Suitability of using coir fibre/polymeric composite for the design of liquid storage tanks. Mater Des 36:847–853
35. chanakan A, Charoenvaisarocha, Jongjit H, Joseph K (2009) Materials and mechanical properties of pretreated coir-based green composites. Compos B 40:633–637
36. Athijayamani A, Thiruchitrambalam M, Natarajan U, Pazhanivel B (2009) Effect of moisture absorption on the mechanical properties of randomly oriented natural fibres/polyester hybrid composite. Mat Sci Eng A 517:344–353
37. Siddiquee (2014) Investigation of an optimum method of biodegradation process for jute polymer composites. Am J Eng Res 3(1):200–206
38. Shinji O (2008) Shinji Ochi, Mechanical properties of kenaf fibres and kenaf/PLA composites. Mech Mat 40(4–5):446–452
39. Oksman K, Skrifvars M, Selin JF (2003) Natural fibres as reinforcement in polylactic acid (PLA) composites. Comp Sci Tech 63:1317–1324

40. Petinakis E, Yu L, Simon G, Dean K (2013) Natural fibre bio-composites incorporating poly (lactic acid). In: Masuelli MA (ed) Fiber reinforced polymers—the technology applied for concrete repair, Web of Science, pp 41–59
41. Hu R, Lim JK (2007) Fabrication and mechanical properties of completely biodegradable hemp fibre reinforcd polylactic acid composites. J Compos Mat 41:1655–1669
42. Singh S, Mohanty AK, Sugie T, Takai Y, Hamada H (2008) Renewable resource based biocomposites from natural fiber and polyhydroxybutyrate-co-valerate (PHBV) bioplastic. Comp Part A 39(5):875–886
43. Singh S, Mohanty AK (2007) Wood fiber reinforced bacterial bioplastic composites: fabrication and performance evaluation. Comp Sci Tech 67:1753–1763
44. Chen B, Evans JRG (2005) Thermoplastic starch–clay nanocomposites and their characteristics. Carbohydr Polym 61:455–463
45. Carrado KA, Xu L, Seifert S, Csencsits R, Bloomquist CAA (2000) Polymer–clay nanocomposites derived from polymer-silicate gels. In: Pinnavaia TJ, Beall G (eds) Polymer–clay nanocomposites. Wiley, Chichester, pp 54–55

An Overview on Plant Fiber Technology: An Interdisciplinary Approach

Alan Miguel Brum da Silva, Sandra Maria da Luz,
Irulappasamy Siva, Jebas Thangiah Winowlin Jappes
and Sandro Campos Amico

List of Abbreviations

CML	Compound middle lamella
DP	Degree of polymerization
G	Guaiacyl (G)
GAX	Glucuronoarabinoxylans
H	p-hydroxyphenyl
HG	Homogalacturonan
L	Lumen
MET	Transmission electron microscopy
MFA	Microfibrillar angle
ML	Middle lamella
OH	Hydroxyl groups
P	Primary wall
RG-I	Rhamnogalacturonan I
RG-II	Rhamnogalacturonan II
RTM	Resin transfer molding
S	Syringyl
S1	Secondary wall 1
S2	Secondary wall 2
S3	Secondary wall 3
SMC	Sheet molding compound

A. M. B. da Silva · S. C. Amico
PPG3 M, Federal University of Rio Grande do Sul, Porto Alegre, Brazil

S. M. da Luz (✉)
Gama Campus—University of Brasília, Brasília, Federal District, Brazil
e-mail: sandra.unb@gmail.com

A. M. B. da Silva · I. Siva · J. T. W. Jappes
Centre for Composite Materials, Kalasalingam University,
Anand Nagar, Tamil Nadu, India

© Springer Nature Switzerland AG 2019
Inamuddin et al. (eds.), *Sustainable Polymer Composites and Nanocomposites*,
https://doi.org/10.1007/978-3-030-05399-4_34

1 Introduction

Herbaceous plant fibers are known to humanity since pre-historic times. Evidence suggests they were being used for 34,000 years back, where flax fibers were used for fabrication of clothes, ropes, and baskets [34]. Trough ages, the utilization of fibers evolved, and their use expanded for different applications. Around the third millennium BC, bricks were fabricated using wheat straw, flax and papyrus as reinforcement in a mud matrix, allowing for more resistant constructions [19, 59]. During the European expansion, beginning in the fifteenth century, hemp fibers played a critical role in the manufacturing of strong and durable sails and ropes, allowing ships to cover great distances [55].

Fiber technology continues to expand in many different applications even today. From the well known textile industry to high-performance fibers used as reinforcement in polymer composites. The latter being characteristic of our current global scenario, where economic and environmental concerns related to the exhaustive use of fossil fuels are driving a change towards a sustainable and eco-friendly replacement for the currently used synthetic fibers.

With this current research trend on plant fibers for composite materials, different areas are working together, and the literature is getting richer every day. Nevertheless, this richer literature brings some recurring problems, often including a poor understanding of subjects that do not comprise the author's background. Areas presenting this issue often include plant fiber morphology in relation to structure and composition, polysaccharide chemistry and its interactions, turning the contribution of many works restricted to very particular aspects of fiber application.

Thus, the objective of this work is to contribute to the interdisciplinary research of plant fiber for potential use as composite reinforcement. The scope of this paper will be on herbaceous plant fibers (non-wood) and will provide knowledge about morphology, chemical composition, mechanical behavior, and application. The work also aims to overcome the lack of accurate information about those types of fibers, in contrast to the well-established research of wood fibers provided by the paper and forest industry.

2 Biology of Plant Fibers

A common misunderstanding in the literature is regarding the biology of what is called fibers. As a complex organism, plants present different tissues systems. Fibers are a specific type of cells belonging to sclerenchyma tissue, possessing an elongated shape with thick cell walls, often dead at maturity, functioning as a mechanical support for the plant [40].

Botanically speaking, true plant fibers are divided into two main categories, xylary and extraxylary fibers. As the names suggest, xylary fibers are found associated with the conductive tissue xylem; and the extraxylary fibers can be found associated with the phloem or in the cortex. The fibers of monocotyledons are classified as extraxylary whether or not associated with the vascular bundles (xylem and phloem) [20].

For sake of convenience, many fibers are classified by their location in the plant, listed as: bast fibers (jute, flax, hemp, ramie and kenaf), leaf fibers (abaca, sisal and pineapple), seed fibers (coir, cotton and kapok), core fibers (kenaf, hemp and jute), grass and reed fibers (wheat, corn and rice) and all other types (wood and roots) [21].

One should be careful when characterizing plant fibers to follow the proper scientific classification, since many apparent fibers are not fibers at all, like cotton, which are individual epidermal trichomes around the seeds of *Gossypium*, and raffia fibers, which are leaf segments of the *Raphia* palm. What is also common to find in the commercial form of monocotyledons fibers, is an aggregate of fibers and vascular bundles [20, 40].

A critical point in understanding the fiber structure is how to differentiate a single fiber cell, a fiber bundle, and yarns. The natural aggregate of fiber cells, with variable dimensions, constitutes a bundle. The bundle is the most common form of commercially extracted fibers. To become a yarn, a bundle has to be segregated mechanically, sometimes with the aid of chemicals, into single fiber cells, this cells will then be spun into a twisted aggregate, which will display other properties like twist/angle degree (Fig. 1) [42].

Plant fibers are the outer cell wall remaining of dead plant cells. They are arranged in a tube-like manner, where many elongated cells overlap and stack together to form a macroscopic "single fiber" or as explained before, a bundle (Table 1).

All properties exhibited by plant fibers are, therefore, an expression of the properties of their cell walls. So, a careful study of cell wall formation, composition, and interactions inside the fiber bundle are of higher importance.

Plant cell walls are divided into three main regions: middle lamella, primary wall, and secondary wall. Plant cells originate from specialized dividing cells grouped into regions called meristems. In this region, cells will divide to form a new cell, which will then expand until reaching its final size.

In the cytoplasm of the dividing cell, after nucleus division, an equatorial alignment of vesicles originated from the Golgi apparatus, containing hemicelluloses and pectins, will fuse together to form a cell plate, which will extend up to the point where it will connect to the cell wall, separating the cell content into two new cells [15]. Upon cytoplasm division, cellulose microfibrils are synthesized by a protein structures known as "rosettes", which are located in plasma membrane, the newly synthesized cellulose will be deposited in the cell plate and then interact with hemicellulose (forming the cellulose/hemicellulose network) giving origin to the primary wall [7]. The innermost region of the cell plate (free of cellulose), will give origin to the middle lamella, which is pectin-rich at the growth stage.

The fiber cells will grow in the axial direction, achieving an elongated shape. When growth is complete, fiber cells will further reinforce its walls with the

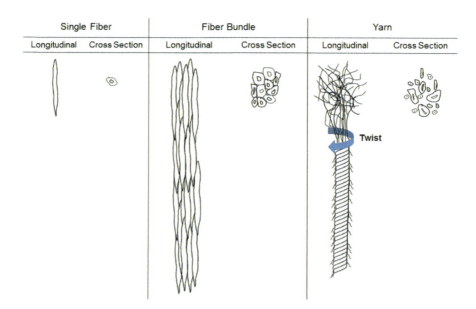

Fig. 1 Schematic of a longitudinal and cross-section view of single fibers (fiber cell), fiber bundles and yarns

Table 1 Dimensions of single fibers and fiber bundles from common herbaceous plants used for fiber extraction [42, 49]

Plant	Fiber structure	Length (mm)	Diameter (μm)
Flax	Single	9–70	5–38
	Bundle	100–1500	40–620
Hemp	Single	5–55	10–51
	Bundle	650–5000	25–500
Jute	Single	2–5	10–25
	Bundle	150–3600	25 – 200
Coir	Single	0.3–1.2	10–20
	Bundle	36–330	50–460

deposition of a secondary wall between the primary wall and the plasma membrane. The secondary wall will compress the outer primary wall to the adjacent middle lamella and reduce the cytoplasm, which at the end of wall deposition, will disappear, leaving a space called lumen. We can easily visualize the transition between the different cell wall regions in a cross-section view (Fig. 2).

Both primary and secondary cell walls present a lamellate structure consisting of progressing deposition of cellulose layers, being the layers adjacent to cytoplasm the most recent ones. Secondary walls are further differentiated in three regions, called S1 (adjacent to the primary wall), S2 and S3 (adjacent to cytoplasm). For fibers, usually, the S2 layer is the most conspicuous layer and can comprise more than 80% of the total thickness of the cell wall, thus dictating much of the fiber properties (Fig. 3).

An Overview on Plant Fiber Technology: An Interdisciplinary ... 981

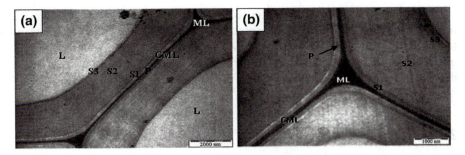

Fig. 2 Transmission electron microscope (TEM) image from an ultrathin cross section of kenaf fiber in low **a** and high **b** magnification (reprinted from industrial crops and products, 31/1, H. P. S. Abdul Khalil, A. F. Ireana Yusra, A. H. Bhat, M. Jawaid, Cell wall ultrastructure, anatomy, lignin distribution, and chemical composition of Malaysian cultivated kenaf fiber, 113–121, Copyright (2010), with permission from Elsevier). Legend: L = Lumen; ML = Middle Lamella; CML = Compound Middle Lamella; P = Primary wall; S1–3 = Secondary wall layers

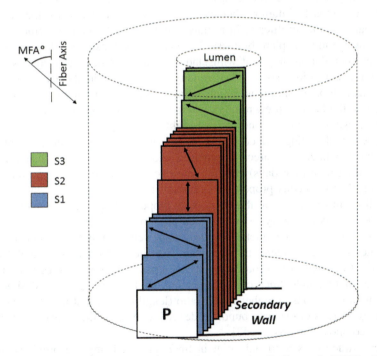

Fig. 3 Lamellar deposition of cell wall layers and average microfibrillar angle (MFA) within the secondary wall. The layers are numbered in order of deposition, being the S1 the first (in contact with the primary wall, P) and S3 the last (lumen side) [45]

During secondary wall deposition, occurs the impregnation of cell corners, middle lamella, primary wall and secondary wall by lignin, giving an extra structural stability and impermeability throughout the whole fiber structure. With lignification, the middle lamella and primary wall can be hard to distinguish, being then called as compound middle lamella (CML) [33].

2.1 Fiber Quality

Plant fibers are a result of a living organism, thus the life cycle of plants will affect how those fibers will form and which properties they will present. Plant health depends on how favourable the environment is and variables like soil composition, climate conditions, water and nutrients available will indirectly affect fiber quality [54].

For practical use of the fibers, they need to be extracted from the rest of the plant. Usually, a degrading process (retting) is used for loosening the fibers from the other plant tissues, thus facilitating fiber extraction. Retting is based on the action of microorganisms that proliferate in a moist condition and secrete specific enzymes that rot away much of plant biomass. Because fibers constitute a less susceptible tissue for degradation they will withstand the whole process. Due to this natural degradation separation, some heterogeneity is expected in the extracted fibers, besides that, climate conditions can also play a role in fiber quality if the same are subjected to field retting (dew retting), where fibers are left on the field (2–3 weeks) for retting using the moist conditions of night dew [65].

Besides field retting, water retting is another typical procedure, plant parts containing the fibers are directly immersed and kept under water, resulting in a shorter retting time (4–6 days) and in a more constant degradation, yielding fibers with more homogeneous properties. Important to notice that an extended retting time will result in damage to the fibers, so a proper retting schedule optimized for each fiber type is necessary to assure good properties [38, 39].

Further refinement of extracted bundles can be achieved by a mechanical process like scutching and combing, by chemicals like the mercerization (NaOH) or by enzymatic treatment. Those process aims to separate bigger bundles into smaller one or even to single fiber cells, depending on the final product desired. For composite applications, a higher aspect ratio (length/width) that maintain the same or better properties of the large bundle is desired to increase the relative contact area with the composite matrix [4, 42].

Fiber extraction is a crucial point in fiber quality, being the previous step to commercial availability, it dictates the final properties of the product before its use. Many drawbacks of using plant fibers for engineering applications, like in composite materials, is due to the high heterogeneity of properties presented in the fibers, thus, reliability should be pursued in areas including fiber extraction and plant cultivation, allowing plant fibers to compete in the market of custom-made synthetic fibers [23, 63].

3 Fiber Chemistry

Cellulose is the main constituent of plant cell walls, its content on fibers varies conform species and values can be found in the range of 43% (coir fibers from coconut) to 92% (cotton fibers) [6]. The presence of cellulose confers high mechanical properties to the cell walls, allowing plants to achieve impressive sizes above ground.

The improved mechanical properties of cellulose found in plant cell walls are the result of its chemical structure, where 6–10 β-(1→4)-glucan chains hydrogen bond to form 2 nm diameter fibrils, this initial structure is directly correlated to the 6–10 enzymatic subunits of the rosette structure. Six of this fibrils, parallel to each other, form a 36 glucan chain microfibril, which possesses a modulus of elasticity in the axial direction of ~ 134 GPa [10]. The length of cellulose microfibrils varies in the order of few micrometres, commonly the degree of polymerization (DP) from the glucose units is used to express microfibril length. The DP found in secondary walls is in the range of 14,000 (~ 7 μm) while the DP in primary walls is often in the order of 8000 (~ 4 μm) [56] (Fig. 4).

Each glucose unit bear three reactive hydroxyl groups (–OH), which give a single cellulose chain a hydrophilic nature, in contrast, when in crystalline microfibrils, hydroxyl groups present in the interior of the structure will not be available for interaction with water molecules, thus crystalline cellulose present a more hydrophobic profile [24].

The highly ordered cellulose microfibril patterns of crystalline and non-crystalline regions are not completely understood and are thought to be either an amorphous shell surrounding a crystalline core, a crystalline and amorphous segments alternating along the axis length of the microfibril, or a combination of both [50, 51]. The amorphous cellulose region is quite similar to highly ordered linear hemicelluloses or with short side chains, probably leading to a gradual nanostructured transition between them [10].

Microfibrils can be composed of two types of cellulose, named cellulose Iα and Iβ. Iα has a single chain triclinic unit cell, whereas cellulose Iβ has two chains monoclinic unit cell (Fig. 5). In both forms, cellulose chains are parallel to each other but in high alkaline concentrations, cellulose fibrils can rearrange in an antiparallel alignment, called cellulose II with the respective forms IIα and IIβ. It is reported that the ratio of cellulose α and β can be influenced by the interaction of cellulose microfibrils with hemicelluloses [29].

Microfibril cellulose is synthesized in spirals around the elongation axis of fibers cells, those spirals stag onto each other and pile up to form layers. The angle formed by those microfibrils spirals in relation to the fiber elongation axis is called microfibrillar angle (MFA) (Fig. 3). The MFA has a direct correlation with mechanical properties of the fiber cell and bundle, since cellulose is a polymer which has the strongest linkages (covalent bonds) oriented along its chain, when a load is applied in the fiber axis direction, microfibrils with lower angles, will result

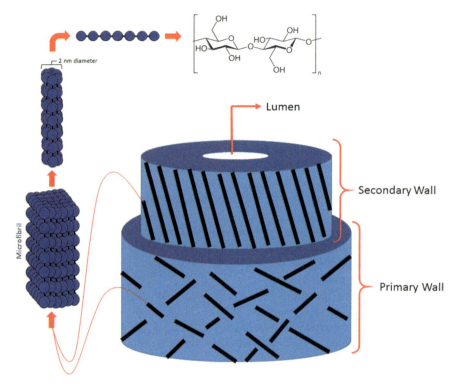

Fig. 4 Cellulose hierarchical structure in the cell wall. The micro-arrangement of cellulose chains is equal for both primary and secondary wall, but the difference in microfibril length and orientation arise at the ultrastructure level

in higher stiffness (Fig. 6). Because the S2 layer is the region with higher cellulose content and lower MFA, it will dictate the mechanical properties of the fiber [9, 10].

Closely associated with cellulose, the heterogeneous group of polysaccharides called hemicelluloses shows a branched β-(1→4) linked backbone of glucose, mannose or xylose. This structure allows the non-covalent interaction by hydrogen bonding between themselves and to cellulose chains, acting as cross-linking polysaccharides [47]. The branched (amorphous) region is the main responsible for fiber moisture absorption [6].

In plant fibers, total hemicellulose content usually lies in the range of 0.25% (coir) to 30% (bamboo) [21], but lack of proper definition of its heterogeneous components is often noticed in fiber characterization works, resulting in scarce information regarding fiber cell wall composition of different species.

The essential difference between cell walls of dicotyledons and monocotyledons (grasses) plants is well known, where hemicellulose polysaccharides are markedly different (Table 2). Regarding cell wall distribution, is known that secondary wall of both monocots and dicots are more abundant in hemicelluloses than the primary wall, probably following the high concentration of cellulose [11].

Fig. 5 Cellulose Iα **a** and Iβ **b** chain arrangement, reducing end of the chain showed in red [61]

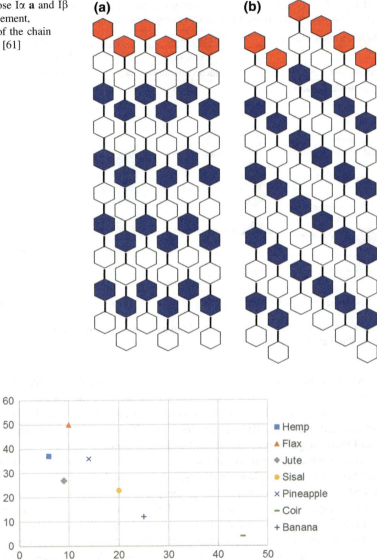

Fig. 6 The relation between MFA and elasticity modulus (values adapted from reviews of [1, 43])

The chemical structure of polysaccharides present in this work was drawn having as basis the standard provided by the Consortium for Functional Glycomics (Fig. 7), which provides a useful method for an easy visualization of polysaccharides components.

Table 2 Main hemicellulose polysaccharides found in dicot and monocot cell walls, other hemicelluloses are also present but in lower concentrations [18, 62]

Wall location Plant type	Primary wall	Secondary wall
Dicots	Xyloglucan	Glucuronoxylan
Monocots	Glucuronoarabinoxylan	Glucuronoarabinoxylan *

*Fewer side chains

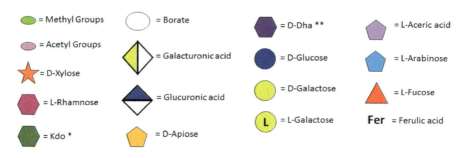

* 2-keto-3-deoxy-d-manno-octulosonic acid   ** 2-keto-3-deoxy-d-lyxo-heptulosaric acid

Fig. 7 Monosaccharide symbols in part from the consortium for functional glycomics. Adapted from Varki et al. [60]

Xyloglucan found in the primary wall of dicots resembles the cellulose structure, where the same β-(1→4)-glucan backbone is present, the addition of α-D-Xylose-(1→6) in the glucan backbone give rise to a branching pattern that alternates with unbranched glucose. Xylose residues can be even further branched with the attachment of galactose, fucose and arabinose units (Fig. 8).

Transitioning to the secondary wall, xylose residues are now linked in a β-(1→4) backbone, known as xylans. In dicots, a frequent xylan modification is an α-(1→2)-linked glucuronic acid, when this is the prevalent substitution, they are known as glucuronoxylans (Fig. 9). Acetylation of glucuronoxylans at position O-3 of xylose residues is also a characteristic feature of this hemicellulose polysaccharide.

Hemicellulose of monocot plants are dominated by xylans in both primary and secondary wall, and they present substitutions with L-arabinose and glucuronic acid, being then known as glucuronoarabinoxylans (GAX) (Fig. 10). A particular feature of monocot GAX is the presence of ferulic acid attached to O-5 of some arabinose residues. They confer a significant cross-link ability between GAX molecules and between GAX and Lignin, possible by dimerization through ester and ether linkages [25, 62]. The typical properties of monocot fibers, like coarse texture and recalcitrance towards lignin extraction, is then directly correlated to feruloylation degree of GAX, which is also believed to serve as a nucleation site for lignin formation [25].

Fig. 8 Xyloglucan polysaccharide structure. A β-(1→4) glucan backbone with xylose side chains

Fig. 9 Glucuronoxylan polysaccharide structure. A β-(1→4) xylose backbone with predominant substitution with glucuronic acid

Fig. 10 Glucuronoarabinoxylan polysaccharide structure. The major hemicellulose polysaccharide structure present in monocots, with a β-(1→4) xylose backbone and characteristic feruloylated arabinose side chains

There is an apparent pattern of hemicelluloses shifting from a glucan backbone in the primary wall to a xylan backbone in the secondary wall, xylan backbones found in the secondary wall are less branched, leading to a more efficient surface with cellulose chains [62].

Pectins are another polysaccharide group present in plant cell walls and are characterized by the presence of galacturonic acid units covalently linked in a repeating backbone or in a dimer repeating backbone together with rhamnose [41]. Regarding the backbone and side chains, pectic polysaccharides can be divided mainly into homogalacturonan (HG), rhamnogalacturonan I (RG-I) and rhamnogalacturonan II (RG-II) (Fig. 11).

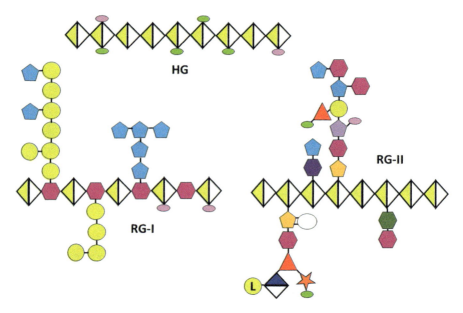

Fig. 11 Pectin major components. Homogalacturonan (HG) is the most abundant form and present a backbone of D-GalA linked in an α- 1,4 configuration, is partially methyl-esterified at O-6 and acetylated at O-2 and O-3 positions. Rhamnogalacturonan I (RG-I) constitute a backbone of [4]-α-D-GalA-(1, 2)-α-L- Rha-(1,]n in which the GalA residues are highly acetylated at O-2 or O-3 positions. Rhamnogalacturonan II (RG-II) have an HG backbone and complex side chains, generally exist in an RG-II borate diester dimer, cross-linking HG in the wall [3]

They can be presented in a range of 0.2–6.6% [40], during fiber cell elongation, pectins are the main constituent of the middle lamella, bonding adjacent cells. It is also present in the primary wall and to a lesser extent in the secondary wall, acting as a cross-linking polymer, connecting hemicelluloses, lignin and proteins [46, 58]. At the end of cell expansion, during deposition of the secondary wall, there is usually a decline in pectin content, where the function of bonding adjacent fiber cells becomes eclipsed by the cementation and densification of middle lamella and primary wall by lignin. In most plant fibers the pectin contribution to fiber properties is not of much importance, but care should be taken in fibers with diminished lignification, like in Flax (*Linum usitatissimum*), where pectin plays a significant role in fiber cells cohesion and consecutively mechanical performance of the fiber bundle [5, 12].

When fiber cells reach their final size, the deposition of the secondary wall begins, reinforcing the current structure. At this point lignification occurs, with lignin precursors synthesized in the cytoplasm and then transported to initiation sites at middle lamella, cell corners, and cell wall, filling voids in the intercellular space and between polysaccharides [14].

The lignification of cell corners, middle lamella, and the primary wall is relatively fast, probably due to the high porosity of those regions [14]. Lignification of the secondary wall is slower and usually completes after deposition of the S3 layer. In fiber cells, because the difference in volume between the compound middle lamella and secondary wall, around 70% of the total lignin is located in the cell wall, but not as concentrated as in the CML. The cohesion provided by lignification of fibers will enhance the load transfer capability between cellulose microfibrils by cross-linking of anchored hemicelluloses, as well as transfer load between fiber cells inside the bundle, through the CML [8].

Lignin is the second most abundant polymer found in nature, second only to cellulose, in non-wood fibers, lignin percentage is found in the range of 0.6% (Ramie) to 45% (Coir) [21] and is composed of three types of phenyl propane precursors, the so-called monolignols (hydroxycinnamyl alcohols) (Fig. 12), p-coumaryl (4-hydroxycinnamoyl), coniferyl (3-methoxy-4-hydroxycinnamoyl) and sinapyl (3,5-dimethoxy-4-hydroxycinnamoyl) alcohols. They differ only in the degree of substitution by methoxyl groups in the aromatic ring at C_3 and C_5 positions. Those monolignols will be initially oxidized by laccases and peroxidases to form a resonance-stabilized phenoxy radical, which will then polymerize into lignin through either a β-O-4, α-O-4, 5-5, β-β, 4-O-5, β-5 or β-1 linkage (Fig. 13) [35].

In the polymerized form, lignin assumes a very complex and branched structure without a repeating backbone like in the other polysaccharides. The aromatic constituents of the monolignols in the polymer are called p-hydroxyphenyl (H), guaiacyl (G) and syringyl (S) moieties [36] (Fig. 12). Generally, for herbaceous plants, all lignin moieties can be found, differently than wood lignin, which mainly consists of G/S and traces of H [13].

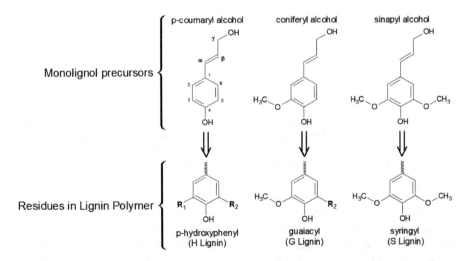

Fig. 12 Hydroxycinnamyl alcohols precursors of H, G and S lignin. R_1 and R_2 can be a hydrogen atom or another lignin molecule

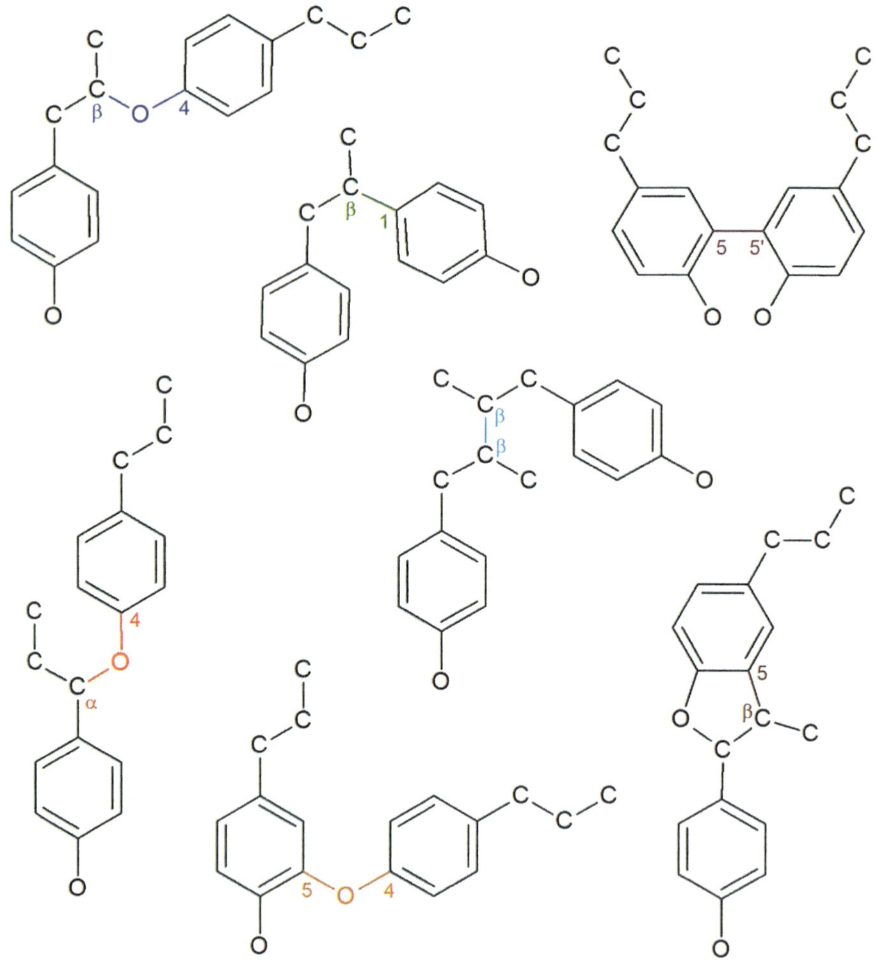

Fig. 13 Chemical bonds in lignin polymeric structure

The proportions of H/G/S moieties vary greatly by species and by region of the cell wall, for example, is know that H lignin content is higher in the middle lamella than in the cell wall. The proportions of those moieties will determine the type of linkage and the degree of branching, which provides a valuable knowledge of the lignin reactivity [17].

The overall chemical composition of plant fibers greatly depends on the part of the plant and age, for instance, fibers extracted from the bottom of the plant will have a higher lignin content compared to fibers from the top, due to the fact that cells from the top are younger and did not have enough time to undergo lignification. This dependence and variation also influence the levels of cellulose, hemicellulose and pectin contents, contributing to fiber heterogeneity [49].

4 Engineering Aspects of Non-wood Fibers in Composite Applications

When talking about plant fibers as an engineering material, a complete knowledge of fiber biology, structure, chemistry, and physical properties is required for effective use. In the view of fibers as reinforcement in composite materials, some critical factors are oblivious for many scholars and researchers. This is often the case with cellulose crystallinity (amorphous material removal), lignin removal and defibrillation.

As mentioned before, cellulose is the main constituent of plant fibers and is arranged in a microfibril structure with crystalline domains. Those domains have excellent mechanical performance due to the extensive hydrogen bonds formed between cellulose chains [26], which suggests that increases in the crystalline content reflect on increases of the mechanical performance of the cell wall and thus the mechanical performance of the entire fiber.

Extensive works have been published in the area of fiber treatment for composite applications, most of them report effects in crystallinity of alkaline treated fibers, which can remove amorphous domains (hemicellulose and lignin), increasing the crystallinity index [53].[1]

The concern is in understanding that the fiber structure behaves as a composite material by itself at two levels, at cell wall, with cellulose as reinforcement in a hemicellulose/lignin matrix, and at bundle level, with secondary cell walls acting as reinforcement in a compound middle lamella matrix (Fig. 14). Utilizing a higher crystalline fiber in a bundle format may not correspond to the total performance which one single fiber could present, due to discontinuity of cellulose-rich secondary walls [8].

To obtain higher length per diameter ratios (which provides excellent performance for fibrous reinforcement elements), a complete delignification of the compound middle lamella is required. With current methods, which involves higher alkaline concentrations at higher temperatures, there is also disruption of lignin present in the cell wall. This results in premature exposure to water sensitive regions of hydrogen bonding between cellulose and hemicellulose cross-linkages, significantly affecting the mechanical performance of the cell wall [48].

For this reason, delignification processes are limited to gentler extractions, preserving the bundle structure in the centre and defibrillating single fibers at the surface (Fig. 15), enhancing surface roughness, which promotes mechanical interlocking with the matrix in the manufactured composite material.

The challenge in the use of herbaceous plant fiber as reinforcement in polymer composites is due to its incompatibility at interface level with the matrix, this zone is where fiber and resin (matrix) are chemically and/or mechanically combined. Interfacial strength plays a significant role in mechanical properties of composites.

---

[1]Relation between amorphous and crystalline regions, which can be calculated from x-ray diffraction analysis, most commonly by Segal's Method [52].

Fig. 14 Composite Bundle. The secondary wall presenting lower MFA and extensive lamella deposition assume a role of discontinuous reinforcement material due to its higher mechanical performance in relation to the compound middle lamella (CML), which acts as a continuous transferring stress matrix

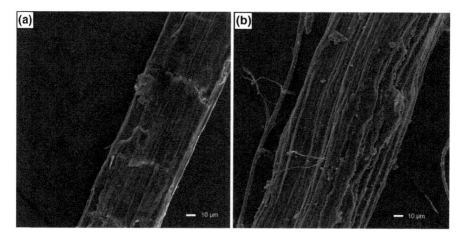

Fig. 15 Curauá (*Ananas Comosus* var. *erectifolius*) fiber bundle untreated (**a**) and NaOH treated (**b**). With the removal of lignin by NaOH, defibrillation of single fiber cells occurs at bundle surface

If the adhesion between phases is weak, then relatively weak load transfer occurs, resulting in a decrease of the mechanical properties [57].

This incompatibility arises from the hydrophilic profile of the chemical constituents of the plant fiber, which are rich in OH groups. To overcome this issue, engineering of the fiber properties can be done using chemical treatments, which can modify fiber chemistry by external agents like sodium hydroxide, acetic anhydride, silanes, etc.

The aforementioned mercerization treatment (NaOH), will interact with OH groups of the fiber through the ionization of the hydroxyl group to alkoxide. The removal of those hydrophilic hydroxyl groups improves the overall moisture resistance of the fiber, but the most desired effect will be the increase in contact surface, facilitating the already mentioned mechanical interlocking with the resin [37, 44].

Similar to mercerization, is the acetylation treatment, which will provide a more extensive hydrophobization of the fiber by means of the introduction of an acetyl functional group at the free hydroxyl groups (Fig. 16), taking out existing moisture, improving compatibility with hydrophobic matrices and bringing dimensional stability to the composite [32].

Another approach for modification of plant fibers is the use of coupling agents, which can provide a bridge at the chemical level between the fiber structure and resin used in the composite, one example of those agents are the silanes.

These chemicals are hydrophilic compounds with different groups attached to silicon, presenting a bi-functional structure, where one end can react with hydroxyl groups of the fiber and the other end interact with the matrix components [30].

Chemical structure of silanes can be generalized as follows:

$$R_{(4-n)} - Si - (R'X)_n (n = 1, 2)$$

Where R is a group that can be hydrolyzed to form a silanol group in aqueous solution and further reacts with hydroxyl groups present in the fiber (Fig. 17), R groups can be chlorine, methoxy, ethoxy, etc. X is organofunctional that is able to react with the matrix, they can be a vinyl, y-aminopropyl, y-methacreloxypropyne, etc. R' is an alkyl bridge (alkyl spacer) which, depending on its length, can influence the hydrolysis of the silane [64].

During hydrolysis, the produced silanol molecules undergo a self-condensation process, where the formation of O–Si–O bonds take place, generating oligomers of increasing size depending on reaction time (ageing). This is important to get a total cell wall modification since, depending on the silanol oligomers size, the penetration in the dense cell wall can be blocked. To avoid this condition, the pH of the

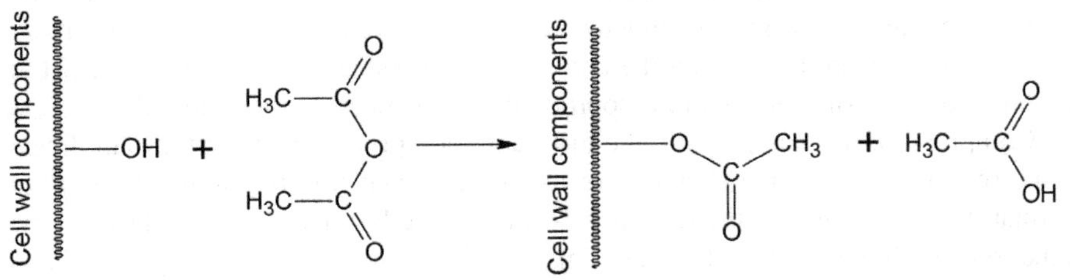

Fig. 16 The schematic reaction of acetic anhydride with hydroxyl groups present in plant fiber cell walls

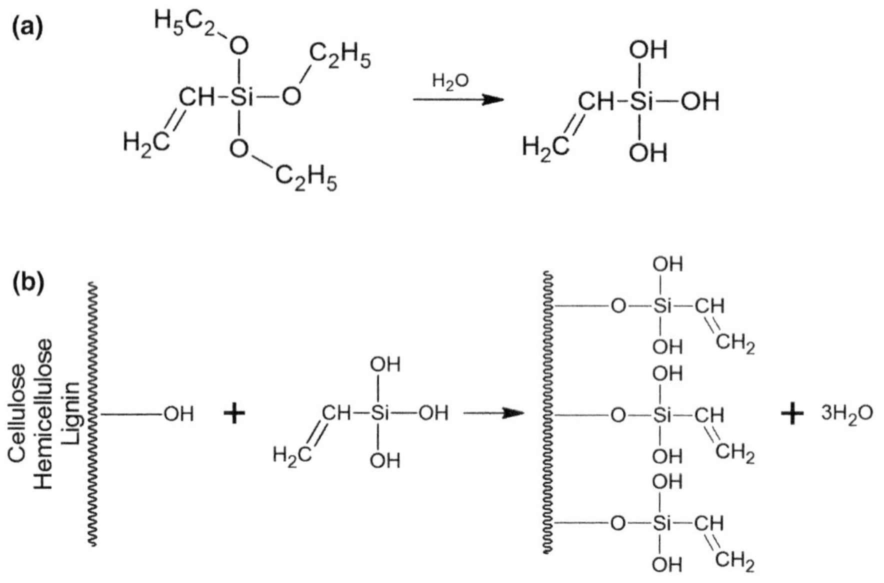

Fig. 17 Silane Modification. a Formation of silanol trough hydrolysis. b The initial interaction of the silanol groups with the OH groups of fiber components is done by hydrogen bonding, but under heating conditions, the hydrogen bonds are converted into primary covalent bonds, releasing water (Grafting process)

solution can be lowered to slow down the condensation reaction, giving enough time for the solution to penetrate into the fiber [28].

Once the proper fiber modification is done, the fiber will bear matrix reactive groups linked to its components, decreasing fiber hydrophilicity, increasing dimension stability and fiber-matrix adhesion via primary covalent bonds, giving molecular continuity across the interface region of the composite [24].

Commercial application of non-wood fibers as reinforcement in polymeric composites is growing every day, notably in the automotive industry [31, 66] where concerns on weight reduction give plant fibers an edge over the more dense glass fibers. Nevertheless, when manufacturing plant fiber-reinforced composites, attention should be taken to consider the composite processing, plant and fiber format to be utilized, each of them is more appropriate for a specific use (Table 3).

Compression moulding is still the most common process method for plant fibers reinforced polymer composites due to its low cost and low requirements on fiber format, being able to use the raw fiber bundles directly after extraction [16]. Some other process like resin transfer moulding (RTM) require certain stability of fibers in a predetermined format, usually utilizing fibers yarns in a textile format (woven or non-woven) [27]. Fibers that permit the spun process to become a yarn can also be used for a continuous process, like pultrusion with thermoplastic resins [22].

Table 3 Composite processing and application of some important plant fibers utilized as reinforcement

Plant	Botanical structure	Comercial format	Composite Processing	Application
Sisal	Perivascular fibre from vascular bundle caps	Bundle, fabrics and Non-woven	Compression moulding	Semi-finished parts in automotive industry; house interior roofing
Hemp	Secondary phloem and pericyclic fibres	Bundle, fabrics and non-woven	Compression moulding	Semi-finished parts in automotive industry
		Fabrics and non-woven	SMC	Bus body component
			RTM	Signposts; forniture; automotive industry
Flax	Pericyclic fibers	Bundle, fabrics and non-woven	Compression moulding	Semi-finished parts in automotive industry
		Fabrics and non-woven	SMC	Spoilers, fenders and funnels
		Fabrics and Non-Woven	RTM	Housing for radar units; speaker boxes; automotive industry
		Yarn	Termoplastic pultrusion	Semi-structural profiles

SMC = Sheet Molding Compound; RTM = Resin Transfer Molding [2, 22, 27]

5 Conclusion

In our continuously connect society, every day more multidisciplinary groups are working on shared issues, which can only improve our complete understanding of the subject. With plant fiber technology, this kind of approach is a must and the demand for eco-friendly and renewable materials in light of climate changes and depletion of fossil fuels makes this topic a global interest.

The present work provided a broader view of the key aspects involving herbaceous plant fiber research, consolidating the biological, chemical and engineering knowledge to this interdisciplinary field of study.

Basic research is still in demand due to the not complete elucidation, especially regarding differences between herbaceous and wood fibers as well as the chemical structure of plant cell walls, which is still being updated as our current technology develops more sophisticated characterization techniques.

Acknowledgements The authors wish to thank the CNPq, Capes and FAP-DF for financial assistance during this work.

References

1. Akin DE, Eder M, Burgert I, Müssig J, Slootmaker T. (2010) What are natural fibres? In: Müssig J (ed) Industrial applications of natural fibres. Wiley, Chichester, UK, pp 11–48. https://doi.org/10.1002/9780470660324.ch2
2. Anandjiwala RD, John M (2010) Sisal—cultivation, processing and products. In: Industrial applications of natural fibres. Wiley, Chichester, pp 181–95. https://doi.org/10.1002/9780470660324.ch8
3. Atmodjo MA, Hao Z, Mohnen D (2013) Evolving views of pectin biosynthesis. Annu Rev Plant Biol 64:747–79. https://doi.org/10.1146/annurev-arplant-042811-105534
4. Baley C, Busnel F, Grohens Y, Sire O (2006) Influence of chemical treatments on surface properties and adhesion of flax fibre-polyester resin. Compos Part A Appl Sci Manuf 37:1626–1637. https://doi.org/10.1016/j.compositesa.2005.10.014
5. Beakou A, Charlet K (2013) Mechanical properties of interfaces within a flax bundle—part II: numerical analysis. Int J Adhes Adhes 43:54–59. https://doi.org/10.1016/j.ijadhadh.2013.01.013
6. Biagiotti J, Puglia D, Kenny JM (2004) A Review on natural fibre-based composites-part I: structure. Proc Prop Vegetable Fibres J Nat Fibers 1(2):37–41. https://doi.org/10.1300/j395v01n02
7. Bidlack J, Malone M, Benson R (1992) Molecular structure and components integration of secondary cell walls in plants. ProcOklaAcadSci 75:51–56
8. Booker RE, Sell J (1998) The nanostructure of the cell wall of softwoods and its functions in a living tree. Eur J Wood Wood Prod 56:1–8. https://doi.org/10.1007/s001070050255
9. Bourmaud A, Morvan C, Bouali A, Placet V, Perré P, Baley C (2013) Relationships between micro-fibrillar angle, mechanical properties and biochemical composition of flax fibers. Ind Crops Prod 44:343–351. https://doi.org/10.1016/j.indcrop.2012.11.031
10. Burgert I, Dunlop JWC (2011) Micromechanics of cell walls. In: Wojtaszek P (ed) Mechanical integration of plant cells and plants, vol 9, 1st edn. Springer, Berlin Heidelberg, pp 27–52. https://doi.org/10.1007/978-3-642-19091-9
11. Caffall KH, Mohnen D (2009) The structure, function, and biosynthesis of plant cell wall pectic polysaccharides. Carbohydr Res 344:1879–1900. https://doi.org/10.1016/j.carres.2009.05.021
12. Charlet K, Béakou A (2011) Mechanical properties of interfaces within a flax bundle—part I: experimental analysis. Int J Adhes Adhes 31:875–881. https://doi.org/10.1016/j.ijadhadh.2011.08.008
13. Dalimova GN, Abduazimov KA (1994) Lignins of herbaceous plants. Chem Nat Compd 30:146–159. https://doi.org/10.1007/BF00629995
14. Donaldson LA (2001) Lignification and lignin topochemistry—an ultrastructural view. Phytochemistry 57:859–73. https://doi.org/10.1016/s0031-9422(01)00049-8
15. Drakakaki G (2015) Polysaccharide deposition during cytokinesis: challenges and future perspectives. Plant Sci 236:177–184. https://doi.org/10.1016/j.plantsci.2015.03.018
16. Drieling A, Müssig J, Graupner N, Müssig J, Piotrowski S, Carus M (2010) Economic aspects. In: Müssig J (ed) Industrial applications of natural fibres. Wiley, Chichester, UK, pp 49–86. https://doi.org/10.1002/9780470660324.ch3
17. Duval A, Lawoko M (2014) A review on lignin-based polymeric, micro- and nano-structured materials. React Funct Polym 85:78–96. https://doi.org/10.1016/j.reactfunctpolym.2014.09.017
18. Ebringerova A, Hromadkova Z, Heinze T (2005) Hemicellulose. Adv Polym Sci 186:1–67. https://doi.org/10.1007/b136816
19. El-Gohary M (2012) The contrivance of new mud bricks for restoring and preserving the Edfa ancient granary–Sohag, Egypt. Int J Conserv Sci 3:67–78
20. Evert RF (2006) Esau's plant anatomy. https://doi.org/10.1002/0470047380

21. Faruk O, Bledzki AK, Fink HP, Sain M (2012) Biocomposites reinforced with natural fibers: 2000–2010. Prog Polym Sci 37:1552–1596. https://doi.org/10.1016/j.progpolymsci.2012.04.003
22. Friedrich K, Evstatiev M, Angelov I, Mennig G (2007) Pultrusion of flax-polypropylene composite profiles. In: Handbook of engineering biopolymers. Carl Hanser Verlag GmbH & Co. KG, München, pp 223–36. https://doi.org/10.3139/9783446442504.007
23. Gandini A, Belgacem MN, Barkoula N-M, Peijs T, Dufresne A, Mosiewicki MA et al (2011) Interface engineering of natural fibre composites for maximum performance, 1st edn. Woodhead Publishing, Cambridge
24. George J, Sreekala MS, Thomas S (2001) A review on interface modification and characterization of natural fiber reinforced plastic composites. Polym Eng Sci 41:1471–1485. https://doi.org/10.1002/pen.10846
25. Grabber JH, Ralph J, Lapierre C, Barrière Y (2004) Genetic and molecular basis of grass cell-wall degradability. I. Lignin-cell wall matrix interactions. Comptes Rendus Biol 327:455–65. https://doi.org/10.1016/j.crvi.2004.02.009
26. Haigler CH (1985) The functions and biogenesis of native cellulose. In: Nevell T, Zeronian S (ed) Cellulose chemistry and its applications. Ellis Horwood Ltd., Chichester, UK, pp 30–83
27. Hänninen T, Hughes M, Baur E, Otremba F, Huber T, Graupner N, et al (2010) Composites. Industrial applications of natural fibres. Wiley, Chichester, UK, pp 381–480. https://doi.org/10.1002/9780470660324.ch19
28. Hill CAS (2006) Wood modification: chemical, thermal and other processes. Wiley, Chichester, UK. https://doi.org/10.1002/0470021748
29. Jarvis MC (2000) Interconversion of the Iα and Iβ crystalline forms of cellulose by bending. Carbohydr Res 325:150–154. https://doi.org/10.1016/S0008-6215(99)00316-X
30. John M, Anandjiwala RD (2008) Recent developments in chemical modification and characterization of natural fiber-reinforced composites. Polym Compos 29:187–207. https://doi.org/10.1002/pc.20461
31. Joshi SV, Drzal LT, Mohanty AK, Arora S (2004) Are natural fiber composites environmentally superior to glass fiber reinforced composites? Compos Part A Appl Sci Manuf 35:371–376. https://doi.org/10.1016/j.compositesa.2003.09.016
32. Kabir MM, Wang H, Aravinthan T, Cardona F, Lau K (2007) Effects of natural fibre surface on composite properties : a review. In: eddBE2011 1st international postgraduate conference engineering design and development built environment sustain, vol 27–29. Wellbeing, pp 94–99
33. Koch G, Schmitt U (2013) Topochemical and electron microscopic analyses on the lignification of individual cell wall layers during wood formation and secondary changes. Plant Cell Monogr 20:41–69. https://doi.org/10.1007/978-3-642-36491-4_2
34. Kvavadze E, Bar-Yosef O, Belfer-Cohen A, Boaretto E, Jakeli N, Matskevich Z et al (2009) 30,000-year-old wild flax fibers. Science 325:1359. https://doi.org/10.1126/science.1175404
35. Laurichesse S, Avérous L (2014) Chemical modification of lignins: towards biobased polymers. Prog Polym Sci 39:1266–1290. https://doi.org/10.1016/j.progpolymsci.2013.11.004
36. Lewis NG (1990) Lignin : occurrence, biogenesis and biodegradation
37. Li X, Tabil L, Panigrahi S (2007) Chemical treatments of natural fiber for use in natural fiber-reinforced composites: a review. J Polym Environ 15:25–33. https://doi.org/10.1007/s10924-006-0042-3
38. Liu M, Fernando D, Daniel G, Madsen B, Meyer AS, Ale MT et al (2015) Effect of harvest time and field retting duration on the chemical composition, morphology and mechanical properties of hemp fibers. Ind Crops Prod 69:29–39. https://doi.org/10.1016/j.indcrop.2015.02.010
39. Martin N, Mouret N, Davies P, Baley C (2013) Influence of the degree of retting of flax fibers on the tensile properties of single fibers and short fiber/polypropylene composites. Ind Crops Prod 49:755–767. https://doi.org/10.1016/j.indcrop.2013.06.012

40. McDougall GJ, Morrison IM, Stewart D, Weyers JDB, Hillman JR (1993) Plant fibers—botany, chemistry and processing for industrial use. J Sci Food Agric 62:1–20. https://doi.org/10.1002/jsfa.2740620102
41. Mohnen D (2008) Pectin structure and biosynthesis. Curr Opin Plant Biol 11:266–277. https://doi.org/10.1016/j.pbi.2008.03.006
42. Müssig J, Haag K (2014) The use of flax fibres as reinforcements in composites. In: Biofiber reinforcements in composite materials
43. Mwaikambo LY (2006) Review of the history, properties and application of plant fibres. African J Sci Technol 7:120–133
44. Mwaikambo LY, Tucker N, Clark AJ (2007) Mechanical properties of hemp-fibre-reinforced euphorbia composites. Macromol Mater Eng 292:993–1000. https://doi.org/10.1002/mame.200700092
45. Niklas KJ (1992) Plant Biomechanics: an engineering approach to plant form and function. The University of Chicago Press, Chicago and London
46. O'Neill MA, Ishii T, Albersheim P, Darvill AG (2004) Rhamnogalacturonan II: structure and function of a borate cross-linked cell wall pectic polysaccharide. Annu Rev Plant Biol 55:109–39. https://doi.org/10.1146/annurev.arplant.55.031903.141750
47. Pauly M, Gille S, Liu L, Mansoori N, de Souza A, Schultink A et al (2013) Hemicellulose biosynthesis. Planta 238:627–642. https://doi.org/10.1007/s00425-013-1921-1
48. Rowell R (2005) Handbook of wood chemistry and wood composites
49. Rowell RM (2008) Natural fibres: types and properties. In: Properties and performance of natural-fibre composites. Elsevier, pp 3–66. https://doi.org/10.1533/9781845694593.1.3
50. Salmén L, Bergström E (2009) Cellulose structural arrangement in relation to spectral changes in tensile loading FTIR. Cellulose 16:975–982. https://doi.org/10.1007/s10570-009-9331-z
51. Salmén L, Burgert I (2009) Cell wall features with regard to mechanical performance. A review. Holzforschung 63:121–129. https://doi.org/10.1515/HF.2009.011
52. Segal L, Creely JJ, Martin AE, Conrad CM (1959) An empirical method for estimating the degree of crystallinity of native cellulose using the X-ray diffractometer. Text Res J 29:786–794. https://doi.org/10.1177/004051755902901003
53. Ben SAEO, Chaabouni Y, Msahli S, Sakli F (2012) Morphological and crystalline characterization of NaOH and NaOCl treated Agave americana L. fiber. Ind Crops Prod 36:257–66. https://doi.org/10.1016/j.indcrop.2011.09.012
54. Singh SR, Kundu DK, Tripathi MK, Dey P, Saha AR, Kumar M et al (2015) Impact of balanced fertilization on nutrient acquisition, fibre yield of jute and soil quality in new gangetic alluvial soils of India. Appl Soil Ecol 92:24–34. https://doi.org/10.1016/j.apsoil.2015.03.007
55. Skoglund G, Nockert M, Holst B (2013) Viking and early middle ages northern Scandinavian textiles proven to be made with hemp. Sci Rep 3:2686. https://doi.org/10.1038/srep02686
56. Somerville C (2006) Cellulose synthesis in higher plants. Annu Rev Cell Dev Biol 22:53–78. https://doi.org/10.1146/annurev.cellbio.22.022206.160206
57. Thakur VK, Thakur MK, Gupta RK (2014) Review: raw natural fiber-based polymer composites. Int J Polym Anal Charact 19:256–271. https://doi.org/10.1080/1023666X.2014.880016
58. Thompson JE, Fry SC (2000) Evidence for covalent linkage between xyloglucan and acidic pectins in suspension-cultured rose cells. Planta 211:275–286. https://doi.org/10.1007/s004250000287
59. Unger F (1866) Botanische Streifzüge auf dem Gebiete der Culturgeschichte—Ein Ziegel der Dashurpyramide in Ägypten nach seinem Inhalte an organischen Einschlüssen. Sitzungsberichte der Kais. Akad. der Wissenschaften Wien Math. Klasse 54:33–62
60. Varki A, Cummings RD, Esko JD, Freeze HH, Stanley P, Marth JD et al (2009) Symbol nomenclature for glycan representation. Proteomics 9:5398–5399. https://doi.org/10.1002/pmic.200900708

61. Viëtor RJ, Mazeau K, Lakin M, Pérez S (2000) A priori crystal structure prediction of native celluloses. Biopolymers 54:342–354. https://doi.org/10.1002/1097-0282(20001015)54:5%3c342:AID-BIP50%3e3.0.CO;2-O
62. Vogel J (2008) Unique aspects of the grass cell wall. Curr Opin Plant Biol 11:301–307. https://doi.org/10.1016/j.pbi.2008.03.002
63. Van de Weyenberg I, Ivens J, De Coster A, Kino B, Baetens E, Verpoest I (2003) Influence of processing and chemical treatment of flax fibres on their composites. Compos Sci Technol 63:1241–1246. https://doi.org/10.1016/s0266-3538(03)00093-9
64. Xie Y, Hill CAS, Xiao Z, Militz H, Mai C (2010) Silane coupling agents used for natural fiber/polymer composites: a review. Compos Part A Appl Sci Manuf 41:806–19. https://doi.org/10.1016/j.compositesa.2010.03.005
65. Yu H, Yu C (2007) Study on microbe retting of kenaf fiber. Enzyme Microb Technol 40:1806–1809. https://doi.org/10.1016/j.enzmictec.2007.02.018
66. Zah R, Hischier R, Leão AL, Braun I (2007) Curauá fibers in the automobile industry—a sustainability assessment. J Clean Prod 15:1032–40. https://doi.org/10.1016/j.jclepro.2006.05.036

Nanocellulose-Reinforced Adhesives for Wood-Based Panels

Elaine Cristina Lengowski, Eraldo Antonio Bonfatti Júnior,
Marina Mieko Nishidate Kumode, Mayara Elita Carneiro
and Kestur Gundappa Satyanarayana

1 Introduction

The use of wood panels is increasing in two ways. The first is by limiting the dimensions of the log diameters, by the anisotropy and other natural defects that solid wood possesses and the second by the search for greater use of the wood [93, 107]. Therefore the production of panels of reconstituted wood represents a rational use of this raw material [18].

There are several types of wood panels, which include: laminated wood panels, agglomerated wood panels or wood fiber panels [53]. It is reported that 416 million

---

E. C. Lengowski
Faculty of Forestry Engineering, Federal University of Mato Grosso (UFMT),
Fernando Corrêa da Costa St, 2367 - Boa Esperança, Cuiabá, MT 78068-600, Brazil
e-mail: elainelengowski@gmail.com; elainecristina@ufmt.br

E. A. Bonfatti Júnior · M. E. Carneiro
Department of Forest Engineering and Technology (DETF),
Federal University of Paraná (UFPR), Av. Pref. LotharioMeissner,
632, Jardim Botânico, Curitiba, PR 80.210-170, Brazil
e-mail: bonfattieraldo@gmail.com

M. E. Carneiro
e-mail: mayaraecarneiro@gmail.com

M. M. N. Kumode
Laboratory of Wood, Pontifical Catholic University, Curitiba, PR, Brazil
e-mail: mnishidate@gmail.com

K. G. Satyanarayana (✉)
PIPE & Department of Chemistry, Federal University of Parana,
Curitiba, Brazil
e-mail: gundsat42@hotmail.com; kgs_satya@yahoo.co.in

K. G. Satyanarayana
Poornaprajna Scientific Research Institute (PPISR), Sy. no. 167, Poornaprajnapura,
Bidalur Post, Devanahalli, Bangalore 562 110, Karnataka, India

© Springer Nature Switzerland AG 2019
Inamuddin et al. (eds.), *Sustainable Polymer Composites and Nanocomposites*,
https://doi.org/10.1007/978-3-030-05399-4_35

m³ of wood panels were produced in 2016, of which 42% were wood panels and the remaining 58% were fiber based panels [35]. It is interesting to note that each panel has an application such as internal or external use, in furniture or civil construction. Depending on the environment to which the panels will be exposed, there are different types of adhesives that should be used, the most common ones being urea formaldehyde for indoor use and low moisture content, while phenol formaldehyde resin for external use [54].

Improvement of adhesive bonding is a routine process in the wood industry [29] as it is one of the key steps in the production of panels. Changes in adhesion to wood are desirable in terms of performance improvement and adhesive economy [43]. Among the several opportunities offered by nanotechnology for the forest products industry [15] the reinforcement of adhesives with nanocellulose has been already identified as an opportunity, which has been explored. This has shown improvement in both the physical and mechanical properties of the panels [43].

Obtaining nanoscale cellulose fibers and its application as reinforcement in the preparation of biodegradable composites as well as nanocomposites has attracted great attention during the last years [101, 116]. This is attributed to the unique properties of nanomaterials such as high aspect ratio, crystallinity and surface area, excellent mechanical properties combined with less weight and biodegradability [29, 76].

With this background, this chapter presents an overview of the use of nanocellulose in wood-based panels, with examples, of the use of different types of nanocelluloses as reinforcement in several types of adhesives in the production of different types of panels. This chapter also presents some important concepts and properties of all the raw materials used, viz., adhesives, wood, and nanocelluloses.

2 Wood-Based Panels

The solid wood presents some disadvantages, because it is a heterogeneous and anisotropic product, i.e., It possesses different physical properties in its tangential, longitudinal and radial axes [13, 39, 108]. It should also be taken into account that the dimensions of the wood pieces limit their use besides the natural defects, such as knots, grain inclination, the percentage of juvenile and adult wood and reaction wood, among others, all of which interfere with the rheological behaviour of the wood [54]. Also, it is reported that many times the mechanical properties of wood are unsatisfactory for certain uses [104]. Because of the above-mentioned limitations of solid wood, reconstituted wood products have been produced by gluing of veener, boards, slabs, particles or fibers, and these elements are joined by adhesive bonds [13]; Industrial Research [78]. With the use of glue utilization of the wood has been increasing because the glue allows the use of pieces of small dimensions to obtain products with greater added value.

Reconstituted products, such as particle board, oriented strand board (OSB, also known as flakeboard) and plywood panels, among others, appear as an alternative to solid wood, rendering improvement in the characteristics of the raw material. This is because, they allow greater homogeneity of the physicochemical properties, dimensional stability, full use of wood and residues, thus contributing to the conservation of forests [18]. However, the quality of the final product depends mainly on the adhesion technology [89].

Reconstituted wood panels can be divided into three types: laminated panels (plywood and Laminated Venner Lumber-LVL), particle board (wafer board and OSB), and fiberboard (Medium Density Fiberboard-MDF, High-Density Fiberboard-HDF, and insulation board). The plywood panels are composed of wood overlapping and bonded with adhesives, mainly phenol-formaldehyde and urea-formaldehyde, under pressure and temperature so that they cross their fibers at an angle of 90° [54, 119]. The wood veneer can be of different thicknesses and are always in odd numbers (Finnish along the length of the part, in which the thickness of the blades should not exceed 6.4 mm—0.25 Forest Industries Federation [33]. According to the Standard Specification for Evaluation of Structural Composite Lumber Products [5], LVL can be defined as a structural compound composed by layers of thin wood assembled with adhesives with wood fibers oriented mainly inches. In the bonding of the LVL panels synthetic adhesives used are resistant to humidity; the most commonly used adhesive being the phenol-formaldehyde [75]. Both plywood and LVL are gaining visibility for their benefits in structural and non-structural use [10, 65, 73]. These are already used in applications typically dominated by steel and concrete [83].

Particleboard wood panels may be defined with randomly arranged small particles, agglutinated using adhesives and glued using heat and pressure [54, 87]. The most used adhesives in the production of panels of particleboard wood are the synthetic ones such as urea-formaldehyde, phenol-formaldehyde and melamine-formaldehyde [44] with phenol-formaldehyde is recommended for external use and the urea-formaldehyde recommended for internal use [44]. The waferboard used as a structural material is produced with larger wafer type particles of square or slightly rectangular shapes, glued with random particle distribution and consolidated through hot pressing [54]. While in the waferboard the particles are arranged randomly, the particles are used as layers in the OSB with perpendicular directions [87]. In view of this arrangement, the superior structural behaviour is exhibited in OSB-type panels having high dimensional stability compared to the waferboard [53, 93]. Accordingly, the OSB wood panels are used for structural applications, considering the evolution of the waferboard differs from its precursor in the direction of the particle [49].

The fiber panels are dry-fabric panels made of lignocellulosic fibers, combined with an adhesive under pressure and temperature [53, 78, 87]. In such type of panels primary adhesion occurs through the interlacing of the fibers and the adhesive properties of some chemical components of the wood [46]. When fabricated with low density, these panels can be used for insulation purposes [60], called 'insulation board', with a mean density between 0.02 and 0.40 g cm^{-3} [54]. Natural fiber insulation is known for the good thermal insulation it promotes, but this material also presents good acoustic insulation [80].

Medium density fiberboard (MDF) panels are normally bonded with urea-formaldehyde adhesive and consolidated with hot pressing [78]. Density of MDF varies from 0.50 to 0.80 g cm^{-3} [7, 54]. This is one of the most well-known wood panels, commonly used as a raw material for furniture, carpentry and building products [70]. Panels having a density varying from 0.80 to 1.10 g·cm^{-3}, called as high-density fiberboard (HDF), are similar to the MDF [107]. These are used as a panel for structural purposes commonly used as a core of laminate flooring [86].

3 Adhesives and Adhesion

3.1 Adhesives

Kinloch [62] defined adhesive as any substance applied to the surface, or both surfaces, of two separate objects that bind them and offers resistance to their separation. On the hand, Peschel et al. [87] added to this concept of the condition of adhesives, these being non-metallic substances with which other materials are solidly bonded together by adhesion and cohesion.

Wood adhesives can be classified according to their origin in natural and synthetic [28, 74]. Natural adhesives can be proteins of animal or vegetable origin, while synthetic ones have the petroleum raw material and, although they resemble the natural adhesives in the physical characteristics, they can be formulated to meet specific requirements and have a higher resistance to humidity [28, 82].

Synthetic adhesives can be classified into two types: Thermoplastic adhesives and thermoset adhesives. The two types differ in their chemical structure and response to heat [81, 88]. Table 1 shows the classification of some adhesives including natural adhesives used in the wood panel industry.

Another type of classification of adhesives that can be made according to their purpose of use involving the environment to which each adhesive would be exposed. Accordingly, Table 2 shows the types of adhesives, the environment in which they are used and the name of adhesives for each of these used in the preparation of wood panels.

Thermoplastic adhesives are liquid adhesives whose aggregate state depends on temperature. The curing and melting are reversible, i.e., if heated after curing they will return to the liquid state since they are adhesives that do not form reticles (net of fibers). These adhesives may also be dissolved in a solvent and then reactivated with solvent evaporation [88]. But, the use of this type of adhesives is limited, i.e., they can only be used in non-structural applications, in low-temperature climate and are not resistant to heat or fire [81]. Of this polyvinyl acetate (PVA) adhesive is the most commonly used in wood glueing [34]. Figure 1 shows the polyvinyl acetate monomer.

PVA is a yellow-white liquid adhesive available in a ready-to-use form and which can be applied directly to the wood and cured at room temperature or through

Table 1 Adhesives used in the wood panel industry

Types of adhesives	Adhesives
Natural	Animal protein derivatives (glutin, casein and albumin)
	Derivatives of vegetable origin (soybean meal)
	Derivatives of starch (wheat flour)
	Cellulose ether
	Natural rubber
Thermoplastics	Polyvinyl acetate
	Polyvinyl/acrylate
	Polyethylene
	Polystyrene
	Synthetic rubber
Thermosets	Urea formaldehyde
	Melanin-formaldehyde
	Phenol-formaldehyde
	Resorcin-formaldehyde
	Tannin-formaldehyde
	Phenol-resorcin-formaldehyde

Table 2 Classification of the use environment of the wood panels according to the type of adhesive (Adapted from [36])

Application area	Name of environment where the adhesive is used	Used adhesive
Structural	Exterior use without any restriction	Phenol-formaldehyde
		Resorcin-formaldehyde
		Phenol-resorcin-formaldehyde
		Polímeros de emulsão/ Isocianato
		Melanin-formaldehyde
	Exterior use with a restriction	Melanin-urea formaldehyde
		Isosyante
		Epoxy
	Interior	Urea-formaldehyde
		Casein
Semi-structural	Exterior use with limitations	Polyvinyl acetate "crosslinking"
		Polyurethane
Non structural	Interior	Polyvinyl acetate (PVAc)
		Construction elastomers
		Contact elastomers
		Hot-melt

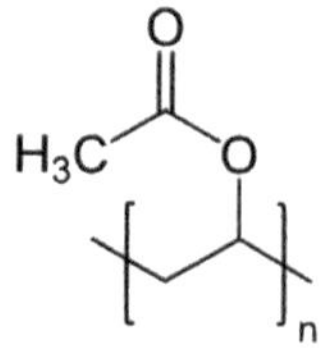

Fig. 1 Polyvinyl acetate (PVA) adhesive monomer. Reproduced from [9] with the kind permission of the publishers

high frequency. After curing this adhesive exhibits high mechanical resistance; however, its use is not recommended in environments with high temperatures and high humidity [36]. Polyvinyl acetate adhesives are fixed by the loss of water mainly by diffusion of water from the adhesive in the wood [34]. This type of adhesive is used for any and all wood glueing operations. Major areas include bonding of corrugated panels, finger-jointing, laminating and assembling [48].

Unlike the thermoplastic adhesives, thermosets are plastics when cured of a soft solid or viscous liquid prepolymer results in a molecule of higher molecular weight and with higher melting point and therefore, will not have the cure reversed by heat [88, 121]. The cure of thermoset adhesives is heat induced, reaching 200 °C. These adhesives generally are stronger than the thermoplastic adhesives and more recommended for high-temperature applications [88] and are more commonly used in wood structures [121]. In spite of a large number of adhesives available for wood panels, the most used are synthetic ones, such as phenol-formaldehyde, urea-formaldehyde, resorcinol-formaldehyde and melamine-formaldehyde [54, 107]. These four adhesives make up approximately 90% of all adhesives used in wood panels and all of these are derived from fossil fuels [14].

Phenol-formaldehyde (PF), applied in a broad spectrum of engineered wood products, is very strong and resistant to dry and humid conditions and exhibits strong adhesion to wood [41, 89, 121]. Figure 2 shows part of a polymer chain of the phenol-formaldehyde adhesive.

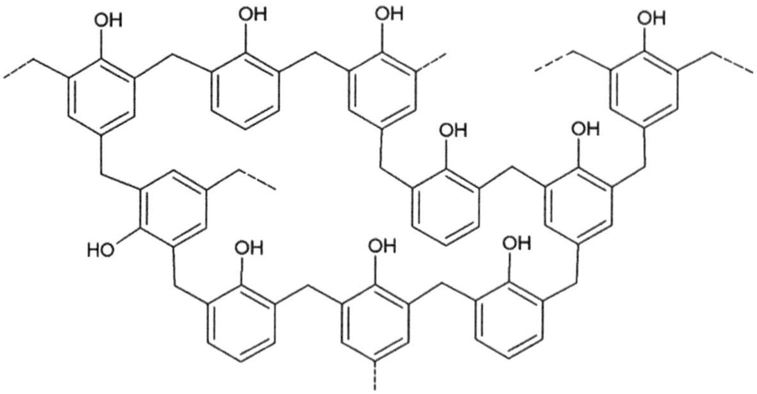

Fig. 2 The molecular structure of the polymer chain of phenol-formaldehyde. Adapted from [124] with the kind permission of the publishers

It can be cured hot or cold; however, for the cold setting process, it is necessary to reduce the pH with the addition of an acid. The PF adhesive is mainly applied to particle or fiberboard, plywood, pressed laminated wood, glued laminated wood, waferboard and OSB [54, 107].

Urea-formaldehyde (UF) is a low-cost, structural and internal-use adhesive [54, 89]. Figure 3 shows part of a polymer chain of the urea-formaldehyde adhesive.

This adhesive is fast hardening, high resistance to dry bonding and presents colourless glue joints. This adhesive can be hardened hot or cold; however, when fast hardening is desired a hardener should be applied [61]. The glue joints of this adhesive are high strength, but brittle and inelastic. Therefore, the stress of the wood, caused by changes in humidity and temperature, impairs glue joints and decreases adhesive performance [41, 121, 74].

Resorcinol-formaldehyde (RF) is a brown, cold-curing, catalyst-requiring adhesive. Figure 4 shows part of a polymer chain of the resorcinol-formaldehyde adhesive. This adhesive is much more reactive than PF [54]. Adhesive bonds of RF

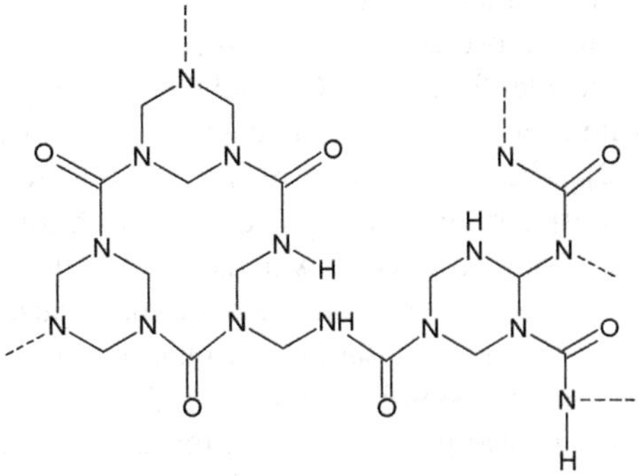

Fig. 3 The molecular structure of the polymer chain of urea-formaldehyde. Adapted from [85], with the kind permission of the publishers

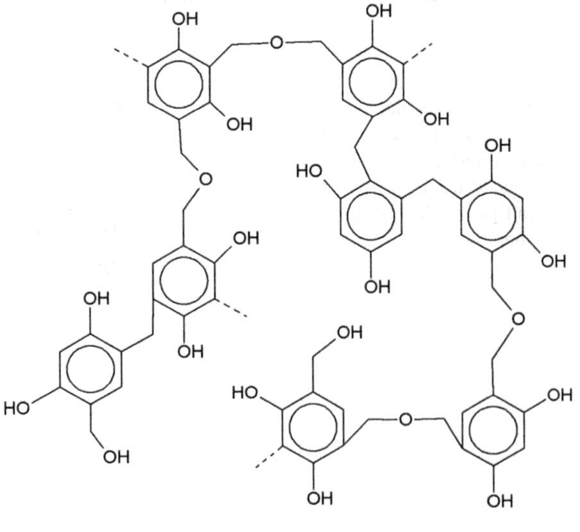

Fig. 4 The molecular structure of the resorcinol-formaldehyde polymer chain. Adapted from [47], with the kind permission of the publishers

are resistant to moisture, boiling water, oil and many other solvents, i.e., an adhesive suitable for exterior use [88]. This adhesive used in the production of rolled beams, shipbuilding, and aviation. However, due to its high cost of production, it is hardly used in pure form. Therefore, it is usually mixed in the same proportion with PF [41, 54, 121].

Melamine-formaldehyde, a product between the condensation of melamine and formaldehyde, is a white-coloured adhesive classified as an intermediate between PF and UF. Figure 5 shows part of a polymer chain of the melamine-formaldehyde adhesive.

RF is a more resistant material than wood [88]. It is a hot curing around (93 °C) structural adhesive Melamine-formaldehyde has the advantages of being more resistant to moisture than UF and curing faster than PF [24, 89]. This can be used outdoors without any restriction [36]. However, its cost is higher than that of these two resins and therefore is commonly used as an additive to improve the performance of UF glue beating, even this form of use is marketed under the nomenclature melamine-urea-formaldehyde [54, 89].

Formaldehyde, present in the four synthetic adhesives mentioned above, is one of the most common chemicals in current use, with the simplest aldehyde of molecular formula H_2CO and the boiling point of -19 °C [97]. It is a colourless gas with pungent, inflammable and highly reactive odour [51] and highly carcinogenic [52]. The use of these resins can lead to the emission of formaldehyde into the atmosphere, generating occupational hazards to the workers involved in the manufacturing process, as well as users of the installations where these panels are used [49]. Exposure to formaldehyde may cause irritation to the mucous membranes of the eyes, nose, nasal cavity, pharynx and larynx, and may also cause drowsiness, nausea, and skin irritation through frequent contact and prolonged exposure [112]. Even coated panels can emit formaldehyde, these emissions are regulated by panel buyers countries [49]. Regulatory standards establish maximum emission limits and analytical methods for gaseous formaldehyde measurements [95].

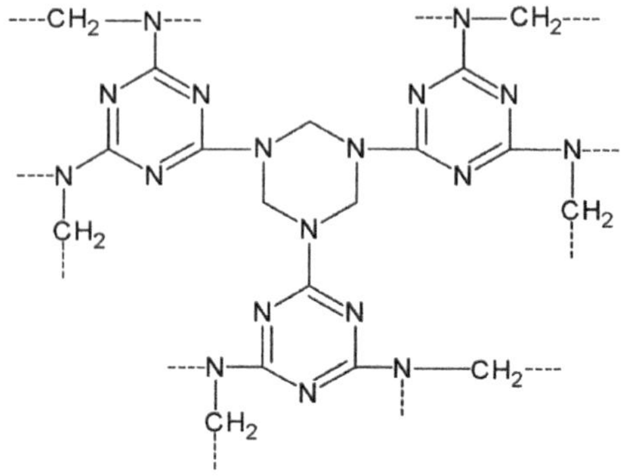

Fig. 5 The molecular structure of the polymer chain of melamine-formaldehyde. Reproduced from [90] with the kind permission of the publishers

3.2 Adhesion

It is understood that adhesion is the force of attraction between molecules of different materials, such as the force between adhesive substance and the junction piece, and cohesion force of attraction between molecules of the same type, such as forces inside a layer of glue [87]. These authors have defined that cohesion is the force of attraction between molecules of the same type, as for example, the forces inside a layer of glue. Therefore, a good adhesive must adhere to the surface of the solid and have sufficient cohesion to ensure the bonding of the solids. According to Schultz and Nardin [100], Petrie [88] and Ebnesajjad and Landroch [27], there are six main adherence theories, viz. mechanical theory, electrical theory also known as electrostatic theory, wetting theory, theory of the diffusion of polymers, also known as diffusion theory, chemical bonding theory and weak boundary layer theory. These are explained below:

- Mechanical theory: According to this, the fluidity and penetration of the adhesive into porous substrates leads to the formation of hooks or a mechanical interlacing of the adhesive tightly attached to the substrate after curing and hardening of the adhesive.
- Electrical theory (Electrostatic theory): According to this, the forces of attraction in adhesion in terms of electrostatic effects at the interface between the adhesive/bonding system is compared to a capacitor, in which the armatures are the two electrical layers formed by the contact of the two substrates. The adhesion results from the forces of attraction developed between the two armatures.
- Wetting theory: According to this, adhesion results from the molecular contact between two materials that develop surface attraction forces. The process of establishing continuous contact between an adhesive and a substrate is called 'wetting', which can also be defined as the adhesion of a liquid to a solid.
- Theory of the diffusion of polymers (diffusion theory): According to this, adhesion occurs through the diffusion of segments of polymer chains. Adhesive forces can be visualized as those produced in mechanical adhesion but on a molecular scale. However, the applications of this theory are also limited. The mobility of long polymer chains is very restricted, severely limiting the molecular interpretation proposed in this theory.
- Chemical bonding theory: According to this theory, adhesion occurs through chemical bonds (covalent and metallic ionic). It is currently believed that adhesion at the interface, from the molecular point of view, is due to the action of secondary forces.
- Weak boundary layer theory: This theory proposes the existence of a finite boundary layer composed of absorbed molecules at the interface, which is different in their constitution from the constituent molecules of the adhesive and the adhesive.

The mechanisms related to the adhesion process and wood panels can be explained by the mechanical theories, chemical adhesion and diffusion of polymers

[37]. In the bonding of porous surfaces, such as wood, the initial process is done mechanically [37]. There is penetration of the adhesive at the cellular scale, filling void intercellular spaces, increasing the bonding durability in the wood, especially when the adhesive is diffused into cellulose and hemicellulose molecules. According to this author, deeper the penetration of the more resistant adhesive, greater is the bond, which may even exceed the resistance of the wood. According to the theory of polymer chain diffusion, adhesion occurs through ionic or covalent primary bonds, and/or by secondary intermolecular forces; however, there is no evidence that primary bonds between wood and adhesive occur [37]. After adhesive penetration into the wood, the adhesion is chemically strengthened by attractive intermolecular forces such as Van der Waal forces, dipole-dipole forces and hydrogen bonding [37]. If the polymer chain extends between the molecules of the wood, the adhesion is reinforced by the diffusion theory. Adhesion is a very complex field beyond the reach of any model or theory. In practice, several adhesion mechanisms can occur simultaneously [100].

3.2.1 Factors Influencing the Adhesion Process

The adequate bonding of wood is directly related to various physical-chemical characteristics of the adhesive. These include viscosity, gelatinization time, solids content and pH, and the intrinsic characteristics of the wood, anatomical, physical, chemical and mechanical properties.

Physico-Chemical Characteristics of the Adhesive

The viscosity is one of the most important properties of an adhesive [25], and can be defined as physical property that characterizes the resistance of a fluid to the flow, high viscosity liquids have low fluidity, such as honey, while those already having low viscosity have high fluidity, such as water. According to Peschel et al. [87], this parameter depends on the temperature, decreasing with the elevation of the temperature of the liquid. According to Iwakiri [54] and Gonçalvez and Lelis [45], when the adhesive viscosity is high, the uniform distribution of the adhesive on the wood is difficult, with insufficient penetration into the wood structure, damaging wetting and leaving a thick tail. However, adhesives with low viscosity have higher penetration and their absorption by the wood is also greater, and in extreme situations, can result in excessive absorption of adhesive by the wood [4, 45, 54, 107]. The gel time is important for the quality of the adhesive since it is related to the maximum admissible viscosity for its application. The gel time is measured in seconds, minutes or hours, and corresponds to the period from the preparation of the adhesive to the application to the hardening "point", or gel phase, when it reaches maximum elasticity [20]. In the industrial scope the gel time is a characteristic foundation, since, from this time on, it is no longer possible to manipulate the resin [22].

In general, the working time of the adhesives should not be very long, as it would require a longer pressing time. However, the short working time results in the difficulty of applying and spreading the adhesive in the wood, due to its rapid polymerization, causing a decrease in the strength of the glue line [22, 54]. The content of solids corresponds to the number of solids contained in the adhesive, which is composed of solid components and volatile liquids. When the panel is subjected to hot pressing, evaporation of the liquid components occurs, which is called "cure", that is, solidification of the adhesive, forming the glue line that is responsible for the bond between the substrates [54]. The pH, hydrogen potential, of an aqueous solution is defined as the concentration of dissociated H^+ ions [97] and its determination is made by direct reading in apparatus called pH meters. In the case of bonding of wood, it is important to consider the influence pH of both wood and resin [54]. According to Wang et al. [114, 115] and Wang et al. [113] the pH of the adhesives should not exceed the range of 2.5–11, because beyond these limits the resin causes degradation of the fibers of the wood.

Intrinsic Characteristics of Wood

The anatomical structure is very diversified, especially in hardwood species, composed of cellular elements that are arranged in various ways to constitute the wood [30]. This cellular organization depends on the botanical species, the age of the plant and the environmental conditions in which it develops [17]. In addition, each cell element has a characteristic of the shape and dimensions being linked to the genotypic characteristics of the species, function of the cellular element and phylogenetic position [110]. The anatomical properties of the wood have a significant influence on the bonding, such as the variability in density and porosity that occurs in early wood (also called spring wood → less dense due to larger cells and thinner cell walls) and latewood (also called summer wood → produced in spring and later), core and sapwood, juvenile and adult wood. Also, the influence of the dimensional instability of the reaction wood, as well as the direction of the grain, in which the penetrability relates to the cutting direction [3] is highlighted.

For the production of particleboards and fiber panels, the wood density is a very important factor since it is related to the compaction ratio of the panel. The compaction ratio and indicates the degree of densification of the wood particles in the panel structure and will affect the properties and qualities of the wood [107]. According to these authors, the compaction ratio should be in the range of 1.3–1.6 so that proper densification and consolidation of the panel in the desired final thickness occurs.

According to Thoemen et al. [107], the characteristics of the growth rings, heartwood and sapwood, tree age, porosity, reaction wood and angle of inclination of the cellular elements are favourably or unfavourably involved, since they may or may not favour the bonding process. In the process of bonding, several adhesion mechanisms occur simultaneously in the wood. The influence of the wood anatomy on the bonding process is related to its structure with respect to the differences of dimensions of the cellular elements, size, disposition, and frequency of the cellular cavities that, in turn, are related to the porosity and permeability of the wood [2]. The interaction between adhesive and substrate occurs mainly by vessels and voids, but there is an effective participation of the rays and, to a lesser extent, the axial parenchyma in this process [2, 103]. Figure 6 shows the glue line adhesive penetration of a plywood panel.

The pH of the wood varies according to the species and is around 3–6, and there may be changes of pH inside a piece of wood as a result of the migration of extractives from lower layers to superficial layers, altering the bonding conditions [54]. The capability is a characteristic of the adhesive and refers to its ability to tolerate contact with more acidic or more basic materials without altering its pH [89]. Some woods may present extractives with pH that inhibit the hardening of the adhesive, impairing the development of resistance and adequate cohesion in the glue line. While in some woods, the pH may favour the pre-hardening of the adhesive, impeding the movement and mobility functions, such as the fluidity, penetration, and wetting of the adhesive in the wood [67]. The most important physical properties of wood in terms of bonding are the density and moisture content of the wood. Antagonistic to porosity and the penetration of adhesives, the density can cause significant effects on adhesion. In low-density woods, there is greater penetration of the adhesive and may result in greater consumption of adhesive. In the case of high-density wood, there are larger dimensional changes

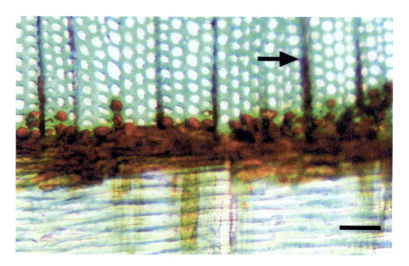

Fig. 6 Penetration of adhesives on the various types of wood cells. Reproduced from [103] with the kind permission of the publishers

resulting from variations in moisture content, generating higher glue line stresses, making the glueing process difficult [54, 77].

The influence on the moisture content of the wood in the cure of the adhesives is related to the amount and the rate of absorption of liquid adhesive by the wood, the lower the moisture content in the wood, the higher the rate of absorption, cure rate and solidification of the wood adhesive [54, 89]. According to Almeida [4], good adhesion between adhesive and wood is obtained provided the moisture content of this adhesive should not exceed 20%. In the wood/glue system, the tensions generated in the bonded product are of extreme importance in the general balance of the resistance. The greater the resistance of the glue line in relation to the strength of the wood, the greater the percentage of rupture or faults in the wood in the interface with the glue line [11]. The tension generated in the glue line is manifested by shear stresses in the plane of the adhesive bond and in the direction perpendicular to it [54].

3.3 Adhesive Additives

In the search for improvements in the adhesion process, specific properties in the panels or reduction of the cost or consumption of adhesives in the industry, additives are commonly used in the tailings formulations. Among the main additives is hardening accelerating agents one generally one comes across. Adhesive curing is influenced by the pH of the environment, in that sense pH modifiers may be employed to promote an acid or alkaline environment, depending on the type of adhesive used. Typically, with UF and melamine-urea-formaldehyde adhesives that cure in an acid environment, ammonium sulfate is used because of the fact that in addition to promoting proper pH for a cure, it leads to the formation of a less hydrolyzable microstructure [84]. For the PF adhesive, sodium hydroxide is used as a hardening accelerator because curing occurs in the basic media [32]. Types of the addition of some reagents and their effects on the final characteristics of paper are listed below:

- Paraffin emulsion: This is used to control the dimensional variation of the panels in the short term, preventing the entry of water into the liquid form by capillarity [19, 117].
- Preservatives: The insecticides and fungicides are commonly used to increase the durability of the panels. Some researchers have already employed the use of nanoparticles loaded with biocides, such as tebuconazole or chlorothalonil, aiming at the slow release of these preservatives [26, 69].
- Filler materials: These are non-stickable materials that are added to the adhesive in order to increase the total volume of the adhesive and reduce the cost without affecting the viscosity of the adhesive. These act as penetration controlling agents, avoiding excessive adhesive penetration in the case of the production of laminated panels; however excess extender may impair adhesion. Kaolin, nut shell flour, coconut husk flour are other commonly used filler materials [54].

Improvements in the rheological properties of adhesives increased mechanical and moisture resistance, reduction of formaldehyde emissions and lower production costs of wood panels are some of the goals that can be achieved with the use of additives. In this context research on the use of nanocelluloses as adhesive additives has shown promising results [6, 38, 111]. These are discussed in the next Section.

4 Nanocellulose

Nanocellulosic materials can be extracted by different methods from different plant biomasses [29, 1]. According to Fujisawa et al. [40], the nanocelluloses can be divided into three groups, viz. Nanocrystalline cellulose (CNC), microfibrillated cellulose (CMF) and nanofibrillated cellulose (CNF). The first CNC is a highly crystalline material with free of defects. This is extracted by the hydrolysis of the amorphous regions present along the axis of cellulosic fibers, by means of a chemical process of acid or enzymatic hydrolysis followed by mechanical agitation of the suspension of nanocrystals in water. The most common nanocellulose production process is by acid hydrolysis, while the most commonly used acid being sulfuric acid. Nitric acid, hydrochloric acid, phosphoric acid and hydrobromic acid can also be used in acid hydrolysis, although on a smaller scale [96, 120].

The second one CMF, obtained by a method of the mechanical disintegration of the cellulosic pulp in water. Finally, the third NCFs are extracted laterally in its nanoscale substructural units (nanofibrils) using combined processes of chemical oxidation with the reagent 2,2,6,6-tetramethylpiperidine-1-oxy, followed of mechanical disintegration in water, or only by the method of mechanical disintegration. It may be noted that the process of obtaining CMF and CNF is similar, differing only in the final dimensions after the processing of the cellulose [99]. According to Samyn et al. [99], the CMF is commonly produced by homogenization, where the fiber shear is performed by a strong pressure drop and impact forces inside the processing chamber. A similar effect has been observed by the use of grinding processing [66], where processed suspensions generally contain a heterogeneous mixture of CMF and CNF which are characterized by different diameters and aspect ratios (length/diameter) [99]. The different aspects of the nanocelluloses described above are shown below in Fig. 7.

When isolated and prepared, CNCs have excellent physicomechanical, optical, magnetic, electrical and conductimetric properties, covering a wide range of uses, different from those obtained by materials seen on a macroscopic scale. The advantages of CNC are related to its properties, such as high mechanical strength and stiffness, low density, durability, uniform size distribution, high specific surface area, low coefficient of thermal expansion, high hydrophilicity, optical transparency and self-molding that enable them to be used in a variety of uses [125, 98]. Due to its crystalline arrangement, this form of nanocellulose has a high mechanical resistance, the modulus of elasticity being estimated between 50 and 145 GPa [106, 64] and comparable to the resistance of extremely rigid materials [29] indicating its

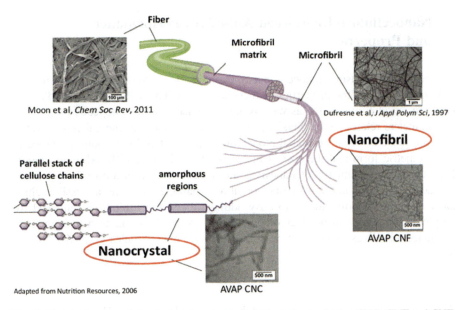

Fig. 7 The mechanism of chemical and mechanical methods for producing CNC, CMF and CNF from cellulose. Reproduced from [79] with the kind permission of the publishers

significant potential as reinforcement material, important, for example, in the automotive industry. However, herein the reported results in the literature are presented in this Chapter.

According to some researchers working on MFCs, they consist of a material obtained by the disintegration of the cellulose subjected to a mechanical process of homogenization, where it is degraded, promoting the exposure and opening of the surfaces previously located inside fibers, fibrils and microfibrils [109] cited by [63, 99, 105]. This process causes an increase in the external surface, allowing a greater area of contact and better bonding between microfibrils, increasing the resistance properties, with a value of modulus of elasticity of 145–150 GPa [55]. On drying, films with lower opacity, high density and transparency would be produced [56]. Reported definition of MFC is fibers with a diameter between 25 and 100 nm, while CNF is nanocelluloses with a diameter between 5 and 30 nm and of a variable length between 2 and 10 μms [96, 101]. Both CMF and CNF have amorphous and crystalline zones composing their structure. On the other hand, CNC refers to cellulose nanoparticles that underwent hydrolysis under controlled conditions and that lead to the formation of structures in the form of small crystalline cylinders and depending on the source of extraction has a diameter of 3–50 nm [101].

Further detailed information on the production of wood pulp and nanocellulose can be found in the following sections: 3.1 Pulping, 4: Cellulose, 5: Nanocellulose, 5.1 Method of NFC and MFC production and 5.2 Method of CNC and MFC production in the chapter on Nanocellulose in paper making in this volume.

5 Nanocellulose-Reinforced Adhesives Performance and Properties

The possibility of using nanocellulose in the adhesive formulation can be a way of promoting gains in the properties of these adhesives. This is because, the adhesion between wood components, as well as among other materials, depends on a series of parameters related to the physicochemical characteristics of the adhesive and the material to be bonded, besides the operational parameters in the bonding process, the geometric form and the size of the pieces to be bonded [29]. Although modifications in the chemistry of adhesives are a path of optimization of mechanical characteristics, the addition of filler or fibrous particles presents a possible alternative route of modification [43]. By adding fillers, the limitations imposed by polymer chemistry can be overcome, and this approach is common in high- performance adhesives, which can be reinforced with nanoparticles [122, 59, 58, 91, 92, 118].

5.1 Effects of the Addition of Nanocellulose on Adhesives

The addition of nanocellulose to the adhesives affects the physicochemical properties of the adhesives, except for the unchanged pH. In general, the percentage increase of nanocellulose in the glue causes an increase in the viscosity and the solids content. Then, the gel time is the property that varies most with this additive, since the partially acidic load in the case of CNC can delay or delay the curing, according to the type of resin to be used.

Damásio et al. [23] evaluated the addition of CNC in the glue mixture of formaldehyde urea glue observed that only the viscosity showed variation, increasing with the increasing percentage of addition in the glue mixture. A similar result has also been reported recently by Ferreira [31], who observed that the increase in the percentage of CNC added (0–8%) in the synthesis of UF adhesives provided an increase in their viscosity, solids content and gelatinization time, in comparison to the synthesis of UF without the addition of CNC. According to the author, the viscosity and solids content gains were marked with a CNC addition up to 6%, although only physical interactions were observed. That is, there was no chemical interaction of the CNC with the other elements present in the adhesive. Similar results were also reported by Liu et al. [68] while evaluating the addition of CNC in lignin PF adhesives. The author has concluded that there was no chemical reaction between the adhesive and CNC since the DSC curves presented only a similar peak at all the compositions and that the addition of the nanocellulose did not affect the energy of the adhesive. Mahrdt et al. [72] found that the addition of CMF in the glue bead of the UF resin caused a delay in the formation of the chemical and mechanical bonding of the resin curing, in addition to increasing the viscosity of the adhesive. However, the addition of CMF allowed better distribution of the adhesive in the wood, with less formation of clots of glue, which do not contribute to the adhesive bond, besides presenting the same penetration in the wood.

Gindl-Altmutter and Veigel [43] contend that the adhesive cure is not excessively prolonged due to the presence of nanocellulose. On the other hand, the severe increase of the viscosity caused by the addition of nanocellulose can represent a serious obstacle for resin spraying and impregnation in the wood. Cardoso et al. [16] have evaluated the addition of CNC produced without the neutralization of the surface charge on the UF glue mixture. They observed that acidic nanocellulose reduced the curing time of the UF resin. Cui et al. [21] produced CNF-reinforced particle boards in tannin-based adhesives. They observed increased gelatinization time and the viscosity of the glue mixture with the addition of CNF (1–3%).

Zhang et al. [123] modified the CNC with 3-aminopropyltriethoxysilane (APTES) and 3-methacryloxypropyltrimethoxysilane (MPS) and evaluated bond strength as well as formaldehyde emission from compensated panels. The results showed higher efficiency for modification with APTES, where 1.5% of modified CNC reduced formaldehyde emission by 53.2% and increased binding resistance by 23.6%, while for MPS modification, the results were 21.3 and 7.0%, respectively.

Ayrilmis et al. [8] used CMF produced from pine sawdust and evaluated the emission of volatile organic compounds (VOC) at different temperatures in LVL panels. The addition of CMF was efficient to reduce the emission up to 35 °C and could be used for furniture for internal use.

Hu et al. [50] have reported that hydrogen bonds between the –OH ends of nanocrystals promote an increase in the frequency and number of hydrogen bonds between UF-nanocrystalline cellulose (UF-CNC) and UF-CNC-wood. In addition, they also promote a higher frequency of the effective hydrogen bonds during the polymerization of the adhesive, allowing gains in resistance of the panels. An American patent [12] claims the use of only nanofibrillated cellulose as an adhesive for the production of particleboard panels. The advantages go beyond the non-emission of VOC, but also advantages in the carbon fixation by the trees producing cellulose used for CNF production.

5.2 Wood Composites with Nanocellulose-Reinforced Adhesives

Damásio et al. [23] have evaluated the addition of CNC in the glue mixture of formaldehyde urea glue. They have observed that the strength of the glue line of dry compaction panels increased with increasing percentage of CNC in the adhesive, where the maximum CNC addition (8%) resulted in an increase of 56% when compared to the control. This shows that the nanocrystals increase the wood-adhesive-wood bonding and interaction [50]. For wet strength, all additions increased this property in relation to the control, but the highest gain occurred with the addition of 2% of nanocellulose. Eichhorn et al. [29] found significant gains when 5% of CNF was added to UF adhesive for the production of bonded joints. In another work, the addition of CNF from 0.5 to 5% UF resin allowed a significant gain in stress and strength until composite failure [29]. According to these authors, the UF adhesive belongs to a class of low-priced, widely used wood adhesives, which are well known for their pronounced fragility and their tendency to develop microcracks that limit their mechanical performance. In addition, the UF adhesive is less moisture-resistant due to the reduction of the molar ratio urea: formaldehyde, leaving free urea groups that bind to water molecules [42]. The addition of CNC contributed to the improvement of the two weaknesses of this type of adhesive, both the resistance of the glue line to the dry and its resistance to moisture [23].

Cardoso et al. [16] have evaluated the addition of CNC in the UF resin and observed a reduction in the swelling in thickness, compared to the panel produced with ammonium sulfate only. However, the water absorption was higher for the panels with the addition of nanocellulose.

Eichhorn et al. [29] have stated that the research group led by Wolfgang Gindl and Josef Keckes has been investigating the reinforcement of adhesives with nanocellulose. To this end, the group tested the addition of CMF and hardwood fibers as resin reinforcement. By adding 5% of untreated pulp fibers and found no significant effect on the value of shear strength of 9.9 MPa. In stark contrast, the addition of 5% nanocellulose, which resulted in a significant increase in shear strength to a value of 13.8 MPa. The researchers justify this increase in the absence of cracks commonly observed in UF glue lines. In addition, they claim that the improvement of the properties of UF resin can open doors to panels of structural uses.

The addition of 2% CNF (m/m) in tannin-based adhesives significantly have been found to improve the mechanical properties of the wood panels produced, since the water absorption did not change significantly [21] In adhesives based on polyurethane or isocyanate, the CNF without chemical modification cannot be dispersed due to the high polarity. For addition of CNF in polyurethane-based resin Richter, et al. [94] did not find significant gains, while some others such as López-Suevos et al. [71] and Kaboorani et al. [57] have observed significantly improved binding strengths and durability of the panels by adding CNF and CNC.

In addition, the CNC increased the dry strength [57] and the CNC increased the resistance to wet and high temperature [71]. Cellulose nanofibrils obtained by high-pressure homogenizer mill added to the UF and melamine urea formaldehyde resins used in the production of agglomerate and OSB showed an increase in their mechanical properties. The improvement was significant mainly for OSB, not as pronounced for MDP [111]. However, in the case of physical properties, there was a reduction in swelling after 24 h in contact with water, and significantly increased internal bonding, flexural strength, and rupture, with the most significant results for OSB panels [111]. This result shows that the combination of larger particles with the enhanced MUF resin contributed to this significant gain. For the addition of CNC as reinforcement in lignin PF resins, Liu and collaborators [68] concluded that the best properties of dry and wet tested glue lines occurred with the addition of 0.25–0.5% CNC.

New MDF panels produced by the addition of 1 and 3% nanocellulose produced from old MDF panels using 8 and 12% new fibers some promising results have been observed [102]. The results showed that values of the highest modulus of rupture (14.47 MPa) and modulus of elasticity (1359.09 MPa) for the panels produced with 12% of glue and 3% of nanocellulose, as well as the highest internal bond strength (0.5 MPa), the lowest swelling in thickness (4.72 and 9.86% after 2 and 24 h on water immersion) and the lowest water absorption (33.11 and 80.63% after 2 and 24 on water immersion).

In the production of particleboard panels and bonded sheet joints, Ferreira [31] has stated that due to the best adhesive and resistance properties observed in the particleboard panels prepared using 4% CNC added UF adhesive, these panels have already been used for applications in glue joints. But, the physical and mechanical results were inferior to those obtained for panels produced with commercial UF adhesive.

In addition to CNF as reinforcement in polyurethane adhesives and water-based polyvinyl acetate (PVAc) adhesives, the good rheological stability of the mixture without the CNF sedimentation after a long preparation time has been reported [94]. However, despite the increase in mechanical properties, the authors believed that new research applying the superficial chemical modification of the CNF should be performed to obtain improvements in properties and justify the industrial applications. Similarly, López-Suevos et al. [71] while producing CNF films with PVAc-latex have observed significant improvements in the storage modulus and thermal properties. Atta-Obeng [6] have found improved shear properties of MDP panels with the addition of MCC in the proportions of 0–10% in PF adhesives; however, the static bending strength was impaired. The authors believe that the presence of MCC resulted in a less pronounced spring back effect during hot-pressing. As pressure decreased, the authors have observed the spring effect occurred resulting in an increase in the thickness of the panel, less interaction of wood particles and adhesive, which resulted in the drop in static bending of the panels.

6 Final Considerations

Cellulose is considered the most abundant renewable polymer on the planet, has many advantages such as biodegradability and low cost and the products obtained from cellulose have wide application, especially in paper production. However, because it is a renewable and widely available resource, there is growing interest in the application of cellulose as an additive in activities with more advanced technologies that use nanotechnology for product development. Within the constant search for better performance of adhesives, the use of nanocelluloses appears as a viable option. The benefits of using nanocelluloses as reinforcements in adhesives for the production of reconstituted wood panels include: the possibility of altering the properties of adhesives, gain in mechanical and physical properties of panels and reduction in formaldehyde emissions by panels using synthetic adhesives. However, despite all the advantages mentioned above, there are still some points to be considered. Therefore, it is concluded that more research needs to be done, either in the application of nanocellulose and its modification in different types of resin, as well as application technologies appropriate to the new conditions of the adhesives.

Acknowledgements At the outset, the authors express their sincere thanks to the Editors of the book (Inamuddin, Sabu Thomas, Raguvendra Mishra and Abdullah M. Asiri), particularly Prof. Inamuddin for inviting us to contribute this Chapter. The authors place on record and appreciate the kind permission given by some of the authors (who have given permission to use their figures), M/s. Elsevier Inc Publishers, IN TECH d.o.o., Rijeka (Croatia), Iran Polymer and Petrochemical Institute with the scientific cooperation of Iran Polymer Society, Royal Society of Chemistry, UK, Chemical Retrieval on the Web (CROW), Springer and Wiley Publishers to reproduce some of the figures from their publications free of charges. One of the authors (KGS) would like to thank the PPISR, Bangalore-India with whom he is associated with presently for their encouragement and interest in this collaboration.

References

1. Abdul KHPS, Tye YY, Leh CP et al (2018) Cellulose reinforced biodegradable polymer composite film for packaging applications. In: Jawaid M, Swain S (eds) Bionanocomposites for packaging applications. Springer, Cham, pp 49–64
2. Albino VCS, Mori FA, Mendes LM (2012) Influence of anatomical features and extractives content wood of *Eucalyptus grandis* w. hill ex maiden in quality bonding. Cienc Florest 22(4):803–811
3. Albuquerque CEC, Latorraca JV (2000) Anatomic features, influence in penetration and adhesion of adhesives. Floresta Ambient 7(1):158–166
4. Almeida VC (2009) Assessment of the potential for the use of tropical wood waste for the production of laterally glued panels—EGP. Federal University of Parana, Thesis
5. American Society for Testing and Materials (2006) ASTM D 5456: standard specification for evaluation of structural composite lumber products. ASTM, West Conshohocken
6. Atta-Obeng E (2011) Characterization of phenol formaldehyde adhesive and adhesive-wood particle composites reinforced with microcrystalline cellulose. Dissertation, Auburn University
7. Ayrilmis N (2007) Effect of panel density on dimensional stability of medium and high density fiberboards. J Mater Sci 42:8551–8557
8. Ayrilmis N, Lee Y-K, Kwon JH et al (2016) Formaldehyde Emission and VOCs from LVLs Produced with Three Grades of Urea-Formaldehyde Resin Modified with Nanocellulose. Build Environ 97:82–87
9. Baghersad S (2016) Coating os silk fabrics by PVA/Ciprofloxain HCl nanofibers for biomedical applications. Iran J Polym Sci Tech 29(2):171–184
10. Baldwin RF, Kurpiel FT, Baldwin RW (2017) Growth and reinvention 2017: a north american perspective on the global wood-based panel industry. Forest Prod J 67(3–4): 144–151
11. Bianche JJ (2014) Wood-adhesive interface and joints' resistance bonded with different adhesives and weight. Federal University of Viçosa, Thesis

12. Bilodeau MA, Bousfield DW (2015) Composite building products bound with cellulose nanofibers. Patent US 20,150,033,983 A1, 05 Feb 2015
13. Buligon EA (2015) Physical and mechanical properties of laminated veneer lumber reinforced gfrp. C Fl 25(3):731–741
14. Campos CI (2005) Physical-mechanical properties of MDF produced with wood fibers from reforestation and alternative adhesives at different levels. University of São Paulo, Thesis
15. Candan Z, Akbulut T (2015) Physical and mechanical properties of nanoreinforced particleboard composites. Maderas Cienc Tecnol 17(2):319–334
16. Cardoso GV, Pereira FT, Ferreira ES et al (2016) Nanocelulose occmo urea-formaldehyde catalyst for the production of agglomerated panels of *Pinus* sp. In: Paper presented at the XV EBRAMEM—Brazilian meeting on timber and timber structures, Brazilian Institute of Wood and Wood, Curitiba, Structures, Curitiba, 9–11 Mar 2016
17. Carlquist S (2001) Comparative wood anatomy. Springer, Berlin
18. Carvalho MZ (2016) Multivariate approach to the behavior of physical-chemical properties and characterization of natural adhesives based on tannins. Federal University of Lavras, Thesis
19. Carvalho L, Martins J, Costa C (2010) Transport phenomena. In: Thoemen H, Irle M, Sernek M (eds) Wood-based panels: an introduction for specialists. Brunel University Press, London, pp 123–295
20. Costa TG (2016) Characterization of synthetic adhesives with addition of silica nanoparticles as reinforcing filler. Federal University of Lavras, Thesis
21. Cui J, Lu X, Zhou X et al (2014). Enhancement of mechanical strength of particleboard using environmentally friendly pine (*Pinus pinaster* L.) tannin adhesives with cellulose nanofibers. Ann For Sci 72(1):27–32
22. Cunha RCB (2016) Implementation of a method for measuring Gel Time of formaldehyde-based resins. Dissertation, Higher Institute of Engineering of Porto
23. Damásio RAP, Carvalho FJB, Carneiro ACO et al (2017) Effect of CNC interaction with urea-formaldehyde adhesive in bonded joints of *Eucalyptus* sp. Sci For 45(113):169–176
24. Diem H, Mathias G, Wagner RA (2012) Amino resins. Ullmann's Encyclopedia of Industrial Chemistry, Wiley-VCH, Weinheim
25. Din Z-U, Xiong H, Wang Z et al (2018) Effects of different emulsifiers on the bonding performance, freeze-thaw stability and retrogradation behavior of the resulting high amylose starch-based wood adhesive. Colloids Surf A 538(5):192–201
26. Ding X, Richter DL, Matuana LM et al (2011) Efficient one-pot synthesis and loading of self-assembled amphiphilic chitosan nanoparticles for low-leaching wood preservation. Carbohydr Polym 86:58–64
27. Ebnesajjad S, Landrock AH (2014) Adhesives technology handbook, 3rd edn. Elsevier, Amsterdã
28. Eckelman CA (1999) Brief survey of wood adhesives. Purdue University Cooperative Extension Service, West Lafayette
29. Eichhorn SJ, Dufresne A, Aranguren M et al (2010) Review: current international research into cellulose nanofibres and nanocomposites. J Mater Sci 45(1):1–33
30. Esteban L, Casasús AG, Oramas CP et al (2003) Wood and its anatomy. Fundación Conde de Valle de Salazar, Madrid
31. Ferreira JC (2017) Synthesis of urea-formaldehyde adhesives with the addition of kraft lignin and nanocrystalline cellulose. Federal University of Viçosa, Thesis
32. Fink J (2013) Reactive polymers Fundamentals and applications—a concise guide to industrial applications, 2nd edn. William Andrew, Norwich
33. Finnish Forest Industries Federation (2002) Handbook of finnish plywood. Kirjapaino Markprint Oy, Lahti
34. Fiorelli J (2002) Use of carbon fibers and glass fibers to reinforce wooden beams. Dissertation, São Paulo University
35. Food and Agriculture Organization (2018) Global production and trade of forest products in 2016. http://www.fao.org/forestry/statistics/80938/en/ Accessed 12 Mar 2018

36. Forestry Products Laboratory (1999) Wood handbook—wood as an engineering material. General Technical Reports FPL-GTR-113. USDA, Forest Service, Madison
37. Forestry Products Laboratory (2010) Wood handbook—wood as an engineering material. General Technical Reports FPL-GTR-190. USDA, Forest Service, Madison
38. Forestry Products Laboratory (2012) Nanocelluloses: potential materials for advanced forest products. In: Proceedings of nanotechnology in wood composites symposium. General Technical Reports FPL-GTR-218. USDA, Forest Service, Madison
39. Fratzl P, Weinkamer R (2007) Nature's hierarchical materials. Prog Mater Sci 52(8):1263–1334
40. Fujisawa S, Okita Y, Fukuzumi H et al (2011) Preparation and characterization of TEMPO-oxidized cellulose nanofibrils films with free carboxyl groups. Carbohydr Polym 84(1):579–583
41. Gardziella A, Pilato LA, Knop A (2000) Phenolic resins: chemistry, applications, standardization, safety and ecology, 2nd edn. Springer, Heidelberg
42. Gavrilovic GI, Neskovic O, Diporovic MM et al (2010) Molar-mass distribution of urea-formaldehyde resins of different degrees of polymerisation by MALDI-TOF mass spectrometry. J Serb Chem Soc 75(5):689–99
43. Gindl-Altmutter W, Veigel S (2015) Nanocellulose-modified wood adhesives. In: Oksman K, Mathew AP, Bismarck A et al (eds) Handbook of green materials. World Scientific Publishing, Hackensack, pp 253–264
44. Gonçalvez FG (2012) Agglomerated panels of *Acacia mangium* wood with urea-formaldehyde adhesives and powdered tannin of Acacia mearnsii bark. Rural Federal University of Rio de Janeiro, Thesis
45. Gonçalvez FG, Lelis RCC (2009) Properties of two synthetic resins after addition of Modified tannin. Floresta Ambient 12(2):01–07
46. Grigsby WJ, Thumm A (2012) The interactions between wax and UF resin in medium density fiberboard. Eur J Wood Wood Prod 70(4):507–517
47. Gupta R, Kandasubramanian B (2015) Hybrid caged nanostructure ablative composites of octaphenyl-POSS/RF as heat Shields. RSC Adv 5:8757–8769
48. Haubrich JL, Gonçalves C, Tonet A (2007) Vinyl adhesives present solutions for wood. Rev Mad 103:66–70
49. Hellmeister V (2017) OSB panel of raft wood residue (Ochroma pyramidale). University of São Paulo, Thesis
50. Hu K, Kulkarni DD, Choi I et al (2014) Graphene-polymer nanocomposites for structural and functional applications. Prog Polym Sci 39(11):1934–1972
51. International Agency for Research on Cancer (2006) Formaldehyde, 2-butoxyethanol and 1-tertbutoxypropan-2-ol. IARC Monogr Eval Carcinog Risks Hum 88:1–478
52. International Agency for Research on Cancer (2012) Chemical agents and related occupations: a review of human carcinogens. IARC Monogr Eval Carcinog Risks Hum 100:1–628
53. Irle M, Barbu C (2010) Wood-based panel technology. In: Thoemen H, Irle M, Sernek (eds) Wood-based panels: an introduction for specialists. Brunel University Press, London, pp 1–94
54. Iwakiri S (2005) Painéis de madeira reconstituída. FUPEF, Curitiba
55. Iwamoto S (2009) Elastic modulus of single cellulose microfibrils from tunicate measured by atomic force microscopy. Biomacromol 10(9):2571–2576
56. Jonoobi M, Mathew AP, Oksman K (2012) Producing low-cost cellulose nanofiber from sludge as new source of raw materials. Ind Crops Prod 40:232–238
57. Kaboorani A, Riedl B, Blanchet P et al (2012) Nanocrystalline cellulose (NCC): a renewable nano-material for polyvinyl acetate (PVA) adhesive. Eur Polym J 48(11):1829–1837
58. Khalili SMR, Jafarkarimi MH, Abdollahi MA (2009) Creep analysis of fibre reinforced adhesives in single lap joints-experimental study. Int J Adhes Adhes 29(6):656–661
59. Khalili SMR, Shokuhfar A, Hoseini SD et al (2008) Experimental study of the influence of adhesive reinforcement in lap joints for composite structures subjected to mechanical loads. Int J Adhes Adhes 28(8):436–444

60. Khedari J, Nankongnab N, Hirunlabh J et al (2004) New low-cost insulation particleboards from mixture of durian peel and coconut coir. Build Environ 39(1):59–65
61. Kim MG (2000) Examination of selected synthesis parameters for typical wood adhesive-type urea-formaldehyde resins by ^{13}C NMR spectroscopy. I. J Appl Polym Sci 75(10):1243–1254
62. Kinloch AJ (1987) Adhesion and adhesives: science and technology. Chapman & Hall, London
63. Kolakovic R, Peltonel L, Laaksonen T et al (2011) Spray-dried cellulose nanofibers as novel tablet excipient. AAPS Pharm Sci Tech 12(4):1366–1373
64. Lahiji RR, Xu X, Reifenberger R, Raman A, Rudie A, Moon RJ (2010) Atomic Force Microscopy Characterization of Cellulose Nanocrystals. Langmuir 26(6):4480–4488
65. Lam F (2001) Modern structural wood products. Prog Struct Eng Mat 3(4):238–245
66. Lengowski EC (2016) Formation and characterization of films with nanocellulose. Federal University of Paraná, Thesis
67. Lima CKP, Mori FA, Mendes LM et al (2007) Anatomic and chemical characteristics of eucalyptus clones wood and its influence upon bonding. Cerne 13(2):123–129
68. Liu Z, Zhang Y, Wang X et al (2015) Reinforcement of lignin-based phenol-formaldehyde adhesive with nano-crystalline cellulose (NCC): curing behavior and bonding property of plywood. Mater Sci Appl 6:567–575
69. Liu Y, Laks P, Heiden P (2002) Controlled release of biocides in solid wood. II. Efficacy against *Trametes versicolor* and *Gloeophyllum trabeum* wood decay fungi. J Appl Polym Sci 86(3):608–614
70. Lubis MAR, Hong MK, Park BD (2017) Hydrolytic removal of cured urea–formaldehyde resins in medium-density fiberboard for recycling. J Wood Chem Technol. https://doi.org/10.1080/02773813.2017.1316741
71. López-Suevos F, Eyholzer C, Bordeanu N et al (2010) DMA analysis and wood bonding of PVAc latex reinforced with cellulose nanofibrils. Cellulose 17(2):387–398
72. Mahrdt E, Pinkl S, Schmidberger C et al (2016) Effect of addition of microfibrillated cellulose to Ureaformaldehyde on selected adhesive characteristics and distribution in particle board. Cellulose 23(1):571–580
73. Meng Q-X, Zhu G-Q, Yu M-M et al (2018) The effect of thickness on plywood vertical fire spread. Procedia Eng 211:555–564
74. Messmer A (2015) Life cycle assessment (LCA) of adhesives used in wood consructions. Master thesis (Ecological System Design), Swiss Federal Institute of Technology Zurich, Swiss, Zurich p 82
75. Molina JC, Calil Neto C, Calil Junior C et al (2013) Evaluation of the behavior of rectangular beams (LVL) with horizontal and vertical lamination. Mad Arq Eng 14(35):1–13
76. Mondragon G, Peña-Rodriguez C, Gonzáles A et al (2015) Bionanocomposites based on gelatin matrix and nanocellulose. Eur Polym J 62:1–9
77. Motta JP, Oliveira JTS, Alves RC (2012) Influence of moisture content on the adhesion properties of eucalyptus wood. Construindo 4(2):96–103
78. National Institute of Industrial Research (2017) The complete technology book on wood and its derivatives. NIIR, Delhi
79. Nelson K, Restina T, Iakovlev M et al (2016) American process: production of low cost nanocellulose for renewable, advanced materials applications. In: Madsen L, Svedberg E (eds) Materials research for manufacturing. Springer Series in Materials Science, vol 224. Springer, Cham
80. Nguyen DM, Grillet A-C, Diep TMH et al (2018) Influence of thermo-pressing conditions on insulation materials from bamboo fibers and proteins based bone glue. Ind Crops Prod 111:834–845
81. Nitthiyah A (2013) Optimization and characterization of melamine urea formaldehyde (MUF) based adhesive with waste rubber powder (WRP) as filler. University Malaysia Pahang, Thesis

82. Olorunnisola AO (2018) Design of wood connections. In: Olorunnisola AO (ed) Design os structural elements with tropical hardwoods. Springer, Berlim, pp 209–236
83. Ozarska B (1999) A review of the utilization of hardwoods for LVL. Woood Sci Technol 33(4):341–351
84. Park B-D, Kang E-C, Park S-B et al (2011) Empirical correlations between test methods of measuring formaldehyde emission of plywood, particleboard and medium density fiberboard. Eur J Wood Wood Prod 69(2):311–316
85. Périchaud AA, Isakakov RM, Kurbatov A et al (2012) Auto-reparation of polyimide film coatings for aerospace applications challenges and perspectives. In: Abadie MJM (ed) High performance polymers—polyimides based—from chemistry to applications. InTech, London, pp 215–244
86. Pervan D (2018) Mechanical locking system for panels and method of installing same. US Patent 2018/0,030,738 A1, 1 Feb 2018
87. Peschel P, Hornhardy E, Nennewitz I et al (2016) Tabellenbuch Holztechnik. Europa-Lehrmittel Nourney, Vollmer GmbH & Co. KG, Haan
88. Petrie EW (2000) Handbook of adhesives andokl sealents, 2nd edn. McGraw-Hill, New York
89. Pizzi A (2015) Synthetic adhesives for wood panels: chemistry and technology. In: Mittal KL (ed) Progress in adhesion and adhesives. Wiley, Hoboken, pp 85–126
90. Polymer Properties Database (2015) Melamine-formaldehyde resins. http://polymerdatabase.com/polymer%20classes/MelamineFormaldehyde%20type.html. Accessed 30 Mar 2018
91. Prolongo SG, Gude MR, Ureña A (2009) Synthesis and characterisation of epoxy resins reinforced with carbon nanotubes and nanofibers. J Nanosci Nanotechnol 9(10):6181–6187
92. Prolongo SG, Gude MR, Ureña A (2010) Rheological behaviour of nanoreinforced epoxy adhesives of low electrical resistivity for joining carbon fiber/epoxy laminates. J Adhes Sci Technol 24(6):1097–1112
93. Ramage MH, Burridge H, Busse-Wicher M et al (2017) The wood from the trees: the use of timber in construction. Renew Sustain Energy Rev 68:333–359
94. Richter K, Bordeanu N, Lópes-Suevos F et al (2009) Performance of cellulose nanofibrils in wood adhesives. In: Schindel-Bidinelli E (ed) Proceedings of the swiss bonding. Rapperswil-Jona, Switzerland, pp 239–246
95. Risholm-Sundman M, Larsen A, Vestin E et al (2007) Formaldehyde emission—comparison of different standard methods. Atmospheric Environ 41(15):3193–3202
96. Rojas J, Bedoya M, Ciro Y (2015) Current trends in the production of cellulose nanoparticles and nanocomposites for biomedical applications. In: Polleto M (ed) Cellulose—fundamental aspects and current trends. InTech, Rijeka, pp 193–228
97. Rumble JR (2018) CRC handbook of chemistry and physics, 98th edn. CRC Press, Boca Raton
98. Salajková M, Berglund LA, Zhou Q (2012) Hydrophobic cellulose nanocrystals modified with quaternary ammonium salts. J Mater Chem 22(37):19798–19805
99. Samyn P, Barhoum A, Öhlund T et al (2018) Review: nanoparticles and nanostructured materials in papermaking. J Mater Sci 53(1):146–184
100. Schultz J, Nardin M (2003) Theories and mechanisms of adhesion. In: Pizzi A, Mittal KL (eds) Handbook of adhesive technology. Marcel Decker, New York, pp 61–75
101. Sehaqui H, Allais M, Zhou Q et al (2011) Wood cellulose biocomposites with fibrous structures at micro-and nanoscale. Compos Sci Technol 71(3):382–387
102. Sheykhi ZH, Tabarsa T, Mashkour M (2016) Effects of nano-cellulose and resine on MDF properties produced from recycled mdf using electrolise method. J Wood Forest Sci Technol 23(3):271–288
103. Singh A, Dawson B, Rickard C et al (2008) Light, confocal and scanning electron microscopy of wood-adhesive interface. Microsc Analy 22(3):5–8
104. Song J, Chen C, Zhu S et al (2018) Processing bulk natural wood into a high-performance structural material. Nature 554:224–228

105. Syverud K, Chinga-Carrasco G, Toledo J et al (2011) A comparative study of *Eucalyptus* and *Pinus radiata* pulp fibres as raw materials for production of cellulose nanofibrils. Carbohydr Polym 84(3):1033–1038
106. Tanpichai S, Quero F, Nogi M et al (2012) Effective young's modulus of bacterial and microfibrillated cellulose fibrils in fibrous networks. Biomacromol 13(5):1340–1349
107. Thoemen H, Irle M, Sernek M (eds) (2010) Wood-based panels: an introduction for specialists. Brunel University Press, London
108. Toquarto S (2002) Random heterogeneous materials. Springer, Berlim
109. Turbak AF, Snyder FW, Sandberg KR (1983) Microfibrillated cellulose, a new cellulose product: Properties, uses and commercial potential. J Appl Polym Sci: Appl Polym Symp 37:815–827
110. Urbinati CV (2013) Influence of anatomical characteristics on cast joints of *Schizolobium parayba* var. Amazonicum (hyber ex. Ducke) barneby (Paricá). Thesis, Federal University of Lavras
111. Veigel S, Rathke J, Weigl M et al (2012) Particle board and oriented strand board prepared with nanocellulose-reinforced adhesive. J Nanomater. https://doi.org/10.1155/2012/158503
112. Veronez D, Farias ELP, Fraga R et al (2010) Potential for occupacional health rish for those teachers, researchers and technical workers of anatomy who are exposed to formaldehyde. InterfacEHS 5(2):63–76
113. Wang XM, Casilla R, Zhang Y et al (2016) Effect of extreme ph on bond durability of selected structural wood adhesives. Wood Fiber Sci 48(4):1–15
114. Wang X, Huang Z, Cooper P et al (2010) The ability of wood to buffer highly acidic and alkaline adhesives. Wood Fiber Sci 42(3):398–405
115. Wang X, Huang Z, Cooper P et al (2013) Effects of pH on lap-shear strength for aspen veneer. Wood Fiber Sci 45(3):294–302
116. Wegner T, Skog KE, Ince PJ et al (2010) Uses and desirable properties of wood in the 21st century. J Forest 108(4):165–173
117. Xu X, Yao F, Wu Q et al (2009) The influence of wax-sizing on dimension stability and mechanical properties of bagasse particleboard. Industrial Crops Produ 29(1):80–85
118. Yoon SH, Kim BC, Lee KH et al (2010) Improvement of the adhesive fracture toughness of bonded aluminum joints using e-glass fibers at cryogenic temperature. J Adhes Sci Technol 24(2):429–444
119. Yuce B, Mastrocinque E, Packianather MS et al (2014) Neural network design and feature selection using principal component analysis and Taguchi method for identifying wood veneer defects. Produm Manufac Res 2(1):291–308
120. Zeni M, Favero D, Pacheci K et al (2015) Preparation of microcellulose (Mcc) and nanocellulose (Ncc) from eucalyptus kraft ssp pulp. Polym Sci 1:1–5
121. Zeppenfeld G, Grunwald D (2005) Klebstoffe in der Holz-und Möbelindustrie. DRW-Verlag, Weinbrenner
122. Zhang Y, You B, Huang H et al (2008) Preparation of nanosilica reinforced waterborne silylated polyether adhesive with high shear strength. J Appl Polym Sci 109(4):2434–2441
123. Zhang H, Zhang J, Song S et al (2011) Modified nanocrystalline cellulose from two kinds of modifiers used for improving formaldehyde emission and bonding strength of urea-formaldehyde resin adhesive. BioResources 6:4430–4438
124. Zhong Y, Jing X, Wang S et al (2016) Behavior investigation of phenolic hydroxyl groups during the pyrolysis of cured phenolic resin via molecular dynamics simulation. Polym Degrad Stab 125:97–104
125. Zhou J, Chen J, He M et al (2016) Cellulose acetate ultrafiltration membranes reinforced by cellulose nanocrystals: preparation and characterization. J Appl Polym Sci 133(39):1–7. https://doi.org/10.1002/app.43946

Nanocellulose in the Paper Making

Elaine Cristina Lengowski, Eraldo Antonio Bonfatti Júnior,
Marina Mieko Nishidate Kumode, Mayara Elita Carneiro
and Kestur Gundappa Satyanarayana

1 Introduction

Paper has been defined as a material having two dimensions, which is produced from an aqueous suspension of fibers, which in turn are 'artificially interlaced and subsequently dewatered through mechanical and thermal processes' [79]. It may be noted that the art of producing paper began more than two millennia ago due to the dire need felt at that time to communicate and record the discoveries in materials

---

E. C. Lengowski
Faculty of Forestry Engineering, Federal University of Mato Grosso (UFMT),
Fernando Corrêa da Costa St, 2367, Boa Esperança, Cuiabá, MT 78068-600, Brazil
e-mail: elainelengowski@gmail.com

E. A. Bonfatti Júnior · M. E. Carneiro
Department of Forest Engineering and Technology (DETF), Federal University
of Paraná (UFPR), Av. Pref. LothárioMeissner, 632, Jardim Botânico, Curitiba,
PR 80.210-170, Brazil
e-mail: bonfattieraldo@gmail.com

M. E. Carneiro
e-mail: mayaraecarneiro@gmail.com

M. M. N. Kumode
Laboratory of Wood, Pontifical Catholic University, Curitiba, PR, Brazil
e-mail: mnishidate@gmail.com

K. G. Satyanarayana
PIPE & Department of Chemistry, Federal University of Parana, Curitiba, Brazil

K. G. Satyanarayana (✉)
Poornaprajna Scientific Research Institute (PPISR), Sy. No. 167, Poornaprajnapura,
Bidalur Post, Devanahalli, Bangalore 562 110, Karnataka, India
e-mail: gundsat42@hotmail.com

© Springer Nature Switzerland AG 2019
Inamuddin et al. (eds.), *Sustainable Polymer Composites and Nanocomposites*,
https://doi.org/10.1007/978-3-030-05399-4_36

that can be transported [53]. Of course, today the paper is being used for a lot of diverse uses, which include printing, hygiene, writing, and packaging.

In a simplified way, modern paper production can be divided into three stages: pulping, bleaching of the pulp and the production of the paper itself. It is possible to disregard the bleaching of the pulp while producing brown pulp, which will be used for papers of packaging applications. One of the earliest milestones for the industrial production of cellulosic pulp was the development of the Kraft process in 1879 [63], starting from this, the paper industry has consistently sought to improve the quality of the paper.

It is interesting to note that like in many fields, the paper industry is characterized by investing a lot in research and development. This includes research in search of new raw materials of the high industrial profile, modification and additives for the Kraft process, reduction or non-use of chlorine compounds in bleaching and, more recently, biorefinery. Van Heiningen [167] defines the term 'biorefinery' as an industry that transforms raw materials from renewable sources, such as sugarcane bagasse, wood, forest residues and black liquor into higher value-added products such as biofuels and biomaterials. In this sense obtaining nanocellulose from biomaterials for the most diverse commercial applications fits very well in this concept.

In fact, the credit goes to Wegner et al. [180], who put the use of nanotechnologies in the forest-based industry as one of the main novelties to be developed in the 21st century. Besides, they also suggested two paths for the application of nanotechnology to forest producers, viz., the first path is for nanotechnologies and nanomaterials developed in other industrial sectors to be adopted and deployed in materials, processes, and products used or produced by the forest-based industry. The second path is the development of completely new materials or product platforms using nanoscale structures and properties derived from wood. Although it was still unknown about the exact economic impacts and opportunities for wood as nanomaterials, but it was expected that all nanomaterials and nano-enabled products would grow to exceed one trillion dollars per annum as technology would be developed in the 21st century [64].

While in the last few decades, traditional uses of paper have been found to decline, other new avenues have opened up during the last decades or so. These include incorporation of nanotechnology since the 1990s into papermaking leading to lower energy costs, development of low-cost products with improved paper quality, biocompatible and flexible with sophisticated functionalities [17, 133]. It may not be exaggerating to state that the use of nanotechnology makes it possible to improve the sustainability of paper making processes for the following reasons:

- More efficient use of resources whereby more resistant papers can be formed with the smaller amount of fibers (by weight);
- Use of secondary materials such as recycled fibers or fibers with inferior properties, which can be converted into nanofibers and used as additives in papermaking, or produce high-quality papers using secondary fibers with the addition of nanofibers;

- Development of new materials such as papers with unique properties, such as films with plastic characteristics, good barrier properties for smart packaging, etc.

Considering the above and published reports on the preparation, characterization and various applications of nanomaterials in general and nanocellulose in particular and the latters' use in paper making, this Chapter will present characteristics of the most used wood in the world for pulp and paper production, main methods of obtaining cellulose in nature, process of bleaching of pulp, paper making, processes to obtain different types of nanocellulose (microfibrillar, nanofiber and cellulose nanocrystals), applications of nanocellulose in the paper making, applications of nanocellulose in paper making through coating and films as well as by nanocellulose-reinforced pulp and the resulting effects of the use of nanocellulose in paper production. The Chapter will also present marketing aspects and possible future opportunities and finally concluding remarks.

2 Wood for Pulp and Paper Production

Although a variety of woods are available, only certain types are used in the paper industry. The basic wood density is considered the most important parameter in its quality evaluation since it has a strong relationship with the other wood properties [61]. Besides, it has a strong effect on the variables of the pulping process and the characteristics of paper pulp [138]. This property is defined by the ratio of the absolute dry weight of the wood to its fully saturated volume. The ideal types of cellulosic pulps for papers for printing and writing and for absorbent papers have been distinguished [134]. According to these authors, the most required criteria for printing and writing papers are lower energy consumption in the mechanical refining, greater specific volume and greater opacity. On the other hand, high capacity for water absorption and increased softness are important in the case of manufacturing of absorbent papers. The authors further state that pulps originating from wood with lower basic densities are ideal for the production of the first type of paper since they have fibers with a smaller thickness and smaller mass per length. For the second type of paper, pulps from denser woods are suitable, because these woods possess fibers with higher thickness than the low-density woods and therefore these would present the greater potential of liquid absorption, and greater mass per length of fibers.

In hardwoods, penetration of liquids occurs rapidly through the vessels, but penetration in the transverse direction practically does not exist. This is because of the fact that the pit membranes of the punctures (Depressions in the secondary cell wall is called 'pit') prevent the passage of the cooking liquor (This is a mixture of chemical reagents-NaOH + Na_2S). It may be noted that in the case of softwood the penetration of the cooking liquor already occurs through the tracheids cells that, unlike the cellular elements of the hardwoods and therefore have good permeability, even in the transverse direction [166]. Considering the dimensions of the chips, it is emphasized that the thickness should be in the range of 4–6 mm, in different sizes so that both impregnation and diffusion are compromised and, consequently, the delignification rate is decreased [103]. The wood is chemically constituted by polymers that perform the structural functions. These are cellulose, hemicelluloses, and lignins [49]. In addition to these structural components present in wood, there are other constituents, which include starches, proteins, pectins beside other substances soluble in water or other organic solvents called extractives or accidental compounds of wood. From the chemical point of view, the amount and type of lignin directly interfere with the conditions of the pulping process. In general, lignin is classified according to the relative amount of the monomers guaiacila (G), syringyl (S) and p-hydroxyphenyl (H), derived from coniferyl, synapyl and p-coumarilic alcohols respectively. Structures of these are shown in Fig. 1.

In the *Eucalyptus* wood, lignin is generally formed by the siringila and guaiacila units (SG lignin), while in conifers it is formed by units guaiacila and p-hydroxyphenyl (lignin GH) [15, 49]. The syringyl/guaiac ratio (S/G) significantly affects the degradation and solubilization of lignin from hardwoods, to the point that increasing this ratio allows the use of a lower alkali load, resulting in higher yields [60].

With fibrous structure and length between 2 and 5 mm, the cellulose originated from the conifers, such as species of the genus *Pinus*, is called 'long fiber pulp' and has application in the papers that demand greater resistance, as is the case of paper used in the manufacture of packaging. On the other hand, the short-fiber pulp can range from 0.5 to 2 mm and is produced from hardwoods, such as *Eucalyptus, Acacia, Propulos* and *Betula*, and is used for the production of printing and writing papers and tissue paper for sanitary purposes. Table 1 presents the technological characteristics of the main forest species planted in the world for pulp and paper production.

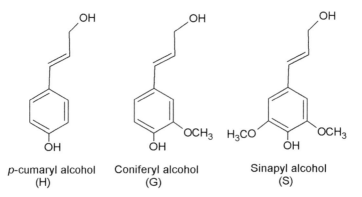

Fig. 1 Precursor alcohols of the phenylprazoid units' p-hydroxyphenyl (H), guaiacyl (G) and syringyl (S). Reproduced from Barbosa et al. [15] with the kind permission of the publishers

Table 1 Characteristics of the most used wood in the world for pulp and paper production

Type of wood	Species	Location	A	BD	TE	TL	HL
Hardwoods [137]	E. grandis x E. urophylla	Brazil	6	0.47	3.1	28.1	68.9
	Eucalyptus globules	Chile	12	0.63	5.6	25.9	68.5
	Eucalyptus nitens	Chile	12	0.52	4.5	27.1	68.4
	Acacia mangiun	Indonesia	6	0.52	5.2	28.0	66.8
	Acacia crassicarpa	Indonesia	6	0.57	4.1	29.4	66.5
	Populus tremulloides	Canada	55	0.37	6.9	22.1	71.0
	Betula pendula	Finland	67	0.50	4.8	17.4	77.8
Softwood [171]	Pinus taeda	Brazil	21	0.43	2.8	26.7	70.5
	Pinus silvestris	Finland	45	0.43	6.4	25.6	68.0

A age in years; BD basic density in g·cm^{-3}, TE total extractives in %; TL total lignin in %; HL holocelullose content in %

3 Paper Making

3.1 Pulping

The production of cellulosic pulp can occur from processes that use different types of energies, viz., mechanical, thermal and chemical or the combination of these. Despite the great diversity of these pulping processes, the alkaline ones have become the main ones used extensively. With a well known and most used worldwide the 'Kraft' process, developed by the German chemist Carl F. Dahl in 1879 [63] and patented in 1884 [150], it is reported that today 90% of the whole cellulose pulp produced in the world comes from this process [111]. The Kraft process was the result of the evolution of the soda process and aimed at dissolving the middle lamella by means of the removal of lignin with consequent individualization of the wood fibers. For this purpose, the wood chips are placed in a digester that is pressurized with the alkaline cooking liquor (NaOH and Na_2S). The hegemony of its use stems from its advantages over other processes and this process is adaptable to different types of lignocellulosic materials, producing high-quality pulp and of high bleachability with high efficiency of recovery of chemical reagents and energy.

It is interesting to note that the increase in the rate of delignification and yield are opposite to each other. This is because the reagents used in the pulping processes are not specific for lignin removal; they also remove carbohydrates, which contribute to a reduction in yield [150]. Advantages, the Kraft process demands precise control of its parameters. There are several parameters that affect the delignification rate, the most important are: type of wood, chip quality, alkaline load, cooking time and temperature.

The impregnation of the cooking liquor in the chips aims to distribute the cooking liquor evenly into the wood [184]. The impregnation consists of two different phases: pore penetration and diffusion [84]. These phases are very important for the efficiency of delignification and are directly related to the quality of wood and chip size.

In pulp obtained by chemical processes, the degree of delignification of the pulp is measured by the kappa number, which expresses the residual lignin present in the pulp after cooking [161] and is dependent on the alkaline load, cooking time and temperature. The alkaline charge is applied proportionally to the amount of wood in the digester by seeking a predetermined target kappa number. According to Almeida [7], the increase of alkaline charge leads to greater delignification with consequent reduction of the kappa number. However, greater delignification promotes the greater generation of fines, possibly resulting from the fragmentation of fibers. Time and temperature have been combined into a single control parameter called 'H factor', which represents the extent of the reaction [172]. Even though different temperatures can be used, delignification can be estimated accurately by 'H factor', provided that the other parameters of the pulping process remain constant. According to Sixta [148] the 'H factor' is defined as follows:

$$H = \int_{t_0}^{t} k_L \cdot dt$$

where k_L is the relative reaction velocity of the pulp. Assuming that the activation energy of the reaction is 134 kJ · mol^1, the H-factor can be expressed as:

$$H = \int_{t_0}^{t} \frac{k_{L(T)}}{k_{100\,°C}} \cdot dt = \int_{t_0}^{t} Exp \cdot \left(43.19 - \frac{16113}{T}\right) \cdot dt$$

where t is time and T is temperature, the above equation is valid for temperatures above 100 °C.

According to Sixta [148], the H factor is the area under the relative reaction velocity curve versus time. This parameter is designed to predict the temperature or cooking time required to reach a given kappa number. The result is valid only when the other cooking conditions, such as the effective alkali concentration and the ratio of liquor to wood, remain constant [148]. Figure 2 shows a graph of a conventional Kraft cooking having a maximum temperature of 170 °C, the heating time of 80 min and time of 60 min at a constant temperature, culminating at an H factor of 1100.

H-factor is a useful process control tool for Kraft pulp industries. Even the most modern plants use this parameter to control the degree of delignification of the pulp [136].

3.2 Bleaching

This process follows the previous one. It should be noted that the pulp obtained as explained in the previous Section is so intense without the modification, which compromises the mechanical strength of the fiber. Accordingly, bleaching process,

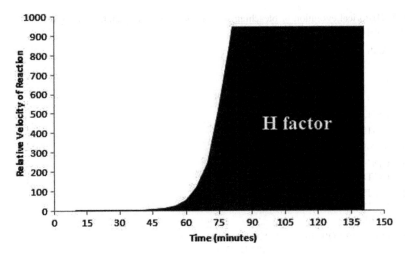

Fig. 2 H-factor plot of a conventional Kraft cooking. Eraldo Antonio Bonfatti Jr. unpublished

which is a chemical process that aims to improve the brightness and cleanliness of cellulosic pulp by removing and/or modifying chromophore and leukoprophic groups are used [55, 72]. After the reactions with the cooking liquor, the lignin in the wood, which is almost colourless, is coloured due to the release of chromophoric groups [55, 72]. Bleaching generally occurs in stages and its progress is always followed by brightness, which is a measure of reflectance of visible blue light at the wavelength of 457 nm, of pulp sheets or paper under standard conditions [31]. The most used method to measure this process is the standard described by ISO 2470: 1999—Paper, board, and pulps. It may be noted that measurement of diffuse blue reflectance factor is expressed in % ISO, while the non-reflective material, absolute black, has a brightness of 0% and a perfect reflectance of light is considered 100%.

Over the years the bleaching chemistry has been changing rapidly, starting with the discovery of bleaching power of chlorine on vegetable fibers, being used in its elemental form (chlorination, C), as calcium hypochlorite (hypochlorination, H) and as chlorine dioxide (dioxidation, D). Table 2 shows the evolution of bleaching sequences over the years. However in the 1990s, due to environmental reasons, the industries had to develop chlorine-free bleaching sequences or sodium hypochlorite [31], as these compounds are the main contributors to the formation of chlorinated organic compounds (Absorbable Organic Halides, AOX) [22]. After this environmental concern, elemental chlorine free (ECF = elemental chlorine free) sequences were created, which is the main technology used today, and totally chlorine free (TCF) sequences, which do not use elemental chlorine or any other chemical reagent that contain chlorine in the molecule. At this stage, oxygen (oxygenation, O), caustic soda (alkaline extraction, E), hydrogen peroxide (perioxidation, P) and ozone (ozonolysis, Z) associated with chelation (Q) and acid hydrolysis (A) are the most commonly used in industries.

Table 2 Historical evolution of bleaching sequences (Prepared by authors using the information from [9, 149])

Time	With element chlorine	ECF	TCF
1880	H		
1910	HEH		
1930	CEH, CEHEH		
1950	CEHDED, CEDED		
1960	(CD)EHDED		
1970	CD(EP)HD(EP)D, OC/DEDED		
1980	O(CD)(EP)D(EP)D, OC/DEODD, OC/D(EO)DD	O/OD(EPO)D(D)	
1985	O(DC)(EO)D(EP)D	O/OD(EPO)D_ND, O/OA/D(EPO)D_ND, O/OD$_{HOT}$(EPO)-D_ND	
1990	O(DC)(EPO)D(EP)D	O/OA/D(EOP)DP	
1995			O/OZ/QPZ/QP, O/OQPQP, O/OQ(PAA)QP, O/OQZQZP, O/OQZQP, O/OAZPZP
2000		O/OD(EPO)D(EP)D, O/AO/D(EPO)D	
2005		O/OZQD(EPO)D, O/AO/D(EPO)D, O/OD(EPO)D(D)	
2010		O/OA(EOP)DP	

ECF elemental chlorine free, *TCF* totally chlorine free

The choice of the bleaching sequence will depend on the kappa number after cooking, the type of raw material used in the pulping process, the end use of the bleached pulp and the desired final brightness. As the bleached pulps present on the market have an average brightness of 90% ISO.

3.3 Drying

After bleaching, the cellulosic pulp will be in aqueous suspension, with a consistency of 10–12%. This needs to be transformed, into cellulose bales with final humidity of 10% for the purposes of commercialization and transportation. It may be noted that this step of drying the bleached pulp is the final stage of the manufacturing process of the bleached cellulosic pulp. The process consists of the wet process step followed by the forming step, where the cellulose sheets are formed and finally, the dewatering step. The gradual removal of mater happens through the use of force of gravity, heating, and vacuum. It should be noted that in these steps the pulp is arranged on a permeable forming screen which is also responsible for conducting the pulp along the dryer [48]. The final step of drying the pulp is the removal of the water by the compressor rolls in the pressing step, with water not being withdrawn in the forming step would be removed. Then, a stronger pulp sheet is cut, packed and finally transported.

3.4 Paper Production

3.4.1 Preparation of the Cellulosic Pulp

It may be noted that when there is the production of integrated pulp and paper, the cellulosic pulp will be transported by piping. In pulp industries, which are separate from that of paper, it is necessary to dry the cellulosic pulp to make the transport for the paper mill. Therefore, if the paper mill is integrated with the pulp mill (normally, both paper and pulp are produced using the same mill), the drying step described above does not happen. In that case, the wet cellulosic pulp is pumped through pipes until the production of paper. If there is no integration between the pulp mill and paper mill, the pulp bales will remain dry and packaged for paper production. The first stage of paper production is the mass preparation. At this stage, for non-integrated paper mills, the cellulose bales are placed in the hydrapulper to be disaggregated and reduce the consistency of the mass. In the case of the integrated industries the cellulose used is already moist and are ready for the next stage, i.e., refining. This is a mechanical treatment in which cellulose pulp fibers are broken into fibrils, thus increasing the surface area and, consequently, the binding capacity between the fibers allowing the formation of a strong network [157]. In this way, the refining allows changing the structure of the fibers of the pulp, resulting in modified properties and increasing the mechanical properties of the finished paper. Figure 3 shows the scanning electron micrograph of the refined bleached cellulosic fibers obtained from *Eucalyptus* wood by one of the authors.

In addition, the energy used in refining causes changes in the fibers, leaving them more prone to collapse during the paper forming process, thereby decreasing the thickness and specific volume of formed sheet. However, refining should not be excessive as it can damage the fibers to the point of reducing the same mechanical properties that have been expected to improve the refining process. It has been recommended that when searching for high brightness pulps, it is necessary to avoid excess during the mechanical refining, as this causes a decrease in the brightness and opacity of the paper [101]. The degree of refining is measured through the Schopper-Riegler grade of drainage, which indicates the ease of the pulp in water shoring and is an important parameter for the evaluation of fiber interweaving—the greater the drainage of the pulp, the lower its capacity to drain water. The most widely used method for determining drainage is described by ISO

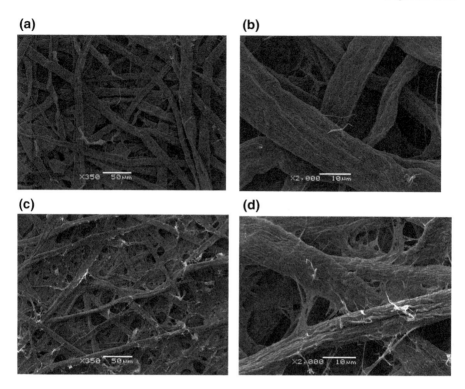

Fig. 3 Surface of a sheet formed by fibers: **a, b**—without refining at two magnifications; **c, d**—with refined fibers at two magnifications as of (**a, b**). Eraldo Antonio Bonfatti Jr—unpublished

5267-1: 1999—Pulps—Determination of drainability—Part 1: Schopper-Riegler method. In order to reach the desired degree of refining, the mass passes through the purification process. This step is aimed at reducing cellulose pulp contamination, and impurities such as plastics, metals, and sand are removed [66] followed by cleaning of pulp to prepare the dough.

In the preparation of the mass, the other components of the paper-making process are added (chemical additives) to improve the mechanical, physical and optical properties of paper as is followed normally in the paper industry [52]. Among the products that may be employed are starches, mineral compounds, vegetable gums, carboxymethylcellulose (CMC) and synthetic polymers [52, 141]. Recently nanocelluloses have also been used as additives in paper production [175].

3.4.2 The Paper Machine

A paper machine consists of different mechanical sections, each of them driven by one motor or an arrangement of one master motor and one or more helping drives, which conventionally speed or torque regulated. Typical sections are represented by

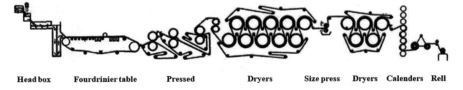

Fig. 4 A schematic drawing of a fourdrinier paper machine. Adapted from Lai [87] with the kind permission from publishers

fourdrinier, press, dryer, calender, and reel [48, 168]. A schematic drawing of a fourdrinier paper machine is shown in Fig. 4.

With the recipe of ready-made paper, the mass is transferred to the head box of the paper machine, where a uniform jet of mass is cast on a constantly moving conveyor belt forming screen flame [48]. At this stage, the first control of the paper thickness takes place; more the dough, the thicker would be the paper. The mass thrown on the mat forms a 5% layer of cellulose and additives and 95% water with water being drawn on the flat table. This stage of the process, known as 'leaf formation', promotes fiber entanglement and gradual water drainage, giving sufficient strength to the paper, so that it can leave the flat table and run through the various cylinders that make up the rest of the process [150].

In this part of the process, the paper passes through hydraulic presses wherein the excess water is removed, increasing the resistance and reducing the thickness of the paper [48, 150]. With the above process steps, the more resistant the paper reaches the drying step, which is promoted by a series of steam-heated cylinders; water would get evaporated from the pressed sheet, leaving it with the required moisture content for its final application [48]. Upon reaching required moisture content, the paper will receive one more layer of the surface additive according to its final use, and after another drying step, it will proceed to the calendar, which will have uniform thickness together with a better surface finish [48]. After this step, the paper is wound in smaller reels and can be commercialized, both in reels and in the form of sheets of sizes standardized for the final consumer.

4 Cellulose

Cellulose is the most abundant organic polymer on the planet and the largest component of plant biomass [89], with an estimated production of 7.5×10^{10} tons per annum [54]. It can be found in pure form, as in cotton, but is commonly found associated with hemicellulose and lignin in the cell wall [35, 89], as in wood, corresponding to approximately 40 to 45% of mass [149]. In addition to plants, it can also be synthesized by bacteria, algae, and fungi, but in lesser amounts [1]. Figure 5 depicts the main routes of obtaining cellulose in nature. Cellulose can also be obtained by synthesis in vitro and should be highlighted with important

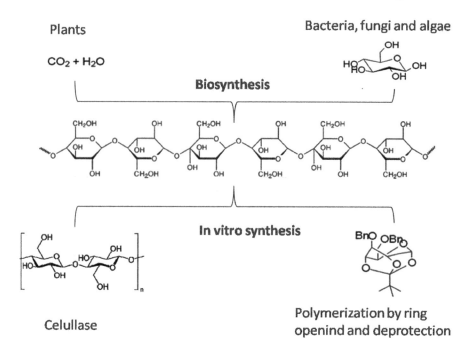

Fig. 5 Main ways of obtaining cellulose

development today [82]. The first report of cellulase-catalyzed cellulose formation was based on cellobiosyl fluoride [83] and the first chemosynthesis was performed through polymerization of substituted D-glucose and with open rings followed by deprotection [108].

Figure 6 shows schematically the ultrastructure. It can be seen from the figure that primary and secondary walls differ in the arrangement of cellulose chains. The secondary wall consists of 3 layers, S1, S2, and S3, and the S3 layer has the lowest cellulose content, being composed mainly of xylan. In the primary wall, the fibers are less ordered and essentially composed of chains in all directions within the plane of the wall. At layer S1 showing the very thin lamellae, the arrangement of the fibrils may be visible as is helical (spiral) in nature with a cross-arrangement in certain species. In layer S2 the cellulose chains are grouped in parallel microfibrils, giving a denser arrangement and aligned with the axis of the fiber. About 40–45% of the dry matter of the secondary wall is composed of cellulose [160].

Cellulose is composed of β-D-anhydroglucopyranose units which bond to each other through the carbons 1–4, forming a basic unit called 'cellobiose', which consists of the binding of two molecules of anhydroglucose [54, 145, 149]. The cellulose chain is linear and high molecular weight, which tends to form hydrogen bonds between the molecules [1].

The degree of polymerization (DP) is up to 20.000; however, it varies widely, and the value is around 10.000 in wood [74]. The hydroxyl groups of the cellulose

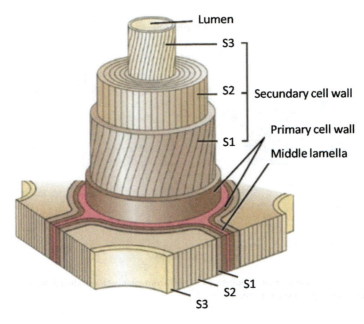

Fig. 6 Ultrastructure of wood. Reproduced from Taiz and Zeiger [160] with the kind permission of the publishers

molecules form hydrogen bonds that may be intramolecular or intermolecular. Their ability to form hydrogen bonds play a major role in leading the crystalline packing which also governs the physical properties of cellulose [74] and are these bonds that make cellulose a stable polymer and appreciated as reinforcement in composites [37, 54].

About 36 individual cellulose molecules are brought together by biomass into larger units known as elementary fibrils or microfibrils, which are packed into larger units called microfibrillated cellulose [54, 89]. The latter are in turn assembled into cellulose fibers. All these are shown in Fig. 7. The diameter of elementary fibrils is about 5 nm whereas the microfibrillated cellulose (also called nanofibrillated cellulose-NFC) has diameters ranging from 20 to 60 nm [6, 89]. The microfibrils are formed during the biosynthesis of cellulose and are several micrometres in length. This microfibrillar aggregates which allow the creation of highly ordered regions (i.e., crystalline) form the core alternate with disordered domains (i.e., amorphous) present at the surface [132]. It is these crystalline regions that are extracted, resulting in nanocrystalline cellulose (NCC). The inter- and intra-molecular interactions networks and the molecular orientations of crystalline regions can vary, giving rise to cellulose polymorphs or allomorphs [23, 89].

As mentioned earlier, the most common way to obtain pulp from wood is through the Kraft chemical pulping process [111], followed by bleaching steps to remove residual lignin on the cellulose surface. This pulp obtained in the paper industry is commonly used to obtain nanocellulose, since pulping and bleaching are

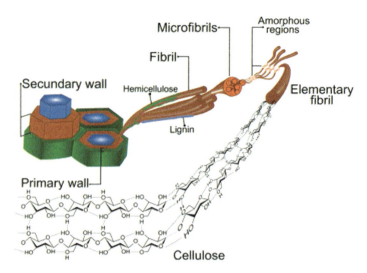

Fig. 7 Hierarchical structure of cellulose extracted from plants. Reproduced from Rojas et al. [125] with the kind permission of the publishers

characterized as pre-treatments necessary to obtain Nanocellulose, whether microfibrillated, nanofibrillated or nanocrystalline [50, 69, 91, 92, 95, 98, 109, 117, 169, 170, 185, 186].

5 Nanocellulose

The term "nanocellulose" refers to cellulosic materials having at least one of their dimensions in nanometer scale. Nanocelluloses can be produced by different methods and from various lignocellulosic sources [1].

According to Fujisawa et al. [46], so far nanocelluloses can be divided into three groups: cellulose nanocrystals (CNC), micro-fibrillated cellulose (CMF) and nano-fibrillated cellulose (CNF). While the first one (CNC) is produced by a chemical process of acid hydrolysis followed by mechanical agitation of the suspension in water, the second (CMF) is obtained by mechanical disintegration of the cellulosic pulp in water and finally third one CNF is prepared using the combination of chemical oxidation followed by mechanical disintegration in water, or only by the mechanical disintegration method. These mechanisms are shown in Fig. 8.

Reported definition of CMF is fibers with a diameter between 25 and 100 nm, while CNF are nanocelluloses with a diameter between 5 and 30 nm and a variable length between 2 and 10 μm [125, 139]. Both CMF and CNF have amorphous and crystalline zones composing their structure.

According to Samyn et al. [133], the microfibrillated nanocelulose is commonly produced by homogenization, where the fiber shear is performed by a strong

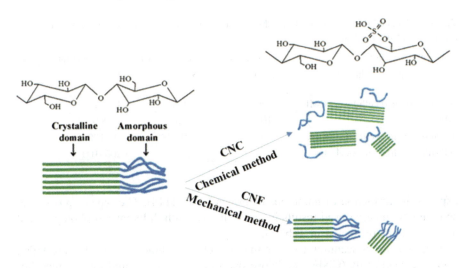

Fig. 8 The mechanism of chemical and mechanical methods for producing CNC and CNF from cellulose. Reproduced from Sofla et al. [151] with the kind permission of the publishers

pressure drop and impact forces inside the processing chamber. A similar effect is observed by the use of grinding process [91], where processed suspensions generally contain a heterogeneous mixture of CMF and CNF which are characterized by different diameters and aspect ratios (length/diameter) [133]. The CMFs are usually characterized by a smaller aspect ratio than the CNF [86, 85, 107, 177]. Depending on the number of processing steps, or passes through the mill, the geometry of the fibers in a suspension is reduced, generating more CNF, which leaves the suspension more homogeneous [91, 133]. Larger nanocellulose suspensions (CMF) present a large tendency to aggregate and flocculate microfibrils, a fact also justified by the surface charge of nanofibrils [91]. Non-uniformities in the suspending media are constructed by the high tendency of aggregation of single cell microfibrils and/or flocculation with larger fibers. According to Kumar et al. [86, 85] the CMF is generally produced by a single mechanical treatment of the cellulosic pulp, while the CNF is produced by mechanical treatment after the chemical pretreatment of the original pulp fibers [86, 85].

On the other hand, CNC refers to cellulose nanoparticles that underwent hydrolysis under controlled conditions and that lead to the formation of structures in the form of small crystalline cylinders [139]. Depending on the source of extraction, these crystallites will have a diameter of 3–50 nm. The CNFs, as well as the CMFs, exhibit zones with high fibrillation intensity due to the shear forces that the fibers undergo in the production process, whereas the nanocrystals are exclusively from the crystalline regions of the cellulose molecule. Nano-fibrillated cellulose has amorphous and crystalline regions that make up its more elongated chain in the longitudinal direction. In this way, the long length of nano-fibrillated cellulose chains associated to its surface containing a wide range of hydroxyl groups, which

Table 3 Nanocellulose derivatives and their dimensions (Reproduced from [75] with the kind permission of the Publishers)

Nanocellulose derivate	Diameter (nm)	Length (nm)	The aspect ratio (L/d)
Microcrystalline cellulose—MCC	>1000	>1000	1
Micro fibrillated cellulose—CMF	10–40	>1000	100–150
Microfibril	2–10	1000	>1000
Cellulose nano crystalline—CNC	2–20	100–600	10–100
TEMPO-oxidized nanocellulose	3–4	>1000	200–100

exposes the formation of numerous hydrogen bonds [114]. The type of processing and the raw material used results in nanocellulose with different morphologies and dimensions, as presented in Table 3.

Figure 9 shows scanning electron micrographs of cellulose microfibrils (CMFs), cellulose nanofibrils (CNFs), cellulose nanocrystals (CNCs) and others microfibrilatted cellulose that can be applied in papermaking, such as MCC and microfibril.

5.1 Method of CNF and CMF Production

As a semi-crystalline polymer, cellulose allows the extraction of nanostructures with different morphological properties (length, diameter, and aspect ratio), depending on mechanical and physical, depending on the extraction method applied [125]. The methods for producing nanocelluloses can be divided into chemical, physical and biological [44]. Some forms of procurement, various types of equipment and also combinations of chemical, enzymatic and/or mechanical treatments have already been tried for the production of nanocelluloses [57, 71, 114, 130]. The nanocellulose can be produced by mechanical methods such as grinding, cryoencation with high-pressure homogenization with liquid nitrogen, steam explosion, high-intensity ultrasound etc. Some pre-treatments may be used prior to mechanical processes to promote the accessibility of the hydroxyl groups, increase the internal surface, alter the crystallinity, break the hydrogen bonds of the cellulose and thus increase the reactivity of the fibers. The pre-treatments are different chemical hydrolysis (alkali or acid) or enzymatic [125].

5.1.1 Mechanical Methods

According to Rojas et al. [125], the mechanical treatments can isolate nanofibers from the primary and secondary cell wall without severely degrading cellulose. It is reported that depending on the types of mechanical treatment and levels of mechanical force used, inter fibrillar hydrogen bonding is broken [70, 123, 131,

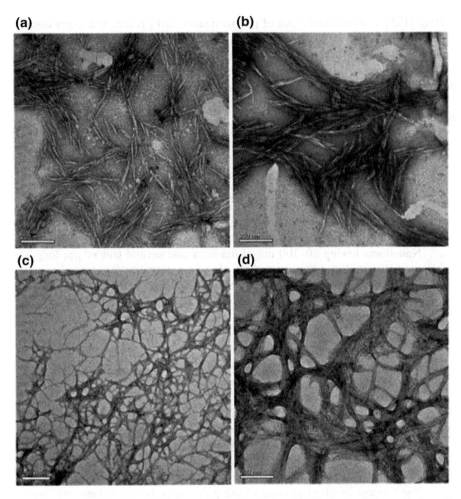

Fig. 9 TEM images of nanocelluloses extracted by: **a**—CNC prepared by sulfuric acid hydrolysis, **b**—cellulose nanocrystals isolated by formic acid hydrolysis, **c**—CNF prepared by 2,2,6,6- tetramethylpiperidine-1-oxyl (TEMPO)-mediated oxidation, and **d**—cellulose nanofibrils fabricated by pulp refining. Reproduced from Liu et al. [97] with the kind permission of the Publishers

179]. For example, microfluidization and high-intensity ultrasonic treatments produce a high shear degree, causing transverse cleavage along the longitudinal axis of the cellulose fibers. This process tends to damage the microfibrillar structure, reducing the molar mass and the degree of crystallinity of cellulose.

The mechanical methods would involve high production costs, besides they being less efficient and requiring higher energy inputs compared to that of chemical methods [96]. In view of these, it is reported that a chemical pretreatment would be necessary, which reduces energy consumption besides obtaining more hydrophobic surface [125]. Further, the degree of polymerization (DP) is reported to get usually reduced from 1200 DP to 1400 DP between 850 and 500 by the mechanical treatment. It may be noted that a high cellulose DP is desirable because of correlation of cellulose with the tensile strength of the nanofiber, which is reported to be at least 2 GPa [26, 118].

The following are some of the mechanical methods used to produce nanocellulose:

(i) High-Pressure Homogenization (HPH): In this process, first known quantity of the cellulose (2–7% w/v) is passed through slurry at high pressure into a vessel through very small spring-loaded valve assembly using low velocity. This is then exposed to a pressure drop to atmospheric condition with the valve opening and closing in a cyclic motion [45]. This method is reported to be an efficient method for refining of cellulosic fbers in view of its high efficiency, simplicity and without requiring any organic solvents [78]. Nanofibers having 20–100 nm of diameter and several tens of μm long are normally produced by this method. However, clogging of the homogenizer, high energy consumption, and mechanical damage of the crystalline micro fibril structure are some of the limitations of this method [96, 174, 178].

(ii) Microluidizer: This method uses the equipment, which consists of an intensifier pump and an interaction chamber. While the first is for increasing the pressure, the second is for defibrillating the fibers using two types of forces, viz., shear and impact against colliding streams and the channel walls [42]. Dimensions of CNFs produced by this process are of several μm long and less than 100 nm [125].

(iii) Grinding: In this method mechanism involved is fibrillation of cellulose using a suitable equipment say, grinder to break the hydrogen bond and the cell wall structure of the cellulose by shearing force besides individualization of pulp to nanoscale fibers [146]. Accordingly, this method uses grinding equipment consisting of a static and rotating grindstone (1400–3000 rpm). The process involves passing of the pulp slurry between these two stones [92, 125]. Accordingly, the cell wall structure would break down by the shear and compression forces, which generate a gel due to the suspension of nanocelluloses. Therefore, a number of cycles to be passed by the pulp/fibers through a grinder or the amount of energy for processing the fibers is important parameters which affect the quality of resultant NFC produced by this method. The diameters of NFCs produced by this process range from about 5–157 nm [56, 155, 182].

(iv) Cryocrushing: This method involves immersion of water swollen cellulosic fibers into liquid nitrogen followed by its crushing by mortar and pestle [45]. Accordingly, in this method, high impact forces would be applied in order to the freeze the cellulosic fibers leading to rupture of cell wall due to the pressure exerted by ice crystals and thus, liberating nanofibers [146]. Nanofibers of soya beanstalks have been produced by this method by cryocrushing and high-pressure defibrillation procedures [175, 176]. Normally, this method produces CNFs with diameters from 30 to 80 nm [3].

(v) Steam explosion: This method is a thermomechanical process. In this method, cellulose is kept at 200–270 °C is exposed to a high pressure of steam maintained between 14 and 16 bars. Then, the steam penetrates the biomass by diffusion for short periods of time between 20 s to 20 min. This is followed by applying sudden decompression (explosion), which would generate shear forces hydrolyzing the glycosidic and hydrogen bonds, between the glucose chains [77, 125]. The diameter of CNFs produced by this methods lies in the range of 10 µm–50 nm [30, 36].

(vi) High-intensity Ultrasonication: This method is a mechanical process wherein oscillating power is used to isolate cellulose fibrils by hydrodynamic forces of ultrasound [28]. According to Rojas et al. [125], the cavitation during the process leads to a powerful mechanical oscillating power. The gas bubbles formed would expand and explode breaking down the cellulose fibers [27]. The diameter of the nanocellulose produced by this method lies in the range of 5–35 nm [57].

It is reported that the mechanical treatment causes changes in fiber structure [32]. The author suggests following four phenomena can be observed due to the defibrillation process. First one is the internal fibrillation (IF), which is difficult to observe by microscopy techniques. Here, loosening of the fiber bundle takes place, which causes swelling and increased fiber flexibility. The swelling of the cellulose increases its accessibility to reagents, and consequently their reactivity. The second effect is the external fibrillation (EF) at the surface of the fiber. This is basically the defibrillation process of the fibrils, but without their complete removal. It may be noted that when these fibrils extend completely from the fiber there is the generation of the nanofibers (CMF), as the third phenomenon showing the structural alteration. And finally, the fourth one involves the dimensional reduction of the fiber itself by mechanical wear through fiber cutting (FC). The entire phenomenon mentioned above can be observed by microscopy techniques as is evident from Fig. 10, which is a transmission electron micrograph of nanocellulose obtained from Lengowski [91].

5.1.2 Electrospinning

This is a method to form the fibers using an electrical rather than a mechanical driving force and is termed as an 'electromechanical' method. Here, the cellulose dispersion is extruded and electrospun under the effect of a high electric field [43], following a 3D spiral trajectory. Once the solvent evaporates, it leaves behind randomly oriented nanofibers in the collector. The CNFs morphology produced by this technical depends on the strength of electric field, solution feed rate and the tip-to-collector distance [125].

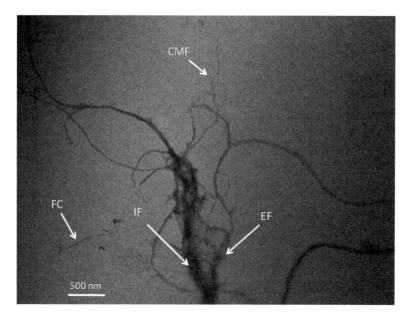

Fig. 10 Effects of refining on the production of nanofibers by the mechanical process. Reproduced from Lengowski [91]

5.2 Methods of CNC and MCC Production

5.2.1 Acid Hydrolysis

The mechanism for obtaining CNC by acid hydrolysis involves the removal of the amorphous regions from the cellulose elementary fibrils by hydrolysis, leaving only the crystalline regions [114]. This cellulose is obtained by cutting the elementary fibrils into small fragments followed by bleaching. Subsequently, CNC is extracted from bleached samples by strong acid hydrolysis under strictly controlled conditions of concentration, temperature, agitation, and time [125]. A typical production process involves acid hydrolysis, washing, centrifugation, dialysis, and sonication to form a suspension followed by drying by freeze-drying or heat-drying [54, 92]. CNC is also known as whiskers or cellulose nanocrystals.

The difference between the production of CNC and MCC by acid hydrolysis lies in the reaction time or in the concentration of the reagent, where less time and lower concentrations are used for MCC compared to those to produce CNC [38, 173, 185, 186]. MCC can be characterized as a white powder of fibrous particles with sizes of about 40 μm with a DP 100–200 and about 80% crystallinity, while the CNC has dimensions of 5–10 nm wide, 100–300 nm long with 90% crystallinity when made from cotton and wood cellulose. On the other hand, other sources like bacteria, algae, and tunicin produce nanocrystals with larger size distributions and dimensions comparable to those of CMF (width: 5–60 nm, length: 100 nm to several μm) [10].

It may also be noted that CNC and MCC are similar to small cylinders or crystalline characters, isolated from acid hydrolysis of the fibers. This is illustrated in Fig. 11, which shows scanning electron micrographs of CNC obtained from Beauvalet [21] (Fig. 11a) and MCC obtained from Thoorens et al. [162] (Fig. 11b).

It may be noted that the hydrolysis processes rely on the fact that the crystalline regions are insoluble in acids under the conditions in which they are employed. This is due to their inaccessibility because of the high organization of the cellulose molecules in their nanostructure. On the other hand, the natural disorganization of the molecules in the amorphous regions favours the accessibility of the acids and consequently the hydrolysis of the cellulose chains present in these regions [132]. Sulfuric and hydrochloric acids are the most commonly used for acid hydrolysis, but phosphoric and hydrobromic acids have also been used [90].

The most commonly used method for the preparation of CNC is acid hydrolysis of cellulosic materials using sulfuric acid (64% w/w). Cellulose nanofibers have also been produced from hardwood by treatment with the 2,2,6,6-tetramethylpiperidine1-oxyl radical in combination with sodium bromide and NaClO [130].

The CNC has a high aspect ratio, a high modulus and good compatibility with matrix materials [127]. Your morphology is the elongated crystalline rodlike shape and has a limited flexibility because it has no amorphous regions. These CNCs have a degree of crystallinity (55–90%). However, it should be noted that the degree of crystallinity, aspect ratio, and morphology depends on the source of cellulosic material and preparation conditions [144]. The colloidal behaviour and superficial charge of CNCs depend on the acid used for their production [90].

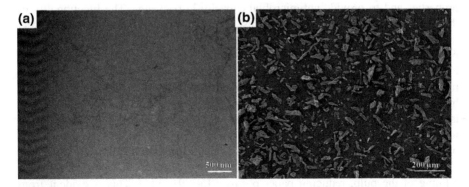

Fig. 11 a—Nanocrystalline cellulose morphology **b**—microcrystalline cellulose. Reproduced from Beauvalet [21] (Fig. 11a) and [162] (Fig. 11b) with the kind permission of the publishers

5.2.2 Enzymatic Hydrolysis

It is well known that enzyme is generally used to modify and/or degrade the lignin and hemicelluloses contents in biomass without altering the cellulose portion. It is known that enzyme helps in the restrictive hydrolysis of several elements or selective hydrolysis of specified components in the cellulosic fibers [71]. These enzymes are produced by cellobiohydrolases. There are two types. The first category is A- and B-type cellulases, which are capable of attacking the crystalline portion of cellulose. On the other hand, the second category is C and D type endoglucanases, which are capable of attacking the disordered structure (amorphous) of cellulose [8]. It is reported that enzymatic methods are highly expensive as these methods take long treatment time for a successful hydrolysis and also due to the isolation process of the enzymes [77]. Actually, this process can be used as a pre-treatment for production of nanocellulose by the mechanical method to reduce the energy consumption to produce CNF.

6 Applications of Nanocellulose in Paper Making

6.1 Nanocellulose-Reinforced Pulp

The development of strength in paper is influenced by several factors, viz., length and strength of fiber used, degree of adhesiveness, fiber-fiber contact area and bonding agents in the formation of dry or moist [11, 12, 81, 159]. Due to the unique properties of nanocellulose, there is a growing tendency to use it as a paper reinforcement additive [112]. Many researchers have been using nanocellulose as an additive and films in paper making, either to improve (i) strength properties, barrier properties in food packaging, paper brightness, printability [80] or (ii) to reduce paper weight without loss in mechanical properties, while improving thermal properties and to provide antimicrobial capacity in packages. The functionality of the nano paper emerges from the intrinsic properties of the nano fibrous network, the additional loading of specific nano materials or the additional deposition and modelling of thin films of nanomaterials on the paper surface [17]. According to Zimmermann et al. [187], reactive sites exposed on the surface of the cellulose micro fiber (CMF) perform the formation of a network of nanofibers due to the hydrogen bonds formed. Because of the nanometer scale, the amount of these bonds is enhanced by the larger contact surface between nano and microfibers. This increases the apparent density of the paper, making it more resistant to the passage of air and humidity besides the gain in the mechanical properties [86, 85, 106, 115].

As an additive, the nanocellulose has a similar effect to that produced by the refining of the pulp, reducing paper porosity [50, 91, 117]. This is evident from Fig. 12, which shows scanning electron micrographs of cellulose sheet with and without any additive highlighting the effect of the addition of CMF to the paper

[91]. It may be noted that Fig. 12a is micrograph of cellulose sheet without any additive, while Fig. 12b, c are micrographs of cellulose sheet with CMF additive and CMF coating. Besides, such addition of CMF to cellulose gives significant gains in the mechanical properties of the paper, whether produced by virgin fibers [117] or by secondary fibers [116]. It is also reported that wet strength is the main functionality for tissue paper, paper towels, cardboard and other papers [129]. These authors have also reported that the cationic polyelectrolyte (Poly Amideamine Epichlorohydrin—PAE) developed in the 1950s has been used as a wet strength additive in the paper making process. In order to increase the mechanical properties of the paper, the use of nanocellulose and additives such as PAE [4, 5] and amphoteric starch [93] have already been reported.

Investigations have also been carried out on the possibility of using CMF and CNF as a reinforcing and filler in a pilot experiment using different degrees of fibrillation and filler mixtures (CMF, CNF, precipitated calcium carbonate, cationic polyacrylamide and two types of starch) [164]. The author has observed that the CNF improved the resistance properties (Scott Index and tensile index) more than the CMF, and also that the CNF associated to the starch presented better resistance to traction being wet. Finally, the author has concluded that in conjunction with CMF or CNF, the filling content could be increased from 30 to 40%, which would imply a potential savings of 3–6.5% compared to conventional sheets. A dose of 6% of applied CNF has been found to promote about 40% energy savings, while the

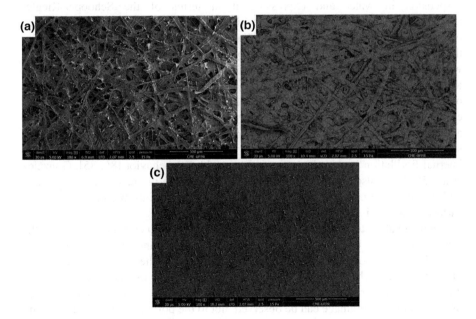

Fig. 12 Effects of nanocellulose on papermaking **a** cellulose sheet without nay additive; **b** cellulose sheet with CMF such additive; **c** cellulose sheet with CMF such coating. Reproduced from Lengowski [91]

12% dose reduced energy expenditure by 85% compared to the control to produce pulps at 35 °SR [34].

Compared with micro-sized cellulose, nanocellulose is more effective as an additive for the paper industry. This is attributed to the interactions between the nanosized elements, which are connected by hydrogen bonds forming a percolated network when the nanocellulose is dispersed in the pulp [3, 73, 75, 76, 80, 183].

Another study has concluded that the drying and wetting process of paper with nanocellulose depends on the amount of added nanocellulose because thenanocellulose clogs the pores and give a greater amount of inter fibrillar connections and preventing fiber-water bonds in the rewetting [120]. It has also been pointed out that 6–12% of CMF could be used in the paper making mix [121], but indices above 5% have been reported to cause a decrease in traction and tear properties [117]. On the other hand, the use of CMF as additives to increase the resistance of thermo-mechanical pulps to produce new types of packaging has also been reported [59]. In addition to improving the mechanical properties, these authors have also observed that the addition of 6 g/m^2 of nanocellulose reduced the permeability and drainage of the cellulosic mass. The enzymatic pretreatment combined with mechanical shear forces at high pressure has been found to significantly improve the mechanical strength of the paper without affecting drainage [50, 51]. In another study Gonzalez et al. [50] have concluded that the porosity and mechanical properties were improved when 9% of nanocellulose was added; however the Shopper Riegler degree (This is the '*degree* of refining of a pulp suspension in water and expressing it in terms of the Schooper-Riegler (SR) number, and to determine the de-watering time') was altered, which becomes a disadvantage in the drainage stage and can cause operational problems in the drying. A similar result was observed by Damásio [34] when 6 and 12% of CNF were added. A decrease in 'freeness' (It is 'a measure of how quickly water is able to drain from a fiber furnish sample. In many cases there is a correlation between freeness values and either (a) a target level of refining of pulp, or (b) the ease of drainage of white water from the wet web, especially in the early sections of a Fourdrinier former') has been observed while using nanofibers from different sources with increased fibrillation due to water imbibed in the cell cavity and also to internal fibrillation itself, which is consistent with SR behavior [86, 85]. González et al. [50] have also confirmed the tendency to increase the resistance to drainage with the increase of the addition of CNF to the pulp, mainly because of the high surface area of this material.

There are also studies which have reported about the influence of drainage by the size of fibers used, ionic strengths, type of polyelectrolyte and pH of the suspension [156, 159]. Both of the above studies have found that the higher the degree of fibrillation, or higher the concentration of CMF in the mixture, the lower the drainage. However, when the mixture of cationic polyelectrolytes and CMF are used, alteration in drainage can be observed, not to the point of damaging the drying process [159].

The addition of nanocellulose makes the sheet drainage slightly impaired. To overcome this limitation, cationic polymers, such as polyacrylamide, have been used as a fixative for the retention of nanocellulose, as well as for better suspension drain ability [99, 181].

In all the above studies where CMF was added to the sheets, an increase in density of the paper was observed. An increase of 4–30% in density for thermo-mechanical pulp sheets containing 4% CMF and a 10% increase in density to 7% CMF in coniferous Kraft pulp leaves have been observed [39, 102].

While these authors observed a 20% increase in density with the addition of 20% CMF, Sehaqui et al. [140] have observed 30–50% increase in density with the addition of 10% homogenized CMF to softwood Kraft pulp sheets.

Optical properties have also been found to be influenced by the addition of nanocellulose [34]. The increase of nanofibers in the paper composition is reported to cause a significant reduction in the light scattering coefficient of the pulp. The opacity also has also been found to decrease in its values due to the decrease of the light scattering coefficient, with the increase of transparency.

6.2 Coating and Films

Effects of the development of nanocellulosic films and surface deposition of nanocellulose films on paper have been studied by many researchers [50, 58, 91, 122, 124, 135, 149, 153, 154, 158]. An increase in surface density, a decrease in water absorption, reduction in permeability and surface porosity has been observed by Sjöström [149] and Lengowski [91]. Increase in surface porosity has already been shown in Fig. 12. On the other hand, the increase in the surface of the fibers with the micro-fibrillation process has been found to favor a greater number of inter fiber bonds due to the greater availability of OH-groups [50], water retention capacity and increase in Schopper-Riegler grade, implying increased drying cost.

Spence et al. [153] have studied the ability of water retention in nano-cellulose films produced from bleached and unbleached pulp from hardwood and longwood. They have observed that water retention was lower for long-fiber films compared to those of short-fibers and non-bleached pulps compared to bleached pulps. They have also evaluated CMF films with different lignin contents and observed that the samples with higher lignin content had higher rates of water vapour transmission. Besides, significantly higher water retention values and larger surface area were observed by these authors in CMF prepared from unbleached hardwood in comparison to other samples. Another study has reported that the water retention for surface depositions of unbleached nanocellulose when the source of moisture was on the opposite side to that of the surface film [91]. The presence of lignin in the production of its CMF films has been found to provide longer, narrower and more connected pores, thus increasing the rate of water transmission [154]. On the other hand, larger and lignified pores have been found to increase the rate of water vapour transmission because of their lower adsorption capacity [62]. It has also been observed that the specific surface area is strongly correlated with the difficulty of removing water content for the pulps, suggesting that water diffusion is more

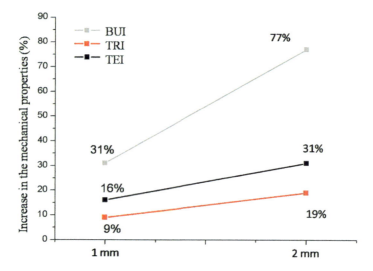

Fig. 13 Increase in the mechanical properties after superficial film deposition on the sheet in relation to a sheet without any additive. BUI = burst index; TRI = tear index; TEI = tension index. 1 mm and 2 mm = deposited moisture film thickness. Reproduced from Lengowski [91]

dependent on pore structure and surface area geometry [154]. On the other hand, Lengowski [91] found that increasing the thickness of the nanocellulosic film on paper caused an increase in mechanical strength (tension, tear and burst index). This is illustrated in Fig. 13, which shows the plots of two different thickness of moisture versus values of tension index, tear index and burst index. On the other hand, this thickness did not influence water absorption properties. While using a bleached and unbleached nano cellulose film, the author observed an increase in the tensile and burst index for bleached CMF deposition, and for the tear, the presence of lignin in CMF caused a gain in this property [91]. These are illustrated in Fig. 14, which shows the plots of mechanical properties with and without any additive on the deposition of surface film (coating) on the sheet. However, the author did not observe any influence of lignin on the thermal stability of the papers. Further, the author has also observed an increase in the mechanical properties of sheets with the addition of CMF (additive) in the production of the sheet compared to that of the sheets without additives. These results are illustrated in Fig. 15, which shows the plots of mechanical properties after surface film deposition on the sheet with and without the addition of CMF [91]. It can be seen from the figure that the gain in the tensile and tear properties are similar to the surface deposition of a nanocellulose film, while the burst index shows a reduction in resistance compared to the use of CMF as an additive. The author observed that the results of the mechanical properties of the papers are enhanced when CMF was used as a reinforcement additive as well as a coating on the paper. It can also be seen that while the values of tensile index, burst index and tear index showed a 134, 50 and 44% increase respectively compared to those values without any addition to the paper [91].

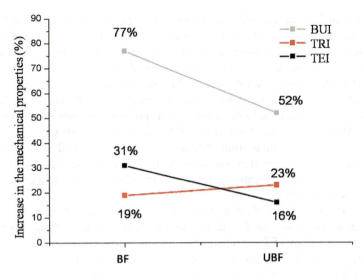

Fig. 14 Increase in the mechanical properties after superficial film deposition (coating) on the sheet in relation to a sheet without any additive. Effect of deposition of unbleached and bleached nanocellulose on mechanical properties of BUI = burst index; TRI = tear index; TEI = tension index. BF = bleached nanocellulose film; UBF = unbleached nanocellulose film. Reproduced from Lengowski [91]

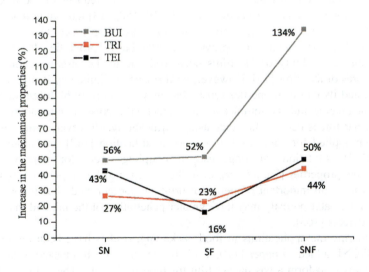

Fig. 15 Increase in the mechanical properties after superficial film deposition on the sheet in relation to a sheet without any additive. Effect of CMF addition, CMF film coating and CMF addition and coating on a paper sheet. SN = sheet with CMF addition; SF = sheet with CMF film coating; SNF = sheet with CMF film coating and addition Reproduced from Lengowski [91]

A dense surface of the films is an important feature, as it is related to the porosity, which determines the barrier properties of the films [122]. The author has used a paper coating of nanocellulose to improve brightness properties, surface roughness, absorption, and permeability. He noticed an improvement in absorption and porosity and a very small impact on the other properties.

For CMF based films the porosity can be modified by drying from different solvents, creating an adjustable feature that provides an advantage over the melt-formed plastics. Henriksson et al. [58] were able to modify the porosity of water-dried CMF-based films from 28% to porosities of up to 40% with dry films from solvents such as methanol and acetone. In addition, when used as a coat of CMF on paper, the air permeability was found to reduce by 10% as a consequence of surface porosity [158].

Sauders et al. [135] evaluated substitution of CMF in different amounts of acetylation. It was observed that there was no improvement in the contact angle (hydrophobicity) from 15% of substitution and increased crystallinity, while there was a decrease in the rupture stress and in the stress carrying capacity of the material. These have been attributed to acetylation delaying the ability of the fibrils to form bonds during the sheet forming process. Another study has not observed any significant reduction of tensile strength and improvement in the hydrophobicity of the papers after CMF acetylation [124].

Similarly, several studies have also been carried out using CMF as a barrier to gases and liquids [13, 16, 47, 65, 91, 100, 110, 113, 147, 158]. The increase in film thickness, the amount of nanocellulose in the mass [110] reduces the permeability to oxygen, water, oils and carbon dioxide [13, 110, 158, 113] indicating that the use of larger thicknesses or greater percentages in mass, simply acts to increase the tortuous path of oxygen and an improvement of the barrier properties. It is believed that the air permeability of the films decreased due to the disconnection of the surface pores of the films [13]. However, performance is limited by its hygroscopic capacity, and its barrier properties loose efficiency at relatively high humidity [13, 113]. When nanocellulosic samples are in contact with moisture, a reduction of the inter fibrillar forces occurs, thus increasing permeability. However, many methods have been applied for nanocellulose surface modification [142], including acetylation [104], silylation, grafting, use of coupling agent for improving the hydrophobic properties [76, 88, 99], or in the presence of lignin [91]. Recognizing that porosity is an important property for printing, another study has reported that papers with greater porosity may have greater penetration of the ink and may cause printing defects [163].

Considering the applications in food packaging, studies have been reported on the CNC/CNF added to paper [119, 128]. In the study by Rampazzo et al. [119] CNC was used to form a gas barrier film for food packaging. They observed very promising results, providing oxygen and carbon dioxide permeability values hundreds of times smaller than those of equal thickness compared to common barrier synthetic polymers over a wide range of temperatures.

On the other hand, Saini et al. [128] in their study added nisin to CNF and evaluated its potential as an antimicrobial agent in food packaging. Addition of

CNF to paper sheets was done in two ways, either directly as an additive or as surface deposition such as coating. The authors found that the latter method (surface deposition) was more efficient than the former (direct addition).

Another property studied is the surface roughness of the composites of CNC-CNF. Damásio [34] produced CNF-CNC nanocomposites by the casting technique, using CNF as the dispersion polymer matrix for different CNC dosages as mechanical reinforcement. The addition of cellulose nanocrystals allowed the reduction of the surface roughness of the nanocomposites produced, increasing the mechanical properties significantly. The incorporation of CNC allowed reduction of the opacity by up to 53%, with consequent gains of transparency of the nanocomposites.

Fang et al. [41] have been reported about its modification by CNF. In this study, CNFs made from TEMPO-oxidized CNFs, where the 2,2,6,6-tetramethylpiperidine-1-oxyl (TEMPO)/NaBr/NaClO was used to modify the surface properties of the pristine wood fibers by selectively oxidizing the C6 hydroxyl groups of glucose produced transparent nanopapers. This nanopaper exhibited ultrahigh optical transparency ($\sim 96\%$) and ultrahigh haze ($\sim 60\%$), thus delivering an optimal substrate design for solar cell devices, and that may influence a new generation of environmentally.

Another field that is being explored is the use of nanocellulose together with starch in the development of biomaterials, looking for the substitution of polymers derived from petroleum. As starch presents a brittle characteristic, it is necessary to use plasticizers that improve its flexibility [14] although the mechanical and barrier properties have been compromised. Several studies have been reported on the possible applications of biofilms with starch in various sectors such as food, agricultural and pharmaceutical and also in other sectors, where biodegradability is required [18, 19, 105]. Some examples of products under study are garbage bags, films to protect food, diapers, and flexible rods; in agriculture as a film in the ground cover and containers for plants and in the production of slow release fertilizers. Also, biofilms made using starch and nanocellulse can also be used in the preparation of capsules, in the release of medicines, in the substitution of styrofoam, in the protection of equipment during transport and in several other applications [40, 126, 152, 165].

Incorporation of different types of nanocellulose (CNF, CMF, and CNC) as a reinforcing agent in starch films has been studied [29]. All the films presented excellent transparency and increased thermal stability in relation to the mechanical properties; however, the best results were for films reinforced with CMF due to the greater aspect ratio exhibited by this type of fiber. One of the main difficulties of the use of nanocelluloses is its tendency to flocculation and agglomeration in the matrix [94]. While using dispersing agents in the composition Campano et al. [24] have observed a greater homogeneity in the matrix. These authors have also evaluated different retention agents in the mixture, which allowed greater retention of the nanoparticles, causing an increase of traction with lower loads. These authors have suggested that further studies should be carried out with retention and dispersing agents with nanocelluloses in order to industrially apply such an additive. In fact,

CNC as a retention agent has already been used for fines in paper production [94]. The highly negative charge of the CNC allowed a strong interaction between cationic polyelectrolytes promoting good drainage and high retention of micro and nanoparticles.

According to Abdul Khalil et al. [2] films based on biodegradable polymer reinforced with nanocellulose present great potential and innumerable benefits for its use in packaging. However, the development of this type of biocomposites is still in the initial stage, and it is necessary to deepen the knowledge in several aspects, such as its quality, cost, and utility.

7 Market and Opportunities

The current demand for sustainability in products is driven by consumers and retailers seeking to differentiate themselves by reducing impacts of product lifecycle on the environment by reducing and minimizing the quantity of packaging and improving the sustainability of their supply chains [25]. The data indicate that this is a growing market with the development of safer and more sustainable products [67]. Also, the cellulose material being a renewable and biodegradable material having low toxicity, this new material can replace petroleum-based packaging, metal components, and other non-renewable materials, which are mostly non-biodegradable and non-renewable. In view of this, cellulose nanomaterials represent an important niche for the design and development of more sustainable products. Nano-cellulose can be used in a variety of products, among the main ones in the cement industry, automotive in the internal and external polymer structure of automobiles, packaging industry whether for food or packaging that demand high strength, waterproof papers, oils and oxygen, special papers, Tissue papers, paper cups, hygienic and absorbent products (disposable diapers), biofilms for plastic replacement, electronics, as well as acting as an excellent stabilizer for suspensions and emulsions such as paints and cosmetics and can be used as a basis for prints 3D of bones and cartilages [68, 143].

According to the "Global Nanocellulose market analysis and trends—Industry forecast to 2020" report, the global nanocellulose market is estimated to reach $295 million by 2020, with a CAGR of 22.15 over the next 5 years due to the expansion of applications and increasing the appeal for green alternatives to oil products. According to the BCC Research Report [20], the global nanocellulose market was $46.8 million in 2014, projected to be $277.7 million in 2019, with a CAGR of 42.8% by 2019. Considering the types of nano-cellulose, cellulose nanofibrillation had a market of 28.2 million in 2014, it is estimated to be 158.3 million in 2019, while nanocrystalline cellulose had a market of $18.0 million in 2014 and is expected to be 116.6 million in 2019.

According to Cowie et al. [33], the annual global market potential for high-volume applications of nanocellulose is estimated at 32.8 million metric tons, based on current markets and middle market penetration estimates. The largest uses for nanocellulose associate a paper and pulp industry are projected to be in packaging coatings (5.3 million metric tons), replacement for the plastic packaging (4.1 million metric tons), plastic film applications (3.3 million metric tons), paper filler (2.4 million metric tons), packaging filler (2.4 million metric tons) and paper coatings (2.2 million metric tons).

8 Final Considerations

The application of nanoscience is an opportunity for greater profitability and autonomy for the pulp and paper industry. This is because it is possible to produce additives and coating for papers using the same raw materials (starch, CMC, synthetic polymers, resins, alkylketene dimer emulsions, alkenyl succinic anhydride) used for paper making without needing other materials to improve the internal bonding of paper fibers, reduce porosity, whiteness, opacity, etc. This opens up possibilities for improving existing products and developing new low-cost, multi-purpose products. There are already different types of nanocellulose that can be used in paper production. However, it is necessary to know the mode of production of the paper and the routes of production of nanocellulose, since the structure or morphology of these different nanocelluloses can be altered according to the routes of and therefore can modify the final properties of the paper. The benefits of nanocelulsoe in pulp and paper industry products include: increased tensile and burst strengths, weight loss, improved barrier properties for oils, oxygen and moisture, better printing surface, optically transparent and/or coloured layer coatings, biodegradability, cost reduction with additives and potentially with drying. The use of nanocelluloses in papermaking presents interesting possibilities, and offers improvements in cost-benefit, energy efficiency and biocompatibility, in addition to generating new products with uses are not available today.

Acknowledgements At the outset, the authors express their sincere thanks to the Editors of the book (Inamuddin, Sabu Thomas, Raguvendra Mishra and Abdullah M. Asiri), particularly Prof. Inamuddin for inviting us to contribute this Chapter. The authors place on record and appreciate the kind permission given by some of the authors (who have given permission to use their figures), M/s. Elsevier Inc Publishers, Springer, Sociedade Brasileira de Química—SBQ, Brazil, www.plantphysiol.org or www.plantcell.org—"Copyright American Society of Plant Biologists." InTech Open Publishers, IOP Publishing and the Vietnam Academy of *Science* and Technology (VAST) to reproduce some of the figures from their publications free of charges. One of the authors (KGS) would like to thank the PPISR, Bangalore-India with whom he is associated with presently for their encouragement and interest in this collaboration.

References

1. Abdul Khalil HPS, Davoudpour Y, Nazrul Islam MD et al (2014) Production and modification of nanofibrillated cellulose using various mechanical processes: a review. Carbohydr Polym 99:649–665
2. Abdul Khalil HPS, Tye YY, Leh CP, Saurabh CK et al (2018) Cellulose reinforced biodegradable polymer composite film for packaging applications. In: Jawaid M, Swain S (eds) Bionanocomposites for packaging applications. Springer, Cham, pp 49–64
3. Abe K, Iwamoto S, Yano H (2007) Obtaining cellulose nanofibers with a uniform width of 15 nm from wood. Biomacromol 8:3276–3278
4. Ahola S, Turon X, Österberg M et al (2008) Enzymatic hydrolysis of native cellulose nanofibrils and other cellulose model films: effect of surface structure. Langmuir 24 (20):11592–11599
5. Ahola S, Salmi J, Johansson L-S et al (2008b). Model films from native cellulose nanofibrils. Preparation, swelling, and surface interactions. Biomacromol 9(4):1273–1282
6. Akil HM, Omar MF, Mazuki AAM et al (2011) Kenaf fiber reinforced composites: a review. Mater Des 32:4107–4121
7. Almeida FS (2003) Influence of alkaline load on the Lo-solids® pulping process for eucalyptus wood. Dissertation, Unisity of São Paulo
8. Anderson SR, Esposito D, Gillette W et al (2014) Enzymatic preparation of nanocrystalline and microcrystalline cellulose. Tappi J 13(5):35–42
9. Andrade M (2011) The fiber line of the future for eucalyptus kraft pulp. In: Paper presented at the 5 th international colloquium on eucalyptus pulp, Federal Uniuversity of Viçosa, Porto Seguro, 8–11 may 2011
10. Angles MN, Dufresne A (2000) Plasticized starch/tunicin whiskers nanocomposites. 1. Structural analysis. Macromol 33(22):8344–8353
11. Ankerfors M, Duker E, Lindstrom T (2013) Topo-chemical modification of fibres by grafting of carboxymethyl cellulose in pilot scale. Nord Pulp Pap Res J 28(1):6–14
12. Ankerfors M, Lindström T, Henriksson G (2013b) Method for the manufacture of microfibrillated cellulose. US patent 8,546,558, 8 Feb 2006,
13. Aulin C, Gallstedt M, Lindstrom T (2010) Oxygen and oil barrier properties of microfibrillated cellulose films and coatings. Cellulose 17:559–574
14. Azeredo HMC (2012) Fundamentals of food stability. EMBRAPA, Brasília
15. Barbosa LCA, Maltha CRA, Silva VL et al (2008) Determination of the siringyl/guaiacyl ratio in eucalyptus wood by pyrolysis-gas chromatography/ mass spectrometry (PY–GC/MS) (PI-CG/EM). Quím Nova 31(8):2035–2041
16. Bardet R, Reverdy C, Belgacem N et al (2015) Substitution of nanoclay in high gas barrier films of cellulose nanofibrils with cellulose nanocrystals and thermal treatment. Cellulose 22 (2):1227–1241
17. Barhoum A, Samyn P, Öhlund T et al (2017) Review of recent research on flexible multifunctional nanopapers. Nanoscale 9:15181–15205
18. Bastioli C (2005) Handbook of biodegradable polymers. Rapra Technology Limited, Shawbury
19. Batista JA, Tanada-Palmu PS, Grosso CRF (2005) The effect of addition of fatty acids on pectin films. Ciênc Tecnol Aliment 25:781–788
20. BCC Research (2015) 'Biomaterial of the Future' Nanocellulose to Send Market Booming with 42.8% CAGR. https://www.bccresearch.com/pressroom/avm/biomaterial-of-the-future-nanocellulose-to-send-market-booming-with-42.8-percent-cagr?fbclid=IwAR2AiLB_BSpIdEPCaGGHaEhupzMsL675cwpyPJZakEJGBj30tUTRBMTJxnc. Accessed 9 Dec 2018.
21. Beuvalet M (2016) Application of cellulose nanomaterials in thermoplastic composites. Univesity of Waterloo, Thesis
22. Bonfatti EA Jr (2013) Oxygen delignification for kraft pulp with high kappa number. Dissertation, Unisity of São Paulo

23. Brinchi L, Cotana F, Fourtunati E et al (2013) Production of nanocrystalline cellulose from lignocellulosic biomass: technology and applications. Carbohydr Polym 94:154–169
24. Campano C, Merayo N, Balea A et al (2017) Mechanical and chemical dispersion of nanocelluloses to improve their reinforcing effect on recycled paper. Cellulose 25(1):269–280
25. Carbon Disclosure Project (2012) CDP supply chain report. https://www.marriott.com/MarriottInternational/CorporateResponsability/Performance_New_2016/SPG_PDFs/CDP-Supply-Chain-Report-2012.pdf. Accessed 6 Jun 2018
26. Chaker A, Mutjé P, Vilar MR et al (2014) Agriculture crop residues as a source for the production of nanofibrillated cellulose with low energy demand. Cellulose 21(6):4247–4259
27. Chen P, Yu H, Liu Y et al (2013) Concentration effects on the isolation and dynamic rheological behavior of cellulose nanofibers via ultrasonic processing. Cellulose 20(1):149–157
28. Cheng Q, Wang S, Rials TG (2009) Poly(vinyl alcohol) nanocomposites reinforced with cellulose fibrils isolated by high intensity ultrasonication. Compos Part A Appl Sci Manuf 40:218–224
29. Cheng G, Zhou M, Wei Y-J et al (2017) Comparison of mechanical reinforcement effects of cellulose nanocrystal; cellulose nanofiber; and microfibrillated cellulose in starch composites. Polym Compos https://doi.org/10.1002/pc.24685
30. Cherian BM, Leão AL, Souza SF, Thomas S, Pothan LA, Kottaisamy M (2010) Isolation of nanocellulose from pineapple leaf fibres by steam explosion. Carbohydr Polym 81:720–725
31. Colodette JL, Santos VLS (2015) General principles of bleaching. In: Colodette JL, Gomes FJB (eds) Cellulose pulp bleaching. Federal University of Viçosa, Viçosa, pp 173–202
32. Coutts RSP (2005) A review of Australian research into natural fibre cements composites. Cem Concr Compos 27(5):518–526
33. Cowie J, Bilek EM, Wegner T et al (2014) Market projections of cellulose nanomaterial-enabled products—part 2: volume estimates. Tappi J 13(6):57–69
34. Damásio RAP (2015) Characterization and nanoscale applications of nanofibrillated cellulose (NFC) and cellulose nanocrystals (CNC). Dissertation, Federal University of Viçosa
35. de Souza e Lima MM, Borsali R (2004) Rodlike cellulose microcrystals: structure, properties, and applications. Macromol Rapid Comm 25:771–787
36. Deep B, Abraham E, Cherian BM et al (2011) Structure, morphology and thermal characteristics of banana nano fibers obtained by steam explosion. Bioresour Technol 102:1988–1997
37. Dufresne A (2008) Processing of polymer nanocomposites reinforced with polysaccharide nanocrystals. Macromolecules 15(8):4111–4128
38. Eichhorn SJ (2011) Cellulose nanowhiskers: promising materials for advanced applications. Soft Matter 7(2):303–315
39. Eriksen Ø, Syverud K, Gregersen Ø (2008) The use of microfibrillated cellulose produced from kraft pulp as strength enhancer in TMP paper. Nord Pulp Pap Res J 23:299–304
40. Fakhouri FM, Fontes LCB, Gonçalves PVM et al (2007) Films and edible coatings based on native starches and gelatin in the conservation and sensory acceptance of Crimson grapes. J Food Sci Technol 27:369–375
41. Fang Z, Zhu H, Yuan Y et al (2014) Novel nanostructured paper with ultrahigh transparency and ultrahigh haze for solar cells. Nano Lett 14(2):765–773
42. Ferrer A, Filpponen I, Rodríguez A et al (2012) Valorization of residual empty palm fruit bunch fibers (EPFBF) by microfluidization: production of nanofibrillated cellulose and EPFBF nanopaper. Bioresour Technol 125:249–255
43. Frey MW (2008) Electrospinning cellulose and cellulose derivatives. Polym Rev 48(2):378–391

44. Frone AN, Panaitescu DM, Donescu D (2011) Some aspects concerning the isolation of cellulose micro-and nano-fibers. Sci Bull B Chem Mater Sci UPB 73(2):133–152
45. Frone AN, Panaitescu DM, Donescu D et al (2011) Preparation and characterization of PVA composites with cellulose nanofibers obtained by ultrasonication. BioResources 6(1):487–512
46. Fujisawa S, Okita Y, Fukuzumi H et al (2011) Preparation and characterization of TEMPO-oxidized cellulose nanofibrils films with free carboxyl groups. Carbohydr Polym 84(1):579–583
47. Fukuzumi H, Saito T, Iwata T et al (2009) Transparent and high gas barrier films of cellulose nanofibers prepared by TEMPO-mediated oxidation. Biomacromol 10(1):162–165
48. Ghosh AK (2011) Fundamentals of paper drying-theory and application from industrial perspective. In: Ahasan A (ed) Evaporation, codensation and heat transfer. InTech, London, pp 535–541
49. Gomide JL, Gomes FJB (2015) Production and composition of unbleached pulps. In: Colodette JL, Gomes FJB (ed) Cellulose pulp bleaching. Federal University of Viçosa, Viçosa, Brazil. pp 59–115
50. Gonzalez I, Boufi S, Pèlach M et al (2012) Nanofibrillated cellulose as paper additive in eucalyptus pulps. BioResources 7(4):5167–5180
51. González I, Vilaseca F, Alcalá M et al (2013) Effect of the combination of biobeating and NFC on the physico-mechanical properties of paper. Cellulose 20(3):1425–1435
52. Gullichsen J, Paulapuro H (2000) Papermaking science and technology: papermaking chemistry. Fapet Oy, Helsinki
53. Gunaratne SA (2001) Paper, printing and the printing press: a horizontally integrative machohistory analysis. Gazette 63(6):459–479
54. Habibi Y, Lucia LA, Rojas OJ (2010) Cellulose nanocrystals: chemistry, self-assembly, and applications. Chem Rev 110(6):3479–3500
55. Hart PW, Rudie AW (2012) The bleaching of pulp, 5th edn. TAPPI Press, Atlanta
56. Hassan ML, Mathew AP, Hassan EA et al (2012) Nanofibers from bagasse and rice straw: process optimization and properties. Wood Sci Technol 46(1):193–205
57. He W, Jiang X, Sun F et al (2014) Extraction and characterization of cellulose nanofibers from Phyllostachys nidularia munro via a combination of acid treatment and ultrasonication. BioResources 9(4):6876–87
58. Henriksson M, Berglund LA, Isaksson P et al (2008) Cellulose nanopaper structures of high toughness. Biomacromol 9:1579–1585
59. Hii C, Gregersen ØW, Chinga-Carrasco G et al (2012) The effect of MFC on the pressability and paper properties of TMP and GCC based sheets. Nord Pulp Pap Res J 27(2):388–396
60. Hinche M, Bassa AGMC, Rottmann W et al (2011) Biotech enhanced levels of syringil lignin improves *Eucalyptus* pulping efficiency. In: Paper presented at the 5th international colloquium on eucalyptus pulp, Federal Uniuversity of Viçosa, Porto Seguro, Brazil. 8–11 May 2011
61. Horáček P, Fajstavr M, Stojanović M (2017) The variability of wood density and compression strength of Norway spruce (*Picea abies*/L./Karst.) within the stem. Beskydy 10 (1–2):17–26
62. Hu Y, Topolkaraev V, Hitner A et al (2000) Measurement of water vapor transmission rate in highly permeable films. J Appl Polym Sci 81(3):1624–1633
63. Hu J, Zhang Q, Lee D-J (2018) Kraft lignin biorefinery: a proposal. Bioresour Technol 247:1181–1183
64. Hullmann A (2006). The economic development of nanotechnology—an indicators based analysis. European Commission, DG Research, Unit "Nano S&T—Convergent Science and Technologies". Staff working paper. http://nanotechnology.cz/storage/nanoarticle_.pdf. Acessed 22 Feb 2018
65. Hult EL, Iotti M, Lenes M (2010) Efficient approach to high barrier packaging using microfibrillar cellulose and shellac. Cellulose 17(3):575–586

66. Höglund H (2009) Mchanical pulping. In: Ek M, Gellerstedt G, Henriksson G (eds) Pulp and paper chemistry and technology, vol 2. Pulping chemistry and technology. Walter de Gruyter GmbH & Co., Berlin, pp 57–89
67. Ianuzzi A (ed) (2012) Greener products: the making and marketing of sustainable brands. CRC Press, Boca Raton
68. International Organization for Standardization (2017) ISO/TC 6: paper, board and pulps
69. Ireana Yusra AF, Juahir H, Firdaus NWNA et al (2018) Controlling of Green nanocellulose fiber properties produced by chemo-mechanical treatment process via SEM, TEM, AFM and image analyzer characterization. J Fundam Appl Sci 10(1s):1–17
70. Isogai A (2013) Wood nanocelluloses: fundamentals and applications as new bio-based nanomaterials. J Wood Sci 59(6):449–459
71. Janardhnan S, Sain M (2006) Isolation of cellulose microfibrils—an enzymathic approach. BioResources 1:176–188
72. Jardim CM, Colodette JL(2015) Pulp chromophoric groups. In: Colodette JL, Gomes FJB (eds) Cellulose pulp bleaching. Federal University of Viçosa, Viçosa, Brazil, pp 203–215
73. Jiang F, Hsieh YL (2014) Super water absorbing and shape memory nanocellulose aerogels from TEMPO-oxidized cellulose nanofibrils via cyclic freezing-thawing. J Mater Chem A 2(2):350–359
74. John MJ, Thomas S (2008) Biofibres and biocomposites. Carbohydr Polym 71:343–364
75. Julkapli NM, Bagheri S (2016) Developments in nano-additives for paper industry. J Wood Sci 62:117–130
76. Kajanto I, Kosonen M (2012) The potential use of micro-and nanofibrillated cellulose as a reinforcing element in paper. J-For 2(6):42–48
77. Kalia S, Boufi S, Celli A et al (2014) Nanofibrillated cellulose: surface modification and potential applications. Colloid Polym Sci 292(1):5–31
78. Keerati-u-rai M, Corredig M (2009) Effect of dynamic high pressure homogenization on the aggregation state of soy protein. J Agric Food Chem 57:3556–3562
79. Keller S (2013) Paper drying in the manufacturing process. In: Banik G, Brückle I (eds) Paper and water, 2nd edn. Butterworth Heinemann, Oxford, pp p173–211
80. Kim BY (2014) Investigation of coating color penetration depending on the properties of base paper. J Korea TAPPI 46(2):16–21
81. Klemm D, Kraner F, Moritz S et al (2011) Nanocelluloses: a new family of nature-based materials. Angew Chem Int Ed 50(2):5438–5466
82. Kobayashi S, Sakamoto J, Kimura S (2001) In vitro synthesis of cellulose and related polysaccharides. Progr Polym Sci 26(9):1525–1560
83. Kobayashi S, Uyama H, Masashi O (2001b) Enzymatic polymerization for precision polymer synthesis. Bull Chem Soc Jpn 74(4):635–613
84. Kolavali R (2013) Diffusion of ions in wood. Thesis, Chalmers University of Technology
85. Kumar V, Bollström R, Yang A et al (2014) Comparison of nano-and microfibrillated cellulose films. Cellulose 21(5):3443–3456
86. Kumar A, Singh SP, Singh AK (2014) Preparation and characterization of cellulose nanofibers from bleached pulp using a mechanical treatment method. Tappi J 13(5):25–31
87. Lai YZ (2012) Wood and wood products. In: Kent J (ed) Handbook of industrial chemistry and biotechnology. Springer, Boston, pp 1057–1115
88. Laine J, Lindström T, Nordmark GG et al (2002) Studies on topochemical modification of cellulosic fibres-part 2. The effect of carboxymethyl cellulose attachment on fibre swelling and paper strength. Nord Pulp Pap Res J 17(1):50–56
89. Lavoine N, Desloges I, Dufresne A et al (2012) Microfibrillated cellulose–its barrier properties and applications in cellulosic materials: a review. Carbohydr Polym 90(2): 735–764
90. Lee KY, Tamelin T, Schulter K (2012) High performance cellulose nanocomposites: comparing the reinforcing ability of bacterial cellulose and nanofibrillated cellulose. ACS Appl Mater Interfaces 4(8):4078–86

91. Lengowski EC (2016) Formation and characterization of films with nanocellulose. Federal University of Paraná, Thesis
92. Lengowski EC, Muñiz GIB, Nisgoski S et al (2013) Cellulose acquirement evaluation methods with different degrees of crystallinity. Sci Forest 41(98):185–194
93. Lengowski EC, Bonfatti EA Jr (2017) Incorporation of amphoteric starch and nanocellulose in paper. In: Paper presented at the 1st semana de aperfeiçoamento em engenharia florestal, 17–24 July 2017. Federal University of Paraná, Brazil, Curitiba city
94. Lenze CJ, Peksa CA, Sun W et al (2016) Intact and broken cellulose nanocrystals as model nanoparticles to promote dewatering and fine-particle retention during papermaking. Cellulose 23(6):3951–3962
95. Li W, Wang R, Liu S (2011) Nanocrystalline cellulose prepared from softwood kraft pulp via ultrasonic-assisted hydrolysis. BioResources 6(4):4271–4281
96. Li J, Wei X, Wang Q et al (2012) Homogeneous isolation of nanocellulose from sugarcane bagasse by high pressure homogenization. Carbohydr Polym 90(4):1609–1613
97. Liu C, Li B, Du H et al (2016) Properties of nanocellulose isolated from corncob residue using sulfuric acid, formic acid, oxidative and mechanical methods. Carbohydr Polym 151:716–724
98. Liu Y, Sui Y, Liu C et al (2018) A physically crosslinked polydopamine/nanocellulose hydrogel as potential versatile vehicles for drug delivery and wound healing. Carbohydr Polym 188:27–36
99. Loranger E, Jradi K, Daneault C (2012) Nanocellulose production by ultrasound-assisted TEMPO oxidation of kraft pulp on laboratory and pilot scales. In: IEEE international ultrasonics symposium, IUS, Taipei, Taiwan, article number 6562112, pp 953–995
100. López-Rubio A, Lagaron JM, Ankerfors M et al (2007) Enhanced film forming and film properties of amylopectin using micro-fibrillated cellulose. Carbohydr Polym 68(4):718–727
101. MacDonald RG (ed) (1968) The pulping of wood, 2nd edn. Mcgraw-Hill Inc., New York
102. Manninen M, Kajanto I, Happonen J et al (2011) The effect of microfibrillated cellulose addition on drying shrinkage and dimensional stability of wood-free paper. Nord Pulp Pap Res J 26(3):297–305
103. Mättänen M, Tikka P (2012) Determination of phenomena involved in impregnation of softwood chips. Part 1: method for calculating the true penetration degree. Nord Pulp Paper Res J 27(3):550–558
104. Mertaniemi H et al (2012) Functionalized porous microparticles of nanofibrillated cellulose for biomimetic hierarchically structured superhydrophobic surfaces. RSC Adv 2:2882–2886
105. Missio AL, Mattos BD, Ferreira DF et al (2018) Nanocellulose-tannin films: from trees to sustainable active packaging. J Clean Prod 2:143–151
106. Mohanty AK, Drzal LT, Misra M (2003) Nano reinforcement of bio-based polymers-the hope and reality. Polym Mater Sci Eng 88:60–61
107. Moon RJ, Martini A, Naim J et al (2011) Cellulose nanomaterials review: structure, properties and nanocomposites. Chem Soc Rev 40:3941–3994
108. Nakatsubo F, Kamitakahara H, Hori M (1996) Cationic ring-opening polymerization of 3,6-Di-O-benzyl-α-D-glucose 1,2,4-Orthopivalate and the first chemical synthesis of cellulose. J Am Chem Soc 118(7):1677–1681
109. Nelson K, Retsina T (2014) Innovative nanocellulose process breaks the cost barrier. Tappi J 13(5):19–23
110. Nygards S (2011) Nanocellulose in pigment coatings: aspects of barrier properties and printability in offset. Dissertation, Linköping University
111. Oliveira RCP, Mateus M, Santos DMF (2018) Chronoamperometric and chronopotentiometric investigation of kraft black liquor. Int J Hydrog Energy. https://doi.org/10.1016/j.ijhydene.2018.01.046
112. Osong SH, Norgren S, Engstrand P (2016) Processing of wood-based microfibrillated cellulose and nanofibrillated cellulose, and applications relating to papermaking: a review. Cellulose 23(1):93–123

113. Österberg M, Vartiainen J, Lucenius J et al (2013) A fast method to produce strong NFC films as a platform for barrier and functional materials. ACS Appl Mater Interfaces 5(11): 4640–4647
114. Pääkkö M, Ankerfors M, Kosonen H et al (2007) Enzymatic hydrolysis combined with mechanical shearing and high-pressure homogenization for nanoscale cellulose fibrils and strong gels. Biomacromol 8(6):1934–1941
115. Podsiadlo P, Choi S-Y, Shim B et al (2005) Molecularly engineered nanocomposites: layer-by-layer assembly of cellulose nanocrystals. Biomacromol 6(6):2914–2918
116. Potulski DC (2016) Influence of nanocellulose on the physical and mechanical properties of primary and recycled paper of Pinus and Eucalyptus. Federal University of Paraná, Thesis
117. Potulski DC, Muñiz GIB, Klock U et al (2014) The influence of incorporation of microfibrillated cellulose on mechanical strength properties of paper. Sci Forest 42(103): 345–351
118. Rahimi M, Behrooz R (2011) Effect of cellulose characteristic and hydrolyze conditions on morphology and size of nanocrystal cellulose extracted from wheat straw. Int J Polym Mater Po 60(8):529–541
119. Rampazzo R, Alkan D, Gazzoti S et al (2017) Cellulose nanocrystals from lignocellulosic raw materials; for oxygen barrier coatings on food packaging films. Packag Technol Sci. https://doi.org/10.1002/pts.2308
120. Rantanen J, Maloney TC (2013) Press dewatering and nip rewetting of paper containing nano-and microfibril cellulose. Nord Pulp Pap Res J 28(4):582–587
121. Rantanen J, Pirttiniemi J, Kuosmanen P et al (2014) Development of a microfibrillated cellulose composite web forming method. In: Paper presented at TAPPI international conference on nanotechnology for renewable materials, TAPPI, Vancouver, 23–26 June 2014
122. Richmond F (2014) Cellulose nanofibers use in coated paper. University of Maine, Thesis
123. Robles NB (2014) Tailoring cellulose nanofibrils for advanced materials. KTH Royal Institute of Technology, Stockholm
124. Rodionova G, Lenes M, Eriksen Ø et al (2011) Surface chemical modification of microfibrillated cellulose: improvement of barrier properties for packaging applications. Cellulose 18(1):127–134
125. Rojas J, Bedoya M, Ciro Y (2015) Current trends in the production of cellulose nanoparticles and nanocomposites for biomedical applications. In: Poletto M, Ornaghi HL Jr (eds) Cellulose—fundamental aspects and current trends. InTech, Rijeka, pp 193–228
126. Róz ALD (2003) The future of plastics: biodegradable and photodegradable. Polymers 13(4):4–5
127. Sacui IA, Nieuwendaal RC, Burnett DJ et al (2014) Comparison of the properties of cellulose nanocrystals and cellulose nanofibrils isolated from bacteria, tunicate, and wood processed using acid, enzymatic, mechanical, and oxidative methods. ACS Appl Mater Interfaces 6(9):6127–6138
128. Saini S, Sillard C, Belgacem MN et al (2016) Nisin anchored cellulose nanofibers for long term antimicrobial active food packaging. RSC Adv 6:12437–12445
129. Saito T, Isogai A (2005) A novel method to improve wet strength of paper. Tappi J 4(3):3–8
130. Saito T, Kimura S, Nishiyama Y, Isogai A (2007) Cellulose nanofibers prepared by TEMPO-mediated oxidation of native cellulose. Biomacromol 8:2485–2491
131. Saito T, Nishiyama Y, Putaux J-L et al (2006) Homogeneous suspensions of individualized microfibrils from TEMPO-catalyzed oxidation of native cellulose. Biomacromol 7(6):1687–1691
132. Samir M, Alloin F, Dufresne A (2005) Review of recent research into cellulosic whiskers, their properties and their application in nanocomposite field. Biomacromol 6:612–626
133. Samyn P, Barhoum A, Öhlund T et al (2018) Review: nanoparticles and nanostructured materials in papermaking. J Mater Sci 53(1):146–184
134. Santos SD, Sansígolo CA (2007) Wood basic density effect of *Eucalyptus grandis* x *Eucalyptus urophylla* clones on bleached pulp quality. Ciên Flor 17(1):53–63

135. Saunders RE, Pawlak JJ, Lee JM (2014) Properties of surface acetylated microfibrillated cellulose relative to intra- and inter-fibril bonding. Cellulose 21(3):1541–1552
136. Segura TES, Santos JRS, Sarto C et al (2016) Effect of kappa number variation on modified pulping of *Eucalyptus*. BioResources 11(4):9842–9855
137. Segura TES, Zanão M, Santos JRS et al (2012) Kraft pulping of the main hardwoods used around the world for pulp and paper production. In: 2012 TAPPI PEERS CONFERENCE, TAPPI Press, pp 1592–1599
138. Segura TES, Silva Júnior FG (2016) Potential of *C.citriodora* for kraft pulp production. TAPPI J 15(3):159–164
139. Sehaqui H, Allais M, Zhou Q et al (2011) Wood cellulose biocomposites with fibrous structures at micro- and nanoscale. Comp Sci Technol 71(3):382–387
140. Sehaqui H, Zhou Q, Berglund L (2013) Nanofibrillated cellulose for enhancement of strength in high-density paper structures. Nord Pulp Pap Res J 28(2):182–189
141. Serviço Nacional De Aprendizagem Industrial (2013) Cellulose. Senai, São Paulo
142. Sharma S, Zhang X, Nair SS et al (2014) Thermally enhanced high performance cellulose nano fibril barrier membranes. RSC Adv 4:45136–45142
143. Shatkin JA, Wegner TH, Bilek EM et al (2014) Market projections of cellulose nanomaterial-enabled products—part 1: applications. Tappi J 13(5):9–12
144. Sinko R, Qin X, Keten S (2015) Interfacial mechanics of cellulose nanocrystals. MRS Bull 40(4):340–348
145. Siqueira G, Bras J, Dufresne A (2009) Cellulose whiskers versus microfibrils: influence of the nature of the nanoparticle and its surface functionalization on the thermal and mechanical properties of nanocomposites. Biomacromol 10(2):425–432
146. Siró I, Plackett D (2010) Microfibrillated cellulose and new nanocomposite materials: a review. Cellulose 17(3):459–494
147. Siró I, Plackett D, Hedenqvist M et al (2011) Highly transparent films from carboxymethylated microfibrillated cellulose: the effect of multiple homogenization steps on key properties. J Appl Polym Sci 119(5):2652–2660
148. Sixta H (2006) Handbook of pulp. Wiley-VCH Verlag GmbH & Co, KGaA, Weinheim
149. Sjöström E (2013) Wood chemistry fundamentals and applications. Academic Press, New York
150. Smook G (2016) Handbook for pulp and paper technologists. TAPPI Press, Atlanta
151. Sofla MRK, Brown RJ, Tsuzuki T et al (2016) A comparison of cellulose nanocrystals and cellulose nanofibers extracted from bagasse using acid and ball milling methods. Adv Nat Sci Nanosci Nanotech 7:035004
152. Souza AC, Benze R, Ferrão ES et al (2012) Cassava starch biodegradable films: influence of glycerol and clay nanoparticles content on tensile and barrier properties and glass transition temperature. LWT J Food Sci Technol 46(1):110–117
153. Spence KL, Venditti RA, Rojas OJ et al (2010) The effect of chemical composition on microfibrillar cellulose films from wood pulps: water interactions and physical properties for packaging applications. Cellulose 17(4):835–848
154. Spence KL, Venditti RA, Rojas OJ et al (2011) A comparative study of energy consumption and physical properties of microfibrillated cellulose produced by different processing methods. Cellulose 18(4):1097–1111
155. Stelte W, Sanadi AR (2009) Preparation and characterization of cellulose nanofibers from two commercial hardwood and softwood pulps. Ind Eng Chem Res 48(24):11211–9
156. Su J, Mosse WKL, Sharman S et al (2013) Effect of tethered and free microfibrillated cellulose (MFC) on the properties of paper composites. Cellulose 20(4):1925–1935
157. Swinehart D (2012) Fundamentals of refining. MeadWestvaco Center for Packaging Innovation, Rayleigh
158. Syverud K, Stenius P (2009) Strength and barrier properties of MFC films. Cellulose 16:75–85
159. Taipale T, Österberg M, Nykänen A et al (2010) Effect of microfibrillated cellulose and fines on the drainage of kraft pulp suspensions and paper strength. Cellulose 17(5):1005–1020

160. Taiz L, Zeiger E (2017) Plant physiology, 6th edn. Sinauer Associates, Sunderland
161. Technical Association of Pulp and Paper Industry (2013) T 236 om-13: kappa number of pulp. TAPPI Press, Atlanta
162. Thoorens G, Krier F, Leclercq B et al (2014) Microcrystalline cellulose, a direct compression binder in a quality by design environment—a review. I J Pharm 473(1–2):64–72
163. Tognetta L, Santos O, Dragoni O et al (2014) Paper. SENAI, São Paulo
164. Torvinen K (2014) Binding fillers for high filler content papers by using CNF/CMF. In: Paper presented at international conference on nanotechnology for renewable materials, TAPPI, Vancouver, 23–26 June 2014
165. Tuovinen L, Peltonen S, Jarvinen K (2003) Drug release from starch-acetate films. J Control Release 91(4):345–354
166. Usta I (2005) A review of the configuration of bordered pits to simulate the fluid flow. Maderas Cien Tecnol 7(2):121–132
167. Van Heiningen ARP (2006) Converting a kraft pulp mill into an integrated forest biorefinery. Pulp Pap Canada 107(6):38–43
168. Valenzuela A, Bentley JM, Lorenz RD (2005) Evaluation of torsional oscillations in paper machine sections. IEEE Trans IndusAppl 41(2):493–501
169. Viana LC, Muñiz GIB, Hein PRG et al (2016) NIR spectroscopy can evaluate the crystallinity and the tensile and burst strengths of nanocellulosic films. Maderas Cienc Tecnol 18(3):493–504
170. Viana LC, Muniz GIB, Magalhaes WLE (2017) Physical and mechanical properties of nano-structed films produced from the unbleached *Pinus* sp. kraft pulp. Sci Forest 45(116): 653–662
171. Vivian MA, Segura TES, Bonfatti Júnior EA et al (2015) Wood quality of *Pinus taeda* and *Pinus sylvestris* for kraft pulp production. Sci Forest 43(105):183–191
172. Vroom KE (1957) The H factor: a means of expressing cooking times and temperatures as a single variable. Pulp Pap Canada 58(3):228–231
173. Wang Y, Cao X, Zhang L (2006) Effects of cellulose whiskers on properties of soy protein thermoplastics. Macromol Biosci 6(7):524–531
174. Wang H, Li D, Zhang R (2013) Preparation of ultralong cellulose nanofibers and optically transparent nanopapers derived from waste corrugated paper pulp. BioResources 8:1374–1384
175. Wang B, Sain M (2007) Isolation of nanofibers from soybean source and their reinforcing capability on synthetic polymers. Compos Sci Technol 67(11–12):2521–2527
176. Wang B, Sain M (2007) Dispersion of soybean stock-based nanofiber in a plastic matrix. Polym Int 56(4):538–546
177. Wang B, Sain M, Oksman K (2007) Study of structural morphology of hemp fiber from the micro to the nanoscale. Appl Compos Mater 14:89–103
178. Wang Y, Wei Y, Li J et al (2013) Homogeneous isolation of nanocellulose from cotton cellulose by high pressure homogenization. J Mater Sci Chem Eng 1(5):49–52
179. Wang Q, Zhao X, Zhu JY (2014) Kinetics of strong acid hydrolysis of a bleached kraft pulp for producing cellulose nanocrystals (CNCs). Ind Eng Chem Res 53(27):11007–11014
180. Wegner T, Skog KE, Ince PJ et al (2010) Uses and desirable properties of wood in the 21st century. J For 108(4):165–173
181. Wågberg L, Decher G, Norgren M et al (2008) The build-up of polyelectrolyte multilayers of microfibrillated cellulose and cationic polyelectrolytes. Langmuir 24:784–795
182. Xie C, Liu Z-M, Wu P et al (2013) Optimization of preparation technology of alkali pretreated reed pulp nano-cellulose. Chem Ind For Prod 33(1):32–36
183. Xu Q, Li W, Cheng Z et al (2014) TEMPO/NaBr/NaClO-mediated surface oxidation of nanocrystalline cellulose and its microparticulate retention system with cationic polyacrylamide. BioResources 9(1):994–1006
184. Xu Y, Yin X, Lin T et al (2018) Silica retention by the addition of sodium metaaluminate during the impregnation stage of bamboo kraft pulping. J Wood Chem Technol 38(1):35–43

185. Zeni M et al (2015) Preparation of microcellulose (Mcc) and nanocellulose (Ncc) from eucalyptus kraft ssp pulp. Polym Sci 1:1–5
186. Zeni M, Favero D, Pacheco K et al (2015) Preparation of microcellulose (Mcc) and nanocellulose (Ncc) from Eucalyptus kraft ssp pulp. Polym Sci 1:1–5
187. Zimmermann T, Bordeanu N, Strub E (2010) Properties of nanofibrillated cellulose from different raw materials and its reinforcement potential. Carbohydr Polym 79:1086–1093

Impact of Nanoparticle Shape, Size, and Properties of Silver Nanocomposites and Their Applications

Arpita Hazra Chowdhury, Rinku Debnath, Sk. Manirul Islam and Tanima Saha

1 Introduction

The genesis of nanotechnology can be traced back to 1959 when in a meeting of the American Physical Society, Richard Feynman first introduced the branch of science [1]. One can define nanotechnology as the production and modification of structures which have at least one dimension less than 100 nm. The birth of nanoscience could be attributed to Michael Faraday as he had reported the intense red colour of stained glass, which originated from small particles of goldin 1857. He also reported that different size of gold particles gave rise to different resultant colours [2]. Recently, nanomaterials have emerged as very important materials in the scientific world as these act as a bridge between bulk materials and isolated atoms and molecules. They increase fractions of atoms at the surface due to their large surface to volume ratio. Nanomaterials have superior properties than the bulk substances due to the following attributes, like mechanical strength, thermal stability, catalytic activity, electrical conductivity, magnetic properties and optical properties etc. The applications of nanomaterials are continuously expanding and their applications in different fields are shown below (Fig. 1).

Arpita Hazra Chowdhury, Rinku Debnath—These authors contributed equally to this manuscript.

A. Hazra Chowdhury · Sk.Manirul Islam (✉)
Department of Chemistry, University of Kalyani, Kalyani, Nadia 741235,
West Bengal, India
e-mail: manir65@rediffmail.com

R. Debnath · T. Saha (✉)
Department of Molecular Biology and Biotechnology,
University of Kalyani, Kalyani, Nadia 741235, West Bengal, India
e-mail: sahatanima@klyuniv.ac.in

© Springer Nature Switzerland AG 2019
Inamuddin et al. (eds.), *Sustainable Polymer Composites and Nanocomposites*,
https://doi.org/10.1007/978-3-030-05399-4_37

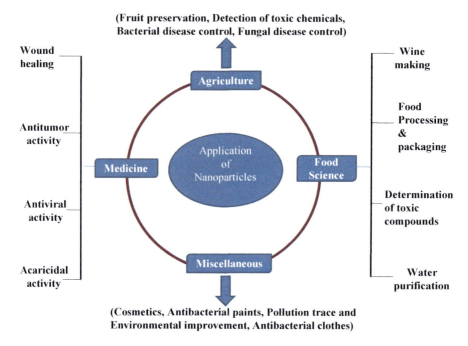

Fig. 1 Schematic illustration of different applications of nanoparticles

Researchers have growing interest in silver nanoparticles (AgNPs) as they have remarkable localized surface plasmon resonance and antimicrobial properties, which render them with unique properties for applications in broad-spectrum surface-enhanced Raman spectroscopy (SERS) [3, 4], as antimicrobial agents [5–7], biological/chemical sensors and biomedicine materials [8–10], biomarkers [11–13] and so on. Size of AgNPs usually varies within 1–100 nm. They are incorporated into industrial applications of catalysis, electronics, and photonics as they have unique electrical, optical and thermal properties. Recently, many synthetic methods and approaches for preparing AgNPs have been reported through physical, chemical, photochemical and biological routes. Every method has its own advantages and disadvantages like costs, scalability, particle sizes and size distribution and so on [14–18].

In recent years, nanocomposite (NC) materials have drawn much attention and interest at industrial and academic level due to their improved properties than single metal nanoparticles. Nanocomposite can be defined as the combination of materials to develop new properties of the materials where one of the materials has a size in the range of 1–100 nm. So, there are two parts to nanocomposite i.e. continuous phase and discontinuous reinforcing phase. Hence, nanocomposites can have a combined or have completely different electrochemical, mechanical, electrical, thermal, optical and catalytic properties of the component materials [19–23].

There can be different phases of nanocomposites such as zero-dimensional (0D) (core-shell), 1D (nanowires and nanotubes), 2D (lamellar) and 3D (metal matrix composites) [9]. Nanocomposite materials have developed as an appropriate replacement to overcome the limitations of microcomposites and monolithic while having synthetic challenges like the control of elemental composition and stoichiometry in the nanocluster phase. Nanocomposite materials have uniqueness in design and property combinations that are not observed in conventional composites and these properties establish the nanocomposites materials as the materials of the 21st century. Even though the first speculation on these properties was reported as early as 1992 [24], the general understanding of them is yet to be established.

The focus of this chapter is on silver nanocomposite systems. These NC materials have promising properties which make them suitable for wide range of structural and functional applications in various fields.

2 Different Synthesis Methods of Silver Nanoparticles

Different synthesis methods and approaches for AgNPs production have been reported by using chemical, physical, photochemical and biological routes. Each method has its own advantages and disadvantages involving costs, scalability, particle sizes and size distribution and so on [14–17, 25].

2.1 Physical Methods

In physical methods, metal nanoparticles are generally synthesized by evaporation-condensation, which is carried out at atmospheric pressure using a tube furnace. The source material is vaporized into a carrier gas, within a boat centred at the furnace. Different kinds of nanoparticles such as Ag, Au, PbS, and fullerene, have been produced using the evaporation/condensation technique [26–28]. Silver nanoparticles have been synthesized by laser ablation of the solution of metallic bulk materials [29, 30]. The most important advantage of laser ablation technique over another conventional method for synthesizing metal colloids is the absence of toxic chemical reagents in solutions. In summary, the physical synthesis utilizes the physical energies for preparing AgNPs with nearly narrow size distribution. The physical approach can produce large quantities of AgNPs in a single process as well as this is also the most effective method to produce AgNPs. On the other hand, primary costs for the investment of equipment have to be kept in mind before implementing such methods.

2.2 Photochemical Methods

This method is based on the light-assisted the reduction of the metal cation M^{n+} to M^0. The mechanism of this method is based on the addition of one or more electrons to a photoexcited species. The aqueous and alcoholic solution of silver perchlorate ($AgClO_4$) was subjected to photoreduction by UV-light irradiation at 254 nm. This photochemical reaction involved electron transfer from a solvent molecule to the electronically excited state of Ag^+ to form Ag^0. In most of the cases, UV excitation is usually required as the metal cations and/or the metal salts absorb only in this region. This method is advantageous in harsh conditions like increased temperatures can be avoidable resulting in effective control of shape and size of Silver nanoparticle (AgNP) [31].

$$Ag^+ + H_2O \xrightarrow{h\nu} Ag^0 + H^+ + HO^\bullet$$

$$Ag^+ + RCH_2OH \xrightarrow{h\nu} Ag^0 + H^+ + R\overset{\bullet}{C}HOH$$

$$Ag^+ + R\overset{\bullet}{C}HOH \xrightarrow{h\nu} Ag^0 + H^+ + RCHO$$

$$nAg^0 \rightarrow AgNP$$

Various photo-induced synthetic processes have been developed recently. Huang and Yang [32] synthesized AgNPs by photoreduction of $AgNO_3$ in layered inorganic clay suspensions. This suspension acts as a stabilizing agent which prevents aggregation of nanoparticles. Light irradiation leads to the disintegration of the AgNPs into a smaller size with a single mode distribution until a relatively stable size and diameter distribution were achieved [32]. However, this method requires high-end equipment and experimental environment.

2.3 Biological Methods

Lately, biosynthetic methods have emerged as a facile and simple alternative to more complex chemical synthetic methods to prepare AgNPs. In biosynthetic methods, natural reducing agents like polysaccharides, biological microorganisms like bacteria, and fungi or plant extracts, i.e. green chemistry are used. This method includes a broad range of natural resources for the synthesis of AgNPs. This method has several advantages over conventional chemical routes of synthesis and it is also an environment-friendly approach as well as a low-cost technique for nanoparticle synthesis. However, the drawback of this method is that it is not easy to synthesize a large quantity of AgNPs by using biological synthesis.

2.3.1 Microbe-Assisted Synthesis

There are two types of microbe-assisted synthesis of AgNPs:

(a) Bacterial Synthesis: Bacteria can produce inorganic materials either intra- or extracellularly. Thus, they act as potential biofactories for the synthesis of nanoparticles such as gold and silver. Specifically, silver is widely known for its biological properties. AgNPs with different shapes and sizes can be effectively synthesized by varying different parameters using different bacteria such as *Klebsiella pneumonia*, *Lactobacillusfermentum*, *Bacillus flexus*, *Escherichiacoli*, and *Enterobacter cloacae* at different pH, temperatures and concentrations of $AgNO_3$ solutions [33–36].

(b) Fungal Synthesis: Researchers have been interested in patenting their research work on the microbial synthesis of nanoparticles. One of these significant works is the synthesis of AgNPs with particle size 5–50 nm, harnessing wet biomass of *Trichoderma reesei* fungus after 120 h of continuous shaking at 28 °C [37]. When antimicrobial properties of AgNPs were tested on *Escherichia Coli* and *Staphylococcus aureus*, it was found that *E. coli* was more susceptible to silver nanoparticles than *S. aureus* [38]. Extracellular nanoparticles were formed by using thermophilic fungus *Humicola* sp. which reacted with Ag(+) ions and reduced the precursor solution [39].

2.3.2 Plant-Mediated Synthesis

Microbe-mediated synthesis requires highly aseptic conditions and their maintenance which lowers the industrial feasibility of this method. Therefore, the plant extract mediated synthesis (Fig. 2) of nanoparticles is potentially advantageous

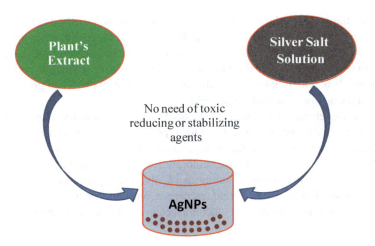

Fig. 2 Schematic diagram showing the one pot plant-mediated synthesis of silver nanoparticles

Table 1 Synthesis of silver nanoparticles using different plant extracts

Plants	Plant's part	Particle size (nm)	Particle shape	References
Averrhoa carambola	Fruit	12–16	Spherical	[41]
Abutilon indicum	Leaf	7–17	Spherical	[42]
Withania somnifera	Leaf	5–30	Spherical	[43]
Eclipta prostrate	Leaf	35–60	Triangles, pentagons, hexagons	[44]
Nelumbo nucifera	Leaf	25–80	Spherical, triangular	[45]
Citrus sinensis	Peel	10–35	Spherical	[46]
Pelargonium graveolens	Leaf	16–20	Spherical	[47]
Tanaetum vulgare	Fruit	10–40	Spherical	[48]
Tea extract	Leaf	20–90	Spherical	[49]
Tribulus terrestris	Fruit	16–28	Spherical	[50]

over microbe assisted synthesis due to the ease of improvement, the less biohazard and elaborate process of maintaining cell cultures [40]. It is the best way to synthesize nanoparticles without using any toxic chemical reducing and stabilizing agents as it provides natural capping agents for the stabilization of silver nanoparticles. A list of silver nanoparticles synthesis using different plant extracts is given in Table 1.

2.4 Chemical Methods

Besides all the methods described earlier, the most common method is the chemical reduction method for nanoparticle synthesis because of its convenience and simple equipment. It is required to control the growth of metal nanoparticles to prepare nanoparticles with a spherical shape and small size with a narrow particle size distribution. It is widely known that chemical reduction method can produce silver nanoparticles at low cost with high yield.

Usually, the chemical synthesis process of AgNPs in solution requires the following three main components: (1) metal precursors, (2) reducing agents and (3) stabilizing/capping agents. Generally, silver nitrate [51–54], silver acetate [55, 56], silver citrate [56–58], and silver chlorate [56, 57, 59] are the most frequently used precursors for the chemical synthesis of silver nanoparticles. Among various reducing agents, borohydride, citrate, ascorbate, and compounds with hydroxyl or carboxyl groups like aldehydes, alcohol, carbohydrates, and their derivatives

[60–63] are the most commonly used reductants. The colloidal solution formation from the reduction of silver salts involves two stages (i) nucleation and (ii) subsequent growth. It is also observed that the size and the shape of synthesized AgNPs are strongly dependent on these two stages. The two phenomenon i.e., nucleation and growth of initial nuclei can be controlled by adjusting different reaction parameters like reaction temperature, pH, precursors, reducing agents (i.e. NaBH4, ethylene glycol, glucose) and stabilizing agents (i.e. PVA, PVP, sodium oleate) [64–66] (Fig. 3).

3 Nanocomposite Systems

Nanocomposite systems have been extensively studied since the 1990s and, it is observed that a steady and continuous increase has taken place in the number of publications on the subject, including time to time reviews. Nanocomposites can be defined as multiphase solid materials in which one of the phases has dimensions of less than 100 nm. There are two parts of NC: (i) continuous phase and (ii) discontinuous reinforcing phase. Nanocomposite materials can be classified into three different categories according to their matrix materials.

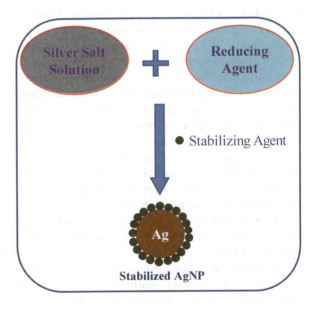

Fig. 3 Schematic diagram showing the one-pot chemical synthesis of AgNPs

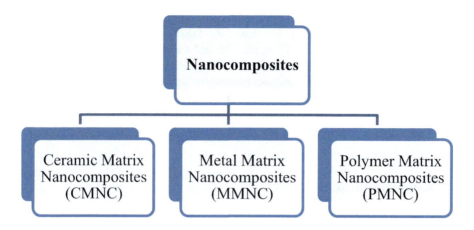

Hence, nanocomposites can have a combined or have markedly different electrical, mechanical, electrochemical, thermal, catalytic and optical properties of the component materials [20–23].

3.1 Silver-Ceramic Matrix Nanocomposites

The synthesis of nanocomposites composed of noble metals (Au, Ag, Pt and Pd, as well as AuAg alloy) and ceramic matrixes such as metal oxides (ZnO, TiO_2, Cu_2O, MnO_2, Fe_2O_3, WO_3, CeO_2 etc) has received considerable attention in recent years for their applications in heterogeneous catalysis, photocatalysis, drug delivery, solar cells, surface enhanced Raman spectroscopy and many other important areas. Now-a-days, among the many nano-catalysts developed, controllable integration of different noble metals (e.g., Au, Ag, Pt, and Pd) and metal oxides (e.g., TiO_2, CeO_2, and ZrO_2) into single nanostructures has become one of the hottest research topics as they not only merge the functions of individual nanoparticles (NPs) but also show a unique combined and synergetic catalytic properties compared to the single-component materials. Generally, these composites are easily prepared by different methods like impregnation [67], co-precipitation [68], deposition-precipitation [69] and many more.

Liu et al. [70] synthesized plasmonic silver nanoparticle incorporated mesoporous metal–oxide (MMO) semiconductors to get increased photocatalysis. Different typical MMO such as TiO_2, ZnO, and CeO_2 semiconductors were synthesized by integrating evaporation-induced self-assembly and in situ pyrolysis of metal salts. Then Ag nanoparticles of different amounts were then loaded in these MMO semiconductors through an efficient photo-deposition process. The Ag nanocrystals were synthesized with sizes of 50–100 nm and then they were embedded in MMO semiconductors (Fig. 4) [70].

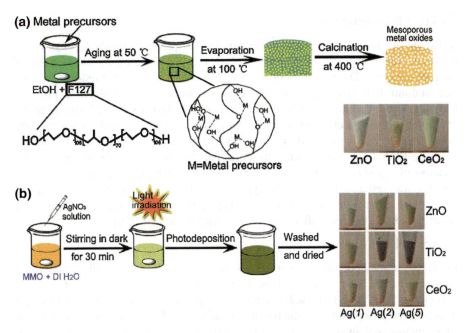

Fig. 4 Schematic diagram showing the synthesis of **a** mesoporous metal oxides (MMO) and **b** Ag/MMO nanocomposites [70]

3.2 Silver-Metal Matrix Nanocomposites

Metal matrix nanocomposites (MMNC) can be described as the materials where nanosized rein-forcement material is implanted in a ductile metal or alloy matrix. Most common metal matrix nanocomposite for silver is gold-silver nanocomposite. Choudhury et al. [71, 72] synthesized Ag—Aunanocomposite substrates by a one-step galvanic replacement reaction from thin films of silver, coated on glass slides. Then there was deposition of metallic layers on the cleaned slides using a vacuum evaporation chamber under high vacuum ($<5 \times 10^{-7}$ Torr). At first, there was an adhesion layer of chromium deposited on the slides, then gold ($\sim$5 nm) and silver ($\sim$600 nm) films were deposited, without breaking vacuum [72].

3.3 Silver-Polymer Matrix Nanocomposites

Recently, researchers in various fields incorporated silver nanoparticles into the polymer matrix to enhance its performance. Polymer materials act not only as an excellent host for incorporating nanoparticles but also terminate the growth of the particles by controlling their nucleation [73]. Silver-polymer nanocomposites can be prepared by using two main approaches.

(a) Insitu Polymerization

In the in situ method, the silver nanoparticles are prepared inside a polymer by dissolving metallic precursor salt in the polymer, followed by chemical reduction of the precursor salt. The reduction of Ag^+ to Ag^0 takes place. Several reducing agents like sodium borohydride, hydrazine etc. are used in the reduction process. Curcumin-loaded chitosan-PVA silver nanoparticles film (CCPSNP) was prepared by adding $AgNO_3$ to chitosan solution. AgNP solution was formed, to which, poly (vinyl alcohol), glutaraldehyde (a crosslinker) and curcumin solution were added [74]. Ag-PVA film was prepared by Porel and his group by mixing an aqueous solution of $AgNO_3$ and poly(vinyl alcohol) (PVA) wherein the silver precursor $AgNO_3$ was reduced by the hydroxyl groups of the PVA macromolecule [75]. Further, Ag-polyaniline nanocomposite was prepared from a mixture of aniline and silver nitrate as precursor after well rinsing with nitric acid [76].

(b) Exsitu Polymerization

In the ex situ method, silver nanoparticles are formed first and then dispersed into a polymer matrix. The prepared nanoparticles show higher dispersibility in the polymer and get long-term stability against aggregation. Sonication provided the dispersion of nanoparticles in the polymer matrix [77]. Thin nanocomposite films of silver nanocrystal in polystyrene matrix were prepared by sonicating polystyrene and silver nanoparticles with toluene for even dispersion of the nanoparticles [78]. Silver embedded mesoporous polyaniline nanocomposite [79] and mesoporous cross-linked polymer (polyacrylic acid) (MCP-1) supported silver nanoparticles [80] were prepared by dispersing mesoporous polymers in TRIS stabilized AgNPs and then stirring it at room temperature for 1 h. At the end of the reaction, black coloured Ag-NPs containing mesoporous polymer mPANI/Ag and mesoporous polyacrylic acid/Ag nanocomposites were obtained (Fig. 5).

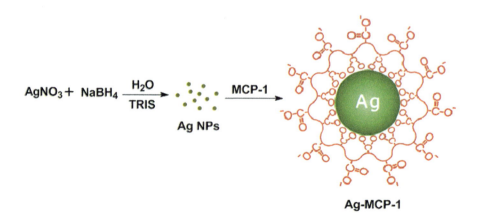

Fig. 5 Schematic diagram showing the synthesis of mesoporouspolyacrylic acid/Ag nanocomposite [80]

4 Applications of Silver Nanocomposites

Recently silver nanocomposites are used extensively as anti-microbial agents in the health industry, food packaging, water treatment, agriculture, winemaking, and textile coatings. They are used as absorbents, photocatalysts, and sensors for detection and removal of environmental pollutants. Silver nanoparticles are also used in cosmetics, personal care products, and electronic devices.

4.1 Medical Field

Silver nanocomposites represent the most important polymer based functionalizing agent among the numerous nanocomposites due to their antimicrobial properties [81, 82]. Two different mechanisms are responsible for antimicrobial activity of silver nanoparticles, (a) adhesion to the cell surface, degradation of lipopolysaccharides, increasing permeability [83] and (b) penetration inside the bacterial cell, DNA damaging [84]. The silver nanoparticles release silver ions, which bind to electron donor groups in biological molecules containing sulfur, oxygen or nitrogen.

Liu et al. [85] demonstrated the doping of TiO_2 with silver that greatly improved photocatalytic bacterial inactivation. Polyethyleneglycol-polyurethane-TiO_2 silver nanocomposites showed antimicrobial activity against *E. coli* and *Bacillus subtilis* [86]. Silver nanoparticle encapsulated porous PMMA (poly methyl methacrylate) spheres have also shown good antimicrobial activity [87, 88]. Silver nanocomposites of POA (porous aluminium oxide) are effective against both Gram-negative and Gram-positive bacteria, *E. coli* and *Staphylococcus epidermidis*. Nadagouda and Varma [89] studied the applications of Ag nanocomposites with biodegradable carboxymethylcellulose in antimicrobial and antifungal coatings and biomedical devices. Zeolite is a porous crystalline material of hydrated sodium aluminosilicate. Silver can electrostatically bind to zeolite with high affinity and form silver-zeolite (SZ). Human saliva contains several kinds of cations which help in the release of silver ions from SZ and under anaerobic conditions they inhibit the growth of several oral bacteria [90]. SZ incorporated mouth rinses, acrylic resins, and tissue conditioners are used in the dental field [91]. Matsuura et al. [92] showed that tissue conditioners containing SZ exhibit the antimicrobial effect against *Candida albicans*, *Pseudomonas aeruginosa* and *Staphylococcus aureus* for a week and the use of SZ incorporated mouth rinse for 5 days showed a reduced plaque score [93].

Recently, synthesis of graphene oxide (GO) metal nanocomposites has gained attention due to excellent biocompatibility and high antibacterial activity of GO. They have possibilities to be used as an antibacterial reinforcement in biomaterials, exploration of the antiseptic properties and cytotoxic activity of GO-containing nanocomposites. Silver nanoparticles functionalized magnetic graphene oxide nanocomposites (GO/Fe_3O_4/PEI/Ag) were synthesized via in situ generate silver nanoparticles on magnetic graphene oxide surface using polyethylenimine as reducing and stabilizing agent. The GO/Fe$_3$O$_4$/PEI/Ag nanocomposites highly enrich the low-abundant glycopeptides from complex biological samples. These nanocomposites might be used as low-abundant disease biomarkers. Dendritic polymer encapsulated silver nanoparticles have antimicrobial activity, besides this, they are also used as markers for cell labelling [94]. Lesniak et al. [95] demonstrated the synthesis of silver/dendrimer nanocomposites that can be used for in vitro cell labelling.

Carbohydrates such as glucose, lactose, and oligosaccharides significantly improve the cytotoxicity and cellular uptake nature of silver nanoparticles. To kill the cancerous cells in phototherapy, the carbohydrate modified silver nanoparticles can be used as a new tool [96]. Silver nanoparticles are used for the identification of calf thymus DNA (*ct*-DNA) due to their strong fluorescence signal [97]. For environmental protection and some disease detection applications such as disease diagnosis, drug screening, epidemic prevention, the oligonucleotide probe bound silver nanoparticles are used as an indicator in electrochemical DNA sensors that are able to pair with the sample DNA sequence. Based on aggregation, induced by sequence-specific hybridization, silver nanoparticles are used in the design of colourimetric assays [98]. In mouse models, silver nanoparticles exhibit anti-inflammatory property by inhibiting the interferon gamma activity and tumour necrosis factor alpha. Hence, they can play an important role in anti-inflammatory therapies [99, 100].

The emerging field of nanomedicine seeks to exploit the novel properties of engineered nanomaterials for diagnostics and therapeutic applications. The engineered nanoparticles are used to carry drug payloads, image contrast agents or gene therapeutics for diagnosing and treatment. Due to the antimicrobial activity of silver nanoparticles, they are now increasingly used in wound dressings, catheters, and various household products. The chitin-AgNP composite scaffolds have good blood clotting ability hence they are widely used in wound healing and tissue engineering applications [101]. In orthopaedic implants, prostheses, vascular grafts and wound dressings, nanosilver reinforced polymers are commonly used. Surgical instruments, contraceptive devices, creams, gels also contain nanosilver [102–104].

Cattles are affected and die by a tick *Rhipicephalus microplus* (Boophilus) leading to economic losses associated with milk, meat, and leather production [105]. Continuous use of acaricides (used for control of ticks) not only contaminates the environment and animal products but also helps in selection of chemical-resistant ticks. Bergeson [106] has reported that more than 100 silver containing pesticides are synthesized due to their antimicrobial and photocatalytic properties. Sodium dodecyl sulfate (SDS) modified photocatalytic TiO$_2$/Ag nanomaterial conjugated with dimethomorph (DMM) are used as nanopesticide in agriculture [107].

4.2 Food Industry

Active packaging is one type of innovative food packaging concept in which active functions like scavenging of oxygen, moisture or ethylene, emission of ethanol and flavours in addition to the antimicrobial activity are involved. Whereas in antimicrobial packaging, to reduce pathogen contamination and extend the shelf-life of food, an antimicrobial substance is included in the packaging system [108]. But the antimicrobial substances are neutralized or diffuse rapidly into the food, so their direct application has limited benefits. Incorporation of the antimicrobial substances into polymers allow their slow release in the packaging system for an extended period and prolong their effect during transport and storage of food [108]. The biopolymer-based films and coatings provide physical protection to foods, improve food quality and enhance the shelf life of food products due to their properties like - acting as barriers against moisture, oxygen, flavour, aroma, and oil [109]. Furthermore, biopolymer films act as antioxidants, antifungal agents, antimicrobials, colours, and other nutrients which are important in the food preservation [110, 111]. Due to these potential applications, biopolymer-based antimicrobial films are used in the food industry including meat, fish, poultry, cereals, cheese, fruits, and vegetables [112–114]. One of the promising candidates in this field is chitosan-based nanocomposite films. Chitosan is nontoxic, biodegradable, and biocompatible so the chitosan-based Ag nanocomposites are used in films and coatings [115, 116]. Silver nanoparticles are cost-effective and have a wide range of applicability than other antimicrobial substances. Hence, they are most commonly used polymer additive for antimicrobial food packaging [108, 117]. Besides the antimicrobial activity, they also extend the shelf life of fruits and vegetables by absorbing and decomposing ethylene from these foodstuffs [118].

Although silver releasing systems are commonly used as food additives in antimicrobial food packaging the single form of silver-releasing system is used in ion-exchange from microporous minerals sector in which silver ions partially replace the naturally occurring sodium ions in clays or other porous minerals [119]. When silver ions come in contact with moisture, they are again substituted by sodium ions present in the release environment and leached from the surface. This is practical, as the release of silver ions will depend on the amount of saline moisture, which is a crucial risk factor for the development of microbes on surfaces. A wide range of polymers and other surfaces are used to incorporate the substituted minerals. They are able to withstand any kind of plastic processing or operating temperature compared to natural antimicrobial substances for their low migration rate and high melting point [120]. Different clays like montmorillonites (MMT) [121] or tobermorites [122], silver zirconium phosphates [123] and silver zeolites [124–128] are used as minerals in this technology. These materials are generally manufactured as a 3–6 μm thick layer containing 1–5% silver content for coating on polymeric or stainless steel surfaces. The food processing equipment like cutlery, cutting boards, countertops, containers, etc. are coated with this type of layer.

Recently silver nanoparticles are marketed as nutrition supplement in form of a colloidal solution. The colloidal silver has high particle surface area for maximum effectiveness. Among the all-metal colloidal mineral supplements, the silver products are most popular [129].

In wine making SO_2 is an important antioxidant that inhibits fungal growth, effects of dissolved oxygen and endogenous enzymes of grape-polyphenoloxidase, tyrosinase and peroxidase [130]. SO_2 also improves the colour and stability of wine during ageing but it produces a negative effect on taste and odour [131–133]. Due to the antimicrobial activity of silver nanomaterials towards a wide range of Gram-negative and Gram-positive bacteria, some fungi and viruses can replace the SO_2. It has been shown that use of colloidal silver complex (CSC) in the production of white and young red wine, displayed very similar chemical and sensory activities as that of SO_2 [134].

Electronic devices

Silver (Ag) is superior in electrical and thermal conductivities among all the noble metals. With increased surface energy, the melting point of the small-sized silver nanoparticle is drastically low which make them useful as conductive fillers in microelectronic materials [135]. The electrical loss is reduced with the lower surface roughness that gives better packing when the electrical conductors are fabricated with a thick film of silver nanoparticles [136]. Silver nanoparticles are used in electro-optical devices and sensors due to their electro reflectance (ER) effect. In this field, silver nanoparticles alter the absorption spectrum of the particle ensemble by changing the stored electronic charge on the particles with 100 times more effective than a bulk metal surface. Absorption spectroscopy directly monitors the changes in the electrostatic charge stored on small metal particles [137].

Silver nanoparticles have the ability to enhance electrical and optical properties of the polymer composites. Due to these superior properties of Ag, various conducting polymers like—polypyrrole, polythiophene, polyaniline based Ag nanocomposites are used in producing newer materials with high conductivity. Ag nanocomposites are produced in combination with silver nanowires (AgNWs) with conducting polymer matrices which are used as conductive filler and thermal interfacial material in sophisticated nanodevices [138–140]. The silver nanoparticle embedded dielectric Teflon matrix nanocomposites showed higher electrical conductivity with an increased film thickness at various silver nanoparticle concentration [141]. Multiwalled carbon nanotube (MWNTs) with high electrical and mechanical properties are used as electrodes. Silver (Ag)/polymer composites prepared by incorporating multiwalled carbon nanotubes (MWNTs) with Ag nanoparticles significantly improve the electrical conductivity [142]. To improve an optical sensor fiber, Ag-doped silica nanocomposite is coated along with bent silica on the surface of it, which is useful to trace ammonia in a gas sample [143]. Silver-poly vinyl alcohol (Ag-PVA) composites are used for light guiding and optical sensing applications [144, 145]. Silver nanoparticles rapidly trap free electrons so they are used in semiconductor applications. TiO_2 is a known photocatalyst capped with the silver nanoparticle to form TiO_2–AgNP nanocomposites which are used for better semiconductor efficiency [143, 146]. Conventional solar cells coated with silicon was not efficient because silicon is a poor absorber of light. To enhance the light trapping efficiency of such solar cells, undercoating silver nanoparticles along with silicon layer was performed [147]. Similarly, in ultra-thin light filters, the silver-Poly-methylmethacrylate-Poly ethylene terephthalate membrane (Ag-PMMA-PET) composite has been used. Li et al. [148] suggested that use of PVP nanofibers-silver nanoparticle composites as a thin layer in organic solar cells increase 19.44% power conversion efficiency.

Glucose oxidase (GOx) immobilized stimuli-responsive silver nanocomposites are used in optical enzyme biosensor for sugar concentration analysis. When a sugar solution like glucose is applied to the surface of optical enzyme biosensor, the interparticle distances of the silver nanoparticle present in the silver nanocomposite are increased and absorbance strength of surface Plasmon resonance is decreased [149]. In a variety of techniques like fluorescent, radiochemical, piezoelectric technology and quartz crystal microbalance, silver nanoparticle probes are also used.

4.3 Water Treatment

As silver has antimicrobial property, Ag nanoparticles and nanocomposites are used in water purification devices to retard the growth of waterborne microorganisms. The Ag-containing nanomaterials can be a more cost-effective way for water treatment than the chemical method. The stabilization and immobilization of AgNPs in polymeric ion-exchange matrices is a promising approach for water treatment processes. Porous silver nanocomposites such as cellulose/Ag nanocomposites [150], chitosan-silver nanocomposites [151], silica silver composite [152] etc., have antibacterial characteristics and are also used in water treatment system. Silver nanoparticle loaded biocompatible and biodegradable polymer, sodium carboxymethyl cellulose (CMC) is used for water treatment application [153]. Zeolite, sand, fibreglass, anion and cation resin loaded silver nanoparticles are used in groundwater purification systems as antibacterial agent [154]. Porous ceramic Ag nanocomposites and thin-film layer containing nanosilver particles are used as an antibacterial substance in water filter [155]. The silver incorporated nanocomposite ceramic membranes exhibit good salt rejection capacity and effective membrane permeability [156]. Ceramic membrane fabricated silver nanoparticles have also been used to prevent biofouling [157]. Ceramic materials, casting with nanoparticles have more nanoscale pore sizes than ceramics with conventional sintering [158]. Silver-decorated ceramic membrane removes all *E. coli* after 24 h of contact time, whereas, bacterial growth was observed on undecorated membranes [159]. Silver nanoparticles incorporated polymeric membranes are also used to mitigate the biofouling and reduce the microbial activity by releasing silver ions from the membrane. The released silver ions lyse the bacterial cell by adhering to the cell and change the permeability of the cell wall [160, 161].

Fe_3O_4-silver nanomaterial is a bifunctional composite that has the superparamagnetic and antibacterial properties against *E. coli*, *Staphylococcus epidermis*, and *B. subtilis*. This nanocomposite easily removes material from water with its superparamagnetic property. To enhance the water treatment efficiency and recyclability, mesoporous polymer nanofiber membranes can be designed with specific pore sizes and desired filtration properties. So, the nanocomposites of super magnetic/silver nanoparticle/polymer nanofiber can be a promising water disinfectant [162, 163].

4.4 Textiles

The silver nanoparticles form bonds with the fibers of different fabrics like nylon, polyester, cotton and produce the nanoengineered fabrics. High surface area relative to the volume of particles that increases their chemical reactivity with the fibers helps the nanoparticles stick to fabrics more permanently. The silver nanoparticle coated fabrics prevent moisture, odour, dirt and have antibacterial activity. Due to these properties, they are used in medical bandages, bed linings and sports socks [164]. For prevention of foot odour, use of silver containing socks has been reported. Polymer nanofiber coated fabric materials are applied in textiles to act as a waterproof textile material. Polymer nanofiber layered fabrics with dirt-proof, stain-proof, and superhydrophobic properties are also available. Perelshtein et al. [165] prepared the silver nanocomposite coated fabric and experimented with its antibacterial activity against *E. coli* and *Staphylococcus aureus*. Silver nanocomposites of Polyvinyl alcohol (PVA) and polyvinyl pyrrolidone (PVP) are the type of polymer nanocomposites that form hydrogen bonds with polar species [166].

4.5 Nanopaints

Polymer matrices give high stability to the silver nanoparticles in polymer-silver nanocomposites. Without significant oxidation and aggregation, the polymer nanocomposites can be stable up to 200 °C. This property enables the production of silver nanoparticle embedded homogeneous paints. The nanopaint is an excellent coating material with outstanding antibacterial properties that can be applied to various surfaces including wood, glass, and polystyrene [167].

4.6 Personal Care Products

Silver nanoparticles used in cleanser soap has been found useful in treating acne due to their bactericidal and fungicidal properties [168]. Nanosilver containing hand wash of 15 mg per litre concentration has been found highly efficient with short exposure time to prevent transmission of infectious diseases [169]. Some toothpaste and tooth creams contain silver nanoparticles which produce a natural tooth enamel like a thin layer in teeth to reduce sensitivity and pain. For imparting freshness to the skin, certain day and night creams contain silver nanoparticles [170]. Zinc oxide (ZnO) or titanium dioxide (TiO$_2$) effectively absorb UV light. Incorporation of silver nanoparticles in ZnO and TiO$_2$ make them small and transparent, enabling them to exclusively absorb UV light excluding the absorption of visible light. Silver nanoparticles are also used in face and body foams, wet wipes, deodorants, lip products etc. [171].

5 Conclusion

The present review focuses on a vivid discussion relating to various synthesis methods and applications of silver nanocomposites. This would help in our understanding and enrich our knowledge on the recent developments in the area. However, further research is needed for nanocomposite synthesis through faster, economical and cheap processes and the extension of their applicability that would result in increased specificity, competence and efficiency. Keeping in view the increasing environmental pollution due to different anthropogenic activities care should be taken to focus on green synthesis methods that would not release any harmful chemicals in the environment. Novel strategies can also be implemented towards the development of recyclable and biodegradable nanocomposite materials preventing their accumulation in the environment after disposal.

References

1. Khademhosseini A, Langer R (2006) Drug delivery and tissue engineering. Chem Eng Prog 102(2):38–42
2. Faraday M (1857) The bakerian lecture: experimental relations of gold (and other metals) to light. Philos Trans R Soc Lond 147:145
3. Konrad MP, Doherty AP, Bell SEJ (2013) Stable and uniform SERS signals from self assembled two-dimensional interfacial arrays of optically coupled Ag nanoparticles. Anal Chem 85:6783–6789
4. Meheretu GM, Cialla D, Popp J (2014) Surface enhanced raman spectroscopy on silver nanoparticles. Inter J Biochemistry Biophysics 2:63–67
5. Franci G, Falanga A, Galdiero S, Palomba L, Rai M, Morelli G, Galdiero M (2015) Silver nanoparticles as potential antibacterial agents. Molecules 20:8856–8874
6. Jana S, Pal T (2007) Synthesis, characterization and catalytic application of silver nanoshell coated functionalized polystyrene beads. J Nanosci Nanotechnol 7:2151–2156
7. Stiufiuc R, Iacovita C, Lucaciu CM, Stiufiuc G, Dutu AG, Braescu C, Leopold N (2013) SER-sactive silver colloids prepared by reduction of silver nitrate with short-chain polyethylene glycol. Nanoscale Res Lett 8:47

8. Evtugyn GA, Shamagsumova RV, Padnya PV, Stoikov II, Antipin IS (2014) Cholinesterase sensor based on glassy carbon electrode modified with Ag nanoparticles decorated with macrocyclic ligands. Talanta 127:9–17
9. Thanha NTK, Green LAW (2010) Functionalisation of nanoparticles for biomedical applications. Nano Today 5:213–230
10. Alon N, Miroshnikov Y, Perkas N, Nissan I, Gedanken A, Shefi (2014) Substrates coated with silver nanoparticles as a neuronal regenerative material. Int J Nanomed 9:23–31
11. Bu Y, Lee S (2012) Influence of dopamine concentration and surface coverage of Au shell on the optical properties of Au, Ag, and AgcoreAushell nanoparticles. ACS Appl Mater Interfaces 4:3923–3931
12. Luo Y, Ma L, Zhang X, Liang A, Jiang Z (2015) SERS detection of dopamine using label free acridine red as molecular probe in reduced graphene oxide/silver nanotriangle sol substrate. Nanoscale Res Lett 10:230
13. Rivero PJ, Urrutia A, Goicoechea J, Matias IR, Arregui FJ (2013) A lossy mode resonance optical sensor using silver nanoparticles-loaded films for monitoring human breathing. Sens Actuators B 187:40–44
14. El-Nour KMM, Eftaiha A, Al-Reda A, Ammar AA (2010) Synthesis and applications of silver nanoparticles. Arabian J Chem 3:135–140
15. Smetana AB, Klabunde KJ, Sorensen CM (2005) Synthesis of spherical silver nanoparticles by digestive ripening, stabilization with various agents, and their 3-D and 2-D superlattice formation. J Colloid Interface Sci 284:521–526
16. Wakuda D, Kim KS, Suganuma K (2008) Room temperature sintering of Ag nanoparticles by drying solvent. Scrip Mater 59:649–652
17. Lee H, Chou KS (2005) Inkjet printing of nanosized silver colloids. Nanotechnology 16:2436–2441
18. Anna Z, Ewa S, Adriana Z, Maria G, Jan H (2009) Preparation of silver nanoparticles with controlled particle size. Procedia Chem 1:1560–1566
19. Twardowski TE (2007) Introduction to nanocomposite materials: properties, processing, characterization. Destech Publications, Incorporated, Lancaster, PA
20. Pina S, Oliveira JM, Reis RL (2015) Natural-based nanocomposites for bone tissue engineering and regenerative medicine: a review. Adv Mater 27:1143–1169
21. Rafiee MA, Rafiee J, Wang Z, Song H, Yu Z-Z, Koratkar N (2009) Enhanced mechanical properties of nanocomposites at low graphene content. ACS Nano 3:3884–3890
22. Mariano M, El Kissi N, Dufresne A (2014) Cellulose nanocrystals and related nanocomposites: review of some properties and challenges. J Polym Sci, Part B Polym Phys 52:791–806
23. Hu H, Onyebueke L, Abatan A (2010) Characterizing and modeling mechanical properties of nanocomposites-review and evaluation. J Miner Mater Charact Eng 9:275
24. Gleiter H (1992) Materials with ultrafine microstructures: retrospectives and perspectives. Nanostruct Mater 1:1–19
25. Anna Z, Ewa S, Adriana Z, Maria G, Jan H (2009) Preparation of silver nanoparticles with controlled particle size. ProcediaChem 1:1560–1566
26. Gurav AS, Kodas TT, Wang LM, Kauppinen EI, Joutsensaari J (1994) Generation of nanometer-size fullerene particles via vapor condensation. J Joutsensaari Chem Phys Lett 218:304–308
27. Kruis F, Fissan H, Rellinghaus B (2000) Sintering and evaporation characteristics of gas-phase synthesis of size selected PbS nanoparticles. Mater Sci Eng B 69:329–334
28. Magnusson MH, Deppert K, Malm JO, Bovin JO, Samuelson L (1999) Gold nanoparticles: production, reshaping, and thermal charging. J Nanoparticle Res 1:243–251
29. Mafune F, Takeda J, Kondow T, Sawabe H (2000) Formation and size control of silver nanoparticles by laser ablation in aqueous solution. J Phys Chem B 104:9111–9117
30. Sylvestre JP, Kabashin AV, Sacher E, Meunier M, Luong JHT (2004) Stabilization and size control of gold nanoparticles during laser ablation in aqueous cyclodextrins. J Am Chem 126:7176–7177

31. Pacioni NL, Borsarelli CD, Rey V, Veglia AV (2015) Synthetic routes for the preparation of silver nanoparticles: a mechanistic perspective. In: Udekwu KI, Alarcón EL, Griffith M (eds) Silver nanoparticle applications: in the fabrication and design of medical and biosensing devices. Springer International Publishing AG, Switzerland, p 13
32. Huang H, Yang Y (2008) Preparation of silver nanoparticles in inorganic clay suspensions. Compos Sci Technol 68:2948–2953
33. Zhang M, Zhang K, De Gusseme B, Verstraete W, Field R (2014) The antibacterial and anti-biofouling performance of biogenic silver nanoparticles by lactobacillus fermentum. Biofouling 30:347–357
34. Priyadarshini S, Gopinath V, MeeraPriyadharsshini N, MubarakAli D, Velusamy P (2013) Synthesis of anisotropic silvernanoparticles using novel strain, bacillus flexus and its biomedical application. Colloids Surf B 102:232–237
35. Gurunathan S, Kalishwaralal K, Vaidyanathan R et al (2009) Biosynthesis, purification and characterization of silvernanoparticles using Escherichia coli. Colloids Surf B Biointerfaces 74:328–335
36. Minaeian S, Shahverdi AR, Nohi AS, Shahverdi HR (2008) Extracellular biosynthesis of silver nanoparticles by somebacteria. J Sci (Islamic Azad University) 17:1–4
37. Vahabi K, Ali Mansoori G, Karimi S (2011) Biosynthesis of silver nanoparticles by fungus, trichodermareesei. Insciences J 1:65–79
38. Ahmad T, Wani IA, Manzoor N, Ahmed J, Asiri AM (2013) Biosynthesis, structural characterization and antimicrobial activity of gold and silver nanoparticles. Colloids Surf B 107:227–234
39. Syed A, Saraswati S, Kundu GC, Ahmad A (2013) Biological synthesis of silver nanoparticles using the fungus Humicola sp. and evaluation of their cytoxicity using normal and cancer cell lines. Spectrochim Acta Part A Mol Biomol Spectrosc 114:144–147
40. Kalishwaralal K, Deepak V, Pandian SRK, Kartikeyan B, Kottaisamy M, Gurunathan S (2010) Biosynthesis of silver and gold nanoparticles using brevibacterium casei. Colloids Surf B Biointerfaces 77:257–262
41. Hazra Chowdhury I, Ghosh S, Roy M, Naskar MK (2015) Green synthesis of water-dispersible silver nanoparticles at room temperature using green carambola (star fruit) extract. J Sol-Gel Sci Technol 73:199–207
42. Ashokkumar S, Ravi S, Kathiravan V, Velmurugan S (2015) Synthesis of silver nanoparticles using A. indicum leaf extract and their antibacterial activity. Spectrochim Acta Part A Mol Biomol Spectroscopy 134:34–39
43. Raut RW, Mendhulkar VD, Kashid SB (2014) Photosensitized synthesis of silver nanoparticles using withania somnifera leaf powder and silver nitrate. J Photochem Photobiol, B 132:45–55
44. Rajakumar G, Abdul Rahuman A (2011) Larvicidal activity of synthesized silver nanoparticles using eclipta prostrata leaf extract against filariasis and malaria vectors. Acta Trop 118:196–203
45. Santhoshkumar T, Rahuman AA, Rajakumar G, MarimuthuS Bagavan A, Jayaseelan C (2011) Synthesis of silver nanoparticles using Nelumbo nucifera leaf extract and its larvicidal activity against malaria and filariasis vectors. Parasitol Res 108:693–702
46. Kaviya S, Santhanalakshmi J, Viswanathan B, Muthumary J, Srinivasan K (2011) Biosynthesis of silver nanoparticles using Citrussinensis peel extract and its antibacterial activity. SpectrochemActa A Mol Biomol Spectrosc 79:594–598
47. Shankar SS, Ahmad A, Sastry M (2003) Geranium leaf assisted biosynthesis of silver nanoparticles. Biotechnol Prog 19:1627–1631
48. Dubey SP, Lahtinen M, Sillianpaa M (2010) Tansy fruit mediated greener synthesis of silver and gold nanoparticles. Process Biochem 45:1065–1071
49. Suna Q, Cai X, Li J, Zheng M, Chenb Z, Yu CP (2014) Greensynthesis of silver nanoparticles using tea leaf extract and evaluation of their stability and antibacterial activity. Colloid Surf A Physicochem Eng Aspects 444:226–231

50. Gopinatha V, Ali MD, Priyadarshini S, Thajuddinb N, MeeraPriyadharsshini N, Velusamy P (2012) Biosynthesis of silvernanoparticles from Tribulus terrestris and its antimicrobial activity: a novel biological approach. Colloid Surf B Biointerface 96:69–74
51. Tolaymat TM, El Badawy AM, Genaidy A, Scheckel KG, Luxton TP, Suidan M (2010) An evidence-based environmental perspective of manufactured silver nanoparticle in syntheses and applications: a systematic review and critical appraisal of peer-reviewed scientific papers. Sci Total Environ 408:999–1006
52. Creighton JA, Blatchford CG, Albrecht MG (1979) Plasma resonance enhancement of Raman scattering by pyridine adsorbed on silver or gold sol particles of size comparable to the excitation wavelength. J Chem Soc Faraday Trans 75:790–798
53. Sui Z, Chen X, Wang L, Chai Y, Yang C, Zhao J (2005) An improved approach for synthesis of positively charged silver nanoparticles. ChemLett 34:100–101
54. Shi Y, Lv L, Wang H (2009) A facile approach to synthesize silver nanorods capped with sodium tripolyphosphate. Mater Lett 63:2698–2700
55. Horiuchi Y, Shimada M, Kamegawa T, Mori K, Yamashita H (2009) Size-controlled synthesis of silver nanoparticles on Ti-containing mesoporous silica thin film and photoluminescence enhancement of rhodamine 6G dyes by surface plasmon resonance. J Mater Chem 19:6745–6749
56. Zielinska A, Skwarek E, Zaleska A, Gazda M, Hupka J (2009) Preparation of silver nanoparticle. ProcChem 1:1560
57. Henglein A, Giersig M (1999) Formation of colloidal silver nanoparticles. Cappingaction of citrate. J Phys Chem B 103:9533–9539
58. Pietrobon B, Kitaev V (2008) Photochemical synthesis of monodisperse size-controlled silver decahedral nanoparticles and their remarkable optical properties. Chem Mater 20:5186–5190
59. Mayer AB, Hausner SH, Mark JE (2002) Colloidal silver nanoparticles generated in the presence of protective cationic polyelectrolytes. Poly J 32:15–22
60. Sivaraman SK, Elango I, Kumar S, Santhanam V (1997) Room-temperature synthesis of gold nanoparticles—size-control by slow addition. CurrSci 7:1055–1059
61. Yoosaf K, Ipe BI, Suresh CH, Thomas KG (2007) In situ synthesis of metal nanoparticles and selective naked-eye detection of lead ions from aqueous media. J PhysChem C 111:12839–12847
62. Chou KS, Lai YS (2004) Effect of polyvinyl pyrrolidone molecular weights on the formation of nanosized silver colloids. Mater Chem Phys 83:82–88
63. Chou KS, Lu YC, Lee HH (2005) Effect of alkaline ion on the mechanism and kinetics of chemical reduction of silver. Mater Chem Phys 94:429–433
64. Chen SF, Zhang H (2012) Aggregation kinetics of nanosilver in different watercondition. Adv Nat Sci Nanosci Nanotechnol 3:035006-1–035006-7
65. Dang TMD, Le TTT, Blance EF, Dang MC (2012) Influence of surfactant on the preparation of silvernanoparticles by polyol method. Adv Nat Sci Nanosci Nanotechnol 3:035004-1–035004-4
66. Patil RS, Kokate MR, Jambhale C, Pawar SM, Han SH, Kolekar SS (2012) One-pot synthesis of PVA-capped silvernanoparticles their characterization and biomedicalapplication. Adv. Nat. Sci.: Nanosci Nanotechnol. 3:015013-1–015013-7
67. Zorn K, Giorgio S, Halwax E, Henry CR, Grönbeck H, Rupprechter G (2011) CO oxidation on technological Pd–Al_2O_3 catalysts: oxidation state and activity. J Phys Chem C 115:1103–1111
68. Guzman J, Carrettin S, Fierro-Gonzalez JC, Hao YL, Gates BC, Corma A (2005) CO oxidation catalyzed by supported gold: cooperation between gold and nanocrystalline rare-earth supports forms reactive surface superoxide and peroxide species. Angew Chem Int Ed 44:4778–4781
69. Wang Y, Van de Vyver S, Sharma KK, Leshkov YR (2014) Insights into the stability of gold nanoparticles supported on metal oxides for the base-free oxidation of glucose to gluconic acid. Green Chem 16:719–726

70. Liu T, Li B, Hao Y, Han F, Zhang L, Hu L (2015) A general method to diverse silver/mesoporous–metal–oxidenanocomposites with plasmon-enhanced photocatalytic activity. Appl Catal B 165:378–388
71. Liu K, Bai Y, Zhang L, Yang Z, Fan Q, Zheng H, Yin Y, Gao C (2016) Porous Au-Ag nanospheres with high-density and highly accessible hotspots for SERS analysis 16:3675–3681
72. Dutta Choudhury S, Badugu R, Ray K, Lakowicz JR (2012) Silver–gold nanocomposite substrates for metal-enhanced fluorescence: ensemble and single-molecule spectroscopic studies. J Phys Chem C 116:5042–5048
73. Li HJ, Zhang AQ, Hu Y, Sui L, Qian DJ, Chen M (2012) Large-scale synthesis and self-organization of silver nanoparticles with tween 80 as a reductant and stabilizer. Nanoscale Res Lett 7:612
74. Vimala K, Yallapu MM, Varaprasad K, Reddy NN, Ravindra S, Naidu NS, Raju KM (2011) Fabrication of curcumin encapsulated chitosan-PVA silver nanocomposite films for improved antimicrobial activity. J Biomater Nanobiotechnol 2:55–64
75. Porel S, Ramakrishna D, Hariprasad E, Gupta D, Radhakrishnan P (2011) Polymer thin film with in situ synthesized silver nanoparticles as a potent reusable bactericide. Curr Sci 101:927–934
76. Wankhade Y, Kondawar S, Thakare S, More P (2013) Synthesis and characterization of silver nanoparticles embedded in polyaniline nanocomposite. Adv Mater 4:89–93
77. Guo Q, Ghadiri R, Weigel T, Aumann A, Gurevich E, Esen C, Medenbach O, Cheng W, Chichkov B, Ostendorf A (2014) Comparison of in situ and ex situ methods for synthesis of two-photon polymerization polymer nanocomposites. Polymers 6:2037–2050
78. Lim MH, Ast DG (2001) Free-standing thin films containing hexagonally organized silver nanocrystals in a polymer matrix. Adv Mater 13:718–721
79. Mandi U, Roy AS, Banerjee B, Islam SM (2014) A novel silver nanoparticle embedded mesoporous polyaniline (mPANI/Ag) nanocomposite as a recyclable catalyst in the acylation of amines and alcohols under solvent free conditions. RSC Adv. 4:42670–42681
80. Mandi U, Roy AS, Kundu SK, Roy S, Bhaumik A, Islam SM (2016) Mesoporouspolyacrylic acid supported silver nanoparticles as an efficient catalyst for reductive coupling of nitrobenzenes and alcohols using glycerol as hydrogen source. J Colloid Interface Sci 472:202–209
81. Panáček A, Kvítek L, Prucek R, Kolář M, Večeřová R, Pizúrová N, Sharma VK, Nevěčná TJ, Zbořil R (2006) Silver colloid nanoparticles: synthesis, characterization, and their antibacterial activity. J Phys Chem B 110(33):16248–16253
82. Kvitek L, Panáček A, Soukupova J, Kolář M, Večeřová R, Prucek R, Holecova M, Zbořil R (2008) Effect of surfactants and polymers on stability and antibacterial activity of silver nanoparticles (NPs). J Phys Chem C 112(15):5825–5834
83. Sondi I, Salopek-Sondi B (2004) Silver nanoparticles as antimicrobial agent: a case study on E. coli as a model for Gram-negative bacteria. J Colloid Interface Sci 275(1):177–182
84. Li Q, Mahendra S, Lyon DY, Brunet L, Liga MV, Li D, Alvarez PJ (2008) Antimicrobial nanomaterials for water disinfection and microbial control: potential applications and implications. Water Res 42(18):4591–4602
85. Liu J, Li X, Zuo S, Yu Y (2007) Preparation and photocatalytic activity of silver and TiO_2 nanoparticles/montmorillonite composites. Appl Clay Sci 37(3):275–280
86. Shah MSAS, Nag M, Kalagara T, Singh S, Manorama SV (2008) Silver on PEG-PU-TiO_2 polymer nanocomposite films: an excellent system for antibacterial applications. Chem Mater 20(7):2455–2460
87. Kong H, Jang J (2008) Antibacterial properties of novel poly (methyl methacrylate) nanofiber containing silver nanoparticles. Langmuir 24(5):2051–2056
88. Lee EM, Lee HW, Park JH, Han YA, Ji BC, Oh W, Deng Y, Yeum JH (2008) Multihollow structured poly (methyl methacrylate)/silver nanocomposite microspheres prepared by suspension polymerization in the presence of dual dispersion agents. Colloid Polymer Sci 286(12):1379–1385

89. Nadagouda MN, Varma RS (2007) Synthesis of thermally stable carboxymethyl cellulose/metal biodegradable nanocomposites for potential biological applications. Biomacromol 8(9):2762–2767
90. Kawahara K, Tsuruda K, Morishita M, Uchida M (2000) Antibacterial effect of silver-zeolite on oral bacteria under anaerobic conditions. Dent Mater 16(6):452–455
91. Casemiro LA, Martins CHG, Pires-de-Souza FDC, Panzeri H (2008) Antimicrobial and mechanical properties of acrylic resins with incorporated silver–zinc zeolite–part I. Gerodontology 25(3):187–194
92. Matsuura T, Abe Y, Sato Y, Okamoto K, Ueshige M, Akagawa Y (1997) Prolonged antimicrobial effect of tissue conditioners containing silver-zeolite. J Dent 25(5):373–377
93. Morishita M, Miyagi M, Yamasaki Y, Tsuruda K, Kawahara K, Iwamoto Y (1998) Pilot study on the effect of a mouthrinse containing silver zeolite on plaque formation. J Clin Dent 9:94–96
94. Aroca RF, Goulet PJ, dos Santos DS, Alvarez-Puebla RA, Oliveira ON (2005) Silver nanowire layer-by-layer films as substrates for surface-enhanced Raman scattering. Anal Chem 77(2):378–382
95. Lesniak W, Bielinska AU, Sun K, Janczak KW, Shi X, Baker JR, Balogh LP (2005) Silver/dendrimer nanocomposites as biomarkers: fabrication, characterization, in vitro toxicity, and intracellular detection. Nano Lett 5(11):2123–2130
96. Oh Y, Suh D, Kim Y, Lee E, Mok JS, Choi J, Baik S (2008) Silver-plated carbon nanotubes for silver/conducting polymer composites. Nanotechnology 19(49):495602
97. Sur I, Cam D, Kahraman M, Baysal A, Culha M (2010) Interaction of multi-functional silver nanoparticles with living cells. Nanotechnology 21(17):175104
98. Vilela D, González MC, Escarpa A (2012) Sensing colorimetric approaches based on gold and silver nanoparticles aggregation: chemical creativity behind the assay. A review. Anal Chim Acta 751:24–43
99. Shin SH, Ye MK, Kim HS, Kang HS (2007) The effects of nano-silver on the proliferation and cytokine expression by peripheral blood mononuclear cells. Int Immunopharmacol 7(13):1813–1818
100. Tian J, Wong KK, Ho CM, Lok CN, Yu WY, Che CM, Chiu JF, Tam PK (2007) Topical delivery of silver nanoparticles promotes wound healing. ChemMedChem 2(1):129–136
101. Wu J, Balasubramanian S, Kagan D, Manesh KM, Campuzano S, Wang J (2010) Motion-based DNA detection using catalytic nanomotors. Nat Comm 1:36
102. Chen X, Schluesener HJ (2008) Nanosilver: a nanoproduct in medical application. Toxicol Lett 176(1):1–12
103. Tolaymat TM, El Badawy AM, Genaidy A, Scheckel KG, Luxton TP, Suidan M (2010) An evidence-based environmental perspective of manufactured silver nanoparticle in syntheses and applications: a systematic review and critical appraisal of peer-reviewed scientific papers. Sci Total Environ 408(5):999–1006
104. Gupta A, Silver S (1998) Molecular genetics: silver as a biocide: will resistance become a problem? Nat Biotechnol 16(10):888
105. Ghosh S, Azhahianambi P, de la Fuente J (2006) Control of ticks of ruminants, with special emphasis on livestock farming systems in India: present and future possibilities for integrated control—a review. Exp Appl Acarol 40(1):49–66
106. Bergeson LL (2010) Nanosilver: US EPA's pesticide office considers how best to proceed. Environ Qual Manage 19(3):79–85
107. Yan J, Huang K, Wang Y, Liu S (2005) Study on anti-pollution nano-preparation of dimethomorph and its performance. Chin Sci Bull 50(2):108–112
108. Quintavalla S, Vicini L (2002) Antimicrobial food packaging in meat industry. Meat Sci 62(3):373–380
109. Wong DW, Camirand WM, Pavlath AE (1994) Development of edible coatings for minimally processed fruits and vegetables. Edible Coat Films Improve Food Qual 65–88
110. Han JH (2005) New technologies in food packaging: overview. Innov Food Packag 3–11

111. Mei Y, Zhao Y (2003) Barrier and mechanical properties of milk protein-based edible films containing nutraceuticals. J Agric Food Chem 51(7):1914–1918
112. Labuza TP, Breene WM (1989) Applications of "active packaging" for improvement of shelf-life and nutritional quality of fresh and extended shelf-life foods. J Food Process Preserv 13(1):1–69
113. Cha DS, Chinnan MS (2004) Biopolymer-based antimicrobial packaging: a review. Crit Rev Food Sci Nutr 44(4):223–237
114. Cagri A, Ustunol Z, Ryser ET (2004) Antimicrobial edible films and coatings. J Food Prot 67(4):833–848
115. Rhim JW, Hong SI, Park HM, Ng PK (2006) Preparation and characterization of chitosan-based nanocomposite films with antimicrobial activity. J Agric Food Chem 54 (16):5814–5822
116. Hu Z, Chan WL, Szeto YS (2008) Nanocomposite of chitosan and silver oxide and its antibacterial property. J Appl Polym Sci 108(1):52–56
117. Appendini P, Hotchkiss JH (2002) Review of antimicrobial food packaging. Innov Food Sci Emerg Technol 3(2):113–126
118. Li H, Li F, Wang L, Sheng J, Xin Z, Zhao L, Xiao H, Zheng Y, Hu Q (2009) Effect of nano-packing on preservation quality of Chinese jujube (Ziziphus jujuba Mill. var. inermis (Bunge) Rehd). Food Chem 114(2):547–552
119. Rai M, Yadav A, Gade A (2009) Silver nanoparticles as a new generation of antimicrobials. Biotechnol Adv 27(1):76–83
120. Simpson K (2003) Using silver to fight microbial attack. Plast Addit Compd 5(5):32–35
121. Praus P, Turicová M, Machovič V, Študentová S, Klementová M (2010) Characterization of silver nanoparticles deposited on montmorillonite. Appl Clay Sci 49(3):341–345
122. Coleman NJ, Bishop AH, Booth SE, Nicholson JW (2009) Ag^+ and Zn^{2+} exchange kinetics and antimicrobial properties of 11Å tobermorites. J Eur Ceram Soc 29(6):1109–1117
123. Cowan MM, Abshire KZ, Houk SL, Evans SM (2003) Antimicrobial efficacy of a silver-zeolite matrix coating on stainless steel. J Ind Microbiol Biotechnol 30(2):102–106
124. Galeano B, Korff E, Nicholson WL (2003) Inactivation of vegetative cells, but not spores, of Bacillus anthracis, B. cereus, and B. subtilis on stainless steel surfaces coated with an antimicrobial silver-and zinc-containing zeolite formulation. Appl Environ Microbiol 69 (7):4329–4331
125. Matsumura Y, Yoshikata K, Kunisaki SI, Tsuchido T (2003) Mode of bactericidal action of silver zeolite and its comparison with that of silver nitrate. Appl Environ Microbiol 69 (7):4278–4281
126. Nakane T, Gomyo H, Sasaki I, Kimoto Y, Hanzawa N, Teshima Y, Namba T (2006) New antiaxillary odour deodorant made with antimicrobial Ag-zeolite (silver-exchanged zeolite). Int J Cosmet Sci 28(4):299–309
127. Akdeniz Y, Ülkü S (2008) Thermal stability of Ag-exchanged clinoptilolite rich mineral. J Therm Anal Calorim 94(3):703–710
128. Gulbranson SH, Hud JA, Hansen RC (2000) Argyria following the use of dietary supplements containing colloidal silver protein. Cutis 66(5):373–374
129. Romano P, Suzzi G (1993) Sulfur dioxide and wine microorganisms 373–393
130. Bakker J, Bridle P, Bellworthy SJ, Garcia-Viguera C, Reader HP, Watkins SJ (1998) Effect of sulphur dioxide and must extraction on colour, phenolic composition and sensory quality of red table wine. J Sci Food Agric 78(3):297–307
131. Blaise A, Bertrand A (1998) Altérations organoleptiques des vins. Oenologie. Fondements Scientifique et Technologiques 1182–1216
132. Stratford M, Rose AH (1985) Hydrogen sulphhide production from sulphite by Saccharomyces cerevisiae. Microbiology 131(6):1417–1424
133. Izquierdo-Cañas PM, García-Romero E, Huertas-Nebreda B, Gómez-Alonso S (2012) Colloidal silver complex as an alternative to sulphur dioxide in winemaking. Food Control 23(1):73–81

134. Umadevi M, Christy AJ (2013) Optical, structural and morphological properties of silver nanoparticles and its influence on the photocatalytic activity of TiO_2. Spectrochim Acta Part A Mol Biomol Spectrosc 111:80–85
135. Chen D, Qiao X, Qiu X, Chen J (2009) Synthesis and electrical properties of uniform silver nanoparticles for electronic applications. J Mater Sci 44(4):1076–1081
136. Jiang H, Moon KS, Li Y, Wong CP (2006) Surface functionalized silver nanoparticles for ultrahigh conductive polymer composites. Chem Mater 18:2969–2973
137. Alshehri AH, Jakubowska M, Młożniak A, Horaczek M, Rudka D, Free C, Carey JD (2012) Enhanced electrical conductivity of silver nanoparticles for high frequency electronic applications. ACS Appl Mater Interfaces 4(12):7007–7010
138. Nam S, Cho HW, Lim S, Kim D, Kim H, Sung BJ (2012) Enhancement of electrical and thermomechanical properties of silver nanowire composites by the introduction of nonconductive nanoparticles: experiment and simulation. ACS Nano 7(1):851–856
139. Yu YH, Ma CCM, Teng CC, Huang YL, Lee SH, Wang I, Wei MH (2012) Electrical, morphological, and electromagnetic interference shielding properties of silver nanowires and nanoparticles conductive composites. Mater Chem Phys 136(2):334–340
140. Lee J, Lee P, Lee HB, Hong S, Lee I, Yeo J, Lee SS, Kim TS, Lee D, Ko SH (2013) Room-temperature nanosoldering of a very long metal nanowire network by conducting-polymer-assisted joining for a flexible touch-panel application. Adv Func Mater 23(34):4171–4176
141. Chapman R, Mulvaney P (2001) Electro-optical shifts in silver nanoparticle films. Chem Phys Lett 349(5):358–362
142. Wei H, Eilers H (2008) Electrical conductivity of thin-film composites containing silver nanoparticles embedded in a dielectric fluoropolymer matrix. Thin Solid Films 517(2):575–581
143. Guo H, Tao S (2007) Silver nanoparticles doped silica nanocomposites coated on an optical fiber for ammonia sensing. Sens Actuators B Chem 123(1):578–582
144. Marques-Hueso J, Abargues R, Canet-Ferrer J, Valdes JL, Martinez-Pastor J (2010) Resist-based silver nanocomposites synthesized by lithographic methods. Microelectron Eng 87(5):1147–1149
145. Ananth AN, Umapathy S, Sophia J, Mathavan T, Mangalaraj D (2011) On the optical and thermal properties of in situ/ex situ reduced Ag NP's/PVA composites and its role as a simple SPR-based protein sensor. Appl Nanosci 1(2):87–96
146. Ghosh S, Das AP (2015) Modified titanium oxide (TiO_2) nanocomposites and its array of applications: a review. Toxicol Environ Chem 97(5):491–514
147. Hutter E, Fendler JH, Roy D (2001) Surface plasmon resonance studies of gold and silver nanoparticles linked to gold and silver substrates by 2-aminoethanethiol and 1,6-hexanedithiol. J Phys Chem B 105(45):11159–11168
148. Li X, Choy WCH, Lu H, Sha WE, Ho AHP (2013) Efficiency enhancement of organic solar cells by using shape-dependent broadband plasmonic absorption in metallic nanoparticles. Adv Func Mater 23(21):2728–2735
149. Endo T, Yanagida Y, Hatsuzawa T (2008) Quantitative determination of hydrogen peroxide using polymer coated Ag nanoparticles. Measurement 41(9):1045–1053
150. Pinto RJ, Marques PA, Neto CP, Trindade T, Daina S, Sadocco P (2009) Antibacterial activity of nanocomposites of silver and bacterial or vegetable cellulosic fibers. Acta Biomater 5(6):2279–2289
151. Vimala K, Mohan YM, Sivudu KS, Varaprasad K, Ravindra S, Reddy NN, Padma Y, Sreedhar B, MohanaRaju K (2010) Fabrication of porous chitosan films impregnated with silver nanoparticles: a facile approach for superior antibacterial application. Colloids Surf B 76(1):248–258
152. Egger S, Lehmann RP, Height MJ, Loessner MJ, Schuppler M (2009) Antimicrobial properties of a novel silver-silica nanocomposite material. Appl Environ Microbiol 75(9):2973–2976

153. Hebeish A, Hashem M, El-Hady MA, Sharaf S (2013) Development of CMC hydrogels loaded with silver nano-particles for medical applications. Carbohyd Polym 92(1):407–413
154. Mpenyana-Monyatsi L, Mthombeni NH, Onyango MS, Momba MN (2012) Cost-effective filter materials coated with silver nanoparticles for the removal of pathogenic bacteria in groundwater. Int J Environ Res Pub Health 9(1):244–271
155. Kim ES, Hwang G, El-Din MG, Liu Y (2012) Development of nanosilver and multi-walled carbon nanotubes thin-film nanocomposite membrane for enhanced water treatment. J Membr Sci 394:37–48
156. Kim DG, Kang H, Han S, Lee JC (2012) The increase of antifouling properties of ultrafiltration membrane coated by star-shaped polymers. J Mater Chem 22(17):8654–8661
157. Taurozzi JS, Arul H, Bosak VZ, Burban AF, Voice TC, Bruening ML, Tarabara VV (2008) Effect of filler incorporation route on the properties of polysulfone–silver nanocomposite membranes of different porosities. J Membr Sci 325(1):58–68
158. Kim J, Van der Bruggen B (2010) The use of nanoparticles in polymeric and ceramic membrane structures: review of manufacturing procedures and performance improvement for water treatment. Environ Pollut 158(7):2335–2349
159. DiGiano FA (2008) In pursuit of innovative membrane technology. In: IWA Membrane Research Conference. University of Massachusetts
160. Lv Y, Liu H, Wang Z, Liu S, Hao L, Sang Y, Liu D, Wang J, Boughton RI (2009) Silver nanoparticle-decorated porous ceramic composite for water treatment. J Membr Sci 331(1):50–56
161. Banerjee M, Mallick S, Paul A, Chattopadhyay A, Ghosh SS (2010) Heightened reactive oxygen species generation in the antimicrobial activity of a three component iodinated chitosan–silver nanoparticle composite. Langmuir 26(8):5901–5908
162. Gong P, Li H, He X, Wang K, Hu J, Tan W, Zhang S, Yang X (2007) Preparation and antibacterial activity of Fe3O4@ Ag nanoparticles. Nanotechnology 18(28):285604
163. Jain P, Pradeep T (2005) Potential of silver nanoparticle-coated polyurethane foam as an antibacterial water filter. Biotechnol Bioeng 90(1):59–63
164. Czajka R (2005) Development of medical textile market. Fibres Text Eastern Eur 13(1):13–15
165. Perelshtein I, Applerot G, Perkas N, Guibert G, Mikhailov S, Gedanken A (2008) Sonochemical coating of silver nanoparticles on textile fabrics (nylon, polyester and cotton) and their antibacterial activity. Nanotechnology 19(24):245705
166. Dallas P, Sharma VK, Zboril R (2011) Silver polymeric nanocomposites as advanced antimicrobial agents: classification, synthetic paths, applications, and perspectives. Adv Coll Interface Sci 166(1):119–135
167. Kumar A, Vemula PK, Ajayan PM, John G (2008) Silver-nanoparticle-embedded antimicrobial paints based on vegetable oil. Nat Mater 7(3):236–241
168. Lohani A, Verma A, Joshi H, Yadav N, Karki N (2014) Nanotechnology-based cosmeceuticals. ISRN Dermatol
169. Prabhu S, Poulose EK (2012) Silver nanoparticles: mechanism of antimicrobial action, synthesis, medical applications, and toxicity effects. Int Nano Lett 2(1):32
170. Gleiche M, Hoffschulz H, Lenhert S (2006) Nanotechnology in consumer products. Nanoforum Rep 1–30
171. Gajbhiye S, Sakharwade S (2016) Silver nanoparticles in cosmetics. J Cosmet Dermatol Sci Appl 6(1):48

Toxicological Evaluations of Nanocomposites with Special Reference to Cancer Therapy

Arpita Hazra Chowdhury, Arka Bagchi, Arunima Biswas and Sk. Manirul Islam

1 Introduction

Nanotechnology is developing a new era with the development of previously unknown materials and creating possibilities having the profound impact on the economic status, environment, and society. The nanotechnology tool is allowing scientists and manufactures to fabricate materials literally molecule-by-molecule. Properties associated with matter, custom design of previously unexplored structures, devices and unique systems with remarkable properties, like considerably increased strength, significantly decreased weight, much increased electrical conductivity or having the capacity to change shape, colour could be harnessed. Their applications in the field of modern medical and biological research are immense. Though researchers have devised various techniques and well defined intricate strategies to deliver poorly-soluble drugs into the infected tissue or cells, challenges remain to design drug delivery in a target-specific manner without causing the negative impact on the normal cells and tissues. The synthesis of nanoparticles and nanocomposites covering a broad range of metal, metal-oxide, and semiconductors to fabricate nanostructures with varying morphology are now being used in various ongoing research to successfully deal with the challenges and overcome obstacles. Moreover, nanoparticles or nanofibers in fabrics enhance various physical resistance, without increasing weight, thickness or stiffness of the fabric. Water filters that are only 15–20 nm wide can sieve very small particles, including virtually all viruses and bacteria. Hence this method is presently being implemented as a cost-efficient, portable water treatment system, improving the quality of drinking

---

A. Hazra Chowdhury · Sk. Manirul Islam (✉)
Department of Chemistry, University of Kalyani, Kalyani Nadia, West Bengal, India
e-mail: manir65@rediffmail.com

A. Bagchi · A. Biswas (✉)
Department of Zoology, University of Kalyani, Kalyani Nadia, West Bengal, India
e-mail: arunima10@gmail.com

© Springer Nature Switzerland AG 2019
Inamuddin et al. (eds.), *Sustainable Polymer Composites and Nanocomposites*,
https://doi.org/10.1007/978-3-030-05399-4_38

water in the developing countries. Not only improving the quality of life, nanoparticles like carbon nanotubes have a number of applications which includes producing strong and lightweight sports good. This proves beyond doubt that nanoparticles have an array of important usages in the modern era. But the goodness of nanoparticles and nanocomposites are not confined in the above- mentioned uses only. They are one of the main targets and hope for the betterment of human health and lifespan in future. The variety of nanotechnology-based platforms has been speculated for use in various biological purposes including improved cancer chemotherapy.

Synthesis of nanocomposites with nanoparticles and various matrix carriers improve the target specificity of the drugs along with their effectiveness in the biological system. The current problem with chemotherapeutic drugs is that they mostly affect the normal tissues along with the cancer ones. The utility of nanocomposites has the potential to address the issue of target non-specificity. Hence, they have gained importance for efficient transport of anticancer agents into the cancerous cells without affecting normal tissues or cells. Unlike the free drugs, which get neutralized in the body within the short time interval, these nanocomposites can also accumulate in the tumours, achieving a cytotoxic load, higher than the rest of the body [1, 2]. The nanocomposites increase the lifetime of a drug preventing their degradation when used in combination. The nanocomposites thus function by increasing the half-life of them in the biological system to a significant amount. Moreover, due to the process of angiogenesis, the accumulation of nanocarriers in the tumour tissues is ensured, which enhance effectivity of these carriers to significantly over free drugs [3, 4]. So, transport of these drugs through the bloodstream is facilitated by increasing half-life of nanocarriers avoiding the action of our body's first line of defence [5].

Different targeting ligands, such as monoclonal antibodies, peptides, antibody fragments, growth factors can be actually tagged to nanocarriers to achieve site-specific active targeting. Moreover, this gives an added advantage to avoid the multiple-drug resistance imparted by passive targeting [6].

But nanocarriers have few drawbacks which are responsible for its clinical failure. After entering into the tumour vasculature, the nanocarriers must reach the cancer cells by overcoming the different barriers. But the endothelial barrier adjacent to cancer cells sometimes act as a real hurdle ensuring the failure of nanocarrier functioning [7]. Moreover, the nanocarriers get attached to the first available receptor, failing to penetrate the other tumours. Also, targeting moieties increase the immunogenicity and plasma protein absorption of the nanocarriers, thereby actually minimizing their half-life in the bloodstream, which was supposed to be enhanced [7], decreasing their targeting capacity.

These shortcomings are needed to overcome to ensure the success of nanocarriers. The drawbacks and limited versatility of a single nanoparticle led to the use of nanocomposites in different fields as they exhibit cumulatively all the properties of their components, thereby increasing the versatility to a large extent. Moreover, they also show increased biocompatibility and high stability both in the environment and in biological systems. Metallic nanocomposites are useful for their multifunctional properties and biocompatibility, keeping their own property unhampered. This is the reason why researchers are showing immense interest in the use if nanocomposites for different biomedical applications including drug delivery, imaging, MRI contrast agent, photothermal ablation agents, photoacoustic imaging contrast agents [8–10].

But the researches on the toxicological evaluation of the nanoparticles and nanocomposites are still fragmentary and even contradictory to each other. But evidence severely suggests that metal nanoparticles like gold nanoparticle are involved in showing toxic effects on cellular levels causing size, shape and surface modifications. Though a small group of scientists claimed the use of gold nanoparticles to be essentially non-toxic, various research groups demonstrated their toxicity. The size of the nanoparticles seemed associated with generating toxicity when studied in cell lines like MCF-7 in a time and dose-dependent manner. The small-sized nanoparticles showed lesser toxicity and lesser accumulation of autophagosomes. These toxic effects might be an issue for using these nanoparticles for biomedical therapies as they might affect the healthy cells along the cancerous ones. Hence researchers are in the process of synthesizing nanocomposites to decrease the toxicity associated with the metal nanoparticles.

1.1 Nanocomposite Systems

Since the 1990s, researchers are showing more interest in nanocomposite systems and as a result, the number of publications, including reviews, is continuously increasing. We can define nanocomposites (NC) as multiphase solid materials in which phases must be present with dimensions of less than 100 nm. Nanocomposite systems have two parts: (i) continuous phase and (ii) discontinuous reinforcing phase. Thus, nanocomposite systems can have a combined or have noticeably different electrical, mechanical, electrochemical, thermal, catalytic and optical properties of the component materials [11–14].

1.2 Synthesis of Nanocomposite Systems: Nanocomposite Materials Are Generally Synthesized Using One of the Two Methods

1.2.1 In Situ Method

An effective and simple way to obtain a nanocomposite is to synthesize the nanoparticle in a matrix by an in situ method. The nanocomposite can be synthesized inside the matrix material in this method, from its corresponding precursors. Therefore, this method follows one-step fabrication of nanocomposites, where prevention of particle agglomeration as well as well spatial distribution in the matrix system both, have occurred simultaneously and these are the best advantages of this method. The disadvantage of this method is that the unreacted reagents of the in situ reaction might influence the properties of the final nanocomposite material.

1.2.2 Ex Situ Method

Ex situ method for synthesizing nanocomposites in another useful method, where the pre-made nanoparticles are directly dispersed into the matrix to form the composite. This method also has advantages and disadvantages. This method is very much advantageous for the large-scale industrial synthesis of nanocomposites than the in situ method, but the major challenge of this method is to prepare highly dispersible and stable nanoparticles. Generally, sonication method is applied to disperse the nanoparticles in the matrix.

1.3 Synthesis of Au/Ag Supported Mesoporous Metal-Oxide Nanocomposites

Mesoporous materials having tunable pore structure and modified framework composition, show various applications in adsorption, catalysis, separation, energy storage, conversion, biological uses etc. [15–17].

Mesoporous silicate particles have drawn a significant interest among the variety of inorganic materials as it possesses ordered porous structure, simple and cost-effective synthetic methods and wide range of applications [18–20] Modification of mesoporous silica surface can be done by various functional groups [21–26] and various metal NPs [27–31] and these modifications made the materials potentially applicable in different biomedical fields [32–34]. It should be noted that the antibacterial activity of Ag NPs depends not only on the size of NPs [35] but also on their shape [36]. Therefore, the main reason for carrying out the antibacterial activity is to synthesize monodispersed stable Ag NPs synthesis with the similar shape of the NPs. Ghosh et al. synthesized mesoporous silica flakes (MSF) using tetraethylorthosilicate (TEOS) as a silica source and CTAB as a structure directing agent in hexane at room temperature. They modified MSF with aminopropyltriethoxyl silane (APTS). The amino group of APTS formed -NH_2CH_2OH group by reacting with formaldehyde and the resulted group acted as reducing as well as the stabilizing agent to form monodisperse Ag NPs [37]. Li et al. [38] synthesized homogenously distributed gold nanoparticles within titania framework via a multi-component assembly approach. In this method, titania, gold building clusters, and surfactant are assembled in a single step process, that is they mixed Pluronic surfactant P123, $TiCl_4$, $Ti(OBu)_4$, and $AuCl_3$ in ethanol. Homogeneous mesostructured nanocomposites were obtained by casting the mixture followed by an ageing process. The surfactant P123 was removed by calcining the sample resulting crystalline mesoporous TiO_2 networks embedding gold nanoparticles. On the other hand, the Ag/mesoporous ZnO nanocomposite was synthesized by microwave irradiation route [39]. The reaction was carried out in an argon atmosphere for 15 min with zinc (II) acetate and silver nitrate as precursor salts to synthesize ZnO and Ag NPs respectively. Briefly, zinc (II) acetate and ethylene glycol were added to the aqueous $AgNO_3$ solution in a round-bottomed flask, fitted to the refluxing system inside the microwave oven. The reaction was conducted for 15 min under argon atmosphere. At the end of the reaction, the powder from the liquid was separated by centrifugation with the mother liquid and then washed with water and ethanol. Then the nanocomposite product was dried overnight under vacuum. Chowdhury et al. synthesized $Ag-TiO_2$ nanocomposite through a green synthetic method. The mesoporous anatase TiO_2 was synthesized by a hydrothermal method where they used titanium (IV) oxysulfate (TIOS) as precursor salt, urea as reducing agent and SDS as the surfactant in aqueous solution. Then the silver nanoparticle doped TiO_2 ($Ag-TiO_2$) was obtained by an impregnation method. In this method, water dispersible Ag NPs were used which were obtained from green carambola extract at pH 10 [40].

Sinha et al. [41] synthesized mesoscopic manganese oxide/gold nanoparticle composites by mixing $Mn(NO_3)_2 \cdot 6H_2O$ salt solution with a solution of NaOH and an aqueous solution of CTAB at pH 8.0. The resulting gel (pH 10.5) was heated in a closed vessel. The solid product was filtered calcined at 500 °C for 4 h and then it was stirred in aqueous H_2SO_4 solution followed by filtering and washing with water, and finally dried at 105 °C to obtain mesoporous MnO_2 sheets. The Au metal was vaporized from an Au disk to create a plasma by using the second harmonic of an Nd:YAG pulsed laser with a pulse width of 7 ns and energy of 1 J pulse. The supports which were prepared as thin sheets, placed in front of the cluster beam. Then each side of the wafer was exposed to the cluster beam for the same time interval and finally, the metal content in the composite was measured by chemical analysis (inductively coupled plasma).

1.4 Synthesis of Au/Ag Supported Graphene Nanocomposites

Since the experimental existence in 2004, graphene attracts the huge attention of the scientific community in almost all fields of material science applications. This credit goes to the extraordinary properties of graphene-like its high surface area (~ 2600 m^2g^{-1}), high thermal and mechanical stability, unique electronic and charge transport properties [42, 43]. Moreover, functionalized graphene nanocomposites show a wide array of applications in different fields such as in chemical and biological sensors, charge storage devices, capacitors, nanoelectronic and nanophotonic devices etc. [44–48]. These applications of graphene like structures may vary depending upon the route of synthesis that includes micro-mechanical exfoliation, Chemical vapour deposition (CVD), chemical reduction of graphene oxide (GO) etc. [42, 49–51]. In order to produce large-scale graphene, the chemical exfoliation of GO to produce reduced graphene oxide (RGO) is one of the most cost effective and efficient pathways [52].

The modified Hummers method is used to synthesize graphene oxide was from natural graphite. The synthesized GO was dispersed in DI water by sonication and $AgNO_3$ was added as the precursor salt to obtain Ag nanoparticles. Finally, the Ag/graphene composite was prepared by adding sodium borohydride ($NaBH_4$) as the reducing agent. The synthetic pathway of silver loaded graphene (Ag/G) is shown in Fig. 1 [53].

Wadhwa et al. demonstrated the synthesis of reduced graphene oxide silver (RGO-Ag) nanocomposite by using microwave irradiation. Modified Hummers method was also applied here to synthesize graphene oxide followed by microwave-assisted the reduction of GO and silver nitrate ($AgNO_3$) by hydrazine hydrate via in situ method [52].

The strong reducing agents like hydrazine and sodium borohydride can be used to synthesize metal nanoparticles very easily, but the main drawback of these reducing agents is that they are toxic and hazardous to the environment. To avoid

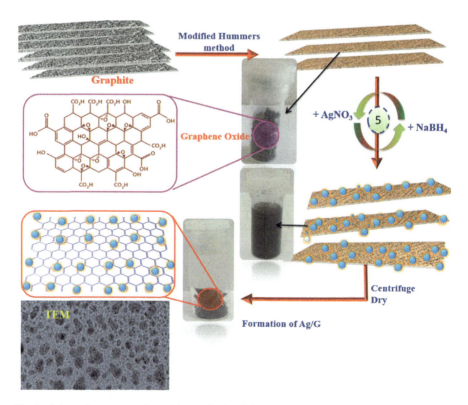

Fig. 1 Schematic representation of the synthesis of silver loaded graphene (Ag/G) composite [53]

such detrimental effects, researchers are recently using a green and inexpensive chemical synthesis approach. silver nanoparticle (AgNPs)–graphene oxide (GO) composite was also synthesized by the green synthetic approach where beta cyclodextrin used as a stabilizing agent and ascorbic acid act as reducing agent [54].

Ju and Chen demonstrated a green and simple in situ approach to synthesize Au nanoparticles on nitrogen-doped graphene quantum dots (Au NPs–N-GQDs). The composite was obtained by simple refluxing of the nitrogen-doped graphene quantum dots and $HAuCl_4 \cdot 4H_2O$ as the precursor salt of Au NPs without using any other reductant and surfactant [55].

1.5 Synthesis of Au/Ag Supported Polymer Nanocomposites

Usually, the nanoparticles tend to aggregate in the polymer matrix and thus the dispersion of nanoparticles in polymer matrices is challenging. This kind of problems of nanoparticles usually results in poor processability of composites and a high defect density [56, 57]. The composite material's physical properties are very much dependable on particle dispersion within the nanocomposite [58]. Toor and Pisano adopted ex situ approach to synthesize nanocomposite material. They prepared PVP coated gold nanoparticles in the form of a dried powder and this powder was dispersed in dimethylformamide (DMF) solvent. On the other hand, the polymer solution was prepared by them where the PVDF in pellets form was mixed with the DMF solvent at 100 °C with continuous stirring. This particle solution was mixed with the polymer solution in various concentration under sonication to prepare the nanocomposite suspension [59] (Fig. 2).

Kanahara et al. [60] prepared amino-terminated polymer particles using the SORP technique. The solution of each polymer in THF was prepared at a certain concentration. Membrane-filtered water was slowly added to the polymer solution in a glass bottle with constant stirring and the resulting mixture was then allowed to stay uncovered at ambient temperature to evaporate the THF. An opaque dispersion of polymer particles in water was obtained after complete evaporation of the THF. Then the aliquot of the aqueous dispersion of polymer particle was mixed with an

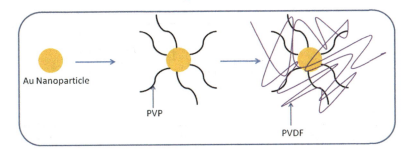

Fig. 2 PVP coated gold nanoparticles are blended with the PVDF polymer

aqueous dispersion of Au NPs and an aqueous PEG solution in a microtube, where PEG was used as a stabilizing agent that can prevent the agglomeration of polymer particles. These composite particles were then separated out by centrifugation followed by washing with water.

Cucurbit [8] uril was used to prepare a gold nanoparticle-polymer composite material, which acts as a supramolecular "handcuff" to grip together with the functionalized gold nanoparticles and acrylamide copolymer. The AuNPs must be functionalized by a water-soluble SAM yet remain accessible for CB [8] host-guest binding, as water solubility is a must for CB [8] ternary system. Water-soluble functionalized-AuNP 3 with a neutral (major) ligand tri (ethylene glycol)-1-butanethiol (EG_3-C_4-SH; 1) and a viologen-containing (minor) ligand, 1-methyl-4,40-bipyridinium-dodecanethiol bisbromide ([MV^{2+}-C_{12}-SH] · $2Br^-$; 2) were prepared by a mixed self-assembled monolayer (mSAM) approach. AuNPs with a diameter of roughly 5 nm were prepared and functionalised with varying ligand mixtures of 1 and 2 leading to the AuNP 3 as depicted in Fig. 3. Another NP control was prepared in a similar manner with a SAM consisting of solely EG_3 (EG_3-AuNP 4) [61] (Fig. 4).

1.6 Synthesis of Au/Ag Supported Dendrimer Nanocomposites

Silver-dendrimer nanocomposites were synthesized by mixing dendrimers and silver nitrate solution to obtain Ag^+/dendrimer complex at pH 7.0. Sodium

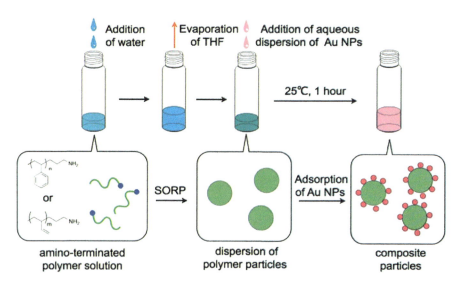

Fig. 3 Schematic illustration of the preparation of composite particles [60]

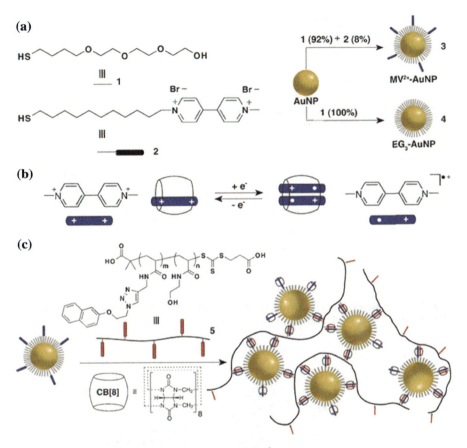

Fig. 4 Schematic representation of **a** preparation of MV^{2+}-AuNP 3 and EG_3-AuNP 4, **b** formation of a 2: 1 $(MV^+)_2$ CB [8] inclusion complex upon reduction and **c** the noncovalent functionalization of MV^{2+}-AuNP 3 with CB [8] and multivalent Np-copolymers 5 [61]

borohydride was added to the mixture as a reducing agent to reduce Ag^+ to Ag^0 and Silver-dendrimer nanocomposite was formed [62]. Zhang et al. showed a simple method of fabrication of thin film composite (TFC) where the silver–polyethylene glycol PEGylated dendrimer nanocomposite is used. They stirred poly (ethylene glycol) methyl ether acrylate (PEGMEA) with the $AgNO_3$ aqueous solution and exposed to the light for several hours to prepare the silver nanocomposite membrane [63] (Fig. 5).

Stable gold-dendrimer nanochains were synthesized in aqueous media without using any templates or organic solvents by regulating the density of dendrimers (Fig. 6). In this approach polyamidoamine (PAMAM) dendrimers self-assembled with gold nanoparticles to obtain one-dimensional nanochains [64].

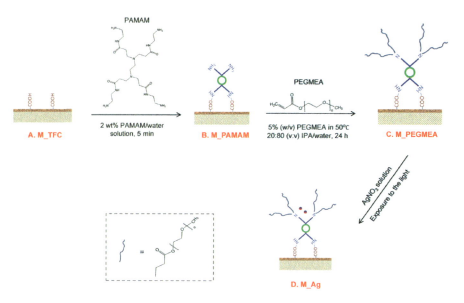

Fig. 5 Schematic diagram of the synthesis of silver–PEGylated dendrimer nanocomposite on the thin film composite membranes [63]

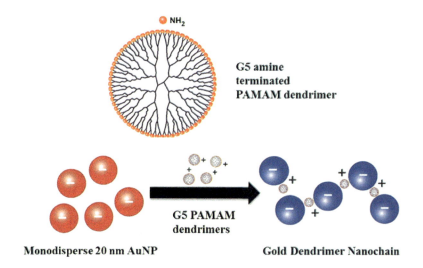

Fig. 6 Schematic representation of the chemical structure of a Generation 5 PAMAM dendrimer and the plausible mechanism of self-assembly of gold nanoparticles (20 nm) with dendrimers to produce electrostatic interactions driven one-dimensional 'nanochains' [64]

2 Applications and Toxicological Evaluations of Gold Nanocomposites

Recently, the use of gold nanoparticles to synthesize different biocompatible nanocomposites has provided various new ways of treatment of different diseases. In cancer, the use of such nanocomposites is becoming popular day by day. In present times, the use of anisotropic gold in the form of nanoparticles has generated much interest amongst scientists all over the world because of the particles' unique properties such as optical, electronic, size- and shape-dependent, and chemical properties, which are completely different from those in bulk and elemental form [65–67]. Moreover, gold nanoparticles (AuNPs) has the potential to act as a photothermal agent [68–70].

2.1 Silica-Based Gold Nanocomposite

The conventional photo-absorbing agents, which are essential for converting radiation energy to heat, have many limitations due to their lack of stability and absorption of radiation. But the use of novel nanostructures has provided the way to overcome such limitations of the conventional agents. The silica-gold-silica (SiO_2–Au–SiO_2) nanocomposite has demonstrated a relatively broad extinction in the NIR region, which is significant for its photothermal effect [71]. Researchers have explored its potential as a photothermal therapy material in vitro in various cell lines of mammalian origin. Researchers have also used Au–SiO_2 nanocomposites for the detection of human ovarian cancer cells (HOC) [72]. By analyzing the optical absorption spectra, it was observed that the treatment of HOC with Au–SiO_2 nanocomposites made them susceptible to change the absorption spectrum in comparison with the control cells. For these reasons, the potential of clinical applications of AuNPs is presently an intense subject for research [73]. Current researchers also indicated that mesoporous silica-coated gold nanocomposites have a strong potential to diagnose and treat breast cancer. They seemed to be potent cytotoxic agents affecting triple negative breast cancer cell line like MDA-MB-231. But toxicological implications of these composites are yet to be evaluated. Questions still remain about whether these can be used in humans with little no side effects on humans and the environment.

2.2 Lipid-Coated Gold Nanocomposite

Liposomal nanoparticles are made up of natural lipids having the potential to encapsulate both water-soluble and insoluble drugs in their core. Their design is for controlled delivery of therapeutic agents enhancing therapeutic efficacy minimizing side effects. Encapsulation or coating of metallic nanoparticles especially Au with lipids is a useful non-covalent approach to stabilize surface chemistry and to increase the compatibility of lipid-coated nanohybrids loaded with drugs at biological level [74–79]. In a study by Kang et al. it has been shown that the administration of docetaxel (DTX) in a lipid bilayer on the nanoparticles leads to the reduction of its side effects, thereby increasing its efficacy. DTX, an anti-cancer drug showed to effectively increase its intracellular delivery and therapeutic efficacy when administered with nanoparticles [80]. The uptake of drugs by the cells was significantly enhanced by this formulation and it also enhanced cytotoxicity compared to uncoated AuNPs and the free drug. The above-mentioned effects were because of cell-cycle arrest in the G_2/M phase of skin melanoma cell line B16F10 and breast cancer cell line MCF-7 cells with the increased population of sub-G_1 phase apoptotic cells. Thus drug-encapsulated lipid-coated nanoparticles have a scope to be used as a promising nanocarrier system for significantly enhanced cancer chemotherapy.

2.3 Manganese Oxide-Based Gold Nanocomposites

Macrophages and monocytes, two of the key components of host response to tumor cells, augments cell proliferation in the tumor microenvironment and in case of infections [81], along with the hypoxic condition, which is also very essential for the survival of tumors and this hypoxic condition can be clearly noticed in case of inflammation or solid tumor formation [82]. Suppression of the hypoxic condition by modulating the signalling pathways of Hypoxia Inducing Factor (HIF) with the use of gold-manganese oxide nanocomposite induces the subset of macrophages to revert back to a cytotoxic and anti-tumorigenic form [83]. This strategy of reversing hypoxic condition of the tumour microenvironment, through their effect on Tumor-Associated Macrophages, can be utilized as a mechanism to combat cancer. Moreover, broad near-infrared absorption of porous gold nanoparticle-manganese monoxide nanocomposites effectively increase the diagnostic time and also provides deeper photoacoustic imaging depth [84], which can be used to perform more accurate MR/Photoacoustic/CT tumour imaging in the human body.

2.4 Chitosan-Based Gold Nanocomposite

A group of researchers used Chitosan, a non-toxic biopolymer, along with gold nanoparticles to develop a nanocomposite that shows properties like the high current response intensity, a high electrocatalytic tendency towards H_2O_2 reduction, high stability, and good biocompatibility. Immunosensors prepared from this gold/chitosan nanocomposite can be used for high-throughput biomedical sensing and clinical applications, such as for the detection of prostate cancer using PSA biomarker, without any sophisticated and complicated fabrication procedure [85]. Nanocomposite, constructed with gold nanoparticles (AuNPs), Carbon nano-onions (CNOs), single-walled carbon nanotubes (SWCNTs) and chitosan (CS) (AuNPs/CNOs/SWCNTs/CS) have been used to develop high-sensitivity electrochemical immunosensor that can detect carcinoembryonic antigens (CEA), which is a clinical tumor marker [86]. Along with the high sensitivity and excellent stability in the biological system, this immunosensor also provides excellent selectivity due to the property of resistance to interference in the presence of other antigens in the serum. Moreover, this platform can be utilized to design various highly selective and sensitive immunosensors to detect important biomarkers such as ciprofloxacin and immunoglobulin A (Fig. 7).

Apart from great application benefits, this type of nanocomposites also offers some health risks. The carbon nanotubes (SWCNTs) used in the above formulation can be hazardous to the people, especially who are producing or handling such nanomaterials. Inhalation of carbon nanotubes has potential to cause inflammation and granuloma formation in the lungs as they can reach the lower respiratory tract and can persist for a year or more. They can also translocate to other organs such as lymph nodes and pleura [87].

Many studies have also shown that nanocomposites composed of gold nanoparticles encapsulated by temperature-sensitive microgel are convenient colloidal systems with trapping capabilities [88, 89]. The biocompatibility of such composites can be used as a system for drug release in low solvent pH, for example in cancer therapy.

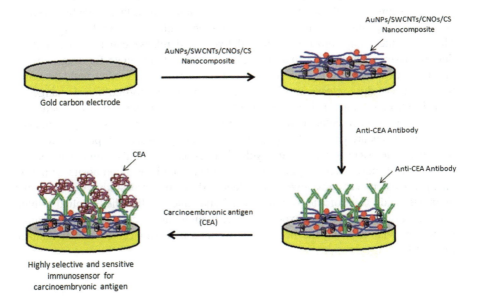

Fig. 7 Fabrication of a highly sensitive and selective immunosensor for carcinoembryonic antigen [86]

2.5 Graphene-Based Gold Nanocomposite

Not only in case of different biomarkers, but many studies have also depicted that these nanocomposites can be very efficient in determining the presence of different components in the environment, especially in the aquatic environment. Researchers have developed gold nanoparticle/Graphene nanocomposite that can be used to determine trace Chromium in water samples [90]. The hexavalent form of chromium acts as a strong oxidizing agent and shows carcinogenic and mutagenic properties, whereas the trivalent form is less toxic and studies have shown that it plays a vital role in many biological processes. So, it is of great importance to determine the level of trace hexavalent chromium in the water bodies to provide control for human and environmental concerns. The gold nanoparticle/Graphene nanocomposite sensors have been used to detect trace hexavalent chromium in the river samples of Indonesia as it shows high stability, high sensitivity, high electrocatalytic activity and low cost of analysis.

Graphene nanocomposites impart toxic effects on human erythrocytes, skin, fibroblasts and on different other cell lines. It is also extensively used in cancer research because of its unique set of characters that provide high mechanical strength and better stability preventing aggregation of the gold nanoparticles [91]. Researchers have constructed biosensors using graphene oxide based AuNP nanocomposite that can detect tumour mutations [92]. Gold nanoparticles were also added as functional agents in N_2- or S- doped graphene sheets (AuNPs-N_2-doped-GN or AuNPs-thiolated GN composite), that shows much enhanced SERS (Surface Enhanced Raman Spectroscopy) attributes on their electro-active surfaces [93, 94]. Along with these efficient diagnostic applications, these graphene oxide/gold nanocomposites have also been proven to be an efficient drug delivery system. Moreover, the whole process of drug delivery and release can be monitored by fluorescent-monitoring [95], making it a more efficient candidate for drug delivery in cancer treatments. Graphene oxide/gold nanocomposite loaded with daunorubicin enhances drug release into cancer cells by inducing morphological changes in cancer cell membrane. This also reduced P-glycoprotein expression and activated apoptosis in cancer cells in both in vitro and in vivo models [96]. Graphene nanocomposites are proposed as potent anti-cancer agents as they produce reactive oxygen species (ROS), induce cell cycle changes and might also initiate apoptosis.

In spite of these useful applications of graphene oxide as a composite with gold nanoparticles, it has some cytotoxic effects. Many in vitro studies reported that graphene oxide is cytotoxic to both normal and cancer cells when applied in high concentration and with long exposure time, though cancer cell lines showed more percentage viability may be because of its inbuilt resistance to cellular damage [97–100].

2.6 Dendrimer Stabilized Gold Nanoparticles

Multifunctional nanocomposites constructed by the researchers using gold nanoparticles stabilized by polyamidoamine (PANAM) dendrimers that can be used for combined detection of tumour cells through many processes such as flow cytometry, confocal microscopy, computed tomography, etc. [101]. These dendrimers are highly branched three-dimensional polymeric macromolecules that have highly configurable architecture. The biocompatibility and pharmacokinetics of this nanoconstruct can be adjusted by tuning the chemical synthesis of the dendrimer. Its high biocompatibility, high drug loading capacity and presence of multiple functional groups on its surface makes it a good candidate for photothermal therapy and targeted cancer therapy. Moreover, its good biodegradability and water solubility augment its use as a carrier for anticancer drugs [102–104]. It was also showed by researchers that incorporation of gold to dendrimer can actually lower the toxicity of dendrimer in a selective manner by modulating the physiochemical parameters of dendrimers.

2.7 Iron Oxide Gold Nanocomposite

Another construct with the gold nanoparticles is the Iron oxide/gold nanocomposite, which has immense importance in theranostics that is both in therapeutics and diagnostics. The flower-shaped iron oxide/gold nanocomposites possess a large number of magnetic domains, leading to enhanced magnetic properties that are helpful in magnetic resonance imaging (MRI) [105]. Not only in MRI, but this nanocomposite is also very useful in computed tomography (CT), Fluorescent optical imaging, hyperthermia and many more diagnostic processes. This nanocomposite has also been used as a carrier for drug delivery for chemotherapy such as cisplatin conjugated nanocomposite.

Several iron oxide-based nanocomposites with gold nanoparticles are under clinical trial to understand their toxicity, but only the dextran-coated superparamagnetic iron oxide is approved for human use by FDA. There are very few researches on the complete toxicological profile of the iron oxide nanocomposites and some of the researches are conflicting with each other. Moreover, some reports suggest superparamagnetic iron oxide be toxic on mouse fibroblast cells whereas reports have also shown that high concentration of this composite failed to show any toxicity.

3 Applications and Toxicological Evaluations of Silver Nanocomposites

The nanoparticles and nanocomposites are popular for use in various fields for various reasons like high surface area-to-volume ratio, increased solubility of the drug and several others. But silver is definitely a suited choice for several others for biocidal properties or microbicidal properties of silver nanoparticles (Table 1) or silver-based nanocomposites [106, 107]. Historically, before penicillin was even discovered, silver was broadly used to combat severe infections, especially for treatment of burns and chronic wounds. Even after the discovery of Penicillin, its use was revitalized in 1968 when silver nitrate was combined with sulfonamide to produce a silver sulfadiazine cream for treating burns [106]. Moreover, antibiotic resistance has imposed a major problem in using the antibiotic drugs available and very recently, silver-based nanocomposites have gained immense importance in instances of infections [108]. Presently, there are a number of medical products available, such as silver-based ointments and bandages that have been proven to be efficiently retarding and preventing bacterial infections [109]. Current researches mainly focus on the improvements in the development of novel silver nanoparticle (Table 1) and composites keeping in mind the wide use and antimicrobial properties of silver. Moreover, researchers are showing more interest towards the exploitation of silver nanoparticle to develop new biologically active materials so that the unique antibacterial properties of silver can be combined with the performance of the biomaterial [110–114]. Silver nanocomposites represent a promising strategy to fight against infections on used medical devices as a problem of proper sterilization as a major cause of hospital deaths in many places around the world. Besides having antibacterial properties, they are also antifungal and antiviral agents. Silver nanoparticles exert cytoprotective effects towards HIV-infected T cells by inhibiting the production of extracellular virions in vitro. They directly interact with the double-stranded DNA of HIV particles. But it is still not known how they affect other viruses. The effects of silver nanoparticles and nanocomposites on fungi are grossly unexplored. Though resistance to existing anti-fungal drugs are less commonly heard and it is not a menace like predominant antibacterial resistance, the long-term concern remains for different types of the antifungal agent as their options are really limited in the present world. Hence, researches are required to develop drugs with novel antifungal mechanisms. Recently, attention has focused on the potential of silver to be used as an antifungal agent, with experimental evidence that silver nanoparticles are capable of exhibiting potent antifungal effects, most likely by destroying the membrane integrity of fungal cells [115–117].

3.1 Graphene Oxide Silver Nanocomposite

The use of silver nanoparticles to develop different nanocomposites is getting very popular among scientists nowadays because of its versatility and high stability. The use of Graphene oxide/silver nanocomposite along with laser exposure (Photodynamic therapy) exhibits a synergistic effect, increasing cytotoxicity to the breast cancer cell lines [118]. This synergistic effect quickly produces reactive oxygen species such as hydroxyl radicals, superoxide ions and singlet oxygen, resulting in oxidative stress and can also include disruption of the cell membrane [119]. These properties of the graphene oxide/silver nanocomposite can be used in

Table 1 Silver nanocomposites and their biological implications

```
                Electrochemiluminiscence immunosensor for trace level
                                    of p53.
                              Antibacterial agent
                           ┌─────────────────────┐
                           │    Silica based     │
   Magnetically controlled │                     │  Graphene oxide       ROS generation.
   antibacterial agent.    │ Iron                │  based                Oxidative stress.
   Uric acid determination │ oxide     AgNPs     │                       Cytotoxic to cancer cells.
   in blood and urine      │ based               │
                           │                     │
                           │   Dendrimer based   │
                           └─────────────────────┘
                              Oxidative stress.
                 Electrochemiluminiscence biosensor for cancer cell.
                       Modulation of renin-angiotensin system
```

future for biomedical applications, especially in targeted cancer therapy. But it should be mentioned that researchers with two lineages of macrophages—a tumour lineage (J774) and peritoneal macrophage collected from Balb/c mouse showed that graphene oxide silver nanocomposite was toxic and induced significant ROS generation compared to silver nanoparticles, though graphene oxide/silver nanocomposites entered less inside cells. Hence the fate of the nanocomposites used should be carefully monitored and is a major concern in developing biocompatible materials.

3.2 Iron Oxide-Based Silver Nanocomposite

The magnetic iron oxide/silver nanocomposites show high anti-bacterial activity, which was tested against *E. coli*. This nanocomposite can also be used as an antibacterial agent which could be magnetically controlled in different biomedical applications. The reason behind it is the fact of the super magnetic properties of the iron oxide nanoparticles are not affected by the modulation of silver ions [120]. A group of researchers has also studied iron oxide-silver oxide quantum dots (QD) decorated cellulose nanofibres as a drug carrier for skin cancer therapy. They introduced two drugs Etoposide and Methotrexate to the melanoma cells in assistance with Fe_3O_4–Ag_2O QD/cellulose nanofibre carrier and showed that the cell viability decreased [121]. This study also indicated that a high number of unloaded nanocomposites were not cytotoxic. Iron oxide-based nanocomposites did not induce any possibility of liver or kidney toxicity. On the other hand, silver nanoparticle alone resulted in increased serum alkaline phosphatase, calcium as well as lymphocyte infiltration in liver and kidney, indicating organ toxicity. These results indicate that in vivo kinetics of nanoparticles are required to be studied to understand their hazards and also nanocomposites might be toxicologically less hazardous than the metal nanoparticle itself.

Moreover, polyaniline (PANI) supported iron oxide/silver nanocomposites is presently the composite adopted to develop a sensor for the tracing and assessing uric acid in human blood and urine sample [122]. High sensitivity, selectivity, and low detection limits augment its potential for various applications.

3.3 Dendrimer-Based Silver Nanocomposites

The silver/dendrimer nanocomposites are of great importance in modern day research. Scientists have already demonstrated several uses of this construct. Xin Jin and group have demonstrated that silver/dendrimer (PAMAM) nanocomposite labelled DNA probe shows high sensitivity and selectivity with significantly low detection limit [123].

Researchers have developed electro-chemiluminescence biosensors for HL-60 cancer cell detection from $g-C_3N_4$ nanosheets and silver-PANAM-luminol nanocomposites, which show great selectivity and low detection limit [124] and has the potential to be used as cell biomarker. 5-fluorouracil loaded silver/PAMAM nanocomposite synergistically induces oxidative stress on cancer cells which were marked by reactive oxygen species and reactive nitrogen intermediate generation, DNA condensation and cytoskeletal compaction, leading to cell blebbing and injury. This also turns on the p53 gene-mediated signalling pathway leading to apoptosis [125].

In addition to these versatile applications of dendrimer-based silver and gold nanocomposites, researchers have also demonstrated the adverse effect of different dendrimers on biological organisms. Researches indicate that the stability of some dendrimers in different physiological conditions varies considerably. In vitro studies in a fish cell line (PLHC-1) have depicted that the PAMAM dendrimer induces toxicity by the generation of reactive oxygen species, which is followed by DNA damage and cell death [126]. In vivo studies have also demonstrated that PAMAM dendrimers induce aggregation of different blood proteins and results in clots in blood vessels [127]. Administration of PAMAM dendrimer in the mouse model induce acute lung failure by modulating the renin-angiotensin system [128]. A considerable number of dendrimers have also been found to be accumulated in some other important organs of the body such as the liver, kidney, heart and in the brain of neonatal rabbit with cerebral palsy. Akhtar and group showed that PAMAM inhibits ERK1/2 and p38 MAPK phosphorylation in both the cortex and medulla region of rat kidney, modulating the MAPK signalling pathway [129].

Even the sub-lethal dose of this dendrimer effects growth and development of zebrafish adversely [130]. So, it is important to do more research on the surface modifications and drug release of such dendrimers for designing a more biocompatible dendrimer construct and make it more suitable for various biological and biomedical applications [131].

3.4 Silica-Based Silver Nanocomposite

Silica-based silver nanocomposites have been extensively used in biomedical fields, especially for developing immunosensors. Researchers have developed an electrochemiluminescence immunosensor for p53 with $Ru(bpy)_3^{2+}$/silver nanoparticles doped silica core-shell nanocomposite (RuAg/SiO2) that shows excellent electrochemiluminescence behaviour with wide linear range, high selectivity, stability and low detection limit [132]. It efficiently detects trace level p53, so can be a very useful tool to be used as a tumour biomarker. Moreover, Yiyan Song and the group have prepared nanocomposite of polydopamine/silver nanoparticle on mesoporous silica (SBA15) that has potential as an antimicrobial agent along with the industrial role as a catalyst [133]. This composite successfully inhibited the growth of *E.coli*, *S. aureus*, and *A. fumigatus*. Mesoporous silica/silver nanocomposite (Ag-SBA-15) also shows high Hg^0 capture capacity with high ability of regeneration and high recyclability, therefore can be used as a catalyst to capture Hg^0 from coal-fired power plant flue gases [134]. These depict the importance of silver/silica nanocomposites in both environmental as well as biomedical applications.

4 Conclusions

In this chapter, different synthetic methods to prepare metal nanocomposites based on recent studies are well described. There is a large scope of future research developing facile, green synthetic route to synthesize metal nanocomposites minimizing the use of hazardous chemical reagents. Synthesis of nanocomposites on a metal base can be done by two different methods: in situ method and ex situ method as described before. Mesoporous metal oxide nanocomposites, silver/gold-supported graphene nanocomposite, silver/gold supported polymer nanocomposites and silver/gold-supported dendrimer nanocomposites are a few varieties of nanocomposites whose synthesis have been discussed in this chapter keeping in mind their wide research usage in causing cell cytotoxicity, in experimental cancer therapy, in antibacterial activity and antifungal activity and in developing immunosensors to name a few. Their roles in the biological system have made it exigent to study and understand the toxicological evaluations of the same in the system as well as to the person exposed to it.

Silver, gold, graphene nanocomposites have shown promising evidence indicating their importance in cancer research and various other fields. They have promising cytotoxic effects on various cancer cells and have potent antibacterial and antifungal activities. But, in spite of their goodness in terms of human healthcare, very few of the researches are actually translating into effective market available drugs and have reached the stage of clinical trial. The reason behind the lag between innovative research to identify new nanocomposites with immense biological potency and the effective market available drug is the dearth of research studies to evaluate the toxicity generated by the composites in the cell system, in animal models and in users who are actually working with the nanocomposites.

It is important to understand the control of the concentrations in using the nanocomposites to have the beneficial effects. Though there are a large number of researches are going on in this field, a systematic in vitro–in vivo extrapolation studies after the application of the nanocomposites is necessary. There are very little information available till to date about the toxic effects of the biomarkers, such as their immunomodulatory effect or ability to alter the genetic expression. In this chapter, the synthesis of nanocomposites relevant to biological research, their wide applications and toxicological evaluations have been discussed with special reference to cancer. But more studies are required on the toxicological implications of the nanocomposites to use them as our friends and not as foes. Intensive toxicological evaluation along with the ongoing research of finding new nanocomposites are required to effectively use nanocomposites in biological systems and as a tool for cancer therapy which will lead to the innovation of modern day target-specific drugs and new arenas in chemotherapy.

References

1. Danhier F, Feron O, Préat V (2010) To exploit the tumor microenvironment: passive and active tumor targeting of nanocarriers for anti-cancer drug delivery. J Control Release 148:135–146
2. Northfelt DW, Martin FJ, Working P, Volberding PA, Russell J, Newman M, Amantea MA, Kaplan LD (1996) Doxorubicin encapsulated in liposomes containing surface-bound polyethylene glycol: pharmacokinetics, tumor localization and safety in patients with AIDS-related Kaposi's sarcoma. J Clin Pharmacol 36:55–63
3. Fang J, Nakamura H, Maeda H (2011) The EPR effect: unique features of tumor blood vessels for drug delivery, factors involved, and limitations and augmentation of the effect. Adv Drug Deliv Rev 63:136–151
4. Maeda H (2001) The enhanced permeability and retention (EPR) effect in tumor vasculature: the key role of tumor-selective macromolecular drug targeting. Adv Enzyme Regul 41: 189–207
5. Morghimi SM, Hunter AC, Murray JC (2001) Long-circulating and target-specific nanoparticles: theory to practice. Pharmacol Rev 53:283–318
6. Matsuo H, Wakasugi M, Takanaga H, Ohtani H, Naito M, Tsuruo T, Sawada Y (2001) Possibility of the reversal of multidrug resistance and the avoidance of side effects by liposomes modified with MRK-16, a monoclonal antibody to P-glycoprotein. J Control Release 77:77–86

7. Lammers T, Kiessling F, Hennink WE, Storm G (2012) Drug targeting to tumors: principles, pitfalls and (pre-) clinical progress. J Control Release 161:175–187
8. Mohammadreza S, Soehnlen ES, Hao J et al (2010) Dual purpose prussian blue nanoparticles for cellular imaging and drug delivery: a new generation of T1-weighted MRI contrast and small molecule delivery agents. J Mater Chem 20(25):5251–5259
9. Liang X, Deng Z, Jing L et al (2013) Prussian blue nanoparticles operate as a contrast agent for enhanced photoacoustic imaging. Chem Commun 49(94):11029–11031
10. Fu G, Feng S, Liu W, Yue X (2012) Prussian blue nanoparticles operate as a new generation of photothermal ablation agents for cancer therapy. Chem Commun 48(94):11567–11569
11. Pina S, Oliveira JM, Reis RL (2015) Natural-based nanocomposites for bone tissue engineering and regenerative medicine: a review. Adv Mater 27:1143–1169
12. Rafiee MA, Rafiee J, Wang Z, Song H, Yu Z-Z, Koratkar N (2009) Enhanced mechanical properties of nanocomposites at low graphene content. ACS Nano 3:3884–3890
13. Mariano M, El Kissi N, Dufresne A (2014) Cellulose nanocrystals and related nanocomposites: review of some properties and challenges. J Polym Sci, Part B Polym Phys 52:791–806
14. Hu H, Onyebueke L, Abatan A (2010) Characterizing and modeling mechanical properties of nanocomposites-review and evaluation. J Min Mater Charact Eng 9:275–319
15. Beck JS, Vartuli JC (1996) Recent advances in the synthesis, characterization and applications of mesoporous molecular sieves. Curr Opin Solid State Mater Sci 1:76–87
16. Davis ME (2002) Ordered porous materials for emerging applications. Nature 417:813–821
17. Liu AM, Hidajat K, Kawi S, Zhao DY (2000) A new class of hybrid mesoporous materials with functionalized organic monolayers for selective adsorption of heavy metal ions. Chem Commun 1145–1146
18. Zhuang TY, Shi JY, Ma BC, Wang W (2010) Chiral norbornane-bridged periodic mesoporous organosilicas. J Mater Chem 20:6026–6029
19. Tsou CJ, Chu CY, Mou CY (2013) A broad range fluorescent pH sensor based on hollow mesoporous silica nanoparticles, utilising the surface curvature effect. J Mater Chem B 1:5557–5563
20. Heidegger S, Gößl D, Schmidt A, Niedermayer S, Argyo C, Endres S, Bein T, Bourquin C (2016) Immune response to functionalized mesoporous silica nanoparticles for targeted drug delivery. Nanoscale 8:938–948
21. Li Z, Barnes JC, Bosoy A, Stoddart JF, Zink JI (2012) Mesoporous silica nanoparticles in biomedical applications. Chem Soc Rev 41:2590–2605
22. Lin YS, Hurley KR, Haynes CL (2012) Critical considerations in the biomedical use of mesoporous silica nanoparticles. J Phys Chem Lett 3:364–374
23. Tao X, Liu B, Hou Q, Xu H, Chen JF (2009) Enhanced accumulation and visible light-assisted degradation of azo dyes in poly (allylamine hydrochloride)-modified mesoporous silica spheres. Mater Res Bull 44:306–311
24. Yuan Q, Chi Y, Yu N, Zhao N, Yan W, Li X, Dong B (2014) Amino-functionalized magnetic mesoporous microspheres with good adsorption properties. Mater Res Bull 49:279–284
25. Huang CH, Chang KP, Oua HD, Chiang YC, Wanga CF (2011) Adsorption of cationic dyes onto mesoporous silica. Microporous Mesoporous Mater 141:102–109
26. Li Y, Zhaou Y, Nie W, Song L, Chen P (2015) Highly efficient methylene blue dyes removal from aqueous systems by chitosan coated magnetic mesoporous silica nanoparticles. J Porous Mater 22:1383–1392
27. Huang RS, Hou BF, Li HT, Fu XC, Xie CG (2015) Preparation of silver nanoparticles supported mesoporous silica microspheres with perpendicularly aligned mesopore channels and their antibacterial activities. RSC Adv 5:61184–61190
28. Tian Y, Qi J, Zhang W, Cai Q, Jiang X (2014) Facile, one-pot synthesis, and antibacterial activity of mesoporous silica nanoparticles decorated with well-dispersed silver nanoparticles. ACS Appl Mater Interfaces 6:12038–12045

29. Liong M, France B, Bradley KA, Zink JI (2009) Antimicrobial activity of silver nanocrystals encapsulated in mesoporous silica nanoparticles. Adv Mater 21:1684–1689
30. Song J, Kim H, Jang Y, Jang J (2013) Enhanced antibacterial activity of silver/polyrhodanine-composite-decorated silica nanoparticles. ACS Appl Mater Interfaces 5:11563–11568
31. Chen CC, Wu HH, Huang HY, Liu CW, Chen YN (2016) Synthesis of high valence silver-loaded mesoporous silica with strong antibacterial properties. Int J Environ Res Pub Health 13:99–112
32. Park JH, Gu L, Maltzahn GV, Ruoslahti E, Bhatia SN, Sailor MJ (2009) Biodegradable luminescent porous silicon nanoparticles for in vivo applications. Nat Mater 8:331–336
33. Tian Y, Qi J, Zhang W, Cai W, Jiang X (2014) Facile, one-pot synthesis, and antibacterial activity of mesoporous silica nanoparticles decorated with well-dispersed silver nanoparticles. ACS Appl Mater Interfaces 6:12038–12045
34. Soto RJ, Yang L, Schoenfisch MH (2016) Functionalized mesoporous silica via an aminosilane surfactant ion exchange reaction: controlled scaffold design and nitric oxide release. ACS Appl Mater Interfaces 8:2220–2231
35. Agnihotri S, Mukherji S, Mukherji S (2014) Size-controlled silver nanoparticles synthesized over the range 5–100 nm using the same protocol and their antibacterial efficacy. Rsc Adv 4:3974–3983
36. Sadeghi B, Garmaroudi FS, Hashemi M, Nezhad HR, Nasrollahi A, Ardalan S, Ardalan S (2012) Comparison of the anti-bacterial activity on the nanosilver shapes: nanoparticles, nanorods and nanoplates. Adv Powder Technol 23:22–26
37. Ghosh S, Vandana V (2016) Nano-structured mesoporous silica/silver composite: synthesis, characterization and targeted application towards water purification. Mater Res Bull 88:291–300
38. Li H, Bian Z, Zhu J, Huo Y, Li H, Lu Y (2007) Mesoporous Au/TiO_2 nanocomposites with enhanced photocatalytic activity. J Am Chem Soc 129:4538–4539
39. Bhattacharyya S, Gedanken A (2008) Microwave-assisted insertion of silver nanoparticles into 3-D Mesoporous zinc oxide nanocomposites and nanorods. J Phys Chem C 112:659–665
40. Hazra Chowdhury I, Ghosh S, Naskar MK (2016) Aqueous-based synthesis of mesoporous TiO_2 and Ag–TiO_2 nanopowders for efficient photodegradation of methylene blue. Ceram Int 42:2488–2496
41. Sinha AK, Suzuki K, Takahara M, Azuma H, Nonaka T, Fukumoto K (2007) Mesostructured manganese oxide/gold nanoparticle composites for extensive air purification. Angew Chem 119:2949–2952
42. Allen MJ, Tung VC, Kaner RB (2010) Honeycomb carbon: a review of graphene. Chem Rev 110:132–145
43. Rao CNR, Sood AK, Subarhmanyam KS, Govindraj A (2009) Graphene: the new two-dimensional nanomaterial. Angew Chem 48:7752–7777
44. Novoselov KS, Geim AK, Morozov SV, Jiang D, Zhang Y, Dubonos SV, Grigorieva IV, Firsov AA (2004) Electric field effect in atomically thin carbon films. Science 306:666–669
45. Novoselov KS, Geim AK, Morozov SV, Jiang D, Katsnelson ML, Grigorieva IV, Dubonos SV, Firsov AA (2005) Two-dimensional gas of massless Dirac fermions in graphene. Nature 438:197–200
46. Stankovich S, Dikin DA, Dommett GHB, Kohlhaas KM, Zimney EJ, Stach EA, Piner RD, Nguyen ST, Ruoff RS (2006) Graphene-based composite materials. Nature 442:282–286
47. Berger C, Song Z, Li X, Wu X, Brown N, Naud C, Mayou D, Li T, Hass J, Marchenkov AN, Conrad AH, First PN, de Heer WA (2006) Electronic confinement and coherence in patterned epitaxial graphene. Science 312:1191–1196
48. Wu J, Pisula W, Mullen K (2007) Graphenes as potential material for electronics. Chem Rev 107:718–747
49. Kim KS, Zhao Y, Jang H, Lee SY, Kim JM, Kim KS, Ahn JH, Kim P, Choi JY, Hong BH (2009) Large-scale pattern growth of graphene films for stretchable transparent electrodes. Nature 457:706–710

50. Acik M, Chabal YJ (2013) A review on thermal exfoliation of graphene oxide. J Mater Sci Res 2:101–112
51. Pei S, Zhao J, Du J, Ren W, Cheng HM (2010) Direct reduction of graphene oxide films into highly conductive and flexible graphene films by hydrohalic acids. Carbon 48:4466–4474
52. Wadhwa H, Kumar D, Mahendia S, Kumar S (2017) Microwave assisted facile synthesis of reduced graphene oxide-silver (RGO-Ag) nanocomposite and their application as active SERS substrate. Mater Chem Phys 194:274–282
53. Saleh TA, Al-Shalalfeh MM, Al-Saadi AA (2018) Silver loaded graphene as a substrate for sensing 2-thiouracil using surface-enhanced Raman scattering. Sens Actuators B 254:1110–1117
54. Dar RA, Khare NG, Cole DP, Karna SP, Srivastava AK (2014) Green synthesis of a silver nanoparticle–graphene oxide composite and its application for As(III) detection. RSC Adv 4:14432–14440
55. Ju J, Chen W (2015) In situ growth of surfactant-free gold nanoparticles on nitrogen-doped graphene quantum dots for electrochemical detection of hydrogen peroxide in biological environments. Anal Chem 87:1903–1910
56. Kim P, Doss NM, Tillotson JP, Hotchkiss PJ, Pan MJ, Marder SR, Li J, Calame JP, Perry JW (2009) High energy density nanocomposites based on surface modified $BaTiO_3$ and a ferroelectric polymer. ACS Nano 3:2581–2592
57. Ehrhardt C, Fettkenhauer C, Glenneberg J, Münchgesang W, Pientschke C, Großmann T, Zenkner M, Wagner G, Leipner HS, Buchsteiner AS, Diestelhorst M, Lemm S, Beige H, Ebbinghaus SG (2013) $BaTiO_3$-P(VDF-HFP) nanocomposite dielectrics – influence of surface modification and dispersion additives. Mater Sci Eng, B 178:881–888
58. Wagener P, Brandes G, Schwenke A, Barcikowski S (2011) Impact of in situ polymer coating on particle dispersion into solid laser-generated nanocomposites. Phys Chem Chem Phys 13:5120–5126
59. Toor A, Pisano AP (2015) Gold nanoparticle/PVDF polymer composite with improved particle dispersion. In: Proceedings of the 15th IEEE international conference on nanotechnology, Rome, Italy
60. Kanahara M, Shimomuraa M, Yabu H (2014) Fabrication of gold nanoparticle–polymer composite particles with raspberry, core–shell and amorphous morphologies at room temperature via electrostatic interactions and diffusion. Soft Matter 10:275–280
61. Coulston RJ, Jones ST, Lee TC, Appel EA, Scherman EA (2011) Supramolecular gold nanoparticle–polymer composites formed in water with cucurbit[8]uril. Chem Commun 47:164–166
62. Jin X, Zhou L, Zhu B, Jiang X, Zhu N (2018) Silver-dendrimer nanocomposites as oligonucleotide labels for electrochemical stripping detection of DNA hybridization. Biosens Bioelectron 107:237–243
63. Zhang S, Qiu G, Ting YP, Chung TS (2013) Silver–PEGylated dendrimer nanocomposite coating for anti-foulingthin film composite membranes for water treatment. Colloids Surf, A 436:207–214
64. Ruiz-Sanchez AJ, Parolo C, Miller BS, Gray ER, Schlegel K, McKendry RA (2017) Tuneable plasmonic gold dendrimer nanochains for sensitive disease detection. J Mater Chem B 5:7262–7266
65. Murphy CJ, Sau TK, Gole AM et al (2005) Anisotropic metal nanoparticles: Synthesis, assembly, and optical applications. J Phys Chem B. 109(29):13857–13870
66. Burda C, Chen X, Narayanan R, El-Sayed MA (2005) Chemistry and properties of nanocrystals of different shapes. Chem Rev 105(4):1025–1102
67. Shaw CP, Fernig DG, Lévy R (2011) Gold nanoparticles as advanced building blocks for nanoscale self-assembled systems. J Mater Chem 21(33):12181–12187
68. Jain PK, Huang X, El-Sayed IH, El-Sayed MA (2007) Review of some interesting surface plasmon resonance-enhanced properties of noble metal nanoparticles and their applications to biosystems. Plasmonics 2(3):107–118

69. Huang X, Jain PK, El-Sayed IH, El-Sayed MA (2008) Plasmonic photother-mal therapy (PPTT) using gold nanoparticles. Lasers Med Sci 23(3):217–228
70. Tong L, Wei Q, Wei A, Cheng JX (2009) Gold nanorods as contrast agents for biological imaging: optical properties, surface conjugation and photothermal effects. PhotochemPhotobiol 85(1):21–32
71. Liang S, Zhao Y, Xu S, Wu X, Chen J, Wu M, Zhao X (2015) A silica-gold-silica nanocomposite for photothermal therapy in near-infrared region 7(1):85-93
72. Mishra YK, Mohapatra S, Avasthi DK, Kabiraj D, Lalla NP, Pivin JC, Sharma H, Kar R, Singh N (2007) Gold–silica nanocomposites for the detection of human ovarian cancer cells: a preliminary study. Nanotechnology 18(34):345606
73. Boisselier E, Astruc D (2009) Gold nanoparticles in nanomedicine: preparations, imaging, diagnostics, therapies and toxicity. ChemSoc Rev 38(6):1759–1782
74. Rasch MR, Rossinyol E, Hueso JL, Goodfellow BW, Arbiol J, Korgel BA (2010) Hydrophobic gold nanoparticle self-assembly with phosphatidylcholine lipid: membrane-loaded and janus vesicles. Nano Lett 10(9):3733–3739
75. Chen Y, Bose A, Bothun GD (2010) Controlled release from bilayer-decorated magnetoliposomes via electromagnetic heating. ACS Nano 4(6):3215–3221
76. Ahmed S, Madathingal RR, Wunder SL, Chen Y, Bothun G (2011) Hydration repulsion effects on the formation of supported lipid bilayers. Soft Matter 7(5):1936–1947
77. Von White G,, Chen Y, Roder-Hanna J, Bothun GD, Kitchens CL (2012) Structural and thermal analysis of lipid vesicles encapsulating hydrophobic gold nanoparticles. ACS Nano 6(6):4678–4685
78. Wijaya A, Hamad-Schifferli K (2007) High-density encapsulation of Fe_3O_4 nanoparticles in lipid vesicles. Langmuir 23(19):9546–9550
79. Xia T, Rome L, Nel A (2008) Nanobiology: particles slip cell security. Nat Mater 7(7):519–520
80. Kang JH, Ko YT (2015) Lipid-coated gold nanocomposites for enhanced cancer therapy. Int J Nanomedicine 10(Spec Iss):33–45
81. Chanmee T, Ontong P, Konno K, Itano N (2014) Tumor-associated macrophages as major players in the tumor microenvironment. Cancer 6(3):1670–1690
82. Lou JJ, Chua YL, Chew EH, Gao J, Bushell M, Hagen T (2010) Inhibition of hypoxiainducible factor-1alpha (HIF-1alpha) protein synthesis by DNA damage inducing agents. PLoS One 5(5):e10522
83. Nath A, Pal R, Singh LM, Saikia H, Rahaman H, Ghosh SK, Mazumder R, Sengupta M (2018) Gold-manganese oxide nanocomposite suppresses hypoxia and augments proinflammatory cytokines in tumor associated macrophages. Int Immunopharmacol 57:157–164
84. Liu Y, Lv X, Liu H, Zhou Z, Huang J, Lei S, Cai S, Chen Z, Guo Y, Chen Z, Zhou X, Nie L (2018) Porous gold nanocluster-decorated manganese monoxide nanocomposites for microenvironment-activatable MR/photoacoustic? CT Tumor Imag 10(8):3631–3638
85. Suresh L, Brahman PK, Reddy KR, Bondili JS (2018) Development of an electrochemical immunosensor based on gold nanoparticles incorporated chitosan biopolymer nanocomposite film for the detection of prostate cancer using PSA as biomarker 112:43–51
86. Rizwan M, Elma S, Lim SA, Ahmed MU (2018) AuNPs/CNOs/SWCNTs/chitosan-nanocomposite modified electrochemical sensor for label-free detection of carcinoembryonic antigen 107:211–217
87. Christou A, Stec AA, Ahmed W, Aschberger K, Amentia V (2016) A review of exposure and toxicological aspects of carbon nanotubes, and as additives to fire retardants in polymers 46(1):74–95
88. Contreras-Caceres R, Sanchez-Iglesias A, Karg M, Pastoriza-Santos I, Perez-Juste J, Pacifico J, Hellweg T, Fernández-Barbero A, Liz-Marzan LM (2008) Encapsulation and growth of gold nanoparticles in thermoresponsive microgel. Adv Mater 20:1666–1670
89. Contreras-Caceres R, Pastoriza-Santos I, Alvarez-Puebla RA, Perez-Juste J, FernandezBarbero A, Liz-Marzan LM (2010) Growing Au/Ag nanoparticles within microgel colloids for improved SERS detection. Chem Eur J 16:9462–9467

90. Sari TK, Takahashi F, Jin J, Zein R, Munat E (2018) Electrochemical determination of Chromium(VI) in river water with Gold nanoparticles-Graaphene nanocomposites modified electrodes 34(2):155–160
91. Yin PT, Kim TH, Choi JW, Lee KB (2013) Prospects for graphene–nanoparticle-based hybrid sensors. Phys Chem Chem Phys 15:12785–12799
92. Benvidi A, Firouzabadi AD, Moshtaghiun SM, Mazloum-Ardakani M, Tezerjani MD (2015) Ultrasensitive DNA sensor based on gold nanoparticles/reduced graphene oxide/glassy carbon electrode. Anal Biochem 484:24–30
93. Yang G, Li L, Rana RK, Zhu JJ (2013) Assembled gold nanoparticles on nitrogen-doped graphene for ultrasensitive electrochemical detection of matrix metalloproteinase-2. Carbon 61:357–366
94. Ju J, Chen W (2015) In situ growth of surfactant-free gold nanoparticles on nitrogen-doped graphene quantum dots for electrochemical detection of hydrogen peroxide in biological environments. Anal Chem 87:1903–1910
95. Wang C, Li J, Amatore C, Chen Y, Jiang H, Wang XM (2011) Gold nanoclusters and graphene nanocomposites for drug delivery and imaging of cancer cells. Angew Chem Int Ed 50:11644–11648
96. Zhang G, Chang H, Amatore C, Chen Y, Jiang H, Wang X (2013) Apoptosis induction and inhibition of drug resistant tumor growth in vivo involving daunorubicin-loaded graphene–gold composites. J Mater Chem B 1:493–499
97. Pinto AM, Gonçalves IC, Magalhães FD (2013) Graphene-based materials biocompatibility: a review. Colloids Surf, B 111:188–202
98. Seabra AB, Paula AJ, de Lima R, Alves OL, Duran N (2014) Nanotoxicity of graphene and graphene oxide. Chem Res Toxicol 27:159–168
99. Chang Y, Yang ST, Liu JH, Dong E, Wang Y, Cao A, Liu Y, Wang H (2011) In vitro toxicity evaluation of graphene oxide on A549 cells. Toxicol Lett 200:201–210
100. Guo X, Mei N (2014) Assessment of the toxic potential of graphene family nanomaterials. J Food Drug Anal 22:105–115
101. Shi X, Wang SH, Van Antwerp ME, Chen X, Baker JR Jr (2009) Targeting and detecting cancer cells using spontaneously formed multifunctional dendrimer stabilized gold nanoparticles 134(7):1373–1379
102. Nanjwade BK, Bechra HM, Derkar GK, Manvi FV, Nanjwade VK (2009) Dendrimers: emerging polymers for drug-delivery systems. Eur J Pharm Sci 38:185–196
103. Cheng Y, Wang J, Rao T, He X, Xu T (2008) Pharmaceutical applications of dendrimers: promising nanocarriers for drug delivery. Front Biosci 13:1447–1471
104. Khan MK, Nigavekar SS, Minc LD, Kariapper MS, Nair BM, Lesniak WG, Balogh LP (2005) In vivo biodistribution of dendrimers and dendrimer nanocomposites—implications for cancer imaging and therapy. Technol Cancer Res Treat 4:603–613
105. Leung KC, Xuan S, Zhu X, Wang D, Chak CP, Lee SF, Ho WK, Chung BC (2012) Gold and iron oxide hybrid nanocomposite materials 41(5):1911–1928
106. Rai M, Yadav A, Gade A (2001) Silver nanoparticles as a new generation of antimicrobials. Biotechnol Adv 27(1):76–83
107. Grishchenko L, Medvedeva S, Aleksandrova G, Feoktistova L, Sapozhnikov A, Sukhov B, Trofimov B (2006) Redox reactions of arabinogalactan with silver ions and formation of nanocomposites. Russian J General Chem 76(7):1111–1116
108. Travan A, Pelillo C, Donati I, Marsich E, Benincasa M, Scarpa T, Semeraro S, Turco G, Gennaro R, Paoletti S (2009) Non-cytotoxic silver nanoparticle—polysaccharide nanocomposites with antimicrobial activity. Biomacromol 10(6):1429
109. Chen JP (2007) Late angiographic stent thrombosis (LAST): the cloud behind the drug—eluting stent silver lining? J Invasive Cardiol 19(9):395–400
110. Kuo PL, Chen WF (2003) Formation of silver nanoparticles under structured amino groups in pseudo—dendritic poly(allylamine) derivatives. J Phys Chem B 107(41):11267–11272
111. Huang H, Yuan Q, Yang X (2004) Preparation and characterization of metal—chitosan nanocomposites. Colloids Surf, B 39(1–2):31–37

112. Fu J, Ji J, Fan D, Shen J (2006) Construction of antibacterial multilayer films containing nanosilver via layer—by—layer assembly of heparin and chitosan—silver ions complex. J Biomed Mater Res, Part A 79(3):665–674
113. Balogh L, Swanson DR, Tomalia DA, Hagnauer GL, McManus AT (2001) Dendrimer—silver complexes and nanocomposites as antimicrobial agents 1(1):18–21
114. Sanpui P, Murugadoss A, Prasad PVD, Ghosh SS, Chattopadhyay A (2008) The antibacterial properties of a novel chitosan—Ag—nanoparticle composite. Int J Food Microbiol 124(2):142–146
115. Kim KJ, Sung WS, Suh BK, Moon SK, Choi JS, Kim JG, Lee DG (2009) Antifungal activity and mode of action of silver nano-particles on Candida albicans. Biometals 22(2):235–242
116. Esteban-Tejeda L, Malpartida F, Esteban-Cubillo A, Pecharroman C, Moya JS (2009) The antibacterial and antifungal activity of a soda-lime glass containing silver nanoparticles. Nanotechnology 20(8):85103
117. Gajbhiye MB, Kesharwani JG, Ingle AP, Gade AK, Rai MK (2009) Fungus—mediated synthesis of silver nanoparticles and their activity against pathogenic fungi in combination with fl uconazole. Nanomedicine 5:382–386
118. Shaheen F, Hammad Aziz M, Fakhar-E-Alam M, Atif M et al (2017) An in vitro study of the photodynamic effectiveness of GO-Ag Nanocomposites against human breast cancer cells 7(11) Pii:E401
119. Gurunathan S, Han JW, Par JH (2015) Reduced graphene oxide–silver nanoparticle nanocomposite: a potential anticancer nanotherapy. Int J Nanomed 10:6257–6276
120. Ghaseminezhad SM, Shojaosadati SA (2016) Evaluation of the antibacterial activity of Ag/Fe_3O_4 nanocomposites synthesized using starch 144:454–463
121. Fakhri A, Tahami S, Nejad PA (2017) Preparation and characterization of Fe_3O_4-Ag_2O quantum dots decorated cellulose nanofibers as a carrier of anticancer drugs for skin cancer 175:83–88
122. Ponnaiah SK, Periakaruppan P, Vellaichamy B (2018) New electrochemical sensor based on a silver-doped iron oxide nanocomposite coupled with polyaniline and its sensing application for picomolar level detection of uric acid in human blood and urine samples
123. Jin X, Zhou L, Zhu B, Jiang X, Zhu N (2018) Silver-dendrimer nanocomposites as oligonucleotide labels for electrochemical stripping detection of DNA hybridization 107:237–243
124. Wang YZ, Hao N, Feng QM, Shi HW, Xu JJ, Che HY (2016) A ratiometric electrochemiluminiscence detection for cancer cells using g-C_3N_4 nanosheet and Ag-PANAM-luminol nanocomposites 77:76–82
125. Matai I, Sachdev A, Gopinath P (2015) Multicomponent 5-fluorouracil loaded PANAM stabilized-silver nanocomposites synergistically induce apoptosis in human cancer cells 3(3):457–68
126. Naha PC, Byrne HJ (2013) Generation of intracellular reactive oxygen species and genotoxicity effect to exposure of nanosized polyamidoamine (PAMAM) dendrimers in PLHC-1 cells in vitro. Aquat Toxicol 132–133:61–72
127. Jones CF, Campbell RA, Brooks AE, Assemi S, Tadjiki S, Thiagarajan G, Mulcock C, Weyrich AS, Brooks BD, Ghandehari H et al (2012) Cationic PAMAM dendrimers aggressively initiate blood clot formation. ACS Nano 6:9900–9910
128. Sun Y, Guo F, Zou Z, Li C, Hong X, Zhao Y, Wang C, Wang H, Liu H, Yang P et al (2015) Cationic nanoparticles directly bind angiotensin-converting enzyme 2 and induce acute lung injury in mice. Part Fibre Toxicol 12:4
129. Akhtar S, Al-Zaid B, El-Hashim AZ, Chandrasekhar B, Attur S, Benter IF (2016) Impact of PANAM delivery systems on signal transduction pathways in vivo: modulation of ERK1/2 and p 38 MAP kinase signaling in the normal and diabetic kidney 514(2):353–363
130. Heiden TC, Dengler E, Kao WJ, Heideman W, Peterson RE (2007) Developmental toxicity of low generation PAMAM dendrimers in zebrafish. Toxicol Appl Pharmacol 225:70–79

131. Naha PC, Mukherjee SP, Byrne HJ (2018) Toxicology of engineered nanoparticles: Focus on poly (amidoamine) dendriers 15(2) pii:E338
132. Wang X, Wang Y, Jiang M, Shan Y, Jin X, Gong M, Wang X (2018) Functional electrospun nanofibers-based electrochemiluminiscence immunosensor for detection of the TSP53 using RuAg/SiO$_2$NPs as signal enhancers 548:15–22
133. Song Y, Jiang H, Wang B, Kong Y, Chen J (2018) Silver-incorporated mussel-inspired polydopamine coatings on mesoporous silica as an efficient nanocatalyst and antimicrobial agent 10(2):1792–1801
134. Cao T, Li Z, Xiong Y, Yang Y, Xu S, Bisson T, Gupta R, Xu Z (2017) Silica-silver nanocomposites as regenerable sorbents for Hg^0 removal of flue gases 51(20):11909–11917

Synthesis, Characterization and Application of Bio-based Polyurethane Nanocomposites

Sonalee Das, Sudheer Kumar, Smita Mohanty and Sanjay Kumar Nayak

List of Abbreviations

PUs	Polyurethane
MMT	Montmorillonite
VOs	Vegetable oils
MDI	Methylene diisocyanate
TDI	Toluene diisocyanate
HDI	Hexamethylene diisocyanate
IPDI	Isophorone diisocyanate
JCO	Jatropha curcas oil
CO	Castor oil
PU/NS	Polyurethane/silica nanocomposites
NS	Nano silica
TGA	Thermogravimetric analysis
DTG	Derivative thermo-gravimetric
DSC	Differential scanning calorimetry
SEM	Scanning electron microscopy
TEM	Transmission electron microscopy
FTIR	Fourier transform infrared spectroscopy
E_a	Activation energy
T_m	Melting temperature
T_g	Glass transition temperature
E'	Storage modulus
E''	Loss modulus
MBPU	Modified bio-based polyurethane
MCO	Modified castor oil
HBPUS	Hyper branched polyurethane
GO	Graphene oxide

S. Das (✉) · S. Kumar · S. Mohanty · S. K. Nayak
Laboratory for Advanced Research in Polymeric Materials,
Central Institute of Plastics Engineering and Technology,
Bhubaneswar 751024, Odisha, India
e-mail: das.sonalee31@gmail.com

RGO	Reduced graphene oxide
f-RGO	Phytoextract-RGO
MWCNTs	Multiwall carbon nanotubes
XRD	X-ray diffraction
BET	Brunauer–Emmett–Teller theory
IPN	Interpenetrating polymer network
EP	Epoxy
PU/EP	Polyurethane/epoxy
DMA	Dynamic mechanical analysis
APTES	Aminopropyltriethoxy silane
WPU	Waterborne polyurethane
SMT	Silylated sodium montmorillonite
SHT	Silylated halloysite nanotubes
POBUA	Palm oil and methylene diisocyanate based polyurethane acrylate
EPOLA	Epoxidized palm oil acrylate
PCL	Poly(e-caprolactone) diol
UPCEA	Poly-(urethane-esteramide)
TiO_2	Titanium dioxide
EDX	Energy dispersive X-ray spectroscopy
WAXD	Wide-angle X-ray scattering
ECNC	*E. Globulus* derived cellulose nanocrystals

1 Introduction

Polyurethane (PU) has been one of the most attractive speciality polymers known for its excellent properties such as scratch resistance, abrasion and chemical resistance. These polymers find extensive applications in many different fields such as foams (flexible, semi-rigid and rigid), elastomers, adhesives, coatings and fibers as shown in Fig. 1.

Basically, these polymers are obtained by the polyaddition reaction of polyol and polyisocyanate as shown in Fig. 2 [1–3].

The properties of the synthesized PU depend upon the molecular weight, the degree of cross-linking, the molar ratio of NCO/OH, effective intermolecular forces and stiffness of different chain segments [4, 5]. In general, the PU chain is composed of a soft segment based on polyol and a hard segment based on diisocyanate and a chain extender [6–8]. The nature of hydrogen bonding (H-bonding) in the rigid segment leads to strong mutual attraction resulting in the formation of micro-domains which can act as physical crosslink providing elastomeric properties to the PU chain segment [7, 8]. The hard segments determine the physical properties such as rigidity, while the soft segment imparts flexibility and elasticity. Recent years have witnessed considerable research effort towards the improvement

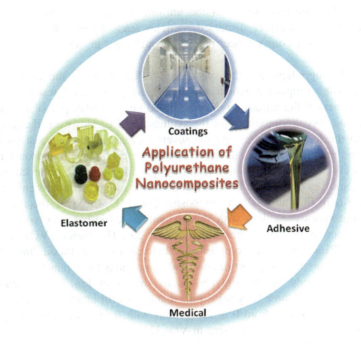

Fig. 1 Applications of polyurethane (PU) in various fields

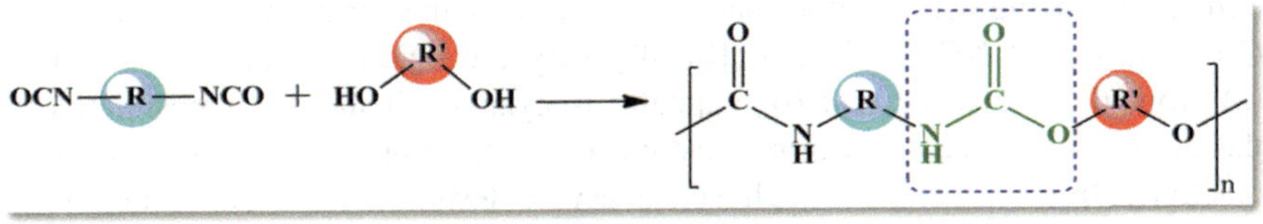

Fig. 2 Synthesis of PUs from diisocyanate and polyol

of the properties of polymer matrix i.e. PU (organic part) through the reinforcement by addition of fillers in the nanometer scale [9, 10]. Nanocomposites represent a unique class of materials that can be described by an ultrafine dispersion of nanometer size fillers in a polymer matrix [11]. The most commonly used nanofillers for PU includes nanosilica, titanium dioxide (TiO_2), organically modified montmorillonite (MMT), graphite, cellulose nanocrystals etc. [12]. The developed PU nanocomposite films using the above nanofillers exhibit good adhesion between the polymer matrix and nanofillers because of the higher surface area and surface-to-volume ratio of the nanoscale building blocks [13]. The introduction of smaller amounts of nanofillers can provide higher thermal, mechanical, optical and flame retardancy properties to the PU composite films as compared to their neat counterparts [14]. Additionally, these fillers are suitable for application in transparent PU coatings owing to the uniform dispersion of the nanoparticles. Moreover, these uniformly dispersed particles leads to no interaction with the incident light. As a consequence, there is no turbidity in the resulting composite material.

The main ingredient for the synthesis of PUs includes the polyol and the polyisocyanate component which are based on petro-based resources. Recent decades witnessed the utilization of petro-based i.e. polyether polyol and polyester polyol as the polyol component for the synthesis of PU. However, the progressive decline in the fossil resources and the severe increase in the oil cost have led to momentous attention towards the development of green PUs based on renewable resources [14, 15]. Vegetable oils (VOs) triglyceride molecule obtained from various plant sources are gaining immense interest as a monomer in a recent decade for the synthesis of PUs and its nanocomposites [16–18]. Moreover, the feasibility of carrying out various reactions with VOs to synthesize bio-based polyol also makes it more attractive for the synthesis of PUs [19, 20]. The various VOs used in the synthesis of PUs and its nanocomposites include castor oil, palm oil, linseed oil etc. Basically VOs are a family part of chemical compounds known as lipids which are predominantly made up of triglycerides as shown in Fig. 3. The triglyceride molecule consists of three fatty acids joined at a glycerol junction. In VOs the fatty acid chain length varies from 0 to 22 carbons with 0 to 3 double bonds per fatty acid. The various fatty acids present in VOs [21].

The other main component used for the synthesis of PU is isocyanate which exists in the form of resonating structure as shown in Fig. 4 [22].

These groups are very reactive, towards nucleophilic agents due to the electron deficiency on the carbon atom and hence most of the reaction occurs across the C = N group of NCO [22, 23]. Broadly they are classified into two group i.e. (a) aromatic isocyanate and (b) aliphatic isocyanate. It has been postulated that the aromatic isocyanate is more reactive than the aliphatic ones with decreased reactivity from primary through secondary to tertiary isocyanate group unless

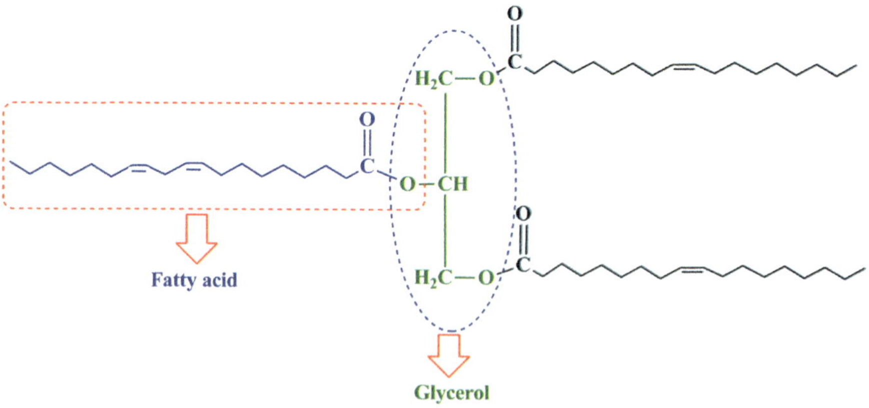

Fig. 3 Triglyceride molecule in vegetable oil

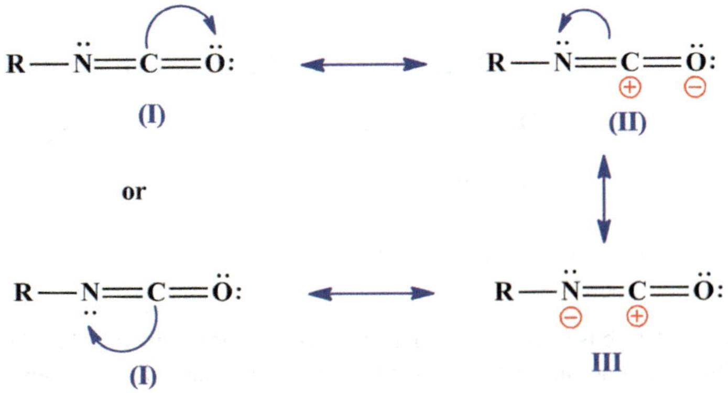

Fig. 4 Resonating structure of isocyanate group

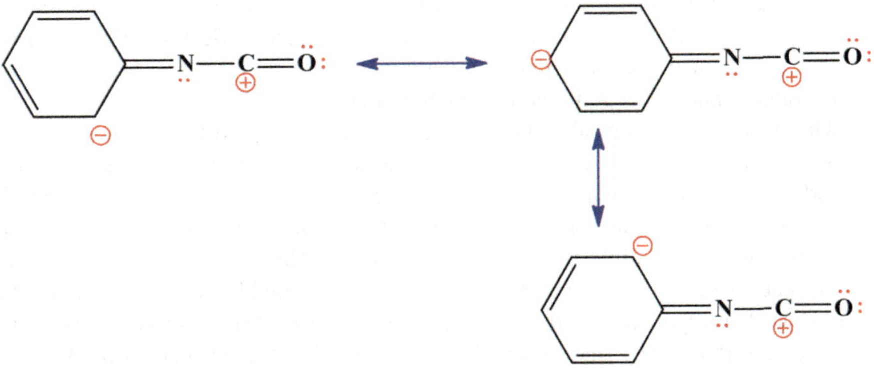

Fig. 5 Resonance in aromatic isocyanate group

catalytic or steric factors results in reversal activity [24]. This can be explained on the basis of the resonating structure wherein, the negative charge on the nitrogen atom is well distributed and stabilized throughout the benzene ring reducing the electron charge on the central carbon atom of the isocyanate when R is aromatic in nature (see Fig. 5) [25].

Thus, the formation of the mesomeric structure in Fig. 5 explains the higher reactivity of aromatic isocyanates such as methylene diisocyanate (MDI) and toluene diisocyanate (TDI) over the aliphatic isocyanate such as hexamethylene diisocyanate (HDI) and isophorone diisocyanate (IPDI). In case of an aromatic isocyanate, the nature of the substituent also determines the reactivity, i.e. electron-attracting substituent in ortho or para position increase the reactivity and electron donating substituent lower the reactivity of isocyanate group [26]. As a general rule, any electron-withdrawing group linked with R will increase the positive charge on carbon, thereby increasing the reactivity of the isocyanate group towards nucleophilic attack while electron donating groups will decrease the reactive of isocyanate

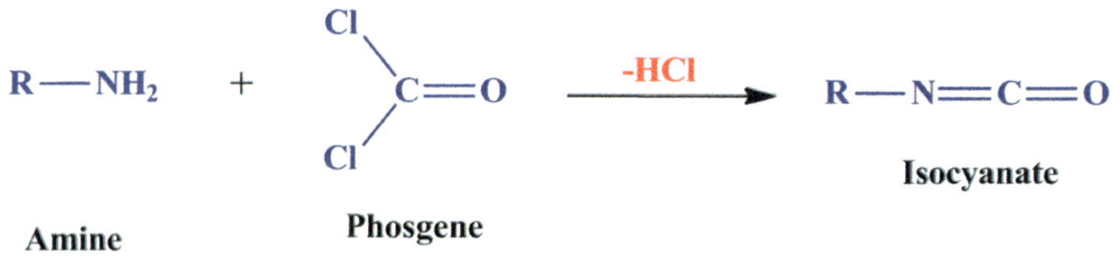

R—NH₂ + Cl₂C=O →(-HCl) R—N=C=O

Amine   Phosgene   Isocyanate

Fig. 6 Synthesis route of isocyanate

groups [22]. The basic synthesis of isocyanate is shown in Fig. 6. The synthesis starts with an amine, aliphatic or aromatic and phosgene. The synthesis route via phosgene was invented in 1884 by Hentschel in which the isocyanate is formed by the elimination of two molecules of hydrogen chloride (HCl).

However, the industrial synthesis of isocyanate through phosgene has to be minimized because of various side reactions that lead to the production of urea and salt complexes. [27]. Moreover, these diisocyanates are toxic and harmful to the environment since they are derived from phosgene and release diamines, free diisocyanate, and hydrogen cyanide on degradation [28, 29].

The diisocyanate normally used in PU synthesis are petroleum-derived as discussed in the above section, but if they are derived from VO sources then this would lead to an increased amount of renewable carbon in such materials. Moreover, the phosgenation synthesis route of isocyanate leads to the emission of carcinogenic products into the atmosphere [28, 29]. Hence, in the current scenario isocyanate based on vegetable oil would be a suitable alternative over the petro-based isocyanate to foster sustainability [30, 31]. Although, extensive research effort is being undertaken for the synthesis of isocyanate from VO precursor but only a few have been used for developing PUs [32, 33].

Thus, it is inevitable from the above discussion that PUs based on bio-based polyol and isocyanate is an essential requirement of the current scenario so as to reduce the excess usage of petro-based products and to formulate and develop environmentally benign materials. However, the PUs synthesized from VO based resources suffer from low mechanical, thermal and chemical properties hence, it is imperative to use nanoparticles for the improvement in the properties of PU matrix [34]. The following section will deal with the synthesis of PUs nanocomposite based on various VOs.

2 Synthesis of Bio-based Polyurethane Nanocomposite from Vegetable Oil

The current trend and present era are inclined towards the development of greener products. Furthermore, as discussed the depletion of fossil fuel has led to rapid research and development towards the synthesis of PUs from renewable resources.

In this concern, vegetable oil like castor oil, jatropha oil, palm oil and soybean oil etc. has led to dramatic utilization for the synthesis of PUs owing to economic, environmental and social advantages. Castor oil (CO) with 92–95% ricinoleic acid is the only commercially available natural polyol with inherent hydroxyl (–OH) group widely used for synthesizing PUs and its nanocomposites [35, 36]. The ricinoleic acid contains a secondary OH group at the 12th carbon position and a double bond at the 9th and 10th carbon. This structural feature distinguishes CO as a candid monomer for synthesizing of PUs amongst all other VOs [35, 36]. Jatropha curcas oil (JCO) is another one attractive VO with high unsaturated fatty acid content and good oxidation stability [37], which makes it a suitable monomer for the synthesis of non-isocyanate polyurethane for various industrial applications [38–42]. Palm oil another vital VO contains 45–60% unsaturated double bonds which can undergo epoxidation to form palm oil-based epoxides which can be used as a green polyol for synthesizing PU [39–43].

However, VO based polyol exhibits disadvantages which include inferior physical properties, poor water resistance, and low thermostability [42–44]. These disadvantages can be addressed by using nanotechnology through the incorporation of nanofillers which can be an effective method for the development of PU properties [45–50]. The following section will deal with the recent trends and development concerning the utilization of various VOs through the exploitation of their structural chemistry for synthesizing different PU nanocomposites using nano-technique.

2.1 Castor Oil Based Polyurethane Nanocomposites

Wang et al. [51] in a recent paper studied about the morphology, thermal and mechanical properties of castor oil-based polyurethane/silica nanocomposites (PU/NS). The NS surface was modified with 3-glycidoxypropyltrimethoxysilane and γ-methacryloxy-propyltrimethoxysilane. Thereafter PU/NS composite was prepared via in situ polymerization technique with castor oil, isophorone diisocyanate and modified NS particle. FTIR spectra indicated a successful blending of NS with silane coupling agents through the appearance of peaks at 2973–2904 cm^{-1} associated with the $-CH_2$-group of 3-glycidoxypropyltrimethoxysilane, at 1636, 1718, 2953, and 2894 cm^{-1} owing to C = C, C = O and C–H stretching vibration of 3-glycidoxypropyltrimethoxysilane group. TGA studies were carried out to investigate the thermal stability of unmodified and modified NS particles. The unmodified NS particles indicate weight loss at 180 °C. However, the modified NS particles indicated two stages of weight loss within 150–600 °C. The higher temperature of weight loss for modified NS particles indicated successful modification with silane coupling agents. Scanning electron microscopy studies indicated the uniform and homogeneous dispersion of modified NS particles within the PU matrix. However, the neat NS particles indicated aggregation with a non-homogenous surface within the PU matrix owing to hydrogen bonding and high specific areas. Moreover, it was observed that 12% loading of modified NS particles within the PU matrix indicated homogenous morphology devoid of any aggregation. The TGA and DTG curve of PU and PU/NS composites indicate three stages of weight loss. The 1st stage was attributed to the decomposition of urethane linkage to form isocyanate, polyol, primary or secondary amine, olefin, and carbon dioxide, the 2nd stage with the decomposition of castor oil segments [52]. The incorporation of NS particles within the PU matrix led to the improvement of $T_{10\%}$ and T_{max1} as compared to neat PU. This was due to o the restriction of PU chain mobility by the NS particles. Moreover, the char residue at 600 °C for PU/NS composite increased to 10.1% in comparison to neat PU. The tensile strength and Young's modulus of the PU/NS composite also increased with an increase in the loading of different NS content. As compared to neat PU, PU/NS composites with 12 wt% loading of neat SiO_2 and modified SiO_2 exhibited 222, 230, and 87% improvement in tensile strength, and 182, 182, and 88% increment in Young's modulus, respectively.

Meera et al. [53] prepared bio-based polyurethane/nanosilica (PU/NS) composite using castor oil (CO) with hexamethylene diisocyanate (HDI) at room temperature. Figure 7 shows the reaction scheme for the synthesis of PU from castor oil and HDI.

The NS particles were modified with silane surface modifiers, i.e., 3-aminopropyltrimethoxysilane for improving the dispersion, compatibility and surface activity of silica particle with PU matrix. The synthesis of PU with castor oil and modified NS particle (i.e. 0.5, 1, 3 and 5 wt%) was carried out through two-step process at 80 °C for 4 h. The prepared nanocomposite films were characterized using Fourier transform infrared spectroscopy (FTIR) to understand the chemical interaction. The thermal stability and glass transition temperature of the films were analyzed using thermogravimetric analysis (TGA) and differential scanning calorimetry (DSC). The dynamic mechanical analysis was done to study the thermo-mechanical properties of the PU nanocomposite films. The morphology of the prepared nanocomposite films was investigated using scanning electron microscopy (SEM) and transmission electron microscopy (TEM). FTIR studies indicate that with an increase in the wt% of NS particles the peak position of urethane (–NH) shifts towards higher wave number i.e. PU-0.5AMS—3332 cm^{-1}, PU-1AMS—3334 cm^{-1}, PU-3AMS—3336 cm^{-1} and PU-5AMS—3338 cm^{-1}. This is due to the presence of strong interaction between NS particles and PU matrix. Moreover, this shift can also be related to the formation of H-bonding between NS particles and PU matrix [54]. However, the peak position of carbonyl peak (–C = O) in PU nanocomposite shifts to lower wave number with an increase in the NS content. These shift in the –NH and –C = O peak confirms the formation of the complete network structure of PU nanocomposite. Moreover, the presence of peaks at 1096 and 774 cm^{-1} for Si–O–Si and O–Si–O also confirms the presence of NS particles within the PU matrix. Neat PU films have low thermal stability and start degrading below 250 °C owing to the presence of labile urethane linkages. On the other hand, the prepared PU nanocomposite films started degradation at around 280 °C due to the presence of NS particles. Neat PU and PU-AMS films showed

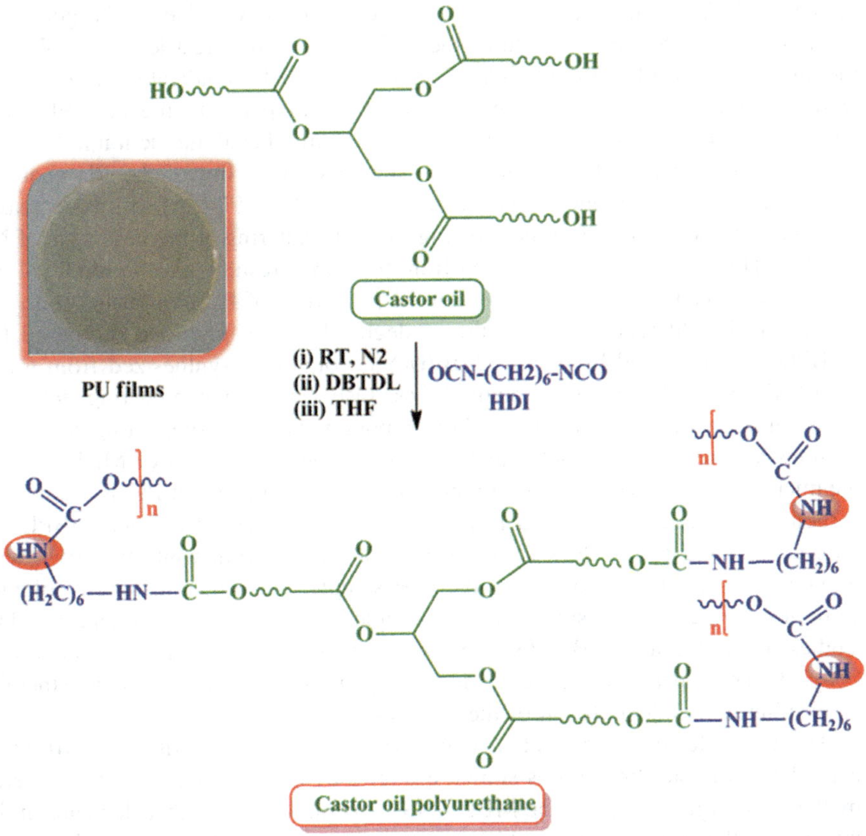

Fig. 7 Reaction scheme for the synthesis of PUs from castor oil and HDI

two-step degradation curves, wherein, the 1st stage at 350 °C was related with the urethane degradation and the 2nd stage between 350 and 500 °C was related with the degradation of polyol. TGA thermogram curve of PU-AMS nanocomposite film indicate improvement in thermal stability and melting temperature (T_m) owing to the presence of silica nanoparticles. It was also observed that the residual weight percentage in case of neat PU was 0.003%, whereas in case of PU-5AMS it was 8.2%. The activation energy (E_a) value of neat PU and was found to be 133 and 139 kJ mol^{-1} whereas, the E_a value of PU-5AMS was found to be between 157 and 166 kJ mol^{-1}. The presence of silica nanoparticles reinforced the PU matrix thereby improving the interfacial interaction and thermal stability of the nanocomposite film. Neat castor oil based PU indicate T_g at −40.1 °C corresponding to a soft segment of the polyol and T_m at 279.1 °C corresponding to the hard segment. On the other hand, the T_g values of PU-0.5AMS, PU-1AMS, PU-3AMS and PU-5AMS was observed at 33.4, 33.7, 32.6 and 29.71 °C respectively. However, the T_m value of PU nanocomposite films was unaltered. The increment in T_g of the PU-AMS nanocomposite films was due to the good dispersion of the NS particles within the PU matrix which restricts its molecular mobility. DMA results of the PU-AMS nanocomposite film indicates an increase in storage modulus (E') and loss modulus (E") as compared to the neat PU films which was due to the reinforcing effect and strong interfacial interaction in between PU and NS particles which restricts the segmental mobility of the PU chain as shown in Fig. 8 [55]. Optical transmittance results of the PU-AMS films decreases with an increase in the NS content owing to light scattering at the interfaces of NS and PU. Thus, the authors concluded from the above results that the modified NS particles can lead to an improvement in the properties of PU composite films.

Das et al. [56] reported about the biodegradability of modified bio-based PU (MBPU) and modified bio-based (MBPU/NS) composite synthesized from transesterified castor oil (MCO) and palm oil based isocyanate by composting technique for a duration of 90 days. The authors reported that the composting technique presents a faster and a cost-effective method for the degradation of MBPUs since the microorganism present in the compost use the ester linkage of PU as a source of carbon and nitrogen. The visual inspection results of neat MBPU and MBPU/NS after composting for 90 days indicated a color change from white to brown with pronounced surface degradation of the later sample. Bacterial and fungal colonies were also formed on the surface of the samples indicating bio-degradation. FTIR studies of MBPU and MBPU/NS composites revealed lowering of urethane carbonyl stretching peaks at 1720 cm^{-1} indicating dissociation of ester linkage thereby confirming the biodegradation of the samples [57, 58].

The probable reason behind this observation was related with the hydrolytic degradation of ester linkages as shown in Fig. 8 by extracellular enzymes secreted by the microorganism [59]. The MBPU/NS films indicated a higher decrease in the intensity of the urethane carbonyl peaks as compared to neat MBPU samples due to its higher hydrophobic behaviour and surface silanol groups that resulted in easy surface adhesion of microorganism [60]. In addition, MBPU/NS surface also indicated higher weight loss as compared to neat MBPU due to the presence of

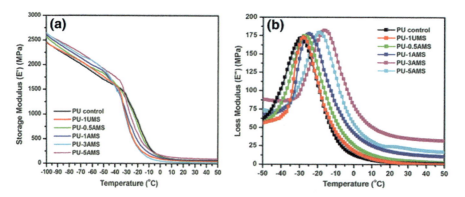

Fig. 8 Storage modulus (E') and loss modulus (E") of PU-nanosilica composite

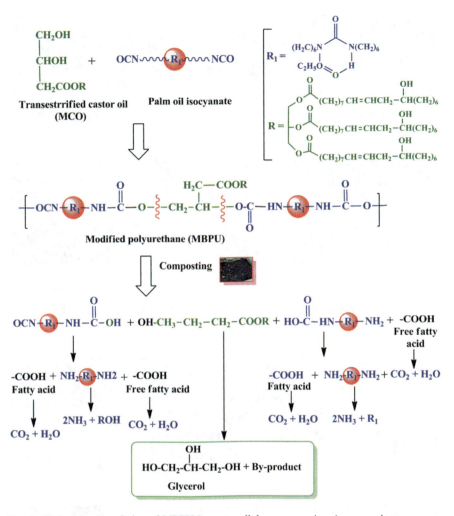

Fig. 9 Hydrolytic degradation of MBPU by extracellular enzymes in micro-organisms

hydroxyl groups on the NS surface that led to attachment of microorganism onto the former surface. SEM micrographs as shown in Figs. 10 and 11 also indicated higher surface degradation of MBPU/NS surface as compared to neat MBPU through the appearance of cracks, holes and fungal mycelia. This was again due to the reasons illustrated earlier. Thus, it can be concluded that the MBPU/NS samples can be a biodegradable as compared to neat PU on account of its structural organization and composition.

Thakur and Karak et al. [61] reported about the preparation of castor oil based-tough hyperbranched polyurethane (HPUs) nanocomposite reinforced with two different types of nanofillers, i.e., reduced graphene oxide (RGO) and phytoextract-RGO (f-RGO). f-RGO was functionalized by reacting RGO with

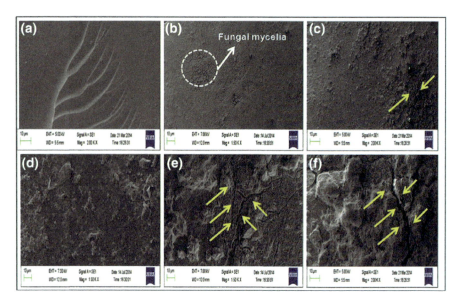

Fig. 10 SEM micrographs of neat MBPU for **a** 0 days, **b** 30 days, **c** 45 days, **d** 60 days and **e** 75 days. Reproduced from Das et al. [56]

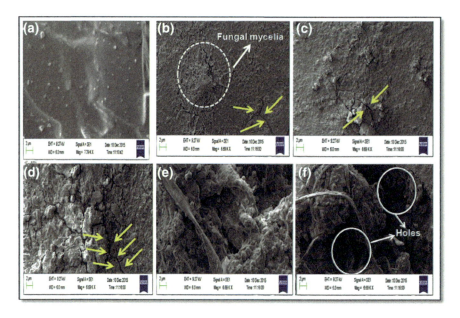

Fig. 11 SEM micrographs of neat MBPU/NS for **a** 0 days, **b** 30 days, **c** 45 days, **d** 60 days and **e** 75 days. Reproduced from Das et al. [56]

toluene diisocyanate (TDI) followed by reaction with 1, 4 butanediol (1, 4 BD). The synthesized HPU/f-RGO nanocomposite showed an improvement in tensile strength to about 525%, modulus to about 42-folds, and toughness to about 18-folds after addition of 2 wt% of f-RGO in HPUs as shown in Fig. 12. Moreover, the elongation at break for the HPU/f-RGO nanocomposite showed an increment from 71% to a maximum of 165%. HPU/f-RGO nanocomposite also exhibited better thermal stability and excellent electrical conductivity to almost 10-fold with 2 wt% loading of f-RGO However, with the same loading RGO nanocomposites, exhibited lower mechanical, electrical and thermal properties as compared to HPU/f-RGO nanocomposites. The authors proposed that the developed HPU/f-RGO nanocomposites can be used for the development of tough, conductive nanocomposites for aerospace and tissue engineering applications.

Zhang et al. [62] reported about the preparation and characterization of in situ polymerized bio-based thermosetting polyurethane/graphene oxide (PU/GO) nanocomposites based on epoxidized soybean oil–castor oil polyol with isophorone diisocyanate (IPDI) as shown in Fig. 13. The functionalization of graphene oxide was carried out through pressure oxidation method followed by reinforcement within the PU matrix. The authors observed similar improvement in the overall properties such as mechanical, thermal and electrical conductivity as reported by Thakur and Karak et al. with a minimal loading of modified graphene oxide. This was due to the strong chemical interaction of urethane group of PU matrix with the hydroxyl (–OH) groups of modified GO.

Ali et al. [63] investigated the synthesis and characterization of polyurethane-multiwall carbon nanotube (PU/MWCNT) nanocomposites based on castor oil via. in situ polymerization technique. The in situ polymerization technique led to the uniform dispersion of the nanoparticles within the PU matrix which is an important criterion in the improvement of nanocomposite properties. MWCNTs are candid and ideal reinforcing material with ordered carbon fibre imparting unique properties to the composite i.e. light weight, stiffness, superconductivity and mechanical

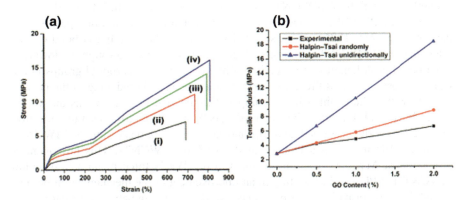

Fig. 12 Tensile test result of hyper branched polyurethane (HPUs) nanocomposite. Reproduced from Karak et al. [61]

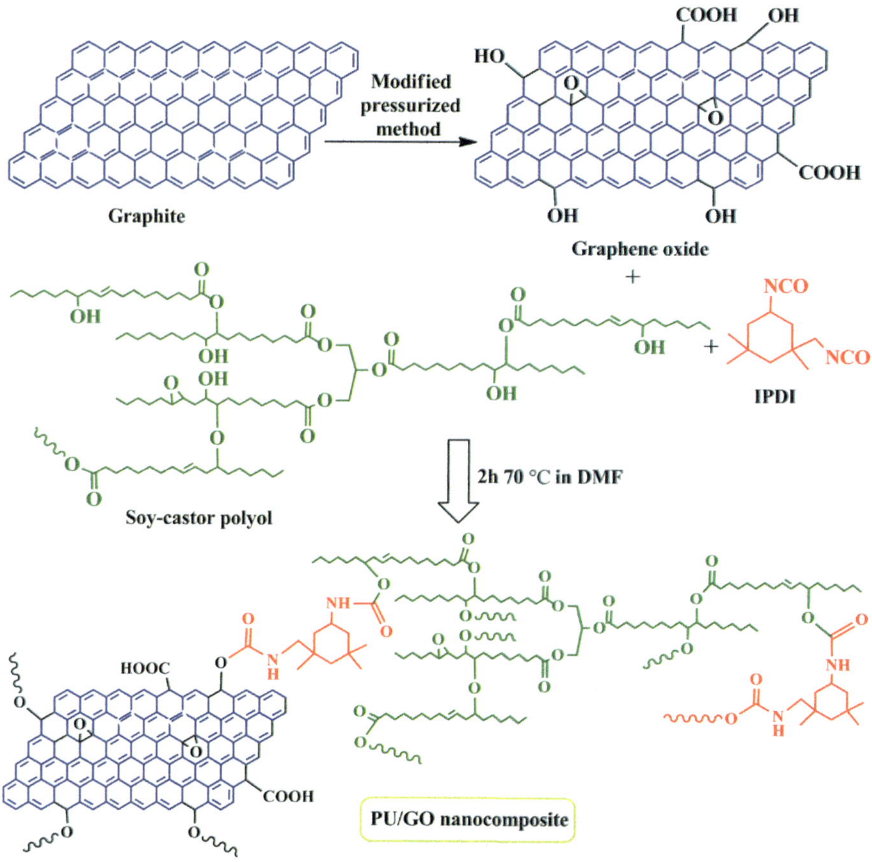

Fig. 13 Reaction scheme for the synthesis of PU/GO nanocomposite

strength. The weight of MWCNTs in castor oil was varied from 0 to 1 wt%. FTIR spectra of the PU/MWCNT nanocomposite indicated broad absorption peak within the range of 1715–1725 cm^{-1} associated with the urethane carbonyl bond of PU. Intercalation of MWCNT within the PU matrix was observed through the lowering of carbonyl peak height and formation of hydrogen bonded carbonyl groups in the nanocomposite. X-ray diffraction (XRD) studies revealed broadening of the peak at $\theta = 21°$ with reduced intensity as compared to the neat PU which was due to the strong interfacial interaction between MWCNTs and PU matrix. Scanning electron microscopy studies revealed uniform and homogeneous dispersion of MWCNTs within the PU matrix at a loading of 0.3 wt% of MWCNTs. This homogenous dispersion of MWCNT was due to the van der Waals force of attraction within MWCNT and PU matrix resulting in an improvement of the mechanical properties. In addition, the formation of H-bonds and covalent bonding between the carboxylic group in MWCNT and PU also contributed towards the observed improvement in mechanical properties. The surface properties of the nanocomposites and neat PU was investigated using the Brunauer–Emmett–Teller (BET) theory. From the theory, it was found that the nanocomposite surface indicated the presence of few pores due to the intercalation of MWCNT within the PU matrix resulting in low permeability.

The incorporation of MWCNTs within the PU matrix had a great impact on the N_2 gas diffusion mechanism. PU/MWCNT based nanocomposites indicated a reduction in gas permeability as compared to neat PU owing to the exfoliation, compatibilization, orientation and re-aggregation of MWCNTs in the PU matrix. Thus, it could be concluded that the incorporation of MWCNTs in the PU matrix led to the improvement in the gas barrier, thermal and mechanical properties. Further, the in situ polymerization technique also led to the homogenous dispersion of MWCNT within the PU matrix resulting in an overall improvement in the properties of the nanocomposite.

Chen et al. [64] synthesized castor oil-based polyurethane/epoxy (PU/EP) interpenetrating polymer network (IPN) reinforced with MWCNTs for damping application. Damping property of a material depends upon its ability to dissipate energy which is directly proportional to the internal friction. Experimental results reveal that the damping ability of PU/EP IPN increases with a loading of 0.1 wt% of MWCNT. This was due to the presence of MWCNT that increases the friction between CNTs and PU resulting in an increment in the rate of dissipating energy. Moreover, the higher surface area and aspect ratio of MWCNT also contributes towards the overall increase in the damping property of the nanocomposites. The above results were also in line with the DMA results indicating improvement in storage modulus (E') of the nanocomposite film with the addition of MWCNTs as shown in Fig. 14. The mechanical properties of the IPN nanocomposite also increased by 30% as compared with the neat counterparts with the loading of 0.1 and 0.7 wt% of MWCNTs respectively. This was due to the strong and large interfacial area between the PU matrix and MWCNT that contributed to the effective load transfer from PU matrix to MWCNT.

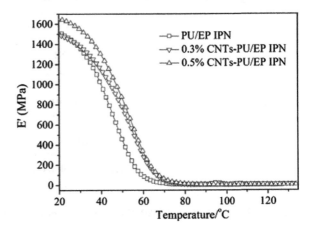

Fig. 14 Variation of Storage modulus (E') in neat PU/IPN and PU/IPN-CNTs. Reproduced from Chen et al. [64]

2.2 Jatropha Oil Based Polyurethane Nanocomposite

Liao et al. [65] studied the surface structure and morphology of waterborne polyurethane/clay nanocomposites (WPU/Clay) prepared via in situ polymerization based on jatropha curcas oil (JCO) and isophorone diisocyanate (IPDI) The author used three different types of nanoclay (i.e. sodium montmorillonite MT, attapulgite AT and halloysite nanotubes HT) whose structures were modified using γ-aminopropyltriethoxy silane (APTES) as shown in Fig. 15.

Thereafter, WPU nanocomposite dispersion was synthesized from JCO, IPDI with three different types of nanoclay as shown in Fig. 16.

FTIR, SEM and TGA studies were carried out to investigate the degree of silylation of the nanoclays. SEM micrographs revealed that the silylated sodium montmorillonite (SMT) nanoclay had distinct and regular layered structure with tight packing. On the other hand, the silylated halloysite nanotubes (SHT) nanoclays indicated smooth surface with a cylindrical shape. FTIR spectra confirmed the successful silylation of the nanoclays with the appearance of a peak at 2930 and 1560 cm^{-1} related to the –CH stretching and deformation vibration of NH$_2$ groups, respectively which was absent in case of unmodified nanoclays. TGA studies indicated two-stage weight losses for unmodified and modified clay respectively wherein, the modified clays indicated higher temperature weight loss as compared to the neat ones. FTIR spectra of WPU nanocomposites based on JCO indicated the appearance of –NH peak at a lower wavelength at 3340 cm^{-1} due to H-bonding. SEM micrographs of the WPU nanocomposites indicated rougher surface due to the presence of clays. However, there was no aggregation of clay nanoparticles within the WPU matrix indicating good compatibility between the WPU and silylated clay nanoparticles. It was observed that WPU/SMT exhibited higher surface roughness as compared to WPU/SAT, WPU/SHT due to the layered

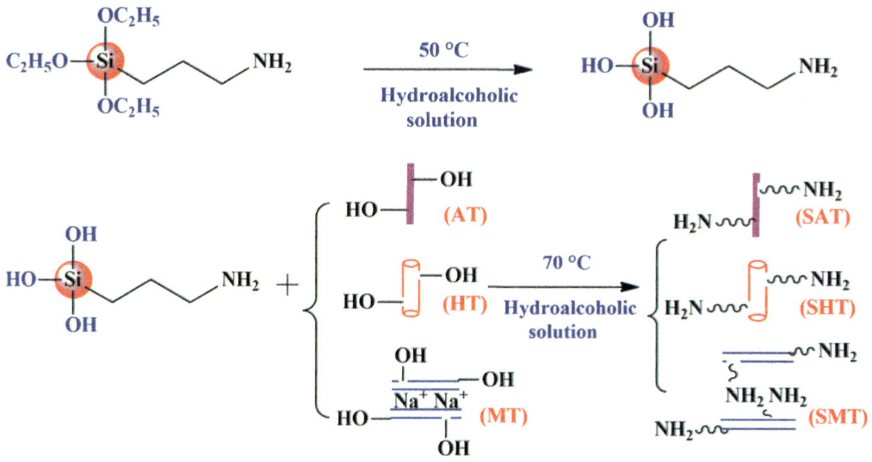

Fig. 15 Scheme showing the APTES modification of three different types of nanoclay

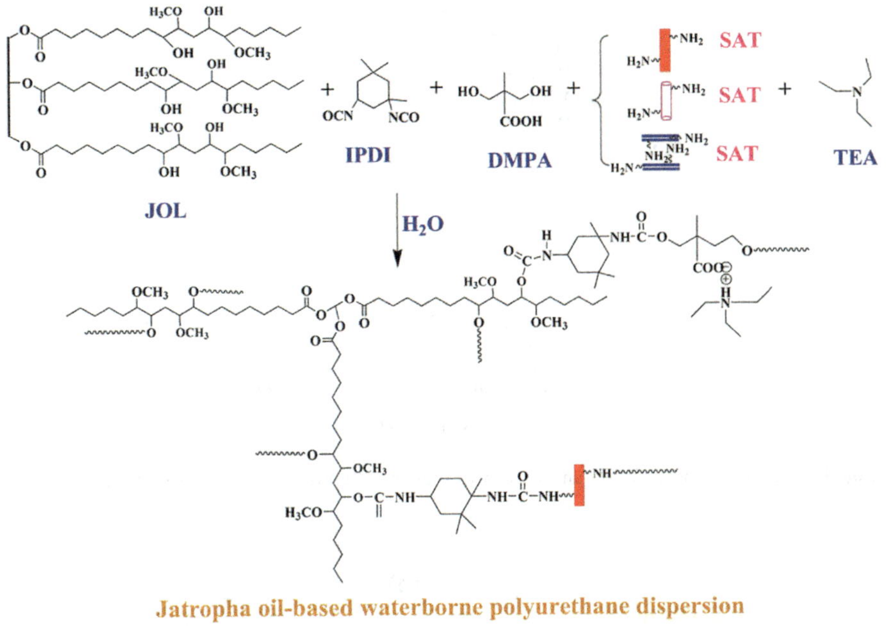

Fig. 16 Preparation of JCO based WPU nanocomposite dispersion. Reproduced from Liao et al. [65]

structure of SMT which led to the restriction of WPU molecular mobility resulting in an increase in the cross-linking density. DMA analyses result indicated strong chemical interactions between clays and WPU matrix thereby increasing the microphase separation WPU/clay nanocomposites. Out of the three nanocomposites, WPU/SHT and WPU/SAT indicated highest microphase separation degree, due to the increase in interaction between the hard segments and soft segments of WPU. The height of tan δ peaks was also found to decrease with the incorporation of nanoclay, which was due to the increase in cross-linking density of PU matrix as shown in Fig. 17.

WPU/SMT indicated the lowest tan δ peak value due to the layered structure of SMT which played an effective role in increasing the cross-linking density and restricting the molecular motion. Tensile test result suggested that the incorporation of nanoclay improved the tensile strength and elongation at break. WPU/SMT and WPU/SAT nanocomposite had the higher tensile strength and lower elongation at break as compared to WPU/SHT. This was due to the presence of higher –NH$_2$ groups and layered structure of SMT resulting in higher cross-linking density and restricting the molecular mobility of the WPU chain as depicted in Fig. 18. As a consequence, the tensile strength increases and the elongation at break decrease for the WPU/SMT nanocomposites. WPU/SAT indicated higher tensile strength due to the formation of more H-bonds that led to strong interfacial interaction between SAT and WPU matrix. On the other hand, reverse phenomenon was observed in

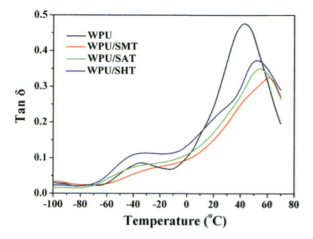

Fig. 17 Tan variation in WPU and its nanocomposite. Reproduced from Liao et al. [65]

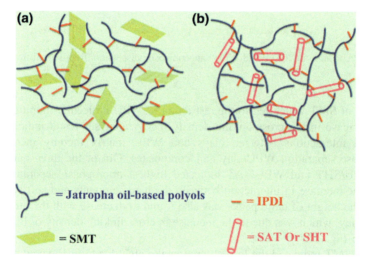

Fig. 18 Model showing the H-bonding interaction between PU and three different types of nanoclay. Reproduced from Liao et al. [65]

case of WPU/SHT which showed higher elongation at break and lower tensile strength. This was due to the higher microphase separation of WPU/SHT which made the soft segments more flexible. Thus, it can be implied from the above results that the properties of WPU/clay nanocomposites depend upon the nature, structure and type of clay.

2.3 Palm Oil-Based Polyurethane Nanocomposites

Dzulkifli et al. [66] studied the synthesis of PU foam from palm oil reinforced with diamine-modified montmorillonite (MMT) nanoclay. MMT was modified with diamino-propane and sodium carbonate and thereafter PU nanocomposite foam was prepared using palm oil, modified MMT and isophorone diisocyanate (IPDI). FTIR spectra indicate the successful modification of MMT by DAP through the appearance of peaks at 3463 cm^{-1}, indicating the successful insertion of DAP into MMT gallery. Presence of peaks at 1712, 1509, and 1216 cm^{-1} corresponding to the bending vibration C = O, N–H, and C–N, respectively, confirmed the successful synthesis of PU foam and PU/MMT foam. X-ray diffraction curves also indicated the successful modification of MMT by DAP through the shifting of the peak from 8.95° for pristine MMT towards the left side at 8.05° for the nanocomposite foam. This shift reveals the achievement of larger interlayer spacing or d-spacing in case of foam nanocomposites. SEM micrographs indicate reduced cell size and exfoliated structure for the PU/MMT composition. This was attributed to the effective dispersion of DAP-MMT within the PU foam matrix resulting in preventing coalescence, and producing smaller and fine cell structure. Thus, it was concluded that the presence of OMMT within the PU foam system led to the overall improvement in the morphology and microstructure of foam.

Adnan et al. [67] studied the development of flexible polyurethane nanostructured biocomposites foams based on palm olein-based polyol with OMMT as nanofiller. FTIR spectra of palm oil-based PU and PU nanocomposites were investigated. It was observed that there was no change in the FTIR spectra of derived PU and PU foam nanocomposites. The broad stretching peak at 3405 cm^{-1} indicates the formation of H-bonded urethane. The band at 2995–2860 cm^{-1} was attributed to the –CH$_2$ stretching vibration and the presence of hydrogen-bonded carbonyl groups was observed at lower wave number at 1733–1731 cm^{-1} for the PU foam nanocomposites. Mechanical test results indicated improvement in tensile strength and tear strength for both petro-based and palm oil-based PU with the loading of 3, 5 and 7 wt% of OMMT as compared to the neat counterparts. This increase in tensile strength and tear strength for both the petro and palm oil-based PU nanocomposites was evident for OMMT loading up to 3 and 5 wt%, beyond which there was an overall decrease in the tear and tensile strength. However, the increment in tensile strength for palm oil-based PU nanocomposites (i.e. 66%) was higher as compared to petro-based PU nanocomposites (33%). Similar result was observed in case of tear strength which increased from 13% in case of petro-based PU nanocomposites to 48% in case of palm oil based PU nanocomposites. This observation was due to the strong H-bond formation between the silicate lamellae of OMMT (mainly silanol, Si–OH and aluminol, Al–OH) with the urethane groups of PU as shown in the Fig. 19.

In addition, the intercalated quaternary ammonium salts of OMMT can act as the "bridge" connecting the MMT layers and the PU chains thereby restricting the molecular mobility [68]. SEM micrographs reveal smaller cell size for the petro and

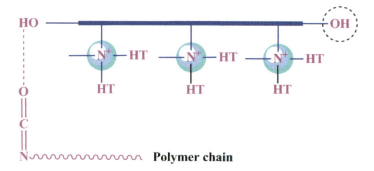

Fig. 19 Mechanism of hydrogen bonding between the PU chain and OMMT

palm oil-based PU foam nanocomposites for a loading of 3, 5 and 7 wt% of OMMT. This was due to the fact that OMMT can serve as a nucleation site for cell formation, as a result, less gas will be available for the growth of the cells. TEM micrographs indicated intercalated OMMT structure within the PU matrix with an average length of 100 nm.

Zaimahwati et al. [69] studied about the palm oil-based polyurethane nanocomposite reinforced with OMMT nanofiller. The palm oil was modified through epoxidation and hydroxylation reaction with acetic acid, hydrogen peroxide and hydrochloric acid to synthesize bio-based polyol. Thereafter PU and its nanocomposite were prepare using the polyol, IPDI and OMMT nanofiller (5 wt%). The iodine value of the palm oil decreased from 56.72 to 14.29 I_2/100 g for palm oil-based polyol after its modification indicating oxidation of double bond. FTIR spectrum of epoxidized palm oil-based polyol indicates a peak at 1050 and 1014 cm^{-1} attributed to the C–O bond of epoxy ring confirming the successful modification of palm oil. TGA results of the synthesized PU/OMMT nanocomposites indicated higher thermal stability as compared to neat PU wherein, the weight reduction of former begins at 150–200 °C with final degradation at of 490 °C. PU/OMMT nanocoating also shows higher gloss as compared to the neat counterparts due to the ability of the OMMT coating surface to reflect light. Hence, the incorporation of OMMT led to the overall improvement of the nanocomposite films.

Salih et al. [70] investigated the thermal and mechanical properties of palm oil and methylene diisocyanate based polyurethane acrylate/clay (POBUA/OMMT) nanocomposites prepared via. in situ intercalative method and electron beam radiation technique. The in situ intercalative polymerization technique is useful for uniform dispersion of the nanofillers within the polymer matrix as shown in Fig. 20. In addition, this technique provides the possibility to polymerize various ranges of thermosets and thermoplastics.

FTIR spectra of the synthesized POBUA nanocomposites indicated the disappearance of NCO peak at 2273 cm^{-1} indicating the complete utilization of –NCO by the –OH group of epoxidized palm oil for the synthesis of urethane linkage of

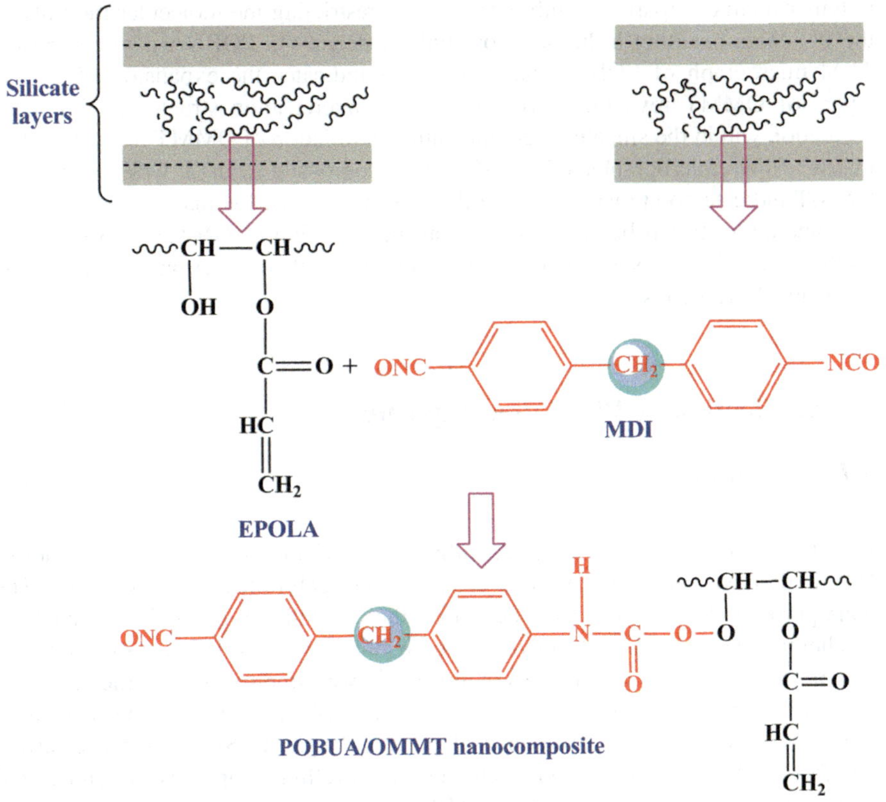

Fig. 20 Synthesis of POBUA/OMMT nanocomposite

PU at 3327 cm^{-1}. The FTIR spectra of POBUA nanocomposite also indicated the appearance of a peak at 1015, 941 and 761 cm^{-1} related to the silicate groups in the OMMT filler indicating that the interaction of layered silicate with PU matrix. TGA thermogram of neat POBUA and its nanocomposite indicate two-stage degradation curves. The first stage of degradation was related with the decomposition of volatile products and the major decomposition which occurred in the second stage, was related to the decomposition of the organic polymer chains. TGA results of the nanocomposite indicate that improved thermostability as compared to neat POBUA due to the presence of OMMT layer creating a protective physical barrier, thereby inhibiting the heat diffusion, and delaying the degradation process. Differential scanning calorimetry (DSC) results indicate higher glass transition temperature (T_g) for the nanocomposites at 61.8 °C as compared to neat PU at 40.5 °C. This was due to the intercalation of POBUA matrix into the silicate layers of the nanoclay that lead to the reduction of rotational and transitional mobility of the polymer chains. Tensile strength results also indicated higher modulus and strength for the POBUA nanocomposites due to the homogenous dispersion of the nanoclay within the polymer matrix, imparting reinforcement, and restricting the molecular mobility of the PU chain resulting in higher cross-linking density in POBUA nanocomposite. TEM micrograph of POBUA nanocomposite indicates the expansion of silicate layers of OMMT revealing formation of the intercalation structure, with some separation within the silicate layers indicating exfoliation of OMMT in the POBUA matrix. SEM micrographs of POBUA nanocomposite with 3 wt% loading of OMMT indicate good dispersion, good adhesion with the PU matrix devoid of any agglomeration. It can be summarized that the inclusion of OMMT into the polyurethane acrylate system led to the improvement of thermal stability and mechanical properties.

3 Application of PU Nanocomposites

3.1 Coatings

Das et al. [33] have studied about the influence of NS inclusion on the properties of MBPU derived from transesterified castor oil and palm oil based isocyanate. The main purpose of the study was to determine the feasibility of palm oil isocyanate as an alternative to petro-based isocyanate for synthesizing MBPU/NS composite. The authors observed that MBPU/NS composite showed improved properties as compared to their neat counterparts due to better cross-linking, H-bonding and phase segregation. Moreover, the presence of polar linkages, O–Si–O bond and strong interfacial interaction led to better coating and swelling properties in synthesized MBPU nanocomposite. Further, the MBPU nanocomposite dispersion was cast onto polycarbonate substrates to determine the adhesive strength, curing properties, contact angle, gloss and abrasion resistance. The authors observed that the MBPU/NS composites derived from transesterified castor oil exhibited faster curing time owing to the presence of primary hydroxyl groups and higher hydroxyl value. The contact angle studies of MBPU/NS coatings indicated hydrophobic behaviour which was due to the H-bonding between the of O–Si–O linkage of NS with the urethane (–NHCOO) linkage of MBPU network that restricts the migration of water as shown in Fig. 21. Further, there was also an overall improvement in the abrasion resistance of the nanocoating as compared to the unreinforced ones owing to higher cross linking and effective bonding of urethane group in MBPU with NS particles. Thus, from the results obtained the authors concluded that palm oil based isocyanate can be used as an alternative to the petro-based isocyanate for the synthesis of green polyurethane nanocomposite.

Xia and Larock [71] studied about the castor oil-based waterborne PU/NS composite prepared through sol-gel process for coating application. This process involved polycondensation and hydrolysis reaction of silicon alkoxides to prepare waterborne PU/NS composite dispersion. Alkoxysilane containing PU was reacted with different weight % of aminopropyltriethoxy silane (APTES) and isocyanate to

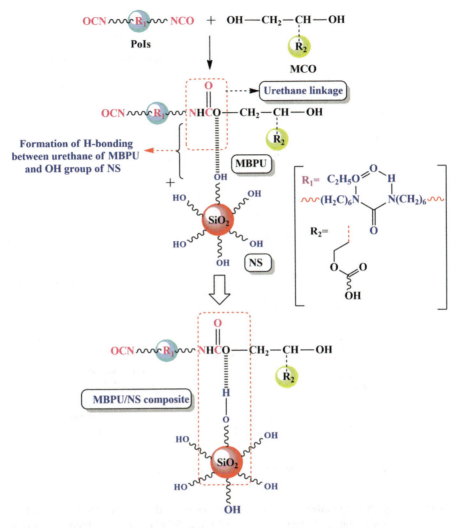

Fig. 21 Possible H-bonding interaction between MBPU and NS particle

form capped PU pre-polymer with core-shell structure as shown in Fig. 22. The sol-gel technique involved chemical cross-linking of PU with NS particle resulting in an increase in the cross-linking density of the nanocomposite films. The authors used rubber elasticity theory to calculate the crosslinking density (v_e) of the nanocomposites and they found that with an increase in NS content from 0 to 2 wt %, v_e increases from 90 to 766 mol m^{-3}. DSC studies also confirmed the increase in cross-linking density with increase in NS loading. It was observed that the PU/NS composites exhibited higher T_g from 18 to 20.9 °C with an increase in NS loading from 0 to 1.5 wt%. The TGA curves indicated three step of degradation.

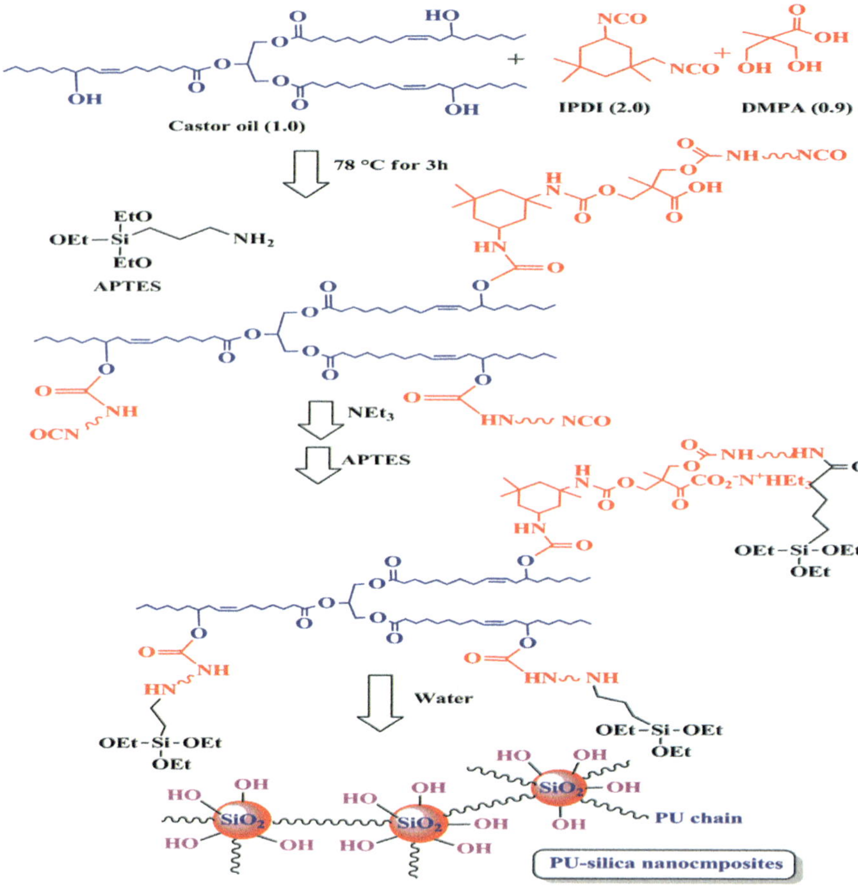

Fig. 22 Synthesis reaction scheme of PU/nanosilica composite

The 1st step of degradation occurred at 150–300 °C due to the dissociation of the urethane bonds, the 2nd step between 300 and 500 °C due to chain scission of castor oil and the last degradation step above 500 °C which was related with the thermo-oxidative degradation of the nanocomposites. With the inclusion of NS particles the temperature corresponding to T_{50} and T_{max} increases due to the increase in cross linking density as depicted in Fig. 23. The presence of NS particles within the PU matrix prohibited the heat and mass transfer thereby reducing the formation of combustible organic components. The topography of the PU-silica nanocomposite films indicated aggregation of APTES with an increase in APTES loading. The authors postulated that the presence of hydrophilic carboxylated groups on the PU chains led to the well dispersion of NS particles within the PU matrix.

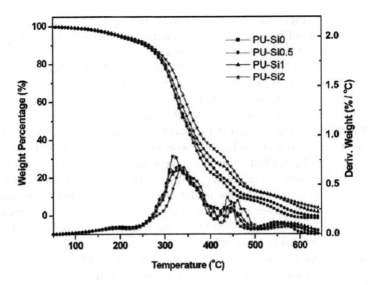

Fig. 23 DTG/TGA thermogram of PU/NS composite. Reproduced from Xia et al. [71]

3.2 Adhesives

Deka et al. [72] studied the adhesive, mechanical, and thermal properties of bio-based hyper-branched polyurethane/clay nanocomposites from nahar seed oil with toluene diisocyanate. The nanoclay (OMMT) wt% within the PU matrix was varied from 1, 2.5 and 5. The main aim of the study was to overcome the poor mechanical strength of hyperbranched polymers (HBPUs) due to the absence of chain coiling and entanglement. HBPU was synthesized using PCL as macroglycol with molecular weight 3000 g mol^{-1} as the long segment with monoglyceride of Mesua ferrea L. seed oil. The obtained HBPU was further modified using epoxy resin and poly (amido amine) hardener to obtain modified hyperbranched polyurethane-epoxy system (MHBPU). The MHBPU was used as the matrix with OMMT as the filler. FTIR spectra of MHBPU and MHBPU/OMMT were identical indicating that OMMT doesn't influence the chemical structure of MHBPU. However, the only difference observed in the spectra of MHBPU/OMMT nanocomposite is the presence of sharp –NH vibration band at 3311 cm^{-1}. This was due to the presence of clay layers that led to restricting the interaction of hard and soft segments. The introduction of clay nanoparticles also led to the shifting of urethane carbonyl peak from 1741 to 1718 cm^{-1} indicating higher H-bonding within the PU matrix and clay in the nanocomposites. X-ray diffraction study was carried out to determine the morphology of clay nanoparticles within the PU matrix. On the basis of Bragg's equation, it was found that the d-spacing of nanocomposites increased from 2.36 (in case of OMMT) to 4.95 nm for MHBPU/OMMT. This indicates exfoliated morphology with the complete insertion of HBPU chains in the gallery spacing of the OMMT. SEM and TEM micrographs also indicate better dispersion of OMMT particles within the MHBPU matrix. This is due to the synergistic actions of both mechanical shearing and diffusion process imparted by mechanical stirring and sonication. Lap shear test was carried out in order to investigate the adhesive strength of the MHBPU and MHBPU/OMMT using different substrates such as plywood, aluminum and polypropylene sheets. Out of all the substrates higher adhesive strength was observed for wood substrates owing to the strong polar interaction of hydroxyl, epoxy, urethane, ether of the cured MHBPU/OMMT system with the hydroxyl groups of wood substrate. Also, the adhesive strength was found to increase with an increase in loading of OMMT wt%. This was due to the exfoliation of clay layers that enhanced the interface interactions via. bridge, loop and tail linkages of PU with the OMMT layers as shown in Fig. 24. This led to the reduction in the amount of voids thereby increasing the length of crack propagation and adhesion strength. Similar observation was seen in case of aluminum substrates due to the strong interactions of polar groups present on the surface of matrix and substrate. The plastic substrates indicated lower adhesive strength owing to its low surface polarity. Thus, the authors concluded that the nanocomposites exhibited high adhesive strength, mechanical properties and thermostability.

Sahoo et al. [73] studied the shear strength of polyurethane/OMMT clay nanocomposites adhesive based on castor oil and palm oil based isocyanate. The clay % was varied from 1 to 5 wt%. The PU adhesive solution was applied onto wood substrates of 0.1 mm thickness and left overnight for curing. The lap shear strength of the neat PU and its nanocomposite was studied as per ATMD 906-82 using a universal testing machine with a loading rate of 600 lb/min. The authors observed that the shear strength values of PU adhesive increase with an increase in clay content. Out of the entire compositions PU/Clay nanocomposite with 3 wt% loading exhibited the highest adhesion strength and was chosen to be the optimized composition. However, PU/Clay nanocomposite with 5 wt% loading showed less shear strength due to the aggregation of nanoclay particles. This aggregation led to the predominance of filler-filler interaction over the polymer-filler interaction. The mode of failure of lap shear was found to be cohesive in nature. Further, a study was carried out wherein, the samples were immersed in water for 24 h. It was observed that water doesn't affect the shear strength even after 10 days of immersion. The authors concluded that the PU/clay nanocomposite exhibited higher adhesion properties due to the homogenous dispersion of the nanoparticles, intercalated structure and good chemical interaction between the clay nanoparticles and PU matrix.

3.3 Medical

Das et al. [74] synthesized sunflower based hyper-branched polyurethane (HBPU) reinforced with Fe_3O_4 nanoparticles via. in situ polymerization technique for designing smart antibacterial biomaterials for biomedical devices and implants.

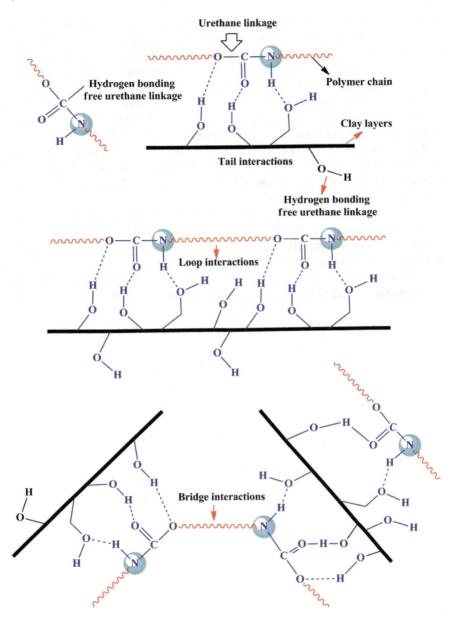

Fig. 24 Model representing the interfacial interaction between MHBPU with OMMT layers

Fe$_3$O$_4$ has various disadvantages which include poor stability, agglomeration, and low performance which can be improved through the incorporation of these nanoparticles in a suitable matrix that can provide strong structural adherence, dispersibility and stability. To overcome the above limitations the author selected

HBPU as a matrix since it provides better biocompatibility, dispersibility, low viscosity, encapsulation and good mechanical properties. The synthesized nanocomposites of Fe_3O_4 with HBPU was named as MHBPU with the loading of Fe_3O_4 of 5, 10 and 15% which can be used as smart materials, shape memory and shape recovery application since its shape can be controlled by using external stimuli such as heat energy and magnetic field, etc. The dispersion of Fe_3O_4 within the HBPU matrix was investigated using SEM and TEM. SEM micrographs indicated uniform and homogeneous dispersion of Fe_3O_4 within the HBPU matrix. A similar observation was seen for TEM image which also confirmed the uniform dispersion of the Fe_3O_4 within the HBPU matrix with an average particle size of 7.65 nm. The uniform stabilization of Fe_3O_4 is because of the chemical interaction with HBPU as shown in Fig. 25.

Mechanical properties i.e. tensile strength of the nanocomposite were also improved as compared to the neat counterparts due to the high surface area of the nanoparticles and strong chemical interaction within Fe_3O_4 and HBPU matrix, facilitating effective stress transfer. However, the elongation at break decreases for the nanocomposites owing to the restriction in HBPU chain mobility. Bio-degradation studies indicated higher degradation rate of the nanocomposites as

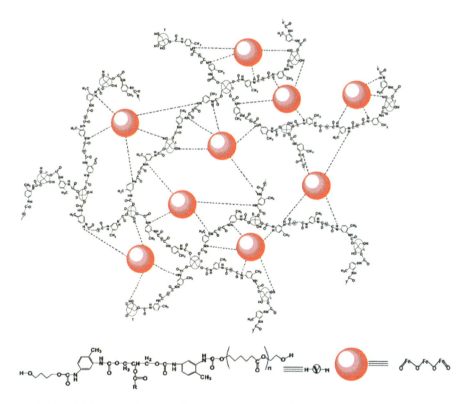

Fig. 25 Model depicting the interaction between HBPU and Fe_3O_4 nanoparticles

compared to HBPUs owing to the structural feature of Fe_3O_4 nanoparticles that led to its easy leaching from the matrix surface. The nanocomposites also exhibited better shape memory behaviour owing to the increase in internally stored energy of the polymer matrix resulting in strong interactions of Fe_3O_4 with the HBPU matrix. The neat HBPU indicated no antibacterial activity whereas; Fe_3O_4 based MHBPU nanocomposites showed zones of inhibition of 13 and 11 mm against *S. aureus* and *K. pneumoniae*, bacteria respectively. The antibacterial activity of the Fe_3O_4 was due to the generation of reactive oxygen species (ROS) that led to the destruction of protein and DNA structure without harming the human cells. The biocompatibility studies using MTT assay indicated good cytocompatibility between the Fe_3O_4 and HBPU matrix. The comparatively better cytocompatibility of the prepared HBPU reinforced Fe_3O_4 nanocomposite was due to the well dispersion of the nanoparticles within the HBPU matrix which imparted better structural support and anchorage substrate to the cells.

Shaik et al. [75] synthesized castor oil based poly (urethane-esteramide) UPCEA/TiO_2 nanocoating with anticorrosive and antimicrobial properties. The author selected TiO_2 nanoparticles since they show excellent biocidal properties and are less volatile in nature. In this study UPCEA was synthesized by the condensation polymerization reaction of N, N-bis (2-hydroxyethyl) castor oil fatty amide with terephthalic acid which was further modified with different percentage of TDI, i.e., 7, 9, 11 and 13% respectively to obtain UPCEA-7, UPCEA-9, UPCEA-11 and UPCEA-13 as depicted in Fig. 26. Thereafter, the synthesized UPCEA was reinforced with TiO_2 nanoparticles with different concentrations, i.e., 0.1 wt%, 0.2 wt%, 0.3 wt%, 0.4 wt%, and 0.5 wt% to obtain the UPCEA/TiO_2 nanocomposite. Out of the entire composition 0.4 wt% of TiO_2 exhibited better dispersion and excellent physicochemical properties. Hence, 0.4 wt% TiO_2 was considered to be the optimum loading and was used for synthesizing UPCEA nanocomposites i.e. UCPEA/TiO_2-7, UCPEA/TiO_2-9, UCPEA/TiO_2-11, and UCPEA/TiO_2-13.

The antibacterial and antifungal activity of UPCEA-13 with 0.4 wt% loading of TiO_2 was studied through agar disc diffusion method. The different bacterial microorganisms used for this study were, *Staphylococcus aureus, Escherichia coli, and Bacillus pasteurii*, and the different fungal strains were *Fusarium solani, Penicillium notatum, and Aspergillus niger*. The bacterial and fungal activity was investigated after 72 h of incubation at 32 °C. The antimicrobial effect of a system relies on the size of nanoparticles so as to impart better dominating attack against the microorganisms. SEM and EDX studies of the coated substrates indicated uniform coating with well trapped TiO_2 nanoparticles within the UPCEA matrix. The coating properties of the nanocomposite on mild steel indicated good adhesion; good scratch hardness with the excellent corrosion resistance properties. Regarding the antimicrobial activity, it was observed that after 48 h of incubation at 27 °C, inhibition zones of UCPEA-13 with 0.4 wt% of TiO_2 versus *Staphylococcus, Escherichia coli, and Bacillus pasteurii* were found to be 14, 25, and 16 mm in diameter, respectively. On the other hand the inhibition zone diameters of UCPEA-13 with 0.4 wt% of TiO_2 against *Fusarium solani, Penicillium notatum, and Aspergillus niger* were found to be 26, 24, and 21 mm, respectively. The above results indicated that the TiO_2 nanoparticles exhibited excellent antimicrobial activity. The TiO_2 nanoparticles could strongly adhere with the electron the donor

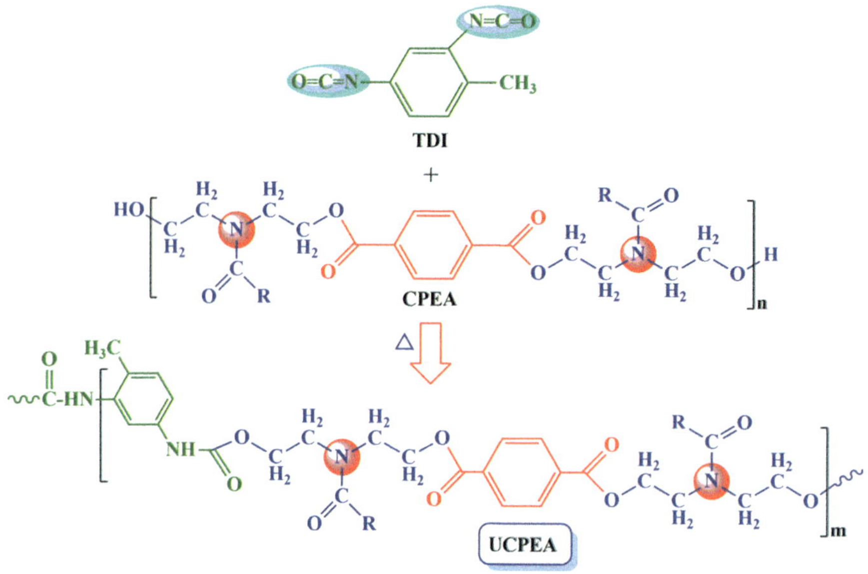

Fig. 26 Synthesis scheme for UPCEA

groups of the biological molecules containing oxygen, nitrogen, and sulphur. As a consequence, the TiO_2 nanoparticles could get uniformly distributed within the cell boundary of microorganism resulting in protection of the rigid outermost cell wall. TiO_2 nanoparticles on interaction with the electron donor groups can produce free radicals and reactive oxygen species that can permanently damage the microbes [76].

3.4 Elastomers

Thakur et al. [77] studied the synthesis and characterization of multi-stimuli responsive smart elastomeric hyperbranched polyurethane/reduced graphene oxide nanocomposites (HBPU/RGO) for shape memory application. The RGO wt% was varied from 0.5, 1.5 and 2.5 respectively. The synergistic effect of HBPU and RGO imparted several advantages which include noncontact stimuli to sunlight, eco-friendly nature and inexpensive practical stimulus. The process of reducing GO has carried out through sonochemical method a promising technology of using ultrasonic irradiation, elevated temperature, high pressure and rapid cooling rates, etc. The sonication method involves the formation of small cavities in the liquid medium. These small cavities implode rapidly generating microscopic shock waves thereby, realising the huge amount of energy in the liquid medium. Hence, the method of ultrasonication prevents the aggregation of GO thereby reducing the reaction within the GO layers. The shape memory behaviour of the nanocomposites was investigated under sunlight, microwave and thermal condition. The shape memory activity of the nanocomposites was found to be excellent with faster shape recovery values when exposed to microwave as compared to sunlight and thermal stimulus as shown in Fig. 27. This was due to the excellent absorbance of RGO towards sunlight and microwave. Moreover, the shape recovery time decreased with an increase in the RGO content in the nanocomposite. This was due to the strong interaction of RGO with HBPU resulting in huge amount of release of elastic strain energy in nanocomposites helping faster shape recovery. The nanocomposites also had excellent thermal stability and mechanical properties with a tensile strength of 27.8 MPa, modulus of 36.3 MPa and toughness of 116 MJ m^{-3} due to the better chemical interaction of RGO with HBPU matrix. Thus, the authors concluded that the resulting nanocomposites exhibited improved thermomechanical and multi-stimuli response shape memory nature.

Ahuja et al. [78] studied the synthesis and characterization of castor oil-based polyurethane nanocomposites elastomer reinforced with organically modified clay (Cloisite 30B/C30B). The high aspect ratio of clay platelet and silicate clay offers huge potential to increase the clay/polymer interfacial area to improve properties, including flame resistance, mechanical properties, gas barrier properties, and thermal stability. Since C30B is hydrophilic in nature it was modified with hydrophobic organic polymers to impart hydrophobic character. This modification results in improving the interfacial adhesion of C30B with the hydrophobic matrix. The PU/C30 B nanocomposites were prepared through ultrasonication method using a high shear mixer with clay % varying from 0 to 5 wt%. TEM indicates the appearance of clear individual clay layers with few inseparable clay platelets and tactoids. The thickness of the platelets was found to be 1–2 nm. It was observed that beyond 3 wt % loading full exfoliated structure was not achieved. WAXD study was used to determine the morphology of the nanocomposites and to distinguish between

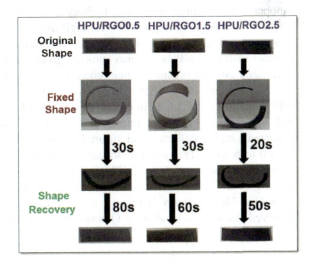

Fig. 27 Shape memory behaviour of synthesized nanocomposites under microwave stimulus. Reproduced from Thakur et al. [77]

ordered (intercalated) and disordered (exfoliated) states of silicates in nanocomposites. In intercalated structure, there is the finite distance within the polymer interlayer with the appearance of new basal reflection related to the interlayer height. On the other hand, in case of exfoliation, there is an increase in the interlayer distance resulting in delamination of the original silicate layers in the polymer matrix. Consequently, there is gradual disappearance of coherent X-ray diffraction from the distributed silicate layers in WAXD curve. WAXD diffraction study of C-30 B indicated a sharp peak at 4.85° corresponding to $d_{(001)}$ diffraction. However, in case of nanocomposites with loading 1 and 2 wt% there is complete absence of $d_{(001)}$ diffraction peak indicating complete exfoliation of the clay platelets. On the other hand, the nanocomposites with C30 B loading beyond 3 wt% indicates broadening of the peak at 4.85° with an increase in basal spacing which was due to the partial exfoliation of clay galleries. The broadening of peak arises due to the due to the partial disruption of parallel stacking of C 30 B. SEM micrographs reveal homogenous dispersion of C 30B within the PU matrix beyond 3 wt%. Above 3 wt % loading of C30B there was an aggregation of nanoparticles due to partial exfoliation as indicated in WAXD studies. FTIR spectra of the PU and its nanocomposite elastomers indicates similar peak confirming the PU clay nanocomposites have similar bands indicating that the PU chains have intercalated into the gallery of layered silicates. It was observed that the tensile strength and modulus increased to 4.49 and 5.88 MPa, respectively, for 4% loading of clay owing to the reinforcement imparted by the dispersed silicate layers in PU matrix. However, beyond 5 wt% of Closite 30B, the tensile strength of the nanocomposite decreased to 3.78 MPa due to agglomeration of the clay. In addition, the elongation at break also increases with the increase of clay content, owing to long fatty acid chains of oil, which imparted high flexibility to the elastomeric film.

Gao et al. [79] synthesized and characterized biocompatible elastomer of waterborne polyurethane (WPU) based on castor oil and polyethylene glycol reinforced with *E. globulus* derived cellulose nanocrystals (ECNc). ECNc nanoparticles obtained from acid hydrolysis of lignocelluloses' have significant properties i.e. high Young's modulus, high aspect ratio, low density, biocompatibility and biodegradability which make it candid reinforcing filler in various polymer matrices. WPUs has also attracted great attention in the recent decade due to reduced volatile organic content emission and environmental safety. Castor oil and polyethylene glycol have been selected as a precursor for the synthesis of WPUs owing to their inherent properties which include bio-degradability, low cost, and easy availability. SEM micrographs of the fractured nanocomposite surface indicated a highly deflected and tortured surface with increase in the ECNc content from 0 to 5 wt%. This was due to the complicated energy dissipating mechanism between the ECNs and the WPU interface. FTIR studies were carried out to investigate the degree of H-bonding by evaluation the –NH (3400–3000 cm^{-1}) and –C = O (1740–1720 cm^{-1}) region of the WPU and its nanocomposites. It was observed that the –NH and –C = O peaks shifts towards lower wave number in case of ECNc based WPU nanocomposites. This finding could be explained on the basis of improvement in the phase segregation degree by ECNc resulting in the microphase segregation of hard segment and a soft segment. As a consequence, the soft segments provide more freedom to –CO and –C–O–C bonds to interact with the –NH of hard segments through H-bonds. The tensile strength of the WPU/ECNc nanocomposites increased from 5.43 to 12.22 MPa with an increase in the ECNc content from 0 to 1 wt%. This was due to the homogenous dispersion of the nanofillers within the WPU matrix as shown in Fig. 28. However, opposite trend was observed with increase in ECNc loading beyond 1 wt% due to the aggregation of nanoparticles. Thus, it can be concluded that ECNc based WPU nanocomposites can be interesting reinforcing material due to its rigid nature and high aspect ratio.

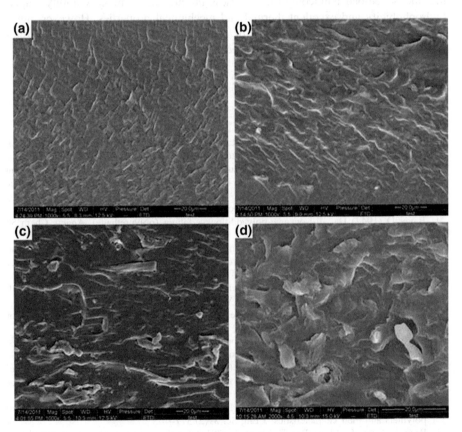

Fig. 28 SEM micrographs of waterborne polyurethane (WPU)/ECNc nanocomposite. Reproduced from Gao et al. [79]

4 Conclusion

Polyurethane nanocomposite from vegetable oil has emerged as versatile materials to overcome the limitation of neat counterparts for wide array of applications. It is imperative from the above literature findings that the properties of bio-based polyurethane nanocomposites can be tailor-made with unique blending with nanoparticles. Also, different techniques such as in situ polymerization, sol-gel technique etc. involved in the dispersion of nanofillers within the polyurethane matrix play a pivotal role in determining the overall properties of the nanocomposite. In addition, the modification of nanoparticles with different surface modifiers has also been beneficial for improving the dispersion of the nanoparticles within the PU matrix. From the above discussion and reports, it is also concluded that the developed nanocomposites possess unique shape memory ability, superhydrophobicity and excellent thermal and mechanical properties. However, much research has to be carried out in the field of polyurethane hybrid nanocomposites from vegetable oil-based polyol and isocyanate respectively.

References

1. Nohra B, Candy L, Blanco JF, Guerin C, Raoul Y, Mouloungui Z (2013) From petrochemical polyurethanes to biobased polyhydroxyurethanes. Macromolecules 46(10):3771–3792
2. Shen L, Haufe J, Patel MK (2009) Product overview and market projection of emerging bio-based plastics PRO-BIP. Report for European polysaccharide network of excellence (EPNOE) and European Bioplastics 243:1–245
3. Berthier JC (2009) Polyurethane PUR. Techniques de lingénieur
4. Bayer O (1947) Das di-isocyanat-polyadditionsverfahren (polyurethane). Angew Chem 59(9):257–272
5. Jincheng W, Shenglin Y, Guang L, Jianming J (2003) Synthesis of a new-type carbonific and its application in intumescent flame-retardant (IFR)/polyurethane coatings. J Fire Sci 21(4):245–266
6. Saunders JH, Frisch KC (1964) Polyurethanes: Chemistry and Technology, Part II. Technology Interscience Publishers, New York
7. Urbanski J, Czerwinski W, Janicka K, Majewska F, Zowall H (1977) Handbook of analysis of synthetic polymers and plastics, UK
8. Durganala S (2011) Synthesis of non-halogenated flame retardants for polyurethane foams. Doctoral dissertation, University of Dayton
9. Pervin F, Zhou Y, Rangari VK, Jeelani S (2005) Testing and evaluation on the thermal and mechanical properties of carbon nano fiber reinforced SC-15 epoxy. Mater Sci Eng, A 405(1–2):246–253
10. Thabet A, Mubarak YA, Bakry M (2011) A review of nano-fillers effects on industrial polymers and their characteristics. J Eng Sci 39(2):377–403
11. Peng L, Zhou L, Li Y, Pan F, Zhang S (2011) Synthesis and properties of waterborne polyurethane/attapulgite nanocomposites. Compos Sci Technol 71:1280–1285
12. Friedrich K, Fakirov S, Zhang Z (eds) (2005) Polymer composites: from nano-to macro-scale. Springer Science & Business Media
13. Pan H, Chen D (2007) Preparation and characterization of waterborne polyurethane/attapulgite nanocomposites. Eur Polym J 43:3766–3772

14. Chang CW, Lu KT (2012) Natural castor oil based 2-package waterborne polyurethane wood coatings. Prog Org Coat 75(4):435–443
15. Silva BB, Santana RM, Forte MM (2010) A solventless castor oil-based PU adhesive for wood and foam substrates. Int J Adhes Adhes 30(7):559–565
16. Xia Y, Larock RC (2010) Vegetable oil-based polymeric materials: synthesis, properties, and applications. Green Chem 12(11):1893–1909
17. Güner FS, Yağcı Y, Erciyes AT (2006) Polymers from triglyceride oils. Prog Polym Sci 31(7):633–670
18. Ronda JC, Lligadas G, Galià M, Cádiz V (2011) Vegetable oils as platform chemicals for polymer synthesis. Eur J Lipid Sci Technol 113(1):46–58
19. Rybak A, Fokou PA, Meier MA (2008) Metathesis as a versatile tool in oleochemistry. Eur J Lipid Sci Technol 110(9):797–804
20. Henna PH, Larock RC (2007) Rubbery thermosets by ring-opening metathesis polymerization of a functionalized castor oil and cyclooctene. Macromol Mater Eng 292(12):1201–1209
21. Liu K, Chapman (1997) Hall Press, New York
22. Sharmin E, Zafar F (2012) Polyurethane: an introduction. INTECH publisher, Croatia
23. Randall D, Lee S (2003) The polyurethanes book. Wiley publishers, New York
24. Szychers M, Szychers (2013) Handbook of polyurethanes. CRC Press Taylor and Francis, Florida
25. Bagdi K (2010) Role of interactions on the structure and properties of segmented polyurethane elastomers
26. Naheed S, Paridah Md, Mohammad J (2014) A review on potentiality of nano filler/natural fiber filled polymer hybrid composites. Polymers 6:2247–2273
27. Fink JK (2017) Reactive polymers: fundamentals and applications: a concise guide to industrial polymers. William Andrew
28. Hojabri L, Kong X, Narine SS (2010) Functional thermoplastics from linear diols and diisocyanates produced entirely from renewable lipid sources. Biomacromol 11(4):911–918
29. Hojabri L, Kong X, Narine SS (2010) Novel long chain unsaturated diisocyanate from fatty acid: synthesis, characterization, and application in bio-based polyurethane. J Polym Sci, Part A: Polym Chem 48(15):3302–3310
30. Wool RP (2014) U.S. Patent No. 8,633,257. U.S. Patent and Trademark Office, Washington, D.C
31. Kyle DR (1993) U.S. Patent No. 5,234,970. U.S. Patent and Trademark Office, Washington, D.C
32. Sahoo S, Kalita H, Mohanty S, Nayak SK (2016) Synthesis of vegetable oil-based polyurethane: a study on curing kinetics behavior. Int J Chem Kinet 48(10):622–634
33. Das S, Pandey P, Mohanty S, Nayak SK (2015) Influence of NCO/OH and transesterified castor oil on the structure and properties of polyurethane: synthesis and characterization. Mater Express 5(5):377–389
34. Kaushik A, Singh P (2005) Synthesis and characterization of castor oil/trimethylol propane polyol as raw materials for polyurethanes using time-of-flight mass spectroscopy. Int J Polym Anal Charact 10(5–6):373–386
35. Mutlu H, Meier MA (2010) Castor oil as a renewable resource for the chemical industry. Eur J Lipid Sci Technol 112(1):10–30
36. Petrović ZS (2008) Polyurethanes from vegetable oils. Polym Rev 48(1):109–155
37. Kumar A, Sharma S (2008) An evaluation of multipurpose oil seed crop for industrial uses (Jatropha curcas L.): a review. Ind Crops Prod 28(1):1–10
38. Segura-Campos MR, Betancur-Ancona D (2016) The promising future of jatropha curcas: properties and potential applications. Nova Science Publishers Incorporated, Hauppauge, NY, USA
39. Abdulla R, Chan ES, Ravindra P (2011) Biodiesel production from jatropha curcas: a critical review. Crit Rev Biotechnol 31(1):53–64

40. Chen CR, Cheng YJ, Ching YC, Hsiang D, Chang CMJ (2012) Green production of energetic jatropha oil from de-shelled jatropha curcas L. seeds using supercritical carbon dioxide extraction. J Supercrit Fluids 66(1):137–143
41. Hazmi ASA, Aung MM, Abdullah LC, Salleh MZ, Mahmood MH (2013) Producing jatropha oil-based polyol via epoxidation and ring opening. Ind Crops Prod 50:563–567
42. Lestari D, Mulder WJ, Sanders JP (2011) Jatropha seed protein functional properties for technical applications. Biochem Eng J 53(3):297–304
43. Pawlik H, Prociak A (2012) Influence of palm oil-based polyol on the properties of flexible polyurethane foams. J Polym Environ 20(2):438–445
44. Pillai PK, Li S, Bouzidi L, Narine SS (2016) Solvent-free synthesis of polyols from 1-butene metathesized palm oil for use in polyurethane foams. J Appl Polym Sci 133(23):1–13
45. Pillai PK, Li S, Bouzidi L, Narine SS (2016) Metathesized palm oil: fractionation strategies for improving functional properties of lipid-based polyols and derived polyurethane foams. Ind Crops Prod 84:273–283
46. Saalah S, Abdullah LC, Aung MM, Salleh MZ, Biak DRA, Basri M, Jusoh ER (2015) Waterborne polyurethane dispersions synthesized from jatropha oil. Ind Crops Prod 64:194–200
47. Aung MM, Yaakob Z, Kamarudin S, Abdullah LC (2014) Synthesis and characterization of Jatropha (Jatropha curcas L.) oil based polyurethane wood adhesive. Ind Crops Prod 60:177–185
48. Gogoi P, Boruah R, Dolui SK (2015) Jatropha curcas oil based alkyd/epoxy/graphene oxide (GO) bionanocomposites: effect of GO on curing, mechanical and thermal properties. Prog Org Coat 84:128–135
49. Zhang L, Zhang H, Guo J (2012) Synthesis and properties of UV curable polyester-based waterborne polyurethane/functionalized silica composites and morphology of their nanostructured films. Ind Eng Chem Res 51(25):8434–8441
50. Hsiao S, Ma CM, Tien H, Liao W-H, Yu-Sheng Wang, Shin-Ming Li, Sheng-Chi Yang Chih-Yu Lin, Ruey-Bin Yang (2015) Effect of covalent modification of graphene nanosheets on the electrical property and electromagnetic interference shielding performance of a water-borne polyurethane composite. ACS Appl Mater Interfaces 7(4):2817–2826
51. Wang C, Xu F, He M, Ding L, Li S, Wei J (2018) Castor oil-based polyurethane/silica nanocomposites: morphology, thermal and mechanical properties. Polym Comp. https://doi.org/10.1002/pc.24798
52. Chattopadhyay DK, Webster DC (2009) Thermal stability and flame retardancy of polyurethanes. Prog Polym Sci 34:1068–1133
53. Meera KMS, Sankar RM, Paul J, Jaisankara SN, Mandal AB (2014) The influence of applied silica nanoparticles on a bio-renewable castor oil based polyurethane nanocomposite and its physicochemical properties. Phys Chem Chem Phys 16(20):9276–9288
54. Liu D, Tian H, Zhang L, Chang PR (2008) Structure and properties of blended films prepared from castor oil-based polyurethane/soy protein derivative. Ind Eng Chem Res 47(23):9330–9336
55. Gu H, Guo J, He Q, Tadakamalla S, Zhang X, Yan X, Huang Y, Colorado HA, Wei S Guo Z (2013) Flame-retardant epoxy resin nanocomposites reinforced with polyaniline-stabilized silica nanoparticles. Ind Eng Chem Res 52(23):7718–7728
56. Das S, Pandey P, Mohanty S, Nayak SK (2017) Evaluation of biodegradability of green polyurethane/nanosilica composite synthesized from transesterified castor oil and palm oil based isocyanate. Int Biodet Biodeg 117(1):278–288
57. Filip Z, Hermann S, Demnerov (2008) FT-IR spectroscopic characteristics of differently cultivated Escherichia coli. Czech J Food Sci 26(6):458–463
58. Fukushima K, Abbate C, Tabuani D, Gennari M, Rizzarelli P, Camino G (2010) Biodegradation trend of poly(e-caprolactone) and nanocomposites. Mat Sci Eng C 30(4):566–574

59. Kim YD, Kim SC (1998) Effect of chemical structure on the biodegradation of polyurethanes under composting conditions 62:343–352
60. Das S, Pandey P, Mohanty S, Nayak SK (2016) Effect of nanosilica on the physicochemical, morphological and curing characteristics of transesterified castor oil based polyurethane coatings. Prog Org Coat 97:233–243
61. Thakur S, Karak N (2013) Bio-based tough hyperbranched polyurethane–graphene oxide nanocomposites as advanced shape memory materials. RSC Adv. 3(24):9476–9482
62. Zhang J, Zhang C, Madbouly SA (2015) In situ polymerization of bio-based thermosetting polyurethane/graphene oxide nanocomposites. J Appl Polym Sci 132(13):1–8
63. Ali A, Yusoh K, Hasany SF (2014) Synthesis and physicochemical behaviour of polyurethane-multiwalled carbon nanotubes nanocomposites based on renewable castor oil polyols. J. Nanomater 2014, Article ID 564384, 9 pages
64. Chen S, Wang Q, Wang T (2012) Damping, thermal, and mechanical properties of carbon nanotubes modified castor oil-based polyurethane/epoxy interpenetrating polymer network composites. Mater Des 38:47–52
65. Liao L, Li X, Wang Y, Fu H, Li Y (2016) Effects of surface structure and morphology of nanoclays on the properties of jatropha curcas oil-based waterborne polyurethane/clay nanocomposites. Ind Eng Chem Res 55(45):11689–11699
66. Dzulkifli MH, Yahya MY, Majid RA (2017) Rigid palm oil-based polyurethane foam reinforced with diamine-modified montmorillonite nanoclay. Mater Sci Eng 204. https://doi.org/10.1088/1757-899x/204/1/012024
67. Adnan S, Maznee TN, Ismail T, Mohd NN, Mariam ND, Nik S, Hanzah NA, Kian YS, Hazimah AH (2016) Development of flexible polyurethane nanostructured biocomposite foams derived from palm olein-based polyol. Adv Mater Sci Eng 2016, Article ID 4316424, 12 pages
68. Nikolaidis AK, Achilias DS, Karayannidis GP (2012) Effect of the type of organic modifier on the polymerization kinetics and the properties of poly(methyl methacrylate)/organomodified montmorillonite nanocomposites. Eur Polym J 48(2):240–251
69. Zaimahwati AH, Rihayat T, Reflianto D, Gea S (2014–2015) The manufacture of palm oil-based polyurethane nanocomposite with organic montmorillonite nanoparticle as paint coatings. Int J Chem Tech Res 7(5):2537–2544
70. Salih AM, Ahmad MB, Ibrahim NA, Mohd Dahlan KZH, Tajau R, Mahmood MH, Wan MdZWY (2014) Thermal and mechanical properties of palm oil-based polyurethane acrylate/clay nanocomposites prepared by in-situ intercalative method and electron beam radiation. AIP Conf Proc 117:1584
71. Xia Y, Larock RC (2011) Preparation and properties of aqueous castor oil-based polyurethane-silica nanocomposite dispersions through a sol-gel process. Macromol Rapid Commun 32(17):1331–1337
72. Deka H, Niranjan K (2011) Bio-based hyperbranched polyurethane/clay nanocomposites: adhesive, mechanical, and thermal properties. Polym Adv Technol 22(6):973–980
73. Sahoo S, Kalita H, Mohanty S, Nayak SK (2018) Shear strength and morphological study of polyurethane-OMMT clay nanocomposite adhesive derived from vegetable oil-based constituents. J Renew Mater 6(1):117–125
74. Das B, Mandal M, Upadhyay A, Chattopadhyay P, Karak N (2013) Bio-based hyperbranched polyurethane/Fe_3O_4 nanocomposites: smart antibacterial biomaterials for biomedical devices and implants. Biomed Mater 8(3):1–12
75. Mohammed RS, Manawwer A, Naser MA (2015) Development of castor oil based poly (urethane-esteramide)/TiO_2 nanocomposites as anticorrosive and antimicrobial coatings. J Nanomater 2015, Article ID 745217, 10 pages. http://dx.doi.org/10.1155/2015/745217
76. Hajipour MJ, Fromm KM, Akbar Ashkarran A (2012) Antibacterial properties of nanoparticles. Trends Biotechnol 30(10):499–511

77. Thakur S, Karak N (2014) Multi-stimuli responsive smart elastomeric hyperbranched polyurethane/reduced graphene oxide nanocomposites. Mater Chem A 2:14867–14875
78. Ahuja D, Kaushik A (2016) Castor oil-based polyurethane nanocomposites reinforced with organically modified clay: synthesis and characterization. J Elastomers Plast 49(4):1–17
79. Peng Gao Z, Jun Zhong T, Sun J, Wang X, Yue C (2012) Biocompatible elastomer of waterborne polyurethane based on castor oil and polyethylene glycol with cellulose nanocrystals. Carbohyd Polym 87(3):2068–2075

Clay Based Biopolymer Nanocomposites and Their Applications in Environmental and Biomedical Fields

K. Sangeetha, P. Angelin Vinodhini and P. N. Sudha

1 Introduction

Clay-based biopolymer nanocomposite was an interdisciplinary subject which holds together the hands of scientists from polymer science, biology, chemistry, physics, materials and biomedical engineering. Collaborators of different discipline would always result in qualitative work and new ideologies which steps into a variety of medical and nonmedical applications. Clay is a natural material which recovers the position against synthetic materials in pharmaceutical technology as it was non-toxic and available plenty in nature. A quality research was going on at present with countless publications and innovative findings dealing with clay reinforced polymeric matrixes were available. The incorporation of nanofiller in the polymeric material will improve the mechanical, barrier and other matrix properties of nanocomposites used for biomedical and environmental applications [40]. This huge number of publications implies the importance of polymer clay nanocomposite as drug delivery system, as hemostasis, in tissue regeneration etc. In this chapter, a detailed discussion on the biomedical and environmental application of clay-polymer nanocomposite was given elaborately with recent findings.

The term clay refers to the group of materials made up of layered silicates or clay minerals with traces of metal oxides and organic matter [51]. They belong to the family of phyllosilicate or sheet silicates made of hydrated alumina–silicates. The basic building unit of clay minerals is composed of tetrahedral silicates and octahedral hydroxide sheets and they were arranged as 1:1 (e.g., kaolinite and serpentine) and 2:1 (e.g., smectite, chlorite, and vermiculite) to give rise to various

---

K. Sangeetha · P. Angelin Vinodhini · P. N. Sudha (✉)
Biomaterials Research Lab, D.K.M. College for Women,
Vellore 632001, Tamilnadu, India
e-mail: drparsu8@gmail.com

© Springer Nature Switzerland AG 2019
Inamuddin et al. (eds.), *Sustainable Polymer Composites and Nanocomposites*,
https://doi.org/10.1007/978-3-030-05399-4_40

classes of the clay mineral. Clay minerals possess specific physicochemical characteristics such as high surface reactivity (adsorption and cation exchange capacity), colloidal and swelling capacity, optimal rheological behaviour, and high water dispersibility, which render them suitable for different biological applications including pharmaceutics, cosmetics, veterinary medicine, biomaterials, and biosensors.

In general, clays were nanometer in size due to this it can be easily incorporated into polymer matrixes. Clay was hydrophilic in nature and it can be readily mixed with hydrophilic polymers like poly(ethylene oxide) [37] and poly(ethylene glycol) [21] to prepare polymer-clay nanocomposite. But in case of hydrophobic polymers, the preparation of nanocomposite with good miscibility was not possible by physical mixing of polymer and clay. In such cases, these clays were converted to organophilic by exchanging the cations present in the clay layer with surfactants such as quaternary alkylammonium and alkylphosphonium ions [25].

The schematic representation of various classes of clay mineral and crystal structure of some commonly used clay in the pharmaceutical application was shown in Figs. 1 and 2 respectively.

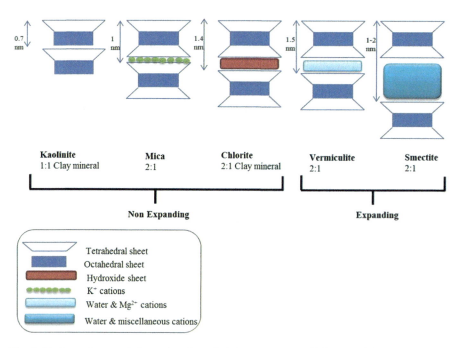

Fig. 1 Various classes of clay minerals and their arrangement [38]. Copyright (2015) Royal Society of Chemistry

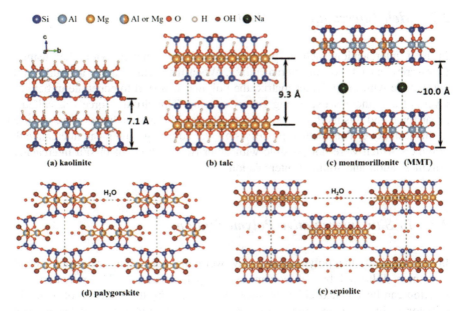

Fig. 2 Crystal structure of pharmaceutical used clays; **a** kaolinite, **b** talc, **c** montmorillonite, **d** palygorskite and **e** sepiolite, where dashed lines represent the unit cell [90]. Copyright (2016) Elsevier

2 Preparation Methods of Polymer Clay Nanocomposites

Solution intercalation, melt mixing and in situ polymerization techniques are the most commonly used methods to prepare polymeric clay nanocomposites [81]. During the preparation, the polymer was intercalated or exfoliated in layered hosts of clay with significant improvement in properties with reduction of component weight.

2.1 Solution Intercalation Method

In solution intercalation method, the polymer was dissolved in a solvent system and the silicate layer was swellable. The layered silicate is first swollen in a solvent, such as water, chloroform, or toluene. During mixing of silicate and polymeric solution, the polymer chains intercalate among themselves and the solvent was displaced within the interlayer of silicate. After the removal of solvent, the intercalated system remained with nanoscale morphology.

2.2 Melt Intercalation Method

In melt intercalation method polymer was directly mixed with clay using a twin-screw extruder or an internal mixer. When the surface layers are sufficiently compatible with the polymer matrix, the polymer will start to penetrate in between layers of clay thereby expanding gallery spacing. This driving force was called as shear force and this method was more commercial because solvents were not used in this technique. This method possesses greater advantage and it can intercalate the polymers which were not possible with other two methods of in situ intercalative polymerization and solution intercalation.

2.3 In Situ Intercalative Polymerization

In this method, the polymer and clay were intercalated by choosing suitable monomers followed by subsequent in situ polymerization. The polymerization reaction can be initiated either by heat or radiation, by the diffusion of a suitable initiator, or by an organic initiator or catalyst fixed through cation exchange inside the interlayer before the swelling step. Here the monomer was used directly as a solubilizing agent for swelling the layered silicate. Subsequent polymerization takes place after combining the silicate layers and monomer, thus allowing the formation of polymer chains between the intercalated sheets [91]. This method was widely used for thermosetting polymer-layered silicate nanocomposites.

The diagrammatic representation of exfoliation and intercalation during formation of polymer clay composite was shown in Fig. 3 and the possible interaction between the clay and polymer matrix as illustrated in Fig. 4.

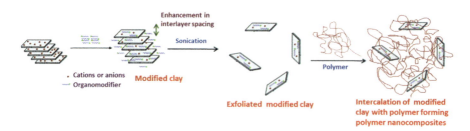

Fig. 3 Schematic diagram showing clay modification and intercalation of polymer to form polymer nanocomposites [51]. Copyright (2015) Elsevier

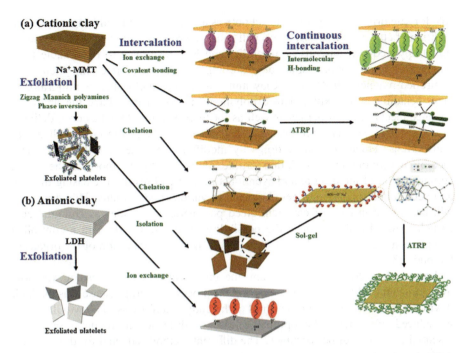

Fig. 4 Schematic representation of possible strategies for the interaction of clay mineral with the polymer [23]. Copyright (2014) Elsevier

3 Biomedical Applications of Polymeric Clay Nanocomposites

Indigenous people all around the world from ancient days have used clay both internally (geophathy) and externally (clay baths) for prevention of illness, physical healing, to improve the general health thereby prolonging their life. Clay therapy was not a new concept or modern healing technology for us, as our ancient cultures and aboriginal tribes were more familiar with clay and they have used for a variety of illness and injuries. Oral consumption of clay will cure the infection, helps in balancing pH of the body, regulate gastrointestinal problems, and counter poisoning by chelation. Some of the clay-polymer composites along with their application in the biomedical field were discussed in this section.

3.1 As Drug Delivery System

In general, usage of a single polymer or clay mineral will not meet all the necessary requirement of drug release and hence the combination of a polymeric material with clay mineral will give adequate support for effective drug delivery. Clay play a vital role for safe drug delivery as it can act both as an excipient and active agent [3]. The addition of clay even lesser than 5% will enhance properties of mechanical strength, thixotropy, reduced gaseous permeability and heat resistance to remarkable level which was highly appreciable for adsorbing the drug and releasing them in a controlled manner [39, 69, 88]. Clay minerals such as halloysite, montmorillonite clay, bentonite were explored in a polymeric matrix as they have the ability to enhance mechanical properties as well as for their potential to act as drug delivery modifiers [61, 68]. The presence of clay mineral in nanocomposite will act as transporting vehicle to deliver drugs by modifying the rate of drug release and thereby improving the dissolution profile of drug [85]. In most of the cases, administration of the drug at a conventional dosage will result in releasing them in an uncontrolled manner and get circulated to various body organs. This drawback will be evaded in case of clay reinforced polymeric composites as colloidal clay particles and biopolymer have the possibility to release the drug in a controlled manner at a specific needed site without affecting the other portion of the organs in the body [86].

For an ideal drug delivery system, the important factor to consider was how much amount of drug was released during the process—if drug release was too high it will lead to adverse effect or harmful to the body and if it was too low will results in reduced efficiency. The optimal release of drug within required time was essential for powerful performance. The different methods adapted for drug release was shown in Fig. 5.

Kohay et al. [50] developed a novel formulation of the organic-inorganic composite in which the drug doxorubicin was incorporated in micelles (M-DOX) of polyethylene glycol-phosphatidylethanolamine (PEG-PE) adsorbed over clay layer of montmorillonite (MMT). On varying the ratio of PEG-PE/MMT, two different formulations were fabricated and named as low, high composite and it was compared with pure doxorubicin. In their work, they have used MCF-7 cells for carrying out bioassay (in vitro study) at a regular time interval of 2 and 6 h and Adriamycin resistant cell line (A2780-ADR) for evaluating cytotoxicity. They reported the following information: In low composite a single layer of polymer and for high composite two layers of polymer were intercalated between the platelets of montmorillonite clay. The release trend followed the order of high formulation > low formulation > DOX/MMT and in MCF-7 cells high formulation exhibits higher cytotoxicity whereas, for Adriamycin resistant cell line (A2780-ADR), low formulation demonstrated the highest cytotoxicity.

Sabbagh et al. [73] synthesized novel halloysite-based chitosan/oxidized starch nanocomposite hydrogel beads by incorporating clay mineral halloysite nanotubes and evaluated the changes in swelling behaviour, thermal properties as well as drug loading/releasing characteristics of hydrogel beads. The embedding of halloysite into hydrogel structure will prolong the release of drug metronidazole (MTZ) from nanocomposite hydrogel beads resulting in controlled drug release.

Lal et al. [53] examined in vitro oral delivery of insulin using montmorillonite poly lactic-*co*-glycolic acid nanocomposites. The prepared composite was evaluated for various parameters such as drug content, physicochemical properties,

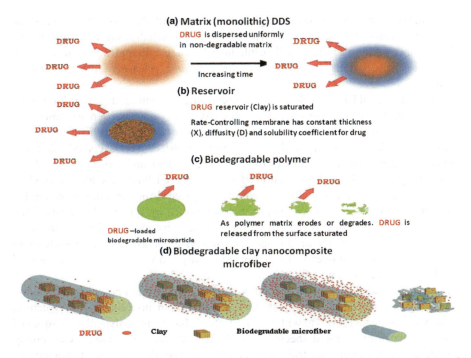

Fig. 5 Different method of drug release based on (a) monolithic, (b) reservoir, (c) degradation and (d) combined a, b and c drug delivery systems. Jafarbeglou et al. [45]. Copyright (2016) Royal Society of Chemistry

in vitro insulin release profile and cytotoxicity. The results from MMT assay reveal the nontoxicity of the formulation and this nanocomposite was effective in releasing insulin without burst effect which will definitely serve as an efficient oral drug delivery vehicle of insulin in near future. Controlled release of strontium ranelate (SRA), a drug for osteoporosis from composite scaffold polycaprolactone-laponite was reported by Nair et al. [62]. By varying Laponite-SRA complex content from 3 to 12 wt% an array of composites were prepared and evaluated in vitro using human osteosarcoma cells. They reported ratio 3 wt% laponite-SRA complex was ideal for drug release which was confirmed using ALP activity.

Khlibsuwan et al. [48] studied the drug release behaviour of spray-dried chitosan-magnesium aluminium silicate (MAS) nanocomposite microparticles, loaded with propranolol HCl. Investigation on the effect of crosslinker (sodium tripolyphosphate) and content of clay (magnesium aluminium silicate) was also reported. A sustained release of propranolol was achieved for all types of microparticles in 0.1 M HCl and phosphate buffer pH 7.4 for several hours. The drug release clearly slowed down at low and neutral pH, whereas the incorporation of 1–3% TPP caused only a slight/moderate decrease in release rate, irrespective of the type of release medium with increasing MAS content. Similarly, in another study, Fan and his coworkers [35] investigated drug release behaviour of diclofenac sodium (DS) using novel sodium alginate/hydroxyapatite/halloysite nanocomposite hydrogel beads. The addition of hydroxyapatite and halloysite nanotube to the biopolymer sodium alginate will result in enhancing entrapment efficiency from 62.85 to 74.63% and thereby controlling the burst release of drug diclofenac sodium.

Bera et al. [10] have developed novel composite by coating montmorillonite clay on the membrane of alginate-Arabic gum (AG) gel and has been examined for intragastric flurbiprofen. The drug release rate was compared with uncoated formulation and they reported that the coated matrices (F-O, coated) depicted improved drug loading ability (DEE, 94.49 ± 1.23%) and slower drug release profile (Q_{8h}, 69.24 ± 0.65%) endowed with improved buoyancy and mucoadhesive suggesting them as excellent drug delivery system. Another study revealing the sustained drug release of clay composite was reported by Ji et al. [46] in which the mechanical strength of gelatin scaffold was enhanced by adding halloysite nanotube and it was loaded with the drug ibuprofen. Halloysite incorporated composite show extended drug release over 100 h compared to only 8 h when it was directly mixed with gelatin (i.e. without halloysite).

Bounabi et al. [16] have developed novel nanocomposite of poly(2-hydroxyethyl methacrylate) (HEMA) with montmorillonite clay and investigated the structural properties of nanocomposite with the addition of clay using analytical techniques of Fourier transform infrared spectroscopy (FTIR), differential scanning calorimetry (DSC) and thermogravimetric analysis (TGA). The drug release behaviour of paracetamol loaded in prepared hydrogel nanocomposite was reported with the following findings: Crosslinking of clay will reduce drug burst rate and release rate of the drug from novel composite will decrease with increase in the content of MMT loading. The composite will delay drug dissolution rate and thereby improve patient compliance through the reduction of multiple dosing.

Huang and his coworkers [43] have prepared chitosan composite hydrogels by adding halloysite nanotubes to the solution of chitosan. They have performed cytotoxicity assay using MC3T3-E1 cells and reported that the composite will support the growth of MC3T3-E1 cells indicating its biocompatibility. The composite hydrogels show a maximum drug entrapment efficiency of 45.7% for doxorubicin (DOX) which is much higher than that of pure chitosan hydrogel (27.5%) suggesting composite hydrogel shows superior drug release efficiency.

Curcumin release from composite carboxymethyl cellulose–montmorillonite clay was carried out by Madusanka and his coworkers. The sustained release rate of more than 60% was achieved from curcumin activated nanocomposite in distilled water (maintained at a pH of 5.4) at 25 °C within 2 h and 30 min. From the results, Madusanka group have reported that curcumin activated carboxymethylcellulose–montmorillonite nanocomposite is an effective curcumin carrier with enhanced solid state properties [59]. Salcedo et al. [74] prepared composites of a modified montmorillonite with chitosan by simple solid-liquid interaction and loaded with drug oxytetracycline. The penetration enhancement properties of nanocarrier towards the drug were evaluated using cell model Caco-2. Presence of clay mineral particles in the polymer matrix resulted in effectively enhanced drug permeation and improved bioavailability in vivo after oral administration of the drug.

Kevadiya and his coworkers [47] examined drug release rate of diclofenac sodium by intercalating "in situ" the anionic drug DC into cationic polymer chitosan in montmorillonite clay. They reported the in vitro cell viability assay of the drug-loaded composite in cancer cell was less toxic than the pristine drug. They suggested applying this formulation as fruitful material in much more applications with different loaded drugs and biomolecules.

Abdeen and his coworkers [1] have prepared nanocomposite for drug carrier by intercalating chitosan into layered silicate of montmorillonite clay and the release rate of drug ibuprofen from nanocomposite was studied. The drug release of ibuprofen was affected by pH value of the medium, drug loading capacity, the percentage of chitosan in nanocomposite and its morphology. They have noted the in vitro drug release at pH of 5.4 and 7.8 which mimic the gradient of pH from the stomach to intestine. On concluding their work, the sustained-drug release was achieved and they have reported the prepared nanocomposite was effective for oral administration as it avoids the interaction between drug and gastric mucosa with prolonged duration of drug activity.

3.2 In Tissue Engineering

Tissue engineering is an emerging field widely employed to develop scaffolds for replacing the damaged tissues and organs [54]. The basic principle in tissue engineering was to fabricate scaffold which should be highly compatible with living tissues and they will help to enhance cell proliferation, adherence and differentiation thereby eventually develop regeneration or growth of new tissues in infected or defected area [5]. In order to design a composite system for replacing tissues, dispersion technology and interfacial interaction between the matrix systems was crucial to attaining. This will be effectively achieved by choosing similar molecular structures so that they will result in composites with improved properties [56].

The presence of clay in the composite will usually enhance cell adhesion and spreading. A possible reason for this increased cell attachment was due to the fact clay will create more focal adhesion sites through reactive functional groups where direct adhesion of cell attachment was taken place [36]. Another interesting reason for enhanced cell proliferation in polymeric-clay composite was due to the creation of particular hydrophobicity/hydrophilicity balance between hydrophobic polymer chains and hydrophilic clay dispersion could directly mediate cell adhesion as well as promote protein adsorption [75]. Afsar and Ghaee [2] also reported a similar result that the presence of clay will enhance cell adhesion and proliferation. Composite of chitosan/alginate/halloysite was prepared and modified with amination. The cytocompatibility was evaluated using Alamar Blue assays by culturing fibroblasts L929 on the scaffolds. They reported that the amination treatment and incorporation of clay will improve cell adherence and proliferation.

Boyer et al. [17] developed a faster gelling composite to heal articular cartilage, a connective tissue using biomaterial assisted cell therapy. Boyer and their coworkers prepared laponite silated hydroxypropylmethyl cellulose hydrogel having the very high mechanical strength the basic criteria for healing cartilage tissues. The in vitro studies shows biocompatibility and nontoxic nature of composite make them a suitable material for treating cartilage defects. An interesting composite work reported more recently was Sr^{+2}-modified chitosan/montmorillonite scaffold fabricated using freeze-drying method. Biocompatibility study was evaluated by employing scaffold in human osteoblasts (hOBs). Cell viability and proliferation were reported from assays of MTT (3-(4,5-dimethylthiasol-2-yl)-2,5-diphenyltetrazolium bromide) and DNA content analysis. Demir et al. [28] have concluded that the composite will possess all desirable physicochemical and biological characteristics which could mimic as an ideal biomimetic template for the repair of defective bone with osteoblasts.

A novel three-dimensional composite scaffold of chitosan/hydroxyapatite/montmorillonite clay was reported by Vyas et al. [87]. The presence of clay will reduce the swelling and degradation rate which was highly required for enhancing the mechanical strength of scaffold. Cell proliferation was tested using MG 63 cell and the cellular response show non-cytotoxic behaviour implying hemocompatibility with more favourable microenvironment for osteoblast cell proliferation. Bhowmick and his colleagues [13] reported an improved osteogenic response of osteoblast cell line MG-63 cells as a direct function of modulating organically modified montmorillonite clay content in composite chitosan/hydroxyapatite-zinc oxide (CTS/HAP-ZnO). An interesting report of a decrease in tensile strength, antibacterial effect and cytocompatibility with the MG-63 cell was observed with composite (in absence of OMMT) compared to composite with organically modified montmorillonite clay. The composite with clay will decrease swelling ability thereby increasing mechanical strength and cell proliferation of composite.

Chappidi and Mills [22] have prepared a novel composite scaffold with a unique combination of poly-glycerol sebacate (PGS), polycaprolactone (PCL) and halloysite clay nanotubes (HNTs). The presence of aluminosilicate clay nanotubes will help to enhance its structure, mechanical properties and have potential to support the development of tissues and their growth. Chitosan-agarose-gelatine nanocomposite porous scaffolds doped with halloysite was fabricated using the freeze-drying method and prepared scaffold was underwent to in vitro, in vivo studies to evaluate its biocompatibility and biodegradability. The relative number of attached cells onto chitosan-agarose-gelatine halloysite scaffold was eventually increased with halloysite content from 3 to 6 wt% which implies good cytocompatibility behavior of the scaffold. The in vivo study was conducted using rats with slight inflammatory effect and the reported results confirm implant was degraded without rejection and full restoration of the blood supply was achieved within six weeks [63]. Bonifacio et al. [15] prepared a novel tri-component hydrogel based on gellan gum, glycerol, halloysite and examined for different soft tissue engineering application. The in vitro results show a good human dermal fibroblasts biocompatibility mimicking the native microenvironment.

Huang et al. [44] prepared composite hydrogels of Sodium alginate (SA)/halloysite nanotubes and examined for tissue engineering application using pre-osteoblast (MC3T3-E1) cell line. The cell adhesion and proliferation were increased in the composite as compared to pure sodium alginate hydrogel revealing the decreased pore size due to halloysite will result in increased cell attachment and low cytotoxicity. Liu et al. [55] fabricated halloysite reinforced alginate composite scaffold by solution mixing and freeze drying. The results from analytical techniques will confirm the interfacial interactions between halloysite, alginate and also the enhancement of cell proliferation and mechanical strength was achieved due to the presence of halloysite. The reported porosity of 93–97% and pore size of around 100–200 μm, confirms the improvement in the number of sites for cell attachment. The cytotoxicity examination using mouse fibroblast cells display better attachment which in turn should be a potential member to apply in tissue engineering.

Olad and Azhar [66] reported novel three-dimensional chitosan-gelatin/nanohydroxyapatite-montmorillonite composite scaffold and examined physicochemical properties. The addition of MMT in the composite will moderate mechanical behaviour, water absorption ability, density, biodegradation, biomineralization suggesting them to apply in bone tissue engineering application in future. A novel nanocomposite was reported by Buffa and his coworkers [19] in which they have prepared composite by loading Sr (II) on halloysite nanotube with the biopolymer matrix of (3-polyhydroxybutyrate-co-3-hydroxyvalerate). The loading of strontium and clay enhance the porosity and mechanical resistivity making them as an effective material to apply for *invitro* studies. The *invitro* study was reported by checking the compatibility of the composite in L929 fibroblast cells and the results proved that it was an effective material to apply for bone regeneration purpose.

3.3 *As Regenerative Repair in Wound Healing*

The largest organ in the human body was skin and it will act as a barrier to harmful mediums and prevent them to enter into the body. Sometimes skin gets wounded due to physical, chemical, mechanical and/or thermal damages. The healing of damaged skin in a natural way was more complex and it will continue for few months to years. To overcome this slow complex process, a new technique "Wound healing" was emerged and has attracted great interest to treat excessive lose of skin. A new combination of materials including natural, synthetic and even combination of both have been used for application of skin regeneration.

One such material was clay and we are familiar with clay from ancient history that our ancestors have widely used clay and clay-based products for skin ailment and for wound dressing. The usage of clay in the medical field was well documented worldwide from early days of mankind and it will continue till today. But after 19th century more scientific inquiry was emerged to analyze chemistry behind its structure, properties and the interaction/mechanism of the healing process.

The intense level of investigation and research was more concentrated in the second half of the 20th century evidencing that these clay-based products are much more compact with skin regeneration. Some of the documented research works were highlighted in the area of wound healer.

Before discussing in detail about the research works in this area, we should aware of the general term "Regenerative medicine" which comprises a broad category involving regeneration of cells, tissues, and organs in order to perform its original function. The incorporation of clay in polymers will result in enhancing mechanical properties, improve cell adhesion, proliferation, differentiation of progenitor cells and also act as biomolecule carrier [27, 36]. A simple representation of clay mineral in regenerative medicine was illustrated in Fig. 6.

An interesting study related with clay-based composite was recently reported by Kurczewska et al. [52] in which they have explained the influence of clay (halloysite nanotubes) in terms of releasing rate and biological activity of dressing material. The release rate of the drug was slowed down by halloysite as its structure allows functionalization of exterior and interior surfaces with organic ligands. This factor will result in reducing release rate as the system would act as a double barrier which in turn can be effective for long-term treatment of wounds.

Kaolin loaded polyurethane hydrogel composite was prepared using simple one-pot synthesis and hemostatic capabilities of composite were compared with commercial hemostatic dressing materials. Blood clotting index was measured using hemolysis assay and the clotting time was measured through rheology by taking Quick Clot Combat Gauze (QCCG) as a reference. Lundin and his coworkers reported that kaolin loaded polyurethane foams of 5 and 10% exhibits enhanced clotting behaviour and was highly recommended for wound dressing

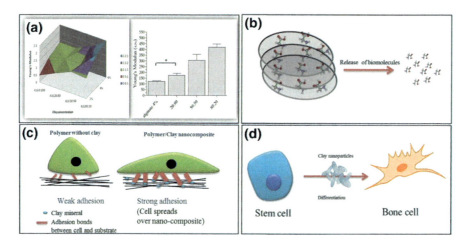

Fig. 6 Application of cationic clay minerals in regenerative medicine: enhancing the mechanical property of polymers (**a**), delivering biomolecules (**b**), improving cell adhesion (**c**), and facilitating the proliferation and differentiation of stem cells to bone cells (**d**). Ghadiri et al. [38]. Copyright (2015) Royal Society of Chemistry

[58]. Yang's research group [89] prepared nanoclay LMSH cross-linked semi-IPN silk sericin/poly(NIPAm/LMSH) nanocomposite hydrogel for wound dressing. The wounded area was treated with prepared hydrogel nanocomposite and recovery rate was monitored at regular interval of time. They reported the nanocomposite has achieved 83% of healing at 6th day followed by complete recovery within 13 days and thereby suggesting this composite as a suitable wound healer.

Solvent casting method was adapted by Devi and Dutta [29] to prepare nanocomposite films of chitosan-bentonite. The nanocomposite has good swelling and excellent antibacterial properties simultaneously, could serve as a barrier against infection, are low-antigenic, could adsorb wound exudates and show high water-vapour permeability. They reported this preliminary study was highly supported for fabricating them as wound care products. Another interesting work on wound treatment was reported by Sirousazar et al. [80]. This research group has formulated a novel wound dressing material by employing freeze-thawed and non-freeze-thawed egg white/PVA/MMT nanocomposite hydrogels. In vitro and in vivo studies were used for evaluating its cytotoxicity and rate of wound recovery. The results of in vitro study reveal the optimum level of biocompatibility was achieved and the in vivo study supports the fast closure of wound edges with the significant rate. Similar technique of freezing-thawing method was used by Noori group [65] to prepare nanocomposite hydrogels of poly (vinyl alcohol)chitosan/montmorillonite. The presence of clay will improve the mechanical properties and swelling behavior of the system. They have reported the addition of 3% of montmorillonite clay will increase tensile modulus to 35% and this enhanced mechanical property make this material highly suitable for wound dressing.

Ambrogi et al. [6] have developed a composite by intercalating chlorhexidine into the montmorillonite-chitosan film. Even though chlorhexidine is an excellent antimicrobial agent the usage of chlorhexidine was limited in wound healing because of its negative aspect of cytotoxicity towards human fibroblasts. This should be overcome on interacting with the chitosan-MMT film at a concentration of 1 and 5% of chlorhexidine resulting in sustained release of drug in long-term purpose with decreased cytotoxicity and can act as good antibiofilm and antimicrobial agents. Nistor et al. [64] have prepared crosslinked collagen/poly (N-isopropyl acrylamide) network embedded with montmorillonite clay and the prepared hybrid hydrogel was targeted for tissue engineering application to cover burned tissues and the healing of damaged skin was monitored. Pseudoplastic behaviour was attained in the hydrogel which was due to the presence of clay which makes them highly suitable for skin healing.

3.4 As Biosensors

Clay-based polymeric biosensors have been escalating as a new generation of biosensors and bioassays with ultra-low limits of detection and enhanced detection sensitivity. For sensor application, conducting polymers was an excellent choice as

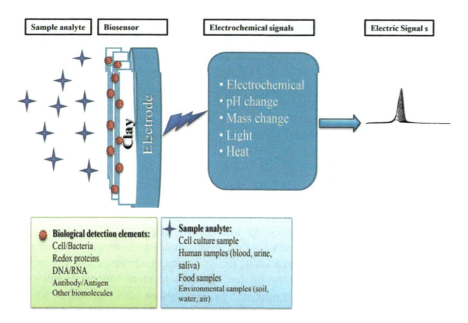

Fig. 7 Representation of a clay-based material in biosensors. Ghadiri et al. [38]. Copyright (2015) Royal Society of Chemistry

they can be easily tuned with different materials like clay to improve its surface properties and immobilization. The sensor is defined as a device that will measure analytical informative signal arises due to a chemical interaction between medium around the analyte and sensor device. If the receptor is a biological module like enzyme or antibody within the cell then the biological mechanism arises was termed as "Biosensor" Fig. 7.

Clay mineral based biosensors were prepared by depositing thin films of clay on the conductive substances [60]. Besombes et al. [11] have developed cholesterol biosensor by using Laponite- poly (amphiphilic pyrrole derivative) matrix. The incorporation of clay will enhance the sensitivity of sense, as this system does not show sensor application in the absence of clay Laponite. This finding suggests the importance of clay in sensing biological substances. Another interesting biosensor was reported by Emre et al. [33] where they developed polymer–clay nanocomposites by grafting polymethylmethacrylate (PMMA) with laponite clay and a conducting polymer; poly(BIPE) [poly(4-(2,3-dihydrothieno[3,4-*b*][1,4]dioxin-5-yl)-7-(2,3-dihydrothieno[3,4-b][1,4]dioxin-7-yl)-2-benzyl-1*H*- benzo [d] imidazole)] was used for detection of glucose. They reported that this system will serve as a proper immobilization platform for enzyme molecules achieving superior biosensor performance with a more proper enzyme deposition and proficient surface chemistry. By this way, an electrochemical signal for glucose detection was amplified.

Sarkar et al. [78] developed chitosan/montmorillonite nanocomposite by depositing it on indium tin oxide coated glass substrate. This fabricated enzymatic biosensor acts as an excellent platform to detect organophosphorous (chlorpyriphos) through measuring immobilization of acetylcholine esterase. Barlas et al. [9] prepared folic acid (FA) modified poly(epsilon-caprolactone)/clay nanocomposite and have tested for selective cell adhesion and proliferation. The results suggest that this material can be adapted for various bio-applications such as 'cell culture on-chip', biosensors and design of tools for targeted diagnosis or therapy.

Advanced plastic medical devices have been prepared using clay mineral-polymeric composites. One such interesting example was the replacement of medical tubing such as catheters prepared by latex rubber and poly(vinyl chloride) was allergic to patients and this will be successfully replaced by non-allergic composite material prepared using polymer polystyrene-b-poly(ethylene-co-butylene)-b-polystyrene mixed with polyolefins and a lesser amount of filler clay mineral [82]. An interesting medical grade nanocomposite used as medical devices ranging from examination gloves and tubings to blood bags and dialysis equipment was prepared by polyvinyl chloride-montmorillonite clay [77].

4 Environmental Applications of Polymeric Clay Nanocomposites

Water, the elixir of life has now become much more important than precious metals like gold, platinum as it was depleted continuously by natural and anthropogenic activities. We may say water as a powerful indicator of sustainability, which in turn indicate the level of social development in a particular community. Water is also an issue that is linked with health, nutrition and many other factors that affect our society including the condition of nature itself! It is not an exaggeration to say that water is life. Several countries all around the world are expected to face severe water crisis by the year 2025 [18]. Contamination and degradation of the aquatic environment were considered as one of the major global concerns of our society. In particular, industrial effluents such as organic and inorganic wastes, heavy metal ions, dyes, aromatic compounds, etc. pose considerable risk to drinking water sources. Among various adsorbents used for remediating effluents, clay was considered as a suitable candidate as its layered structure has the ability to imprison water in between interlayer space resulting in heavy metal adsorption and ion exchange. Hence in this chapter, we have discussed clay based composites as a potential candidate for remediating wastes by adopting different techniques which was considered as a favorable subject for researchers.

4.1 For Heavy Metal Removal

Various scientists all around the world have conducted a huge number of studies to explore adsorptive characteristics of clay and all these experiments will result in concluding the importance of clay as an excellent adsorbent for removing different toxic heavy metal ions from aqueous solutions. This chapter represents literature review of investigations done in the last decade with clay minerals and their modification as composites with a different polymeric combination and their positive results towards removal of heavy metals.

Edathil et al. [30] developed alginate clay composite using different clay minerals of sepiolite, montmorillonite, bentonite clays and a comparative study was reported on the adsorption performance of heavy metal chromium, total organic anions and iron. A maximum adsorption was achieved at 2.0 wt% of filler (clay) for all the combination of reported clay minerals with alginate. The sorption of the contaminant was confirmed from the results of FT-IR and Scanning electron microscopy suggesting that this composite was effective for removal of heavy metal due to the presence reinforcing material clay. Ahmad and Hasan [4] synthesized L-methionine montmorillonite encapsulated guar gum-g-polyacrylonitrile (GPCM) hybrid nanocomposite by free radical graft copolymerization. Adsorption studies were carried out for the removal of metal ions Cu (II) and Pb (II). The obtained results showed that optimum adsorption for both metal ions was achieved at pH of 5 and equilibrium time of 150 min. The adsorption kinetics of both metal ions followed pseudo-second-order model and the adsorption isotherm was well fitted by Langmuir model. The maximum adsorption capacity of the adsorbent was 125.00 mg/g for lead and 90.91 mg/g for copper.

Heydari et al. [41] have developed nanocomposite of β-Cyclodextrin polymeric bentonite clay by emulsion polymerization. The addition of clay enhances thermal stability of composite and it was confirmed by analytical techniques of DSC and TGA. The composite was employed to remove heavy metals such as copper, zinc, and cobalt from drinking water using batch adsorption method. From the calculated values of Kd and percentage of metal adsorption, the sequence followed for metal adsorption onto nanocomposites was found to be $Zn^{2+} > Cu^{2+} > Co^{2+}$. These findings suggest that maximum removal was achieved with an increase in clay content thereby reporting the composite as green absorbent and economically more effective to use on large scale for the purpose to purify drinking water.

Piri et al. [71] synthesized Polyaniline-clay nanocomposite using in situ chemical oxidative polymerization technique and this adsorbent was used to remove metal ions Pb (II) and Cd (II) from aqueous solution. Batch adsorption study was carried out by varying parameters such as pH of the solution, adsorption dose, metal ion concentration and contact time. The obtained experimental data and isothermal kinetic studies were well inserted for pseudo- second kinetic and Temkin model confirming a companionship of physical and chemical adsorption. They also reported that the preparation method "in situ chemical oxidative polymerization" was a new technique for lead removal and it was economic to employ for large-scale removal of metal ions. Onyango et al. [67] used two different techniques such as batch adsorption and fixed bed column for removal of heavy metal hexavalent chromium using polypyrrole modified montmorillonite composite. For the batch study, Langmuir model was well fitted for the composite to remove chromium with maximum adsorption capacity of 119.34 mg/g at 298 K. In fixed bed column study process parameters such as initial Cr (VI) concentration, bed mass and flow rate were considered. Onyang and his coworkers reported maximum sorption rate was achieved at a lower flow rate and initial Cr (VI) concentration and high bed mass.

Rafiei et al. [72] used the combination of polyacrylic acid-organobentonite to prepare nanocomposite by successive intercalation of cetyltrimethyl ammonium as a surfactant and it was employed to remove Pb (II) ions from aqueous solutions. The interaction of polyacrylic acid in clay was confirmed from XRD diffractogram. The sorption study was performed using batch adsorption and it was compared to untreated bentonite clay. They have reported maximum sorption of 93 mg/g was achieved which was twice as that of untreated bentonite having removal rate of 52 mg/g.

Choudhury et al. [24] synthesized nanocomposites of bentonite clay-hydroxyapatite at three different pH of 3, 7 and 10 which was further cross-linked by glutaraldehyde to increase its thermal stability. These entire composite were utilized to remove toxic lead (II) ion. The sorption capacity was analyzed by mathematical and statistical optimizing technique (response surface methodology). They reported maximum sorption was observed for composite prepared at pH of 7 by fitting Langmuir isotherm model with sorption capacity of 346 mg/g at 30 °C and the mechanism was well fitted with pseudo-second-order kinetics indicating the coexistence of both physisorption and chemisorption. Around 99% of removal was reported with this composite indicating the excellency of prepared material for lead removal.

4.2 For Dye Removal

With the fast development of industry, water pollution has become a leading global risk factor for human health [14]. The disposal of dyes without proper treatment will pollute both surface and groundwater as they are non-biodegradable and it will disrupt the growth of living organisms thereby causing illness, disease, and even human death. Some of the commonly reported dyes include methylene blue, bromophenol blue dye, congo red, methylene violet etc.

Liu et al. [57] developed a composite bead with the combination of carboxymethyl cellulose/k-carrageenan/activated montmorillonite for removing cationic methylene blue dye and reported the extent of adsorption by studying its kinetics and isotherms. The composite optimum ratio was found to be 1:1:0.4 for CMC:kC:AMMT with the removal rate of 98% of dye and follows Langmuir isotherm (R^2 0.999) fitted with pseudo-second-order kinetics. Fabryanty and his workers [34] prepared clay based composite using bentonite, alginate in three different compositions. These composites were tested for the removal of crystal violet dye using batch adsorption study. The adsorption of crystal violet dye by composite follows both Langmuir (q_{max} 601.9339 mg/g) and Freundlich (K_f 36.3399 $(mg/g)(L/mg)^{-n}$) model and the kinetic was well fitted with pseudo-second-order getting higher R^2 value of 0.9916 and reported the adsorption of dye by composite was controlled by chemisorption.

A composite bead of Chitosan/KSF-montmorillonite of weight ratio 1–25% w/w was reported recently by Pereira and his coworkers [70]. The adsorption capacity of the composite was tested by evaluating removal of remazol blue dye under batch operations by varying parameters: pH (2–8), contact time (0–660 min) and dye concentration (100–1600 mg L^{-1}). The results revealed effective removal of dye onto composite by following Langmuir model suggesting monolayer adsorption of dye onto composite. Uyar et al. [84] have successfully removed methylene blue dye using a composite of alginate–clay quasi-cryogel beads. They developed this composite by following a novel technique of cryogelation by deep-freezing the alginate beads at $-21\ °C$. The removal of dye was conducted using batch adsorption method and adsorption of methylene blue follows endothermic physical adsorption and well fitted with pseudo second-order kinetics.

Azha and his research group [8] reported industrial dye removal using composite prepared by coating commercial grade acrylic polymer emulsion (APE) and bentonite on cotton cellulosic fiber (CCF). This composite was effective to remove around 95% of cationic dye pollutant within 30 min by following pseudo-second-order kinetics. Hossein and Neda [42] developed hydrogel nanocomposite of acrylic acid and 2-acrylamido-2-methylpropane sulfonic acid monomers in presence of montmorillonite (MMT) by using ammonium persulfate as an initiator and methylene bisacrylamide as a crosslinker. The prepared composite was tested for methylene blue dye adsorption by varying the parameters of agitation time, MMT content, pH, initial dye concentration, adsorbent dose, and temperature. The sorption of dye will fit well with Redlich–Peterson model with the kinetic of pseudo-second-order.

El-Zahhar et al. [32] prepared polymeric clay nanocomposite was prepared using in situ polymerization by incorporating kaolinite clay into poly(acrylamide co-acrylic acid) with cross-linker. This composite was employed for removal of bromophenol blue (BPB) dye. The sorption study was conducted with the effect of parameters including time of contact, dye concentration and temperature. They reported that the dye removal was fitted with Lagergren (pseudo first order reaction rate) with free energy change ($\Delta G°$) value of -110.323 kJ/mol indicating the feasibility and spontaneous nature of sorption reaction. In another study, El-Zahhar [31] reported polymeric clay nanocomposite of polyacrylonitrile (PAN)-kaolin for the removal of methylene blue dye and chromium from aqueous solution. The adsorption capacity was calculated and reported as 127 mg/g of chromium and 68 mg/g methylene blue dye.

Bhattacharyya and Ray [12] synthesized hydrogel using chitosan and the copolymer of acrylic acid and acrylamide by free radical polymerization. This prepared hydrogel was filled with micro and nano-sized bentonite clay. The removal rate of dyes malachite green and methyl violet from water was reported. The percentage of dye removal was maximum for malachite green. The adsorption data were fitted with 1st and 2nd order kinetics and the combined Langmuir–Freundlich isotherm.

Darei et al. [26] prepared novel thin film composite by coating chitosan/organoclay on commercial polyvinylidene fluoride (PVDF) microfiltration membrane. Two differently modified nanoclay (Cloisite 15A and 30B) were introduced into chitosan aqueous solution with varying content ranged between 0.5 and 2 wt% in final dope solution. The prepared membranes were evaluated for dye removal and the results reported were as follows: (a) Membrane coated with Cloisite 15A/chitosan was more efficient for methylene blue removal from water. (b) Membrane-coated with Cloisite 30B/chitosan was effective for acidic orange 7 dye at acidic pHs. Anirudhan et al. [7] removed cationic dyes such as malachite green (MG), methylene blue (MB) and crystal violet (CV) from aqueous solutions at pH of 4.8 using nanocomposite of immobilized-amine-modified bentonite–polyacrylamide (HA-Am-Bent–PAA) nanocomposite. The removal rate was reported as 98% for malachite green, 97% for methylene blue and 94% for crystal violet. Saravanan et al. [76] have synthesized blend using chitin, bentonite clay and the resulted blend was crosslinked with glutaraldehyde. The blend was employed to remove heavy metals copper and chromium from dye effluent collected from Ambur industrial, India. The adsorption of both the metal ions follow Freundlich isotherm suggesting this blend is more effective and cost-effective to apply in a large-scale application.

4.3 For General Wastewater Treatment

The disposal of untreated effluent to the environment without proper removal of waste hazardous material will result in contaminating groundwater and various open sources of water such as a lake, pond and river. This will leads to change the colour of water bodies, increase the microbial growth so that these water bodies cannot be used for drinking as well as domestic household purposes. To overcome this issue clay composite were employed and some of the studies were discussed here. Unuabonah et al. [83] have prepared chitosan modified hybrid clay composite by combining clay, chitosan, $ZnCl_2$ and commercial Alum. This composite was applied for microbial removal from contaminated water. Within reasonable time complete adsorption of bacteria was reported. Anionic micropollutants from wastewater effluent were removed by treating them with composite prepared from quarternized polyvinylpyridinium-co-styrene (QPVPcS) to montmorillonite (MMT). Diclofenac (anionic pollutant) was effectively removed by the composite and filtration process was carried out by column study. The kinetic and the adsorption at equilibrium of diclofenac to composite was calculated using Langmuir equation. The results indicate effective removal was attained by the composite [49]. A similar micropollutant removal of diazinon was demonstrated by protonated poly (4-vinyl-pyridine-co-styrene) (HPVPcoS) and montmorillonite (MMT) clay using column study. This composite was able to remove with the efficiency of 100% having values of affinity and capacity coefficients K_L 5.6×10^4 L/mol and Q_{max} of 5.8×10^{-4} mol/g respectively [79].

An interesting composite material prepared using in situ polymerization was reported by Bunhu et al. [20]. Lignocellulose-clay nanocomposites were prepared by Bunhu and his group by varying the weight ratio of montmorillonite clay to lignocellulose ranged from 1:9 to 1:1 by grafting them with a novel combination of Poly(methyl methacrylate), Polymethacryloxypropyltrimethoxysilane and Poly (methacrylic acid). All these monomers were coupled with lignocellulose and clay. The presence of clay will increase the thermal stability of composite and they suggest it was highly suitable for wastewater treatment in future studies.

5 Future Challenges

Like most of the newly emerging disciplines, the area of research dealing with polymer-clay nanocomposite also has opportunities and challenges. The most difficult task for researchers to deal with polymer clay nanocomposite was not only to prepare them with superior performance but also to check biocompatible nature of the resulting material as it should possess acceptable biological function. In general, polymer nanocomposite with nontoxic performance in vitro may not necessary to perform biocompatible in vivo. Hence on developing clay reinforced polymer composite for biomedical application includes necessary requirement to succeed both in vitro as well as in vivo. Another interesting challenge for researchers to deal with polymer-clay nanocomposite was to understand the fundamental interactions between the polymer matrix and inorganic clay filler as it plays a primary role in controlling the structure-property relationship.

6 Conclusion

We hope this chapter was informative and useful for the researchers who want to use clay minerals in combination with polymer as potential adsorbent and as effective material to apply in various biomedical applications. We have discussed more recent papers which will highlight the advancement of clay based composites in both medical and non-medical field. As stated in the introduction, these materials have likely been in use for quite some time already, but as the chemist and materials scientist become better at designing the system through fundamentals, new products and applications utilizing this technology will grow in number and capability. Though an attempt to address the health and environmental application of polymeric clay composites has been taken in this chapter, there is yet lot more to be done in this field. We address the future challenges in this chapter hence it was necessary to provide continuous effort on the theoretical as well as applied work of modified clay based polymer nanocomposites globally both in academia and industry.

References

1. Abdeen R, Salahuddin N (2013) Modified chitosan-clay nanocomposite as a drug delivery system intercalation and in vitro release of ibuprofen. J Chem 2013:1–9
2. Afsar HA, Ghaee A (2016) Preparation of aminated chitosan/alginate scaffold containing halloysite nanotubes with improved cell attachment. Carbohyd Polym 151:1120–1131
3. Aguzzi C, Cerezo P, Viseras C, Caramella C (2007) Use of clays as drug delivery systems: possibilities and limitations. Appl Clay Sci 36(1–3):22–36
4. Ahmad R, Hasan I (2017) L-methionine montmorillonite encapsulated guar gum-g-polyacrylonitrile copolymer hybrid nanocomposite for removal of heavy metals. Groundwater Sustain Develop 5:75–84
5. Ahsan SM, Thomas M, Reddy KK, Sooraparaju SG, Asthana A, Bhatnagar I (2018) Chitosan as biomaterial in drug delivery and tissue engineering. Int J Biol Macromol 110:97–109
6. Ambrogi V, Pietrella D, Nocchetti M, Casagrande S, Moretti V, De Marco S, Ricci M (2017) Montmorillonite-chitosan-chlorhexidine composite films with antibiofilm activity and improved cytotoxicity for wound dressing. J Colloid Interface Sci 491:265–272
7. Anirudhan TS, Suchithra PS, Radhakrishnan PG (2009) Synthesis and characterization of humic acid immobilized-polymer/bentonite composite and their ability to adsorb basic dyes from aqueous solutions. Appl Clay Sci 43:336–342
8. Azha SF, Shahadat M, Ismail S (2017) Acrylic polymer emulsion supported bentonite clay coating for the analysis of industrial dye. Dyes Pigm 145:550–560
9. Barlas FB, AgSeleci D, Ozkan M, Demir B, Seleci M, Aydin M, Tasdelen MA, Zareie HM, Timur S, Ozcelik S, Yagci Y (2014) Folic acid modified clay/polymer nanocomposites for selective cell adhesion. J Mater Chem B 2:6412–6421
10. Bera H, Reddylppaguntu S, Kumar S, Vangala P (2017) Core-shell alginate-ghatti gum modified montmorillonite composite matrices for stomach-specific flurbiprofen delivery. Mater Sci Eng, C 76:715–726
11. Besombes J-L, Cosnier S, Labbe P, Reverdy G (1995) Improvement of the analytical characteristics of an enzyme electrode for free and total cholesterol via laponite clay additives. Anal Chim Acta 317(1–3):275–280
12. Bhattacharyya R, Ray SK (2014) Micro- and nano-sized bentonite filled composite superabsorbents of chitosan and acrylic copolymer for removal of synthetic dyes from water. Appl Clay Sci 101:510–520
13. Bhowmick A, Banerjee SL, Pramanik N, Jana P, Gnanamani A, Das M, Paban Kundu P (2018) Organically modified clay supported chitosan/hydroxyapatite—zinc oxide nanocomposites with enhanced mechanical and biological properties for the application in bone tissue engineering. Int J Biol Macromol 106:11–19
14. Bolisetty S, Mezzenga R (2016) Amyloid-carbon hybrid membranes for universal water purification. Nat Nanotechnol 11(4):365–371
15. Bonifacio MA, Gentile P, Ferreira AM, Cometa S, De Giglio E (2017) Insight into halloysite nanotubes-loaded gellan gum hydrogels for soft tissue engineering applications. Carbohyd Polym 63:280–291
16. Bounabi L, Mokhnachi NB, Haddadine N, Ouazi F, Barille R (2016) Development of poly (2-hydroxyethyl methacrylate)/clay composites as drug delivery systems of paracetamol. J Drug Deliv Sci Technol 33:58–65
17. Boyer C, Figureiredo L, Pace R, Lesoeur J, Rouillon T, Visage CL, Tassin JF, Weiss P, Guicheux J, Rethore G (2018) Laponite nanoparticle-associated cellulose as an injectable reinforced interpenetrating network hydrogel for cartilage tissue engineering. Acta Biomater 65:112–122
18. Bremere I, Kennedy M, Stikker A, Schippers J (2001) Growth of 200% predicted for desalination in waterscarce countries by 2025. Int Desalination Water Reuse Q 11(2):7–12
19. Buffa SD, Bonini M, Ridi F, Severi M, Losi P, Volp S, Al Kayal T, Soldani G, Baglioni P (2015) Design and characterization of a composite material based on Sr (II)-loaded clay nanotubes included within a biopolymer matrix. J Colloid Interface Sci 448:501–507

20. Bunhu T, Chaukura N, Tichagwa L (2016) Preparation and characterization of polymer-grafted montmorillonite-lignocellulose nanocomposites by in situ intercalative polymerization. J Appl Chem 2016:1–8
21. Chang CW, Van Spreeuwel A, Zhang C, Varghese S (2010) PEG/clay nanocomposite hydrogel: a mechanically robust tissue engineering scaffold. Soft Matter 6(20):5157–5164
22. Chappidi DY, Mills DK (2016) Tissue engineering nanoclay composite scaffolds composed of poly-glycerol sebacate and poly-caprolactone. In: 32nd southern biomedical engineering conference
23. Chiu C-W, Huang T-K, Wang Y-C, Alamani BG, Lin J-J (2014) Intercalation strategies in clay/polymer hybrids. Prog Polym Sci 39:443–485
24. Choudhury PR, Mondal P, Majumdar S (2015) Synthesis of bentonite clay based hydroxyapatite nanocomposites cross-linked by glutaraldehyde and optimization by response surface methodology for lead removal from aqueous solution. RSC Adv 5:100838–100848
25. Chrzanowski W, Yunsun Kim S, Abou Neel EA (2013) Biomedical applications of clay. Aust J Chem 66:1315–1322
26. Darei P, Siavash Madeeni S, Salehi E, Ghaemi N, Ghari HS, Khadivi MA, Rostami E (2013) Novel thin film composite membrane fabricated by mixed matrix nanoclay/chitosan on PVDF microfiltration support: preparation, characterization and performance in dye removal. J Membr Sci 436:97–108
27. Dawson JI, Kanczler JM, Yang XBB, Attard GS, Oreffo ROC (2011) Skeletal regeneration: application of nanotopography and biomaterials for skeletal stem cell based bone repair. Adv Mater 23:3304–3308
28. Demir AK, Elcin AE, Elcin YM (2018) Strontium-modified chitosan/montmorillonite composites as bone tissue engineering scaffold. Mater Sci Eng, C 89:8–14
29. Devi Nirmla, Dutta Joydeep (2017) Preparation and characterization of chitosan-bentonite nanocomposite films for wound healing application. Int J Biol Macromol 104:1897–1904
30. Edathil AA, Pal P, Banat F (2018) Alginate clay hybrid composite adsorbents for the reclamation of industrial lean methyldiethanolamine solutions. Appl Clay Sci 156:213–223
31. El-Zahhar AA (2015) A polymer-organoclay nanocomposite for simultaneous removal of Chromium(vi) and organic dyes. Eur Chem Bull 4:10–12
32. El-Zahhar AA, Awaad NS, El-Katori E (2014) Removal of bromophenol blue dye from industrial waste water by synthesizing polymer-clay composite. J Mol Liq 199:454–461
33. Emre FB, Kesik M, Kanik FE, Akpinar HZ, Alsan-Gurel E, Rossi RM, Toppare L (2015) A benzimidazole-based conducting polymer and a PMMA–clay nanocomposite containing biosensor platform for glucose sensing. Synth Met 207:102–109
34. Fabryanty R, Valencia C, Soetaredjo FE, Nyooputro J, Santoso SP, Kumiawan A, Ju Y-H, Ismadji S (2017) Removal of crystal violet dye by adsorption using bentonite—alginate composite. J Environ Chem Eng 5(6):5677–5687
35. Fan L, Zhang J, Wang A (2013) In situ generation of sodium alginate/hydroxyapatite/halloysite nanotubes nanocomposite hydrogel beads as drug-controlled release matrices. J Mater Chem B 45:6261–6270
36. Gaharwar AK, Mihaila SM, Swami A, Patel A, Sant S, Reis RL, Marques AP, Gomes ME, Khademhosseini A (2013) Bioactive silicate nanoplatelets for osteogenic differentiation of human mesenchymal stem cells. Adv Mater 25(24):3329–3336
37. Gaharwar AK, Kishore V, Rivera C, Bullock W, Wu CJ, Akkus O, Schmidt G (2012) Physically crosslinked nanocomposites from silicate-crosslinked PEO: mechanical properties and osteogenic differentiation of human mesenchymal stem cells. Macromol Biosci 12 (6):779–793
38. Ghadiri M, Chrzanowski W, Rohanizadeh R (2015) Biomedical applications of cationic clay minerals. RSC Adv 5:29467
39. Gunister E, Pestreli D, Unlu CH, Atici O, Gungor N (2007) Synthesis and characterization of chitosan–MMT biocomposite systems. Carbohydr Polym 67:358–365

40. Han C, Zhao A, Varughese E, Sahle-Demessie E (2018) Evaluating weathering of food packaging polyethylene-nano-clay composites: release of nanoparticles and their impacts. NanoImpact 9:61–71
41. Heydari A, Khoshnood H, Sheibani H, Doostan F (2017) Polymerization of β-cyclodextrin in the presence of bentonite clay to produce polymer nanocomposites for removal of heavy metals from drinking water. Polym Adv Technol 28(4):524–532
42. Hossein H, Neda K (2015) Removal of cationic dyes by poly(AA-co-AMPS)/montmorillonite nanocomposite hydrogel. Desalination Water Treat 1–12
43. Huang B, Liu M, Zhou C (2017) Chitosan composite hydrogels reinforced with natural clay nanotubes. Carbohyd Polym 175:689–698
44. Huang B, Liu M, Long Z, Shen Y, Zhou C (2017) Effects of halloysite nanotubes on physical properties and cytocompatibility of alginate composite hydrogels. Mater Sci Eng C 70 (1):303–310
45. Jafarbeglou M, Abdouss M, Shoushtari AM, Jafarbeglou M (2016) Clay nanocomposites as engineered drug delivery systems. RSC Adv 6:50002–50016
46. Ji L, Qiao W, Zhang Y, Wu H, Miao S, Cheng Z, Gong Q, Liang J, Zhu A (2017) A gelatin composite scaffold strengthened by drug-loaded halloysite nanotubes. Mater Sci Eng C 78:362–369
47. Kevadiya BD, Rajkumar S, Bajaj H (2015) Application and evaluation of layered silicate–chitosan composites for site specific delivery of diclofenac. Biocybernetics Biomed Eng 35 (2):120–127
48. Khilbsuwan R, Siepmann F, Siepmann J, Pongjanyakul T (2017) Chitosan-clay nanocomposite microparticles for controlled drug delivery: effects of the MAS content and TPP crosslinking. J Drug Deliv Sci Technol 40:1–10
49. Kohay H, Izbitski A, Mishael YG (2015) Developing polycation-clay sorbents for efficient filtration of diclofenac: effect of dissolved organic matter and comparison to activated carbon. Environ Sci Technol 49(15):9280–9288
50. Kohay H, Izbitski A, Mishael YG (2017) PEG-PE/clay composite carriers for doxorubicin: effect of composite structure on release, cell interaction and cytotoxicity. Acta Biomater 55:443–454
51. Kotal M, Bhowmick AK (2015) Polymer nanocomposites from modified clays: recent advances and challenges. Prog Polym Sci 51:127–187
52. Kurczewska J, Pecyna P, Ratajzak M, Gajecka M, Schroeder G (2017) Halloysite nanotubes as carriers of vancomycin in alginate-based wound dressing. Saudi Pharm J 25(6):911–920
53. Lal S, Perwez A, Rizvi MA, Datta M (2017) Design and development of a biocompatible montmorillonite PLGA nanocomposites to evaluate in vitro oral delivery of insulin. Appl Clay Sci 147:69–79
54. Langer R, Vacanti JP (1993) Tissue engineering. Science 260(5110):920–926
55. Liu M, Dai L, Shi H, Xiong S, Zhou C (2015) In vitro evaluation of alginate/halloysite nanotube composite scaffolds for tissue engineering. Mater Sci Eng C 49:700–712
56. Liu M, Zheng H, Chen J, Lim S, Huang J, Zhou C (2016) Chitosan-chitin nanocrystal composite scaffolds for tissue engineering. Carbohyd Polym 152:832–840
57. Liu C, Omer AM, Ouyan X (2018) Adsorptive removal of cationic methylene blue dye using carboxymethyl cellulose/k-carrageenan/activated montmorillonite composite beads: isotherm and kinetic studies. Int J Biol Macromol 106:823–833
58. Lundin JG, McGann CL, Daniels GC, Streifel BC, Wynne JH (2017) Hemostatic kaolin-polyurethane foam composites for multifunctional wound dressing applications. Mater Sci Eng C 79:702–709
59. Madusanka N, Nalin de Silva KM, Amaratunga G (2015) A curcumin activated carboxymethyl cellulose–montmorillonite clay nanocomposite having enhanced curcumin release in aqueous media. Carbohyd Polym 134:695–699
60. Mallouk TE, Gavin JA (1998) Molecular recognition in lamellar solids and thin-films. Acc Chem Res 31(5):209–217

61. Massaro M, Colletti CG, Noto R, Riela S, Poma P, Guernelli S, Parisi F, Milioto S, Lazzara G (2015) Pharmaceutical properties of supramolecular assembly of co-loaded cardanol/triazole-halloysite systems. Int J Pharm 478:476–485
62. Nair BP, Sindhu M, Nair PD (2016) Polycaprolactone-laponite composite scaffold releasing strontium ranelate for bone tissue engineering applications. Colloids Surf B 143:423–430
63. Naumenko EA, Guryanov ID, Yendluri R, Lvov YM, Fakhrullin RF (2016) Clay nanotube—biopolymer composite scaffolds for tissue engineering. Nanoscale 8:7257–7271
64. Nistor MT, Vasile C, Chiriac AP (2015) Hybrid collagen-based hydrogels with embedded montmorillonite nanoparticles. Mater Sci Eng C 53:212–221
65. Noori MK, Hassan ZM (2015) Nanoclay enhanced the mechanical properties of Poly(Vinyl Alcohol)/Chitosan/Montmorillonite nanocomposite hydrogel as wound dressing. Procedia Mater Sci 11:152–156
66. Olad A, Azhar FF (2014) The synergetic effect of bioactive ceramic and nanoclay on the properties of chitosan–gelatin/nanohydroxyapatite–montmorillonite scaffold for bone tissue engineering. Ceram Int 40(7):10061–10072
67. Onyango MS, Mbakop S, Leswifi TY, Mthombeni NH, Setshedi K (2016) Application of polymer-natural clay composite in water treatment. In: Proceedings of the 2016 annual conference on sustainable research and innovation
68. Park JK, Choy YB, Oh JM, Kim JY, Hwang SJ, Choy JH (2008) Controlled release of donepezil intercalated in smectite clays. Int J Pharm 359:198–204
69. Perez CJ, Alvarez VA, Vazquez A (2008) Creep behavior of layered silicate/starch-polycaprolactone blends nanocomposites. Mater Sci Eng A 480:259–265
70. Pereira FAR, Sousa KS, Cavalcanti GRS, Franca DB, Quieroga LNF, Santos IMG, Fonseca MG, Jaber M (2017) Green biosorbents based on chitosan-montmorillonite beads for anionic dye removal. J Environ Chem Eng 5(4):3309–3318
71. Piri S, Zanjani ZA, Piri F, Zamani A, Yaftian M, Davari M (2016) Potential of polyaniline modified clay nanocomposite as a selective decontamination adsorbent for Pb(II) ions from contaminated waters; kinetics and thermodynamic study. J Environ Health Sci Eng 14:1–20
72. Rafiei HR, Shirvani M, Ogunseitan OA (2016) Removal of lead from aqueous solutions by a poly(acrylic acid)/bentonite nanocomposite. Appl Water Sci 6(4):331–338
73. Sabbagh N, Akbari A, Arsalani N, Eftekhari-Sis B, Hamishekar H (2017) Halloysite-based hybrid bionanocomposite hydrogels as potential drug delivery systems. Appl Clay Sci 148:48–55
74. Salcedo I, Sandri G, Aguzzi C, Bonferoni C, Cerezo P, Sanchez-Espejo P, Viseras C (2014) Intestinal permeability of oxytetracycline from chitosan-montmorillonite nanocomposites. Colloids Surf B 117:441–448
75. Samani F, Kokabi M, Soleimani M, Valojerdi MR (2010) Fabrication and Characterization of electrospun fibrous nanocomposite scaffolds based on poly(lactide-co-glycolide)/poly(vinyl alcohol) blends. Polym Int 59:901–909
76. Saravanan D, Hemalatha R, Sudha PN (2011) Synthesis and characterization of crosslinked chitin/bentonite polymer blend and adsorption studies of Cu (II) and Cr (VI) on chitin
77. Sarfraz A, Warsi MF, Sarwar MI, Ishaq M (2012) Improvement in tensile properties of PVC-montmorillonite nanocomposites through controlled uniaxial stretching. Bull Mater Sci 35(4):539–544
78. Sarkar T, Narayanan N, Solanki PR (2017) Polymer-clay nanocomposite-based acetylcholine esterase biosensor for organophosphorous pesticide detection. Int J Environ Res 11(5–6):591–601
79. Shabtai IA, Mishael YG (2017) Catalytic polymer-clay composite for enhanced removal and degradation of diazinon. J Hazard Mater 335:135–142
80. Sirousazar M, Jahani-Javanmardi A, Kheiri F, Hassan ZM (2016) In vitro and in vivo assays on egg white/polyvinyl alcohol/clay nanocomposite hydrogel wound dressings. J Biomater Sci 1–30
81. Styan KE, Martin DJ, Poole-Warren LA (2008) In vitro fibroblast response to polyurethane organosilicate nanocomposites. J Biomed Mater Res A 86(3):571–582

82. Tiggemann HM, Tomacheski D, Celso F, RIbeiro VF, Bachtigall SMB (2013) Use of wollastonite in a thermoplastic elastomer composition. Polym Testing 32(8):1373–1378
83. Unuabonah EI, Adewuyi A, Kolawole MO, Omorogie MO, Olatunde OC (2017) Disinfection of water with new chitosan-modified hybrid clay composite adsorbent. Heliyon 3(8):e00379
84. Uyar G, Kaygusuz H, Erim FB (2016) Methylene blue removal by alginate–clay quasi-cryogel beads. React Funct Polym 106:1–7
85. Viseras C, Cerezo P, Sanchez R, Salcedo I, Aguzzi C (2010) Current challenges in clay minerals for drug delivery. Appl Clay Sci 48:291–295
86. Viseras C, Aguzzi C, Cerezo P, Bedmar MC (2008) Biopolymer—clay nanocomposites for controlled drug delivery. Mater Sci Technol 24(9):1020–1026
87. Vyas V, Kaur T, Thirugnanam A (2017) Chitosan composite three dimensional macrospheric scaffolds for bone tissue engineering. Int J Biol Macromol 104:1946–1954
88. Wu T-M, Wu C-Y (2006) Biodegradable poly(lactic acid)/chitosan-modified montmorillonite nanocomposites: preparation and characterization. Polym Degrad Stab 91:2198–2204
89. Yang C, Xue R, Zhang QS, Yang S, Liu P, Chen L, Wang K, Zhang X, Wei Y (2017) Nanoclay cross-linked semi-IPN silk sericin/poly(NIPAm/LMSH) nanocomposite hydrogel: an outstanding antibacterial wound dressing. Mater Sci Eng C 81:303–313
90. Yang J-H, Lee J-H, Ryu H-J, Elzatahry AA, Alothman ZA, Choy J-H (2016) Drug–clay nanohybrids as sustained delivery systems. Appl Clay Sci 130:20–32
91. Zanetti M, Lomakin S, Camino G (2000) Polymer layered silicate nanocomposites. Macromol Mater Eng 279(1):1–9

Thermal Behaviour and Crystallization of Green Biocomposites

Vasile Cristian Grigoras

List of Abbreviations

APTES	3-aminopropyltriethoxysilane
Ac-CNC	Acetate cellulose nanocrystals
ACNC	Acetylated cellulose nanocrystals
ATBC	Acetyltributyl citrate ATBC
A-sisal	Alkali treated sisal fibers
BC	Bacterial cellulose
BCN	Bacterial nanocellulose
BF	Bamboo fibers
BPU	Biobased polyurethane
CA	Citric acid CA
CE	Cellulose
CF	Cellulose fibers
ChNC	Chitin nanocrystals
CO	Cotton
CNC	Cellulose nanocrystals
CNCSD	Conventional spray dried cellulose nanocrystals
CNCFD	Freeze dried cellulose nanocrystals
C18-g-CNC	Cellulose nanocrystals grafted with long alkyl chain (C18)
CNF	Cellulose nanofibers
CNC-g-PLLA	Poly(L-lactide)-grafted-cellulose nanocrystals
CNW	Cellulose nanowhiskers
C_p	Heat capacity
ΔC_p	Heat capacity step (at glass transition)
ΔE_a	Activation energy
DDMSiCl	N-dodecyldimethylchlorosilane
DMA	Dynamic mechanical analysis
DSC	Differential scanning calorimetry

V. C. Grigoras (✉)
"Petru Poni" Institute of Macromolecular Chemistry, 700487 Iassy, Romania
e-mail: crgrig@icmpp.ro

© Springer Nature Switzerland AG 2019
Inamuddin et al. (eds.), *Sustainable Polymer Composites and Nanocomposites*,
https://doi.org/10.1007/978-3-030-05399-4_41

DTG	Derivative thermogravimetry
DTA	Differential thermal analysis
ENR	Epoxidized natural rubber
EVA	Ethylene vinyl alcohol
GC	Green composite
GLU	Glutaraldehyde
Gly	Glycerol
GTA	Glycerol triacetate
GMA	Glycidyl methacrylate
GPS	3-glycidoxypropyltrimethoxy silane
dH/dt	Enthalpy variation in time
HAlk	Alkalinized hemp fibers
HCE	Hydrolysed cellulose
HF	Hemp fibers
HW	Hard wood
ΔH	Enthalpy
ΔH_m	Melting enthalpy
ΔH_c	Crystallization enthalpy
KF	Kenaf fibers
Kraft	Bleached kraft softwood
LA-CNC	Lactate cellulose nanocrystals
LDI	Lysine-based diisocyanate
LDPE	Low density polyethylene
MA	Maleic anhidride
MBC	Modified bamboo cellulose
MC	Microcrystalline cellulose
MFC	Microfibrilated cellulose
MRSF	Modified rice straw fibers
NBSK	Black spruce and northern bleached softwood kraft
ODI	Octadecyl isocyanate
PBA	Poly(butyl acrylate)
PBAT	Poly(butylene adipate-co-terephthalate)
PBI	4-phenylbutyl isocyanate)
PBSu	Poly(butylene succinate)
PCL	Poly(ε-caprolactone)
PFA	Polyfurfuril alchohol
PHB	Poly(hydroxy butyrate)
PHBV	Poly(hydroxy butyrate-co-valerate)
PL	Plastified lignin
PLA	Poly(lactic acid)
PLA-g-CNC	Poly(lactic acid)-grafted-cellulose nanocrystals
PLA-g-MA	Poly(lactic acid-grafted-maleic anhydride)
PLLA	Poly(L-lactide)
PLM	Polarizing light microscopy

PPC	Poly(propylene carbonate)
PP	Poly(propylene)
PP-g-MA	Poly(propylene-grafted-maleic anhydride)
PU	Polyurethane
PVC	Poly(vinyl chloride)
PVA	Poly(vinyl alchohol)
PVAc	Poly(vinyl acetate)
RF	Ramie fibers
RH	Rice husk
RS	Rice straw
SEBS	Styrene-ethylene-butadiene-styrene
SEBS-g-MA	Maleic anhydride-grafted-styrene-ethylene-butadiene-styrene
SCF	Standard size cellulose fibers
SPA	Anhydride plasticized soy protein
S-sisal	Sylane treated sisal fibers
SW	Softwood
TAC	Triacetate citrate TAC
TDI	Toluene isocyanate TDI
TGA	Thermogravimetric analysis
T_c	Crystallization temperature
$T_{c(onset)}$	Crystallization onset temperature
T_{cc}	Cold crystallization temperature
T_g	Glass transition
T_m	Melting temperature
T_m^o	Equilibrium melting point
T_{max}	Temperature of maximum decomposition rate
TPS	Thermoplastic starch
ΔS	Entropy
U-sisal	Untreated sisal fibers
χ_c	Crystallinity index
σ_e	Fold surface free energy

1 Introduction

A general thermal behaviour of the green composites (GCs) can provide important results which if are well understood, represent the key factor in the choice of the material applications range. In this regard, speciality literature discussed most important aspects generally related to the glass transition range, melting or crystallization behaviour of the GC either to the range of temperatures that cause degradation of the organic structure of matrix or natural fillers. Reported literature results show that the temperature generally not only degrades the material structure but affects the most properties of GC, too.

Thermal properties of a GC are commonly determined by the physical and chemical characteristics of main constituents—natural fibers and polymer matrix, but interactions between them represent a decisive role. A gap between the filler and the matrix means that the heat is less conducted through the material, but a good interfacial bonding, as well as a better bonding, is related with a modification of GC thermal properties. An efficient way to improve the physical properties of GC consists in the chemical modification of natural fillers surfaces in order to facilitate interactions at the interface with the polymer matrix [1].

The most important results related to the thermal behaviour of various GCs, available by literature from last years will be reviewed in this chapter. Specifically, the literature data concerned to both on temperature range defined by thermal stability as well as in the temperature range where thermally degradation occurs will be analyzed. The main topics related to the effects of the interface between natural fillers and the polymeric matrix on the thermal behaviour will be highlighted. In this regard, the modifications induced by fillers loading on glass transition, melting and crystallization behaviour on the one hand, as well modifications induced on thermal degradation or on degradation mechanism, on the other hand, will be discussed.

2 Thermal Analysis as an Analytical Method of Green Composites Characterization

Thermal analysis unifies a lot of well-defined experimental procedures and tests; the main purpose of them consist on the qualitative and quantitative evaluation of the major changes induced on physicochemical processes and structure of the material that is subject to a temperature change. Actually, the temperature is known as fundamental state variable parameter that influences chemical reactions, physical properties or structural transformations of a material. Considering this, it can support as a general concept that any scientific and experimental method through which can be measured the effects produced by the temperature variations can be considered as thermal analysis. Thermal properties of a material can be analyzed by several methods, but in this chapter, we refer just to differential scanning calorimetry (DSC) and thermogravimetric analysis (TGA) and derivative mass loss (DTG). For a well understanding of GC thermal behaviour, a short review of these methods will precede the presentation of experimental data.

2.1 Differential Scanning Calorimetry

DSC measures the heat effects associated with phase transitions and chemical reactions function of temperature or time in a dynamic or isothermal program, respectively. In this method (Fig. 1), the heat flow difference between the sample and a reference at the same temperature is recorded as a function of temperature [2].

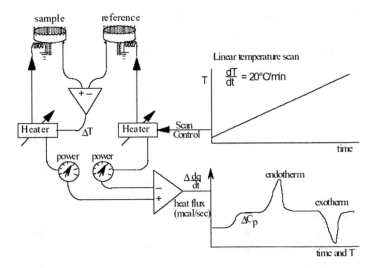

Fig. 1 Schematic representation of DSC physical principle. Adapted from Mathot [2]

The reference is an inert material such as alumina, or just an empty aluminium pan. The physical principle of this method assumes the measuring of the heat flow through material subjected on heating or cooling at a constant rate (in dynamic measurements) or at a constant temperature for a time (in isothermal measurements). In the endothermic process (e.g. most phase transitions), the heat is absorbed and normally the heat flow through the sample is higher compared with the reference: hence enthalpy variation in time (dH/dt) is accepted to have a positive value. The opposite situation happens in the exothermic process where the heat is released and dH/dt is negative as well. Reported papers from the literature that contain data provided by DSC analysis show many results, but the most relevant parameters which intensively discussed and describe a general material thermal behaviour are:

- the temperature for thermal transitions: T_g, T_c, and T_m;
- enthalpy ΔH and entropy ΔS variation assigned to thermal transition;
- heat capacity C_p and step of the heat capacity at glass transition ΔC_p;
- crystallinity index: χ_c (%) (assuming that ΔH_m is known);
- heat absorbed or evolved during cure reactions or decomposition process;
- sub-glass or solid state transitions (rarely, but relevant in some studies);
- crystallization kinetics (heat evolved during isothermal crystallization).

Because DSC method is intensively used in calorimetry and polymer science, a series of physical processes and chemical reaction investigated are listed in Table 1 [3].

Table 1 Physical processes in materials evaluated by the DSC method

Process/transition	Exothermal	Endothermal	Parameter
Solid solid transition	▲	▼	T
Glass transition			T, ΔC_p
Crystallization	▲		T_c, ΔH_c
Melting		▼	T_m, ΔH_m
Vaporization		▼	ΔH
Sublimation		▼	ΔH
Adsorbtion	▲		ΔH
Desorbtion		▼	ΔH
Absorbtion		▼	ΔH
Dehydration		▼	ΔH
Solid solid reaction	▲	▼	ΔH
Solid-liquid reaction	▲	▼	ΔH
Solid-gas reaction	▲	▼	ΔH
Oxidative degradation	▲		ΔH
Reduction		▼	ΔH
Oxidation	▲		ΔH
Polymerisation	▲		ΔH
Crosslinking	▲		ΔH
Catalytic reaction	▲		ΔH
Combustion	▲		ΔH

Adapted from Turi [3]

2.2 Thermogravimetric Analysis

The thermogravimetric analysis involves the heating of a sample at a constant rate up to the completed decomposition and monitoring the weight and rate of mass loss function of temperature. It is commonly used to monitor the polymer degradation reactions. TGA only requires a sensitive method for monitoring weight and the sample temperature. Ideally, the sample chamber should have a controllable environment/atmosphere to monitor the degradation process under various conditions (e.g. in the presence of oxygen or under dry nitrogen). The operation principle of TGA instrument is shown in Fig. 2 [3].

In the most accurate TGA experiments, the temperature is set to a constant value and then the sample weight has monitored the function of time. Thus, the results give quantitative information about the rate of composite degradation process at that temperature value. Repeating this experiment for several temperatures will give information about the temperature range under which this material can be used. Usually, these experiments take a long time and useful results have been obtained if the degradation occurs at a slow rate. The sample seems to resist at higher temperatures less than it would actually survive under long-term exposure at lower

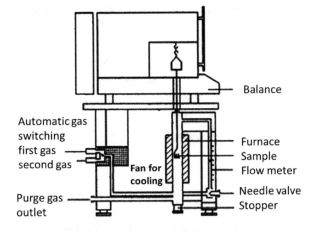

Fig. 2 Schematic drawn of TGA instrument. Adapted from Turi [3]

temperatures, thus the results obtained will be very sensitive to scanning rate. The sensitivity is large because degradation rates may be slow compared to typical scanning rates such as 10–20 °C/min. By coupling the degradation products resulted from a TGA experiment to the inlet channel of the other instruments such as a mass spectrometer, infrared spectrometer or to a pyrolysis instrument, the obtained data can be useful in the study of polymer degradation mechanisms. Thus, the coupled instruments help to identify the structure of degradation products. By a combination of the structural information with those provided by kinetics analysis, a complete picture of the degradation process results.

TGA experiments provided valuable results for:

- determination of thermal stability of a composite matrix;
- the compositional analysis of composite;
- study of thermo-oxidative degradation processes;
- determination of volatile products (paraffinic, aromatic, plasticizer, solvents, moisture, cross-linking agents and unreacted accelerators) from composites;
- determination of carbon black content;
- kinetic studies used for assignment of the thermal degradation mechanism of the composite.

The poor thermal resistance of GCs shows significant disadvantages of these materials since, depending on the application, GCs must pass regulatory fire tests. For example, the thermal degradation of natural fibers begins at approximately 200 °C, what substantially limits the number of suitable polymers used as a matrix for GCs fabrication. The thermal properties of a GC are mainly governed by the characteristics of natural fibers, but they are also dependent on the matrix polymer and the interactions between these main constituents. Figure 3 shows the stages of a typical combustion in GCs.

The interface between the matrix and the reinforcement fiber in green composites is responsible not just for mechanical properties, but also for the thermal

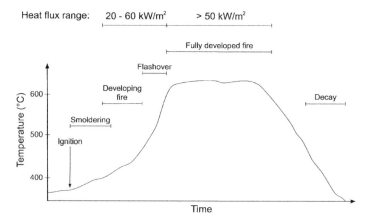

Fig. 3 Stages of combustion in green composites. Reproduced from Väisänen et al. [4]

stability of the material subjected to heat. The thermal stability of composite is enhanced by improving this interface between polymer matrix and fibers, meaning a higher energy is required to separate the constituents. Unlike to this reality, a poor interface in composite decrease the thermal resistance to heat, and finally the thermal stability is reduced [4].

An important issue on thermal degradation of a material consists of the atmosphere which influences the thermogravimetric results leading to a clearly different behaviour. Usually, the experiments can be conducted both in an inert atmosphere (helium or nitrogen) as well as in oxidative atmosphere (air or oxygen). Further, the literature recorded many works in this field that present different results from TGA analysis conducted at different flows of gases (nitrogen). By using an inert atmosphere, results of cellulose thermal degradation showed that the main degradation mechanism (a single DTG peak) was assigned to the formation of macromolecules containing rings with double bonds. When an oxidative atmosphere was used, the main peak in DTG was partially overlapped with the exothermic peak from DTA, assigned to the oxygen reaction with the cellulose. Accordingly, the main DTG peak was shifted to lower temperatures in the oxidative atmosphere as compared to the inert one [5].

3 Glass Transition and Physical Ageing of Green Biocomposites

The glass transition temperature—namely T_g represents the temperature when amorphous (noncrystalline) polymers are converted from a brittle, glass-like form to a rubbery, flexible form.

This is not a true phase transition, but one that involves a change in the local degrees of freedom, assigned to change in heat capacity (C_p) of the composite. Above the glass transition temperature, certain segmental motions of the polymer are comparatively unhindered by the interaction with neighbouring chains. Below the glass transition temperature, such motions are greatly hindered, and the relaxation times associated with these hindered motions are usually long compared to the duration of the experiment. The operative definition of glass transition temperature is that value, with a tolerance of few degrees, to which the specific heat, coefficient of thermal expansion, free volume and the dielectric constant (in the case of a polar polymer) all change rapidly. Since the mechanical behaviour of polymers markedly changes at the glass transition temperature, it is an important characteristic of every polymer. In the DSC experiment, the glass transition shows a step in the baseline, indicating a change in the heat capacity of the composite. No enthalpy is associated with this transition (reason due it is also called second-order transition); therefore, the effect in a DSC curve is slight, but observable and measurable.

A series of factors can affect the behaviour of composites in glass transition region imposed by the chemical structure of polymer matrix (main chain flexibility, stereoregularity, lateral voluminous groups, crosslinking, molecular mass) or by the effects of plasticizers or crystallinity. But a substantial number of relevant works from speciality literature contains interesting results about the interface effects between the polymer matrix and natural fillers on glass transition temperature; therefore we refer to these in the next paragraph.

Recent work recorded the effect of the interface on specific heat capacity and glass transition of poly(butylene succinate)-hemp GCs (PBS-hemp) [6]. Here is highlighted the role of specific intermolecular interactions present at the interface on the values of specific heat capacity data as resulted from DSC. The maximum of derivative of specific heat capacity signal from DSC is assigned to glass transition temperature as is known. In this study, the presence of two amorphous phases—mobile one in bulk and rigid one at interface explained the differences degree mobility of PBS amorphous segments. The rigid amorphous phase limited the relaxation of chain segments inducing a broadening of glass transition region. The effect of oil palm fibers on glass transition temperature of biocomposites was reviewed [7]. A higher content of oil palm in phenol-formaldehyde matrix increased the void formation, facilitating thus the chain mobility at lower temperatures that decreased the glass transition temperature. The opposite effect of oil palm was recorded in the polyurethane-based composites or in polyvinyl chloride (PVC)-epoxidized natural rubber composites, where the glass transition temperature increased with the fiber content. In addition, the chemical treatment (benzoylation) of fibers reduced the glass transition temperature due to a plasticization effect of fibers that diffused or dissolved into the PVC matrix.

Cuinat-Guerraz et al. [8] showed in their work that by a hydrothermal ageing treatment of the bio-epoxy/flax and the polyurethane/flax composites, the glass transition temperature was changed. There was remarked an important reduction of T_g value (15.1 ± 3.2%) for the bio-epoxy/flax, but smaller one (10.7 ± 5.8%) for the polyurethane/flax composites. This reduction was explained by authors based on the water absorption at the interface, which acts as a plasticizer. Once again, the key role played by the interface in the polymeric biocomposites on the relaxation of polymer segments in the matrix was highlighted [9]. They showed that the chemical treatment of ramie fibers at the surface was proven to be effective for the improvement of poly(lactic acid) PLA-ramie composites. In Fig. 4, the results obtained from DSC analysis are presented.

Ramie fibers have surface chemically treated with alkali, 3-aminopropyltriethoxy silane (silane 1) and γ-glycidoxypropyltrimethoxy silane (silane 2). In this way, the interfacial adhesion with PLA matrix was enhanced, resulting in a T_g value increase such that the mechanical properties of composites were improved. In a more recent work same author assigned the decrease of glass transition temperature in PLA matrix-ramie fibers composite with the increase of chains mobility [10]. Authors found that if PLA-g-MA (poly lactic acid-g-maleic anhydride copolymer) was used as the matrix, the chain mobility of PLA molecule was improved, and accordingly, the glass transition shifted to lower temperatures (Fig. 5). More than that, the ramie fibers increased the amorphous fraction in the composite during cooling, due to a decrease of PLA chain regularity by the ramie fibers, which finally reduced crystallization ability. Restriction of PLA chain mobility was highlighted when was studied the PLA-banana fiber biocomposites [11]. Glass transition increased in these materials at lower fiber content, while the increase of fiber loading caused a poor adhesion of banana fiber on PLA matrix.

The most important issue related to the study of the GCs thermal behaviour in the glass transition region is to analyze the effect of various cellulose particles at the

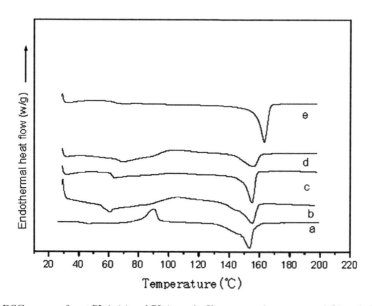

Fig. 4 DSC curves of neat PLA (**a**) and PLA-ramie fiber composites: untreated (**b**) and alkali (**c**), silane 1 (**d**) or silane 2 (**e**) respectively. Reproduced from Yu et al. [9]

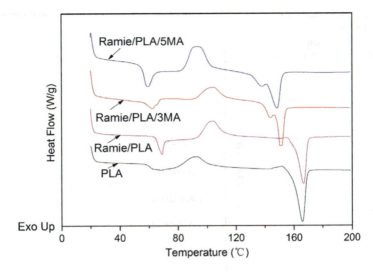

Fig. 5 DSC thermograms of neat PLA, ramie/PLA without and with different MA content. Reproduced from Yu et al. [10]

nanoscale range. Different cellulose nanoparticles as nanowhiskers, nanofibers or nanocrystals isolated from various natural fibers, showed an important effect on polymeric matrix chain mobility as literature results related. From these works, only a few most representative were reviewed in this chapter.

Thermal analysis of GCs based on cellulose nanowhiskers (CNW) using PLA as a matrix, was performed [12]. The results provided by DSC analysis showed a slight increase of T_g values for PLA-CNW-S composites (obtained by solution casting technique) compared with T_g values of neat PLA. These findings were explained by the addition of CNW-S that reduced PLA chain flexibility. In addition, the mobility restriction of PLA chains was related to the formation of hydrogen bonds between OH groups of PLA and CNW-S, which increased the energy needed for occurring of glass transition in this material.

Nanocellulose particles isolated from bagasse by acid hydrolysis were used for the preparation of nanocomposite films based both on PVA as well on crosslinked PVA (10% wt. solid PVA) with glyoxal [13]. As expected, the glass transition temperatures in nanocomposites with crosslinked PVA increased due to a restriction imposed by crosslinks of nanocellulose particles.

Cellulose nanocrystals isolated from many natural fibers were inserted into a biodegradable polymer matrix and studied by many authors. A relevant example is a work in which the cellulose nanocrystals extracted from cotton seeds were acetylated [14]. The polylactic acid was reinforced with these acetylated cellulose nanocrystals (ACN) at different contents. For lower ACN contents (up 2%), the glass transition temperature increased as Fig. 6 shown. obtained results were explained by the presence of the interactions between the acetylated surface of cellulose nanocrystals and PLA matrix, which imposed a restriction on the polymer

Fig. 6 The DSC curves of PLA and PLA-ACN nanocomposites contains ACN. Reproduced from Lin et al. [14]

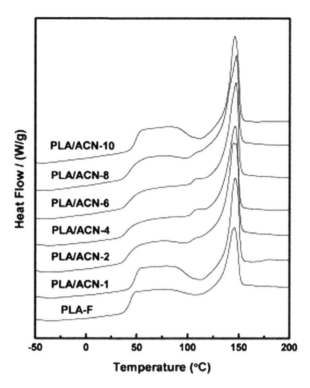

chain mobility. If ACN content increases at 4–6%, a decreasing of glass transition temperatures were explained by means of a high number of interactions between crystalline and amorphous phases from nanocomposite. If a more content of ACN was added in excess (8–10%) to PLA matrix, the self-aggregation occurred between nanofillers such that restrain the PLA chain mobility in the amorphous phase, accordingly the glass transition temperature increased again in these nanocomposites. In case of CG preparation based on cellulose nanocrystals has proven to be a real success by using PLA as a matrix [15]. Here was used the neat and grafted cellulose nanocrystals onto PLLA chains to preparing PLLA-nanocomposites. Thermal behaviour of these materials studied by DSC analysis is shown in Fig. 7. The results denoted that glass transition temperature strongly was decreased (around of 5 °C) in PLLA grafted with cellulose nanocrystals (CNC-g-PLLA). The glass transition increased in PLLA matrix when it was reinforced only with neat cellulose nanocrystals, but much more when PLLA matrix is was reinforced with grafted cellulose nanocrystals. The confinement effect in nanocomposites induced by a very high number of cellulose nanocrystals surfaces was claimed by authors in this study. This effect was countered by the grafting of cellulose nanocrystals which limited the increase of glass transition temperature in PLLA matrix.

The effect of restriction imposed by the cellulose nanocrystals on chain mobility of polypropylene carbonate (PPC) matrix was demonstrated [16]. Data obtained

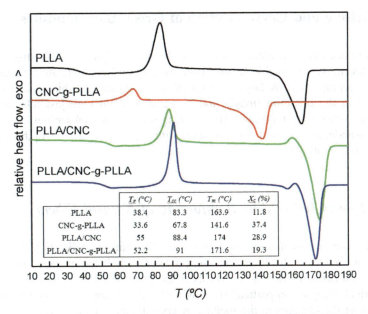

Fig. 7 DSC curves of quenched samples from the melt of PLLA, CNC-g-PLLA nanohybrid, PLLA/CNC nanocomposite and PLLA/CNC-g-PLLA nanocomposite. Reproduced from Lizundia et al. [15]

from DSC analysis and sustained by dynamic mechanical analysis (DMA) results, clearly showed an increase of the glass transition temperature with cellulose nanocrystals content. The thermoplastic and plasticized starch (with glycerol) used as a matrix in the preparation of the CG was reinforced with cellulose nanocrystals extracted from hemp fibers [17]. Heterogeneity of plasticized starch induced a separation between glycerol-rich phases and starch-rich continuous phases, hence the recording a two glass transitions by DSC analysis was expected for these materials. The low glass transition temperature corresponding to glycerol-rich phases (around of −58 °C) did not change with cellulose nanocrystals content on matrix. The intermolecular interactions between starch phases and cellulose nanocrystals limited the starch chain mobility and the glass transition temperature increased from 43.3 to 48.7 °C. Plasticized starch was used also as a matrix for incorporation of cellulose nanocrystals isolated from ramie fibers [18]. The loading of this filler produced the same effect on behaviour in glass transition of the GC: T_g values of rich starch phases increased from 26.8 to 55.7 °C. This result involved the occurrence of intermolecular interactions between starch and stiff cellulose crystallites, which reduced the molecular chains flexibility of starch matrix. In a most recent study [19], the results from thermal analysis of pectin-cellulose nanocrystals films showed that the molecular mobility of amorphous pectin chains did not change by incorporation of cellulose nanocrystals, thus the glass transition temperatures were insignificantly changed.

4 Melting and Crystallization of Green Biocomposites

The interface between natural fillers and different polymer matrix structures represent the most responsible factor in the melting and crystallization behaviour of green biocomposites. A large number of works related to this topic have been reported in the literature (maybe too exceedingly), but here just relevant results from last years were presented. However, in this chapter subsection, many results were considered pertinent for discussing this issue, so these will be presented separately as follows.

4.1 Effects Induced by Reinforcing of Natural Fibers

A great influence of natural fibers in the melting and crystallization behaviour of CG is attributed to the interfacial morphology, which has a direct influence on mechanical performances or wettability of the material. Numerous results provided by thermal analysis (in particularly DSC), claims the importance of this interface, confirming the changes in the melting or crystallization behaviour.

From many others natural fibers, bamboo fibers (BF) are a typical case. A comprehensive work reviewed the literature data related to composite loaded with this fiber [20]. Data resulted from DSC analysis of crystallization process in a various polymer matrix, sustained the conclusions that BF acts as nucleation agents even as a source of so call β-nucleators. After chemical treatment of these surface fibers with alkali, new bonds were generated, inducing the improved thermal properties of composites. Thus, the adhesion degree between fibers and matrix was enhanced and the crystallinity degree increase, for example. In the same work, some results were related to the crystallization behaviour: a different temperature of the crystallization peak temperature at a certain cooling rate has been found directly proportional to the content of BF loaded in GC. This remark was assigned to a heterogeneous nucleation effect of BF which increased the crystallization ability of polypropylene (PP) matrix. In the CG based on recycled PP, the melting temperature lightly decreased, but crystallization rate increased less than in neat PP composites. The crystallinity in this matrix was obtained at a lower cooling rate, thus the nucleation role played by fibers has been proved once again. Thermal behaviour of polylactic acid (PLA) and polybutylene succinate (PBSu) matrices reinforced with BF was investigated [21]. Here, the effect of lysine-based diisocyanate (LDI) used as a coupling agent in crystallization and melting of biocomposites was obvious as results presented in Table 2 are shown. Also, the influence of LDI content in composites crystallization temperature and crystallization enthalpy is relevant. Additions of LDI decreased the enthalpy of melting too, but not the melting temperature. Bamboo fibers increased the crystallization temperature meaning that the crystallization rate increased in nonisothermal processes. Actually, the urethane bonds between BF and polymer matrix promoted the

Table 2 Effect of LDI on crystallization and melting behavior of the PLA and PBS/BF (70/30) composites [21]

Sample	LDI content (%)	T_c (°C)	ΔH_c (J/g)	T_m (°C)	ΔH_m (J/g)
PLA	–	112.8	33.2	165.3	41.1
PLA/BF composite	0	115.2	23.5 (33.6)	162.3	27.4 (39.1)
	0.11	118.4	21.5 (30.7)	162.7	23.5 (33.6)
	0.33	118.7	20.2 (28.9)	161.8	23.6 (33.7)
	0.65	119.4	16.4 (23.4)	162.2	23.1 (33.0)
	1.30	120.2	14.5 (20.7)	161.6	22.8 (32.6)
PBS	–	68.1	60.1	112.1	65.3
PBS/BF composite	0	74.6	40.8 (58.2)	112.0	42.3 (60.4)
	0.11	78.0	37.8 (54.0)	112.1	35.8 (51.1)
	0.33	80.2	38.1 (54.4)	112.3	29.4 (42.0)
	0.65	82.2	37.3 (53.3)	112.9	24.5 (35.0)
	1.30	82.6	31.2 (44.6)	112.1	21.7 (31.0)

( ) Values divided by the weight proportion of polymers

nucleation effect. The crystallization enthalpy decrease involved the limited mobility of polymer matrix chains by LDI addition. The strong interfacial interaction between the polymer matrix and BF yielded a confinement of polymer chain orientation which explained the decrease of melting enthalpy, as authors sustained.

Interesting results were obtained by an extensive study of isothermal crystallization kinetics on bamboo cellulose/poly(ε-caprolactone) (MBC/PCL) composites [22]. Data were extracted from differential scanning calorimetry (DSC) study at different crystallization temperatures (T_c). For a detailed patterning of the isothermal crystallization process, it was applied the Avrami model to the data provided from DSC experiments, and parameters describing the isothermal crystallization process were computed. According to the values of these parameters, MBC reduced the crystallization half-time of composites as compared to the crystallization process of neat PCL. This work pointed out the quality of nucleating agent of the MBC fillers in PCL matrix. The *Avrami exponent* **n** linearly increased with crystallization temperature and recorded values in the range of 1.81–2.70 (most between 2 and 3) for all composites. Actually, these data explained a morphology evolution in these composites: gradual crystals growth from two-dimensional to a spherical three-dimensional one. Effect of the MBC loading consisted of an important increase of the crystallization kinetic constant **K** and improvement of the overall crystallization rate of PCL. These results supporting further the role played by MBC as nucleating agents. Following Hoffman–Weeks theory and after a more DSC data processing, authors of this work provided the values of equilibrium melting point ($T_m^°$). The results showed an increase of $T_m^°$ with MBC content in the composites. Based on these results, the values of the spherulite growth rate and folding-surface free energy of PCL/MBC composites have been found. Both parameters values have increased in the presence of MBC, proving the mobility constraints imposed by fibers on the PCL chains in the inter-spherulitic regions.

Thermal behaviour of GCs containing kenaf fibers (KF) was reviewed [23] and in one of the papers presented here, thermal analysis results showed wide endothermic DSC peaks present in the thermograms of the modified kenaf-PLA composites, due to the presence of moisture. Kenaf fibers were chemically treated with silanes, the most effective treatment being with 3-glycidoxypropyltrimethoxy silane (GPS). This has a great effect on the material plasticization, most of the provided data indicating the presence of moisture which reduced the mechanical performances. If observe the data from this work presented in Table 3, the crystallization process as investigated by DSC showed a strong impact of the KF on recrystallization, melting and morphology development.

Data reveal a melting temperature decrease as compared to the blend, but enthalpy values increase according to DSC results. Author remarked the modification of the onset recrystallization temperature with kenaf fiber length, highlighting the effect of fibers loading on crystallization rate increase in the polymer matrix. The non-isothermal crystallization behaviour of PHBV matrix in kenaf fiber composite model, bulk and compatible PHBV (with maleic anhydride) was investigated [24]. Data analysis resulted from DSC analysis followed the kinetic models of Avrami, Jeziorny, and Mo's analysis as well as Kissinger approach. Based on Avrami analysis, the author concluded that KF and the introduction of maleate groups onto PHBV chains did not significantly affect the crystallization kinetics of PHBV matrix in the model, bulk and compatibilized systems. Neither the processing conditions (for example different cooling rates at 20–5 °C/min), practically did not have an important influence on the crystallization process. The approximate crystallization behaviour of polymer resin in model and bulk composites could be considered as an important result in composite processing. Relevant results were obtained when the crystallization behaviour of poly(L-lactic acid) PLLA based composites prepared with KF, rice straw (RS) and two pigments was investigated [25]. This study revealed the importance of these fillers in nucleation ability of PLLA matrix in the composites. Cooling from the melt at 10 °C/min of neat PLLA induced a consistent amorphous phase. Instead, the crystallinity index increased in a composite containing KF, but much more in those containing RS in same conditions. The data resulted from analysis of the isothermal crystallization showed that all samples had a maximum rate of crystallization around of 105 °C. This rate was found about three times faster for the composite containing only KF, KF with red pigment and RS, compared with those containing neat PLLA. Both the

Table 3 DSC data for PLA-kenaf fiber composites [23]

Sample	T_g (°C)	T_m (°C)	ΔH_m (J/g)	X_c
PLA	60	160.3	–	–
PLA:Kenaf (5:5)	56.9	164.2	19.2	20.5
PLA:Kenaf (5:5) GPS (1%)	58.7	168.6	22.8	24.3
PLA:Kenaf (5:5) GPS (3%)	59.3	168.1	23.1	24.6
PLA:Kenaf (5:5) GPS (5%)	59.1	167.3	23.3	24.9

heterogeneous nucleation as well as tri-dimensional spherulitic growth rate increased the value of *Avrami exponent* **n** (at 3.1) for sample PLLA-RS-yellow. An interesting behaviour was observed on subsequent heating after isothermal crystallization of the samples. The analysis of melting temperatures and crystallinity as a function of T_c showed three distinct regions corresponding to the values around 90, 110 and 130 °C, respectively. There was involved a reorganization of PLLA chains during the melting process; the metastable crystals were formed in isothermal process melt and recrystallized into a more stable crystal followed by a melting of the latter. In another work, thermal behaviour of composites consisting of polylactic acid—modified rice straw fibers (PLA-MRSF) and poly-butyl acrylate (PBA) was investigated by DSC analysis [26]. Obtained results support that the presence of PBA increased the cold crystallization temperature (T_{cc}) and reduced the crystallinity in PLA matrix. In addition, the MRSF loading increased the PLA crystallinity matrix. Instead, both T_g and T_m values of composites have not changed as compared with the pure PLA. Accordingly, higher PBA content was not a suitable compatibilizer between PLA and MRSF as this work sustained.

In composites based on polycaprolactone (PCL) containing rice husk (RH), the DSC heating scans show in Fig. 8 a continuous decrease of the melting peak temperature with the increase of RH content [27]. Furthermore, the RH content decreased the crystallinity extent of PCL phase in PCL-RH eco-composites, while the results recorded lower crystallinity index values for all samples compared with the neat PCL. By RH loading in composite, the PCL chains mobility was limited and the packing affected, thus resulting in a less perfect crystalline structure

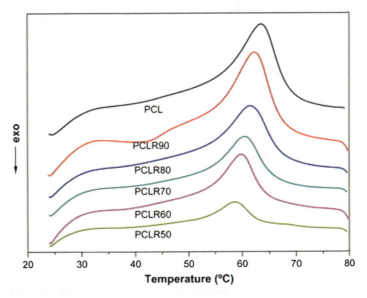

Fig. 8 DSC heating scans at 10 °C/min of PCL/RH ecocomposites. Reproduced from Zhao et al. [27]

compared with neat PCL. The conclusion of this work sustained that a reducing of melting temperature and a free volume of PCL phase led to a depression in the crystallinity.

Ramie is another natural fiber used in composite manufacturing. Its content affects the thermal behaviour of melting and crystallization, too. In order to investigate the thermal and mechanical properties, the different content of poly (butylene adipate-co-terephthalate) (PBAT) was incorporated in same PLA matrix [28]. Results show in Fig. 9 a gradual decrease of T_m values and crystallinity index (χ_c) in ramie/PLA composites with PBAT content increasing. This behaviour was caused by a decrease in the PLA molecular chain regularity. In other work, the same author analyzed the effect of ramie fibers on polylactic acid-grafted-maleic anhydride/ramie (PLA-g-MA/ramie) composites [29]. Here, similar results have been obtained, but during the cooling process, the amorphous fraction in the polymer matrix seems to be changed by the ramie fibers. Once more, the decrease of the PLA-g-MA molecular chains regularity and addition of ramie fibers explained the χ_c decrease in ramie-PLA composites.

Coir used as other natural fiber, has an opposite effect in PLA based biocomposites if it is treated in alkali solution (NaOH) [30]. The number of H-bonds on fiber surface has increased by fiber treatment, inducing a better interface and the crystallinity index increased in these materials, which finally improved the mechanical properties. Coir fibers substantially decreased the cold crystallization temperature (T_{cc}) of biocomposites based on PLA matrix, confirming their nucleating agent role, which accelerated the crystallization process in PLA matrix.

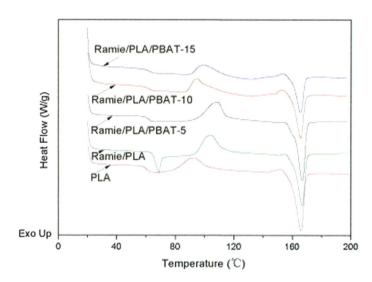

Fig. 9 DSC scans of neat PLA, ramie/PLA without and with different PBAT content. Reproduction from Yu and Li [28]

Some natural fibers like sisal, after a chemical treatment, exhibit transcrystallinity effects on PLA matrix [31]. Both untreated sisal fibers (U-sisal) as well as treated with alkali (A-sisal) and silane (S-sisal) was loaded into PLA matrix for the evaluation of crystallinity and isothermal crystallization behaviour using polarized optical microscopy. Results denoted that all type of sisal fibers had a nucleating ability, influencing transcrystallization in PLA matrix. More than, after computed the interfacial free energy functions for PLA/U-sisal, PLA/A-sisal and PLA/S-sisal composites, author supposed that the difference between these fibers has little or no influence to the nucleation ability in PLA matrix.

Natural fibers extracted from pineapple leaves have proven to be a less efficient filler when was reinforced in neat PP matrix or even in a matrix of PP grafted with maleic anhydride (PP-g-MA) [32]. Thermal analysis performed by DSC revealed a decrease of the composite melting temperature, caused by the interface between fibers and polymer matrix. Further, fibers showed a heterogeneous nucleation effect of the crystallization process, which caused the formation of small size crystals containing defects. Combined with transcrystallinity effects, this is the reason for a limited crystallinity fraction in these materials. The same behaviour was remarked after analyzing the thermal properties of LDPE based composites containing pineapple leaf fibers [33]. The DSC thermograms showed a melting enthalpy decreasing with the fibers incorporation in the LDPE matrix. Sustained by XRD data, these results could be explained by a limited molecular diffusion and finally packing of LDPE chains in the crystal growth process. The crystallinity extent in these composites increased just thanks to the nucleation effect induced by fibers. Slightly higher values of melting enthalpy were obtained for composites containing long fibers as compared with those containing short fibers. Author supposed that the effects of strong the interfacial interaction between thermoplastic matrix and long fiber that confined polymer chain orientation were involved. By comparison, this work concludes that different lengths of fibers did not have an important effect on melting and crystallization of PP matrix.

Recent work presents the behaviour of the GC containing agave fibers in poly-hydroxybutyrate (PHB) and poly-hydroxy butyrate-co-valerate (PHBV) matrix [34]. DSC results showed that the melting temperature did not modify, as well as crystallinity index had a standard deviation less than 5%. Results confirmed that the nucleation was only improved by fibers addition; in the crystallization process, the agave fibers restricted the motion of the chain which should have some effect on overall crystallinity.

Green composites based on PLA matrix filled with fibers extracted from black spruce and northern bleached softwood kraft (NBSK) was obtained through melt compounding and injection [35]. Results denoted again the role of nucleating agents played by cellulosic fibers in the PLA crystallization process.

Investigation of the nonisothermal crystallization kinetic in these composites sustained an increase of crystallization temperature and crystallinity, and a crystallization half-time decrease, as Fig. 10 is shown. Further results relieved the effect of pressurized N_2 gas during DSC analysis which affected the foaming of cellulosic fiber composites based on PLA. A further increase of crystallinity and a

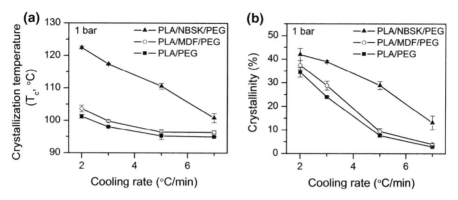

Fig. 10 a Crystallization temperature variation and **b** degree of crystallinity of the PLA materials measured at various cooling rates. Reproduced from Ding et al. [35]

morphology improving were observed in these composites, but the N_2 gas had not any effect on crystallization temperature or half-time crystallization.

Nowadays flax fibers are well-known as the most natural technical fibers. Chemical treatment of flax fiber surface increased the interface adhesion with the different biodegradable polymer matrix. For the first time was proceeded to the alkalinization of the flax fibers, and then treated with organo silane by an optimized procedure [36]. The significant effects of treated fibers were remarked on cold crystallization process of heating: untreated flax fibers acted as nucleating agents and better promoted cold crystallization upon heating at their surface. By alkalization and organo silane treatment of flax fibers, PLA chains became entangled and covalently bonded over the fiber surface. Thus chains became less mobile to have required mobility on crystallization process, and the kinetic was reduced. The compatibilization of PP matrix by addition of grafted glycidyl methacrylate on PP (PP-g-GMA), the styrene-ethylene-butadiene-styrene block copolymer (SEBS) and SEBS-g-GMA improved the interface between hemp fibers in composites [37]. The melting and crystallization behaviour investigated by DSC analysis denoted a general decrease of PP matrix crystallization temperature with the addition of SEBS copolymers as compared to non-compatibilized PP matrix. Thanks to a poor miscibility with the polyolefin matrix, the segregation of the SEBS on the fiber surface occurred. As a result, a flexible interphase layer around the fibers has been produced, reducing thus the nucleating effect on the PP crystallization as authors sustained. On the contrary, the composite containing PP-g-GMA showed a crystallization temperature similarly to that observed in binary PP-g-GMA/Hemp composites. The interaction between fibers and polymer matrix restricted the chain mobility and a reduction of the dynamic in crystallization process was observed in case of compatibilized systems. Other fibers that do not promote the nucleation in these composites hindered the crystallization due to an induced confinement effect, according to authors.

Soy protein flour plasticized with an anhydride (SPA) was blended with polylactic acid (PLA) in order to obtain an eco-friendly composite [38]. After subjected to a biodegradation process under specific conditions, the thermal behaviour of samples was investigated by DSC analysis. The results showed a compatibilization between SPA and PLA in melt region, SPA acting as a plasticizer in this blend; both the glass transition temperatures and melting temperatures decreased with soy protein content increase. Soy protein incorporation in the blend has an important effect on acceleration of the biodegradation rate of PLA compared with pure PLA. The effect became visible with the increase of the degradation time in soil: the glass transition temperatures increased in the early stages of biodegradation and then decreased. This behaviour could be explained by the amorphous phase decrease in PLA by degradation process, resulting in a crystalline phase increase, as well as rigidity increase. When degradation process was extended for a long time, the raised depolymerisation in the blends reduced glass transition temperatures.

Known as one of the most abundant biodegradable polymer founded in many renewable resources, starch is widely used in preparing the biodegradable materials. PLA is the most efficient biodegradable polymer which can be combined with starch. The thermal behaviour of such material has been reported by a study of the isothermal crystallization kinetics in composites of thermoplastic starch/poly(lactic acid) (TPS/PLA) [39]. (DSC) provided data for samples crystallized at different temperatures (T_c). Following a kinetic-based on the Avrami theory, the process of isothermal crystallization outlined a complete imagining of the crystallization process dynamic. The obtained results from data analysis could be used to compute parameters like the spherulite growth rate, overall crystallization rate, and activation energy (ΔE_a). In case of TPS/PLA composites, the values of these parameters were really affected by the incorporation of TPS. The crystallinity in PLA-based composites was improved by TPS which acts as nucleating agent. The computed values of fold surface free energy (σ_e) according to Lauritzen-Hoffman kinetic theory for PLA blends showed a higher value in blends as compared with neat PLA. This finding was explained by a higher work of chain folding (q) needed in the crystallization process, due to a constraint imposed by TPS on the PLA chain mobility in the composite melt. Later was investigated the crystallization behaviour of thermoplastic starch/poly-ε-caprolactone (TPS/PCL) composites by DSC analysis [40]. The no integer values founded for *Avrami exponent* **n** (between 1.8 and 2.6) showed that TPS acted as a nucleating agent, and the mechanism consisted of a mix of thermal and athermal nucleation processes. However, these processes were followed by a gradual growth of crystals from two to three dimensions. As results showed, the decrease of crystallization rate from Avrami analysis has resulted in a crystallinity index reduction, effect sustained also by XRD data.

Biodegradable PCL has been used to prepare biocomposites by melt-mixing with 5 and 15 wt% of cotton (CO), cellulose (CE) and hydrolyzed-cellulose (HCE) [41]. The purpose was to study the influence of types and content of lignocellulose fillers on the morphology, crystallization behaviour and thermal properties of these biodegradable eco-composites. Obtained results have sustained the decrease of PCL crystallinity for a lower content of CO and CE matrix, while the

HCE content had no effect. By increasing the filler content, the crystallinity degree of the matrix decreased to less extent, regardless of which filler type was used. The calculated values of the theoretical melting point were clearly reduced in these biocomposites and were assigned to heterogeneous nucleation sites occurred at the lower content of CO and CE. Data resulted from crystallization kinetic investigation denoted a reduction of induction and half-crystallization times at low fillers loading, but this effect was attenuated at higher filler contents. Cellulose fibers isolated from hardwood (HW), softwood (SW), and bleached kraft softwood (Kraft) pulp was used in producing and characterization of fully biodegradable natural PLA composites [42]. Thermal behaviour of these materials on heating denoted a general increase of the cold crystallization temperatures (T_{cc}) with the pulp fibers addition as data from Table 4 show. When more fibers were loaded, T_{cc} values decreased, that was explained by the author by a heterogeneous nucleation at lower temperatures induced by pulp fibers in PLA matrix. The double melting peak present in neat PLA became more pregnant after the pulp fibers addition.

Furthermore, pulp fibers promoted the recrystallization at a higher temperature: as data from Table 4 show, the melting enthalpy for all composites increased at the higher temperature as the fiber content increased from 30 to 40% (w/w). Pulp fibers did not increase the maximum crystallinity index in composites despite the nucleation effects induced.

4.2 Effect of Micro and Nanocellulose Loading

Different procedures applied for extraction of cellulose from various natural fibers have allowed being obtained particles with different dimensions and orientation, which once they were loaded in a GC, had a significant effect on crystallization and melting behaviour.

First literature results related to the effect produced by microfibrillated cellulose (MFC) on thermal behaviour of the green composites were reported [43]. In this connection, nanocomposites based on PLA matrix and MFC were prepared in two different states (fully amorphous and crystallized) to investigate the thermal and

Table 4 DSC results of neat PLA and pulp fiber-reinforced PLA composites [42]

Sample	T_g (°C)	T_{cc} (°C)	ΔH_{cc} (J/g)	ΔH_m (J/g)	X_{sample} (%)	X_{max} (%)
PLA	60.7	125.3	35.7	37.0	1.3	39.4
PLAHW30	62.9	121.2	25.8	36.5	0.8	40.5
PLAHW40	62.9	114.0	23.0	23.9	0.9	42.5
PLASW30	62.8	119.1	27.2	27.8	0.7	42.4
PLASW40	63.1	115.5	22.1	22.8	0.8	40.6
PLAKraft30	62.9	112.2	26.1	26.6	0.5	40.5
PLAKraft40	62.8	110.0	22.3	22.7	0.4	40.4

mechanical properties. The presence of MFC accelerated the PLA matrix crystallization as DSC measurements revealed. Results confirmed that the particles of MFC clearly increased the nucleation in PLA matrix, and thus cold crystallization occurred at lower temperatures. The overall crystallinity increased in such composites improved thus their mechanical properties (storage modulus). Soon after this result, the same author prepared composited based on partially crystallized PLA matrix containing MFC, in order to reduce the manufacturing time of PLA parts [44]. Experimental data indicated that by using different annealing times (at 80 °C) has been prepared composites having different crystallinity degree. These results showing actually a possibility of replacing of fully neat crystallized PLA matrix with partially crystallized PLA. Later, MFC has been used to prepare fully biodegradable composites based on polyvinyl alcohol [45]. Thermal behaviour of these materials studied by DSC method indicated an increase of the glass transition temperature along with a reduction of melting temperature and crystallinity due to crosslinking of PVA matrix in MFC-PVA composites. Results showed also that the crystallization process in PVA matrix was prevented by nano- and micro-fibrils of MFC.

Known as efficient nanofiller used in the GCs preparation, cellulose nanowhiskers (CNW) has been isolated from oil palm and then incorporated in PLA matrix at the different content [46]. Addition of CNW in PLA matrix reduced the PLA chains mobility, consequently, glass transition temperature in these nanocomposites slightly increased, as DSC results sustain. On the other hand, CNW induced a decrease of T_{cc} values of PLA-CNW as compared to the neat PLA; this indicated a faster nucleation induced by CNW which acts as a nucleating agent for PLA. Values of melting temperatures slightly increased due to the perfection of crystalline morphology, finally leading to better mechanical performances of these materials.

The effect of cellulose fibers standard size (SF) compared with the cellulose nanofibers (CNF) induced in the crystallization process was established [47]. In this report, after studying isothermal crystallization by optical microscopy of PLA composites, authors did not observe an important effect of SF or CNF particles on crystallization process. Results sustained that spherulites developed in PLA matrix have similar sizes, concluding that CNFs doesn't have influence in the nucleation process. These findings were supported by results obtained from DSC analysis.

Using DSC results, the effect induced by interactions between the cellulose nanofibers (CNF), cellulose nanocrystals (CNC) and PU matrix based on PCL diol, both for the unfilled and filled samples was investigated [48]. Values of melting temperatures, fusion enthalpy and crystallinity index extracted for the unfilled PU film and nanocomposite materials reinforced with CNF or CNC are listed in Table 5. Nucleation effect induced by cellulose nanofillers slightly increased the crystallinity index in all nanocomposites as compared to the neat PU matrix, the effect induced by CNF being stronger. Presence of cellulosic fillers at the higher content (over 5 wt% of CNC and 7.5 wt% of CNF) imbedded the PCL crystallites growth, as result of the crystallinity index decreased in the nanocomposites.

Table 5 DSC data for PU (CNC/CNF) composites [48]

Sample	T_m (°C)	ΔH_m (J/g)	X_c (%)
PU 0%	49.4	60.4	38.5
PU/CNF 2.5%	50.1	63.1	41.7
PU/CNF 5%	52.2	65.2	44.6
PU/CNF 7.5%	50.7	59.3	45.2
PU/CNF 10%	51.0	57.5	40.0
PU/CNC 2.5%	49.0	61.8	40.7
PU/CNC 5%	48.9	64.2	42.9
PU/CNC 7.5%	49.9	57.4	40.2
PU/CNF 10%	50.1	49.8	39.7

A good dispersion and reinforcement of CNF fillers at nanoscale level improved the crystallinity levels in PCL matrix, due to a number of interactions that increased through physical bonding. The numerous hydroxyl groups (OH) on the CNF surfaces interacted with PU chains, providing a higher thermodynamic compatibility between components.

Biodegradable nanocomposites were obtained by loading polylactic acid (PLA) matrix with cellulose nanofibers (CNF) having diameters ranging from 11 to 44 nm [49]. The nanocomposites have been evaluated in detail by DSC analysis. The main purpose of this study was to investigate the influence of treated (with 3-aminopropyltriethoxysilane) and untreated nanofibers on the thermal properties of PLA matrix. Results confirmed that the addition of cellulose nanofibers shifted the cold crystallization peak at lower temperatures and broadened, as compared to the cold crystallization of neat PLA. This finding supported the nucleation role played by cellulose nanofibers (CNF) on PLA crystallization process. Furthermore, the silane treatment of nanofibers shifted to higher values and sharpened the cold crystallization peak of PLA/CNFS, as compared with the neat PLA. The CNF fillers increased the crystallinity index in PLA nanocomposites, but CNFS didn't have the same effect. Silane treatment induced a stronger adhesion between the PLA matrix and the CNFS, thus reducing the crystallization ability at fiber surface. The surface of CNF fibers has been modified by esterification to improve the dispersion and interfacial adhesion with PLA matrix in order to prepare green nanocomposites [50]. Loading of treated CNF in PLA matrix caused a slight increase in glass transition and melting temperatures. However, the nanocomposite films do not show a significant difference in thermal behaviour as compared with unfilled PLA film. Cellulose nanofibers have proven to be more effective when is dispersed in plasticized PLA matrix. Bio-nanocomposites based on CNF and PLA matrix plasticized with glycerol triacetate (GTA) recorded better thermal properties [51]. A visual monitoring of CNF diffusion in the PLA matrix using optical microscopy showed no phase separation occurred in the nanocomposites. A slight decrease in the T_m of PLA matrix, from 174.5 to 169.1 °C resulted due to the addition of 20 wt% of GTA to neat PLA. Thus, GTA plasticized the PLA matrix and increased chain mobility, accordingly the crystallinity index increased from 23 to 60%.

Results sustained that plasticizer and CNF dispersion enhanced the nucleation as well as the crystallization process of PLA matrix. DSC analysis on cooling at 20 °C/min of these nanocomposites highlighted the melt crystallization peak of PLA matrix containing CNF while melting was not visible for plasticized PLA. An opposite behaviour was observed by the further addition of 1 wt% of CNF to the plasticized PLA: crystallinity index of the PLA nanocomposite slightly dropped from approx. 60–55%. This result has been explained rather by an impingement phenomenon in the bulk crystallization process due to CNF than the induced nucleating effect. In another work, the same author prepared GCs based on plasticized PLA matrix with triacetate citrate (TAC) and reinforced with cellulose nanocrystals (CNC) or chitin nanocrystals (ChNC) [52]. The higher aspect ratio of ChNC as compared with CNC explained a high efficiency of these particles in the nucleation and crystallization processes. Accordingly, nanocomposites with better mechanical properties were obtained by using ChNC.

Cellulose nanofibers surface modified with oleic acid have been used for the preparation of nanocomposites based on PLA matrix [53]. Hydroxyl groups substituted by acyl groups on the CNFs surface, esterified CNF increasing thus the nanofibers hydrophobicity such that finally the compatibilization with PLA matrix was improved. Increasing the CNFs content (from 4 to 8%) on PLA matrix enhanced the crystallization process in nanocomposites, as DSC results sustained.

The results confirmed that more dense and perfect crystals had been developed in PLA matrix due to the ability of modified CNFs to modify PLA chain orientation. The regularity needed for crystallization process was increased, more than, a nucleation process induced by these fibers further increased the crystallinity index in these materials. These assumptions were sustained by DSC thermograms on cooling as shown in Fig. 11. The higher content (12%) of modified CNF added to PLA matrix caused fillers agglomeration and viscosity increased during nanocomposites preparation. This obstructed the crystallization process, and crystallinity index in samples decreased again.

Another strategy for preparation of PLA GCs based on nanocellulose fillers consist on the exploitation either the miscibility of the PLA matrix with a PVAc (acting as a dispersion medium for the nanocrystals) and/or the chemical modification of PLA and CNFs by radical grafting with glycidyl methacrylate (GMA) [54]. DSC analysis shows that the phase behaviour and the crystallization process of PLA in composites were clearly affected by the functionalization and the cellulose nanofibers content. The melting temperatures of the nanocomposites decreased with PVAc amount increase and the cold crystallization peak is almost not noticeable, as DSC results confirm. Functionalized CNFs cause the strong interactions at the polymer-filler interface and higher miscibility between PVAc and PLA, preventing the crystallization process in these nanocomposites. Crystallinity index in nanocomposites decreased due to hindrance caused by PVAC chains on nucleation and growth PLA crystals, despite to CNFs added to the matrix. Isothermal DSC thermograms showing a reduction of PLA overall crystallization rate from the melt in the presence of PVAc and CNC.

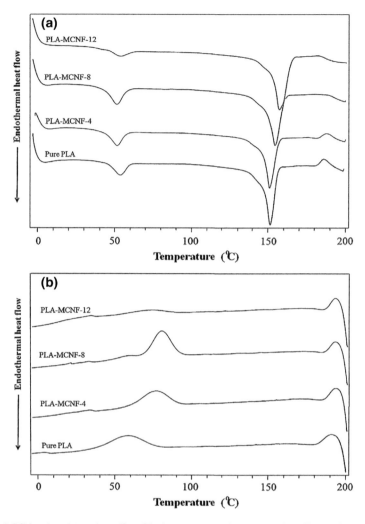

Fig. 11 DSC heating (**a**) and cooling (**b**) thermograms of the pure PLA film and PLA–MCNF nanocomposites. Reproduced from Almasi et al. [53]

The crystallization process of PLLA has a special interest in industrial processing; the using of cellulose nanocrystals (CNC) as a bio-based nucleating agent was proved an efficient way. The CNC surface was chemically treated with silane (*n*-dodecyldimethylchlorosilane) to improve nucleation efficiency of PLLA on both nonisothermal as well as isothermal crystallization [55]. Obtained partial silanized particles (SCNC) had 15 nm width and 200–300 nm length. By addition just of 1% SCNC, the crystallization rate of PLLA was strongly increased, as compared with CNC untreated particles where did not have such strong effects. The highly dispersed treated nanocrystal phase has combined with the larger specific surface area for crystallite nucleation in PLLA/SCNC composites as compared with PLLA/CNC. The length of cellulose nanocrystals has been reduced by silylation, thus high aspect ratio of nanocrystals has been modified, and nanospheres were obtained instead of rods. This has a real effect on composite structuration at the nanoscale level, significantly improving the crystallinity index and mechanical performances of CG. Valuable results were obtained with CNC modified with long alkyl chain grafted (C18-g-CNC) or PLA grafted with CNC (PLA-g-CNC) [56]. Crystallization behaviour of bio-nanocomposites based on PLA/natural rubber blends filled with unmodified and modified CNC was evaluated. Obtained results showed an influence of the modified CNC on crystallization process, improving the nucleation and increasing cold crystallization temperatures. Natural rubber was proven to have more affinity toward C18-g-CNC as against to PLA-g-CNC, a result which explained the reduction of nucleation efficiency in PLA matrix by C18-g-CNC. The crystals with reduced dimensions were developed, changing thus the nanocomposite morphology, and author reported good mechanical properties. Using hydrolyzed CNCs by treatment with sulfuric acid (CNC-S) and hydrochloric acid (CNC-H), a better dispersion in poly(3-hydroxybutyrate-co-3-hydroxyvalerate) (PHBV) matrix was obtained [57]. Following this way, the reinforcement at nanoscale of PHBV matrix was enhanced by the presence of both types of CNCs particles. Comparing the reinforcing effects, in Fig. 12a DSC results confirmed the stronger effect on PHBV morphology produced by the CNC-H particles, especially at higher content loading (12%). As conclusions of this study sustain, the larger aspect ratio, higher crystallinity and especially no residual acid groups exhibited by CNC-H particles have contributed to the thermal behaviour. The effect was raised by the intermolecular hydrogen bonding interactions, increased on such materials, leading to crystallite perfection, and the crystallization rate increased and finally improved the mechanical properties. Variation of T_c and $T_{c(onset)}$ temperatures with both CNCs shown in Fig. 12b, sustained that in the CNC-H/PHBV nanocomposites higher T_c and the smaller $T_{c(onset)} - T_c$ can be achieved, which means a stronger nucleation effect produced by CNC-H as compared to CNC-S.

Nonisothermal crystallization performed by DSC analysis and isothermal crystallization studied by PLM of PLA nanocomposites loaded with conventional spray-dried CNC (CNCSD) and the freeze-dried CNC (CNCFD) were investigated [58]. A better porosity of PLA matrix was obtained by dispersion of the CNCSFD as compared with CNCFD particles. The nucleation effect induced by CNCSFD particles had strong affect on crystalline nucleation and crystallization rates of PLA. Taking into account this result, PLA based composites showing better mechanical performances were obtained if CNCSFD particles were used.

An efficient way to synthesise nanocomposites containing CNCs consisted on in situ ring-opening polymerization of L-lactide in the presence of CNC [59]. Due to the bimodal molecular weight distribution of free PLLA, the CNC-PLLA nanomaterials showed different thermal behaviour, both for glass transition as well as for cold crystallization and melting. Consequently, by blending of CNC-g-PLLA with low molecular weight and free PLLA homopolymer, resulted in a hydrophobic and homogenous material that was not able to develop any crystallinity. Blends of

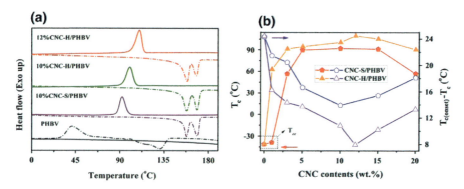

Fig. 12 DSC curves (**a**), the crystallization temperature (T_c) and $T_{c(onset)} - T_c$ (**b**) for neat PHBV and the nanocomposites with various CNC contents. Reproduced from Yu et al. [57]

CNC-g-PLLA with higher molecular weights PLLA allowed the crystals growth in these nanocomposites. Thus, T_{cc} and T_m values increased with PLLA molecular weight; nanomaterial containing the highest molecular weight showed the highest crystallinity index as results confirmed.

Addition of CNC modified with surfactants (s-CNC) and silver (Ag) nanoparticles in PLA matrix denoted different thermal behaviour in the PLA-CNC films [60]. An increase in crystallinity index was recorded for nanocomposites containing a binary system of s-CNC-PLA, as thermal analysis showed. The nucleation was enhanced in these nanocomposites by a good dispersion of s-CNC in the polymer matrix as favoured by the presence of a surfactant. Instead, lower crystallinity index resulted in a binary system containing Ag nanoparticles because the nucleation was not improved. The presence of both s-CNC and Ag nanoparticles in PLA matrix didn't improve enough the crystallinity index in the nanocomposites; as expected their mechanical performances were reduced as well.

For the first time, the CNC incorporation in poly(butyleneadipate-co-terephthalate) pure matrix (PBAT) was studied [61]. Functionalization of CNCs with an aliphatic and aromatic isocyanate (octadecyl and 4-phenylbutyl isocyanate), and then addition to PBAT at 5 or 10% content, yielded nanocomposites having same thermal behaviour. Temperatures of crystallizations from the melt (T_c) increased with treated or untreated CNC amount. Under quiescent conditions, the CNC surfaces acted actually as nucleation centres for PBAT matrix, despite chemical treatment with isocyanates. Just a slow decrease of crystallinity index was observed on nanocomposites samples containing CNCs functionalized with 4-phenyl butyl isocyanate due only to processing technique used, as author sustained.

Interesting results reported in literature highlighted the improving crystallinity in a composite matrix through surface grafting reaction of CNCs with PHBV [62]. Toluene isocyanate (TDI) has been used as a coupling agent in grafting reaction, so by monitoring TDI/PBHV content on copolymers, the length and density of grafted chains on CNCs surfaces were controlled. Thermal behaviour investigation of grafted CNCs by DSC analysis indicated important differences both on cooling and subsequent second heating, as a function of TDI/PBHV fractions. Results shown in Fig. 13 seem to preserve the initial morphology of PBHV in resulted copolymers.

The changes in crystallization temperatures as well in cold crystallization behaviour became visible for copolymers containing a high density of grafted PHBV chains (PHCN8 and PHCN10 from Fig. 13b). The double melting peak

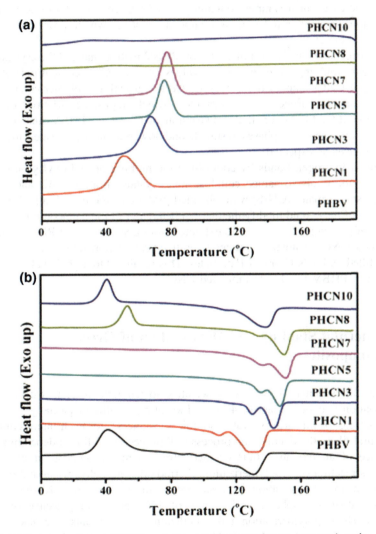

Fig. 13 DSC traces of neat PHBV and the resulting copolymers prepared under various TDI/PHBV fractions obtained during the first cooling (**a**) and second heating (**b**) scans at 10 °C/min. Reproduced from Yu and Qin [62]

observed in Fig. 13a in the second heating cycle denoted a possibility of melting transition and the crystallinity control in copolymers by modifying grafting density or side chain lengths of PHBV.

To produce electrospun bionanocomposites films, the PLA-PHB blend was plasticized with acetyltributyl citrate (ATBC) and reinforced with CNCs [63]. Due to the effect of plasticization, PLA-PHB-ATBC matrix showed increased T_{cc} and ΔH_{cc} values as the content of CNCs increased based on the rearrangement of PLA segments during crystallization. Both melting enthalpy, as well as crystallinity index, increased in a sample containing less CNC content (1%) due to a better dispersion of nanoparticles in a composite matrix, which promoted a higher nucleation effect.

Heterogeneous nucleation of CNCs on PHBV matrix changed the crystallization behaviour, implying an increase of T_c values, at lower CNC content [64]. A higher loading of CNCs amount (over 4%) limited the growth of polymer crystals due to an increase of hydrogen bonds formation, which imposed restrictions on PHBV chain mobility. Moreover, if the dispersion of CNCs on PHBV matrix was not homogenous, separate phases resulted, and lowered the composites mechanical performances as expected.

Tuning hydrogen bonds by controlling the number of hydroxyl groups of cellulose nanocrystals so-named cellulose nanocrystals citrate (CN-C) and cellulose nanocrystal formates (CN-F) were obtained [65]. Once reinforced in PBHV matrix, these nanofillers could modify the thermal behaviour of the nanocomposites. Lower values of parameter $T_{c\,(onset)} - T_c$ extracted from DSC results for PHBV filled with untreated CNCs denoted a greater overall crystallization rate as compared with those filled with CN-C or CN-F particles. This result led to high T_c values observed in case of PHBV filled with untreated CNCs.

5 Thermal Stability and Degradation of Green Composites

Natural fillers generally were very sensitive to temperature changes; the thermal degradation is expected above 400 °C. Excepting volatile or partially stable components (e.g. pectin, waxes, water, soluble substances), for each major component of natural fibers the degradation process will proceed as follows: depending on the presence of an oxidative (air) or inert atmosphere, decomposition of cellulose involves reaction of depolymerization, thermoxidation, dehydration and glycosans formation [5]. Dehydration process occurs in a range of 210–260 °C in a non-oxidative atmosphere, or in range of 160–250 °C in presence of oxygen (or air). The depolymerization and volatilization of glucosans take place at about 310 °C in an inert atmosphere, but maximum rate increases at 350 °C (oxygen) or 375 °C (helium). The hemicellulose components decompose at relatively low temperatures (159–175 °C), this process being preceded by the cellulose decomposition, but its effect is proportionally limited by the content in the fiber. Instead, lignin decomposition process consists on three steps: below 220–250 °C side chains are broken and split, followed by the formation of free radicals between 300 and 400 °C and a series of condensation reactions above 400 °C.

Understanding of this item has a major importance on the practical application of composites since any mechanical performance was generally altered by temperature increase. Considering that some issues related to the degradation process of natural particles, organic structures may impose limitations on different applications of material obtained. As in precedent section of this chapter, the most important results related to the thermal degradation of GCs from last years reported in the literature will be presented both for natural fibers as well as for cellulose or starch nano-particulate fillers.

5.1 Thermogravimetry of Green Composites Containing Natural Fibers

A considerable number of natural fibers have been used in composite processing, but here will be overviewed only results related to composites containing common natural fibers. For each relevant fiber type used, one paragraph will contain and discussed results provided by one or more works related to this issue.

The TG/DTG analysis of common lingo-cellulose fibers denotes similar results that could be related to the thermal decomposition of their main constituents. As thermogravimetric parameters from Table 6 suggest, the TG curve contains three stages of weight loss: the first one up to about 200 °C, is ascribed to a maximum weight loss of 10%; this step is followed by a second stage up to about 500 °C, where the loss is more than 70 wt%. In the final third stage, about 20 wt% of the mass is a loss, allowing the extending range of TG/DTG test up to 800 °C. The maximum rate of thermal decomposition was displayed in DTG curves as the main peak, while the components of the fibers could have assigned like a shoulder or tail peaks in a DTG curve. The values shown in Table 6 are similar to others reported in the literature for distinct natural fibers but may vary depending on treatments

Table 6 Thermogravimetric parameters of common natural fibers [66]

Natural fiber	1st stage weight loss (%)	1st stage DTG peak (°C)	2nd stage onset T_0 (°C)	2nd stage weight loss (°C)	2nd stage DTG shoulder (°C)	2nd stage DTG main peak (°C)	2nd stage DTG tail (°C)	3rd stage weight loss (°C)
Jute	8	60	260	89	290	340	470	3
Sisal	9	52	250	76	275	345	465	15
Wood	2	107	290	85	270	367	400	13
Cotton	4	55	265	91	280	330	410	5

applied to the fiber. According to other results, after water loss in the first stage (DTG peak), the thermal degradation of main lignocellulose constituents of the fiber began at the onset of the second stage [66]. The cellulose decomposition was ascribed by the main DTG peak of this stage, while the hemicellulose and the end of lignin decomposition were represented by shoulder peak or by the tail peak, respectively. After the third stage, the residual weight was related to char or to other products resulted during decomposition reactions.

An important standpoint in discussions of related literature results is represented by the effect of different atmospheres on the thermogravimetric analysis of natural fibers. Actually, may be used in two distinct atmospheres, inert (helium and nitrogen) and oxidative (air and oxygen). Moreover, different flows of these gases conduct heat at different rates, such that the thermograms obtained in nitrogen may be significantly different from those obtained in helium. In an inert atmosphere, the thermal degradation of cellulose resulted in the main DTG peak associated with the formation of macromolecules containing rings bearing double bonds [5]. This peak was partially overlapped in an oxidative atmosphere due to oxygen reaction with the cellulose (exothermic peak corresponding to this reaction). Consequently, the main DTG peak was shifted to lower temperatures in that atmosphere as compared to the inert one.

Various thermal behaviours of GCs containing bamboo fibers were related to the assignment of an enhanced degree of adhesion between the matrix and bamboo fibers, because a stronger adhesion meaning a better thermal stability [20]. Biocomposites based on PLA, PBS reinforced with bamboo fiber (BF) were investigated [21]. Results of this study denoted a decrease of thermal degradation temperature of both composites as compared with those of pure polymer matrix. By using lysine-based diisocyanate (LDI) as a coupling agent between fibers and polymers matrixes, the thermal degradation temperature increased in composites. Authors sustained an increase of molecular weight due to crosslinking reactions between PLA or PBS matrix and BF, which could induce an increase of thermal degradation temperatures.

Coir fibers (CF) treated by washing with water, alkali treatment (mercerization) or bleaching filled a blended matrix consisted of starch/ethylene vinyl alcohol copolymers in order to prepare biocomposites [67]. Fibers treatments considerably affected the thermal behaviour, namely increased the thermal degradation temperatures, as TGA data sustained. These modifications were attributed to the removal of some easily hydrolyzed substances, which decomposed earlier than the major components, cellulose, and lignin. This could increase the thermal stability in the second step of degradation process as compared with composites filled with untreated fibers. Interesting results reported on thermal behaviour of PLA based biocomposites containing both untreated as well as alkali treated coir fiber were obtained [30]. By alkalization process, the number of H-bonds at surface increased, improving the adhesion at the interface between coir fibers and PLA matrix. The TGA results, consisting of both TGA and DTG curves, are shown for untreated and treated coir fibers in Figs. 14 and 15 respectively.

The chemical treatment of coir fibers seems to have a significant increase in thermal stability, increasing the maximum degradation temperatures in the composite. Removal of hemicellulose during alkali treatment could explain this finding. Nevertheless, due to the embedding of coir fibers in PLA matrix, the thermal degradation temperatures decreased and the thermal stability of biocomposites decreased irrespective of the fiber treatment as results show on Figs. 14 and 15. The author concluded that fiber loading was responsible for the composite thermal stability decrease, imposed by the lower degradation temperature of coir fibers. The effect of lignin loading in PP/CF composites both on presence as well as in absence

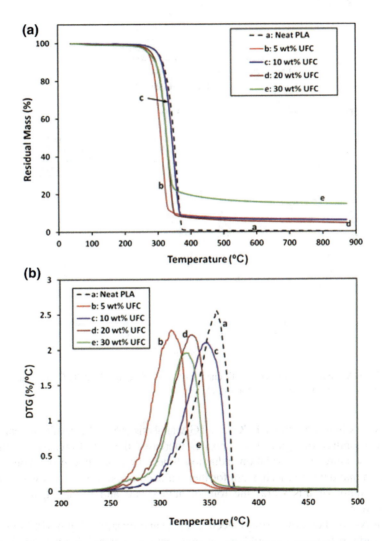

Fig. 14 TGA curves (a) and (b) DTG curves of neat PLA and PLA/UFC (untreated) coir fibre biocomposites. Reproduced from Dong et al. [30]

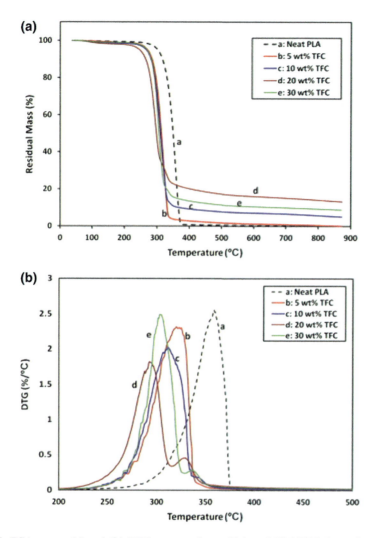

Fig. 15 TGA curves (**a**) and (**b**) DTG curves of neat PLA and PLA/TFC (treated) coir fibre biocomposites. Reproduced from Dong et al. [30]

of MA was investigated by DSC and TGA analysis [68]. The results revealed that the incorporation of lignin in composites increased the initial thermal decomposition temperatures and oxidation induction times. Two possible protections played by lignin that delayed mechanism for thermal decomposition process was involved by the author of this work: the role of antioxidant and barrier against thermal degradations.

Matrix-based on polyfurfuryl alcohol (PFA) and plastified lignin (PF) renewable from various biomasses could be an eco-friendly material having good performances for various applications [69]. These two components were blended in two different ratios, then polymerized and investigated on their behaviour in the thermo-oxidative degradation process; results were compared to those of neat materials. As this work sustained, strong interactions between PL and PFA in the blended matrix acted as protection against chains internal scission and depolymerization during thermal degradation. The neat PL was clearly decomposed at a lower temperature than PFA/PL blend, as results from TG/DTG analysis sustained. Later, PFA bioresin was used as a matrix for preparation of fully green composite containing kenaf fibers (KF) [70]. Thermogravimetric analysis of composites showed a decrease of temperature corresponding to maximum decomposition rate T_{max} (from 466 to 458 °C) with KF content increase from 5 to 20 wt%. This behaviour was explained by increasing of thermally unstable non-cellulosic structures come from the fiber content in GCs. Retention of about 45% char residue at 800 °C was due to a higher thermostability of PFA matrix in the composite.

Thermal stability of composites containing natural fibers increases generally due to a chemical treatment applied to fibers. Waxes and non-cellulosic components were removed from surfaces, increasing thus the interaction with the polymer matrix. Thermogravimetric and pyrolysis analysis investigations bring out such results on KF/epoxy composites containing both treated as well as untreated fibers [71]. By alkalization (with NaOH) of KF, the level of moisture content with 3% decreased, which influenced the weight loss behaviour, as well as their corresponding composites. A high level of moisture content means more voids and hemicelluloses on composites, which increase thermal conductivity with thermal stability decrease. Consequently, by treated KF addition in epoxy composites, the thermal stability of composites has improved as well as its charring capability. The alkalization process reduced the char content after thermal decomposition of these KF/epoxy composites. Loading of composites based on PVC/TPU matrix with KF enhanced their thermal stability, but in composites based on the biodegradable matrix (PLA), KF induced a decrease of thermal degradation temperatures due to PLA depolymerization as in another study was showed [23]. Silanization of KF fibers induced an increase of thermal stability in these composites.

The thermal degradation of GC containing alkalinized hemp fibers (HAlk) revealed different behavior when were loaded in the different matrix [72]. The degradation temperature increased from 351 °C in neat PP matrix containing 25 wt % hemp fibers, to 376 °C for PP/HAlk and up to 391 °C for PP-SEBS-g-MA/HAlk. This behavior was explained by the barrier effect induced by hemp fibers against degradation process, which increased in the ternary system due to a better adhesion of interface imposed by MA groups. A different behaviour was reported for composites based on MAPP matrix filled with silane treated hemp fibers, which denoted a decreased thermal stability as compared with those filled with untreated fibers [73]. Results obtained by thermogravimetric analysis sustained decreasing values of the temperature up to 25% at 10% weight loss (T_{10}). The weight loss was less than 2% at 240 °C, due to silane treatment of hemp fibers.

Besides the applications of natural fibers on building structures and automotive area, literature does not report too many results regarding the use of the flax fibers (FF) in GCs processing. A representative study where a green composite based on bio-based resin reinforced with flax and basalt fibers was investigated [74]. Results from TGA analysis showed a lower thermal stability if it contained neat FF due to a higher amount of lignocellulosic content, as compared to that contained basalt fibers in addition.

Thermal behaviour of PLA composites containing ramie fibers (RF) showed increased values of thermal degradation temperatures if the RF surface was treated as compared with those which was reinforced with untreated RF [9]. Results were explained by the chemical bonds between PLA matrix and ramie fiber after treatment that could enhance the interfacial adhesion. It's worthy to note that the alkali treatment of RF induced lower thermal degradation temperature in composites as compared with the composites with silane-treated fiber. This result revealed a worsening of interface bonds between alkalinized RF and PLA composites as compared with the interface of silane treated RF and PLA. Later, it was founded that in case of composites based on PLA matrix reinforced with RF compatibilized with PLA-g-MA [10] or blended with PBAT [28] thermal stability increased.

Microcrystalline cellulose (MCC) renewed from oil-palm biomass has been used as filler for PLA matrix [46]. Results based on TG/DTG data (T_{on}, T_{10}, and T_{50}) showed an improving the thermal stability of PLA/MCC composites compared to pure PLA. Due to an intrinsic flame resistant property, the higher content of cellulose I on MCC was related with the highest char residue (2.4%) displayed for the PLA/MCC 5% composites as compared to pure PLA, or composites containing 1 and 3% MCC. Data from TGA and DTG curves generally denoted a higher thermal stability for all PLA/MCC composites as compared with pure PLA, due to MCC incorporation.

Cassava and pineapple flour have been used as fillers in PLA bio-composites, improving interfacial adhesion when maleated polylactic acid (MAPLA) was added for compatibilization [75]. Both fillers have reduced the degradation temperature of PLA but de-starched cassava flour denoted higher thermal stability due to its higher lignin content compared with pineapple flour. An important result it was that the thermal degradation temperature was increased by adding MAPLA.

Effects of alkaline and silane surface treatments of wheat or rice husk on thermal degradation of PLA biocomposites was observed [76]. Both alkaline treated husks (RA and RG), as well as silane treated husks (RNA and RNG), increased thermal degradation temperatures in composites, due to an increased number of chemical bonds which enhanced interfacial adhesion. As results show in Fig. 16, the PLA biocomposites with untreated husks have lower thermal stability as compared with those filled with treated husks.

However, results from this figure denote lower thermal degradation temperature for GCs containing alkaline treated rice husks as compared with those containing silane treated husks. This study sustained that the interface bonds between alkaline treated husks and PLA were weaker than the interface bonds between silane-treated husks and PLA.

An eco-friendly polymer obtained by graft polymerization of *Hibiscus sabdariffa* fibers on poly butylacrylate (PBA) has thermal behaviour investigated by the thermogravimetric analysis (TGA) [77]. Results showed an increase of the thermal

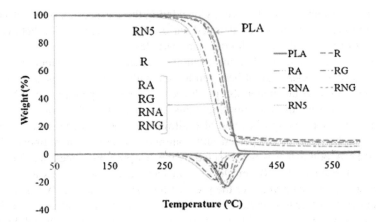

Fig. 16 Thermogravimetric curves of the neat PLA and biocomposites with untreated and treated rice husk. Reproduced from Thao Tran et al. [76]

stability in this material explained by a decay induced by the late decomposition of PBA. Generally, in the synthesis of grafted polymers crosslinked and entangled type of networks could be produced, which subjected to degradation temperatures imbedded char diffusion on the surface, and thus inhibit the degradation process.

Cellulose fiber (CF) extracted from *Grewia optiva* reinforced GC containing blend films of corn starch/poly(vinyl alcohol) (St/PVA) [78]. Films were plastified with citric acid (CA) and crosslinked with glutaraldehyde (GLU). Thermogravimetric analysis of these GC films revealed that for *Grewia optiva* fibers, processes like depolymerization, dehydration and glucosan formation were occurred in the temperature range of 26.0–190 °C. The degradation temperatures of the composites fall between the degradation temperatures for the St/PVA blend films and the CF. Results indicated that CF fibers retarded the dynamic of composites degradation process. The increase of chemical bond strength at the interface between CF and matrix increased the resistance against thermal attack, improving thus thermal properties of the green composite.

5.2 Thermogravimetry of Green Composites Containing Cellulosic Nanoparticles

Thermal stability of cellulose can be differently approached if results are discussed taking in account the nanoscale dimensions of particles. A higher value of specific surface area for a cellulosic nanoparticle increase the thermal conductivity, the degradation temperatures occurring at lower values, despite to a degradation mechanism similar to hemicellulose. But, once embedded in a biodegradable polymer matrix, celluloses nanoparticles thermal degrade at higher temperatures due to a lower thermal conductivity. Considering these aspects, a series of valuable results reported in the literature denoted various thermal degradation behaviours of composites reinforced with cellulose nanoparticles.

Thermal stability of CNCs, for example, depending on polymer matrix containing these particles. Nanofibers reinforced with CNC were used as fillers in sustainable composite materials produced via electrospinning of PVA [79]. Thermogravimetric analysis investigations revealed the effect of CNC on the thermal stability of composites and also allowed to obtain deeper insights into the interactions between the dispersed and continuous phases of nanofibers. The stability of PVA polymer was not affected by electrospinning process as TGA results sustained. Presence of CNC in electrospun nanofibers appeared as a peak in the first-order derivative curve, over the preexisting ones assigned to PVA matrix. This peak shifted to lower temperatures due to the accelerated thermal decomposition of CNCs in PVA melt, attributed to the degradation products that catalyzed the overall degradation process. Treatment of CNCs by hydrolyzation with sulfuric acid (CNC-S) and hydrochloric acid (CNC-H) promoted a better dispersion in poly(3-hydroxybutyrate-co-3-hydroxyvalerate) matrix (PHBV) [57]. Thermogravimetric analysis (TGA) of composites shown in Fig. 17 reveals a higher degradation temperature for CNC-H as compared with CNC-S where surface contained sulfated groups, which significantly lowered the degradation temperature. However, the thermal stability of both nanocomposites was higher than neat PHBV. Increased content of CNC-S in PHBV matrix resulted in a poor thermal stability of nanocomposites, an opposite effect being recorded for PBHV loaded with CNC-H where higher thermal stability was achieved. This study concluded that the number of intermolecular hydrogen bonding interactions between CNCs and PHBV (higher for CNC-H as compared with CNC-S) has a decisive role on thermal stability of composites in addition to a good reinforcement of CNC particles in the PBHV matrix.

The thermal stability of cellulose nanowhisker/bio-based polyurethane (CNW/BPU) composites were investigated indicating a clear increase in thermal degradation temperatures with CNW content [80]. The computed values of activation energies for thermal decomposition of these composites indicated that BPU matrix was decomposed more difficult due to the incorporation of CNW. The analysis of parameters following the thermal decomposition kinetics of the BPU/m-CNW composites also sustained the increasing of the thermal stability of composites due to the incorporation of CNW. Cellulose nanocrystals were modified by acid hydrolysis (HCl or H_2SO_4) and esterified using the modified Fischer method in order to obtain acetate cellulose nanocrystals (Ac-CNCs) and lactate cellulose nanocrystals (LA-CNCs) [81]. Then, these modified CNCs reinforced PLA matrices through melt blending procedure. Under the inert atmosphere, all nanocomposites showed similar thermal stability, but under oxygen, the thermal stability decrease in order: LA-CNCs > AA-CNCs > HCl-CNCs > H_2SO_4-CNCs. Improved thermal stability in nanocomposites containing LA-CNCs and AA-CNCs particles was probably caused by a good dispersion that increased compatibility with PLA matrix, as author sustained.

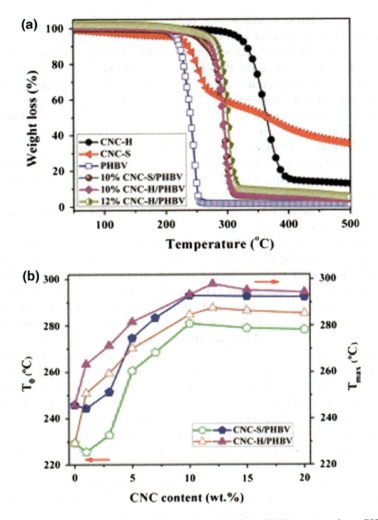

Fig. 17 TGA analysis (a), T_0 and T_{max} (b) as a function of the CNC contents of neat PHBV and the nanocomposites with CNC-S and CNC-H. Reproduced from Yu et al. [57]

Starch matrix has poor thermal stability, but if it is chemically modified following various ways this property is much improved. Thermogravimetric analysis (TGA) was performed on fully biocomposites of glycerol plasticized waxy maize starch with and without hydrolyzed starch nanocrystals [82]. The lower thermal stability reported for the biocomposite mentioned in this work was assigned by a closer association of glycerol with starch nanoparticles. Results showed that the effect of the addition of starch nanoparticles in the unplasticized matrix compared to the one induced in the glycerol plasticized material do not change the thermal stability. This work assumes a decrease of thermal degradation temperature for hydrolyzed starch nanocrystals, which act as flame retardants due to char formation.

Starches from different vegetal sources (tuber, cereal, and legume) were plasticized with glycerol and reinforced with cellulose nanocrystals [83]. In all cases, thermal degradation of TPS films filled with nanocellulose started at higher temperatures than unfilled TPS. This result showed a good thermal stability for TPS matrix containing large starch-rich domains as much as the extent of plasticization (high amylopectin starches) increased. GCs were obtained by reinforcing the glycerol plasticized corn starch matrix with cellulose nanofibrils (CNF) extracted from wheat straw (using steam explosion) [84]. The results from TGA and DSC experiments indicated an interaction between CNF and glycerol, which produced the reduction in onset of degradation temperatures as compared with the pure matrix. TGA investigations revealed that thermal stability of these GCs increased after high shear mechanical treatments. This behaviour was explained by removing of non-cellulosic material which improved crystallinity after treatments, increasing thus the thermal stability.

Bacterial cellulose (BC), when was combined with a biodegradable polymer matrix, generated a fully biodegradable composite with promising applications. The addition of low amounts (1 and 5%) of vegetal, BC fibers and glycerol (plasticizer) to the thermoplastic starch matrix (TPS) resulted in a green composite with better mechanical properties [85]. The results from TGA analysis showed that thermal stability of composites filled with vegetal cellulose and BC slightly increased (with 9 and 7 °C, respectively). The higher stability of cellulose fillers and the excellent compatibility between the two carbohydrate components of composites explained these findings. Very interesting results recently reported were related to preparing of resistant starch/pectin (RS/P) free-standing films reinforced with bacterial nanocellulose (BCN) for biomedical use [86]. Presence of BCN or CNF at 0.5% amount in nanocomposites films produced an almost similar mass loss, as resulted from TG curves. The strong interaction between matrix and both reinforcement phases (BCN or CNF) became clear at an increased concentration (from 1.0 to 3.0%), which promoted the better thermal stability in these composites. The thermal degradation process of nanocellulose (depolymerization and decomposition process of dehydrocellulose) doesn't really affect the mass loss of nanocomposite films with higher amounts of reinforcements.

6 Conclusions

Recent studies focused on the thermal behaviour of GCs presented in this chapter revealed the most important issues about glass transition, melting and crystallization on the one hand, and about thermal degradation and stability range on the other hand. Almost all reviewed works were concerned with the significance of the interface between natural fillers and various polymeric matrices because it plays the most important role in the applications area. The binding between GC components at interface decided their thermal behaviour in any temperature range. The results have been separately presented in this chapter, for each range, in order of temperature increase.

Several studies, since 2006 have been discussed results related to glass transition range which is considered as the important parameter to define the mechanical performances of GCs. For example, when hemp was reinforced on PBSu matrix or banana fibers were reinforced on PLA matrix at any content without any chemical modification, the T_g value of resulted GCs did not show important changes, but its range was broadened. Authors remarked just a slightly decreased T_g value for CG based on phenol formaldehyde and filled with oil palm which explained a possible heterogeneity generated during processing. Decreasing of T_g value by plasticization (bio-epoxy or PU/flax) or due to some confinement effects at fiber interface with polymer matrix also were involved.

Instead, a real improvement in the mechanical performances of GCs was obtained by various chemical treatments applied to the natural fibers or fillers used as reinforcements. For example, T_g value of GC based on PLA matrix containing chemically treated ramie increased, improving thus the mechanical properties or decreased worsening the interfacial adhesion.

As a new finding in the area of GC production, CNC (or CNW) unmodified as a function of loaded content reduced the polymer matrix chains mobility at the interface (in case of PLA, PPC, TPS) and increased the T_g values, or have no any effect (case of pectin). In some situations, due to a chemical treatment (acetylation) applied to the CNC or by grafting (to PLLA) before adding to polymer matrix (PLA), T_g value increased, the effects being proportional to the modified CNC content.

From the cited literature, it seems that only a chemical treatment or any other modifications of fillers that change adhesion at the interface will have an important effect on T_g value, so it could be controlled to obtain the desired mechanical performance when GCs are prepared.

The changes in the thermal behaviour of GC were well presented also when the melting and crystallization region of polymer matrices was studied by DSC analysis. Different natural fillers isolated from various natural resources were considered in almost all presented works as nucleation agents in the crystallization process, which could change the morphology at the interface. Consequently, the resulted modifications of the T_c, T_m, ΔH_m, ΔH_c or χ_c values in polymer matrices were clearly visible in DSC analysis. The overall crystallization rate and crystallinity index value increased due to fibers loading in the follow GC: ramie, coir, rice straw—PLA, kenaf—PLLA, bamboo—PP-g-MA, bamboo—PLA/PBS or rice straw—PLLA. Instead, these values decreased in kenaf—PLA, rice husk—PC, rice straw—PLA/PBA and generally in case of chemically treated fibers. In some situations, fibers like cellulose pulp extracted, sisal, pineapple leafs or agave had reduced or had no effect on GC crystallinity. A special behaviour had TPS, which increased the crystallinity in PLA but decreased in PCL matrix.

Different ways of crystals growth and morphology development induced in polymer matrices by fillers have been determined after evaluation of the polymer matrix crystallization kinetic using DSC method, under both isothermal as well as non-isothermal conditions. By loading of neat fibers, the crystallization kinetic was improved by the bamboo fibers in PCL matrix, kenaf fibers, and rice straw in PLLA, black spruce and northern bleached softwood kraft in PLA, but has no any effect or reduced the crystallization rate when these were chemically treated (ex. sisal—PLA).

Reduced dimensions of cellulose nanofillers extracted from various bioresources had a strong effect on crystallinity and morphology of GCs. The nucleation process was generally increased by these fillers type, and the growth of crystals was dependent on particular conditions. Thus, the crystallinity was improved in PLA based composites containing MFA, CNW or ChNC, and in the PU matrix filled with CNF. Also, in plasticized PLA or PLA/PHB blends, the crystallization was improved by CNF and CNC, respectively. Chemical treatments (hydrolization, esterification, and surfactant) or freeze drying applied on CNC surfaces had increased the crystallization rate in PLA matrix. Decreasing of crystallinity was reported by loading of MFA or silanized CNF in PVA, CNC-g-PLA in PVAc, CNC in PBHV, and by grafting PBHV chains onto CNC surface.

All these results mean that the GC morphology can be tuned to control crystals dimensions, obtaining thus desired mechanical properties (e.g. small crystals—good mechanical properties), as a function of the application area. To obtain GCs with good thermal and mechanical properties, at least two conditions must be satisfied to control the crystallization process: good surface treatment (fillers or/and polymer matrix) and well dispersion of fillers in the polymer matrix (to prevent heterogeneity).

In the most situations, the natural fibers improved the thermal stability of GCs, as much as fiber content increased. A better thermal stability was recorded by the bamboo fibers loaded in PLA or PBS, coir fibers reinforced in EVA/TPS blend, PLA, PP-g-MA, cassava and pineapple flour loaded in PLA or by the plasticized lignin filled with PFA. The degradation temperature was increased when the chemical treatment applied to the fibers enhanced adhesion with the polymer matrix. Thus, the scientific literature has registered results related to the alkalinized fibers as in kenaf—epoxy or PVC/TPU, hemp—PP/SEBS-g-MA and ramie—PLA-g-MA composites or to silanized (rice husk or kenaf) fibers loaded in PLA matrix. On the other hand, a decreased thermal stability was reported for kenaf (or ramie)—PLA, silanized hemp—PP-g-MA and flax (or kenaf)—PFA green composites.

Cellulose pulp fibers increased the thermal degradation temperatures in corn starch/PVA blend based composites, or in case of MCC reinforcing PLA matrix. Also, this property was enhanced in CNW—BPU and in BC—TPS (TPS/pectin blend), but a chemical treatment of cellulose nanoparticles improved thermal degradation furthermore. Thus, it was worth to notice the results that confirmed a thermal stability enhancement for alkalinized CNF—TPS, hydrolyzed CNC—PBHV (or PLA) and hydrolyzed starch nanocrystals—TPS.

Beside the interface between natural fillers (fibers or nanoparticles), a well dispersion in the processing step represent the key factors in a green composite preparation. Following these findings, the thermal properties of sustainable materials will be improved, providing thus desired and good mechanical performances.

References

1. Das O, Bhattacharyya D, Sarmah AK (2016) Sustainable ecocomposites obtained from waste derived biochar: a consideration in performance properties, production costs, and environmental impact. J Clean Prod 129:159–168
2. Mathot VBF (ed) (1994) Calorimetry and thermal analysis of polymers. Carl Hanser Verlag, München
3. Turi EA (ed) (1997) Thermal characterization of polymeric materials. Academic Press, New York
4. Väisänen T, Das O, Tomppo L (2017) A review on new bio-based constituents for natural fiber-polymer composites. J Cleaner Prod 149:582–596
5. Monteiro SN, Calado V, Rodriguez RJS et al (2012) Thermogravimetric behavior of natural fibers reinforced polymer composites—an overview. Mat Sci Eng A 557:17–28
6. Signori F, Pelagaggi M, Bronco S et al (2012) Amorphous/crystal and polymer/filler interphases in biocomposites from poly(butylene succinate). Thermochim Acta 543:74–81
7. Shinoj S, Visvanathan R, Panigrahi S et al (2011) Oil palm fiber (OPF) and its composites: a review. Ind Crops Prod 33:7–22
8. Cuinat-Guerraz N, Dumont M-J, Hubert P, (2016) Environmental resistance of flax/bio-based epoxy and flax/polyurethane composites manufactured by resin transfer moulding. Compos Part A 88:140–147
9. Yu T, Ren J, Li S et al (2010) Effect of fiber surface treatments on the properties of poly(lactic acid)/ramie composites. Compos Part A 41:499–505
10. Yu T, Jiang N et al (2014) Study on short ramie fiber/poly(lactic acid) composites compatibilized by maleic anhydride. Compos Part A 64:139–146
11. Shih YF, Huang CC (2011) Polylactic acid (PLA)/banana fiber (BF) biodegradable green composites. J Polym Res 18:2335–2340
12. Haafiz M, Hassan A, Khalil A et al (2016) Exploring the effect of cellulose nanowhiskers isolated from oil palm biomass on polylactic acid properties. Int J Biol Macromol 85:370–378
13. Mandal A, Chakrabarty D (2014) Studies on the mechanical, thermal, morphological and barrier properties of nanocomposites based on poly(vinyl alcohol) and nanocellulose from sugarcane bagasse. J Ind Eng Chem 20:462–473
14. Lin N, Huang J, Chang PR et al (2011) Surface acetylation of cellulose nanocrystal and its reinforcing function in poly(lactic acid). Carbohydr Polym 83:1834–1842
15. Lizundia E, Vilas JL, León LM (2015) Crystallization, structural relaxation and thermal degradation in poly(l-lactide)/cellulose nanocrystal renewable nanocomposites. Carbohydr Polym 123:256–265
16. Hu X, Xu C, Gao J et al (2013) Toward environment-friendly composites of poly(propylene carbonate) reinforced with cellulose nanocrystals. Compos Sci Technol 78:63–68
17. Cao X, Chen Y, Chang PR et al (2008) Green composites reinforced with hemp nanocrystals in plasticized starch. J Appl Polym Sci 109:3804–3810
18. Lu Y, Weng L, Cao X (2006) Morphological, thermal and mechanical properties of ramie crystallites-reinforced plasticized starch biocomposites. Carbohydr Polym 63:198–204
19. Chaichi M, Hashemi M, Badii F et al (2017) Preparation and characterization of a novel bionanocomposite edible film based on pectin and crystalline nanocellulose. Carbohydr Polym 157:167–175

20. Abdul Khalil HPS, Bhat IUH, Jawaid M et al (2012) Bamboo fibre reinforced biocomposites: a review. Mater Des 42:353–368
21. Lee SH, Wang S (2006) Biodegradable polymers/bamboo fiber biocomposite with bio-based coupling agent. Compos Part A 37:80–91
22. Liu H, Huang Y et al (2010) Isothermal crystallization kinetics of modified bamboo cellulose/PCL composites. Carbohydr Polym 79:513–519
23. Ramesh M (2016) Kenaf (*Hibiscus cannabinus* L.) fibre based bio-materials: a review on processing and properties. Prog Mat Sci 78–79:1–92
24. Buzarovska A, Bogoeva-Gaceva G, Grozdanov A et al (2007) Crystallization behavior of poly(hydroxybutyrate-co-valerate) in model and bulk PHBV/kenaf fiber composites. J Mater Sci 42:6501–6509
25. Dobreva T, Perena JM, Perez E et al (2010) Crystallization behavior of poly(L-lactic acid)-based ecocomposites prepared with Kenaf fiber and rice straw. Polym Compos 31(6):974–984
26. Qin L, Qiu J, Liu M et al (2011) Mechanical and thermal properties of poly(lactic acid) composites with rice straw fiber modified by poly(butyl acrylate). Chem Eng J 166:772–778
27. Zhao Q, Tao J, Yam RCM et al (2008) Biodegradation behavior of polycaprolactone/rice husk ecocomposites in simulated soil medium. Polym Degrad Stab 93:1571–1576
28. Yu T, Li Y (2014) Influence of poly(butylenes adipate-co-terephthalate) on the properties of the biodegradable composites based on ramie/poly(lactic acid). Compos Part A 58:24–29
29. Yu T, Jiang N, Li Y (2014) Study on short ramie fiber/poly(lactic acid) composites compatibilized by maleic anhydride. Compos Part A 64:139–146
30. Dong Y, Ghataura A, Takagi H et al (2014) Polylactic acid (PLA) biocomposites reinforced with coir fibres: evaluation of mechanical performance and multifunctional properties. Compos Part A 63:76–84
31. Wang Y, Tong B, Hou S et al (2011) Transcrystallization behavior at the poly(lactic acid)/sisal fibre biocomposite interface. Compos Part A 42:66–74
32. Biswal M, Mohanty S, Nayak SK (2009) Influence of organically modified nanoclay on the performance of pineapple leaf fiber-reinforced polypropylene nanocomposites. J Appl Polym Sci 114:4091–4103
33. Chollakup R, Tantatherdtam R, Ujjin S et al (2011) Pineapple leaf fiber reinforced thermoplastic composites: effects of fiber length and fiber content on their characteristics. J Appl Polym Sci 119:1952–1960
34. Torres-Tello EV, Robledo-Ortíz JR, González-García Y et al (2017) Effect of agave fiber content in the thermal and mechanical properties of green composites based on polyhydroxybutyrate or poly(hydroxybutyrate-co-hydroxyvalerate). Ind Crops Prod 99:117–125
35. Ding WD, Jahani D, Chang E et al (2016) Development of PLA/cellulosic fiber composite foams using injection molding: crystallization and foaming behaviors. Compos Part A 83:130–139
36. Le Moigne N, Longerey M, Taulemesse J-M et al (2014) Study of the interface in natural fibres reinforced poly(lactic acid) biocomposites modified by optimized organosilane treatments. Ind Crops Prod 52:481–494
37. Pracella M, Chionna D, Anguillesi I et al (2006) Functionalization, compatibilization and properties of polypropylene composites with Hemp fibres. Compos Sci Technol 66:2218–2230
38. Yang S, Madbouly SA, Schrader JA et al (2015) Characterization and biodegradation behavior of bio-based poly(lactic acid) and soy protein blends for sustainable horticultural applications. Green Chem 17:380–393
39. Cai J, Liu M, Wang L et al (2011) Isothermal crystallization kinetics of thermoplastic starch/poly(lactic acid) composites. Carbohydr Polym 86:941–947
40. Cai J, Xiong Z, Zhou M et al (2014) Thermal properties and crystallization behavior of thermoplastic starch/poly(ε-caprolactone) composites. Carbohydr Polym 102:746–754
41. Luduena L, Vázquez A, Alvarez V (2012) Effect of lignocellulosic filler type and content on the behavior of polycaprolactone based eco-composites for packaging applications. Carbohydr Polym 87:411–421

42. Du Y, Wu T, Yan N et al (2014) Fabrication and characterization of fully biodegradable natural fiber-reinforced poly(lactic acid) composites. Compos Part B 56:717–723
43. Suryanegara L, Nakagaito AN, Yano H (2009) The effect of crystallization of PLA on the thermal and mechanical properties of microfibrillated cellulose-reinforced PLA composites. Compos Sci Technol 69:1187–1192
44. Suryanegara L, Nakagaito AN, Yano H (2010) Thermo-mechanical properties of microfibrillated cellulose-reinforced partially crystallized PLA composites. Cellulose 17:771–778
45. Qiu K, Netravali AN (2012) Fabrication and characterization of biodegradable composites based on microfibrillated cellulose and polyvinyl alcohol. Compos Sci Technol 72:1588–1594
46. Haafiz MKM, Hassan A, Zakaria Z et al (2013) Properties of polylactic acid composites reinforced with oil palm biomass microcrystalline cellulose. Carbohydr Polym 98:139–145
47. Kowalczyk M, Piorkowska E, Kulpinski P et al (2011) Mechanical and thermal properties of PLA composites with cellulose nanofibers and standard size fibers. Compos Part A 42:1509–1514
48. Benhamou K, Kaddami H, Magnin A et al (2015) Bio-based polyurethane reinforced with cellulose nanofibers: a comprehensive investigation on the effect of interface. Carbohydr Polym 122:202–211
49. Frone AN, Berlioz S, Chailan JF et al (2013) Morphology and thermal properties of PLA—cellulose nanofibers composites. Carbohydr Polym 91:377–382
50. Abdulkhani A, Hosseinzadeh J, Ashori A et al (2014) Preparation and characterization of modified cellulose nanofibers reinforced polylactic acid nanocomposite. Polym Test 35:73–79
51. Herrera N, Mathew AP, Oksman K (2015) Plasticized polylactic acid/cellulose nanocomposites prepared using melt-extrusion and liquid feeding: mechanical, thermal and optical properties. Compos Sci Technol 106:149–155
52. Herrera N, Salaberria AM, Mathew AP et al (2016) Plasticized polylactic acid nanocomposite films with cellulose and chitin nanocrystals prepared using extrusion and compression molding with two cooling rates: effects on mechanical, thermal and optical properties. Compos Part A 83:89–97
53. Almasi H, Ghanbarzadeh B, Dehghannya J (2015) Novel nanocomposites based on fatty acid modified cellulose nanofibers/poly(lactic acid): morphological and physical properties. Food Pack Shelf Life 5:21–31
54. Mariano P, Minhaz-Ul H, Debora P (2014) Morphology and properties tuning of PLA/cellulose nanocrystals bionanocomposites by means of reactive functionalization and blending with PVAc. Polymer 55:3720–3728
55. Pei A, Zhou Q, Berglund LA (2010) Functionalized cellulose nanocrystals as biobased nucleation agents in poly(L-lactide) (PLLA)—crystallization and mechanical property effects. Compos Sci Technol 70:815–821
56. Bitinis N, Fortunati E, Verdejo R et al (2013) Poly(lactic acid)/natural rubber/cellulose nanocrystal bionanocomposites. Part II: properties evaluation. Carbohydr Polym 96:621–627
57. Yu HY, Qin ZY, Liu L et al (2013) Comparison of the reinforcing effects for cellulose nanocrystals obtained by sulfuric and hydrochloric acid hydrolysis on the mechanical and thermal properties of bacterial polyester. Compos Sci Technol 87:22–28
58. Kamal MR, Khoshkava V (2015) Effect of cellulose nanocrystals (CNC) on rheological and mechanical properties and crystallization behavior of PLA/CNC nanocomposites. Carbohydr Polym 123:105–114
59. Miao C, Hamad WY (2016) In-situ polymerized cellulose nanocrystals (CNC)-poly(L-lactide) (PLLA) nanomaterials and applications in nanocomposite processing. Carbohydr Polym 153:549–558
60. Fortunati E, Armentano I, Zhou Q et al (2012) Multifunctional bionanocomposite films of poly(lactic acid), cellulose nanocrystals and silver nanoparticles. Carbohydr Polym 87:1596–1605
61. Morelli CL, Belgacem MN, Branciforti MC et al (2016) Supramolecular aromatic interactions to enhance biodegradable film properties through incorporation of functionalized cellulose nanocrystals. Compos Part A 83:80–88

62. Yu HY, Qin ZY (2014) Surface grafting of cellulose nanocrystals with poly (3-hydroxybutyrate-co-3-hydroxyvalerate). Carbohydr Polym 101:471–478
63. Arrieta MP, López J, López D et al (2016) Biodegradable electrospun bionanocomposite fibers based on plasticized PLA–PHB blends reinforced with cellulose nanocrystals. Ind Crops Prod 93:290–301
64. Malmir S, Montero B, Rico M et al (2017) Morphology, thermal and barrier properties of biodegradable films of poly (3-hydroxybutyrate-co-3-hydroxyvalerate) containing cellulose nanocrystals. Compos Part A 93:41–48
65. Yu HY, Yao JM (2016) Reinforcing properties of bacterial polyester with different cellulose nanocrystals via modulating hydrogen bonds. Compos Sci Technol 136:53–60
66. Monteiro SN, Calado V, Rodriguez RJS et al (2012) Thermogravimetric stability of polymer composites reinforced with less common lignocellulosic fibers—an overview. J Mater Res Technol 1(2):117–126
67. Rosa MF, Chiou BS, Medeiros ES et al (2009) Effect of fiber treatments on tensile and thermal properties of starch/ethylene vinyl alcohol copolymers/coir biocomposites. Bioresour Technol 100:5196–5202
68. Morandim-Giannetti AA, Agnelli JAM, Lanças BZ et al (2012) Lignin as additive in polypropylene/coir composites: thermal, mechanical and morphological properties. Carbohydr Polym 87:2563–2568
69. Guigo N, Mija A, Vincent L et al (2010) Eco-friendly composite resins based on renewable biomass resources: polyfurfuryl alcohol/lignin thermosets. Eur Polym J 46:1016–1023
70. Deka H, Misra M, Mohanty A (2013) Renewable resource based "all green composites" from kenaf biofiber and poly(furfuryl alcohol) bioresin. Ind Crops Prod 41:94–101
71. Azwa ZN, Yousif BF (2013) Characteristics of kenaf fibre/epoxy composites subjected to thermal degradation. Polym Degrad Stab 98:2752–2759
72. Elkhaoulani A, Arrakhiz FZ, Benmoussa K et al (2013) Mechanical and thermal properties of polymer composite based on natural fibers: moroccan hemp fibers/polypropylene. Mater Des 49:203–208
73. Panaitescu DM, Vuluga Z, Ghiurea M et al (2015) Influence of compatibilizing system on morphology, thermal and mechanical properties of high flow polypropylene reinforced with short hemp fibers. Compos Part B 69:286–295
74. Bakare FO, Ramamoorthy SK, Åkesson D et al (2016) Thermomechanical properties of bio-based composites made from a lactic acid thermoset resin and flax and flax/basalt fibre reinforcements. Compos Part A 83:176–184
75. Kim KW, Lee BH, Kim HJ et al (2012) Thermal and mechanical properties of cassava and pineapple flours-filled PLA bio-composites. J Therm Anal Calorim 108:1131–1139
76. Thao Tran TP, Bénézet JC, Bergeret A (2014) Rice and Einkorn wheat husks reinforced poly (lactic acid) (PLA)biocomposites: effects of alkaline and silane surface treatments of husks. Ind Crops Prod 58:111–124
77. Thakur VJ, Thakur MT, Gupta RK (2013) Development of functionalized cellulosic biopolymers by graft copolymerization. Int J Biol Macromol 62:44–51
78. Priya B, Gupta VK, Pathania D (2014) Synthesis, characterization and antibacterial activity of biodegradablestarch/PVA composite films reinforced with cellulosic fiber. Carbohydr Polym 109:171–179
79. Peresin MS, Habibi Y, Zoppe JO et al (2010) Nanofiber composites of polyvinyl alcohol and cellulose nanocrystals: manufacture and characterization. Biomacromolecules 11:674–681
80. Park SH, Oh KW, Kim SH (2013) Reinforcement effect of cellulose nanowhisker on bio-based polyurethane. Compos Sci Technol 86:82–88
81. Spinella S, Lo Re G, Liu B et al (2015) Polylactide/cellulose nanocrystal nanocomposites: efficient routes for nanofiber modification and effects of nanofiber chemistry on PLA reinforcement. Polymer 65:9–17
82. Garcia NL, Ribba L, Dufresne A et al (2011) Effect of glycerol on the morphology of nanocomposites made from thermoplastic starch and starch nanocrystals. Carbohydr Polym 84:203–210

83. Montero B, Rico M, Rodríguez-Llamazares S et al (2017) Effect of nanocellulose as a filler on biodegradable thermoplastic starch films from tuber, cereal and legume. Carbohydr Polym 157:1094–1104
84. Kaushik A, Singh M, Verma G (2010) Green nanocomposites based on thermoplastic starch and steam exploded cellulose nanofibrils from wheat straw. Carbohydr Polym 82:337–345
85. Martins IMG, Magina SP, Oliveira L et al (2009) New biocomposites based on thermoplastic starch and bacterial cellulose. Compos Sci Technol 69:2163–2168
86. Meneguin AB, Cury BSF, Dos Santos AM (2017) Resistant starch/pectin free-standing films reinforced with nanocellulose intended for colonic methotrexate release. Carbohydr Polym 157:1013–1023

Eco-friendly Polymer Composite: State-of-Arts, Opportunities and Challenge

V. S. Aigbodion, E. G. Okonkwo and E. T. Akinlabi

1 Introduction

Polymer matrix composites (PMCs) have come a long way in becoming a key player in the world of high performing engineering materials. With a wide array of applications due to its properties, these classes of materials are becoming more popular than their metallic and ceramic counterparts and as such replacing then in various applications especially where high strength to weight ratio and low density is a priority [1]. Composites like PMCs being multiphase materials exhibit properties that are dependent on the constituents; resin and fiber/particulate (reinforcement/filler). The resin helps to re-distribute stress and holds the reinforcement together while the reinforcement (fiber/particulate) provides the necessary load carrying operation and hence the strength and stiffness of the composite. Moreover, composites can always be tailored to the choice of the producer [2]. However, most popular polymers are synthetic and coupled with decreasing petroleum resources and environmental policies, they areas such costly and non-biodegradable under standard conditions. Addition of environmental favourable and renewable source of reinforcement to these polymers have not only helped to shift the overall outlook but also give it a new and wider prospect. Aside from the ecological benefits of using these materials, overall cost implication and health problems are also eliminated [3].

As at 2016, an estimated 300 million tons of plastics were produced with the majority of it finding ways into water bodies like rivers, oceans etc. It is been

V. S. Aigbodion (✉) · E. G. Okonkwo
Department of Metallurgical and Materials Engineering,
University of Nigeria, Nsukka, Nigeria
e-mail: victor.aigbodion@unn.edu.ng

V. S. Aigbodion · E. T. Akinlabi
Department of Mechanical Engineering Science, University of Johannesburg,
P.O. BOX 524, Auckland Park, South Africa

© Springer Nature Switzerland AG 2019
Inamuddin et al. (eds.), *Sustainable Polymer Composites and Nanocomposites*,
https://doi.org/10.1007/978-3-030-05399-4_42

estimated that by 2050, there will be more plastics in the ocean than fishes.[1] The menace doesn't only extend to this as poor waste management culture in most developing countries have also led to mismanagement of recyclable portion of these class of wastes. With the rising volume of plastics dumps all around the world, these properties come in handy with regard to reduction in carbon emission and greenhouse gases as is the case due to incineration or burning of these materials, and as such reclamation of lands used for landfills; environmental pollution.

The birth of eco-friendly composites can be likened to man's continual quest for newer breeds of materials crisscrossing from the time of reinforcing bricks with straw to this day where nanoscience and technology has brought in the era of nanocomposites that uses Nano-sized materials like nanoclays, nanotubes, nanofibers, and nanosheets etc. as fillers/reinforcement. This can also be associated to technological breakthroughs that have seen the significant improvement in properties of some polymers thus opening up more areas of application like in the food packaging industry that uses PLA, biomedical, agricultural, construction sector, transport and even in textiles industries [4, 5]. This quest has also led to the reinvestigation of old processes giving rise to newer products [5]. Though ecological challenges seem to be the driving point behind it, dwindling nature of other sources of materials like petroleum has also led to a turnaround in material research. Man's rate of consumption of non-renewable natural resources is ever increasing due and outways the rate at which the earth produces them [6]. Therefore the only option is the search for the sustainable and renewable source of materials and stimulates ambitious policies for a significant increase in resource efficiency, particularly through technical change and innovation. Moreover, annual quantity of agrowastes and natural fibers produces ranks in billions of tons with majority burnt on-field [7]. This can expand rural agriculture-based economics by opening up new market [8]. According to Rana S. and Fangueiro R., cultivation of one hectare of hemp on a hectare area of land causes the absorption of approximately 2.5 tons of atmospheric carbon dioxide during a vegetative season while jute absorbs 2.4 tons of carbon dioxide. On the other hand, production of one ton of polypropylene emits about 3 tons of carbon dioxide into the atmosphere [9]. Thus wastes or materials which are both biodegradables can be used to reduce the cost of production of composites without impairing their properties.

1.1 What Are Eco-friendly Composites (EFC)

According to Adeosun et al., composites that are both biodegradable and renewable are termed eco-friendly composites [10]. This is because they can be easily disposed of without harm to the environment. Asokan et al. [11] sees them as

---

[1]FEATURE: UN's mission to keep plastics out of oceans and marine life. https://news.un.org/en/story/2017/04/556132-feature-uns-mission-keep-plastics-out-oceans-and-marine-life. Accessed on January 10, 2018.

biocomposites since they are manufactured from materials like natural fibers and agro-wastes. According to Mitra bio-composites covers composite materials where at least one of the component is bio-based. These include petroleum-derived polymers like epoxy, polyester etc. reinforced with natural fibres like jute, flax etc.; bio-polymer like polylactic acid (PLA) reinforced by bio-fibers like coir and bio-polymers reinforced with synthetic fibers like glass or carbon [1]. However, a composite that combines both natural fibers/particulates like agro-waste and natural resins are termed green composite because both the resin and reinforcement decompose by the action of micro-organisms. As such it is pertinent to point out that though all the aforementioned composites are friendlier to the environment than the conventional composites and as such can be termed eco-friendly, a subtle demarcation can be seen based on their constituent. Hence eco-friendly polymer matrix composites can either bio-based green since some bio-based polymer matrix composite requires the introduction of antioxidants to help facilitate degradation [12, 13] and hence are called oxo-degradable polymers [13]. Nevertheless, biocompatible materials for the production of EPCs can either be natural or synthetic and as such can be used as either the reinforcement or matrix of a composite [14]. Synthetic degradable polymers can be divided into condensation polymers (polyesters, amides, polyureas, polyurethanes etc.), addition polymers, water-soluble polymers (polyvinyl alcohol, polyethylene glycol) and blends of natural and synthetic polymers (starch and PLA based blends). Decomposition is basically by hydrolysis, microbial action, photo or thermos-oxidation etc. [15]. Natural occurring biocompatible polymers can be divided into agro-polymers and biopolymers. Agro-polymers are polymers derived from biomass products and include polysaccharides like starch, cellulose, chitin, etc. and their respective derivatives, protein like soya protein, casein, zein, gelatin etc., lipids like triglycerides and its derivatives. Biopolymers, on the other hand, include polymers from the following sources.

I. Synthesis of bio-derived monomers. E.g. PLA,
II. Extraction using micro-organisms. E.g. Polyhydroxyalkanoate (PHA)
III. Synthesis of synthetic monomers. E.g. Polycaprolactone (PCL), Polyesteramides (PEA), aliphatic copolymers like PBSA, aromatic copolymers like PBAT [5, 16].

On the other hand, most fillers used in the production of eco-friendly composites are derived from natural fibers/particulates like Nanosilica, graphene, nanoclays, carbon nanotube, nanocellulose etc. Cellulose and starch-based fibers are the quite prominent because of their availability, cost and high specific strength. Nanotechnology has also seen to the production of these fibers at the nano-sized level. For instance, nanocellulose (NC) can be categorized into three subcategories based on cellulose source and on the production methods, cellulose nanocrystals (CNC), nanofibrillated cellulose and bacterial nano-cellulose. CNCs are highly crystalline [16].

1.2 Why Eco-friendly Composites and Trends

Like earlier stated, eco-friendly Polymer composites have caught the attention of researchers due to some of their daring properties especially with regards to degradation [17]. However, the increase in the research into it can be attributed to various reasons. According to [17], the drive is because of increased drift towards maximum utilization of renewable natural resources whereas [18–20] sees it be as a result of increased environmental awareness. Ease of availability and being economical as compared to conventional composites produced wholly with synthetic material was also opinioned by [21]. Thus from the ongoing, eco-friendly composites have not only come into enhance the already existing products by emulating nature's way of producing material but also to expand on it.

One attractive feature of EPC is that it can turn what is regarded as waste to a very useful material without harm to the environment; a natural way of converting waste to wealth at little or no cost but with many benefits. For example wood flour, rice husk, egg shells and various agro and industrial wastes like sawdust [22–26] have been successfully turned into useful products with many improved properties and at no harm to the environment. Moreover, with the global demand for wood and competition from the pulp and paper industry, use of residues like stalks of cereal is no longer an option but a necessity [27]. A growing class of eco-friendly composite is the wood plastic composite. This class of composites is produced by combining thermoplastics with wood/natural fiber [28] or by melt blending polymer/wood flour or powder mix. These class of composites requires less maintenance and as such offers alternative to wood products. Moreover, they can be made into complex shapes [29]. Aigbodion and co-worker have used many natural fillers in the production of eco-composites among them are (Figs. 1 and 2):

Also, the possibility of hybridization has also increased the range of properties that can be obtained from this class of composite. According to [32] hybrid

Fig. 1 Polymer composites produced with Bagasse [30]

Fig. 2 Polymer composites produced with Breadfruit shell [31]

composites are obtained by two or more different fibers in a common matrix. Thus it allows for modifying the properties of the composite so as to meet the required need. Various works on hybrid eco-friendly polymer composites have shown improved properties. Ravindran et al. [28] used coconut shell powder and wood dust as filler for polyester resin and observed that tensile, flexural and shear strength of the hybrid composites were better than the mono-reinforced composite. The similar result has been observed in the works done by [20, 25, 33]. On the other hand, hybridization also helps in reducing the overall cost of producing a synthetic composite by balancing out some of the components with some renewable and easily available ones. Kasiviswanathan et al. [34] worked on the impact of stacking of natural fibers and a synthetic fiber as reinforcement for polyester resin. Sisal, banana, and E-glass fiber were used and the fibers were stacked alternatively. Although the presence of glass fiber improved the mechanical of the composite, that of the ones with more of sisal and banana fiber were not far off showing that they can serve as a possible replacement. Udhayasankar and Karthikeyan [35] highlighted some of the uses or applications of natural fiber composites.

Nevertheless, the developmental trend of EPCs is ever blazing. Nanoclays are known for their high aspect ratio and intercalative/exfoliation behaviour. The behaviour of nanoclay in a PMC is dependent on the type of clay, pre-treatment method, mode of dispersion of the filler and the resin itself [36]. Addition of nanoclays to EPCs not only leads to improved mechanical properties but higher thermal stability [37]. Rahmat et al. [8] from their work on the addition of nanoclay to rice husk reinforced polypropylene composite observed an improved tensile strength and modulus compared to the rice husk—polypropylene composite. Ashori A. et al. also found that introduction of nano-silica in rice husk and beach bark-Propylene composite enhanced the physical and mechanical properties like flexural strength whereas the tensile properties suffered due to the high weight fraction of the fiber incorporated [38]. Nanocellulose fibers, whiskers etc. have all led to improved optical, electrical, magnetic properties etc. of the produced EPC thus opening new grounds for engineering applications.

1.3 Properties of Eco-friendly Composites

Eco-friendly polymer composites are known to offer very enticing benefits. Besides their ability to degrade without damage to the environment, they are also known for their high strength to weight ratio, low density, excellent corrosion resistance [39]. Aside from that, its constituents are abundant, cheap and renewable [40], thus there is an assured supply means. Some of the properties of eco-friendly composites can be summarized in Table 1.

1.4 Challenges in the Processing of Eco-friendly Composites

Although eco-friendly composites posit material for the future most are lignocellulosic which are highly polar in nature. Therefore coupling with nonpolar matrix leads to poor adhesion at the interface and hence properties that are lower than expected. Addition of hydrophilic fibers to hydrophobic matrix reduces the mechanical strength of the composite formed due to incompatibility between the fiber and matrix. Likewise, nonuniform dispersion and agglomeration of nanoparticle lead to a discrepancy in result [8]. To combat this modification of the surface of the fibres have been seen to be a plausible solution. Therefore some research into biobased composite has been done with the aim of finding the right surface treatment so as to enable proper compatibility. This has been done using varied means ranging from chemical treatment (mercerization, silane treatment, acetylation etc.), physical methods (stretching, calendering, thermo treatment), electric discharge (corona treatment, cold plasma etc.) [42].

The collection, storage, transportation and economics of production is another defying factor that hinders the use of cereal by-product, straw, shell and other agricultural residues in composites [43]. However, research focused on the use of these materials have also followed a developmental market, which has led to the birth of a new market opportunity for these surplus inexpensive field crop husks.

Other challenges include the fact that most reinforcement used in producing eco-friendly polymer composites are difficult to get in the continuous fibrous state.

Table 1 Properties of eco-friendly composites [35, 41]

• Renewable source of reinforcement/matrix	• Reduced dermal and respiratory irritation
• Biodegradable	• Acoustic absorption
• Low density	• Enhanced energy recovery
• Good stiffness and strength	• Reduced tool wear
• No off-gassing of toxic compounds	• Reduced fogging behaviour
• Less spintering	• Favourable processing properties

Moreover fiber quality of plant-based fibers is dependent on many factors including, the age of the plant, retting method etc. [44–46]. The final properties of composite materials depend on fibre properties (morphology, surface chemistry, chemical composition and crystalline contents) as well as matrix properties (nature and functionality) [43]. Bledzki A. J. et al. compared the properties of by-products like barley husk, coconut shell and softwood as reinforcement for thermoplastics like polyethylene. Morphological analysis showed that coconut shell had a relatively smooth morphology. Water absorption behaviour showed that softwood absorbed the highest quantity of water in the first few days. This can be ascribed to the hemicellulose content of the fillers. Barley husk showed best tensile and impact strength whereas coconut-shell shell exhibited the best elongation at break. Addition of coupling agent (MA-PP) led to increased tensile properties and a minimal change in impact strength [43].

1.5 Opportunities

EFCs promises sets of materials with excellent environmental compatibility. Aside from serving as a possible alternative to varying already existing materials, they also provide a chance for the better utility of assumed waste. One of the class of eco-friendly composites that have been gaining relevance is the wood polymer composite which can be a good alternative to wood and as such help to relieve the pressure on the forestry industry. This class of EFC has found application as structural parts for low-cost buildings, cabinet, and upholstery etc. Incorporation of nanoclays has also led to the improved thermal stability of polymer matrix composites without affecting their ecofriendliness thus offering an opportunity for use at higher temperatures [47].

Incorporation of other nanoparticles like nanocellulose in biodegradable polymers like PLA has made it possible for applications in areas like tissue engineering, seizure for wounds etc. in food packaging industry these class of materials are of high demand due to their biodegradability, ability to form barriers to gases and repel water. Other areas of application of EFCs include as biosensors, tissue engineering etc. [48].

Likewise, polymer blend which has been identified as one of the ways to curtail the inherent poor properties of some biodegradable polymers. PLA which is a thermoplastic is known for its brittleness, poor crystallinity and poor impact strength. However blending with other polymers like PBS, rubber, Polypropylene has helped to improve its properties. Polymer blends have found application in the textile industry amongst other engineering applications.

2 Processing of Eco-friendly Composites

The trend in design and processing of polymer composites is one that has continued to evolve, the fact that can be attained to the development of new reinforcement and a better understanding of the effect of the interaction of the matrix and reinforcement. Traditionally, polymers are classified as either thermosets or thermoplastics. Thermoplastics are known for having branched chains. Notable among them are PEEK, PP. PA-6, PPS etc. and are known for varied application in aerospace and automobile industry due to properties like high chemical resistivity, fracture toughness and crack growth resistance. Though they can be reshaped via heating and cooling, they have lower stiffness and strength compared to thermosets. Thermosets, on the other hand, are made by crosslinking or chain extension of monomer chains mostly under low heat and pressure in a processing called curing. They are characterized by rigid three-dimensional structures and high molecular weight and normally decompose before melting, unlike thermoplastic. This can be via poly-condensation, copolymerization or homo-polymerization with a catalyst as in the case of epoxy. Popular thermosets include epoxy, acrylic, polyester, polyurethanes, vinyl ester etc. and have found application as electrical insulators, waterproof coating, circuit board, car parts etc.

Most of the polymers mentioned above are synthetic polymers though when mixed with natural fibers or agro-wastes form eco-friendly composites. Nevertheless, naturally occurring polymers also can also be classified as based on the subunit (mer group) as seen in synthetic polymers. PLA is a biodegradable aliphatic thermoplastic polyester, polyhydroxyalkanoate is also a polyester. These polymers are (bio)degradable polymers due to the potentially hydrolysable ester bonds and relatively short aliphatic chains present in the macromolecules [49]. Other biopolymers include naturally produced polymers like poly-3-hydroxybutyrate (PHB), polyhydroxyvalerate (PHV) and polyhydroxyhexanoate (PHH); synthetic ones like Polybutylene succinate (PBS), polycaprolactone (PCL); Cellulose esters like cellulose acetate and nitrocellulose and their derivatives; starch and its derivates among others. Nevertheless, the chemistry of the matrix and fiber plays an important role in the processing of composite and as such requires a quick preview.

2.1 Chemistry of Eco-friendly Composites: Matrix and Fillers for Production of EFC

Research has shown that the behaviour of composites is dependent on the integrity of the bond between the matrix and reinforcement. Eco-friendly composites are made up of materials with differing chemical make—up and hence exhibits differing behaviour. Thus for a proper understanding of this material, it is pertinent to look at the chemistry of the components which are the matrix and reinforcement. The matrix is one of the most important parts of a composite. Aside from distributing stress and acting as a bonding agent, it also shields the reinforcement from the external environment. In eco-friendly PMCs, the matrix can either be synthetic or natural polymers. Natural polymers can be derived from renewable sources or from petroleum. However, some synthetic based polymers can be derived from renewable sources For instance polyamide can be synthesized from castor oil. One of the most widely used matrices is Poly(lactic acid (PLA). It can be produced by different methods like direct polycondensation. Other methods include azeotropic condensation polymerization, solid state polymerization, coordination-insertion mechanism which involves the use of catalysts like metallic alkoxides. PLA has high transparency and elastic modulus can be thermoplastically processed like conventional plastics and has been widely used in the development of disposable products, such as disposable cutlery, cups, and films. Its brittleness at room temperature and hydrolysability have limited its application. Another important natural occurring eco-friendly polymer used as the matrix is starch. Starch a semi-crystalline biopolymer is a branched homopolymer of glucose, with α-(1→4) linear links and α-(1→6) branched links and is the main carbohydrate reserve in roots, tubers, seeds, fruits, cereals etc. It is normally made up of amylose and amylopectin and depending on the source of origin, the granules can vary in size, shape, chemical composition and structure [50]. Starch in granular form has limited processability and as such is often blended with other polymers. Chitin and its derivative chitosan is another important and widely used natural polymer. This polymer is derived from the exoskeleton of insects, crustacean, insects, arthropods. Chitin has limited solubility in diluted acidic aqueous solutions unlike chitosan.

However due to its, availability, low cost, high biocompatibility, biodegradability, antimicrobial property, ease of chemical modification and excellent film-forming ability. This polymer also possesses properties, including its high viscosity, charge distribution, and release mechanisms, making it particularly suitable as a carrier this polymer has wide range application from the biomedical to industrial areas [50, 51]. Aside from cellulose, poly(hydroxyl butyrate) (PHB); a naturally occurring polyester produced by numerous bacteria in nature as an intracellular reserve of carbon or energy; guar gum, non-ionic polysaccharides another natural occurring polymer used as matrix. Just like natural occurring polymer matrices, the biodegradable polymer matrix can be derived from synthetic sources like fossil fuel via processes like polycondensation. Poly(ε-caprolactone) (PCL) is a biodegradable and biocompatible polymer manufactured by the ring-opening polymerization of ε-caprolactone (CL). PCL has a flexible chain, exhibits a high elongation at break, low modulus and the low melting point which is its major drawback. However, it is normally blended with other polymers to improve stress crack resistance, dyeability, and adhesion. Poly(butylene succinate) (PBS) is another biodegradable aliphatic thermoplastic polyester produced through the condensation polymerization. It has been blended with PLA and has found applications in the textile sector. Others include Poly[(butylene succinate)-co-adipate] (PBSA) an environmentally friendly biodegradable thermoplastic polyester made of butylene succinate adipate random copolymer, poly(butylene adipate-co-terephthalate) (PBAT), an aliphatic-aromatic liner random copolyester synthesized by polycondensation reaction of 1,4-butanediol in the presence of adipic and terephthalic acids etc. [51].

Reinforcement is generally responsible for strengthening the composite and as such to improve the mechanical properties of the composite. Thus it can be said that composites are designed in such a way that the loads applied can be supported by the reinforcements. Reinforcements can be present in different formats e.g. spheroids, spherical, short or continuous fiber, whiskers, platelet etc. However, the most common form of reinforcements is fibers and particulates.

Fibers are classified as materials whose length is much longer than its width. They are generally known for their strength and stiffness; properties which have been taken advantage of in fiber reinforced composites. Fibers can be natural or synthetic. Synthetic fibers can be organic or inorganic based on having carbon basis. Natural fibers can be categorized based on sources: animal, plant (lignocellulosic) and mineral fiber as shown in Fig. 3.

Plant-based fibres are classified according to the part of the plant where they are obtained such as leaf, seed/fruit, stem and bast. Properties of natural fibres vary considerably depending on the fibre diameter, structure, degree of polymerization, crystal structure and source, whether the fibres are taken from the plant stem, leaf or seed, and on the growing conditions. Generally, plant fibres consist of cellulose,

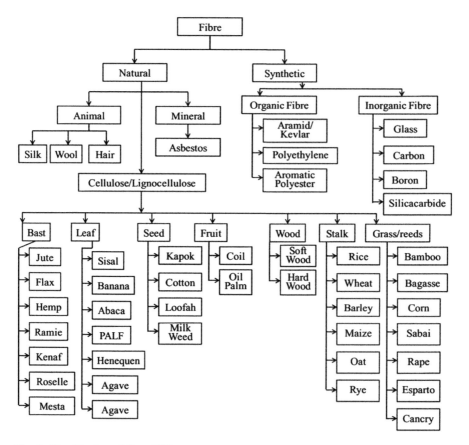

Fig. 3 Classification of fibers [52]

hemicellulose, lignin, pectin and waxes. Cellulose is a linear condensation polymer consisting of D-anhydroglucopyranose units joined together by b-1,4-glycosidic linkages. The molecular structure of cellulose, which is responsible for its supramolecular structure determines many of its chemical and physical properties. In the fully extended molecule, the adjacent chain units are oriented by their mean planes at the angle of 180° to each other. Thus, the repeating unit in cellulose is the anhydrocellobiose unit, and the number of repeating units per molecule is half a degree of polymerization which may be as high as 14,000 in the native cellulose [53]. It provides strength, stiffness and structural stability of the fiber [54, 55]. Reinforcing efficiency of natural fiber is depends upon the nature of cellulose and its crystallinity. Depending on the type of fibre, cellulose has its own cell geometry which is responsible for the determination of mechanical properties of plant fibres [53, 56] (Fig. 4).

Hemicellulose occurs mainly in the primary cell wall and has branched polymers containing five and six carbon sugars of varied chemical structures. It differs from cellulose in three aspects firstly it contains several sugar units, exhibits a considerable degree of chain branching containing pendent side groups give rise to its ion crystalline nature and its degree of polymerization (DP). In the case of hemicelluloses it is 50–30 but in cellulose is 10–100 times more than that of hemicelluloses. Hemicellulose is very hydrophilic, soluble in alkali and easily hydrolyzed in acids [58]. Hemicellulose found in the natural fibers is believed to be a compatibilizer between cellulose and lignin [53].

Lignin is a complex hydrocarbon polymer with both aliphatic and aromatic constituents and it is totally insoluble in most of the solvents and can't be broken down into monomeric units. It is considered to be a thermoplastic polymer having a glass transition temperature of around 90 °C and melting temperature of around 170 °C. It is totally amorphous and hydrophobic in nature. It is not hydrolyzed by acids, but soluble in hot alkali, readily oxidized and easily condensable with phenol [55, 58] hence it often used to partially replace phenol in the phenolic and lignophenolic pre-polymers to produce composites with thermoset matrices. Lignin keeps the water in fibers, acts as a protection against biological attack and as a stiffener to give stem its resistance against gravity forces and wind [53].

Pectin is a collective name for heteropolysaccharides. They provide flexibility to plants. Pectin structure is complex and their side chains are often cross-linked with the calcium ions and arabinose sugars [58]. Pectin is soluble in water only after a

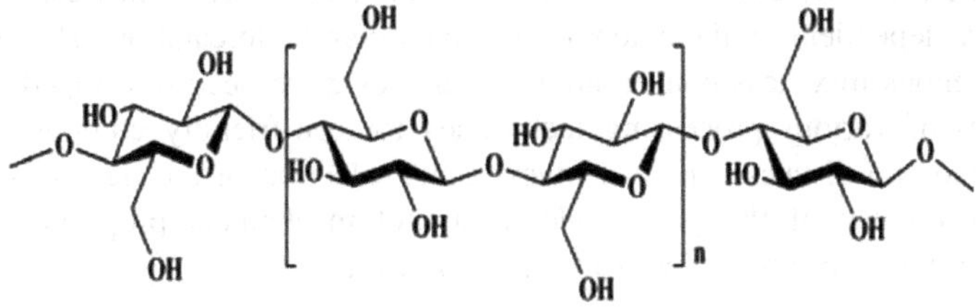

Fig. 4 Chemical structure of cellulose [57]

partial neutralization with alkali or ammonium hydroxide. Pectin is normally removed so as to separate cellulose from hemicellulose in stems. Waxes make up the last part of fibers and they consist of different types of alcohols [58]. They are insoluble in water as well as in several acids. The lignin, hemicelluloses and pectin provide the adhesive to hold the cellulose framework structure of the fiber together. Natural fibers are also composed of a small amount of organic (extractives) and inorganic (ash) components. Organic extractives are responsible for the colour, odour and decay resistance, and inorganic matters enhance abrasive nature of the fiber [55]. It is pertinent to point out that cellulose is one of most common biopolymers on earth and has found applications in various industries.

Animal-based fibers are derived from the hair, wool, and silk comprising mainly of protein and silk. Hairs and wools are made of protein and got from creatures e.g. Sheep's downy, goat hair, horse hair, alpaca hair, and so forth. Silk fiber is the filaments gathered from the dry saliva of bugs or crawling creatures cocoons. Avian strands are the fiber from fowls [59].

With regards to particulates, shells like an eggshell, periwinkle shell with calcium base are very common. Agrowastes like rice husk, bagasse ash, husks, wood dust etc. are on the side with regard to having a cellulose base. For use as particulate reinforcement is also clays. Clays are one of the most important minerals in the earth and have found varied applications in many sectors [60]. Although they are known to be hydrophilic, Nanotechnology has made it possible for a modified class of clay called nanoclay which have been found to improve quality of the product, save cost and doesn't harm the environment [61].

2.2 Various Processing Methods

Processing of PMCs entails a blend of both thermal and or mechanical process. However, a processing method to be used depends on the type of resin (thermoplastic or thermoset), nature of reinforcement (fiber or particulate) and even the size of the reinforcement (micro or nano-sized) [62]. In some processes, the matrix and reinforcement are combined in the mould whereas in some the reinforcement is incorporated into the matrix in a pre-moulding operation which is later used in moulding the composite part [62]. Generally, the extent of dispersion of a filler in a matrix determines the properties. Unlike in micro-sized particles where interfacial adhesion shows the extent of wettability, nano-sized particulates especially nanoclays are dependent on the extent of exfoliation and intercalation. Thus to achieve a homogenous mix, less conservative techniques have been developed. The characteristics of various processing techniques for eco-friendly composites will be highlighted and discussed here. However, the selection of the right process based on the geometry of the part, scale, cost and mechanical properties is key in achieving optimum properties from the composite.

Hand lay up

This is one of the most popular fabrication technique. Known for its low cost compared to other techniques, it involves the manual application of a laminate ply or woven fibers before the resin is poured to wet the fibers. A roller is used to ensure the uniform spread of the resin on the fibers and remove entrapped air before the composite is allowed to cure. Though it offers a cheap method of producing composites especially for academic researchers, its industrial application is minimal because of low production rate and limited volume fraction of the reinforcement that it can accommodate [63]. Also mixing of the resin is dependent on the operator's skill and this method is mostly used for resin with a low viscosity like epoxy, polyester, phenol etc. Many works on natural fiber reinforced polymer matrix composites have been done using this method.

Spray up method

In spray up method, chopped fibers and resin are made to mix at the tip of a spray nozzle before falling into an already prepared mould. Just like in hand lay-up, rollers are used to remove entrapped air. Spray up technique offers an efficient way of producing thermosetting PMC as it can accommodate more variety part sizes as well as the higher volume fraction of reinforcement compared to hand lay-up. But the skill of operator/worker plays a vital role in determining the final quality of the composite part produced [63].

Bag moulding

This involves preparing continuous prepregs which area continuous sheet of fibers already preimpregnated with a layer of the thermoset resin, staking up precuts of the prepregs in a mould, covering with a thin polymer film and curing in an autoclave or a press at a higher temperature. In this method, pressure and vacuum are used to remove excess resins, consolidate the layers of the prepregs while the resin cures. Though a slow and labour-intensive method, it is used in producing parts with accurate fiber orientation, minimal void and controlled fiber volume fraction [62].

Filament winding

Known for producing hollow structural parts using thermosetting resins, it entails pulling a band of continuous fibers through a tank filled with catalyzed resin which will be wound round a rotating mandrel. The resin-coated fiber band is also traversed back and forth along the length of the rotating mandrel to create a helical winding pattern. The winding angle can be varied by controlling the mandrel speed and the traversing rate of the fiber band. The part is cured in an oven, and the mandrel is removed to create a hollow shape. In some applications, such as oxygen tanks, the mandrel is not removed after curing, and it becomes a part of the structure. One of the primary advantages of filament winding is the ability to control resin usage and minimal fiber is used. However, the difficulty in laying fiber exactly along the length of the component and high cost of mandrels especially in producing large components makes it less economical [62, 63].

Pultrusion

This is a process of producing long, straight composite part using continuous fibers. This method of pulling continuous fibers through a tank containing already catalyzed thermosetting resin then through a heated die where the resin-coated fibers are cumulated to form the shape being produced. Curing takes place as they move along the length of the die. For industrial application, parameters like the method of preheating, die temperature, pulling speed has to be set down. Pultrusion provides a fast and economical way of impregnating and curing fibers using an enclosed impregnation area thus limiting volatile emission. On the other hand, it is limited to the production of constant or nearly constant components and cost of heating the dies cannot be overlooked. However, it can be used in producing natural fiber reinforced polymers composites [63].

Compression moulding

Here the composite materials are laid between two moulds before heat and pressure are applied for curing to take place. Varieties of this method exist including bulk moulding compound (BMC), thick moulding compound (TMC), sheet moulding compound (SMC) and wet lay-up compression; all depending on the type of moulded materials [63]. It is known for the rapid production of large quantities of complex parts, part design flexibility; however, the tooling cost is high due to high pressure involved though it can be used to produce parts with the good surface finish.

Resin transfer moulding

This is a very popular liquid composite moulding method. Here dry fiber preform is placed in the mould and thermosetting resin already mixed with a catalyst or curing agent is injected into the mould under pressure. As the prepolymer resin enters the mould, it displaces air and wets the dry fiber with curing occurring at room or elevated temperature. Compared to bag moulding and compression moulding, tooling cost is low as the resins are transferred under low pressure. However improper impregnation of parts can occur leading to increased cost due to the scrapping of the parts. A variant of this method called structural reaction injection moulding (SRIM) is used in cases of resins like polyurethanes due to the high reactivity of the chemicals used in making the matrix.

However, due to the introduction of phases with varying morphologies and at the same time need for proper dispersal, new methods have also come up. Some of these include extrusion followed by injection moulding, melt extrusion, melt compounding followed by compression moulding, direct melting, one step in situ solution polymerization, injection moulding and solution casting after gelatinization [64]. Extrusion is a process of converting a raw material into a product of uniform shape and density by forcing it through a die under controlled condition. It is basically used for thermoplastic composites. It can modify the thermal stability, mechanical strength, elongation, adhesive strength and other mechanical properties of polymers [65]. A variant of extrusion method called solid-state extrusion process which involves solid deformation of polymers which makes it possible to highly oriented structures in quite substantial cross sections for engineering and scientific studies have been studied by [66] where a haul-off is used to pull the composite through the die. Injection moulding is one of the most widely used polymer matrix composite processing techniques. Here the filler used is either short fibers or particulates. In this method, polymer granules and reinforcement are fed into a heated rotating barrel through a hopper. Heat in the barrel melts the polymer pellets while the shearing action of the rotating barrel mixes the polymer melt and filler. The molten composite is then injected into a mould. One of the widely used types of the extruder is the twin screw extruder. However other parameters like speed, temperature etc. affect the properties of the mould. Bledzkiet al. used this method to fabricate NFC composite [43].

Intercalation method is a top-down approach that is based on the exfoliation property of layered silicates like montmorillonite, mica and other nanoclays. Here clays already modified with organic surfactants like amino acid, imidazolium, and phosphonium salts are used as inorganic fillers by intercalating an organic compound into the interlayer space of the silicate. Treatment of the silicate is done to make it be hydrophobic enough to properly mix properly with the matrix. Intercalation can be direct (mechanical technique), in situ (chemical technique) or via melt compounding [67]. Nevertheless one of the major problems with this technique is incompatibility [48]. Unlike in the normal intercalative method, In situ polymerization method or in situ intercalative polymerization method involves polymerization reaction. Here nanoparticles are mixed with a monomer or monomer solution and the polymerization reaction is allowed to take place. Proper dispersion of the particulate in the monomer is very important in this method and as such surface modification is often carried out [67]. This method allows surface modification on the nanofillers without drying. It is one of the most successful methods of producing polymer nanocomposites. Examples include Polyamide 6—clay nanocomposite [48].

Solution polymerization involves exfoliating layered nanoparticles like silicates into single layers using a solvent in which the polymer is soluble. Predispersed in a solution of the polymer, allows the fillers to be finely dispersed in the polymer matrix after which the solvents are later removed by evaporation. However certain polymers are not soluble in conventional low boiling point solvent which limits the application of this method to some polymer matrix [67]. Nevertheless, use of ultrasonic waves can be employed if there is low viscosity. Melt compounding or blending involves mixing the nanofiller with a polymer melt. Unlike solution polymerization, no solvent is required however the level of intercalation depends on the compatibility between the matrix and filler. Melt compounding depends on the shear can be less effective than in situ polymerization in producing an exfoliated nanocomposite. Thus for better exfoliation ultrasonic mixing which provides high shear force has been proposed. However, due to the weakening of this force when the viscosity is high, it would be difficult to achieve uniform dispersion of the nanofillers into the polymer melts on a large scale [67]. This is one area where solution polymerization has an upper hand [48]. Other methods include emulsion polymerization which has the advantage of no environmental concern as in solution polymerization but involves the use of surfactant and stirring, ultrasound irradiation etc.

Although various techniques exist, new ones continue to evolve with respect to the manufacturing of composites. Tanahashi [67] developed a new method of fabricating nanocomposites that are based on the conventional simple melt compounding technique and does not require surface modification. The concept is based on the fracture strength of agglomerates of the dispersed particles and involves preparing agglomerates with a porous structure, before melt compounding. Nevertheless, the key to successful dispersion lies in having agglomerates with a low strength which can be easily broken down by shear stress induced during melt compounding. With respect to the environment, this method was seen to be environmentally friendly since it doesn't require chemicals used in the in situ and sol-gel method, is more suitable for large-scale production since its more compatible to industrial processes like extrusion and injection moulding. Moreover, it allows for the use of polymers not suitable for in situ polymerization, sol-gel and solution mixing [67]. In likewise manner new concepts like automated tape lay-up (ATL) and automated fiber placement (AFP) which entails the introduction of computer-guided robotic are also enroute. However it is pertinent to point out that each method is unique with its shortfalls and advantages, integration can also be done to achieve better results. For instance in polymer nanocomposites, better intercalation has been achieved using an extruder with the polymer in a melt. Moreover, the condition of the reinforcement (natural fibers and particulates) should always be taken into account when considering the method of choice.

2.3 Effect of Processing Technique

Literature reviews have shown that processing techniques affect the properties of the composite. Diez E. A. while using extrusion method to produce a semiconductive polymeric composite observed that extrusion temperature and screw speeds did not have a much significant effect on the electrical properties of the composite. But with respect to screw type, use of conventional screw gave better result compared to barrier screw [68]. Processing conditions like draw ratio, temperature, speed have been observed to affect the properties of composites produced by extrusion [66, 69]. Cai et al. from their work on the effect of temperature and draw ratio on the mechanical and morphological properties of a wood-polymer composite made with polypropylene and wood flour observed that density, tensile strength and tensile modulus increased with draw ratio. This was seen to be due to a high degree of orientation and morphology achieved due to use of solid state extrusion. Elongation at the break, on the other hand, decreased with increasing draw ratio which was attributed to increase in crystallinity. With regards to temperature, density and tensile modulus decreased with increasing die temperature whereas elongation at break and tensile strength increased [66].

Feldmann M. et al. studied the influence of process parameters on the mechanical properties of a bio-based polymer composed produced via extrusion. Polyamide from castor oil was used as matrix while chopped man-made cellulose was used as reinforcement. Thermogravimetric analysis shows different screw configurations and processing temperatures have an influence on the mass loss of the compounds. Also, different screw configurations and temperature settings led to minor deviations in the mechanical properties and the morphological structure of the bio-compounds [70].

Kadam and Mhaske [71] also looked at the effect of reprocessing via extrusion on the properties of nylon 6/talc nanocomposite and observed that reprocessing of the composites up to three times leads to a decrease in mechanical properties. However, with regard to the concentration of nanotalc, the mechanical properties increased showing that the decrease in mechanical properties of the composite was due to degradation of nylon 6 as was also confirmed by thermal and rheological analysis. Poletto M. looked at the effect of processing condition on cellulose fiber reinforced polystyrene composite. The result of his study showed that extruding at 400 RPM improves the mechanical and dynamic properties of a composite due to a reduction in fibre size, increasing the superficial area of the fiber and as such more contact area with matrix [72]. Reprocessing has also been observed to affect the properties of composites especially when recycling is of primary concern.

Nevertheless, the effect of processing parameters also seems to be a function of the polymer involved. Peinado V. et al. found out from their work on the mechanical and rheological properties of reinforced PLA that reprocessing and recycling via extrusion of this bio-based material doesn't have a significant effect on the mechanical properties. A number of extrusions were also seen to not have any effect on the flexural and tensile modulus of the virgin matrix and composite. Rheological studies also showed that on extrusion, the viscosity of natural PLA decreased as expected due to chain degradation especially above 190 °C. However, the composite having nanoclay showed high viscosity values [69].

Although processing parameters affects the interfacial adhesion, addition of compatibilizers do affect the end product as observed by Gunning et al. [65] who explored the effect of compatibilizer content on the mechanical properties of a bioplastic like polyhydroxybutyrate (PHB) produced via hot melt extrusion. The result showed that addition of compatibilizers to PHB resin reinforced with hemp, jute and lyocell fibres respectively increases resistance to water absorption, improves fiber dispersion, melt flow index and improved flexural modulus. Likewise some processing techniques like film stacking requires pressure, the effect of pressing parameters like pressure and pressing time have been seen to have significant effect on properties of eco-friendly composites [73] from the study on PLA/Flax fibre, PLA/Jute fibre and PLA/cotton fibre composite prepared via film stacking showed PLA/Flax fiber showed the best combination of tensile and flexural strength. All in all with higher pressure, and longer pressing time, the adhesion between the PLA and the fibers is improved which causes significant increase on the mechanical properties [74] also observed that in a polyester-kenaf fiber composite, pressure was the only production condition that had significant impact on the mechanical and water absorption behavior of the composite.

Dispersion method has been seen to affect the mechanical behaviour of composites especially ones with nanoparticles as observed by Agubra et al. [75]. Karripal et al. [36] used ultrasonification method to disperse nanoclay particles in a hybrid composite made of epoxy, glass fibre and nanoclay. Using nanoclay ranging of 0, 2, 3, 5 and 6 wt%, they observed that tensile and flexural strength, as well as the moduli, increased up to 5 wt% fraction. This was attributed to improvement in the dispersion of nanoclay and interfacial adhesion between the matrix and nanoclay which helped in restricting chain mobility. Interlaminar shear strength (ILSS) and hardness of the composite also increased up to 5 wt% before decrease of which was attributed to lesser uniformity in the dispersion at higher weight fraction.

Morphological analysis revealed higher fiber pullout at higher weight fraction which can be due to agglomeration of the nanoclay particles and as such improper wetting. DSC scan, on the other hand, revealed a lower glass transition temperature above 2 wt% nanoclay which can be due to the alteration of chain kinetics due to interaction with the particles. Khoo et al. [16] studied the properties of poly (lactic acid) (PLA)/CNC nanocomposite prepared using solution casting technique. Cellulose nanocrystals (CNC) were synthesized by acid hydrolysis of microcrystalline cellulose (MCC) powder. Morphological analysis using Energy-filtered transmission electron microscopy (EFTEM) studies showed that the CNC exhibited needle-like structure (approximately 10–20 nm in width and 250–300 nm in length), which is a typical measurement found in wood-based nanocellulose. DSC analysis showed that CNC (up to 5 wt%) is capable of acting as nucleating agent for PLA whereas TGA analysis showed that the of decomposition temperatures PLA/CNC nanocomposites were higher than that of pure PLA. Zaini A. S. S. M. et al. evaluated the effect of UV radiation on the mechanical properties of wood polymer composite produced using rice husk, waste fibre and polypropylene. The result shows that with increasing number of hours of exposure to UV radiation, the compressive strength and impact energy decreased whereas density remained barely the same [29].

3 Challenges

3.1 Drawbacks in the Processing of EFCs

The difference in hydrophilicity between polymers and reinforcement has been a core challenge in producing polymer matrix composite. This hydrophilicity results in incompatibility with the hydrophobic polymer matrix. Hydrophilicity of natural fibres indicates the high moisture absorption of the fibres which is the main reason of the weak adhesion to hydrophobic matrices and this causes the produced composites to fail in wet conditions through surface roughening by fibre swelling or delamination. Moisture present during manufacturing will lead to poor processability and low mechanical performance of the composite. Natural fibers also have the tendency to form aggregates during processing, poor moisture resistance, inferior fire resistance, lower durability, variation in quality and price, and difficulty in using established manufacturing process. The major cause for this drawback is the presence of hydroxyl and other polar groups in natural fibres which makes them hydrophilic in nature.

Subsequent to this the majority of natural fibres have low degradation temperatures (<200 °C) which are inadequate for processing with thermoplastics with processing temperatures higher than 200 °C. Interfacial treatments can improve this condition either through surface treatments, resins, additives, or coatings. Nourbakish et al. [38] observed that even though the eco-friendly composites showed good mechanical properties, water intake was also high which was attributed to the high content of the lignocellulosic materials present in the composite. absorbed by the cellulosic material in the composite. As mentioned, the water absorption of composites is due to the hydrogen bonding of the water molecules to the free hydroxyl groups present in the cellulosic cell wall materials and the diffusion of water molecules into the filler-matrix interface. Additionally, large numbers of porous tubular structures present in fibers accelerate the penetration of water by the capillary action [38]. Modification of the surface seems like a viable option as shown by various studies. Silane treatment is used to stabilize polymer composites reinforced with natural fibres by treating the fibres to resist water leaching. Here silicon is accumulated in the cell lumina and bordered pits of fibres thus plugging pathways for penetration of water [76].

Acrylation pretreatment of fibers provides covalent bonds across the interface. Through such treatment, the surface energy of the fibers was increased, thereby providing better wettability and high interfacial adhesion. Pretreatments with permanganate are conducted by using a different concentration of potassium permanganate ($KMnO_4$) solution. Plasma treatment is another effective method to modify the surface of natural polymers without changing their bulk properties. Microwave treatment, corona plasma are some of the common plasma treatment methods. The type of ionized gas and the length of exposure influenced the modification of the wood and synthetic polymer surfaces. Most of the chemical treatments have been found to decrease the fiber strength due to breakage of the bond structure, and disintegration of the noncellulosic materials but silane and acrylation treatment lead to the strong covalent bond formation and the strength was enhanced marginally.

Cruz and Fangueiro [77] in his review of the effect of different modification method on the properties of eco-friendly composite showed that modification improves the properties of the composite by creating sites for proper interfacial adhesion. Chern et al. [78] in their work on the effect of addition of hydrophilic nanoclay on oil palm mesocarp fiber-reinforced polylactic acid/polycaprolactone blend observed that addition of the hydrophilic nanoclay not only led to band shift as shown by FTIR test but improved tensile, flexural and impact strength thus suggesting that it can act as a compatibilizing agent.

Thermogravimetric analysis showed that improved thermal stability which was attributed to the ability of clay to hinder permeability of volatile degradation. Also, an improved glass transition temperature was also observed due to the incorporation of clay. However soaking time, the concentration of solution etc. helps to determine the success of this method. Longer soaking time leads to fiber degradation and as such reduction in mechanical properties of the composite. Use of compatibilizer/ coupling agents has also been tried as an alternative means of improving the adhesion between the fiber and matrix [43]. Observed that addition of MA-PP to barley husk, softwood and coconut shell improved the tensile strength. Zaaba et al. [79] in their work on modification of peanut shell powder with polyvinyl alcohol showed that the modified peanut shell powder showed better interfacial adhesion to the matrix than the unmodified one.

Also, the cost of using chemicals to improve the surface of natural fibres have been highlighted as one of the shortfalls of using natural fibres. However [65] found that use of cost of using compatibilizers like is substantially when compared to the virgin matrix, but same cannot be said of another method of improving the interfacial adhesion. Lower thermal stability of most natural fibers has made processing of EFCs difficult as high processing temperature leads to fiber degradation. This is supported by morphological and thermal analysis of most EFCs. The high variability in diameter and length of fibers is another major drawback. This can be as a result of the environment where the fibre is obtained from.

3.2 Work on Eco-friendly Composites and Effect on Properties

Composites generally are materials where the addition of reinforcement to a matrix helps in bringing out a new class of material with more improved properties. In composites, PMCs have come out as the most prevalent [80] due to the ability to combine materials from different sources and morphologies i.e. particulates and fibers, natural and synthetic, micro, micron and nano to produce a more superior material. Use of different kinds of reinforcement in PMCs have been studied for decades ranging from oxides like Al_2O_3, MgO, SiO_2, Fe_2O_3; fibers like PET fibers; carbides like SiC etc. [81]. However, most of the studies started with the use of synthetic reinforcement but the incorporation of green fillers in a bid to create a more environmentally friendly material has also been seen to be able to perform comfortably well in various regards thus pushing for a total replacement of synthetic materials with green based ones. With regards to tribology, Hassan [82] showed that addition of eggshell particulates to polyester improved both the compressive strength and hardness of the composite (Figs. 5, 6 and 7).

Fig. 5 Eggshell samples. **a** Uncarbonized, **b** uncarbonized (ground), **c** carbonized, **d** carbonized (ground) [82]

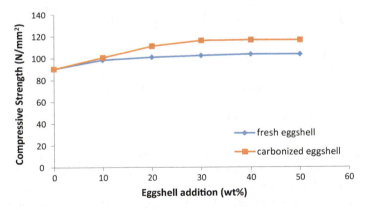

Fig. 6 Variation of compressive strength of polyester/eggshell particulate composites with wt% eggshell addition [82]

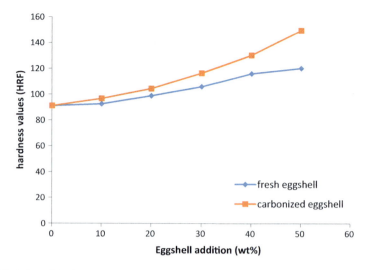

Fig. 7 Variation of hardness value of polyester/eggshell particulate composite with wt% eggshell addition [82]

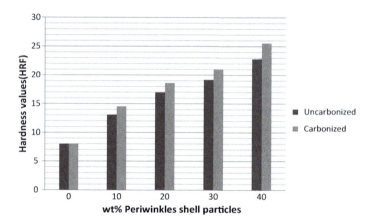

Fig. 8 Variation of hardness values of the composite with periwinkles particles addition [81]

Aigbodion et al. [30] show that incorporation of Baggase ash in recycled low-density polyethylene improved the wear resistance. Similarly, Aigbodion and Asuke observed that carbonization of agro-wastes like periwinkle shell particles before use as reinforcement also affects properties like wear resistance and hardness where the carbonized particles showed better hardness and wear resistance [81] (Figs. 8 and 9).

Likewise [83] showed that with increasing weight fraction of particulates of orange and pomegranate, the composite showcased better wear and hardness result. Atuanya et al. [84] also observed that addition of bean pod particulates to

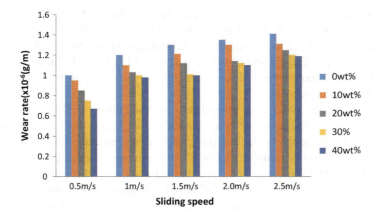

Fig. 9 Variation of wear rate with sliding speed for carbonized periwinkles shell particles [81]

low-density polyethylene led to an increased hardness, flexural and tensile strength. The similar result has also been observed by [31, 85, 86]. Other agro-wastes like a chicken feather, hooves and horns have shown improved dielectric and flame resistance in resins like polyester, and epoxy [87–89].

According to Omah et al. [90], the addition of cassava cortex to polyester improved the dielectric properties. A similar result has also been observed by Kiew et al. [88]. Use of natural fibers like jute, flax, hemp etc. has also shown the tremendous result. From the review carried out by [91, 92], the addition of these fibers have not only helped to increase the mechanical properties but also reduce cost due to the decreased wearing of tools, reduced dermal and respiratory irritation, enhanced energy recovery [93]. Mohanta N. and Acharya S. K. showed that incorporation of *Luffa Cylindrica* fibres in epoxy resin increased the tensile, flexural, impact and interlaminar shear strength of the composite [19] Sound damping properties of composites have improved due to use of natural fibres [94].

Particle size has been observed to play a major role in the properties of composites. Ameh et al. [95] studied the effect of particle size on properties of date palm seed reinforced polyester composite. Optimum tensile strength was observed at the lowest particle size whereas highest hardness was showcased by the composite with highest weight fraction and particle size. Also, water absorption behaviour was observed to decrease with decreasing particle size. Shokoufa et al. [96] worked on the effect of the addition of nanoclay to wheat stalk flour reinforced polypropylene composite. Improved flexural strength was observed and was seen to be due to the high apparent coefficient of nanoclay particles which helps to enhance the interface between the two phases. Singla et al. [97] experimented on the possibility of using lignin powder as reinforcement for producing a bio-based composite. PLA was used as matrix while the weight fraction of lignin was varied. Tensile properties were seen to increase on the addition of the lignin powders which was also supported by the SEM micrograph which showed significant interaction between the components. Impact strength decreased due to coarse and non-uniform shaped lignin particles which cause stress concentration in the PLA matrix or because of intrinsic brittleness of PLA and rigidity of lignin particles. Nevertheless, DSC and FTIR analysis showed the existence of intermolecular interaction with the peaks of hydroxyl groups of lignin and carbonyl groups of PLA in the FTIR spectra shifting toward lower wavenumbers thus indicating the presence of hydrogen bonding interaction between them.

Natural rubber (NR) and epoxidized natural rubber (ENR) are renewable resources that exhibit a unique combination of toughness, flexibility, biocompatibility and biodegradability. Its low cost makes it was an alternative to improve the toughness of brittle thermoplastics like PLA. Pongtanayuta et al. [98] compared the morphology, crystallization behaviour, thermal stability and mechanical properties of PLA/NR blends to that of PLA/ENR blend. Morphological examinations showed a coarse structure which is due to partial compatibility between the blends. However, NR at 10% weight gave optimum property though, at a high content of NR, tensile properties did suffer. Addition of ENR led to a reduced crystallization ability, thermal resistance and tensile properties of the blend. The rubber particles behave as stress concentrators enhancing the fracture energy absorption of brittle polymers and ultimately results in a material with improved toughness. The rubber particles behave as stress concentrators enhancing the fracture energy absorption of brittle polymers and ultimately results in a material with improved toughness.

Even hybrid of these materials has not also fallen short of expected properties. Hybrid composite of coconut coir/chicken feather showed improved flexural strength [99]. Saikishore et al. [100] also observed that addition of chicken feather improves impact strength and hardness of the composite. Njoku et al. [101] showed that compatibility between sisal and periwinkle in sisal/periwinkle/E-glass fiber/polyester composite led to improved tensile strength.

Polymer blends have been seen as one of the ways of improving the properties of biodegradable composites. Blending polymer is a simple technique to enhance the property of the pure polymer. The benefits of blending include providing new materials with desired properties at low cost, quick formulation changes, plant flexibility, high productivity and reduction of the number of grades that need to be manufactured and stored [102]. Hassan et al. [103] also looked at PLA/PBS blend produced via extrusion. Thermal analysis shows that thermal stability of the blends was higher than that of pure PLA and the weight loss of PLA/PBS (40/60 wt%) was lower than neat polymers. DSC thermograms of blends indicated that the thermal properties of PLA did not change noticeably when blended with PBS. DMA analysis shows that storage modulus of the blend was lowered which indicated the increase of molecular mobility by adding PBS due to lower glass transition. Increasing PBS content led to a decrease in tensile strength because PLA has a higher tensile strength than PBS and also due to poor stress transfer across the phase of each polymer. Elongation was increased with increased PBS whereas fracture behaviour changed from brittle to ductile as PBS increased. However, the impact strength of the blends was higher than neat PLA [104]. Improvement in impact strength might be due to the high flexibility of PBS. All in all PLA and PBS have shown partial miscibility [105]. The tensile strength and modulus of blends decreased with the increasing PBS content. But impact strength has improved about two times compared to pure PLA. Darie-Nita et al. [106] blended Poly(lactic acid) (PLA) and poly(butylene succinate) (PBS) using a twin screw extruder. Poly (butylene succinate) (PBS) is a biodegradable polymer, which derived from petroleum resources. PBS has good as well as elasticity and crystallization thus the addition of PBS could enhance the elasticity of PLA/PBS blend fibers. Tensile properties of the blend showed good tensile modulus and strength when adding PBS up to 30 wt%. it was also observed that increasing the PBS contents in the blends made it difficult for spinning Nevertheless PLA/PBS blends of 90:10 ratio could be spun and successfully collected at the take-up speed of 50 m/min and can find application in textile industry.

However, the properties of polymer blend are limited by immiscibility between neat polymers. To solve this problem, compatibilizers are often used as additives to improve the compatibility of immiscible blends. The compatibilization could achieve optimization of the interfacial tension, stabilize the morphology and enhance adhesion between the phases in the solid state [102]. The effect of polypropylene/poly(lactic acid) weight ratios on the properties of blend films compatibilized with polypropylene grafted-maleic anhydride has been investigated. FTIR spectra confirmed the interaction between compatibilizer and polymers whereas morphological analysis showed distinct phases of polypropylene and poly (lactic acid). Increasing of PLA content led to a decrease in melting temperature and crystallinity. Tensile increased with increasing PLA content, while elongation at break was drastically decreased however the blend proved to be a better barrier for oxygen than for water vapour [102].

Use of self-reinforcing polymers have also been looked at self-reinforcement is a situation whereby the same polymer is used as both matrix and reinforcement. Jia et al. [105] compared properties of PLA self-reinforced (PLA-SR) to that of PLA-PBS blend. The composites were made by film stacking and subsequent hot pressing. Tensile tests show that the tensile strength and elastic modulus of the PLA SR composite are higher than those of the PLA–PBS composite by more than 40%. This is due to the PLA film itself having better tensile properties than the PBS film and also due to the stronger interfacial adhesion between the PLA fibre and the PLA. The toughness of the PLA–PBS composite is more than three times higher than that of the PLA SR composite because matrix cracking and fibre/matrix debonding dissipate energy to give a more progressive fracture. This work goes on to show that the stiffness or toughness of a composite can be tailored to specific needs by selecting fibre and matrix constituents. Therefore if damage resistance and tolerance are required, the PLA–PBS composite is more preferable whereas, for stiffness and strength, the PLA SR composite should be offered a better option [105].

4 Current Opportunities

High demand has led to a rise in production capacity for eco-friendly polymers and its composites. The natural fibers such as flax or sisal have relatively high stiffness and are low in cost. Agrowastes like rice husk, bagasse, egg shells, coconut shells are relatively abundant. Annually, billions of tons of agro wastes and natural fibers which would have been a major source of income are burnt on the field producing all to take optimum advantage of the constituents in order to produce superior and a more economical material [7]. Similarly biodegradable polymers market is expanding year by year due to the demand from various sectors. As at 2015, the production capacity of biodegradable polymers is expected to be approximately 2.8 million ton/year [50].

4.1 Future Opportunities in EFC

The interest in biopolymers has continued to grow. In the food packaging industry, use of eco-friendly plasticizers like epoxidized soya bean oil, lactide, PLA oligomer, poly(ethylene glycol) (PEG 2000 or PEG2) poly(ethylene glycol) (PEG 4000) and for PLA films processing has been studied [106]. Wood-based fillers is another area that is bound to continue growing. Past decade has seen a rapid growth in the wood plastic industry. This is due to low cost and reinforcing capacity [107]. The main application areas of wood flour filled composites are the automotive and building industries in which they are used in structural applications as fencing, decking, outdoor furniture, window parts, roofline products, door panels, etc. There are environmental and economic reasons for replacing part of the plastics with wood but the wood could also work as reinforcement of the plastics. The elastic modulus of wood fibres is approximately 40 times higher than that of polyethylene and the strength about 20 times higher [107].

Continual improvement in the understanding of the chemistry behind the interplay between the reinforcement and matrix have continued to open more ground for an increase in the areas of application of various eco-friendly materials. However, the challenges for the future are in terms of processing, characterization and the mechanisms governing the behaviour of these advanced materials. The primary determinant of the performance of a composite material is the interface between the reinforcing fibre/particulate and matrix.

5 Conclusions

According to food and agriculture organization 2014 handbook, apart from enteric fermentation; a major contributor to greenhouse gas emission comes from rice cultivation, crop residue, manure management and burning crop residue [108].

Therefore in the bid to maintain sustainable development of environmentally compatible composites at low cost, the study of this class of materials has continued to pull in increasing research interest with amazing results recorded for various applications in many areas. Natural fibers especially the lignocellulosic ones are known for their hydrophilic. However, surface modification and blending have proved to be a way of modifying the properties. Nanotechnology has also shown to be a bright burner in the drive towards a sustainable environment by providing an avenue for environmentally friendly nanofillers to be incorporated as fillers.

All in all, EFCs promises a set of sustainable materials for the future. Though the tensile strength of composites is dependent on matrix properties whereas modulus is dependent on filler properties, use of environmentally friendly materials for both can allow for proper tailoring of the properties of the composite so as to meet both environmental policy regulation and customer satisfaction.

Our ability to solve the engineering challenges of today and tomorrow is still being driven by the cost and availability. One needs to overcome this challenge to achieve the required degrees of success. Our understanding of how materials interact, how they can be reliably integrated with combinations in functional requirements have to be explored. The challenges, as well as the opportunities, will be in the areas of **materials development, synthesis, fabrication, integration, characterization** and **validating the recycling technologies**, which of course aids capitulate a lot of potential in Innovating new ideas.

References

1. Mitra BC (2014) Environment friendly composite materials: biocomposites and green composites. Defence Sci J 64(3):244–261
2. Layth M, Ansari MNM, Pua G, Jawaid M, Islam MS (2015) A review on natural fiber reinforced polymer composite and its applications. Int J Polym Sci 1–15. http://dx.doi.org/10.1155/2015/243947
3. Swaroop KV, Vinod NR, Rupendra M (2017) Numerical and experimental analysis of a natural fiber reinforced composites. Int J Mech Prod Eng 5(11):25–27
4. Anne B (2011) Environmental-friendly biodegradable polymers and composites. In: Integrated waste management, vol I. http://www.intechopen.com/books/integrated-wastemanagement-volume-i
5. Azwa ZN, Yousif BF, Manalo AC, Karunasena W (2013) A review on the degradability of polymeric composites based on natural fibres. Mater Des 47:424–442
6. Material resources, productivity and the environment: key findings. www.oecd.org. Accessed 5 Feb 2018
7. Islam MS, AhmadMB Hasan M, Aziz SA, Jawaid M, Haafiz MKM, Zakaria SAH (2015) Natural fiber-reinforced hybrid polymer nanocomposites: effect of fiber mixing and nanoclay on physical, mechanical and biodegradable properties in hybrid nanocomposite. BioResources 10(1):1394–1407
8. Rahmat MB, Ab-Wahid WF, Ahmad M (2015) Effect of nanoclay on tensile strength of wood plastic composite made from malaysian rice husk and polypropylene. Int J Mech Prod Eng 3(10):61–63

9. Rana S, Fangueiro R (eds) (2016) Fibrous and textile materials for composite applications. Springer, Berlin
10. Adeosun SO, Lawal GI, Balogun SA, Akpan EI (2012) Review of green polymer nanocomposites. J Miner Mater Charact Eng 11(4):385–416
11. Asokan P, Firdoous M, Sonal W (2012) Properties and potential of bio-fibres, bio-binders, and bio-composites. Rev Adv Mater Sci 30:254–261
12. Abilash N, Sivapragash M (2013) Environmental benefits of ecofriendly natural fiber reinforced polymeric composite materials. Int J Appl Innov Eng Manage 2(1):53–59
13. Vroman I, Tighzert L (2009) Biodegradable polymers. Mater 2:307–344. https://doi.org/10.3390/ma2020307
14. El-Sherbiny IM, Ali IH (2015) Eco-friendly electrospun polymeric nanofibers-based nanocomposites for wound healing and tissue engineering. In: Thakur VK, Thakur MK (eds) Eco-friendly polymer nanocomposites processing and properties. Springer, New Delhi, pp 399–431
15. Chen HN (2012) An overview of degradable polymers. https://doi.org/10.1021/bk-2012-1114.pr002
16. Khoo RZ, Ismail H, Chow WS (2016) Thermal and morphological properties of poly (lactic acid)/nanocellulose nanocomposites. Procedia Chem 19:788–794. https://doi.org/10.1016/j.proche.2016.03.086
17. Kocak D, Merdan N, Yuksek M, Sancak E (2013) Effects of chemical modification on mechanical properties of Luffa cylindrica. Asian J Chem 25(2):637–641
18. Mohanta N, Acharya SK (2013) Tensile. Flexural and interlaminar shear properties of luffa cylindrical fibre reinforced epoxy composites. Int J Macromol Sci 3(2):6–10
19. Pai AR, Jatap RN (2015) Surface morphology and mechanical properties of some unique natural fiber reinforced polymer composites—a review. J Mater Environ Sci 6(4):907–917
20. Panneerdhass R, Baskan R, Rajkumar K, Gnanavebabu A (2014) Mechanical properties of chopped randomly oriented epoxy—luffa fiber reinforced polymer composite. Appl Mech Mater 591:103–107
21. Gupta G, Gupta A, Dhanola A, Raturi A (2016) Mechanical behavior of glass fiber polyester hybrid composite filled with natural fibers. IOP conference series: materials science and engineering. https://doi.org/10.1088/1757-899x/149/1/012091
22. Hassan SB, Oghenevweta JE, Aigbodion VS (2012) Morphological and mechanical properties of carbonized waste maize stalk as reinforcement for eco-composites. Compos B 43:2230–2236
23. Chen RS, AhmadS, Gan S (2016) Characterization of rice husk incorporated recycled thermoplastic blend composites. Bioresources 11(4):8470–8482
24. Safwan MM, Lin HO, Akil HM (2013) Preparation and characterization of palm kernel shell/polypropylene biocomposite and their hybrid composite with Nanosilica. BioResources 8(2):1539–1550
25. Karthik R, Sathiyamurthy S, Jayabal S, Chidambaram K (2014) Tribological behaviour of rice husk and egg shell hybrid particulated coir-polyester composites. IOSR J Mech Civil Eng 75–80. Retrieved from http://www.iosrjournals.org/iosr-jmce/papers/NCCAMABS/Volume-3/39.pdf
26. Prabhu R, Amin AK, Dhyanchandra A (2015) Development and characterization of low cost polymer composites from coconut coir. Am J Mater Sci 5(3C):62–68
27. Ashori A, Nourbakhsh A (2010) Bio-based composites from waste agricultural residues. Waste Manage 30:680–684
28. Ravindran D, Sornakumar T, Prithvirajadurai DS, Varadharajan V (2015) Development of hybrid coconut shell powderwood dust polyester resin based composites. Int J Appl Mech Prod Eng 1(7):1–4
29. Zaini ASSM, Rus ZAM, Rahman NA, Jais FHM, Fauzan MZ, Sufian NA (2017) Mechanical properties evaluation of extruded wood polymer composites. In: 4th international conference on the advancement of materials and nanotechnology (ICAMN IV 2016), AIP conference proceedings 1877, 060005. pp 2–9. https://doi.org/10.1063/1.4999884

30. Aigbodion VS, Hassan SB, Agunsoye OJ (2011) Effect of bagasse ash reinforcement on dry sliding wear behaviour of polymer matrix composites. Mater Des 33:322–327
31. Atuanya CU, Aigbodion VS, Nwigbo SC (2014) Experimental study of the thermal and wear properties of recycled polyethylene/breadfruit seed hull ash particulate composites. Mater Des 53:65–73
32. Fragassa C, Santulli C, Pavlović A, ŠljivićM (2015) Improving performance and applicability of green composite materials by hybridization. Contemp Mater 35–43. https://doi.org/10.7251/comen1501035f
33. Muthukumar S, Lingadurai K (2014) Investigating the mechanical behaviour of coconut shell and groundnut shell reinforced polymer composite. Glob J Eng Sci Res 1(3):19–23
34. Kasiviswanathan S, Santhanam K, Kumaravel A (2015) Evaluation of mechanical properties of natural hybrid fibers, reinforced polyestercomposite materials. Carbon Sci Tech 7/4:43–49 [CST-161-7-4]
35. Udhayasankar R, Karthikeyan B (2015) A review on coconut shell reinforced composites. Int J ChemTech Res CODEN (USA): IJCRGG 8(11):624–637
36. Karippa JJ, Murthy HNN, Rai KS, Sreejith M, Krishna M (2011) Study of mechanical properties of epoxy/glass/nanoclay hybrid composites. J Compos Mater 1–7
37. Meziane O, Bensedira A, Guessoum M, Haddaoui N (2016) Polypropylene-modified kaolinite composites: effect of chemical modification on mechanical, thermal and morphological properties. J Fundam Appl Sci 8(2):494–509
38. Nourbakhsh A, Baghlani FF, Ashori A (2011) Nano-SiO_2 filled rice husk/polypropylene composites: physico-mechanical properties. Ind Crops Prod 33:183–187
39. Chauhan S, Bhushan RK (2017) Study of polymer matrix composite with natural particulate/fiber in PMC: a review. Int J Adv Res Ideas Innovations Technol 3(3):1168–1179
40. Arpitha GR, Sanjay MR, Yogesha B (2014) Review on comparative evaluation of fiber reinforced polymer matrix composites. Adv Eng Appl Sci Int J 4(4):44–47
41. Chandramohan D, Marimuthu K (2011) A review on natural fibers. IJRRAS 8(2):194–206
42. Hashim MY, Roslan MN, Amin AM, Zaidi AMA, Ariffin S (2012) Mercerization treatment parameter effect on natural fiber reinforced polymer matrix composite: a brief review. World Acad Sci Eng Technol 6:1382–1388
43. Bledzki AK, Mamun AA, Volk J (2010) Barley husk and coconut shell reinforced polypropylene composites: the effect of fibre physical, chemical and surface properties. Compos Sci Technol 70:840–846
44. Rozyanty AR, Firdaus MYN, Liew TZ, Yunus NFM (2015) Kenaf-unsaturated polyester composite: the effect of different retting process of kenaf bast fiber on the mechanical properties. Mater Sci Forum 819:256–261
45. Dass PM, Akinterinwa A, Adamu JN, Abba S (2015) The influence of different retting processes on the strength of fibres obtained from *Poliostigma raticulatum*, *Grewia mollis*, *Cissus populnea* and *Hibiscus sabdariffa*. Environ Nat Resour Res 5(4):41–45
46. Gopu RN, Singh A, Zimniewska M, Raghavan V (2013) Comparative evaluation of physical and structural properties of water retted and non-retted flax fibers. Fibers 1:59–69. https://doi.org/10.3390/fib1030059
47. Jawaid M, Tahir PM, Saba N (eds) (2017) Lignocellulosic fibre and biomass-based composite materials: processing, properties and applications. Woodhead Publishing, UK
48. Thakur VK, Thakur MK (eds) (2015) Eco-friendly polymer nanocomposites: chemistry and application. Springer, New Delhi
49. Rydz J, Sikorska W, Kyulavska M, Christova D (2015) Polyester-based (bio)degradable polymers as environmentally friendly materials for sustainable development. Int J Mol Sci 16(1):564–596
50. Ray SS (2013) Environmentally friendly polymer nanocomposites: types, processing and properties. Woodhead Publishing, New Delhi
51. Shukla SK, Mishra AK, Arotiba OA, Mamba BB (2013) Chitosan-based nanomaterials: a state-of-the-art review. Int J Biol Macromol 1–13. http://dx.doi.org/10.1016/j.ijbiomac.2013.04.043

52. Saba N, Tahir PM, Jawaid M (2014) A review on potentiality of nano filler/natural fiber filled polymer hybrid composites. Polymers 6(8):2247–2273. https://doi.org/10.3390/polym6082247
53. Thomas S, Paul SA, Pothan LA, Deepa B (2011) Natural fibres: structure, properties and applications. In: Kalia S, Kaith BS, Kaur I (eds) Cellulose fibers: bio- and nano-polymer composites. Springer, Berlin
54. Shehu U, Audu H, Nwamara MA, Ade-Ajayi AF, Shittu UM, Isa MT (2014) Natural fibre as reinforcement for polymers: a review. SPJTS 2(1):238–253
55. Kabir MM, Wang H, Aravinthan T, Cardona F, Lau KT (2011) Effects of natural fibre surface on composite properties: a review. pp 94–99
56. Kalia S, Dufresne A, Cherian BM, Kaith BS, Averous L, Njuguna J, Nassiopoulos E (2011) Cellulose-based bio- and nanocomposites: a review. Int J Polym Sci 1–36. https://doi.org/10.1155/2011/837875
57. Fan M, Dai D, Huang B (2012) Fourier transform infrared spectroscopy for natural fibres. In: Salih S (ed) Fourier transform—materials analysis. ISBN 978-953-51-0594-7. www.interchopen.com/books/fourier-transform–materials-analysis/Fourier-Transform-Infrared-Spectroscopy-for-natural-fibres. Accessed 5 Feb 2018
58. Kumar R, Obrai S, Sharma A (2011) Chemical modifications of natural fiber for composite material. Der Chemica Sinica 2(4):219–228
59. Anurag T (2016) Study of musa acuminata fibre reinforced composite—a review. Int J Res Aeronaut Mech Eng 4(4):29–42
60. Srinivasan R (2011) Advances in application of natural clay and its composites in removal of biological, organic, and inorganic contaminants from drinking water. Adv Mater Sci Eng 1–17. https://doi.org/10.1155/2011/872531
61. Hamid E, Raji M, Bouhfid R, Qaiss AEK (2016) Nanoclay and natural fibers based hybrid composites: mechanical, morphological, thermal and rheological properties. In: Jawaid M, Qaiss AEK, Bouhfid R (eds) Nanoclay reinforced polymer composites, engineering materials. Springer, Singapore
62. Mallick PK (2018) Processing of polymermatrix composites. CRC Press, Boca Raton
63. Zin MH, Razzi MF, Othman SLK, Abdan K, Mazlan N (2016) A review on the fabrication method of bio-sourced hybrid composites for aerospace and automotiveapplications. IOP conference series: materials science and engineering. https://doi.org/10.1088/1757-899x/152/1/012041
64. Dong P, Prasanth R, Xu F, Wang X, Li B, Shankar R (2015) Eco-friendly polymer nanocomposite—properties and processing. In: Thakur VK, Thakur MK (eds) Eco-friendly polymer nanocomposites processing and properties. Springer, New Delhi
65. Gunning M, Geever LM, Killion JA, Lyons JG, Higginbotham CL (2014) Effect of compatibilizer content on the mechanical properties of bioplastic composites via hot melt extrusion. Polym Plast Technol Eng 53:1223–1235
66. Cai J, Jia M, Xue P, Ding Y, Zhou X (2013) The effect of processing conditions on the mechanical properties and morphology of self-reinforced wood-polymer composite. Polym Compos 1567–1574. https://doi.org/10.1002/pc.22553
67. Tanahashi M (2010) Development of fabrication methods of filler/polymer nanocomposites: with focus on simple melt-compounding based approach without surface modification of nanofillers. Materials 3:1593–1619. https://doi.org/10.3390/ma3031593
68. Díez EA (2014) Effect of extrusion on the electrical, mechanical and rheological properties of an ethylene butylacrylate/carbon black/graphite nanoplatelets nanocomposite. Diploma work, Department of Materials and Manufacturing Technology, Chalmers University of Technology, Gothenburg, Sweden

69. Peinado V, Castell P, García L, Fernández A (2015) Effect of extrusion on the mechanical and rheological properties of a reinforced poly(lactic acid): reprocessing and recycling of biobased materials. Materials 8:7106–7117. https://doi.org/10.3390/ma8105360
70. Feldmann M, Heim HP, Zarges JC (2015) Influence of the process parameters on the mechanical properties of engineering biocomposites using a twin-screw extruder. Compos Part A 1–7. http://dx.doi.org/10.1016/j.compositesa.2015.03.028
71. Kadam PG, Mhaske ST (2014) Effect of extrusion reprocessing on the mechanical, thermal, rheological and morphological properties of nylon 6/talc nanocomposites. J Thermoplast Compos Mater 1–19. https://doi.org/10.1177/0892705714551591
72. Poletto M (2016) Polystyrene cellulose fiber composites: effect of the processing conditions on mechanical and dynamic mechanical properties. Rev Mater 21(3):552–559
73. Hajba S, Tábi T (2014) Development of natural fibre reinforced poly(lactic acid) biocomposites. In: ECCM16—16th European conference on composite materials, Seville, Spain, 22–26 June 2014, pp 1–8
74. Rassmann S, Reid RG, Paskaramoorthy R (2010) Effects of processing conditions on the mechanical and water absorption properties of resin transfer moulded kenaf fibre reinforced polyester composite laminates. Compos Part A 4(11):1612–1619
75. Agubra VA, Owuor PS, Hosur MV (2013) Influence of nanoclay dispersion methods on the mechanical behavior of E-glass/epoxy nanocomposites. Nanomaterials 3:550–563. https://doi.org/10.3390/nano3030550
76. Kalia S, Kaith BS, Kaur I (2009) Pretreatments of natural fibers and their application as reinforcing material in polymer composites—a review. Polym Sci Eng 49(7):1253–1272. https://doi.org/10.1002/pen.21328
77. Cruz J, Fangueiro R (2016) Surface modification of natural fibers: a review. Procedia Eng 155:285–288
78. Chern CE, Nor AI, Norhazlin Z, Ariffin H, Yunus WMZW, Then YY (2014) Enhancement of mechanical and dynamic mechanical properties of hydrophilic nanoclay reinforced polylactic acid/polycaprolactone/oil palm mesocarp fiber hybrid composites. Int J Polym Sci. http://dx.doi.org/10.1155/2014/715801
79. Zaaba NF, Ismail H, Jaafar M (2014) The effects of modifying peanut shell powder with polyvinyl alcohol on the properties of recycled polypropylene and peanut shell powder composites. BioResources 9(2):2128–2142
80. Kumar TV, Chandrasekaran M, Padmanabhan S (2017) Characteristics and mechanical properties of reinforced polymer composites. ARPN J Eng Appl Sci 12(8):2450–2454
81. Asuke F, Aigbodion VS (2016) Experiment numerical study of dry sliding wear behavior of epoxy/periwinkles shell particulate composites. J Chin Adv Mater Soc 1–17. https://doi.org/10.1080/22243682.2015.1124736
82. Hassan SB, Aigbodion VS, Patrick SN (2012) Development of polyester/eggshell particulate composites. Tribol Ind 34(4):217–225
83. Kadhum AAU, Mohammed AA (2016) Investigation the effect of natural materials on wear and hardness properties of polymeric composite materials. Iraqi J Mech Mater Eng 16(4):369–372
84. Atuanya CU, Aigbodion VS, Obiorah SO (2015) Evaluation of the mechanical properties of recycled low-density polyethylene/bean pod particulate bio-composites. J Chin Adv Mater Soc 3(4):345–358. https://doi.org/10.1080/22243682.2015.1081077
85. Sarki J, Hassan SB, Aigbodion VS, Oghenevweta JE (2011) Potential of using coconut shell particle fillers in eco-composite materials. J Alloy Compd 509:2381–2385
86. Aigbodion VS, Atuanya CU, Igogori EA, AndIhom P (2013) Development of high-density polyethylene/orange peels particulate bio-composite. Gazi Univ J Sci 26(1):107–117
87. Subramani T, Krishnan S, Ganesan SK, Nagarajan G (2014) Investigation of mechanical properties in polyester and phenylester composites reinforced with chicken feather fiber. Int J Eng Res Appl 4(12):93–104

88. Kiew KS, Rahman MR, Hamdan S, Talibb ZA (2013) Maleic anhydride modified unsaturated polyester composites reinforced with chicken feather fiber: dielectric and morphological study. World Appl Sci J 25(6):899–907. https://doi.org/10.5829/idosi.wasj.2013.25.06.1347%5b
89. Oladele IO, Omotoyimbo JA, Ayemidejor SH (2014) Mechanical properties of chicken feather and cow hair fibre reinforced high density polyethylene composites. Int J Sci Technol 3(1):66–72
90. Omah AD, Okorie BA, Omah EC, Ezemal C, Aigbodion VS, Orji UU (2017) Experimental correlation between varying cassava cortex and dielectric properties in epoxy/cassava cortex dielectric particulates composites. Part Sci Technol. https://doi.org/10.1080/02726351.2017.1307888
91. Tushar S, Shirish P, Vikram D, Acharya R (2015) Natural fiber reinforced polymer composite material—a review. IOSR J Mech Civil Eng 142–147
92. Madhusudhan T, Swaroop GK (2016) A review on mechanical properties of natural fiber reinforced hybrid composites. Int Res J Eng Technol (IRJET) 3(4):2247–2251
93. Nitin S, Singh VK (2013) Mechanical behaviour of walnut reinforced composite. J Mater Environ Sci 4(2):233–238
94. Tao Y, Li P, Cai L (2016) Effect of fiber content on sound absorption, thermal conductivity and compression strength of straw fiber filled rigid polyurethane foams. BioResources 11(2):4159–4167
95. Ameh AO, Isa MT, Sanusi I (2015) Effect of particle size and concentration on the mechanical properties of polyester/date palm seed particulate composites. Leonardo Electron J Practices Technol 26:65–78
96. Shokoufa N, Nourbakhsh A, Ashkan G, Talaeipour M, Habibollah KE (2013) Investigating the effect of nanoclay on polypropylene-made cellulose composite. Res J Appl Sci Eng Technol 6(21):4022–4029. https://doi.org/10.19026/rjaset.6.3505
97. Singla RK, Maiti SN, Ghosh AK (2016) Crystallization, morphological, and mechanical response of poly(lactic acid)/lignin-based biodegradable composites. Polym Plast Technol Eng 55(5):475–485. https://doi.org/10.1080/03602559.2015.1098688
98. Pongtanayuta K, Thongpina C, Santawiteeb O (2013) The effect of rubber on morphology, thermal properties and mechanical properties of PLA/NR and PLA/ENR blends. Energy Procedia 34:888–897. https://doi.org/10.1016/j.egypro.2013.06.826
99. Alagarsamy SV, Sagayaraj AVS, Vignesh S (2015) Investigating the mechanical behaviour of coconut coir—chicken feather reinforced hybrid composite. Int J Sci Eng Technol Res (IJSETR), 4(12):4215–4221
100. Saikishore T, Rao PP, Reddy MCS (2017) Synthesis & investigation of the mechanical behaviour of luffa, groundnut shell, chicken feather and cowdung fibers reinforced epoxy composites. IJSRD Int J Sci Res Dev 5(4):42–46
101. Njoku RE, Obayi CS, Nnamchi PS (2011) Hybrid effect on the mechanical properties of sisal fiber and E-glass fiber reinforced polyester composites. Niger J Technol 30(3):97–103
102. Nalin P, Suppakula P, Atong D, Pechyen C (2014) Blend of polypropylene/poly(lactic acid) for medical packaging application: physicochemical, thermal, mechanical, and barrier properties. Energy Procedia 56:201–210
103. Hassan E, Wei Y, Jiao H, Muhuo Y (2013) Dynamic mechanical properties and thermal stability of poly(lactic acid) and poly(butylene succinate) blends composites. J Fiber Bioeng Inform 6(1):85–94. https://doi.org/10.3993/jfbi03201308
104. Jompanga L, Thumsorna S, Onb JW, Surinb P, Apawet C, Tirapong C, Narin K, Narongchai OC, Srisawata N (2013) Poly(lactic acid) and poly(butylene succinate) blend fibers prepared by melt spinning technique. Energy Procedia 34:493–499
105. Jia W, Gong RH, Soutis C, Hogg PJ (2014) Biodegradable fibre reinforced composites composed of polylactic acid and polybutylene succinate. Plast Rubber Compos 43(3):82–88. https://doi.org/10.1179/1743289813Y.0000000070

106. Darie-Nita RN, Vasile C, Irimia A, Lipsa R, Rapa M (2016) Evaluation of some eco-friendly plasticizers for PLA films processing. J Appl Polym Sci 1–11. https://doi.org/10.1002/app.43223
107. Vignesh J, Selvam CM (2015) Experimental evaluation of wood dust particulate reinforced polymer composites. IRACST Eng Sci Technol Int J (ESTIJ) 5(4):226–229
108. FAO Statistical Yearbook 2014 Africa: Food and Agriculture (2014) Food and agriculture organization of the United Nations Regional Office for AfricaAccra

Synthesis, Characterization, and Applications of Hemicelluloses Based Eco-friendly Polymer Composites

Xinwen Peng, Fan Du and Linxin Zhong

Abbreviations

EVOH	Ethylene vinyl alcohol
PVDC	Polyvinylidene chloride
3D	Three-dimensional
AGU	Anhydroglucose units
AcGGM/GGM	*O*-acetyl galactoglucomannans
DMF	*N*, *N*-dimethylformamide
DMA/LiCl	*N*, *N*-dimethylacetamide/lithium chloride
DMSO/THF	Dimethyl sulfoxide/tetrahydrofuran
DMAP	4-dimethylamino pyridine
NBS	*N*-bromosuccinimide
TEA	Triethylamine
DS	Degree of substitution
MSA	Methane sulfonic acid
[BMIM]Cl	1-butyl-3-methylimidazolium chloride
IL	Ionic liquid
LC	Lauroyl chloride
LH	Lauroylated hemicelluloses
HFIP	Hexafluoroisopropanol/1, 1, 1, 3, 3, 3-hexafluoro-2-propanol
CDI	*N*, *N'*-carbonyldiimidazole
SET-LRP	Single-electron-transfer mediated living radical polymerization
AcGGM-SH	Thiolated *O*-acetyl galactoglucomannan
PEG-MA	Polyethylene glycol monomethacrylate
ETA	2, 3-epoxypropyltrimethylammonium chloride
QH	Quaternized hemicelluloses

X. Peng (✉) · F. Du · L. Zhong (✉)
State Key Laboratory of Pulp and Paper Engineering, South China
University of Technology, 381 Wushan Road, Guangzhou 510640, China
e-mail: fexwpeng@scut.edu.cn

L. Zhong
e-mail: lxzhong0611@scut.edu.cn

© Springer Nature Switzerland AG 2019
Inamuddin et al. (eds.), *Sustainable Polymer Composites and Nanocomposites*,
https://doi.org/10.1007/978-3-030-05399-4_43

MMT	Montmorillonite
NaH	Sodium hydride
BnGGM	Benzyl galactoglucomannan
TBAI	Tetrabutylammonium iodide
CHMAC	3-chloro-2-hydroxypropyltrimethylammonium chloride
GTMAC	Glycidyltrimethylammonium chloride
METAC	[2-(methacryloyloxy) ethyl] trimethylammonium chloride
HPMA	2-hydroxypropyltrimethylammonium
DME	1, 2-dimethoxyethane
PHL	Pre-hydrolysis liquor
GTMAC	Glycidyltrimethylammonium chloride
METAC	[2-(methacryloyloxy) ethyl] trimethylammonium chloride
MeGlcp-Xylan	O-acetyl-4-O-methylglucuronoxylan
WH	Wood hydrolysate
AG	Arabinogalactan
EDC/NHS	N-ethyl-N'-(3-dimethylamino)propyl carbodiimide hydroxide/ N-hydroxysuccinimide
TA	Tyramine
HRP	260 purpurogallin unit/mg solid
DMT-MM	4-(4, 6-dimethoxy-1, 3, 5-triazin-2-yl)-4-methylmorpholinium chloride
CuAAC	Copper(I)-catalyzed azide-alkyne cycloaddition
AX	Arabinoxylan
AGX	Arabinoglucuronoxylan
[emim][Me$_2$PO$_4$]	1-ethyl-3-methylimidazolium dimethyl phosphate
[DBNH][OAc]	1, 5-diazabicyclo[4.3.0]non-5-enium acetate
[Amim]$^+$Cl$^-$	1-allyl-3-methylimidazolium chloride
XylC6N$_3$	Di-O-(6-azidohexanoyl)-xylan
PLLA	Poly(L-lactide)
PMDETA	N, N, N', N', N''-pentamethyldiethylenetriamine
LLA	L-lactide
TBD	Triazabicyclodecene
PLA	Polylactide
AN	Acrylonitrile
MA	Methyl acrylate
AM	Acrylamide/acrylic amide
DMC	Methacryloyloxy ethyl trimethyl ammonium chloride
APMP	Alkaline peroxide mechanical pulping
MMA	Methyl methacrylate
NIPAM	N-isopropyl acrylamide
GMA	Glycidyl methacrylate
GM	Galactomannan
QCM-D	Quartz crystal microbalance with dissipation
TEMPO	2, 2, 6, 6-tetramethylpiperidine-1-oxyl

Cy	Cysteine
LOD	Limit of detection
AgNPs	Silver nanoparticles
PMP	Polymeric magnetic microparticles
MP	Magnetic microparticles
CMH	Carboxymethyl functionalized hemicellulose/carboxymethyl hemicellulose
Pd NPs	Palladium nanoparticles
XH	Xylan-type hemicelluloses
CKGM	Carboxymethyl Konjac glucomannan
CS	Chitosan
BSA	Bovine serum albumin
WVP	Water vapor permeability
OP	Oxygen permeability
PVA	Polyvinyl alcohol
HPKO	Hydrogenated palm kernel oil
HLBs	Hydrophilic-lipophilic balances
LDPE	Low-density polyethylene
DMA	Dynamic mechanical analysis
NCH	Chitin nanowhiskers
BH	Bleached hemicelluloses
BAH	Acetylated bleached hemicelluloses
NCC	Nanocrystalline cellulose
CNCC	Cationically modified NCC
HC/SB	Hemicelluloses/sorbitol
GTMAC	Glycidyltrimethylammonium chloride
HL	Hemicellulose/lignin
NFC	Nanofibrillated cellulose
MFC	Microfibrillated cellulose
CNFs	Cellulose nanofibers
CNT	Carbon nanotube
κ-car/LBG	κ-carrageenan/locust bean
GA	Gum arabic
SA	Stearyl acrylate
SM	Stearyl methacrylate
EB	Electron beam
PLGA	Poly(lactic-co-glycolic acid)
TFAA	Trifluoroacetic anhydride
PET	Polyethylene terephthalate
CHPS	3-Chloro-2-hydroxypropyl sulfonic acid
SCHMAC	(S)-(-)-(3-chloro-2-hydroxypropyl)-trimethylammonium chloride
CHPMAC	3-chloro-2-hydroxypropyl-trimethylammonium chloride

Ra	Roughness value
Seq	Equilibrium swelling ratio
HEMA	2-hydroxyethyl methacrylate
HEMA-Im	2-[(1-imidazolyl)formyloxy]ethyl methacrylate
AnMan5A	Enzyme β-mannanase
M-AcGGM	Methacrylated AcGGM
CM-AcGGM	Maleic anhydride-modified M-AcGGM
AA	Acrylic acid
CA	Citric acid
SHP	Sodium hypophosphite
NIPAAm	N-isopropylacrylamide
MBA	N, N'-methylenebis-acrylamide
DMAP/NMP	2, 2-dimethoxy-2-phenylacetophenone/N-methyl pyrrolidone
ACX	Acylated xylan
Hce-MA/AHC	Acylated hemicellulose
LCST	Lower critical solution temperature
APS/TEMDA	Ammonium persulfate/N, N, N', N'-tetramethyl-ethane-1, 2-diamine
MeDMA	[2-(methacryloyloxy) ethyl] trimethylammonium chloride
ECH	Epichlorohydrin
GDEP	Glow discharge electrolysis plasma
MFRHH	Magnetic field-responsive hemicelluloses-based hydrogel
SRs	Swelling ratios
ECH	Electrically conductive hydrogels
ECHH	Electrically conductive hemicellulose hydrogel
AP	Aniline pentamer
C-AcGGM	Carboxylated AcGGM
AT	Aniline tetramer
SRHMGs	Stimuli-responsive hemicellulose microgels
CMCH	Carboxymethyl chitosan-hemicellulose
CHNT	Carboxymethyl chitosan-hemicellulose network
SDS	Sodium dodecyl sulfate
DTPA	Diethylene triamine pentaacetic acid
DHC	Dialdehyde hemicelluloses
CNF	Cellulose nanofibrils
CNC	Nanocrystalline cellulose
NFC	Nanofibrillated cellulose
IPNs	Interpenetrating polymer networks
MA-CMC	Methacrylated carboxymethylcellulose
SWH	Softwood hemicellulose hydrolysate
kC-xylan-PVP	Kappa-carrageenan/xylan/polyvinylpyrrolidone
KPS	Sodium persulphate

PEG	Poly(ethylene glycol)
PEG-PPG-PEG	Poly(ethylene glycol)-b-poly(propylene glycol)-b-poly(ethylene glycol)
IA	Itaconic acid
PAA	Poly(amidoamine)
GO	Graphene oxide
PAM	Polymerized acrylamide
MW-CNTs	Multiwall carbon nanotube
MB	Methylene blue
PEGDE	Polyethylene glycol diglycidyl ether

1 Introduction

During the last few decades, synthetic polymers and polymer-based materials have become indispensable materials in people's lives. The polymer composites are utilized in a very wide range of applications from household to aerospace, from agriculture to medicine or pharmacy. However, due to the growing awareness on environment and human health as well as the scarcity and increase in the price of the fossil resources, tremendous attention and efforts have been given to the environmentally friendly materials. Production of environmentally friendly biomaterials from renewable biomass resources is a critical route to solve the environmental pollution and the upcoming depletion of petrochemical resources. In the past decade, various biomass resources, such as cellulose [1], lignin [2, 3], hemicelluloses [4, 5], chitosan [6, 7], starch [8, 9], psyllium [10], have been used to fabricate biomaterials and biocomposites that have potential applications in various fields. Hemicelluloses are the secondly abundant sustainable polymers on the earth which possessing 25–30% of the total weight in the wood plant, and will not compete with food supply for the production of biopolymers [5, 11–14]. Recently, the global number of trees was estimated to be ~3.04 trillion, and thus generating ~3900 million m^3 of wood per year and providing large amounts of plant materials including cellulose, hemicelluloses, and lignin [12, 13]. The hemicelluloses-based polymer can be cast into films and take a substitute to traditional non-degradable and high-cost petroleum-based products such as ethylene vinyl alcohol (EVOH) or polyvinylidene chloride (PVDC). Additionally, it also can be fabricated into three-dimensional (3D) hydrogel networks and used in adsorbing the heavy metals [15–20] and methylene blue [21, 22], drug delivery [23–26], food package [27–29] and paper surface engineering [30]. However, hemicelluloses have some inherent disadvantages, for instance, brittleness, water/moisture sensitivity, high water vapour permeability, and so on. Physical, chemical and enzymatic modifications of hemicelluloses by either bulk or surface modification are essential ways to settle these problems because of abundant free hydroxyl groups and carboxyl distributed along the backbone and side chains of hemicelluloses [5]. Up to now, various chemical methods, such as esterification, etherification, ionization, amination, amidation, acetylation, grafting copolymerization, crosslinking, blending, have been explored to modify hemicelluloses and fabricate functional materials. This chapter gives an overview of the recent advances in the synthesis of various polymer composites based on hemicelluloses such as modified hemicellulose-based materials, hemicellulose-based particles, films, and gels, as well as their potential applications.

1.1 Occurrence of Hemicellulose

Hemicelluloses are the most abundant plant polysaccharides next to cellulose, which form the hydrophilic component of the cell wall in the wood plant. There are several methods applied to isolate hemicelluloses from the wood materials such as acid pretreatment [31], alkaline extraction [32], liquid hot-water extraction [33], steam treatment [34], ionic liquid extraction [35–37], hydrogen peroxide extraction [38], wet-oxidation [39], microwave treatment [40], and organic solvent treatment [41]. Among these methods, acid pretreatment and alkaline extraction are the most important methods for hemicelluloses isolation. However, high cost and environmental pollution make acid treatment, not a good choice. And more importantly, approximately 100% of hemicelluloses are broken down into monomeric sugars and sugar degradation products, including weak acids, furan derivatives, and phenolics during acid treatment. Therefore, alkaline extraction is extensively applied and well-studied in hemicelluloses extraction due to the integrity of extracted hemicelluloses [11, 42, 43]. Of hemicelluloses, modification of xylan draws a special attention, due to its abundance and widely available as compared with other hemicelluloses. Galactglucomannan is another important hemicellulose to develop novel hydrogel, film laminates for oxygen barriers, cationized materials, and multifunctional macroinitiator for single-electron transfer-mediated living radical polymerization [44].

1.2 Structure of Hemicellulose

Hemicelluloses are non-crystalline polysaccharides bearing different anhydroglucose units (AGU) in their chains or branches and have an average degree of polymerization ranging from 100 to 200. Figure 1 gives the structure of the most important AGU units of hemicelluloses [42, 45], to a lesser extent, 4-*O*-methyl-D-glucuronic acid, D-galacturonic acid, L-rhamnose, L-fucose, and a variety of *O*-methylated neutral sugars also exist in hemicellulose [46]. Moreover, The AGU repeating units can be *O*-acetylated at the C-2 and/or C-3 positions. It is well known that hemicelluloses belong to a highly heterogeneous group of noncellulosic

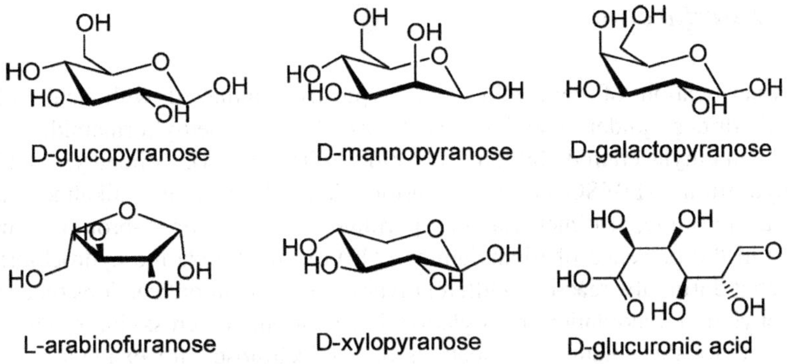

Fig. 1 The main constituents of hemicelluloses (Reproduced from [45] with permission)

polysaccharides containing xylans, mannans and glucomannans, xyloglucans, and β-(1→3,1→4)-glucans [12]. Additionally, hemicellulosic compositions, structures and amounts vary from different sources of different biomass resources. Xylan is the most common type of hemicelluloses in hardwood and gramineous plants. Xylan from many plant materials are heteropolysaccharides with homopolymeric backbone chains of 1, 4-linked β-D-xylopyranose units. Besides xylose, xylan may contain arabinose, glucuronic acid or its 4-O-methyl ether, and acetic, ferulic, and p-coumaric acids. The backbone consists of O-acetyl, α-L-arabinofuranosyl, α-1, 2-linked glucuronic or 4-O-methylglucuronic acid substituents. The frequency and composition of branches are dependent on the source of xylan, thus xylan can be categorized as linear homoxylan, arabinoxylan, glucuronoxylan, and glucuronoarabinoxylan, whereas, softwood hemicelluloses contain mostly O-acetyl galactoglucomannans (AcGGM). AcGGM consists of a backbone chain containing β-(1→4)-linked mannose and glucose units partially substituted with α-(1→6)-linked galactose units [42, 47].

2 Synthesis and Characterization of Modified Hemicelluloses (Zero-Dimensional)

Hemicelluloses have an abundance of free hydroxyl groups distributed along the backbone and side chains and are, therefore, ideal candidates for chemical functionalization. Researchers have explored these options using various techniques such as esterification, etherification, ionization, amination, amidation, acetylation, fluorination, sulfation, benzylation or grafting methods [5]. Chemical modification of wood hemicelluloses offers numerous possibilities to control and tailor the properties of hemicelluloses, leading to conductive, stimuli-responsive, more elastic, and more hydrophobic or thermoplastic products [12].

2.1 Esterification

The esterification of hemicellulose is typically conducted with acid chlorides or anhydrides under solutions such as *N*, *N*-dimethylformamide (DMF), *N*, *N*-dimethylacetamide/lithium chloride (DMA/LiCl), dimethyl sulfoxide/tetrahydrofuran (DMSO/THF), or various ionic liquids and alkaline catalysts such as pyridine, 4-dimethylamino pyridine (DMAP), or *N*-bromosuccinimide (NBS) in the presence of triethylamine (TEA) to neutralize the hydrochloric acid generated during the reaction. Different types of esterification reactions (acetylation, propionylation, oleoylation, lauroylation, benzylation, and cross-linking) have been used to tailor the potential applications of hemicelluloses, for example improving the hydrophilicity, thermal stability, thermoplastic property of hemicelluloses, as well as their solubility in organic solvents for a good use in heavy metal ions adsorption, drug delivery and food packaging [11, 12].

The esterification of hemicelluloses always happens with acid chlorides or anhydrides having short alkyl chains. In order to improve the esterification of hardwood xylan with a high degree of substitution (DS), two different approaches were investigated. The acetylation reaction of xylan-type hemicelluloses can be carried out with acetic anhydride in a homogeneous solution of DMF/LiCl in the presence of DMAP as a catalyst. This reaction can also be performed under heterogeneous conditions in the absence of organic solvent and use methane sulfonic acid (MSA) or DMAP as a catalyst. The highest DS of 1.6 and maximal yield of 85% can be achieved under the screened optimal reaction conditions. The results show that the amount of MSA has a significant influence on DS and yield. In addition, the esterification of propionic and hexanoic anhydrides was also conducted to obtain hydrophobic xylan esters with low DS values [48]. Homogeneous esterification of xylan-rich hemicelluloses with maleic anhydride in 1-butyl-3-methylimidazolium chloride ([BMIM]Cl) ionic liquid by using LiOH as a catalyst was performed by Peng et al. [49] (Fig. 2). The novel functional biopolymer has carbon-carbon double bond and carboxyl groups, and the DS values range from 0.095 to 0.75 can be achieved. The influences of the molar ratio of maleic anhydride toanhydroxylose unit in xylan-rich hemicelluloses, reaction temperature, time, and dosage of LiOH on the DS value of hemicellulosic derivatives were investigated. Most notably, the DS increased to 0.72 when 0.02 g LiOH was applied, and the thermal stability of the new polymer was lower than that of the native xylan-rich hemicelluloses.

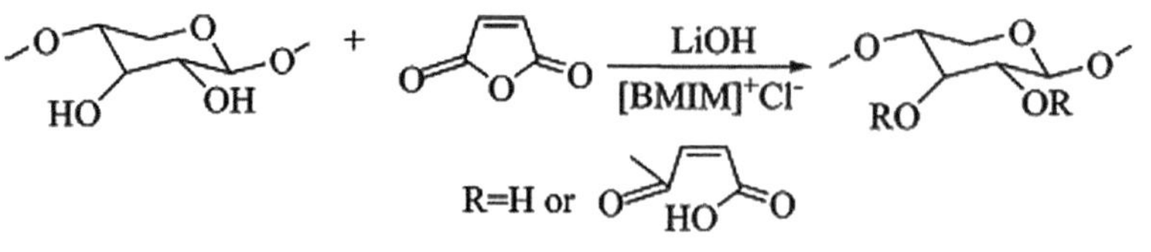

Fig. 2 Reaction scheme of homogeneous esterification of XH with MA in [BMIM]Cl ionic liquid (Reproduced from [49] with permission)

Zhang et al. [50] performed the esterification of hemicelluloses with butyryl chloride in [BMIM]Cl ionic liquid (IL) using TEA as a neutralizer, as shown in Fig. 3. The as-prepared butyrylated hemicellulose was characterized by FT-IR, TGA, ^{1}H, and ^{13}C NMR spectroscopies, and the results indicate that the DS of the products changed with the reaction conditions such as molar ratio of the substrates, temperature and reaction time. The highest DS of 1.89 can be obtained at the preferred reaction condition. Furthermore, the hydrophobicity and thermal stability of the butyrylated hemicelluloses increased with the increasing DS value.

Hemicelluloses modified with acid chlorides or anhydrides having long alkyl chains were also been investigated. Sun et al. [51] obtained stearoylated hemicelluloses with different DS values (0.20–1.71) by a rapid reaction of wheat straw derived hemicelluloses with stearoyl chloride in homogeneous DMF/LiCl system with DMAP as a catalyst and TEA as an acid acceptor. The concentrations of stearoyl chloride and TEA, as well as the reaction temperature, have different effects on the DS. The results also demonstrated that TEA was better than pyridine as an acid acceptor. Under optimum conditions, the highest DS value of 1.71 can be obtained, and approximately 92% of the free hydroxyl groups in native hemicelluloses are stearoylated with a minimal degradation ($\leq 8\%$). The thermal analysis shows an increase in the thermal stability of the stearoylated hemicelluloses. Another ester with a low DS has been synthesized by oleoylation of sugarcane bagasse hemicelluloses using NBS as a catalyst in the DMF/LiCl system [52]. The thermal stability of the oleoylated hemicelluloses decreases slightly upon chemical modification, but no significant further decrease in thermal stability was observed for the derivatives with a DS $\geq$ 0.29. Such low-DS polymers have great application in food packaging. Similarly, lauroylated hemicelluloses (LH) can be prepared in homogeneous [BMIM]Cl IL without catalyst [53]. The relationships between reaction parameters including the molar ratio of lauroyl chloride (LC) to anhydroxylose unit in hemicelluloses, reaction temperature, and reaction time with the DS value of hemicellulose are studied. The highest DS value of 1.82 can be achieved in the optimal condition. A significant degradation of the derivative occurs with the increase in molar ratio, reaction time, and temperature, and thus the thermal stability of the derivative is lower that of the native hemicelluloses. The synthesis of poplar wood hemicellulose-based esters containing different groups (including short and long alkyl chains) are similar to Sun et al. [51]. The GPC

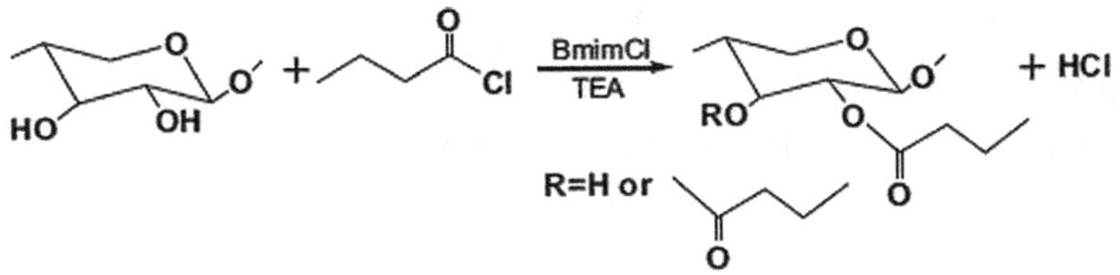

Fig. 3 Butyrylation of hemicelluloses (Reproduced from [50] with permission)

analysis reveals no significant degradation for the polymers during the reaction process [54, 55].

In some cases, xylan esters can be electrospun into nanofibers, the mechanical properties of the esters are also investigated. A series of xylan esters with varying alkyl chain lengths (C2–C12) are synthesized by heterogeneous and homogeneous reactions. The thermal stability and solubility in CHCl$_3$ increase after esterification. The film forming performance of the new polymers is highly dependent on the molecular weight, while the tensile strength and Young's modulus decrease with the increase of alkyl chain length and conversely in the hydrophobicity and elongation at break of the xylan esters. Especially, the high-molecular-weight xylan esters in hexafluoroisopropanol (HFIP) can be electrospun into nanofibers [56].

Hemicelluloses-based esters can be used as drug carriers. Xylan ibuprofen esters with a high drug loading were synthesized via the activation of the ibuprofen carboxylic acid with N, N'-carbonyldiimidazole (CDI) in DMSO [57]. For the sake of enhancing its hydrophobicity, sulfation of xylan ibuprofen esters was carried out in the SO$_3$/DMF system (Fig. 4). The xylan ibuprofen esters form spherical nanoparticles with a size ranging from 328 to 473 nm (PDI = 0.123–0.255) after dialyzed in DMA; the mean diameter, however, reduces from 342 to 162 nm (PDI = 1.04) after sulfation (Fig. 5). In addition, preliminary stability tests indicate that hydrolytic stability decreases with the introduction of sulfate groups.

There is an increasing need for barrier coatings in food packaging to prevent oxygen, carbon dioxide, water and toxic substances. O-acetyl-galactoglucomannan (GGM) is the main hemicellulose in softwoods and has been widely applied in the synthesis of biopolymers. For the environmental consideration, a way to prepare

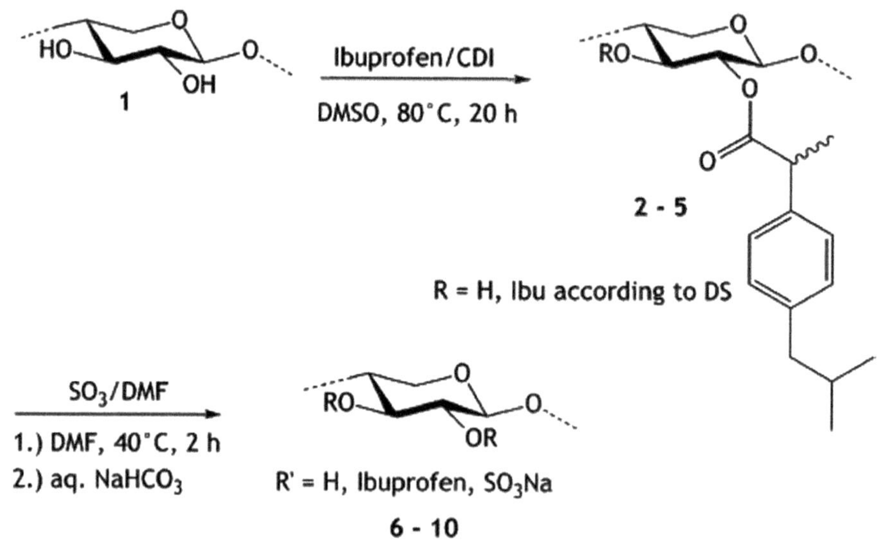

Fig. 4 Synthesis of xylan ibuprofen esters and sulfated xylan ibuprofen esters (Reproduced from [57] with permission)

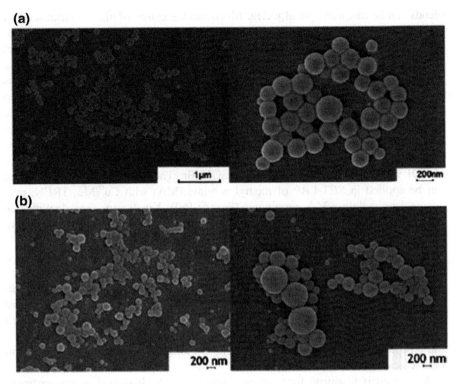

Fig. 5 SEM images of nanoparticles of xylan ibuprofen esters (upper) and sulfated xylan ibuprofen esters (beneath) prepared by dialysis: **a** $DS_{Ibu} = 0.79$, **b** $DS_{Ibu} = 1.24$ (*upper*); **a** $DS_{Sulfate} = 0.15$, **b** $DS_{Sulfate} = 0.25$ (*beneath*) (Reproduced from [57] with permission with permission)

GGM ester has been proposed by Kisonen et al. [58]. In their work, GGMs were esterified with phthalic and benzoic anhydrides, respectively. The results showed that GGM ester has a better water vapour resistance and grease barrier than native GGM when coated onto carton board with a thin coating (1–3 g/m²). Penetrating 0.1% rapeseed oil with a 2.4 g/m² coating thickness cost up to 54 h and the lowest water vapor transmission value was 39 g/m²/24 h with 9.7 g/m² coating, which indicates a significant decrease in grease barrier in comparison to the uncoated carton board. Moreover, the high-molar-mass GGM coating performed better than the low-molar-mass GGM coatings. To sum up, GGM and GGM esters provide a sustainable and safe conception for barrier coatings in food packaging.

Buchanan et al. [59] suggested that arabinoxylan acetate/propionate/butyrate can be rapidly prepared using MSA as a catalyst. The average molecular weight of the esters was ∼500,000, and the glass transition temperatures (T_g's) range from 61 to 138 °C and increase with the decrease of DS, the thermal stability is enhanced in comparison to the parent arabinoxylan but undergoes a rapid degradation when the temperature was above 225 °C. Surprisingly, arabinoxylan acetate/cellulose acetate blends can be cast into optically clear films, and the clarity of films is influenced by solvent, which provides a possibility of incorporating these arabinoxylan derivatives and other polysaccharide esters into the mixture for preparing novel composites.

Single-electron-transfer mediated living radical polymerization (SET-LRP) has appeared as an effective strategy to achieve living radical polymerization and produce graft copolymers with brush-like architecture without the use of any toxic catalyst. This method is appealing for the design of hybrid materials from macromolecules such as polymers based on sugar moieties, known as polysaccharides. An esterification between AcGGM and α-bromoisobutyric acid with the assistance of CDI yields a macroinitiator ester AcGGM-Br. The macroinitiator as a catalyst can be applied in SET-LRP of methyl acrylate (MA) with Cu^0/Me_6-TREN, producing hemicellulose/MA copolymer (Fig. 6) [60]. Kinetic analyses demonstrate the living property of the SET-LRP process together with a high conversion of up to 99.98%, offering a brush-like structure of the hybrid copolymers.

The synthesis of hydroxycinnamic acid xylan esters from oat spelt arabinoxylan/birchwood glucuronoxylan in the presence of LiCl/DMA and pyridine/DMAP was reported by Wrigstedt et al. [61]. Owing to the antioxidative nature of ferulic and sinapic acids, the esterified xylan polymers remained antioxidative activity and exhibited a better lipid antioxidation when comparing with the native oat spelt and birch wood xylans. Further studies show that ferulic acid glucuronoxylan esters and ferulic acid xylan esters were more efficient antioxidants than arabinoxylan and sinapic acid-modified arabinoxylan.

Thiolation is a way to form an ester bond with the hydroxyl groups in hemicelluloses. Maleki et al. [62] proposed a one-pot procedure to synthesize thiolated O-acetyl galactoglucomannan (AcGGM-SH) by AcGGM-mediated nucleophilic ring-opening of γ-thiobutyrolactone and the activation of the hydroxyl groups in hemicellulose (Fig. 7). ^{1}H, ^{13}C NMR spectroscopy and Ellman's reagent assay confirmed that the thiol groups were successfully incorporated into the backbone of hemicellulose. Interestingly, three AcGGM hydrogels were fabricated by using three different methods named thiol-ene click reaction, thiol-Michael addition and disulfide bond formation reaction, respectively. Besides, some polymers such as polyethylene glycol monomethacrylate (PEG-MA) were grafted onto AcGGM-SH by a thermally induced thiol-ene click reaction. Hence, the thiolation of AcGGM is indeed a good choice to design hemicellulose-based esters and hydrogels by employing "click" chemistry.

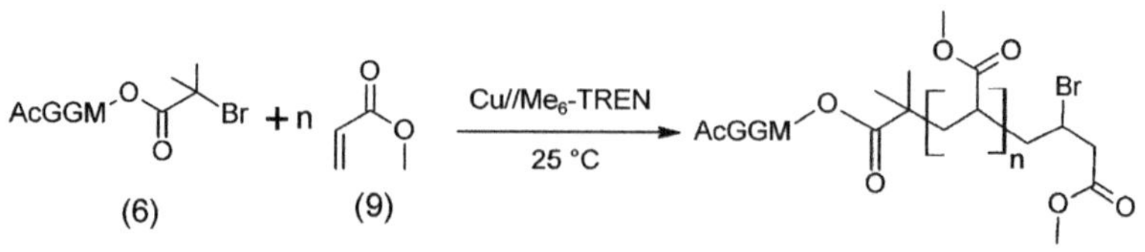

Fig. 6 The SET-LRP of MA (9) with α-Briba-AcGGM (6) as a macroinitiator (Reproduced from [60] with permission)

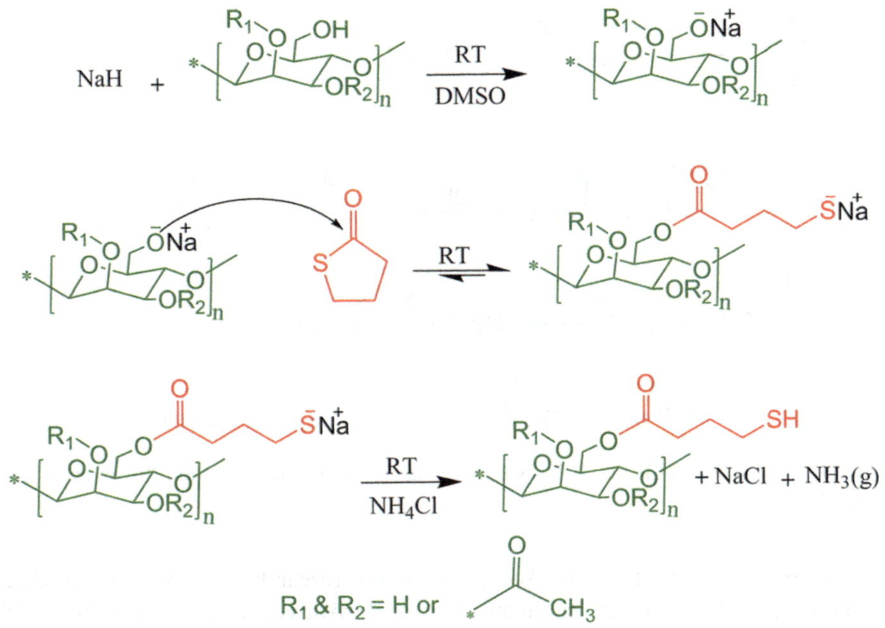

Fig. 7 One-pot thiolation of AcGGM in DMSO (Reproduced from [62] with permission)

2.2 Etherification

Etherification is extensively studied by researchers for chemical modification of hemicelluloses, and many reactions including methylation, alkylation, benzylation and cationic moieties are usually introduced by this way. This reaction occurs between hydroxyl and alkylating agent such as acrylamide, alkyl halides (chlorides, bromides, and iodides), alkyl sulfonates, and epoxides by using alkaline (e.g. sodium hydroxide) as a catalyst under the organic (e.g. DMSO, ethanol, ethanol/ acetone, isopropanol, butanol, and DMF) or inorganic (e.g. water) system [5]. The ether bonds are more stable than ester bonds and not easy to be hydrolyzed in an alkaline environment. What's more, the solubility and film formability of hemicelluloses are improved through etherification. However, In order to reduce the degradation of hemicelluloses, heterogeneous reaction systems like water/organic solvents are required [12]. Ren et al. [63] synthesized novel hemicellulosic derivatives through five different heterogeneous/homogeneous systems by etherification with acrylamide in the presence of sodium hydroxide as a catalyst (Fig. 8). The as-prepared hemicellulose derivatives containing carbamoylethyl and carboxyethyl achieved the highest DS value of 0.58 under the optimal condition. Especially, reaction medium has a significant influence on the extent of etherification in terms of solvent polarity and stereochemistry. The reaction efficiency increased as the polarity of the solvent decreased. In addition, the thermal stability

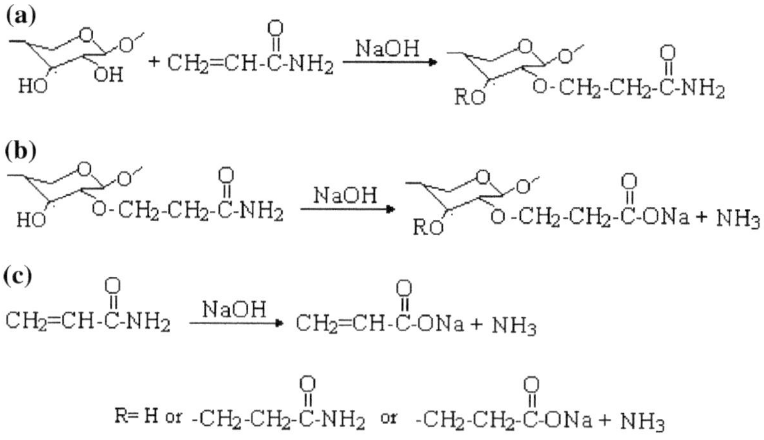

Fig. 8 Etherification of hemicelluloses with acrylamide in alkaline condition (Reproduced from [63] with permission)

decreased after etherification. According to the research described above, Peng et al. [64] synthesized identical hemicellulosic derivatives with a higher DS of 0.92 in butanol/water media. The molecular weights of the products were lower than that of the native hemicellulose due to the significant degradation of hemicellulose during alkaline activation in the etherification reaction. Furthermore, the new derivatives exhibited a less elastic behaviour as compared with the native hemicellulose due to the influences of molecular weight and the functional groups on the structure of macromolecule chains.

Etherification of hemicelluloses with 2, 3-epoxypropyltrimethylammonium chloride (ETA) can be carried out under sodium hydroxide/water media, obtaining quaternized hemicelluloses (QH). Afterwards, smooth and heat-resistant films basing on QH and montmorillonite (MMT) platelets were prepared by a vacuum-filtrated technique. It turns out to be the electrostatic and hydrogen-bonding interactions between the electropositive QH and the exfoliated anionic MMT that contribute to the film-forming process. The excellent thermal property provides a great potential application for the nanocomposite films in flame retardant [65, 66]. By using similar condition, cationic galactomannan/xylan can also be prepared (Fig. 9). The desired DS of 1.3 could be obtained with a maximal grafting rate of 48 and 64% of mass yield at the optimal condition [67]. In a similar way, Ren et al. [68] synthesized a cationic hemicellulose from sugarcane bagasse hemicellulose. ^{13}C NMR spectra justified that the etherification reaction mainly occurred to C-3 position of hemicelluloses. In addition, the molecular weight of the new products reduced sharply from 28,890 (native hemicellulose) to 15,520 g mol^{-1} (DS = 0.33), which is attributed to the degradation of hemicellulose (especially when prolonging the time of alkaline activation at higher temperature).

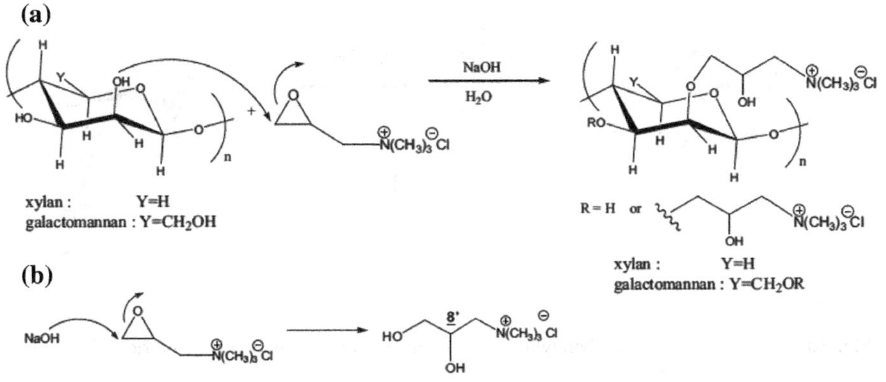

Fig. 9 a Cationisation of the hemicelluloses and b concurrent reaction forming the by-product (Reproduced from [67] with permission)

Methylation is a typical example of etherification, Fang et al. [69] prepared a methylated hemicellulose by using methyl iodide as an alkylating reagent and sodium hydride (NaH) as a catalyst in DMSO. The reaction mechanism involves methyl sulfinyl anion capturing a proton from the hemicellulose to form the polyalkoxide. The conversion of methylation was high up to 90% with a DS value of 1.7; an endothermic degradation and an obvious increase in thermal stability were observed due to the existence of methyl after methylation.

Hydroxyalkylated xylans were produced by the etherification of xylan and epoxides (propylene oxide, butyl glycidyl ether, and allyl glycidyl ethers) under an alkaline condition. The DS values of the functionalized xylans varied from 1.5 to 0.2. Moreover, the xylan derivatives exhibited a significant oxygen barrier and surface strength as well as low mineral oil migration, which indicates a promising application in barrier coatings for packaging and pigment coating binders for printing papers [28].

Williamson benzylation was used for preparing hydrophobic benzyl galactoglucomannan (BnGGM). In this research, the benzylation of AcGGM hemicellulose was carried out using benzyl chloride as a benzylation reagent and tetrabutylammonium iodide (TBAI) as a phase transfer catalyst in water/sodium hydroxide system, and then surface modification was conducted by plasma treatment followed by styrene addition or vapor-phase grafting of styrene, which was followed by lamination [70]. The vapour-phase-grafted films showed a better tolerance toward to humidity than the plasma-treated ones. The contact angle and oxygen permeability measurements indicate that the films were not only water-resistance but also had excellent oxygen barrier and moisture tolerance properties. Therefore, the surface grafting and lamination methods seem to be a good choice for obtaining good barrier properties even in high humidity environment. Another similar example about hemicelluloses benzylation is the benzylation of wheat straw hemicelluloses in an ethanol/distilled water system (Fig. 10). The low DS value ranging from 0.09 to 0.35 was dependent on the volume ratio of

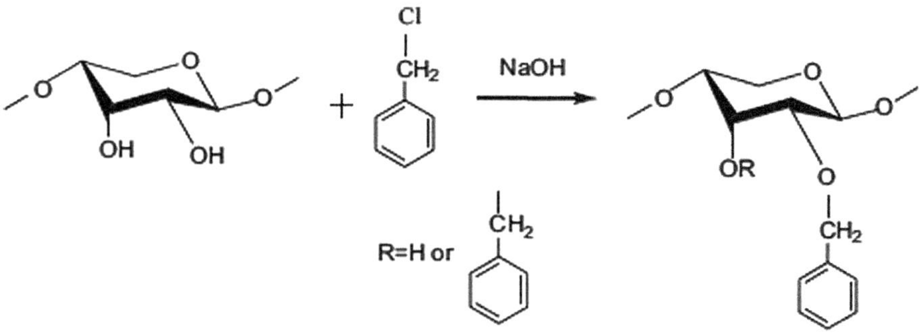

Fig. 10 Reaction scheme of benzylation of hemicelluloses with benzyl chloride (Reproduced from [71] with permission)

ethanol/water, the molar ratio of sodium hydroxide or benzyl chloride to anhydroxylose unit in hemicellulose, reaction temperature and reaction time. FT-IR and ^{13}C NMR spectra identified that benzyl groups were incorporated onto the backbone of hemicellulose, the introduction of benzyl groups increased the thermal stability as well as the hydrophobic property of the new materials [71].

Thiols are very useful groups in developing new materials and bio-inspired polymers. By combining the etherification and thiol-ene reaction, novel thioether xylans were synthesized with a protected thiol that can be stored at atmosphere. At first, allyl groups were introduced in the backbone of xylans by etherification with allyl chloride in aqueous alkaline condition at 40 °C, giving etherified xylans with a DS of up to 0.49. Afterwards, the allyl groups in xylans were reacted with different thiols via thiol-ene reaction by using potassium peroxydisulfate as a radical initiator and water as the solvent, producing novel thiol-, amine- or amino acid functionalized xylans. The thiol-functionalized xylans have broad applications such as preparing hydrogel scaffolds, cross-linking foams by a thiol-thiol oxidative coupling reaction, and modifying filter paper surface by the simple dipping method [72].

The cationization of hemicelluloses was carried out by using 3-chloro-2-hydroxypropyltrimethylammonium chloride (CHMAC), glycidyltrimethylammonium chloride (GTMAC), [2-(methacryloyloxy) ethyl] trimethylammonium chloride (METAC), or ETA as a quaternary nitrogen source in heterogeneous (e.g. H_2O) or homogeneous system (e.g. DMSO) [73]. As described in etherification [65, 67, 68], cationic 2-hydroxypropyltrimethylammonium (HPMA) xylans based on birch wood xylan and ETA were synthesized in 1, 2-dimethoxyethane (DME)/H_2O system. In this research, adjusting the molar ratio of HPMA and anhydroxylose units of xylans led to the altering of DS in cationic xylans and the highest DS was calculated to be 1.64. Furthermore, the HPMA xylans can be cast into films and had a polyionic structure with positive charged ammonium group (Fig. 11) [74]. In a similar case, a cationic ETA-GGM polyelectrolyte was synthesized and subsequently re-acetylated to obtain enhanced thermal stability and hydrophobic property [75]. Higher reaction efficiency and lower degradation were achieved by using

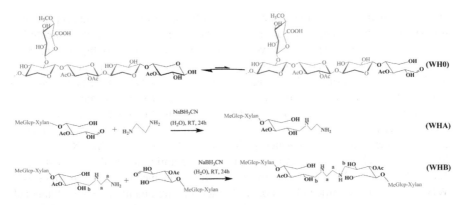

Fig. 11 (*i*) MeGlcp-Xylan reducing-end hemiacetal in equilibrium with the ring-opened aldehyde, (*ii*) MeGlcp-Xylan amine functionalization, (*iii*) head-to-head chain extension of MeGlcp-Xylan (Reproduced from [79] with permission)

THF/H_2O or DMSO/H_2O as a reaction media in comparison to that of only with plain water. As reported, the cationised GGM can be potentially utilized as polyelectrolyte layers in films and coatings.

Liu et al. exploited hemicelluloses that isolated from pre-hydrolysis liquor (PHL) to render cationic copolymers under the glycidyltrimethylammonium chloride (GTMAC)/NaOH system [73]. The cationization parameters were investigated, the result indicates the charge density and DS value of the cationic hemicelluloses increased with the increasing dosage of NaOH from 1 to 5% (wt.), but rapidly decreased when the amount of NaOH exceeded 5% (wt.). By copolymerization of xylan and [2-(methacryloyloxy) ethyl] trimethylammonium chloride (METAC) in the presence of $K_2S_2O_8$, a cationic xylan copolymer was produced. The as-prepared xylan-METAC copolymer has an application as a flocculant for dye removal from wastewater [76]. In Kong's research, a higher DS of cationic xylan was obtained by using a semi-dry process compared to that of in the wet process [77]. The GPC analysis shows a substantial degradation of the xylan copolymers occurred during the cationization under alkaline conditions. Conversely, cationic ETA-hemicellulose copolymers with low average DS value were prepared in homogeneous DMSO/H_2O media without significant degradation in alkaline media [78].

2.3 Amination

A reductive amination is a common approach used in polysaccharide chemistry to immobilize different types of amino-functionalized moieties entities to saccharide structures by using the ring-opened aldehyde at the reducing end of the hemicellulosic chain as a reactive center for attaching amino coupling agents. This reaction is conducted in aqueous media and under mild conditions, and thus represents a robust and highly selective method for the chemical modification of hemicellulose [12]. The hemicellulosic derivative exhibits an enhancing amphiphilic property with increasing molecular weight after reducing amination, which can be utilized as reinforce way for films and coatings. In the report by Dax et al. [79], reductive amination was conducted between native GGM and a series of amino-functional fatty acids in the presence of $NaBH_3CN$, yielding block-structure amphiphilic GGM derivatives which were only water-soluble [80]. In another sample, a chain-extended O-acetyl-4-O-methylglucuronoxylan-rich (MeGlcp-Xylan) wood hydrolysate (WH) was prepared through reductive amination with the assistance of $NaBH_3CN$ (Fig. 12). Especially, the formability and mechanical performances of WH films were significantly improved after the treatment, and simultaneously reduced the need of co-components. In addition, the produced films exhibited an excellent oxygen barrier performance. Therefore, reductive amination could be applied for increasing the molar mass of a low-molecular-weight xylan with mono and bifunctional amines on its reducing end and has a potential utilization in films and coatings [81].

Ehrenfreund-Kleinman et al. [82] synthesized arabinogalactan (AG)-based sponges by reductive amination on the reducing end of the oxidized AG with different amines in the presence of $NaBH_4$. The as-prepared sponges exhibited high swelling and had a possible use in cell growth in tissue engineering. Particularly,

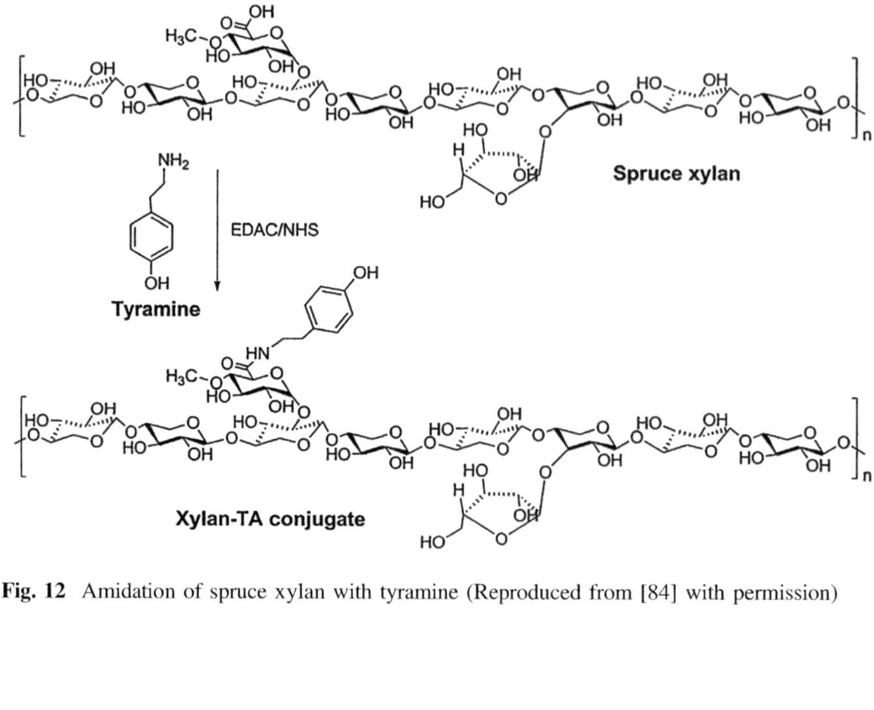

Fig. 12 Amidation of spruce xylan with tyramine (Reproduced from [84] with permission)

AG-chitosan sponges showed good biocompatibility with an inflammatory response confined to the implant site which decreased with time.

2.4 Amidation

Amidation of hemicellulose with amine-functionalized compounds can be defined as the coupling of amines to carboxylic acids through amide bonds and is commonly activated by N-ethyl-N'-(3-dimethylamino)propyl carbodiimide hydroxide/N-hydroxysuccinimide (EDC/NHS) (carbodiimide-mediated amidation) in aqueous media [83]. Kuzmenko et al. [84] reported the amidation of glucuronic acid groups from xylan backbone and tyramine (TA) with the activation of EDC/NHS (Fig. 13) followed by enzymatic crosslinking of the phenol-containing TA-xylan conjugate in the presence of HRP/H_2O_2, and fabricating a 3D hydrogel in a negligible time (20 ± 5 s) for cell encapsulation and vivo delivery. The spruce xylan-based hydrogel exhibits an obvious increase in the storage module and excellent swelling stability, which provides a great potential application in tissue engineering. In another case, oxidized GGM (containing uronic acid) was functionalized with arginine or 1, 6-diaminohexane by using the same carbodiimide-mediated amidation. This amidation is performed after the oxidation of hydroxyl groups in GGMs [83]. Except for EDC, 4-(4, 6-dimethoxy-1, 3, 5-triazin-2-yl)-4-methylmorpholinium chloride (DMT-MM) was also employed as a superior acid-amine coupling reagent for the amidation of an acidic xylo-oligosaccharide fragment from 4-O-Methylglucuronoxylan hemicellulose without the tight control of PH. As a result, amidation of C-6 acids was performed efficiently by using the coupling reagent DMT-MM and formed an amide linkage. A triazine side product which hindered C-1 amidation, however, was observed. Furthermore, the modified monomers were polymerized via copper(I)-catalyzed azide-alkyne cycloaddition (CuAAC) "click" reaction, and thus forming a soft gel. This pathway provides a rarely used system of DMT-MM with preferable performance compares to EDC under certain conditions [85].

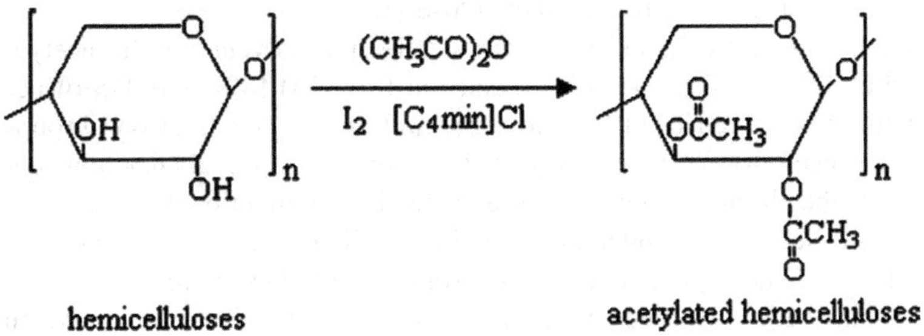

Fig. 13 Acetylation of wheat straw hemicelluloses (Reproduced from [89] with permission)

2.5 Acetylation

Hemicellulose acetylation is an important chemical modification that introduces acetyl groups to the hemicellulose chain through esterification reactions as mentioned above. This reaction was carried out in non-water medium such as ionic liquids (ILs), DMA/LiCl, DMF, in the presence of DMAP, pyridine or NBS as catalysts. The thermal stability, mechanical properties, oxygen transmittance and hydrophobicity of the acetylated hemicellulose have been investigated by altering the reaction conditions [11].

Fundador et al. [86] acetylated xylan from hardwood with acetic anhydride or propionic anhydride in homogenous DMA/LiCl system by the catalysis of pyridine, obtaining xylan acetates and xylan acetate propionates with varied DS values. The structure and DS of the modified xylans were elucidated by ^{1}H, ^{13}C NMR spectroscopy, and two-dimensional (2D) NMR techniques. The results show equal reactivities at the C-2 and C-3 of xylose during the acetylation process. Moreover, the thermal stability and solubility in chloroform of xylan were improved after acetylation. In this study, xylan acetate (DS = 2.0) nanofibers with a diameter of 163–429 nm were formed by an electrospinning method and using HFIP as a solvent. Mechanical tests of xylan acetate propionate films showed that the tensile strength and elongation at break increased as the DS decreased from 1.1 to 0.6. Another acetylation of wheat straw hemicellulose was reported by Sun et al. under the same condition mentioned above, except for using DMAP as a catalyst. Both the yield and DS of the acetylated hemicellulose were increased with temperature (from 60 to 85 °C) and reaction time (from 12 to 60 h); and a high DS value of 1.49 could be obtained under the optimal condition (85 °C, 60 h) with 80% of the free hydroxyl groups being acetylated. The GPC analysis revealed a sharp decomposition when the reaction time was 72 h at 85 °C, resulting in a low molecular-average weight of 22,890 g mol^{-1}. Moreover, the chemical modification leads to a decreased thermal stability [87]. Due to the toxicity of pyridine and the high cost of DMAP, a new catalyst NBS was utilized for the acetylation of sugarcane bagasse hemicellulose with acetic anhydride under an almost solvent-free system. The overall yield varied from 66.2 to 83.5% and DS ranged from 0.27 to 1.15 by altering the reaction temperature and duration. Furthermore, the results of TG-DSC indicate that the thermal stability of the modified hemicellulose was higher than that of the native hemicellulose [88].

ILs are considered to be a green and effective solvent for the acetylation of hemicelluloses by using iodine as a catalyst (Fig. 14) [89]. The DS of acetylated hemicelluloses ranged from 0.49 to 1.53, and about 83% hydroxyl groups in the native hemicelluloses were acetylated under the optimal reaction condition. Moreover, the thermal stability was enhanced after functionalization. In order to achieve full acetylation under mild conditions, Stepan et al. [90] reacted rye arabinoxylan (AX) and spruce arabinoglucuronoxylan (AGX) with acetyl chloride and acetic anhydride in two new ILs, respectively. The first system was 1-ethyl-3-methylimidazolium dimethyl phosphate ([emim][Me$_2$PO$_4$]) with CHCl$_3$

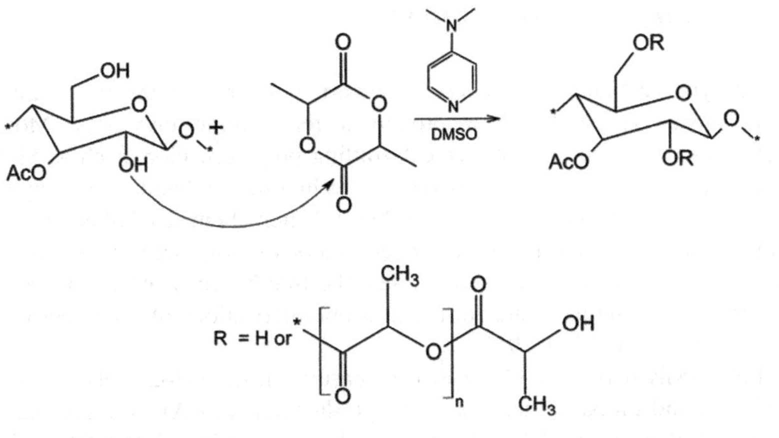

Fig. 14 Synthetic pathway for the grafting of L-lactide oligomers from the acetylated galactoglucomannan (AcGGM) backbone (Reproduced from [96] with permission)

as a co-solvent, and the other was 1, 5-diazabicyclo[4.3.0]non-5-enium acetate ([DBNH][OAc]) without co-solvent. The complete acetylation was achieved within 5 min in both reaction systems under the optimal condition. FT-IR and ^{1}H NMR spectroscopy confirm the full acetylation of xylan. GPC result shows no significant degradation occurred during acetylation of rye AX but a minor depolymerization took place in the case of spruce AGX. Both of the acetylated xylans maintained a relatively high molecular weight. In addition, the solubility of the new polymers in $CHCl_3$ and dimethyl carbonate allowing it casting into transparent films.

Acetylation has a significant influence on the solubility, water content and thermal properties of the modified hemicelluloses. The differences have been well investigated by the reaction between aspen glucuronoxylan and acetic anhydride in formamide/pyridine system [91]. For instance, the water content in humidity strongly decreased after acetylation. The fully acetylated glucuronoxylan was soluble only in aprotic solvents, such as chloroform and DMSO, while the native glucuronoxylan was partially soluble in hot water. The decomposition temperatures for a 10% weight loss were 283 and 335 °C for non-acetylated glucuronoxylan and the acetylated sample, indicating an improving thermal stability after acetylation. DSC analysis indicated that glucuronoxylan with a DS of 1.2 had a significant glass transition temperature during 160–200 °C and could form films under pressure, but not the case in non-acetylated samples.

A new method for synthesizing acetylated hemicelluloses was developed in 1-allyl-3-methylimidazolium chloride ([Amim]$^+$Cl$^-$) ionic liquid without using catalyst [92]. The DS ranged between 0.03 and 1.25 and increased with the reaction time and temperature. The properties of the as-prepared hemicellulosic derivative such as solubility, equilibrium water content, thermal stability were also significantly changed. In conclusion, eco-friendly ionic liquids are excellent solvents to generate value-added hemicellulose acetate with high DS and can be used in various industrial applications.

2.6 Grafting Copolymerization

Graft polymerization is an effective method to overcome the disadvantages of hemicelluloses and tailor their properties for specific end uses. Most graft copolymers are formed by free radical grafting polymerization, such as SET-LRP, which is a water-tolerating polymerization technique conductible at benign conditions with the catalysis of Cu(0). In addition, "click chemistry" offers a promising pathway for the synthesis of graft-copolymers of polysaccharides. Among the radical polymerizations of hemicelluloses, the free hydroxyl groups in hemicelluloses are used as initiator sites to perform polymerizations of a number of monomers to form grafts [5, 93–95].

Graft copolymers XylC6N$_3$-g-PLLAs basing on di-O-(6-azidohexanoyl)-xylan (XylC6N$_3$) and Propargyl-terminated poly(L-lactide) (PLLA) were prepared in the presence of N, N, N', N', N''-pentamethyldiethylenetriamine (PMDETA) and copper (I) bromide via click chemistry. The results indicate that the grafted PLLA side-chains act as an internal plasticizer for xylans [94]. Similarly, L-lactide (LLA) oligomers were functionalized with wood hydrolysate derived hemicellulose-rich fractions by ring opening graft polymerization (Fig. 15), the as-prepared copolymers can be applied as a compatibilizer for hydrolysate/PLLA film [96]. Persson et al. [97] developed biodegradable PLA-g-xylan copolymers

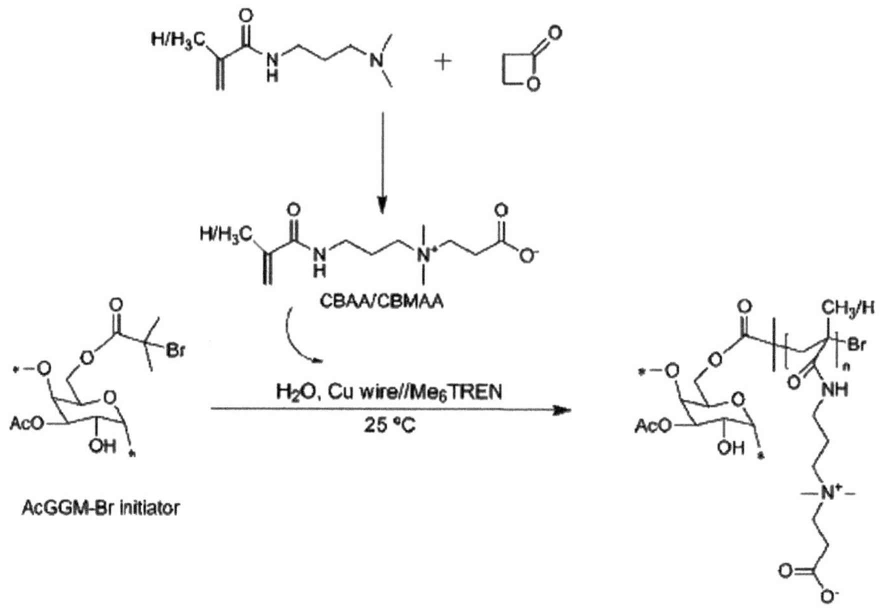

Fig. 15 Schematic outline of the pathways for grafting poly(CBAA-3) and poly(CBMAA-3) from the AcGGM-Br macroinitiator via SET-LRP under mild conditions (Reproduced from [100] with permission)

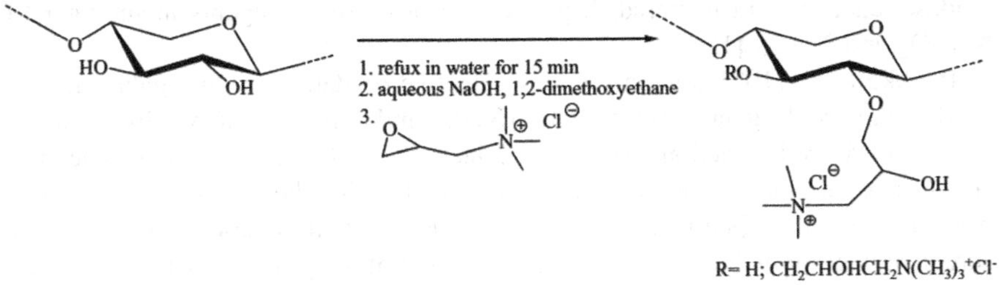

Fig. 16 Reaction scheme of the cationization of xylan with EPTA (Reproduced from [74] with permission)

with different branch length in the catalysis of triazabicyclodecene (TBD). The obtained polymers with long polylactide (PLA) branch exhibited better mechanical and thermal properties than the polymers with short PLA branch.

Acrylonitrile (AN) or methyl acrylate (MA) can be grafted onto hemicelluloses containing lignin. The graft polymerization of hemicellulose with low and high lignin content can be carried out by using ceric ammonium nitrate and $FeSO_4/H_2O_2$ redox system as an initiator, respectively [98]. Taking $FeSO_4/H_2O_2$ as an initiator, Dong et al. [93] synthesized copolymers by grafting acrylamide (AM) and methacryloyloxy ethyl trimethyl ammonium chloride (DMC) on the hemicelluloses from alkaline peroxide mechanical pulping (APMP) effluent. The resulting copolymers show enhanced physical strength and water resistance on corrugated paper, which is attributed to the functional groups (–OH and –NH$_2$) and the covalent bonds formed between the hemicellulose and the graft copolymer.

According to the previous description [60], a series of hemicellulose-based graft-copolymers with tunable hydrophilicity were synthesized via SET-LRP process, including methyl methacrylate (MMA)/hemicellulose, N-isopropyl acrylamide (NIPAM)/hemicellulose, and acrylamide (AcAm)/hemicellulose copolymers [99]. In Edlund's report, AcGGM-graft-GMA copolymers with a conversion of high up to 80% were successfully prepared by the SET-LRP procedure of AcGGM-Br and glycidyl methacrylate (GMA). A new class of zwitterionic polymers were prepared by grafting poly(carboxybetaine) onto AcGGM-Br via the same technique (Fig. 16) [100]. In another case, grafting copolymers composed of methylated xylan and polystyrene were synthesized by O'Malley's group, and the mechanism involved the reaction of "living" polystyrene with the methyl glucuronate moiety and generation of well-defined comb-like structures [101]. Therefore, SET-LRP is regarded as a powerful method for the design of new hemicellulosic copolymers.

2.7 Oxidation

Oxidation is a reaction that transforms the alcohol groups of hemicellulose to aldehydes or carboxylic acids in the presence of specific enzymes such as galactose oxidase, catalase and horseradish peroxidase or oxidizing agents including KI/I_2, $NaClO_2$ and $NaIO_4$ [12].

Parikka et al. [102] studied the regioselective oxidation of guar galactomannan (GM), tamarind galactoxyloglucan (XG), and spruce GGM by galactose oxidase-catalyzed reaction to form galactoaldehydes and further selectively chemical oxidation to obtain galacturonic acids with the assistance of I_2/KI or $NaClO_2$ (Fig. 17). Quartz crystal microbalance with dissipation (QCM-D) measurement was applied to determine the interaction of the products with cellulose, the result showed that the chemo-enzymatically oxidized galacturonic polysaccharides with an unmodified backbone had a better ability to interact with cellulose than 2, 2, 6, 6-tetramethylpiperidine-1-oxyl (TEMPO)-oxidized products. Similarly, a series of polysaccharides containing terminal galactose were oxidized by galactose oxidase, catalase and horseradish peroxidase. The oxidations result in significant changes in properties of different polysaccharides. For instance, tamarind xyloglucan formed a gel and larger particles were dispersed in the solution of spruce galactoglucomannan after oxidation [103, 104].

TEMPO oxidation can specifically convert the primary alcohols on C-6 to carboxylate groups. Song et al. [105] reported a TEMPO/$NaClO_2$ oxidized β-$_D$-glucan and the product was applied as an enhanced bonding agent for papermaking. A significant enhancement in tensile strength and folding endurance of paper could be observed when using oxidized β-$_D$-glucan a strengthening agent. Xylan can be oxidized by $NaIO_4$ to introduce dialdehydes and hemiacetal bonds formed between the aldehyde groups and hydroxyl groups of the modified xylan during the freeze-casting process, obtaining cross-linking hydrogels with the reinforce of nanocrystalline cellulose [106]. The strength of the gels decreased with the increasing degree of oxidation due to the change in xylan molecular rigidity. Another aldehyde functionalized xylan MGX was prepared by the oxidation of $NaIO_4$ [107]. It is found that periodate oxidation is specific to vicinal diols for the formation of aldehydes at low periodate concentration and no depolymerization occurs at 0.05 $NaIO_4$/xylose ratio. The effect of different oxidants on the degradation rate of oxidized arabinogalactan-chitosan sponges was proposed by Ehrenfreund-Kleinman et al. [108]. The result indicates the degradation of sponges oxidized with sodium chlorite was faster than that of the sample with sodium periodate.

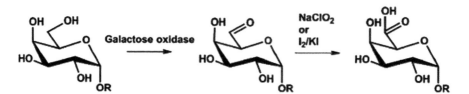

Fig. 17 Oxidation of the galactosyl units of polysaccharides to galacturonic acid ($R = backbone$ $of the polysaccharide$) (Reproduced from [102] with permission)

3 Hemicellulose-Based Particles (Zero-Dimensional)

Hemicelluloses possess many hydroxyl groups and reducing aldehyde groups, and exist as helical chains or random-coil chains in aqueous solution, which provides a possibility to act as a green reducing agent and stabilizing agent by capping metallic nanoparticles in the special structure [109]. In addition, hemicelluloses are widely used as a substrate or coatings for nanoparticles and have an application in target drug delivery and chemical catalysis.

By using biopolymer xylan as a stabilizing and reducing agent, Luo et al. [110] prepared highly stable xylan/Au nanoparticles (AuNPs) composites with the highly sensitive sensing of cysteine. The obtained nanoparticles could distinguish cysteine (Cy) among dozens of kinds of amino acids with a limit of detection (LOD) of 0.57 μm, exhibiting a potential application for Cys detection in real biological samples. Similarly, well-distributed spherical silver nanoparticles (AgNPs) could be rapidly synthesized by using bamboo hemicellulose as stabilizer and glucose as reducer with the assistance of microwave [111]. The effects of reaction parameters on the shape of AgNPs were investigated, the result revealed that the average particle size of AgNPs increased with $AgNO_3$ concentration. The coexistence of silver Ag(0) and Ag(I) was found by XPS analysis. In another case, xylan/AgNPs composites were developed via tollens reaction by reduction of $[Ag(NH_3)_2]^+$ under microwave irradiation (Fig. 18) [109]. The nanoparticles showed a high selectivity and sensitivity for Hg^{2+} detection with a low LOD of 4.6 nm, indicating a suitable detector for harmful heavy metal detection. Xylan-coated magnetic microparticles were produced by emulsification of Fe^{2+}/Fe^{3+} suspension and xylan solution, followed by interfacial cross-linking in the presence of terephthaloyl chloride [112]. The obtained polymeric magnetic microparticles (PMP) can undergo not to be dissolved at gastric PH but not of that in magnetic microparticles (MP) without the xylan coating, which demonstrates the xylan coating did shield magnetite from the gastric pH.

In Wu's research, carboxymethyl functionalized hemicellulose (CMH) was used as a substrate for the deposition of palladium nanoparticles (Pd NPs) to provide a novel heterogeneous catalyst CMH-Pd(0) by using ethanol as a solvent and an in situ reducing agent. The CMH-Pd(0) was a very active and stable biobased catalyst for Heck coupling reaction and could be easily recovered and reused at least 5 times [113]. Chen et al. [114] used xylan-type hemicelluloses (XH) supported terpyridine-palladium (II) nanoparticles to catalyze Suzuki-Miyaura reaction with a high yield up to 98%. In another study by Chen et al. [115], PdNPs@XH catalyst was synthesized with the same deposition-precipitation method and showed excellent catalytic activity in the Suzuki, Heck, and Sonogashira coupling reactions. These works broaden the applications of hemicelluloses in green catalysis.

The BSA-loaded CKGM-CS nanoparticle basing on carboxymethyl Konjac glucomannan (CKGM) and chitosan (CS) was formed by electrostatic interactions and was applied for the encapsulation of bovine serum albumin (BSA). The mean

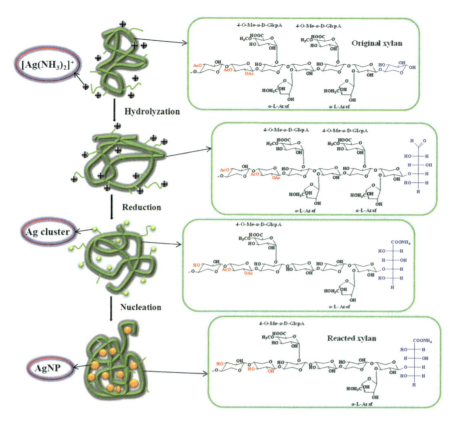

Fig. 18 The possible formation course of xylan-AgNPs composites (Reproduced from [109] with permission)

size of the nanoparticles increased with an increase in either the CKGM or CS concentration. Particularly, the concentration of CS can effectively dominant the zeta potential measurements, which was attributed to its cationic nature [116]. As observed, the nontoxic CKGM-CS nanoparticles have a promising application as green drug carriers. Xylan furoate pyroglutamate was prepared by esterification between xylan and pyroglutamic acid/furan-2-carboxylic acid in the presence of CDI [117]. Small-size spherical nanoparticles (60–85 nm) yielded by simple dialysis of the obtained esters in DMSO. Garcia et al. [118] employed an original method by neutralizing the alkaline xylan solution with HCl or acetic acid to form micro and nanoparticles. The result demonstrated that the size of xylan particles was dependent on xylan concentration and nanoparticles with 100–900 nm could be obtained only in a low xylan content.

4 Hemicelluloses-Based Films and Coatings (Two-Dimensional)

Hemicelluloses have inherent properties including low oxygen permeability, water solubility, and biodegradability. The low oxygen permeability makes hemicelluloses have a promising application in barrier films and coatings, notably in food-packaging. Considering the low cost and environmental friendliness of hemicelluloses, they are good substitutions for the conventional package materials, for example, aluminium foil, EVOH or PVDC. However, the disadvantages of water/moisture sensitivity, high water vapour permeability (WVP) and brittleness hinder their better use in films and coatings. Thus, chemical or enzymatic modifications of hemicellulose are required to alter their physicochemical properties by grafting, crosslinking, blending and other methods, and finally improve the oxygen barrier, WVP and mechanical properties of native hemicelluloses for special use [12]. The oxygen barrier properties can be quantified by oxygen permeability (OP) and expressed in cm^3 $\mu m/m^2$ d kPa, while WVP is usually determined by the cup test in the order of g m^{-1} Pa^{-1} h^{-1} or g m^{-1} Pa^{-1} s^{-1}.

4.1 Blending

Hemicelluloses-based films can be formed by organic and inorganic additives including emulsifiers (sucrose esters, palmitic acid, etc.) [119–121], plasticizers (glycerol, sorbitol, xylitol) [122, 123], polyvinyl alcohol (PVA) (44), MMT [27, 65, 124], chitosan [27, 125], cellulose derivatives [126–132], alginate [122], carbon nanotube [133], and other materials [134–137].

The emulsifiers have a significant effect on the stabilization for the structure of the film especially during drying, and can strongly influence the barrier and mechanical properties of films. In Phan's research, hydrogenated palm kernel oil (HPKO) and sucroesters with varied hydrophilic-lipophilic balances (HLBs) were successively added to AX/glycerol solution, and films were formed after stirring, homogenizing and drying. The film with 2.5% SP10 shows a 30% decrease in WVP value when comparing to lipid-free AX film. The presence of SP10 also improved the moisture resistance of film due to its low melting point and hydrophobic property. However, SP30, SP40, and SP70 do little effect on the water absorption rate and the contact angle of films [121]. Another two similar reports about the AX-lipid-based edible films elucidated the influences of drying temperature and lipid type on film structure and properties, [119, 120]. Increasing drying temperature induced a gain of mechanical performances (elongation) of film containing lipid and an apparent "bilayerlike" structure that led to the reduction of moisture transfer. The lowest WVP and a smallest mean diameter (0.54 µm) were observed in HPKO-AX emulsion films. Additionally, water contact angle measured on the HPOK-AX film was comparable to those observed for low-density polyethylene

(LDPE) films (>90°). However, only triolein-AX films gained elongation after emulsification. As illuminated, HPKO is a good emulsifier to enhance water resistance of AX-based film [120].

A plasticizer was required to avoid brittleness of hemicellulosic films, and the most commonly used plasticizers are propylene glycol, glycerol, sorbitol, and xylitol. AcGGM was physically blended with alginate or carboxymethyl cellulose in the presence of plasticizer (glycerol, sorbitol or xylitol), and films were formed by solution-casting technique. All films have been evaluated by dynamic mechanical analysis (DMA) equipped with a humidity scan, Ox-Tra® Mocon and DSC to measure their storage modulus, oxygen permeability and thermal properties. It was found that the incorporation of plasticizer resulting in a low O_2 permeability [≤ 4.6 (cm^3 μm)/(m^2 d kPa)], high mechanical toughness, and flexibility within the AcGGM-based film. Therefore, AcGGM film is an excellent candidate for making new renewable barrier materials for food packaging [122]. Without alginate or carboxymethyl cellulose, corn hull arabinoxylan films with a plasticizer (glycerol, propylene glycol, or sorbitol) had lower WVPs (0.23 ~ 0.43 × 10^{-10} g m^{-1} Pa^{-1} s^{-1}) and may be attributed to the antiplasticization effect. Moreover, the blend films had a tensile strength of 10–61 MPa, modulus of 365–1320 MPa, and elongation ranging from 6 to 12%. Interestingly, grapes coated with arabinoxylan/sorbitol films achieved a decrease of 41% in weight loss rates after 7 days [123].

Reinforcing agents such as inorganic clays possess the flame retardant property and other superior physical properties. MMT is commonly used to incorporate with biopolymers in film casting through a green and simple paper-marking method. The heat-resistance hybrid films (Fig. 19) were prepared from QH and MMT with different ratios [65]. The film forming mechanism is the electrostatic and hydrogen-bonding interaction between QH and MMT nanoplatelets. The thermal stabilities of the hybrid films were higher than those of QH due to the addition of MMT and the proper proportion of QH and MMT. In a study by Chen et al. [124], QH/MMT/PVA or QH/MMT/chitin nanowhiskers (NCH) hybrid films with different proportions were formed by vacuum filtration. Compared with QH-MMT film, the films containing PVA or NCH exhibited an improved mechanical strength, thermal stability, and oxygen barrier property, as well as a higher optical transparency.

Chitosan (CS) is an electropositive biopolymer due to the presence of $-NH_2$ groups and could generate electrostatic interactions with anionic groups in acidic media. Together with biodegradability and biocompatibility, chitosan is an excellent film-forming candidate. When chitosan was introduced into QH/MMT matrix, producing nanocomposite films with nacre-like structure and excellent mechanical properties [27]. UV-vis test shows that the opacity was increased with the increment of MMT amount. With the addition of small amount of CS, QH-MMT-CS film showed a high tensile strength of 57.8 MPa, which was 30.2% higher than QH-MMT composite film. The new prepared films also possess a good thermal stability, lower oxygen permeability and preferable water vapour permeability. Basing on the above results, these hemicelluloses-based nanocomposite films can be good candidates for non-renewable films and have a great potential application

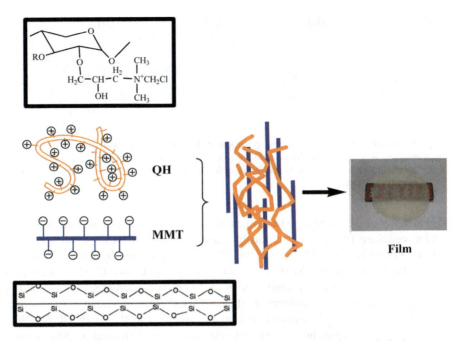

Fig. 19 The QH-MMT film forming process (Reproduced from [65] with permission)

in packaging. By introducing chitosan and glycerol, bagasse hemicelluloses films were prepared with varied hemicellulose concentrations, chitosan/glycerol amount and the drying temperature [125]. Due to the highly hydrophilic nature of glycerol, the WVP of hybrid films rose as the increasing glycerol amount and decreased by 48% as the drying temperature increased from 25 to 55 °C. Besides, the tensile strength and the elongation were changed significantly in the same range of temperature. In summary, drying temperature has a noticeable effect on the mechanical properties of these films.

In recent years, incorporation of biodegradable reinforcements such as celluloses and other polymers has been extensively used for producing composite films with enhancing mechanical performances and good barrier property. Xylan from corn cob was blended with two kinds of cellulose and elaborating two composite films by solvent casting [126]. One film was hydrophilic and fabricated from bleached hemicelluloses (BH), unmodified cellulose and glycerol (as a plasticizer). The other film was hydrophobic and composed of acetylated bleached hemicelluloses (BAH) and acetylated cellulose, as shown in Fig. 20. The hydrophobic film had a higher maximum weight loss temperature than the hydrophilic one, revealing a higher thermal stability for the hydrophobic film. In addition, hydrophobic film (a Young's modulus $\sim$2300 MPa, a tensile strength of 44.1 MPa, and a strain at break of 5.7%) showed better mechanical properties than the hydrophilic films (a Young's modulus of 3 MPa, a strength of 3.3 MPa, and a strain at break of 5.3%), indicating acetylated cellulose did not alter its reinforcing potential.

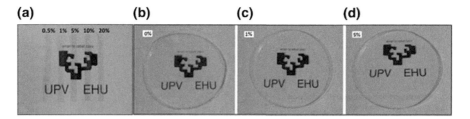

Fig. 20 Optical images of films: **a** test specimens of hydrophilic films with different percentages of nanofiber, **b** hydrophobic films without reinforcement, **c** hydrophobic films with 1% of acetylated cellulose and **d** hydrophobic films with 5% of acetylated cellulose (Reproduced from [126] with permission)

Huang et al. [128] reported a comparative study on the influence of nanocrystalline cellulose (NCC) or cationically modified NCC (CNCC) to the prepared hemicelluloses/sorbitol (HC/SB) films. NCC was cationized by glycidyltrimethylammonium chloride (GTMAC). The presence of NCC and CNCC remarkably improved the mechanical properties of HC/SB films with increased elastic modulus and higher tensile stress. Moreover, HC/SB/CNCC films exhibited better thermal stability and a relatively smooth surface as compared to that with HC/SB/NCC films. For the first time, hemicellulose/lignin (HL) film plasticized by 30% sorbitol was reinforced with different loadings of nanocelluloses (ACNF, CNC, OCNF, ACCNF). The film exhibited significant improvement in flexibility, transparency, mechanical and moisture barrier performances. For instance, the film with ACNF enhanced Young's modulus and tensile strength up to 319 MPa and 266%, respectively; while a film with CNC20 possessed a lower MVP and a less moisture content as compared to other films. This work shows a great feasibility of turning agricultural residues into high value-added film materials [127]. Kisonen et al. [129] took the good affinity between GGM and nanofibrillated cellulose (NFC) to cast films and coated with either succinic esters of GGM [GGM-Su1 (DS = 1.0) and GGM-Su2 (DS = 1.6)] or native GGM. The contact angle measurement revealed that NFC-GGM film double-coated with GGM-Su2 had better a hydrophobic characteristic than other films; the lowest OP value of 0.1 [(cm^3 µm) (m^2 d kPa)] and an improvement in barrier property and stiffness were achieved by the addition of GGM-Su2 on NFC-GGM substrate. Furthermore, all films were demonstrated to be grease impenetrable even at high temperature. Similarly, NFC/glucomannan films could be fabricated by taking sorbitol as a plasticizer. The incorporation of NFC improved the tensile and bursting strength of the films. Notably, the films can be gravure printed with solvent-based ink using a gravure K-proofing press due to its relatively high surface energy [130]. In another case, microfibrillated cellulose (MFC) was mixed with GGM to form composite films with the addition of glycerol. The reinforcement of MFC improved the mechanical properties and decreased the moisture uptake of the films at low glycerol content (5–15%) [131]. Peng and co-workers synthesized nanocomposite films basing on xylan-rich hemicelluloses and cellulose nanofibers (CNFs) in the presence of sorbitol. The strong hydrogen bonding between CNF and xylan allowed good interfacial adhesion between xylan matrix and CNF, which results in a high mechanical strength and increased thermal stability [132].

For the purpose of gaining conductivity of hemicelluloses, carbon nanotube (CNT) is utilized as a filler material. The prepared CNT-hemicellulose films were cast by either spinning or drop-drying method. The CNT-hemicellulose films showed excellent conductive properties with a bulk conductivity of up to 2000 S cm^{-1} [133].

Hemicelluloses also can be utilized as reinforcement. Hemicellulose from autohydrolysis process was incorporated with κ-carrageenan/locust bean (κ-car/LBG) system in the presence of 0.30% (w/w) glycerol to prepare polymer-blend films. The addition of hemicellulose resulted in a higher tensile strength and a lower moisture content but did not lead to an obvious change in WVP and elongation at break [136].

Without any addition, glucose and hemicelluloses-based films were prepared with cellulosic and hemicellulosic portions from alkaline pretreatment of cotton stalks at three different temperatures. The result showed that pretreatment temperature did significant influence on the yield of cellulose and hemicellulosic portion and the physical properties. The optimal pretreatment temperature was determined to be 60 °C, at which could form an integrated film with a proper content of lignin, and the film had better oxygen permeability and moderate mechanical properties. Increasing pretreatment temperature to 90 °C, however, was detrimental to the mechanical properties of the films (Fig. 21) [134]. By using autohydrolysis and alkaline extraction of hemicelluloses as mentioned above, free-standing films could be prepared from the hemicellulosic fractions with different xylan/GGM ratios. The rapeseed xylan-GGM films exhibited strain-to-break values >60% without any plasticizers [137]. Xylan could be used as an additive for production of biodegradable wheat gluten film. Mechanical properties changed not only due to the PH of suspension and drying temperature, but also were attributed to the type and composition of xylan [135]. Hemicelluloses can be applied as coating materials

Fig. 21 Films obtained from hemicelluloses isolated at the end of pretreatments conducted at 25, 60 and 90 °C (Reproduced from [134] with permission)

with gum arabic (GA) for food microencapsulation. The synthesis of microcapsules by encapsulating fish oil with the emulsions of GA, GA-HC and HC as the coating materials using spray dryer method was reported by Tatar et al. [29] According to SEM observation, the as-produced microcapsules were surface smooth with no cracks and splits. The viscosity increased with the addition of HC, which resulted in the coalescence of oil particles and large amounts of oil particles with the largest size. Finally, the combination of GA and HC lowered the cost of the coating material.

4.2 Cross-Linking

Cross-linking is a good way to increase the molecular weight of polymers. By using the laccase-catalyzed cross-linking method, aromatic moieties were successfully bounded to wood hemicelluloses. Subsequently, carboxymethylcellulose sodium salt and glycerol were mixed into the cross-linking hemicelluloses and cast into films. The cross-linking sample with high molecular weight (high-MW) hemicellulosic fractions formed a film that gave better mechanical properties and lowers oxygen barrier when compared with those with low-MW hemicellulosic fractions, confirming that molecular weight could significantly affect the strength and barrier properties of films [138].

4.3 Grafting

Grafting is an effective method for modifying biopolymers and their special use. Hemicellulose-lipid-based films were prepared by grafting various omega-3 fatty acids onto arabinoxylans (AXs) polymeric chains in the presence of glycerol, using cold oxygen plasma associated with electron beam (EB) irradiation treatment. Among these films, the linseed oil-based edible film had the lowest WVP (1.09×10^{-10} g m^{-1} s^{-1} Pa^{-1}), which was 50% lower than the unmodified film. Surface hydrophobicity of fish oil-based and marine oil-based films increased and were almost comparable to those of LDPE films [139]. The same technique was also applied in preparing AX-based films by grafting stearyl acrylate (SA) and stearyl methacrylate (SM). The AX-based films with functional acrylates had contact angles of superior to 110°, which was much higher than the original film (70°), indicating an improved hydrophobicity after treatment. Additionally, a decrease of 24% in the WVP occurred after modification. Combining the oxygen plasma and electron beam (EB) irradiation technique with copolymer grafting is demonstrated to be a good way to improve barrier properties of films for better use in packaging [140]. By using plasma treatment, β-(1→3) (1→6)-glucan was grafted onto the poly(lactic-co-glycolic acid) (PLGA) film, resulting in an increasing hydrophobicity and inducing cell affinity onto the polymer surface. The obtained plasma/β-glucan PLGA film facilitated the cell proliferation procedure and was expected to be applied in skin tissue engineering [141].

4.4 Other Modifications

Modifications such as oxidation, amination, fluorination and etherification are extensively employed to obtain hemicelluloses-based films. Hydrophobic plastic films from maize bran heteroxylan were prepared by periodic oxidation with sodium periodate, followed by reductive amination with dodecylamine in the presence of sodium cyanoborohydride. The DS values ranged from 0.5 to 1.1, and a higher DS resulted in an increase in elongation and a decrease in the elastic modulus [142]. Gröndahl et al. [143] proposed a method to produce hydrophobic AX-based films by gas-phase surface fluorination with trifluoroacetic anhydride (TFAA) (Fig. 22). IR spectra confirmed the presence of fluorinated moiety only on the film surface. A hidden increase of water contact angle for AX film with 7% fluorine (70°) compared to an untreated film (30°), except that, a decrease in the equilibrium moisture content took place from 18 to 12% after the treatment. The relatively modest increase in the hydrophobicity may be attributed to a partial hydrolysis of the trifluoroacetate groups in contact with the water droplet.

In a research by Laine et al. [28], xylan extracted from bleached birch kraft pulp were hydroxyalkylated by propylene oxide or glycidyl ethers and produced barrier coatings in the presence of sorbitol/glycerol with or without citric acid. The data shows a considerable WVP and the best OP of hydroxypropylated xylan coating was nearly one-third of those for polyethylene terephthalate (PET) coating. The modified films also exhibited oil and grease resistance. Thus, the results elucidate its value-added applications in packaging and pigment coating binders for printing papers. In another case, xylan was functionalized with 3-Chloro-2-hydroxypropyl sulfonic acid (CHPS), (S)-(-)-(3-chloro-2-hydroxypropyl)-trimethylammonium chloride (SCHMAC) or 3-chloro-2-hydroxypropyl-trimethylammonium chloride (CHPMAC)/CHPS, respectively. Afterwards, these samples were applied for films casting. The products were water soluble and reproducible. Quaternized films had a

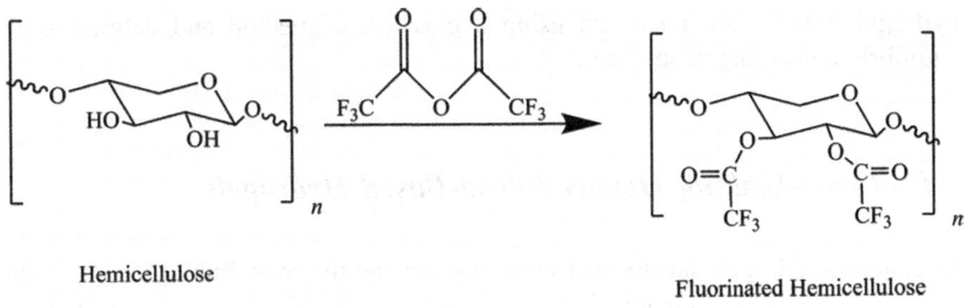

Fig. 22 Fluorination of hemicelluloses (Reproduced from [143] with permission)

supreme tensile strength (64.3 MPa) and hydroxypropyl sulfonate xylan films gained a higher Young's modulus (3350 MPa). Furthermore, the xylan-quaternized/sulfonated film was surface smooth but fractured with a higher thickness of 234 ± 18 µm as observed by SEM. The mechanical properties gave values that are close to the average value of the two derivatives [144]. As previously described [70], the benzylated AcGGM-based film was manufactured to be a potential oxygen and moisture barrier. In order to investigate superstructural features of the hemicellulose-based film, xylan-based furan-2-carboxylic acid ester was prepared by mixing furan-2-carboxylic/CDI/DMSO solvent with xylan/DMSO solution, and the obtained xylan furoate was cast into the film. Porous structure with a macropore size of smaller than 250 nm and a meshwork-like pore-structure in the upside within the film could be observed. In addition, a roughness value (R_a) of 150 nm and a maximum height (R_{max}) of 1500 nm were determined [145]. Conclusively, research in the area of hemicellulosic films and coatings is ongoing and seems to hold great promise for future practical applications.

5 Hemicellulose-Based Hydrogels (Three-Dimensional)

Hydrogels are polymeric three-dimensional networks that have a high capacity of adsorbing a large amount of water or biological fluids and swell to many times comparing with their dry mass while maintaining structural integrity [43]. From the biocompatible and biodegradable point of view, natural polysaccharides and especially hemicelluloses are extensively used in fabricating hydrogels. Hemicellulose hydrogels, however, generally suffer from poor mechanical performances, less stretchable for special use, and thus hindering their applications. To overcome these drawbacks, the chemical modification has been made to prepare many types of highly stretchable and tough hemicellulose-based hydrogels [146]. Given to their favorable properties such as high stretchability, responsiveness, swelling/deswelling, and good mechanical performances, hemicellulose-based hydrogels have been employed in many fields such as waste treatment [15–20, 147], dye adsorption [21, 22, 148, 149], drug delivery and release [23–26, 150–152], tissue engineering [153], biodetector [154], biosensor [24, 25, 151, 155–160], conductive polymers [153, 161, 162] and so forth. The swelling ratios of the hydrogel samples are measured using a gravimetric method and defined as the equilibrium swelling ratio (Seq).

5.1 Cross-Linking Hemicellulose-Based Hydrogels

In general, hydrogels are formed either by covalently cross-linking or physically cross-linking, such as chemical treatment [163, 164], heating, or microwave irradiation. Hemicellulose-based cross-linking polymers offer resistances to thermal

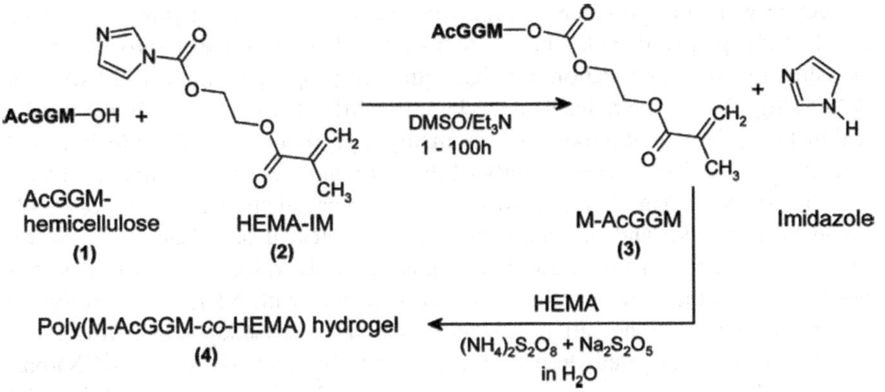

Fig. 23 Synthetical route applied for the preparation of poly(M-AcGGM-co-HEMA) hydrogels (Reproduced from [150] with permission)

degradation, cracking by liquids and other harsh environments [11]. In a research by Andersson et al. [152], AcGGM-based hydrogels were prepared by three steps. Firstly, 2-hydroxyethyl methacrylate (HEMA) was mixed with CDI in anhydrous $CHCl_3$ and obtained 2-[(1-imidazolyl)formyloxy]ethyl methacrylate (HEMA-Im), followed by HEMA-Im covalently coupled to AcGGM in DMSO with the catalysis of triethylamine; and finally hydrogels were synthesized by the cross-linking of HEMA with the modified AcGGM in the presence of ammonium peroxodisulphate and sodium pyrosulfite as catalysts and a simultaneous incorporation of BSA. The DS value of HEMA (DS_{HEMA}) ranged between 0.1 and 0.36, and BSA release from hydrogels was reduced with an increase of DS_{HEMA}. The addition of enzyme β-mannanase (AnMan5A) enhanced the release of BSA up to 95% within 8 h as compared with that of 60% without a catalyst. In a similar case, Voepel et al. [150] designed neutral hydrogels composed of methacrylated AcGGM (M-AcGGM) and HEMA, and ionic hydrogels basing on maleic anhydride-modified M-AcGGM (CM-AcGGM) and HEMA through radical-initiated polymerization and cross-linking method (Fig. 23). The tests elucidated the drug release rate and swelling ratio of the neutral hydrogels decreased with an increase in the degree of methacrylation. The ionic hydrogels, however, showed quicker drug release kinetics and higher swelling capabilities than the neutral gels, especially at neutral conditions. As expected, the release speed was lowered under acidic media due to the protonation of carboxylic functionalities. According to the above descriptions, these novel hemicellulose-hydrogels have future prospects in oral drug release.

Several monomers have been used to prepare hemicellulose-based hydrogels, including acrylic acid (AA), citric acid (CA), maleic acid, methacrylic acid. For example, hemicelluloses were plasticized and cross-linked by CA with or without sodium hypophosphite (SHP) as a catalyst and formed hydrogels using reactive extrusion. The highest drug release can be obtained in an alkaline medium due to a higher swelling ratio of hydrogels. Additionally, the concentration of CA and the molecular weight of polymers can influence the degradation of hydrogels [23]. Gao et al. [24] prepared xylan-based temperature/PH sensitive hydrogels by the crosslinking copolymerization of xylan with N-isopropylacrylamide (NIPAAm) and AA using N, N'-methylenebis-acrylamide (MBA) as a cross-linker and 2, 2-dimethoxy-2-phenylacetophenone/N-methyl pyrrolidone (DMAP/NMP) as a UV initiating agent. The swelling ability of the hydrogel was greatly influenced by the amounts of NIPAAm, AA and MBA. A drug encapsulation efficiency of 97.60% could be achieved, and the cumulative release rates of acetylsalicylic acid were 90.12 and 26.35% in the intestinal and gastric fluids, respectively. Besides, MTT assay revealed the biocompatibility of the hydrogel with NIH3T3. Therefore, the novel hydrogels have potential applications as oral acetylsalicylic acid drug carriers for the intestinal-targeted drug delivery. Similarly, acylated xylan (ACX)-based magnetic Fe_3O_4 nanocomposite hydrogels (ACX-MNP-gels) were fabricated by copolymerization of ACX with acrylamide and NIPAAM [154], as shown in Fig. 24. Especially, the as-prepared magnetic hydrogels exhibited excellent catalytic activity and gave a sensitive response to H_2O_2 even at a concentration of 5×10^{-6} mol L^{-1}. Therefore, the magnetic hydrogels can be utilized as a detection tool in the field of biotechnology and environmental chemistry.

By taking the UV photo-crosslinking method as described above, Yang et al. [159] synthesized honeycomb-like HC-based hydrogel by photo-cross-linking acylated hemicellulose (Hce-MA) with NIPAAm in NMP. The lower critical

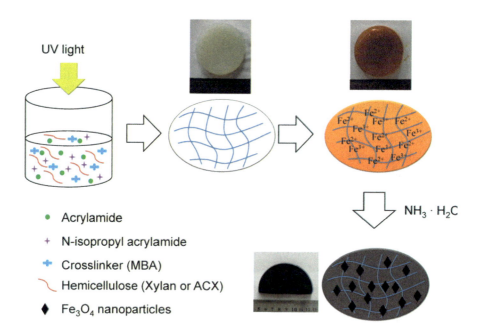

Fig. 24 Schematic process for the preparation of magnetic Fe_3O_4 nanocomposite hydrogels (Reproduced from [154] with permission)

solution temperature (LCST) of the hydrogel increased with Hce-MA and the equilibrium swelling ratio was to rest with the environment temperature, which indicated the temperature sensitivity of the hydrogels. Novel hydrogels basing on wheat straw hemicelluloses were formed by taking AA as a monomer and MBA as a cross-linker under the $K_2S_2O_8/Na_2SO_3$ system and used for acetylsalicylic acid and theophylline release. Equilibrium swelling ratio of the hydrogel can be affected by pH values [25]. Cross-linked with the same monomer AA, hemicelluloses from the spent liquor were utilized to prepare hydrogel by using ammonium persulfate/N, N, N', N'-tetramethyl-ethane-1, 2-diamine (APS/TEMDA) as a redox initiator system. The as-prepared hydrogels gave a rapid response to pH, salt and ethanol, and showed excellent mechanical properties with the highest compressive strength of 105.1 and strain at break reach of 34.8% [155].

By graft copolymerization of AA, acrylic amide (AM) with hemicellulose in the presence of APS/NaHSO$_3$, hydrogels with excellent water absorbency (1128 g/g) were synthesized. The results showed that the undulant surface and broad network structure were responsible for its excellent water absorbency [154]. In a research by Peng et al. [158], ionic hydrogels basing on xylan were prepared by free radical graft copolymerization of AA and xylan. The prepared ionic hydrogels exhibited different swelling degrees to pH, salt, and organic solvents, having potential applications in separation and drug release systems, as well as removing heavy metal ions [18].

Cationic hemicelluloses-based hydrogels were fabricated by Dax et al. [16]. The procedure can be concluded as: firstly, GGM was modified with GMA in the presence of DMAP and obtained GGM-MA, followed by cross-linking GGM-MA with [2-(methacryloyloxy) ethyl] trimethylammonium chloride (MeDMA) as a monomer. The resulting hydrogels can effectively remove arsenate and chromate ions from aqueous solutions. According to the previous description [26, 65, 155], CQH was synthesized by crosslinking quaternized hemicelluloses with epichlorohydrin (ECH), followed by graft copolymerization with AA and providing hemicelluloses-based hydrogels [165]. The biocompatibility test confirms the nontoxicity of the as-prepared hydrogels, which can allow cell growth and have a potential application in functional biomaterials.

By using thiol-ene click reaction with MBA, thiol-Michael addition to MBA, and disulfide bond formation, three AcGGM-hydrogels were synthesized. The as-prepared hydrogels do not dissolve in water when prolonging the immersion time over 54 h, demonstrating the successful formation of a covalently bonded three-dimensional network [62]. Glow discharge electrolysis plasma (GDEP) treatment can provide energetic species such as HO·, H·, and HO$_2$· to induce a chemical reaction in aqueous solution. GDEP treatment was applied for the preparation of dual sensitivity reed hemicellulose-based hydrogels. Surprisingly, all hydrogels have the same phase-transition temperatures of approximately 33 °C under different discharge voltages. However, under the discharge voltage of 600 V, the hydrogels become more sensitive to temperature and pH and thus obtaining a higher deswelling ratio [160]. By using the same technique, Zhang et al. [147] developed a hydrogel from reed hemicelluloses for the adsorption of heavy metal

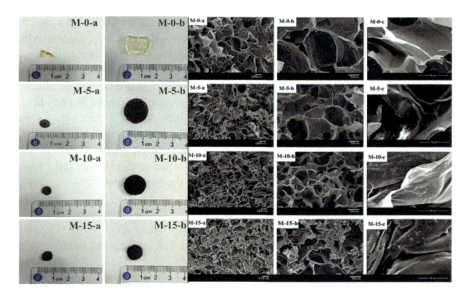

Fig. 25 Photographs of hemicellulose hydrogel (M-0) and MFRHHs (M-5, M-10, and M-15) in the dry state (M-0-a, M-5-a, M-10-a, and M-15-a) and swollen state (M-0-b, M-5-b, M-10-b, and M-15-b) (*left*). Corresponding SEM images ×100 (**a**), ×300 (**b**) and ×2000 (**c**) magnification (*right*) (Reproduced from [26] with permission)

ions. The hydrogel retains high reusability for the adsorption of metal ions after 8 repeated adsorption/desorption cycles. Zhao and co-workers synthesized a magnetic field-responsive hemicelluloses-based hydrogel (MFRHH) with excellent BSA adsorption and controlled release. The fabricating process can be described as the cross-linking between AcGGM and ECH in the presence of $FeCl_3 \cdot 6H_2O$ and $FeCl_2$ ($Fe^{3+}:Fe^{2+} = 2:1$) under basic media. The magnetizations of the MFRHHs increased as the increasing Fe_3O_4 nanoparticles content, the equilibrium swelling ratios (SRs) were 12.1, 7.4 and 2.4 for the M-5, M-10, and M-15 hydrogels, respectively, lower than that of M-0 (hydrogel without Fe_3O_4 nanoparticles) (SR = 27.0). Therefore, the MFRHH has great potential applications in controlled drug delivery and magnetically assisted bioseparation [26]. Figure 25 shows the MFRHH in different states and the macrostructure of MFRHHs with different Fe_3O_4 nanoparticles contents.

5.2 Conductive Hemicellulose-Based Hydrogels

Conductive polymers attract much attention as humans relying on various electrical equipment such as cellphone, computer, car and daily necessities. Conductive polymers such as polyaniline, polypyrrole and polythiophene are extensively applied in the microelectronics industry, including battery technology, photovoltaic devices, light-emitting diodes, and electrochromic displays as well as biomaterial field [161]. Electrically conductive hydrogels (ECH) combine the unique advantages of conductive polymers and hydrogels, but the non-degradability of ECH greatly limits their application. Therefore, conductive hydrogels basing on natural polymers such as hemicelluloses are synthesized to solve this problem due to the good compatibility of hemicellulose and polymers [153]. Zhao et al. [161] prepared an electrically conductive hemicellulose hydrogel (ECHH) by cross-linking AcGGM with epichlorohydrin and conductive aniline pentamer (AP) in basic media under ambient condition (Fig. 26). Increasing the AP content led to an enhancement in thermal stability of ECHH and a reduction in equilibrium swelling ratio. Simultaneously, conductivity increased from 9.05×10^{-9} to 1.58×10^{-6} S/cm. In the second study by Zhao et al. [162], free-standing ECHHs were obtained

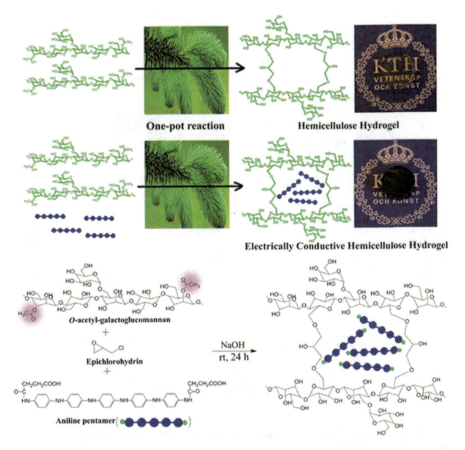

Fig. 26 Schematic synthesis of ECHHs using epichlorohydrin as a cross-linker in basic media (Reproduced from [161] with permission)

by a copolymerization of carboxylated AcGGM (C-AcGGM) with GMA and subsequently covalently coupling with aniline tetramer (AT). The swelling ratio of the hydrogels decreased with an increasing AT content; and particularly, the conductivity increased by two orders of magnitude as the AT content changed from 10 to 40% (w/w).

A series of stimuli-responsive hemicellulose microgels (SRHMGs) were reported by Zhao et al. [151]. AcGGM as a matrix was cross-linked with different functional materials (including poly(acrylic acid), AP, and iron) during spray drying, giving SRHMGs with response to pH, electrochemical stimuli, magnetic field, or dual-stimuli (Fig. 27). Comparing with single-stimuli hydrogel, the one-pot reaction products had more practical applications, such as controlled drug release, magnetic resonance imaging, biosensors, electronic devices, and tissue engineering.

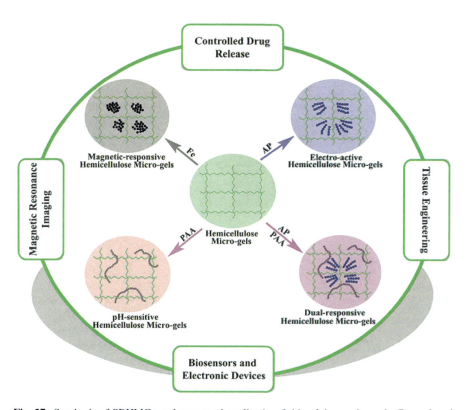

Fig. 27 Synthesis of SRHMGs and suggested application fields of these microgels (Reproduced from [151] with permission)

5.3 Hemicellulose-Polymer Composite Gels (Hydrogels and Aerogels)

To overcome the drawbacks of Hemicellulose hydrogels such as weak thermal stability, brittle properties and weak mechanical properties, compatible ingredients such as clay, chitosan, cellulose derivatives, carbon materials, polyvinyl alcohol and other polymers are utilized to the synthesis of composite hydrogels with enhancing functional properties [166].

Chitosan is also a renewable biopolymer possessing many functional groups such as N-acetyl groups, reactive hydroxyl and amino groups, and has been widely used to prepare natural hydrogels. Hemicellulose hydrogels can be cross-linked and/or reinforced by chitosan to improve their mechanical performances [15, 20]. In the study of Gabrielii et al. [166], xylan-containing glucuronic acid functionalities and chitosan were dissolved in acidic condition to form hydrogels. The gelation mechanism was the complexation between glucuronic acid functionalities in xylan and amino groups in chitosan. The hydrogels were sensitive to pH and responded in a reversible manner to the stimuli. In a similar way, $-NH_2$ groups in chitosan were cross-linked with carboxylic acid groups in hemicellulose derivative obtained from esterification of hemicellulose with citric acid in SHP, and producing an elastic, highly porous and durable hemicellulose citrate-chitosan hydrogel [167]. The aerogel can adsorb up to 100 g of a saline solution and 80 g of water per gram of the material, respectively, which can be used to reduce the overall salinity of the water. Wu et al. [20] demonstrated a cross-linking carboxymethyl chitosan-hemicellulose (CMCH) hydrogel with highly macroporous structure, pH-sensitivity and highly efficient adsorption with metal ions. Furthermore, CMCH could be reused without significant loss of the adsorption capacity. In another work by Wu and co-workers, TiO_2 nanoparticles were incorporated into the carboxymethyl chitosan-hemicellulose network (CHNT) with the assistance of sodium dodecyl sulfate (SDS). Due to the favourable chelating groups existing in its structure, the hydrogels not only had the capacity of removing heavy metal but also exhibited good regeneration of loaded metal ions with EDTA [19]. Since diethylene triamine pentaacetic acid (DTPA) has a high-affinity binding with hemicellulose in the catalysis of SHP, the hemicellulose-DTPA-chitosan hydrogel can be synthesized after crosslinking DTPA-hemicellulose and chitosan [15]. The resulting foam can be used for water desalination and the maximum salt uptake is nearly 0.30 g/g.

Guan et al. [168] reported a novel dialdehyde hemicelluloses (DHC)/chitosan/ Ag composite hydrogel possessing antimicrobial activity against microbes. DHC was obtained by oxidation of hemicelluloses with $NaIO_4$, and the hydrogel was formed by a reduction of silver ions within the cross-linked DHC/chitosan hydrogels, as shown in Fig. 28. The swelling degree of the hydrogel decreased sharply after the addition of silver ions. The composite hydrogel seems to be a potential antimicrobial material and can be used to treat accessible wounds to prevent or kill the existing infection.

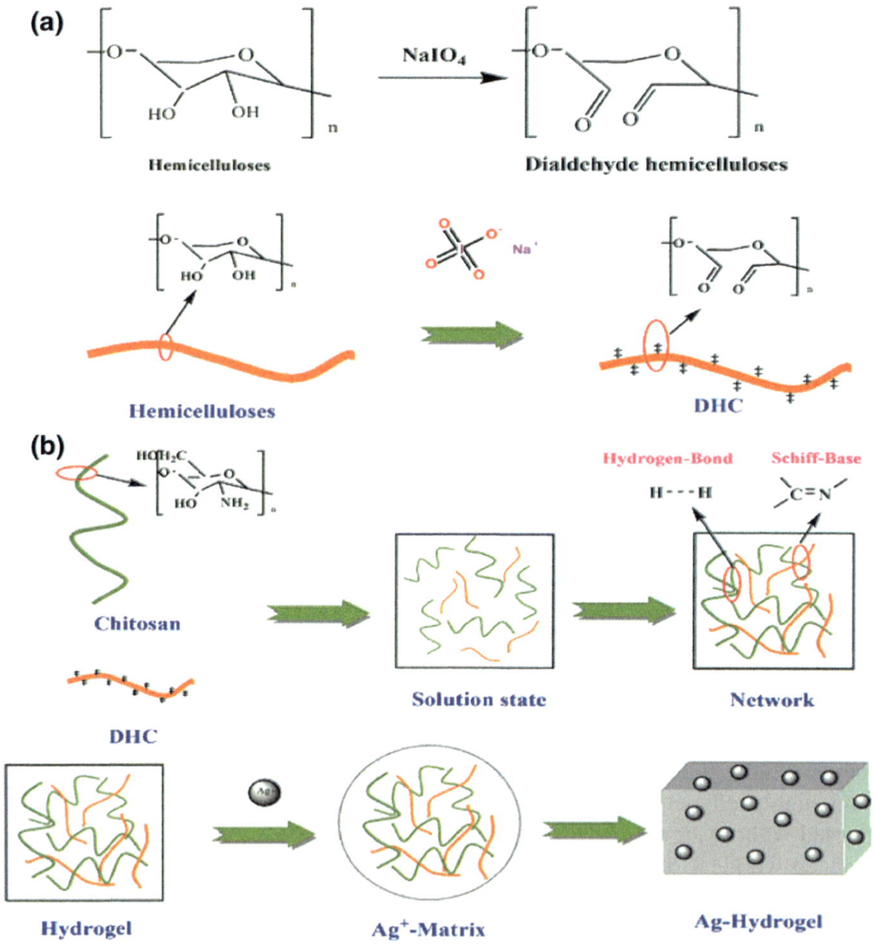

Fig. 28 The mechanism for the synthesis of (DHC)/chitosan/Ag hydrogel (Reproduced from [168] with permission)

In another work by Guan et al. [169], hemicelluloses, PVA, and chitin nanowhiskers were mixed with a mass ratio of 1:1:1 to form a hydrogel in water by using the freeze-thaw technique. FT-IR and NMR spectra confirmed that the repeated freeze-thaw cycles induced the physical crosslinking. The mechanical properties were significantly enhanced by increasing the content of chitin nanowhiskers. As a conclusion, this physical method is a good way to prepare hydrogels with good mechanical properties [169, 170]. Cellulose whiskers were coated to HEMA modified hemicellulose and formed a hydrogel by in situ radical polymerization of HEMA. The resulting hydrogels exhibited a rubber-like behaviour with enhancing toughness, increasing viscoelasticity, and improving recovery behaviour [171].

Identically, nanocellulose such as cellulose nanofibrils (CNF) and nanocrystalline cellulose (CNC) also can be applied as a reinforcing material for hemicelluloses-based gels. Cellulose nanofibrils (CNF)-GGM sponge-like aerogels was elucidated by Alakalhunmaa et al. [172], as shown in Fig. 29. The aerogels could adsorb water up to 37 times of their initial weight and possessed reversible sponge capacity, which can be used as a substitute of petroleum-based materials in food-packaging. Dax et al. [173] designed a hydrogel basing on nanofibrillated cellulose (NFC) and GGM-MA. The electrostatic attraction between the anionic charges in NFC and the quaternary ammonium groups in the polymer chains resulted in the successful connection of NFC to the polymer and thus enhancing the modulus of the hydrogel. Moreover, the hydrogels revealed a high adsorption capacity to chromate ions. Xylan/CNC hydrogels were successfully synthesized by cross-linking the oxidized xylan with CNC during freeze-casting [106]. Within this hydrogel, hemiacetal bonds formed between aldehyde groups and hydroxyl groups during solidification/sublimation process, which was determined by NMR spectra. Thus, freeze-casting/cross-linking method could enable the fabrication of nanoreinforced biopolymer-based hydrogels with tailor-made architectures.

Interpenetrating polymer networks (IPNs) are unique "alloys" of cross-linked polymers in which at least one network is synthesized and/or cross-linked in the presence of the other. IPN hydrogels are endowed with improving responsiveness and mechanical performance, as well as fast adsorption of ionic species like dyes and heavy metal ions. Lately, biopolymer-based IPN hydrogels have been widely reported, especially hemicelluloses-based IPN hydrogels [174, 175]. Malaki et al. [176] fabricated a semi-IPNs hydrogel by through cross-linking methacrylated carboxymethylcellulose (MA-CMC) with AA as co-monomer and MBA as cross-linker in the presence of softwood hemicellulose hydrolysate (SWH), as shown in Fig. 30. The semi-IPN hydrogels demonstrated a highly porous structure and appreciable mechanical performance, and the swelling ratio was similar to the single SWH network. In Malaki's another research, full IPNs hydrogels were prepared from AcGGM via free radical polymerization and a thiol-ene click reaction. The as-prepared IPNs hydrogel showed a faster swelling rate and a higher shear storage modulus (35–40 times higher) than the corresponding single network

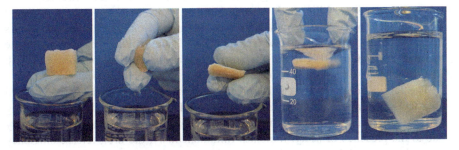

Fig. 29 The repeatable liquid water absorption capacity of the aerogels that maintained their structure and shape in water (Reproduced from [172] with permission)

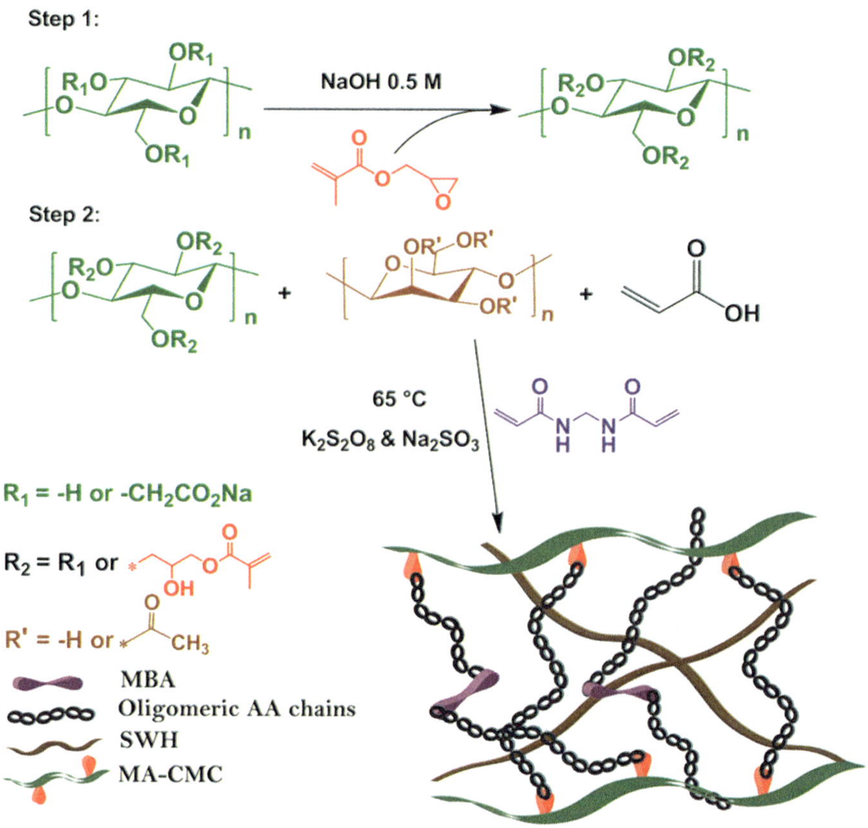

Fig. 30 Schematic outline of the synthesis of SWH-based semi-IPN hydrogels (Reproduced from [176] with permission)

of AcGGM after click reaction [177]. Meena and co-workers reported a hydrogel basing on kappa-carrageenan/xylan/polyvinylpyrrolidone (kC-xylan-PVP) blend, using sodium persulphate (KPS) as a water-soluble initiator under microwave irradiation. The resulting hydrogels also possessed a semi-IPN structure with enhancing swelling ability and water holding capacity as compared to untreated blends [178]. To sum up, the radical graft copolymerization [179] of hemicelluloses offers new opportunities to derive functional hemicellulose-based composite hydrogels; and these IPN hydrogels are preferable to serve as membranes [180], adsorbents and supports and relieve the pressure of the environment.

The fabrication of xylan (with or without acetyl moieties)/poly(HEMA)-based hydrogels is similar to the previous reports [152]. Surprisingly, the resulting hydrogels demonstrated to be stiffer and possessed a lower water swelling capacity due to the presence of acetyl moieties. Simultaneously, the drug release efficiency was enhanced and the hydrogels could act as cargo carriers to deliver an anticancer drug [181]. Basing on the studies described above [152, 181], an elastic, soft, and water-swellable hydrogel named as hemicellulose/HEMA and hemicellulose/poly (ethylene glycol) (PEG) was synthesized [182, 183]. The swelling behaviour of the hydrogels were comparable to the pure poly(2-hydroxyethyl methacrylate) (PHEMA) hydrogels, providing a suitable way for the preparation of novel polymeric structures. Utilizing CuAAC, thermoresponsive hydrogels basing on birchwood xylan were prepared by introducing reactive azide groups on the backbone of xylan via etherification and crosslinking the azide groups with poly(ethylene glycol)-b-poly(propylene glycol)-b-poly(ethylene glycol) (PEG-PPG-PEG) in the presence of propargyl bromide and NaH. The hydrogel showed a reversible swelling ability at low temperature and deswelling performance at high temperature. The compressive modulus of the hydrogel increased at 7 °C and its stiffness decreased at 70 °C [157]. Xylan-MA/PVA blends with different ratios of maleic anhydride (MA) and PVA were heated and then cross-linked to generate hydrogels under acidic media. Xylan-MA was the product of esterification between xylan and MA. The swelling and strength behaviours of the new hydrogel were rest with the content of PVA and MA [184]. Another PVA-enhanced temperature- and pH-sensitive hemicellulosic hydrogel was prepared by grafting MA onto hemicellulose, followed by copolymerization of the obtained acylated hemicellulose (AHC) with NIPAAm and itaconic acid (IA) and incorporated with PVA. The resulting hydrogel with enhancing compressive strength was proved to be biocompatible and the salicylic acid could release rapidly in the simulated gastric fluid at the initial time. As observed, the hydrogels have an extensive application in terms of controllable drug delivery [156]. Novel poly(amidoamine) (PAA)/hemicellulose hydrogel was prepared by graft polymerization of methacrylic acid onto the AcGGM backbone in the presence of acrylamide end-capped PAA oligomers as cross-linkers and $Na_2S_2O_5/(NH_4)_2S_2O_8$ as initiator [17]. The as-prepared hydrogel gained a high swelling degree, low storage moduli, and a high adsorption capacity of various heavy metal ions, and can be used for the treatment of highly contaminated wastewaters [185].

Graphene oxide (GO) and CNT have emerged as carbonaceous materials with the high specific surface area and excellent mechanical strength. GO and CNT is reliable and promising physical fillers for the preparation of composite hydrogels with enhanced mechanical strength [186]. Kong et al. [146] established a drawing procedure that GO was incorporated into polymerized acrylamide (PAM)/carboxymethyl hemicellulose (CMH) solution, followed by ionic crosslinking with Al^{3+} to obtain PAM/GO/Al-CMH nanocomposite hydrogel. The hydrogen bonding between GO and polymer chains and the network formed by Al^{3+} ionically cross-linking CMH led to a decrease in the swelling capacity and an enhancing mechanical property. Figure 31 shows the gelation procedure of PAM/GO/Al-CMH hydrogel and its mechanical property was depicted in Fig. 32.

Another organic-inorganic hybrid hydrogel composed of hemicellulose-g-poly (methacrylic acid) and multiwall carbon nanotube (MW-CNTs) was prepared for the removal of methylene blue. The cross-linking was initiated by $(NH_4)_2S_2O_8/Na_2SO_3$ after adding the cross-linker MBA and the monomer methacrylic acid [149]. The adsorption kinetics of methylene blue (MB) followed the

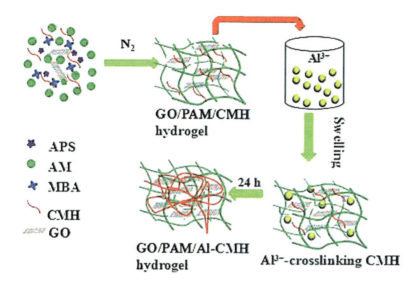

Fig. 31 Formation of GO/PAM/Al-CMH nanocomposite hydrogels (Reproduced from [146] with permission)

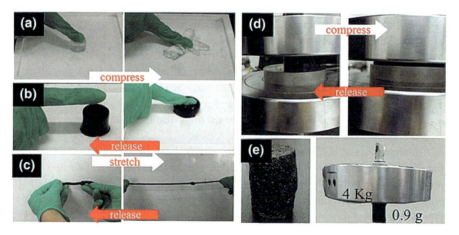

Fig. 32 Mechanical properties of the PAM hydrogel and the GO/PAM/Al-CMH hydrogels (Reproduced from [146] with permission)

pseudo-second-order kinetic model and the mechanism of adsorbing methylene blue is depicted in Fig. 33. A superabsorbent basing on MWCNT-xylan composite and poly(methacrylic acid) was synthesized by the same method described above, and also exhibited high removal rate for MB [148].

Clay can act as a physical crosslinker and provides physically crosslinking during the gelation process of hybrid hydrogels. Cheng et al. [22] developed a facial

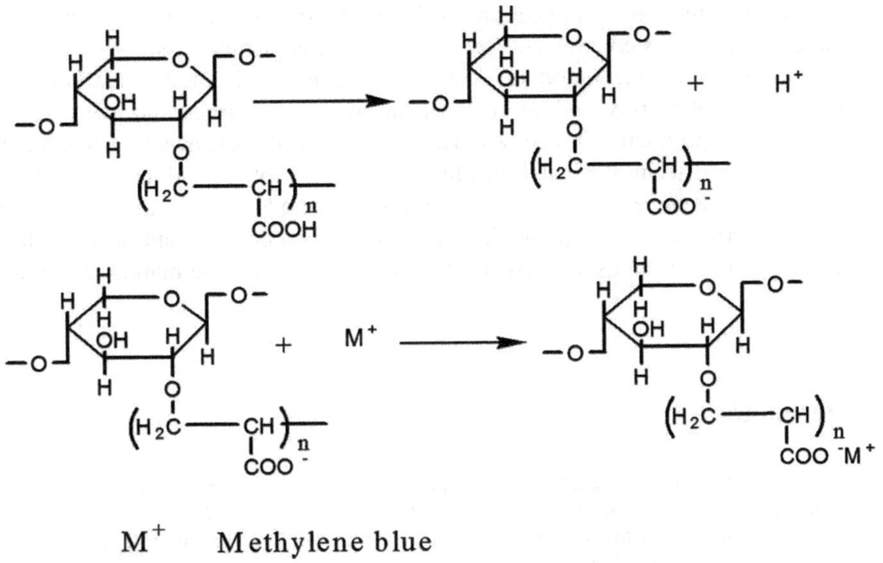

Fig. 33 The possible adsorption mechanism of methylene blue on hydrogels (Reproduced from [149] with permission)

route to prepare hemicellulose/clay hydrogels in the presence of polyethylene glycol diglycidyl ether (PEGDE). The non-covalent interactions between hemicellulose and clay hindered the penetration of water into the hydrogel network, leading to a decrease in swelling ratio. This hybrid hydrogel, however, increased the adsorption of methylene blue by approximately 50 mg/g as compared with the hydrogel without clay. In a study by Sun et al. [21], $CaCO_3$ was used as porogen in the fabrication of stimuli-responsive (pH/salt) porous hydrogels basing on HC and poly(sodium acrylate), the obtained HC-g-poly(sodium acrylate) hydrogels showed a high adsorption capacity of methylene blue, which is appropriate for wastewater application.

6 Summary and Outlook

Hemicelluloses as renewable resources have a lot of favourable inherent properties such as hydrophilicity, biodegradability, biocompatibility, low cost, and non-toxicity, as well as good barrier properties and mechanical properties by using appropriate chemical modification or physical treatment. Therefore, hemicelluloses-based materials hold great potentials in various applications such as packaging, water treatment, and biomedical field, which can form a viable substitute for fossil-based materials to some extent.

However, only a few applications of hemicellulose-based materials are so far explored and suggested. Societal awareness of sustainability issues is a strong driver for the implementation of such biobased materials on the market, still, commercial applicability and success depend also on the price and efficiency of extraction and production processes. There is a huge market awaiting hemicelluloses. In hence, technological breakthroughs are urgent, including the scale-up production of hemicelluloses, new fabrication methods of functional materials, unique performances and applications of various composites, and so on. These technology breakthroughs will trigger the soaring hemicellulose market in the near future.

References

1. Thakur VKT, Hakur MK (2014) Processing and characterization of natural cellulose fibers/thermoset polymer composites. Carbohydr Polym 109(13):102–117
2. Thakur VK, Thakur MK (2015) Recent advances in green hydrogels from lignin: a review. Int J Biol Macromol 72:834
3. Thakur VK, Thakur MK, Raghavan P, Kessler MR (2014) Progress in green polymer composites from lignin for multifunctional applications: a review. ACS Sustain Chem Eng 2 (5):1072–1092
4. Mikkonen KS (2013) Recent studies on hemicellulose-based blends. Composites and nanocomposites. Springer, Berlin, pp 313–336
5. Hansen NM, Plackett D (2008) Sustainable films and coatings from hemicelluloses: a review. Biomacromolecules 9(6):1493–1505
6. Shukla SK, Mishra AK, Arotiba OA, Mamba BB (2013) Chitosan-based nanomaterials: a state-of-the-art review. Int J Biol Macromol 59(4):46
7. Thakur VK, Thakur MK (2014) Recent advances in graft copolymerization and applications of chitosan: a review. ACS Sustain Chem Eng 2(12)
8. Rahmat AR, Wan AWAR, Sin LT, Yussuf AA (2009) Approaches to improve compatibility of starch filled polymer system: a review. Mater Sci Eng C 29(8):2370–2377
9. Avérous L, Halley PJ (2009) Biocomposites based on plasticized starch. Biofuels Bioprod Biorefin 3(3):329–343
10. Thakur VK, Thakur MK (2014) Recent trends in hydrogels based on psyllium polysaccharide: a review. J Clean Prod 82(22):1–15
11. Farhat W, Venditti RA, Hubbe M et al (2017) A review of water-resistant hemicellulose-based materials: processing and applications. Chemsuschem 10(2):305–323
12. Ibn Yaich A, Edlund U, Albertsson AC (2017) Transfer of biomatrix/wood cell interactions to hemicellulose-based materials to control water interaction. Chem Rev 117(12):8177–8207
13. Thakur VK, Thakur MK (2015) Eco-friendly polymer nanocomposites. Advanced structured materials. Springer, India
14. Iwata T (2015) Biodegradable and bio-based polymers: future prospects of eco-friendly plastics. Angew Chem Int Ed Engl 54(11):3210–3215
15. Ayoub A, Venditti RA, Pawlak JJ, Salam A, Hubbe MA (2013) Novel hemicellulose-chitosan biosorbent for water desalination and heavy metal removal. ACS Sustain Chem Eng 1(9):1102–1109
16. Dax D, Chavez MS, Xu C et al (2014) Cationic hemicellulose-based hydrogels for arsenic and chromium removal from aqueous solutions. Carbohydr Polym 111:797–805

17. Ferrari E, Ranucci E, Edlund U, Albertsson AC (2015) Design of renewable poly (amidoamine)/hemicellulose hydrogels for heavy metal adsorption. J Appl Polym Sci 132 (12):41695
18. Peng XW, Zhong LX, Ren JL, Sun RC (2012) Highly effective adsorption of heavy metal ions from aqueous solutions by macroporous xylan-rich hemicelluloses-based hydrogel. J Agric Food Chem 60(15):3909–3916
19. Wu S, Kan J, Dai X et al (2017) Ternary carboxymethyl chitosan-hemicellulose-nanosized TiO_2 composite as effective adsorbent for removal of heavy metal contaminants from water. Fibers Polym 18(1):22–32
20. Wu SP, Dai XZ, Kan JR, Shilong FD, Zhu MY (2017) Fabrication of carboxymethyl chitosan–hemicellulose resin for adsorptive removal of heavy metals from wastewater. Chin Chem Lett 28(3):625–632
21. Sun XF, Gan Z, Jing Z et al (2015) Adsorption of methylene blue on hemicellulose-based stimuli-responsive porous hydrogel. J Appl Polym Sci 132(10):41606
22. Cheng HL, Feng QH, Liao CA et al (2016) Removal of methylene blue with hemicellulose/ clay hybrid hydrogels. Chin J Polym Sci 34(6):709–719
23. Farhat W, Venditti R, Mignard N et al (2017) Polysaccharides and lignin based hydrogels with potential pharmaceutical use as a drug delivery system produced by a reactive extrusion process. Int J Biol Macromol 104:564–575
24. Gao C, Ren J, Zhao C et al (2016) Xylan-based temperature/pH sensitive hydrogels for drug controlled release. Carbohydr Polym 151:189–197
25. Sun XF, Wang HH, Jing ZX, Mohanathas R (2013) Hemicellulose-based pH-sensitive and biodegradable hydrogel for controlled drug delivery. Carbohydr Polym 92(2):1357–1366
26. Zhao W, Odelius K, Edlund U, Zhao C, Albertsson AC (2015) In situ synthesis of magnetic field-responsive hemicellulose hydrogels for drug delivery. Biomacromolecules 16(8): 2522–2528
27. Chen GG, Qi XM, Guan Y et al (2016) High strength hemicellulose-based nanocomposite film for food packaging applications. ACS Sustain Chem Eng 4(4):1985–1993
28. Laine C, Harlin A, Hartman J et al (2013) Hydroxyalkylated xylans—their synthesis and application in coatings for packaging and paper. Ind Crops Prod 44:692–704
29. Tatar F, Tunç MT, Dervisoglu M, Cekmecioglu D, Kahyaoglu T (2014) Evaluation of hemicellulose as a coating material with gum arabic for food microencapsulation. Food Res Int 57:168–175
30. Shen J, Fatehi P, Ni Y (2014) Biopolymers for surface engineering of paper-based products. Cellulose 21(5):3145–3160
31. Nguyen QA, Tucker MP, Keller FA, Eddy FP (2000) Two-stage dilute-acid pretreatment of softwoods. Appl Biochem Biotechnol 84–86(1–9):561–576
32. Egüés I, Sanchez C, Mondragon I, Labidi J (2012) Effect of alkaline and autohydrolysis processes on the purity of obtained hemicelluloses from corn stalks. Biores Technol 103 (1):239–248
33. Hasegawa I, Tabata K, Okuma O, Mae K (2004) New pretreatment methods combining a hot water treatment and water/acetone extraction for thermo-chemical conversion of biomass. Energy Fuels Am Chem Soc J 18(3):755–760
34. And MP, Zacchi G (2003) Extraction of hemicellulosic oligosaccharides from spruce using microwave oven or steam treatment. Biomacromolecules 4(3):617
35. Froschauer C, Hummel M, Iakovlev M et al (2013) Separation of hemicellulose and cellulose from wood pulp by means of ionic liquid/cosolvent systems. Biomacromolecules 14(6):1741–1750
36. Mesbah M, Shahsavari S, Soroush E, Rahaei N, Rezakazemi M (2018) Accurate prediction of miscibility of CO_2 and supercritical CO_2 in ionic liquids using machine learning. J CO_2 Utilization 25:99–107
37. Razavi SMR, Rezakazemi M, Albadarin AB, Shirazian S (2016) Simulation of CO_2 absorption by solution of ammonium ionic liquid in hollow-fiber contactors. Chem Eng Process 108:27–34

38. Gould JM (1984) Alkaline peroxide delignification of agricultural residues to enhance enzymatic saccharification. Biotechnol Bioeng 26(1):46–52
39. Schmidt AS, Thomsen AB (1998) Optimization of wet oxidation pretreatment of wheat straw. Biores Technol 64(2):139–151
40. Li H, Qu Y, Yang Y, Chang S, Xu J (2016) Microwave irradiation—a green and efficient way to pretreat biomass. Biores Technol 199:34–41
41. Chum HL, Johnson DK, Black S et al (1988) Organosolv pretreatment for enzymatic hydrolysis of poplars: I. Enzyme hydrolysis of cellulosic residues. Biotechnol Bioeng 31(7):643–649
42. Saha BC (2003) Hemicellulose bioconversion. J Ind Microbiol Biotechnol 30(5):279–291
43. Hu L, Du M, Zhang J (2018) Hemicellulose-Based hydrogels present status and application prospects: a brief review. Open J Forestry 08(01):15–28
44. Uraki Y, Koda K (2015) Utilization of wood cell wall components. J Wood Sci 61(5):447–454
45. Gandini A (2011) The irruption of polymers from renewable resources on the scene of macromolecular science and technology. Green Chem 13(5):1061
46. Cunha AG, Gandini A (2010) Turning polysaccharides into hydrophobic materials: a critical review. Part 2. Hemicelluloses, chitin/chitosan, starch, pectin and alginates. Cellulose 17(6):1045–1065
47. Thomas S, Visakh PM, Mathew AP (2013) Advances in natural polymers. Advanced structured materials, vol 18. Springer, Berlin, pp 216–217
48. Belmokaddem FZ, Pinel C, Huber P, Petit Conil M, Perez DDS (2011) Green synthesis of xylan hemicellulose esters. Carbohydr Res 346(18):2896–2904
49. Peng XW, Ren JL, Sun RC (2010) Homogeneous esterification of xylan-rich hemicelluloses with maleic anhydride in ionic liquid. Biomacromolecules 11(12):3519–3524
50. Zhang LM, Yuan TQ, Xu F, Sun RC (2013) Enhanced hydrophobicity and thermal stability of hemicelluloses by butyrylation in [BMIM]Cl ionic liquid. Ind Crops Prod 45:52–57
51. Sun RC, Fang JM, Tomkinson J (2000) Stearoylation of hemicelluloses from wheat straw. Polym Degrad Stab 67(2):345–353
52. Sun XF, Sun RC, Sun JX (2004) Oleoylation of sugarcane bagasse hemicelluloses using N-bromosuccinimide as a catalyst. J Sci Food Agric 84(8):800–810
53. Wang HT, Yuan TQ, Meng LJ et al (2012) Structural and thermal characterization of lauroylated hemicelluloses synthesized in an ionic liquid. Polym Degrad Stab 97(11):2323–2330
54. Sun R, Fanga JM, Tomkinson J, Hill CAS (1999) Esterification of hemicelluloses from poplar chips in homogenous solution of N, N-dimethylformamide/lithium chloride. J Wood Chem Technol 19(4):287–306
55. Sun RC, Fang JM, Tomkinson J, Geng ZC, Liu JC (2011) Fractional isolation, physico-chemical characterization and homogeneous esterification of hemicelluloses from fast-growing poplar wood. Paper Chem 44(1):29–39
56. Fundador NGV, Enomoto-Rogers Y, Takemura A, Iwata T (2012) Syntheses and characterization of xylan esters. Polymer 53(18):3885–3893
57. Daus S, Heinze T (2010) Xylan-based nanoparticles: prodrugs for ibuprofen release. Macromol Biosci 10(2):211–220
58. Kisonen V, Xu C, Bollström R et al (2014) O-acetyl galactoglucomannan esters for barrier coatings. Cellulose 21(6):4497–4509
59. Buchanan CM, Buchanan NL, Debenham JS et al (2003) Preparation and characterization of arabinoxylan esters and arabinoxylan ester/cellulose ester polymer blends. Carbohydr Polym 52(4):345–357
60. Voepel J, Edlund U, Albertsson AC, Percec V (2011) Hemicellulose-based multifunctional macroinitiator for single-electron-transfer mediated living radical polymerization. Biomacromolecules 12(1):253–259
61. Wrigstedt P, Kylli P, Pitkanen L et al (2010) Synthesis and antioxidant activity of hydroxycinnamic acid xylan esters. J Agric Food Chem 58(11):6937–6943

62. Maleki L, Edlund U, Albertsson AC (2015) Thiolated hemicellulose as a versatile platform for one-pot click-type hydrogel synthesis. Biomacromolecules 16(2):667–674
63. Ren JL, Peng F, Sun RC (2008) Preparation of hemicellulosic derivatives with bifunctional groups in different media. J Agric Food Chem 56(23):11209–11216
64. Peng X, Ren J, Sun R (2011) An efficient method for the synthesis of hemicellulosic derivatives with bifunctional groups in butanol/water medium and their rheological properties. Carbohydr Polym 83(4):1922–1928
65. Guan Y, Zhang B, Tan X et al (2014) Organic-inorganic composite films based on modified hemicelluloses with clay nanoplatelets. ACS Sustain Chem Eng 2(7):1811–1818
66. Rezakazemi M, Sadrzadeh M, Mohammadi T, Matsuura T (2017) Methods for the preparation of organic-inorganic nanocomposite polymer electrolyte membranes for fuel cells. In: Inamuddin D, Mohammad A, Asiri AM (eds) Organic-inorganic composite polymer electrolyte membranes. Springer International Publishing, Cham, pp 311–325
67. Bigand V, Pinel C, Da Silva Perez D et al (2011) Cationisation of galactomannan and xylan hemicelluloses. Carbohydr Polym 85(1):138–148
68. Ren JL, Sun RC, Liu CF (2007) Etherification of hemicelluloses from sugarcane bagasse. J Appl Polym Sci 105(6):3301–3308
69. Fang JM, Fowler P, Tomkinson J, Hill CAS (2002) Preparation and characterisation of methylated hemicelluloses from wheat straw. Carbohydr Polym 47(3):285–293
70. Hartman J, Albertsson AC, Sjöberg J (2006) Surface- and bulk-modified galactoglucomannan hemicellulose films and film laminates for versatile oxygen barriers. Biomacromolecules 7(6):1983
71. Ren JL, Peng XW, Zhong LX, Peng F, Sun RC (2012) Novel hydrophobic hemicelluloses: synthesis and characteristic. Carbohydr Polym 89(1):152–157
72. Pahimanolis N, Kilpelainen P, Master E, Ilvesniemi H, Seppala J (2015) Novel thiolamine- and amino acid functional xylan derivatives synthesized by thiolene reaction. Carbohydr Polym 131:392–398
73. Liu Z, Ni Y, Fatehi P, Saeed A (2011) Isolation and cationization of hemicelluloses from pre-hydrolysis liquor of kraft-based dissolving pulp production process. Biomass Bioenergy 35(5):1789–1796
74. Schwikal K, Heinze T, Ebringerová A, Petzold K (2005) Cationic xylan derivatives with high degree of functionalization. Macromol Symp 232(1):49–56
75. Kisonen V, Xu C, Eklund P et al (2014) Cationised O-acetyl galactoglucomannans: synthesis and characterisation. Carbohydr Polym 99:755–764
76. Wang S, Hou Q, Kong F, Fatehi P (2015) Production of cationic xylan-METAC copolymer as a flocculant for textile industry. Carbohydr Polym 124:229–236
77. Kong WQ, Ren JL, Wang S, Li MF, Sun RC (2014) A promising strategy for preparation of cationic xylan by environment-friendly semi-dry oven process. Fibers Polym 15(5):943–949
78. Ren JL, Peng F, Sun RC et al (2008) Synthesis of cationic hemicellulosic derivatives with a low degree of substitution in dimethyl sulfoxide media. J Appl Polym Sci 109(4):2711–2717
79. Ibn Yaich A, Edlund U, Albertsson AC (2015) Enhanced formability and mechanical performance of wood hydrolysate films through reductive amination chain extension. Carbohydr Polym 117:346–354
80. Dax D, Eklund P, Hemming J et al (2013) Amphiphilic spruce galactoglucomannan derivatives based on naturally-occurring fatty acids. BioResources 8(3):3771
81. Daus S, Elschner T, Heinze T (2010) Towards unnatural xylan based polysaccharides: reductive amination as a tool to access highly engineered carbohydrates. Cellulose 17 (4):825–833
82. Ehrenfreund-Kleinman T, Gazit Z, Gazit D et al (2002) Synthesis and biodegradation of arabinogalactan sponges prepared by reductive amination. Biomaterials 23(23):4621–4631
83. Leppänen AS, Xu C, Eklund P et al (2014) Targeted functionalization of spruce O-acetyl galactoglucomannans—2,2,6,6-tetramethylpiperidin-1-oxyl-oxidation and carbodiimide-mediated amidation. J Appl Polym Sci 130(5):3122–3129

84. Kuzmenko V, Hagg D, Toriz G, Gatenholm P (2014) In situ forming spruce xylan-based hydrogel for cell immobilization. Carbohydr Polym 102:862–868
85. MacCormick B, Vuong TV, Master ER (2018) Chemo-enzymatic synthesis of clickable xylo-oligosaccharide monomers from hardwood 4-O-methylglucuronoxylan. Biomacromolecules 19(2):521–530
86. Fundador NGV, Enomoto-Rogers Y, Takemura A, Iwata T (2012) Acetylation and characterization of xylan from hardwood kraft pulp. Carbohydr Polym 87(1):170–176
87. Sun RC, Fang JM, Tomkinson J, Jones GL (1999) Acetylation of wheat straw hemicelluloses in N, N-dimethylacetamide/LiCl solvent system. Ind Crops Prod 10 (3):209–218
88. Sun XF, Sun RC, Zhao L, Sun JX (2010) Acetylation of sugarcane bagasse hemicelluloses under mild reaction conditions by using NBS as a catalyst. J Appl Polym Sci 92(1):53–61
89. Ren JL, Sun RC, Liu CF, Cao ZN, Luo W (2007) Acetylation of wheat straw hemicelluloses in ionic liquid using iodine as a catalyst. Carbohydr Polym 70(4):406–414
90. Stepan AM, King AWT, Kakko T et al (2013) Fast and highly efficient acetylation of xylans in ionic liquid systems. Cellulose 20(6):2813–2824
91. Gröndahl M, Teleman A, Gatenholm P (2003) Effect of acetylation on the material properties of glucuronoxylan from aspen wood. Carbohydr Polym 52(4):359–366
92. Ayoub A, Venditti RA, Pawlak JJ, Sadeghifar H, Salam A (2013) Development of an acetylation reaction of switchgrass hemicellulose in ionic liquid without catalyst. Ind Crops Prod 44:306–314
93. Dong L, Hu H, Yang S, Cheng F (2014) Grafted copolymerization modification of hemicellulose directly in the alkaline peroxide mechanical pulping (APMP) effluent and its surface sizing effects on corrugated paper. Ind Eng Chem Res 53(14):6221–6229
94. Enomoto-Rogers Y, Iwata T (2012) Synthesis of xylan-graft-poly(L-lactide) copolymers via click chemistry and their thermal properties. Carbohydr Polym 87(3):1933–1940
95. Edlund U, Albertsson A-C (2014) A controlled radical polymerization route to polyepoxidated grafted hemicellulose materials. Polimery 59(01):60–65
96. Saadatmand S, Edlund U, Albertsson A-C (2011) Compatibilizers of a purposely designed graft copolymer for hydrolysate/PLLA blends. Polymer 52(21):4648–4655
97. Persson J, Dahlman O, Albertsson AC (2012) Birch xylan grafted with pla branches of predictable length. Bioresources 7(3):3640–3655
98. Fanta GF, Burr RC, Doane WM (1982) Graft polymerization of acrylonitrile and methyl acrylate onto hemicellulose. J Appl Polym Sci 27(11):4239–4250
99. Voepel J, Edlund U, Albertsson A-C (2011) A versatile single-electron-transfer mediated living radical polymerization route to galactoglucomannan graft-copolymers with tunable hydrophilicity. J Polym Sci Part A Polym Chem 49(11):2366–2372
100. Edlund U, Rodriguez-Emmenegger C, Brynda E, Albersson A-C (2012) Self-assembling zwitterionic carboxybetaine copolymers via aqueous SET-LRP from hemicellulose multi-site initiators. Polym Chem 3(10):2920
101. O'Malley JJ, Marchessault RH (1966) Characterization of graft copolymers of methylated xylan and polystyrene. J Phys Chem 70(10):3235–3240
102. Parikka K, Leppanen AS, Xu C et al (2012) Functional and anionic cellulose-interacting polymers by selective chemo-enzymatic carboxylation of galactose-containing polysaccharides. Biomacromolecules 13(8):2418–2428
103. Parikka K, Leppanen AS, Pitkanen L et al (2010) Oxidation of polysaccharides by galactose oxidase. J Agric Food Chem 58(1):262–271
104. Leppanen AS, Xu C, Parikka K et al (2014) Targeted allylation and propargylation of galactose-containing polysaccharides in water. Carbohydr Polym 100:46–54
105. Song X, Hubbe MA (2014) TEMPO-mediated oxidation of oat beta-D-glucan and its influences on paper properties. Carbohydr Polym 99:617–623
106. Kohnke T, Elder T, Theliander H, Ragauskas AJ (2014) Ice templated and cross-linked xylan/nanocrystalline cellulose hydrogels. Carbohydr Polym 100:24–30

107. Chemin M, Rakotovelo A, Ham-Pichavant F et al (2016) Periodate oxidation of 4-O-methylglucuronoxylans: influence of the reaction conditions. Carbohydr Polym 142:45–50
108. Ehrenfreund-Kleinman T, Domb AJ, Golenser J (2003) Polysaccharide scaffolds prepared by crosslinking of polysaccharides with chitosan or proteins for cell growth. J Bioact Compatible Polym 18(5):323–338
109. Luo YQ, Shen SQ, Luo JW, Wang XY, Sun RC (2015) Green synthesis of silver nanoparticles in xylan solution via Tollens reaction and their detection for Hg^{2+}. Nanoscale 7(2):690–700
110. Luo Y, Shen Z, Liu P, Zhao L, Wang X (2016) Facile fabrication and selective detection for cysteine of xylan/Au nanoparticles composite. Carbohydr Polym 140:122–128
111. Peng H, Yang A, Xiong J (2013) Green, microwave-assisted synthesis of silver nanoparticles using bamboo hemicelluloses and glucose in an aqueous medium. Carbohydr Polym 91(1):348–355
112. Silva AK, da Silva EL, Oliveira EE et al (2007) Synthesis and characterization of xylan-coated magnetite microparticles. Int J Pharm 334(1–2):42–47
113. Wu CY, Peng XW, Zhong LX, Li XH, Sun RC (2016) Green synthesis of palladium nanoparticles via branched polymers: a bio-based nanocomposite for C–C coupling reactions. RSC Adv 6(38):32202–32211
114. Chen W, Zhong LX, Peng XW, Lin JH, Sun RC (2013) Xylan-type hemicelluloses supported terpyridine–palladium(II) complex as an efficient and recyclable catalyst for Suzuki-Miyaura reaction. Cellulose 21(1):125–137
115. Chen W, Zhong LX, Peng XW et al (2014) Xylan-type hemicellulose supported palladium nanoparticles: a highly efficient and reusable catalyst for the carbon-carbon coupling reactions. Catal Sci Technol 4(5):1426–1435
116. Du J, Sun R, Zhang S et al (2004) Novel polyelectrolyte carboxymethyl konjac glucomannan-chitosan nanoparticles for drug delivery. Macromol Rapid Commun 25(9):954–958
117. Heinze T, Petzold K, Hornig S (2008) Novel nanoparticles based on xylan. Cellul Chem Technol 41(1):13–18
118. Garcia RB, Nagashima T Jr, Praxedes AKC et al (2001) Preparation of micro and nanoparticles from corn cobs xylan. Polym Bull 46(5):371–379
119. Phan The D, Debeaufort F, Péroval C et al (2002) Arabinoxylan-lipid-based edible films and coatings. 3. Influence of drying temperature on film structure and functional properties. J Agric Food Chem 50(8):2423–2428
120. Péroval C, Debeaufort F, Despré D, Voilley A (2002) Edible arabinoxylan-based films. 1. Effects of lipid type on water vapor permeability, film structure, and other physical characteristics. J Agric Food Chem 50(14):3977–3983
121. Phan TD, Péroval C, Debeaufort F et al (2002) Arabinoxylan-lipids-based edible films and coatings. 2. Influence of sucroester nature on the emulsion structure and film properties. J Agric Food Chem 50(2):266–272
122. Hartman J, Albertsson A-C, Lindblad MS, Sjöberg J (2006) Oxygen barrier materials from renewable sources: material properties of softwood hemicellulose-based films. J Appl Polym Sci 100(4): 2985–2991
123. Zhang P, Whistler RL (2004) Mechanical properties and water vapor permeability of thin film from corn hull arabinoxylan. J Appl Polym Sci 93(6):2896–2902
124. Chen GG, Qi XM, Li MP et al (2015) Hemicelluloses/montmorillonite hybrid films with improved mechanical and barrier properties. Sci Rep 5:16405
125. Liu YX, Sun B, Wang ZL, Ni YH (2016) Mechanical and water vapor barrier properties of bagasse hemicellulose-based films. Bioresources 11(2):4226–4236
126. Gordobil O, Egues I, Urruzola I, Labidi J (2014) Xylan-cellulose films: improvement of hydrophobicity, thermal and mechanical properties. Carbohydr Polym 112:56–62
127. Hu S, Gu J, Jiang F, Hsieh YL (2016) Holistic rice straw nanocellulose and hemicelluloses/lignin composite films. ACS Sustain Chem Eng 4(3):728–737

128. Huang B, Tang Y, Pei Q et al (2017) hemicellulose-based films reinforced with unmodified and cationically modified nanocrystalline cellulose. J Polym Environ
129. Kisonen V, Prakobna K, Xu C et al (2015) Composite films of nanofibrillated cellulose and O-acetyl galactoglucomannan (GGM) coated with succinic esters of GGM showing potential as barrier material in food packaging. J Mater Sci 50(8):3189–3199
130. Ma RX, Pekarovicova A, Fleming III PD, Husovska V (2017) Preparation and characterization of hemicellulose-based printable films. Cellul Chem Technol 51(9–10):939–948
131. Mikkonen KS, Stevanic JS, Joly C et al (2011) Composite films from spruce galactoglucomannans with microfibrillated spruce wood cellulose. Cellulose 18(3):713–726
132. Peng XW, Ren JL, Zhong LX, Sun RC (2011) Nanocomposite films based on xylan-rich hemicelluloses and cellulose nanofibers with enhanced mechanical properties. Biomacromolecules 12(9):3321–3329
133. Shao D, Yotprayoonsak P, Saunajoki V et al (2018) Conduction properties of thin films from a water soluble carbon nanotube/hemicellulose complex. Nanotechnology 29(14):145203
134. Bahcegul E, Toraman HE, Ozkan N, Bakir U (2012) Evaluation of alkaline pretreatment temperature on a multi-product basis for the co-production of glucose and hemicellulose based films from lignocellulosic biomass. Bioresour Technol 103(1):440–445
135. Kayserilioğlu BŞ, Bakir U, Yilmaz L, Akkaş N (2003) Use of xylan, an agricultural by-product, in wheat gluten based biodegradable films: mechanical, solubility and water vapor transfer rate properties. Bioresour Technol 87(3):239–246
136. Ruiz HA, Cerqueira MA, Silva HD et al (2013) Biorefinery valorization of autohydrolysis wheat straw hemicellulose to be applied in a polymer-blend film. Carbohydr Polym 92 (2):2154–2162
137. Svard A, Brannvall E, Edlund U (2015) Rapeseed straw as a renewable source of hemicelluloses: extraction, characterization and film formation. Carbohydr Polym 133: 179–186
138. Oinonen P, Areskogh D, Henriksson G (2013) Enzyme catalyzed cross-linking of spruce galactoglucomannan improves its applicability in barrier films. Carbohydr Polym 95(2): 690–696
139. Péroval C, Debeaufort F, Seuvre A-M et al (2003) Modified arabinoxylan-based films. Part B. Grafting of omega-3 fatty acids by oxygen plasma and electron beam irradiation. J Agric Food Chem 51(10):3120–3126
140. Peroval C, Debeaufort F, Seuvre AM et al (2004) Modified arabinoxylan-based films grafting of functional acrylates by oxygen plasma and electron beam irradiation. J Membr Sci 233(1–2):129–139
141. Lee SG, An EY, Lee JB et al (2007) Enhanced cell affinity of poly(D, L-lactic-co-glycolic acid) (50/50) by plasma treatment with β-(1→3) (1→6)-glucan. Surf Coat Technol 201(9–11):5128–5131
142. Fredon E, Granet R, Zerrouki R et al (2002) Hydrophobic films from maize bran hemicelluloses. Carbohydr Polym 49(1):1–12
143. Gröndahl M, Gustafsson A, Gatenholm P (2006) Gas-phase surface fluorination of arabinoxylan films. Macromolecules 39(7):2718–2721
144. Šimkovic I, Gedeon O, Uhliariková I, Mendichi R, Kirschnerová S (2011) Positively and negatively charged xylan films. Carbohydr Polym 83(2):769–775
145. Hesse S, Liebert T, Heinze T (2005) Studies on the film formation of polysaccharide based furan-2-carboxylic acid esters. Macromol Symp 232(1):57–67
146. Kong W, Huang D, Xu G et al (2016) Graphene oxide/polyacrylamide/aluminum ion cross-linked carboxymethyl hemicellulose nanocomposite hydrogels with very tough and elastic properties. Chem Asian J 11(11):1697–1704
147. Zhang W, Liang Z, Feng Q et al (2016) Reed hemicellulose-based hydrogel prepared by glow discharge eletrolysis plasma and its adsorption properties for heavy metal ions. Fresenius Environ Bull 25(6):1791–1798

148. Jing Z, Zhang G, Sun X-F, Shi X, Sun W (2014) Preparation and adsorption properties of a novel superabsorbent based on multiwalled carbon nanotubes-xylan composite and poly (methacrylic acid) for methylene blue from aqueous solution. Polym Compos 35(8):1516–1528
149. Sun XF, Ye Q, Jing Z, Li Y (2014) Preparation of hemicellulose-g-poly(methacrylic acid)/carbon nanotube composite hydrogel and adsorption properties. Polym Compos 35(1):45–52
150. Voepel J, Sjöberg J, Reif M et al (2009) Drug diffusion in neutral and ionic hydrogels assembled from acetylated galactoglucomannan. J Appl Polym Sci 112(4):2401–2412
151. Zhao W, Nugroho RW, Odelius K et al (2015) In situ cross-linking of stimuli-responsive hemicellulose microgels during spray drying. ACS Appl Mater Interfaces 7(7):4202–4215
152. Alexandra AR, Ulrica E, John S, Ann-Christine A, Henrik S (2008) Protein release from galactoglucomannan hydrogels: influence of substitutions and enzymatic hydrolysis by mannanase. Biomacromolecules 9(8):2104–2110
153. Guo B, Glavas L, Albertsson A-C (2013) Biodegradable and electrically conducting polymers for biomedical applications. Prog Polym Sci 38(9):1263–1286
154. Dai QQ, Ren JL, Peng F et al (2016) Synthesis of acylated xylan-based magnetic Fe_3O_4 hydrogels and their application for H_2O_2 detection. Materials (Basel) 9(8):3–16
155. Du J, Li B, Li C et al (2016) Tough and multi-responsive hydrogel based on the hemicellulose from the spent liquor of viscose process. Int J Biol Macromol 88:451–456
156. Liu S, Chen F, Song X, Wu H (2016) Preparation and characterization of temperature- and pH-sensitive hemicellulose-containing hydrogels. Int J Polym Anal Charact 22(3):187–201
157. Pahimanolis N, Sorvari A, Luong ND, Seppala J (2014) Thermoresponsive xylan hydrogels via copper-catalyzed azide-alkyne cycloaddition. Carbohydr Polym 102:637–644
158. Peng XW, Ren JL, Zhong LX, Peng F, Sun RC (2011) Xylan-rich hemicelluloses-graft-acrylic acid ionic hydrogels with rapid responses to pH, salt, and organic solvents. J Agric Food Chem 59(15):8208–8215
159. Yang JY, Zhou XS, Fang J (2011) Synthesis and characterization of temperature sensitive hemicellulose-based hydrogels. Carbohydr Polym 86(3):1113–1117
160. Zhang W, Zhu S, Bai Y et al (2015) Glow discharge electrolysis plasma initiated preparation of temperature/pH dual sensitivity reed hemicellulose-based hydrogels. Carbohydr Polym 122:11–17
161. Zhao W, Glavas L, Odelius K, Edlund U, Albertsson A-C (2014) Facile and green approach towards electrically conductive hemicellulose hydrogels with tunable conductivity and swelling behavior. Chem Mater 26(14):4265–4273
162. Zhao W, Glavas L, Odelius K, Edlund U, Albertsson A-C (2014) A robust pathway to electrically conductive hemicellulose hydrogels with high and controllable swelling behavior. Polymer 55(13):2967–2976
163. Rezakazemi M, Shahidi K, Mohammadi T (2012) Sorption properties of hydrogen-selective PDMS/zeolite 4A mixed matrix membrane. Int J Hydrogen Energy 37(22):17275–17284
164. Rezakazemi M, Shahidi K, Mohammadi T (2012) Hydrogen separation and purification using crosslinkable PDMS/zeolite A nanoparticles mixed matrix membranes. Int J Hydrogen Energy 37(19):14576–14589
165. Qi XM, Chen GG, Gong XD et al (2016) Enhanced mechanical performance of biocompatible hemicelluloses-based hydrogel via chain extension. Sci Rep 6:33603
166. Gabrielii I, Gatenholm P (2015) Preparation and properties of hydrogels based on hemicellulose. J Appl Polym Sci 69(8):1661–1667
167. Salam A, Venditti RA, Pawlak JJ, El-Tahlawy K (2011) Crosslinked hemicellulose citrate-chitosan aerogel foams. Carbohydr Polym 84(4):1221–1229
168. Guan Y, Chen J, Qi X et al (2015) Fabrication of biopolymer hydrogel containing Ag nanoparticles for antibacterial property. Ind Eng Chem Res 54(30):7393–7400
169. Guan Y, Bian J, Peng F, Zhang XM, Sun RC (2014) High strength of hemicelluloses based hydrogels by freeze/thaw technique. Carbohydr Polym 101:272–280
170. Guan Y, Zhang B, Bian J, Peng F, Sun R-C (2014) Nanoreinforced hemicellulose-based hydrogels prepared by freeze-thaw treatment. Cellulose 21(3):1709–1721

171. Karaaslan MA, Tshabalala MA, Yelle DJ, Buschle-Diller G (2011) Nanoreinforced biocompatible hydrogels from wood hemicelluloses and cellulose whiskers. Carbohydr Polym 86(1):192–201
172. Alakalhunmaa S, Parikka K, Penttilä PA et al (2016) Softwood-based sponge gels. Cellulose 23(5):3221–3238
173. Dax D, Bastidas MSC, Honorato C et al (2015) Tailor-made hemicellulose-based hydrogels reinforced with nanofibrillated cellulose. Nord Pulp Pap Res J 30(3)
174. Dragan ES (2014) Design and applications of interpenetrating polymer network hydrogels. A review. Chem Eng J 243:572–590
175. Myung D, Waters D, Wiseman M et al (2008) Progress in the development of interpenetrating polymer network hydrogels. Polym Adv Technol 19(6):647–657
176. Maleki L, Edlund U, Albertsson A-C (2016) Green semi-IPN hydrogels by direct utilization of crude wood hydrolysates. ACS Sustain Chem Eng 4(8):4370–4377
177. Maleki L, Edlund U, Albertsson AC (2017) Synthesis of full interpenetrating hemicellulose hydrogel networks. Carbohydr Polym 170:254–263
178. Meena R, Lehnen R, Saake B (2013) Microwave-assisted synthesis of kC/Xylan/PVP-based blend hydrogel materials: physicochemical and rheological studies. Cellulose 21(1):553–568
179. Rezakazemi M, Sadrzadeh M, Matsuura T (2018) Thermally stable polymers for advanced high-performance gas separation membranes. Prog Energy Combust Sci 66:1–41
180. Rezakazemi M, Ebadi Amooghin A, Montazer-Rahmati MM, Ismail AF, Matsuura T (2014) State-of-the-art membrane based CO_2 separation using mixed matrix membranes (MMMs): an overview on current status and future directions. Prog Polym Sci 39(5):817–861
181. Fonseca Silva TC, Habibi Y, Colodette JL, Lucia LA (2011) The influence of the chemical and structural features of xylan on the physical properties of its derived hydrogels. Soft Matter 7(3):1090–1099
182. Söderqvist Lindblad M, Albertsson A, Ranucci E, Laus M, Giani E (2005) Biodegradable polymers from renewable sources: rheological characterization of hemicellulose-based hydrogels. Biomacromolecules 6(2):684
183. Lindblad MS, Ranucci E, Albertsson AC (2001) Biodegradable polymers from renewable sources. New hemicellulose-based hydrogels. Macromol Rapid Commun 22(12):962–967
184. Tanodekaew S, Channasanon S, Uppanan P (2006) Xylan/polyvinyl alcohol blend and its performance as hydrogel. J Appl Polym Sci 100(3):1914–1918
185. Azimi A, Azari A, Rezakazemi M, Ansarpour M (2017) Removal of heavy metals from industrial wastewaters: a review. ChemBioEng Rev 4(1):37–59
186. Rezakazemi M, Zhang Z (2018) 2.29 desulfurization materials A2. In: Ibrahim D (ed) Comprehensive energy systems. Elsevier, Oxford, pp 944–979

Self-healing Bio-composites: Concepts, Developments, and Perspective

Zeinab Karami, Sara Maleki, Armaghan Moghaddam and Arash Jahandideh

Abbreviations

BG	Bioglass
CNCs	Cellulose nanocrystals
CB	Cucurbit uril
DA	Diels-Alder
ELP	Elastin-like polypeptides
GO	Graphene oxide
MSP	Metallo-supramolecular polymer
MWCNTs	Multi-wall carbon nanotubes
NR	Natural rubber
PDAP	Polydopamine
PU	Polyurethane
PCL	Poly(ε-caprolactone)
Ag NWs	Silver nanowires
UV	Ultraviolet radiation
UPy	Ureidopyrimidinone

1 Introduction

As humans seek immortality, they also like their products to be able to be used infinitely, but man-made materials, such as polymer composites, are vulnerable to damage, failure, and degradation. Defects from deep within the structure, and thus, detecting such defects and repairing them is hardly feasible [1]. On the other hand, biological systems, such as muscles or bones, can be repaired after being exposed to excessive loads through complex mechanisms. Accordingly and inspired by nature,

---

Z. Karami (✉) · S. Maleki · A. Moghaddam · A. Jahandideh
Iran Polymer and Petrochemical Institute (IPPI), P.O. Box 14965-115, Tehran, Iran
e-mail: Z.Karami@ippi.ac.ir; Rana.Karami@gmail.com

© Springer Nature Switzerland AG 2019
Inamuddin et al. (eds.), *Sustainable Polymer Composites and Nanocomposites*,
https://doi.org/10.1007/978-3-030-05399-4_44

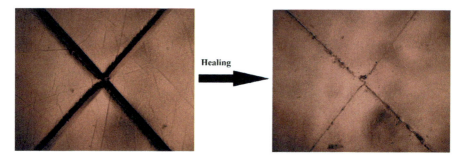

Fig. 1 Healing of a polymer coating through thermal treatment (retro diels-alder and diels-alder reactions)

synthetic healable systems have been developed. Synthetic self-healing materials are smart systems which have the ability to heal and retain the properties, similar to living organisms, after the occurrence of a damage [2].

Different types and applications of polymers including thermoplastics, thermosets, elastomers [3–8], shape memory and supramolecular polymers, coatings, and polymer composites have been reported in the field of self-healing materials, as these materials are used excessively in everyday and industrial applications [9, 10]. Figure 1 shows a healing process of a polymer coating via a thermal treatment.

In this chapter, the basics and fundamentals of healing are explored. The following sections of the chapter deal with self-healing biocomposites, and their types and applications are explained. Finally, a scope into the future of these biocomposites is discussed.

1.1 Fundamentals of Self-healing

The self-healing materials are divided into extrinsic and intrinsic systems. In extrinsic self-healing systems, healing agents such as microcapsules are embedded in the matrix, but this approach is of some drawbacks; for instance, the healing agent gets consumed during healing in the damaged region [11]. For intrinsic self-healing systems, the non-covalent or dynamic covalent chemistries control the healing process and efficiency. Unlike extrinsic healing materials, these types of healable materials can be used repeatedly in healing cycles [12].

Moreover, self-healing mechanisms based on the bonding can be categorized into two groups: covalent bonds, and non-covalent bonds. Indeed, the formation or cleavage of covalent or non-covalent bonds in the polymer materials can result in healing property in the presence or absence of an external stimulus [13]. The reactions which consequently lead to the self-healing of the polymers include covalent bonds, such as the formation of cyclic structures [14], and non-covalent bonds, meaning supramolecular chemistry (H-bonding, ionic interactions, and π–π stacking) [15, 16].

1.2 Biocomposites: Substitutes for Fossil-Based Composites

The ever-changing economic developments leading to increasing oil prices, global warming and endless trash being produced, have led to more light being shed on the urgency of formulating novel composites with sustainable sources, and even better, biodegradable ones [17]. Besides, with the development of polymers and composites fields and petrochemicals being the main supply for these fields, concerns about a fossil fuel-deprived future has grown. Regarding the economic and environmental issues arisen, the production of polymers derived from renewable and sustainable sources have become a necessity [18, 19]. Thus, for instance, petrochemical-based composites can be replaced with biocomposites.

A bio-based composite can consist of a bio-based reinforcing agent, matrix, both or even other ingredients in the formulation such as vegetable oils [20, 21]. Sustainable matrixes include furan derivatives, gelatin, chitin and chitosan, poly (lactic acid) (PLA), poly(lactic-co-glycolic acid) (PLGA), casein, alginate, proteins and their derivatives, reinforcing agents include cellulose and jute fibers, and other ingredients include linseed, Tung, and neem oil, to name but a few [22–32].

As mentioned earlier, based on the active agent and the degree of damage, different mechanisms and transitions can affect the healing property of the materials [33]. In following sections, the mechanisms by which the healing in biocomposites occur are discussed regarding the type of healing, more specifically, based on the bonds and transitions.

2 Self-healing Biocomposites Based on Non-covalent Bonding (Supramolecular)

Features including reversibility, directionality, and sensitivity make the supramolecular chemistry attractive, particularly for self-healing materials. In contrast to covalent bonding, supramolecular networks can remodel rapidly and reversibly from fluid-like to solid-like plastic networks [34, 35]. Therefore, exploiting supramolecular chemistry in biocomposite matrixes can lead to healable biocomposites. The self-healing in these biocomposites are based on hydrogen bonding, metal-ligand coordination, $\pi-\pi$ stacking, ionic interactions and macrocyclic host-guest interactions. Table 1 summarizes the general mechanisms of self-healing for supramolecular polymers. the supramolecular self-healing biocomposites have great potentials to be used in various applications, namely fields of biomaterials, wastewater treatment, and smart materials [36–38].

Table 1 General mechanisms of self-healing for supramolecular materials

Type of interaction	Mechanism
Hydrogen bonding	It is sensitive to pH or temperature changes. For example, at ultraviolet radiation exposure, UPy motifs cleave and so the properties of these materials, such as molecular weight and viscosity, decrease and as a result, the defects heal quickly and efficiently [45]
Metal–ligand coordination	It is a temperature-sensitive complex. It has to be noted that heating leads to possible de-bonds in metal–ligand motifs, which consequently reduce the molecular weight of polymers, as well as the viscosity, and thereby, the healing of the mechanical damages would be facilitated [49]
π–π stacking interaction	It is a temperature-sensitive complex [52]
Ionic interactions	The ionic cross-links easily re-form and rearrange, which facilitate the self-healability [54]
Host–guest interactions	Macrocyclic host–guest interactions can be incorporated in a biocomposite in order to impart self-healability [56]

2.1 Self-healing of Biocomposites on the Basis of Hydrogen Bonding

Directionality and affinity features have made hydrogen bonding (H-bonding) very attractive. Besides, these features have endowed remarkable mechanical strength to the systems with hydrogen bonding. Method of selecting the hydrogen bonding motifs, such as ureidopyrimidinone (UPy, a motif with strong tendency to form H-bonds through a quadruple array of H-bonding donors and acceptors, as shown in Fig. 2), can be affected by the properties of polymers. It should be noted that these motifs are easily introduced into polymer chains either as end-groups or as pending groups [39–41]. UPy end-functionalized supramolecular polymers behave similarly to conventional polymers with an immense dependency on physical and mechanical properties to the temperature [42, 43]. UPy-monomers coupled with the thermally responsive polymer can terminate the formation of thermo-regulated self-healing

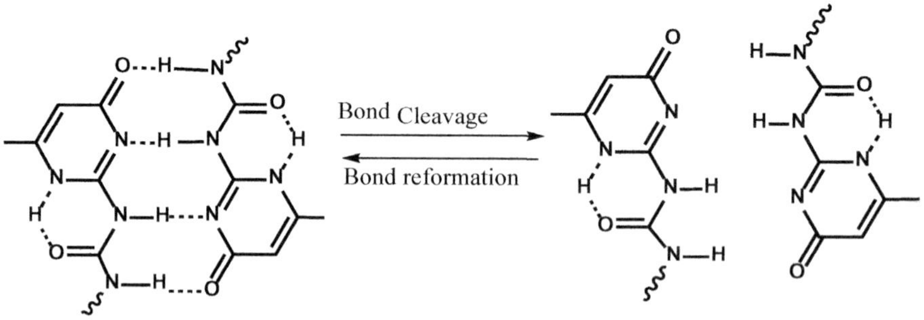

Fig. 2 Multiple H-bonds cleaving and reforming in an Upy motif as a moiety that can cause the healability [39]

polymers. In the supramolecular polymers, using materials with high segmental mobility (which act as the soft segment with low Tg), connected to the segments with the H-bonding motifs (which play the role of the hard segment with high Tg), endow a self-healing property to the system which is sensitive to the changes in temperature or pH [12]. It is also believed that the hydrophobic interactions associated with the hard segments (UPy-groups) lead to the phase separation, and therefore self-healing can be a feature in supramolecular polymers [44]. As aforementioned, H-bonding can be utilized in biocomposites to impart self-healability [45].

For example, the synthesis of light-healable nanocomposites, based on a telechelic poly(ethylene-co-butylene) (functionalized with H-bonding of UPy) and UPy-functionalized cellulose nanocrystals is reported. Under ultraviolet radiation exposure, the UPy motifs of these materials were excited by the absorbed energy that was converted into heat. This phenomenon resulted in the temporary cleavage in the H-bonding motifs, and consequently, a reversible drop in the properties of the supramolecular polymers, such as molecular weight and viscosity. As a result, defects healed quickly and efficiently, even at a filler content of up to 20 w/w%. It must be noted that healing through this method has been performed under heating, and this method is sensitive to the temperature applied [45].

2.2 Self-healing of Biocomposites on the Basis of Metal-Ligand Coordination

The optical and photochemical properties, along with reversibility and tenability (which have been achieved by incorporation of different metal ions and ligand substitutes), have made metal-ligand coordination particularly attractive. In these supramolecular polymers, incorporating ligands into metal-crosslinked polymers leads to the formation of a temperature-sensitive complex in which variations in the temperature may decouple metal ions from the ligand, resulting in bonding and de-bonding [46]. When the metallo-supramolecular polymers are exposed to the ultraviolet (UV) light, the metal-ligand motifs could be electronically excited, and the absorbed energy could be converted into heat. This, in turn, results in the de-bonding of metal-ligand motifs (Fig. 3), which consequently reduces the molecular mass of the polymers as well as the viscosity, and thereby, healing of the mechanical damages would be facilitated [47, 48].

An example for utilizing metal-ligand coordination in self-healing biocomposites discussed the synthesis of the light-healable nanocomposites, including cellulose nanocrystals (CNCs) and a metallo-supramolecular polymer (MSP) based on poly(ethylene-co-butylene) that was end-functionalized with 2,6-bis (1-methylbenzimidazolyl) pyridine ligands and Zn(NTf2) 2. These nanocomposites were able to absorb UV radiations and convert them to heat, and thus, the metal−ligand motifs dissociated. As a result, small defects could be filled by liquefying the material. When the UV light was switched off, the MSP reassembled, and the

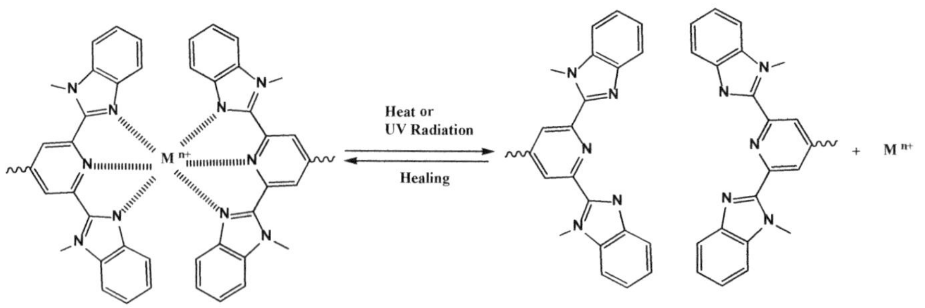

Fig. 3 A representative scheme of self-healing of MSP containing metal–ligand coordination [46]

initial properties were restored. Incorporating CNCs into the MSP matrix improved the strength and stiffness-from 52 and 1.7 MPa for the neat polymer to 135 and 5.6 MPa in 10% w/w CNCs, respectively [49].

2.3 Self-healing of Biocomposites on the Basis of π–π Stacking Interaction

When end-caped π-electron deficient groups interact with other π-electron-rich aromatic backbone molecules, stacking interactions form. In these polymers, tuning T_g facilitates the fabrication of self-healable supramolecular polymers at a wide temperature range (∼50–100 °C). It has also been reported that within one supramolecular network, the intermolecular H-bonding can be combined with π–π Stacking, and consequently, thermal healable networks containing urethane and urea groups as a spacer can be achieved [50, 51].

To endow heal ability to biocomposites by π–π Stacking interactions, a supramolecular healable nanocomposite was prepared, via blending π–π interactions (between a π-electron rich pyrenyl end-capped oligomer) and a chain-folding oligomer, containing pairs of π-electron poor naphthalene-diimide units, with CNCs as the reinforcing agent. The authors prepared and studied a series of nanocomposites, employing different CNCs wt% (from 1.25 to 20.0 wt%) in the healable supramolecular polymeric matrix, via solvent casting followed by compression moulding. The authors studied the healing behaviour of nanocomposites at elevated temperatures (∼85 °C). The results showed that the healing rate decreased by an increase in the CNC content. The best combination of healing efficiency and mechanical properties was obtained when employing 7.5 wt% CNC in the nanocomposite, which resulted in a rehealable composite (100% healability at 85 °C within 30 min). As a result, they showed enhanced mechanical properties for the supramolecular nanocomposites compared to the unreinforced polymer, while efficient thermal healing is still possible [52].

2.4 Self-healing of Biocomposites on the Basis of Ionic Interactions

In this group of supramolecular polymers and polymer composites, the formation of the ionomer leads to the formation of the final polymer network, caused by the ionic interactions. For instance, Xu et al. [53] worked on a self-healable material, exploiting the controlled peroxide-induced vulcanization to generate the physical ionic crosslinks, via polymerization of zinc dimethacrylate in natural rubber (NR). It has been noted that the rubber with covalent cross-linking has higher strengths and modulus, although these types of cross-links induce lower levels of mobility to the rubber chains. The restricted mobility of the rubber chains, in turn, hampered the healability of the system in case of the mechanical damage. On the other hand, NR chains in the ionic supramolecular network had good flexibility and mobility. The ionic cross-links easily reconstructed and rearranged, which facilitated the self-healability [53, 54].

Moreover, studied the self-assembly of chitosan chains with graphene oxide (GO) nano-sheets was studied, where GO worked as the two-dimensional cross-linker due to its multifunctional groups on both sides. The gel-formation process includes two steps: first, chitosan interacts with GO via electrostatic interactions. Due to the complex intra-/inter-molecular hydrogen bonding, the chitosan chains would be in the compressed state. After heating, the hydrogen bonding interactions (among chitosan chains) become weak and the free motion of chitosan chain and GO nano-sheets increase, resulting in the more interactions among them. As a result, the chains stretched have a higher chance to interact with other GO sheets and supramolecular hydrogels of chitosan and GO could be prepared by controlling the concentration of GO, the ratio of chitosan/GO, and the temperature. They also found that at a high GO concentration, healable hydrogels can be prepared at room temperature. However, at lower GO concentrations, the supramolecular hydrogels formed only at an elevated temperature (i.e. 95 °C) [36].

2.5 Self-healing of Biocomposites on the Basis of Macrocyclic Host–Guest Interactions

Macrocyclic host-guest interactions have been employed over the past three decades. Typically, this kind of interactions is formed where a guest molecule is locked within the cavity of the host. It should be noted that the host molecule typically owns external features (that interact with the solvent), and internal features that often cause the formation of the interaction between a 'guest' through either a specific shape or a favourable environment to the host molecules [55]. Macrocyclic host-guest interactions can be incorporated in a biocomposite in order to impart self-healability.

For example, a nanocomposite hydrogel was prepared from brush polymer-modified CNC, as hard domains, and 'soft' polymeric domains, that were bound together by cucurbit [8] uril (CB [8]) supramolecular crosslinks, which form dynamic host-guest interactions. The resulting supramolecular nanocomposite hydrogels showed three important properties: (I) high storage modulus (G' > 10 kPa), (II) rapid sol-gel transition (<6 s), and (III) rapid self-healing (even upon aging for several months) due to the balanced colloidal reinforcement, as well as the selectivity and dynamics of the CB [8] three-component supramolecular interactions [56].

3 Self-healing Biocomposites Based on Covalent Bonding

For self-healing biocomposite based on covalent bonds, the most important covalent bonding, which is commonly used for the preparation of the self-healing polymers and polymer composites is [2+4] cycloaddition, known as DA reaction [14]. Although this reaction was discovered in the year 1950 by Diels and Alder, just in the last decade was it employed as the cross-linking mechanism in healable polymers for healing applications [57]. DA reaction includes a cycloaddition reaction between a conjugated diene such as a furan group and a dienophile such as a maleimide group to form a substituted cyclohexene known as DA adducts [58] (Fig. 4). The formation of reversible adducts is responsible for imparting self-healability [14, 59, 60]. In *retro* DA reactions, the diene and dienophile moieties disconnect. Subsequently, at lower temperatures, the covalent bonds (DA bonds) reform, and the crack would be completely repaired [61].

Furan compounds involved in these reactions are of bio-based materials, which subsequently improve the sustainability of the final product [59]. The synthesis of partially bio-based healable composites based on DA reactions has been reported in the literature [57, 62–66].

Another kind of covalent bonding used for self-healable biocomposites is Schiff-base interaction. A Schiff-base dynamic covalent bond is formed as a result of a reaction between an aldehyde or ketone with primary amines, in which the C=O group is replaced by the C=N-R group, where R may be any alkyl or any aryl group. Schiff-bases which contain aryl substituents are substantially more stable

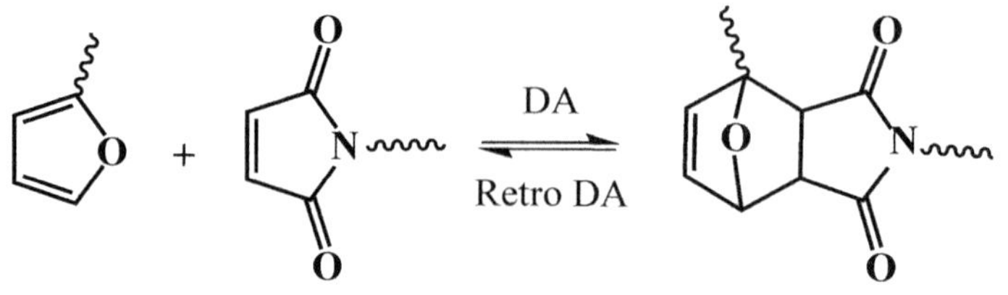

Fig. 4 A representative scheme of reversible DA reaction

and synthesized more readily, while those which contain alkyl substituents are relatively unstable. Schiff-bases of aliphatic aldehydes are relatively unstable, while those of aromatic aldehydes contain effective conjugations and are more stable [67, 68].

3.1 DA Based Self-healing Nanocomposites

Providing DA and retro DA reaction conditions for a nanocomposite matrix endow self-heal ability to the fabricated nanocomposites. These nanocomposites would have potential applications in electronics, as the conductivity of the broken electric circuit can be completely restored [64].

The first example of partially bio-based composites containing nanomaterial, based on DA chemistry was reported by Wu et al. [64], where a composite was prepared from DA polyurethane covalently connected to the functionalized graphene nano-sheets. This flexible composite displayed excellent mechanical properties in addition to its infrared (IR) laser self-healing properties. The reduced graphene oxide could absorb the IR light and convert it into heat promptly, resulting in a local increase in the temperature. The results showed that mechanical properties could be retained after healing via 1 min IR laser irradiation at 980 nm; in terms of Young's modulus, break strength, and break elongation, the healing efficiencies were 100, 96, and 97%, respectively. The healing was also visually confirmed via SEM analysis [64].

The synthesis of a healable composite, made of amino-functionalized multi-wall carbon nanotubes (MWCNTs)/epoxy, was reported based on the DA network. Crosslinking the matrix included two steps; first, the epoxy resin was reacted with furfurylamine or the amino-functionalized MWCNTs; subsequently, the substituent furan groups were reacted with bismaleimide groups through DA reaction, resulting in the formation of the DA network [69]. In this composite, the photothermal conversion of CNTs triggers both DA and retro DA reactions [70]. Both heating and IR irradiation methods can be successful in terms of healing. Nonetheless, heat-triggered self-healing procedure results in an unwanted dissociation of other parts of the specimen, which consequently results in wasting the energy and deformation of the sample [71]. For the heating method, cracks were repaired after 1 min at 120 °C and completely healed after 5 h (which is rather a long time). In irradiation method, the cracks were healed within 30 s of irradiation. It should be noted that the repairing time and efficiency (determined based on the ratio of tensile strengths of the healed and pristine sample) is adjustable by an increase of NH_2-MWCNTs content and also a reduction in the repairing time [69].

In the mobile phone industry where the devices are subjected to service for prolonged periods, healability is of great importance, especially for screens. Such screens are intended to be transparent. In this case, DA polymers, capable of self-healing, are suitable candidates to give repairable conductive screens [72].

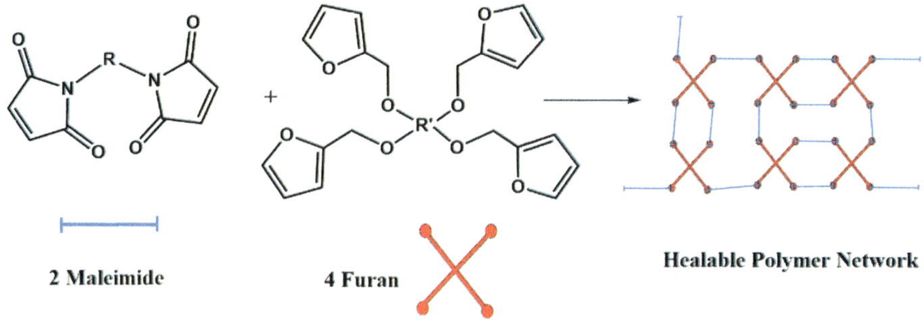

Fig. 5 Monomers and healable network used for conductive screens [65, 72]

Two studies have been reported by Pei et al. [65, 72], concerning the healable conductive screens. In these studies, the simultaneous utilization of silver nanowires (Ag NWs) network and a DA-based polymer resulted in a transparent conductor with desired mechanical properties [65]. The resulted capacitive touch screen sensors have the ability to mend their functions by being exposed to 80 °C for 30 s, which consequently provides the conditions for the efficient retro DA reaction, and thus, further healing of the surface cracks is possible [72]. Figure 5 represents the schematics of the DA healable network.

Cracks healed rapidly and efficiently due to the reformation of DA polymer matrix. In fact, reformation causes the silver nanowires to re-assemble and form an integrated network. The conductivity of the healed samples was measured and it was 97% of the pristine ones [65]. According to the SEM micrographs, the healing occurred through two pathways: (a) the DA reaction of the DA polymer network, i.e. the substrate, and (b) the reformation of the AgNW network. Healed samples possessed 86% of the mechanical strength of the pristine ones. Results also showed that the healing process can be repeatedly accomplished in the same location, without a significant decrease in the surface conductivity [65].

3.2 DA Based Self-healing Biocomposites Containing Fibers

DA reaction may be employed in fiber-reinforced composites for several reasons: (a) it can improve the adhesion of the reinforcement to the polymer matrix, (b) it facilitates the load transfer from the polymer matrix to the reinforcement material, (c) it endows healability to the composite, and (d) it extends the fatigue lifetime of the specimen [57].

Chemical sizing of reinforcements can improve the interfacial adhesion and durability of a composite [73]. One of the chemical sizing methods of fibers (in order to prepare a healable composite) can be maleimidation of fibers [57, 59].

For instance, furan-functionalized epoxy-amine thermosetting matrixes and maleimide-functionalized glass fibers were employed to fabricate a reversible DA composite by Peterson et al. [57]. The authors reported that the DA reaction occurring at room temperature resulted in DA adducts formation, which cleaved from the furan and maleimide moieties at temperatures higher than 90 °C.

Based on the results of the single-fiber micro-droplet pull-out testing, which is a common test method for investigating the mechanical properties of a single fiber [74], healing in the specimens at the interfaces was calculated to be $\sim 41\%$. Herein, the system was capable of being repaired, for up to five healing cycles [57].

Similar fiber sizing method was employed for carbon fibers. After sizing carbon fibers by maleimide, the DA bonds formed at the interphase between carbon fiber surface and the epoxy matrix. These DA bonds endow an interfacial self-healing to the carbon/epoxy composite. The furan groups were dispersed in the matrix by blending the epoxy resin with furfuryl glycidyl ether as an active bio-based epoxy monomer. Both epoxies were then reacted with isophorone diamine to obtain an epoxy-amine matrix. During the healing process, the unreacted or cleaved furan groups from retro DA reaction can participate in the DA reaction with the maleimide groups to reform the fiber-matrix interphase [59].

3.3 DA Based Self-healing Biocomposites Containing Encapsulated Maleimides

Encapsulation of diene and dienophiles to prepare self-healable biocomposites has been rarely considered in the literature, probably since the repeatability is low [12]. One of the few studies in this field is presented by Prartama et al. [66]. In brief, they synthesized a self-healable epoxy-amine thermoset, via encapsulation of multi-maleimides. After the rupture of the microcapsules, the multimaleimides were capable of reacting with the available furans in the matrix. According to the results of the mechanical tests, the healing extent was reported to be 71% [66].

3.4 Self-healing of Schiff-Base Biocomposites

The formation of a Schiff-base from an aldehyde or ketones is a reversible reaction, and generally takes place under acidic or basic catalysis, or upon heating [67, 68]. Schiff-base chemistry as a reversible covalent bond can be utilized to prepare self-healing biocomposites.

For example, Zhang et al. [75] prepared a magnetic self-healing hydrogel by mixing ferrofluid (chitosan-modified Fe_3O_4 nanoparticles) with telechelic difunctional poly(ethylene glycol). The hydrogel showed substantial healability potential, and in fact, healed itself automatically with no external stimulus, a sign of excellent self-healing capability. Authors showed that by a magnet, the magnetic self-healing hydrogel can be remotely operated to pass through a narrow channel with an obstacle in the middle. They also showed that the hydrogel changed its shape and squeezed into the narrow channel, engulfed the glass obstacle in the middle of the channel, and finally, passed through the channel completely after ∼30 min. Although the hydrogel changed its shape during this process, it still maintained its integrity. This experiment demonstrated the excellent cooperation between magnetic and self-healing features of the hydrogel, suggesting the potential drug delivery application of this novel material [75].

In another work, Zeng et al. [76] developed a novel in-situ forming organic/inorganic composite hydrogel with dynamic aldimine crosslinks (Schiff-base interactions), based on elastin-like polypeptides (ELP) and bioglass (BG). For this purpose, they first synthesized ELP containing either primary amines or carboxylic acid functional groups, and chemically modified carboxylic acids to create ELP with aldehyde functional groups. Then, the organic and inorganic components were prepared to create ELP/BG composite hydrogels. The self-healing ability of ELP/BG hydrogels was investigated and the results showed that the dynamic nature of Schiff-base reaction had endowed self-healing properties to the ELP/BG hydrogels. The authors showed that imine bonds, or more specifically, aldimine bonds are reversible, and this property makes them suitable to design self-healing hydrogel system due to constant bond association and dissociation [76–78].

4 Self-healing of Microcapsule-Based Biocomposites

One of the most common methods to endow self-healing properties to different matrixes with no intrinsic healing property is nano-/micro-capsulation. In this case, functional microcapsules (as healing agents inside the coating matrixes) are embedded in the matrix as a preventative method. For achieving self-healing through this route, microcapsules contain reactants for the polymerization of a material, usually similar to the matrix. Catalysts or monomers to carry out polymerization are also dispersed in the matrix (Fig. 6a). As soon as the matrix is damaged, the microcapsules would be ruptured, and due to the release of healing agents (Fig. 6b), the defect can be repaired locally (Fig. 6c) [79–82]. To prepare microcapsulated polymer composites, methods including interfacial, in-situ, and mini-emulsion polymerizations, solvent evaporation, and sol-gel reactions have been reported [83].

Linseed, Tung, neem, and coconut oil have been reported in different papers to be used for microcapsulation leading to the self-healing of the composites [84–87]. Namely, linseed oil was encapsulated in phenol-formaldehyde dispersed in the epoxy matrix through in-situ polymerization. Due to high unsaturated ester content of linseed oil, it tends to polymerize when exposed to oxygen, after being released from the ruptured capsules [88]. An important contributing factor in improving the self-healing of microcapsule embedded composites is the appropriate adhesion

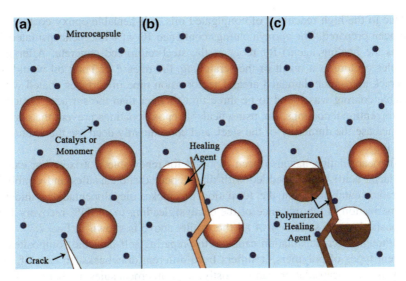

Fig. 6 Self-healing in microcapsulated materials; **a** a crack forms in the matrix, **b** microcapsules get ruptured and the healing agent is released, and **c** the polymerization is carried out to fill the crack [99]

between the microcapsules and the matrix. When the morphology of the capsules is rough, they adhere to the matrix easier through an anchoring mechanism and also break faster upon the failure of the matrix [89].

Another example for the effect of adhesion was through utilizing interfacial polymerization to prepare poly(amidoamine) embedded with linseed oil-filled polyuria microcapsules, where amino groups present on the shell of the capsules enhance the adhesion caused by the polarity [90]. Similar to linseed oil, Tung oil has been used in paints and coatings. It is primarily consisted of a glyceride of eleostearic acid, a conjugated triene, making Tung oil highly unsaturated. It can polymerize through oxidation. For instance, it was encapsulated in poly(urea–formaldehyde) in an in-situ method, and dispersed in epoxy and promoted the self-healing [31].

In another study, soy protein was reported to be encapsulated in PLGA using emulsification solvent evaporation method. For this composite, soy protein released upon the rupture of microcapsules and worked as the cross-linker for the glycolic acid present in the matrix leading to a 48% efficiency of healing [25].

Not only is the microcapsulation method used for imparting self-healing to polymer matrixes, but also as the rust formation and induction of internal stresses (which leads to the development of the cracks) are severe concerns for steel-reinforced concretes [91], embedding epoxy-coated rebar, as a physical barrier, inside the concrete matrix to prevent or delay the corrosion is a good protective method. However, the cracks may form during the transportation or handling of epoxy thermosets coatings, causing negative effects on the protective properties [92].

Due to the high reactivity of conjugated oils, self-healable epoxy coatings have also been prepared. The epoxy coatings contained 10 wt% microencapsulated Tung oil (as the healing agent) for rebar in the steel-reinforced concrete. After microcapsules rupture as a result of the scratching, the release of Tung oil would repair the crack across the damaged area. The corrosion time of surfaces coated with the healable coating was found to be three times longer than the surfaces coated with the conventional coatings. So, using healable epoxy coatings prevent the corrosion, and increase the durability of the steel, and consequently decrease the maintenance costs [93].

Microcapsulation has also been reported for renewable matrixes. For example, poly(lactic acid) was embedded with dicyclopentadiene capsules and Grubbs' 1st generation catalyst. Upon the formation of cracks, dicyclopentadiene underwent polymerization in the presence of the catalyst, leading to an 84% recovery for the composite [94]. A summary of all the results above is shown in Fig. 7.

Although microcapsulation can lead to preparing self-healable composites from different types of polymers, it suffers from different drawbacks. In case of special catalysts, microcapsulation can be costly, large microcapsules are hard to disperse in certain matrixes, small microcapsules may not rupture upon the formation of cracks in the matrix leading to the embedment of capsules to be fruitless, and also, the healing agents deplete after healing in a region, and therefore self-healing can only occur once [95, 96].

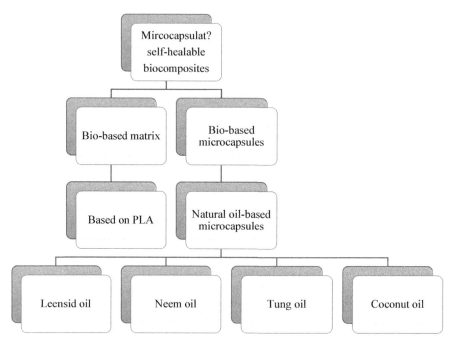

Fig. 7 Microcapsulated self-healable composites based on bio-based matrix or microcapsules [84–87, 94]

5 Self-healing of Biocomposites on the Basis of Melting-Recrystallization Cycles

In addition to physical and chemical bonding, physical transitions can result in self-healing in biocomposites. For example, in order to prepare a multifunctional composite, PDAP particles as the filler (5 wt%) and poly(ε-caprolactone) (PCL) as matrix were used. PCL was grafted on the PDAPs surface to ensure the fillers have strong interfaces with the matrix [97]. PDAP particles are bio-based particles obtained from animal or plant-based dopamine [98].

As heating leads to deformation in the shape of the composite, the light-induced healing of PDAP/PCL composites was used via the melting-recrystallization mechanism. This mechanism constitutes two steps: (a) melting of crystals, which happens due the presence of the hot-spots with temperatures up to melting temperature of the matrix, and subsequent diffusion of polymer chains through the cracks, and (b) recrystallization of the fused chains as a result of cooling to crystallization temperature [40]. Comparing the tensile properties of the pristine and the healed composites showed that the mechanical properties of healed composites increased slightly in terms of tensile strength and modulus, while elongation at break was decreased, suggesting that the healed composite is comparable with the pristine composite in terms of tensile properties [97].

6 Summary and Outlook

It is well known that most polymeric composites consist of petrochemical-based materials. Although the properties of synthetic materials can be more tunable, considering the declining fossil sources, global warming, and other environmental concerns, it is dire to outline a future where composites containing petrochemical- and non-sustainable-based materials consist the majority of goods. Also, inspired by nature and the mechanisms mother Erath utilizes to repair bio-based materials which can be re-used after failure, non-disposable goods are preferred at households, medical field, and industry. Accordingly, researchers have been attempting to prepare composites comprised of sustainable resources which are also self-healable. These materials can be self-healable by intrinsic or extrinsic mechanisms. The intrinsic self-healable composites can heal based on their chemistry; DA and retro DA reactions, hydrogen bonds, supramolecular polymers and non-covalent bonds can endow composites intrinsic self-healing. As for extrinsic self-healing, polymerization needs to be carried out inside the composite upon the failure; the materials needed for the polymerization are embedded in the composite using different methods, for example, microcapsulation. Upon the formation of cracks, the parcels dispersed in the matrix rupture and the materials filled in them react with materials contained in the matrix to polymerize and heal the cracks.

For future studies, researchers can look forward to improving the self-healing efficiency of composites produced, preparing composites based on intrinsic self-healing pathway consisted from other renewable polymers, increasing the healing cycles for extrinsic-based composites, and for all composites, decreasing the costs.

References

1. Yuan YC, Yin T, Rong MZ, Zhang MQ (2008) Self healing in polymers and polymer composites. Concepts, realization and outlook: a review. Express Polym Lett 2:238–250
2. Ghosh SK (2009) Self-healing materials: fundamentals, design strategies, and applications. In: Ghosh SK (ed) Self-healing materials, 1st edn. Wiley-VCH, Weinheim, pp 1–28
3. Rezakazemi M, Dashti A, Asghari M, Shirazian S (2017) H2-selective mixed matrix membranes modeling using ANFIS, PSO-ANFIS, GA-ANFIS. Int J Hydrogen Energy 42:15211–15225. https://doi.org/10.1016/j.ijhydene.2017.04.044
4. Rezakazemi M, Shahidi K, Mohammadi T (2012) Sorption properties of hydrogen-selective PDMS/zeolite 4A mixed matrix membrane. Int J Hydrogen Energy 37:17275–17284. https://doi.org/10.1016/j.ijhydene.2012.08.109
5. Rezakazemi M, Shahidi K, Mohammadi T (2015) Synthetic PDMS composite membranes for pervaporation dehydration of ethanol. Desalin Water Treat 54:1542–1549. https://doi.org/10.1080/19443994.2014.887036
6. Rezakazemi M, Vatani A, Mohammadi T (2016) Synthesis and gas transport properties of crosslinked poly(dimethylsiloxane) nanocomposite membranes using octatrimethylsiloxy POSS nanoparticles. J Nat Gas Sci Eng 30:10–18. https://doi.org/10.1016/j.jngse.2016.01.033
7. Rezakazemi M, Vatani A, Mohammadi T (2015) Synergistic interactions between POSS and fumed silica and their effect on the properties of crosslinked PDMS nanocomposite membranes. RSC Adv 5:82460–82470. https://doi.org/10.1039/C5RA13609A
8. Rezakazemi M, Shahidi K, Mohammadi T (2012) Hydrogen separation and purification using crosslinkable PDMS/zeolite A nanoparticles mixed matrix membranes. Int J Hydrogen Energy 37:14576–14589. https://doi.org/10.1016/j.ijhydene.2012.06.104
9. Rezakazemi M, Sadrzadeh M, Matsuura T (2018) Thermally stable polymers for advanced high-performance gas separation membranes. Prog Energy Combust Sci 66:1–41. https://doi.org/10.1016/j.pecs.2017.11.002
10. Rezakazemi M, Sadrzadeh M, Mohammadi T, Matsuura T (2017) Methods for the preparation of organic-inorganic nanocomposite polymer electrolyte membranes for fuel cells
11. Thakur VK, Kessler MR (2015) Self-healing polymer nanocomposite materials: a review. Polym (United Kingdom) 69:369–383. https://doi.org/10.1016/j.polymer.2015.04.086
12. Su CC, Chen JS (2017) Self-healing polymeric materials. Key Eng Mater 727:482–489. https://doi.org/10.4028/www.scientific.net/KEM.727.482
13. Burattini S, Greenland BW, Chappell D et al (2010) Healable polymeric materials: a tutorial review. Chem Soc Rev 39:1973–1985. https://doi.org/10.1039/b904502n
14. Kloxin CJ (2013) Reversible covalent bond formation as a strategy for healable polymer networks. Heal Polym Syst 62–91. https://doi.org/10.1039/9781849737470-00062
15. Hart LR, Harries JL, Greenland BW et al (2013) Healable supramolecular polymers. Polym Chem 4:4860. https://doi.org/10.1039/c3py00081h
16. Herbst F, Döhler D, Michael P, Binder WH (2013) Self-healing polymers via supramolecular forces. Macromol Rapid Commun 34:203–220. https://doi.org/10.1002/marc.201200675
17. Mülhaupt R (2013) Green polymer chemistry and bio-based plastics: dreams and reality. Macromol Chem Phys 214:159–174

18. Yang Z, Hollar J, He X, Shi X (2011) A self-healing cementitious composite using oil core/silica gel shell microcapsules. Cem Concr Compos 33:506–512. https://doi.org/10.1016/j.cemconcomp.2011.01.010
19. Fornasiero P, Graziani M (2007) Renewable resources and renewable energy: a global challenge, 2nd edn. CRC Press, Boca Raton
20. Iqbal HMN, Kyazze G, Tron T, Keshavarz T (2014) "One-pot" synthesis and characterisation of novel P(3HB)–ethyl cellulose based graft composites through lipase catalysed esterification. Polym Chem 5:7004–7012. https://doi.org/10.1039/c4py00857j
21. Srubar WV, Pilla S, Wright ZC et al (2012) Mechanisms and impact of fiber-matrix compatibilization techniques on the material characterization of PHBV/oak wood flour engineered biobased composites. Compos Sci Technol 72:708–715. https://doi.org/10.1016/j.compscitech.2012.01.021
22. Yang B, Zhang Y, Zhang X et al (2012) Facilely prepared inexpensive and biocompatible self-healing hydrogel: a new injectable cell therapy carrier. Polym Chem 3:3235–3238
23. Ding F, Shi X, Wu S et al (2017) Flexible polysaccharide hydrogel with pH-regulated recovery of self-healing and mechanical properties. Macromol Mater Eng 302:1–9. https://doi.org/10.1002/mame.201700221
24. Desai KGH, Schwendeman SP (2013) Active self-healing encapsulation of vaccine antigens in PLGA microspheres. J Control Release 165:62–74. https://doi.org/10.1016/j.jconrel.2012.10.012
25. Kim JR, Netravali AN (2016) Self-healing properties of protein resin with soy protein isolate-loaded poly(d, l-lactide-co-glycolide) microcapsules. Adv Funct Mater 26:4786–4796. https://doi.org/10.1002/adfm.201600465
26. Yabuki A, Sakai M (2011) Self-healing coatings of inorganic particles using a pH-sensitive organic agent. Corros Sci 53:829–833. https://doi.org/10.1016/j.corsci.2010.11.021
27. Palin D, Wiktor V, Jonkers HM (2016) A bacteria-based bead for possible self-healing marine concrete applications. Smart Mater Struct 25:1–6. https://doi.org/10.1088/0964-1726/25/8/084008
28. Merindol R, Diabang S, Felix O et al (2015) Bio-inspired multiproperty materials: strong, self-healing, and transparent artificial wood nanostructures. ACS Nano 9:1127–1136. https://doi.org/10.1021/nn504334u
29. Abilash N, Sivapragash M (2011) Assesment of self healing property in hybrid fiber polymeric composite. Int J Eng Sci 3:5430–5436
30. Hatami Boura S, Peikari M, Ashrafi A, Samadzadeh M (2012) Self-healing ability and adhesion strength of capsule embedded coatings—micro and nano sized capsules containing linseed oil. Prog Org Coatings 75:292–300. https://doi.org/10.1016/j.porgcoat.2012.08.006
31. Samadzadeh M, Boura SH, Peikari M et al (2011) Tung oil: AN autonomous repairing agent for self-healing epoxy coatings. Prog Org Coatings 70:383–387. https://doi.org/10.1016/j.porgcoat.2010.08.017
32. Chaudhari AB, Tatiya PD, Hedaoo RK et al (2013) Polyurethane prepared from neem oil polyesteramides for self-healing anticorrosive coatings. Ind Eng Chem Res 52:10189–10197. https://doi.org/10.1021/ie401237s
33. Wu DY, Meure S, Solomon D (2008) Self-healing polymeric materials: a review of recent developments. Prog Polym Sci 33:479–522. https://doi.org/10.1016/j.progpolymsci.2008.02.001
34. Aida T, Meijer EW, Stupp SI (2012) Functional supramolecular polymers. Science (80-.) 335:813–817
35. Schmuck C, Wienand W (2001) Self-complementary quadruple hydrogen-bonding motifs as a functional principle: from dimeric supramolecules to supramolecular Polymers. Angew Chemie Int Ed 40:4363–4369. https://doi.org/10.1002/1521-3773(20011203)40
36. Han D, Yan L (2014) Supramolecular hydrogel of chitosan in the presence of graphene oxide nanosheets as 2D cross-linkers. ACS Sustain Chem Eng 2:296–300. https://doi.org/10.1021/sc400352a

37. Lin L-J, Larsson M, Liu D-M (2011) A novel dual-structure, self-healable, polysaccharide based hybrid nanogel for biomedical uses. Soft Matter 7:5816. https://doi.org/10.1039/c1sm05249g
38. Wang L, Zhang X, Xiong H, Wang S (2010) A novel nitromethane biosensor based on biocompatible conductive redox graphene-chitosan/hemoglobin/graphene/room temperature ionic liquid matrix. Biosens Bioelectron 26:991–995. https://doi.org/10.1016/j.bios.2010.08.027
39. Beijer FH, Sijbesma RP, Kooijman H et al (1998) Strong dimerization of ureidopyrimidones via quadruple hydrogen bonding. J Am Chem Soc 120:6761–6769. https://doi.org/10.1021/ja974112a
40. Söntjens SHM, Sijbesma RP, Van Genderen MHP, Meijer EW (2000) Stability and lifetime of quadruply hydrogen bonded 2-Ureido-4[1H]-pyrimidinone dimers. J Am Chem Soc 122:7487–7493. https://doi.org/10.1021/ja000435m
41. Bosman AW, Sijbesma RP, Meijer EW (2004) Supramolecular polymers at work. Mater Today 7:34–39. https://doi.org/10.1016/S1369-7021(04)00187-7
42. Phadke A, Zhang C, Arman B et al (2012) Rapid self-healing hydrogels. Proc Natl Acad Sci 109:4383–4388. https://doi.org/10.1073/pnas.1201122109
43. Feldman KE, Kade MJ, De Greef TFA et al (2008) Polymers with multiple hydrogen-bonded end groups and their blends. Macromolecules 41:4694–4700. https://doi.org/10.1021/ma800375r
44. Cui J, del Campo A (2012) Multivalent H-bonds for self-healing hydrogels. Chem Commun 48:9302. https://doi.org/10.1039/c2cc34701f
45. Biyani MV, Foster EJ, Weder C (2013) Light-healable supramolecular nanocomposites based on modified cellulose nanocrystals. ACS Macro Lett 2:236–240
46. Schubert US, Eschbaumer C, Hien O, Andres PR (2001) 4′-Functionalized 2,2′:6′,2″-terpyridines as building blocks for supramolecular chemistry and nanoscience. Tetrahedron Lett 42:4705–4707. https://doi.org/10.1016/S0040-4039(01)00796-1
47. Kersey FR, Loveless DM, Craig SL (2007) A hybrid polymer gel with controlled rates of cross-link rupture and self-repair. J R Soc Interface 4:373–380. https://doi.org/10.1098/rsif.2006.0187
48. Gohy JF, Lohmeijer BGG, Schubert US (2002) Reversible metallo-supramolecular block copolymer micelles containing a soft core. Macromol Rapid Commun 23:555–560. https://doi.org/10.1002/1521-3927(20020601)23:9<555::aid-marc555>3.0.co;2-k
49. Coulibaly S, Roulin A, Balog S et al (2014) Reinforcement of optically healable supramolecular polymers with cellulose nanocrystals. Macromolecules 47:152–160. https://doi.org/10.1021/ma402143c
50. Burattini S, Colquhoun HM, Fox JD et al (2009) A self-repairing, supramolecular polymer system: healability as a consequence of donor–acceptor π–π stacking interactions. Chem Commun 6717. https://doi.org/10.1039/b910648k
51. Burattini S, Greenland BW, Merino DH et al (2010) A healable supramolecular polymer blend based on aromatic π-π stacking and hydrogen-bonding interactions. J Am Chem Soc 132:12051–12058. https://doi.org/10.1021/ja104446r
52. Fox J, Wie JJ, Greenland BW et al (2012) High-strength, healable, supramolecular polymer nanocomposites. J Am Chem Soc 134:5362–5368. https://doi.org/10.1021/ja300050x
53. Xu C, Cao L, Lin B et al (2016) Design of self-healing supramolecular rubbers by introducing ionic cross-links into natural rubber via a controlled vulcanization. ACS Appl Mater Interfaces 8:17728–17737. https://doi.org/10.1021/acsami.6b05941
54. Sordo F, Mougnier SJ, Loureiro N et al (2015) Design of self-healing supramolecular rubbers with a tunable number of chemical cross-links. Macromolecules 48:4394–4402. https://doi.org/10.1021/acs.macromol.5b00747
55. Cragg Peter J (2010) Supramolecular chemistry, from biological inspiration to biomedical applications, 1st edn. Springer, New York

56. McKee JR, Appel EA, Seitsonen J et al (2014) Healable, stable and stiff hydrogels: combining conflicting properties using dynamic and selective three-component recognition with reinforcing cellulose nanorods. Adv Funct Mater 24:2706–2713. https://doi.org/10.1002/adfm.201303699
57. Peterson AM, Jensen RE, Palmese GR (2011) Thermoreversible and remendable glass-polymer interface for fiber-reinforced composites. Compos Sci Technol 71:586–592. https://doi.org/10.1016/j.compscitech.2010.11.022
58. Surajmal M (2013) Introduction to diels alder reaction, its mechanism and recent advantages: a review. Indo Am J Pharm Res 3:3192–3215
59. Zhang W, Duchet J, Gérard JF (2014) Self-healable interfaces based on thermo-reversible diels-alder reactions in carbon fiber reinforced composites. J Colloid Interface Sci 430:61–68. https://doi.org/10.1016/j.jcis.2014.05.007
60. Barthel MJ, Rudolph T, Crotty S et al (2012) Homo- and diblock copolymers of poly(furfuryl glycidyl ether) by living anionic polymerization: toward reversibly core-crosslinked micelles. J Polym Sci, Part A: Polym Chem 50:4958–4965. https://doi.org/10.1002/pola.26327
61. Gandini A, Hodge P (1998) Application of the diels—alder reaction to polymers bearing furan moieties. 2. Diels—alder and retro-diels—alder reactions involving furan rings in some styrene copolymers. Macromolecules 1:314–321
62. Wu S, Li J, Zhang G et al (2016) High mechanical strength and high dielectric graphene/polyuthane composites healded by near infrared laser. In: 2016 17th international conference on electronic packaging technology ICEPT 2016, pp 157–161. https://doi.org/10.1109/icept.2016.7583110
63. Li Q-T, Jiang M-J, Wu G, Chen L, Chen S-C, Cao Y-X, Wang Y-Z (2017) Photothermal conversion triggered precisely targeted healing of epoxy resin based on thermo-reversible dielsalder network and amino-functionalized carbon nanotubes. ACS Appl Mater Interfaces 9:20797–20807
64. Wu S, Li J, Zhang G et al (2017) Ultrafast self-healing nanocomposites via infrared laser and their application in flexible electronics. ACS Appl Mater Interfaces 9:3040–3049. https://doi.org/10.1021/acsami.6b15476
65. Gong C, Liang J, Hu W et al (2013) A healable, semitransparent silver nanowire-polymer composite conductor. Adv Mater 25:4186–4191. https://doi.org/10.1002/adma.201301069
66. Pratama PA, Sharifi M, Peterson AM, Palmese GR (2013) Room temperature self-healing thermoset based on the diels-alder reaction. ACS Appl Mater Interfaces 5:12425–12431. https://doi.org/10.1021/am403459e
67. Xavier A, Srividhya N (2014) Synthesis and study of Schiff base ligands 7:6–15
68. Hussain Z, Khalaf M, Adil H et al. Metal complexes of Schiff's bases containing sulfonamides nucleus. Res J Pharm Biol Chem Sci 7:1008–1025
69. Li QT, Jiang MJ, Wu G et al (2017) Photothermal conversion triggered precisely targeted healing of epoxy resin based on thermoreversible diels-alder network and amino-functionalized carbon nanotubes. ACS Appl Mater Interfaces 9:20797–20807. https://doi.org/10.1021/acsami.7b01954
70. Yang Y, Pei Z, Zhang X et al (2014) Carbon nanotube–vitrimer composite for facile and efficient photo-welding of epoxy. Chem Sci 5:3486–3492. https://doi.org/10.1039/C6SC90083F
71. Wu M, Li Y, An N, Sun J (2016) Applied voltage and near-infrared light enable healing of superhydrophobicity loss caused by severe scratches in conductive superhydrophobic films. Adv Funct Mater 26:6777–6784. https://doi.org/10.1002/adfm.201601979
72. Li J, Liang J, Li L et al (2014) Healable capacitive touch screen sensors based on transparent composite electrodes comprising silver nanowires and a Furan/Maleimide diels-alder cycloaddition polymer. ACS Nano 8:12874–12882
73. Wu HF, Dwight DW, Huff NT (1997) Effects of silane coupling agents on the interphase and performance of glass-fiber-reinforced polymer composites. Compos Sci Technol 57:975–983. https://doi.org/10.1016/S0266-3538(97)00033-X

74. Hodzic A, Kalyanasundaram S, Kim JK et al (2001) Application of nano-indentation, nano-scratch and single fibre tests in investigation of interphases in composite materials. Micron 32:765–775. https://doi.org/10.1016/S0968-4328(00)00084-6
75. Zhang Y, Yang B, Zhang X et al (2012) A magnetic self-healing hydrogel. Chem Commun 48:9305. https://doi.org/10.1039/c2cc34745h
76. Zeng Q, Desai MS, Jin HE et al (2016) Self-healing elastin-bioglass hydrogels. Biomacromol 17:2619–2625. https://doi.org/10.1021/acs.biomac.6b00621
77. Wei Z, Yang JH, Liu ZQ et al (2015) Novel biocompatible polysaccharide-based self-healing hydrogel. Adv Funct Mater 25:1352–1359. https://doi.org/10.1002/adfm.201401502
78. Haldar U, Bauri K, Li R et al (2015) Polyisobutylene-based pH-responsive self-healing polymeric gels. ACS Appl Mater Interfaces 7:8779–8788. https://doi.org/10.1021/acsami.5b01272
79. Yuan YC, Rong MZ, Zhang MQ et al (2008) Self-healing polymeric materials using epoxy/Mercaptan as the healant. Macromolecules 41:5197–5202. https://doi.org/10.1021/ma800028d
80. Porter R, MIale JB (1984) Extended control of marine fouling—formulation of a microencapsulated liquid organometallic biocide and vinyl rosin paint. Appl Biochem Biotechnol 9:439–445. https://doi.org/10.1007/bf02798398
81. Yeom CK, Kim YH, Lee JM (2002) Microencapsulation of water-soluble herbicide by interfacial reaction. II. Release properties of microcapsules. J Appl Polym Sci 84:1025–1034. https://doi.org/10.1002/app.10383
82. Wei H, Wang Y, Guo J et al (2015) Advanced micro/nanocapsules for self-healing smart anticorrosion coatings. J Mater Chem A 3:469–480. https://doi.org/10.1039/C4TA04791E
83. Zhu DY, Rong MZ, Zhang MQ (2015) Self-healing polymeric materials based on microencapsulated healing agents: from design to preparation. Prog Polym Sci 49–50:175–220. https://doi.org/10.1016/j.progpolymsci.2015.07.002
84. Wang H, Zhou Q (2018) Evaluation and failure analysis of linseed oil encapsulated self-healing anticorrosive coating. Prog Org Coatings 118:108–115. https://doi.org/10.1016/j.porgcoat.2018.01.024
85. Li H, Cui Y, Wang H et al (2017) Preparation and application of polysulfone microcapsules containing tung oil in self-healing and self-lubricating epoxy coating. Colloids Surfaces A Physicochem Eng Asp 518:181–187
86. Marathe R, Tatiya P, Chaudhari A et al (2015) Neem acetylated polyester polyol—renewable source based smart PU coatings containing quinoline (corrosion inhibitor) encapsulated polyurea microcapsules for enhance anticorrosive property. Ind Crops Prod 77:239–250
87. Ataei S, Khorasani SN, Torkaman R et al (2018) Self-healing performance of an epoxy coating containing microencapsulated alkyd resin based on coconut oil. Prog Org Coatings 120:160–166. https://doi.org/10.1016/j.porgcoat.2018.03.024
88. Jadhav RS, Hundiwale DG, Mahulikar PP (2011) Synthesis and characterization of phenol-formaldehyde microcapsules containing linseed oil and its use in epoxy for self-healing and anticorrosive coating. J Appl Polym Sci 119:2911–2916. https://doi.org/10.1002/app.33010
89. Suryanarayana C, Rao KC, Kumar D (2008) Preparation and characterization of microcapsules containing linseed oil and its use in self-healing coatings. Prog Org Coatings 63:72–78. https://doi.org/10.1016/j.porgcoat.2008.04.008
90. Tatiya PD, Hedaoo RK, Mahulikar PP, Gite VV (2013) Novel polyurea microcapsules using dendritic functional monomer: synthesis, characterization, and its use in self-healing and anticorrosive polyurethane coatings. Ind Eng Chem Res 52:1562–1570. https://doi.org/10.1021/ie301813a
91. Subramanian N (2013) Understanding corrosion and cathodic protection of reinforced concrete structures
92. Lau K, Sagüés AA, Powers RG (2007) Long-term corrosion behavior of epoxy coated rebar in Florida Bridges. Houston, TX

93. Chen Y, Xia C, Shepard Z et al (2017) Self-healing coatings for steel-reinforced concrete. ACS Sustain Chem Eng 5:3955–3962. https://doi.org/10.1021/acssuschemeng.6b03142
94. Wertz JT, Mauldin TC, Boday DJ (2014) Polylactic acid with improved heat deflection temperatures and self-healing properties for durable goods applications. ACS Appl Mater Interfaces 6:18511–18516. https://doi.org/10.1021/am5058713
95. Trask RS, Williams HR, Bond IP (2007) Self-healing polymer composites: mimicking nature to enhance performance. Bioinspiration and Biomimetics 2. https://doi.org/10.1088/1748-3182/2/1/p01
96. Samadzadeh M, Boura SH, Peikari M et al (2010) A review on self-healing coatings based on micro/nanocapsules. Prog Org Coatings 68:159–164. https://doi.org/10.1016/j.porgcoat.2010.01.006
97. Xiong S, Wang Y, Zhu J et al (2016) Poly(ε-caprolactone)-grafted polydopamine particles for biocomposites with near-infrared light triggered self-healing ability. Polymer (Guildf) 84:328–335. https://doi.org/10.1016/j.polymer.2016.01.005
98. Ju KY, Lee Y, Lee S et al (2011) Bioinspired polymerization of dopamine to generate melanin-like nanoparticles having an excellent free-radical-scavenging property. Biomacromol 12:625–632. https://doi.org/10.1021/bm101281b
99. Coope TS, Mayer UFJ, Wass DF et al (2011) Self-healing of an epoxy resin using Scandium (III) Triflate as a catalytic curing agent, pp 4624–4631. https://doi.org/10.1002/adfm.201101660

Chemical Modification of Lignin and Its Environmental Application

Zhili Li, Yuanyuan Ge, Jiubing Zhang, Duo Xiao and Zijun Wu

1 Introduction

Lignin constitutes one of the main component of lignocellulosic biomass (15–30% by weight), which is second only to cellulose in mass on the earth. Lignin acts as the essential glue that gives plants their structural integrity and resistance against microbial, chemical attack and prevents other outside stresses from destroying the structure of plant cell walls [1]. Lignin is a complex and recalcitrant phenolic macromolecule composed of three phenylpropane units: p-hydroxyphenyl (H), guaiacyl (G), and syringyl (S), cross-linked by β-o-4, α-o-4, β-β, and 5-5′ bonds etc. (as depicted in Fig. 1). Among them, the main linkage is the β-o-4 bond, about 40–60% of all inter-unit linkages in lignin belong to this bond [2]. The component and structural characterizations of lignin are quite different even it is separated from the same plant. It is to say, the lignin obtained from a plant is a mixture of different lignin polymers which is dependent on the source plant, species, and growing ages of the plant, such as softwood, hardwood, and grass [3].

A large quantity of technical lignin was produced as a main component in black liquor by the pulp and papering industry, which is a big threat to the environment [4]. It is estimated that the global annual production of technical lignin in pulp making is *ca.* 70 million tons, much of which is consumed as a low-value fuel. Although there are some other applications, such as a binder or dispersing agent, no large-scale application has so far been found [4]. Lignin has significant potential as a source for the production of the bio-renewable polymer [5, 6]. The main active sites of lignin are comprised of phenolic or alcoholic hydroxyl groups, which accounts for its

---

Z. Li (✉) · Y. Ge (✉) · J. Zhang · D. Xiao · Z. Wu
School of Chemistry and Chemical Engineering, Guangxi University,
100 Daxuedong Road, Nanning 530004, China
e-mail: lizhili@gxu.edu.cn

Y. Ge
e-mail: geyy@gxu.edu.cn

© Springer Nature Switzerland AG 2019
Inamuddin et al. (eds.), *Sustainable Polymer Composites and Nanocomposites*,
https://doi.org/10.1007/978-3-030-05399-4_45

Fig. 1 **a** The phenylpropane type units: p-hydroxyphenyl (H), guaiacyl (G), and syringyl (S); and **b** β-o-4, **c** α-o-4, **d** β-β, and **e** 5-5′ linkages in lignin

reactivity, hydrophilicity as well as other chemical and physical attributes of lignin [2]. Recently, due to the unique polyphenol structure, chemical stability and wide availability, different kinds of adsorbents, particularly for dyes and heavy metals removal from wastewater are potentially obtainable from lignin [7].

Water pollution by different kinds of contaminants is currently serious concerned [8–13]. A large amount of organic, inorganic, and biological compounds have been reported as water contaminants. For example, the dyes, particularly azo dyes are recalcitrant molecules and resistant to aerobic digestion and are also stable to oxidizing agents that makes them being hard to treat [14]. The heavy metals, lead (Pb), mercury (Hg), cadmium (Cd), chromium (Cr) and arsenic (As), etc., are notorious with high toxicity and carcinogenicity. These toxic pollutants, even with a trace amount in water, can pollute water resources and transfer throughout the food chain to accumulate in animals and mankind, causing various diseases and disorders [15]. Therefore, the pollutants should be removed before its discharge into the environment.

Several physical and chemical treatment techniques have been reported for the removal of pollutants from wastewater, including precipitation, oxidation, biodegradation, ion exchange, adsorption [16], membrane separation [17–19], coagulation and flocculation, flotation, electrochemical methods. These methods still suffer from the limitations of low effectiveness and high cost.

Adsorption using cost-effective adsorbents is now considered as an efficient, convenient and economical method for wastewater treatment, due to the feasibility in design and operation, effectiveness in treating and the recyclability of the adsorbents [20, 21]. Activated carbon (AC) is being recognized as one of the most popular adsorbents used in the removal of contaminants from water. However, AC is of high capital cost due to the prolong production process [22]. In recent years, the development of various environmentally friendly adsorbents from biomass, biopolymers, especially the natural-occurring polymers, including cellulose, lignin, and chitosan, etc., has been intensified [23–25]. It could be seen from the literature that the adsorption of dyes and heavy metals using lignin is one of the most reported methods for the removal of contaminants from water [26–28].

Unfortunately, original lignin is less competitive in adsorption applications due to the low adsorption capability, and lignin does not own specific selectivity for a certain kind of pollutant from complex polluted water [29, 30]. The goal for the lignin modification for pollutants adsorption is to promote the adsorption capability, including adsorption capacity, stability, selectivity and recyclability [15]. The ease of separation from wastewater after the operation and the cost-effectiveness of the adsorbents should also be considered. The modification of lignin could be achieved by various methods including crosslinking, hybridization, hydrogen bonding formation, condensation, grafting and copolymerization [31]. Therefore, the main objective of this article is to collect and compile the most recent reported literature regarding the modified lignin for adsorption applications and to discuss the advantageous and disadvantageous issues of the resulting lignin as an adsorbent to remove o different environmental pollutants from the aqueous phase.

2 Modified Lignin for Dyes Adsorption

Dyes contaminated wastewater is hard to treat because of their inert properties and the residual trace amount of dyes in water. Recently, adsorption techniques using modified lignin have been widely reported in removing dyes as a promising wastewater treatment process. The dyes adsorption capacities of the modified lignin-based adsorbents are tabulated in Table 1. And the modified lignin-based materials for removing dyes will be fully discussed in the following section.

Table 1 Reported modified lignin for adsorption of dyes in water

Modified lignin	Dye	T (°C)/pH	Q_m (mg/g)	Reference
Chitosan-alkali lignin	RBBR	27/5.9	111.11	[32]
ALiCE	MB	20/7.0	36.25	[33]
Chitin/lignin hybrid	DB71	20–50/2.4–8.4	40.0	[28]
CAML	C-3R	RT/4.0–10	99.3%	[34]
	MO		67.0%	
LBF	RB	–/2.0–10	93%	[34]
MLS	MB	40/–	31.23	[35]
	RB		17.62	
Oxidized lignin	EV	30/7.0	70–80 wt%	[36]
	BB		80–95 wt%	
ALR	CV	25/3.0–12	150.4	[37]
Lignin-g-p(AM-co-NIPAM)/MMT	MB	25/1.0–11.0	9646.9	[38]
LS-g-AA	MB	30/3.0–8.0	2013	[39]
LPUF	MG	25–65/2.0–9.0	80	[40]
AAL	MB	30/5.0	63.3	[41]
Lignin sulfonate polymer	MG	30/7.0	60.2	[42]
LSMMs	MG	24/3.0–7.0	150.3	[43]
The hydrogel of acylated hemicelluloses, acrylic acid and lignosulfonate	MB	30–70/7.0–10	2691	[44]
Hydrogel of kraft lignin-N-isopropyl acrylamide	MB	15–45/1.0–11	–	[45]
CML-Al	PB	25/2.0	73.52	[46]
CML-Mn			55.16	
CML-Fe	BR-2	25/2.0	73.6	[47]

Nair et al. [32] reported a novel chitosan-alkali lignin composite prepared from chitosan and alkali lignin (as shown in Fig. 2). The weak interactions between β-1,4-glycosidic linkage, amide and hydroxyl groups of chitosan, and ether, an aromatic ring and hydroxyl groups of alkali lignin, impart enhanced surface and chemical properties to the composite than chitosan and alkali lignin. The composite with 50:50 chitosan: alkali lignin exhibited maximum adsorption (111.11 mg/g) of Remazol Brilliant Blue R (RBBR) compared to chitosan (76.92 mg/g), which showed that the composite exhibits 33% improvement in the maximum adsorption amount. Batch adsorption of RBBR on the composite followed the Langmuir equation, while the dynamic adsorption followed the pseudo-second-order equation. A lignin-chitosan pellet (ALiCE) was prepared for adsorption of methylene blue (MB) [33]. The results indicated the ALiCE had a maximum adsorption capacity of 36.25 mg/g for MB according to the fitting of the Langmuir equation ($R^2 = 0.997$), yielding. The adsorption kinetic data could be fitted well by the pseudo-second-order-model [33]. A chitin/lignin hybrid biosorbent was prepared

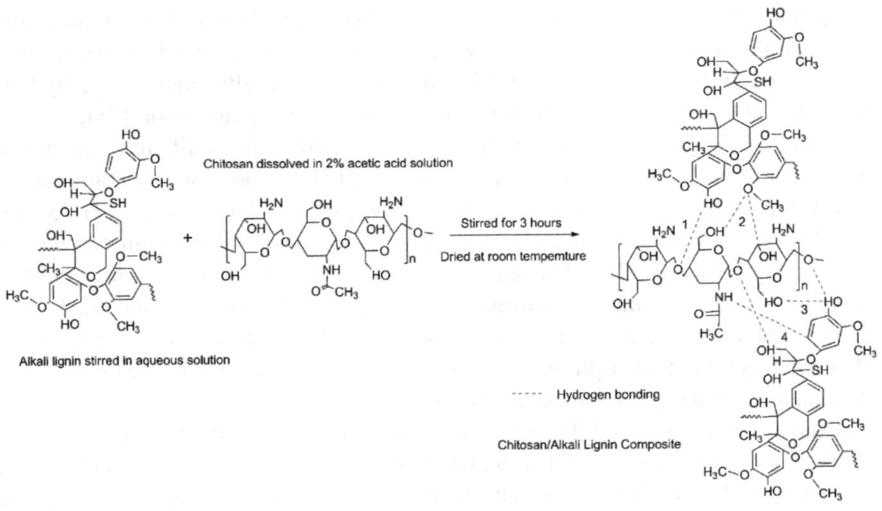

Fig. 2 An illustrative diagram of chitosan-alkali lignin composite. Reprinted with permission [32]. Copyright 2014 Elsevier

and used to adsorb C.I. Direct Blue 71 (DB71) from water [28]. The adsorption capacities of DB71 by the chitin and chitin/lignin hybrid biosorbent were 30.7 and 40 mg/g, respectively, which indicated that the hybrid chitin/lignin material has a stronger adsorptive affinity for DB 71 than the chitin. The adsorption of DB 71 onto the hybrid could be described by the Freundlich equation and the pseudo-second-order equation, respectively. The optimum solution pH for dye removal was in the range 2.4–8.4. The adsorption of DB 71 by the chitin/lignin hybrid was spontaneous and endothermic in nature [28].

Lou et al. [34] developed a ternary graft copolymer based on chitosan, acrylamide, and lignin (CAML) by a microwave-assisted method (as shown in Fig. 3). The copolymer was loosely aggregated powder with a particle size range of 1–3 μm. Adsorption experiment showed that the CAML (chitosan:acrylamide:lignin = 1:1:1) exhibited maximum removal efficiency of 99.3 and 67.0% for reactive orange C-3R and methyl orange, respectively. The CAML had a wide suitable pH range, although the removal efficiency was slightly higher under lower pH values. The removal mechanism was combined charge neutralization and bridging effects [34]. Guo et al. [48] prepared a lignin-based flocculant (LBF) via grafting

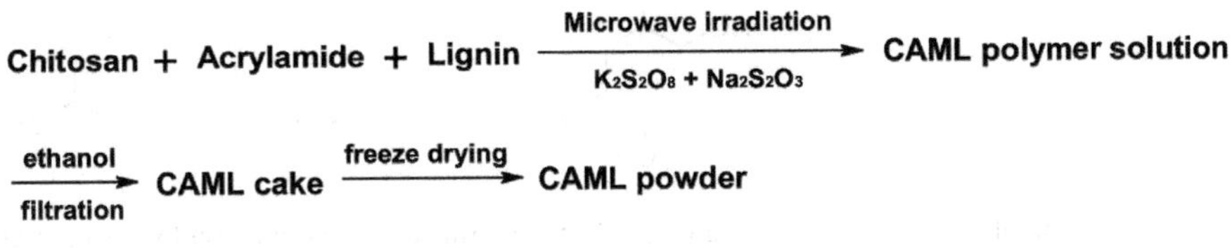

Fig. 3 Synthetic process of CAML. Reprinted with permission [34]. Copyright 2018 Elsevier

dimethyldiallylammonium chloride and acrylamide onto lignin. Adsorption results demonstrated that a high removal efficiency of 93% was achieved for the reactive dye by LBF with the addition of PAC. The high removal efficiency of dye by LBF was because of the electrostatic attraction effect and bridging action [48].

Li et al. [35] prepared magnetic lignin spheres (MLS) from different organosolv lignins with maleic anhydride and Fe_3O_4 nanoparticles. The adsorption amounts of methylene blue and Rhodamine B by the MLS from larch lignin (31.23, and 17.62 mg/g, respectively) were higher than that from poplar lignin (25.95 and 15.79 mg/g, respectively). The adsorption kinetics and isotherm were could be described by the pseudo-second-order equation and Langmuir equation, respectively. Moreover, the MLS from larch and poplar lignin had a good recyclability, after three cycles of adsorption-desorption, the removal efficiencies for the dyes still remained at 98 and 96%, respectively [35].

Couch et al. obtained [36] oxidized lignin products from softwood lignin by using HNO_3 (as shown in Fig. 4). The products were used to remove ethyl violet (EV) and basic blue (BB) dyes from simulated water. The results showed that the dyes removal efficiency was 70–80 wt% for EV and 80–95 wt% for BB within the dyes concentrations of 50 and 400 mg/L. The dye removal was pH and ionic strength dependent. Feng et al. [41] prepared a methylene blue (MB) adsorbent from acetic acid lignin (AAL) via deacetylation in NaOH aqueous solution followed by fractionation in methanol. The maximum adsorption capacity of MB reached to 63.3 mg/g by the AAL. In addition, the adsorption of MB was pH and dosage dependent.

Xu et al. [37] developed an acrylic-lignosulfonate resin (ALR) from calcium lignosulfonate and acrylic acid. The resin had a high surface area of 190.55 m^2/g with a porous structure in an average pore diameter of 11.34 nm. The maximum adsorption capacity of crystal violet (CV) by the ALR was 150.40 ± 4.80 mg/g at 25 °C. The kinetic and equilibrium data could be fitted by the pseudo-second-order equation and Freundlich equation, respectively. In addition, the calculation of thermodynamic parameters demonstrated the adsorption of CV on ALR was exothermic and spontaneous in nature.

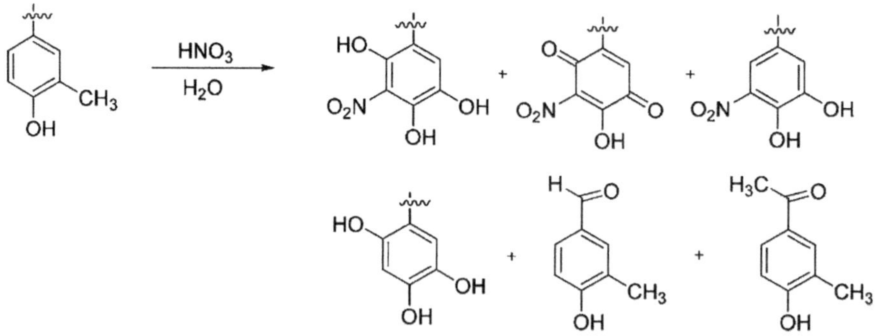

Fig. 4 Scheme of oxidized softwood lignin products. Reprinted with permission [36]. Copyright 2016 American Chemical Society

Wang et al. [38] prepared a hybrid hydrogel from lignin grafted with acrylamide and N-isopropyl acrylamide as well as montmorillonite (lignin-g-p(AM-co-NIPAM)/MMT, as shown in Fig. 5). The prepared hydrogel presented thinner pore walls, good thermal stability and strong mechanical strength due to the existence of montmorillonite. The hydrogel showed an excellent removal efficiency for methylene blue in water, as indicated by the maximum adsorption capacity of 9646.92 mg/g. The adsorption was dependent on the pH and temperature. Dynamic adsorption could be described by the pseudo-second-order equation, while both of the Langmuir and Freundlich equations could describe the equilibrium adsorption. Furthermore, the hydrogel showed an excellent reusability within five adsorption-desorption cycles.

Yu et al. [39] also prepared an LS-g-AA hydrogel by grafting of acrylic acid on lignosulfonate backbone using N,N'-methylene-bis-acrylamide and laccase/t-BHP (tert-butyl hydroperoxide) as cross linker and initiator respectively (Fig. 6). The successful grafting of the monomer was confirmed using FTIR. The prepared hydrogel showed a high adsorption capacity of 2013 mg/g for methylene blue (MB) dye in water. Besides, excellent reusability was shown by the LS-g-AA hydrogel with adsorption capacity of 1757 and 1681 mg/g for 3 and 4 cycles, respectively.

Kumari et al. [40] reported a pine needle lignin-based polyurethane foam (LPUF) as an adsorbent of dyes removal from water. The experiment results showed that the LPUF was effective in removing a cationic dye, malachite green (MG) with a maximum adsorption capacity of 80 mg/g, other than an anionic dye, methyl orange (MO), from the water. The adsorption kinetics and isotherms could

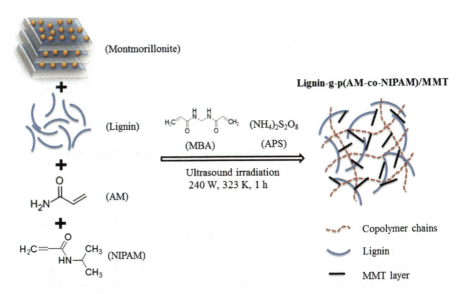

Fig. 5 Diagram of synthesis of lignin-g-p(AM-co-NIPAM)/MMT. Reprinted with permission [38]. Copyright 2017 Elsevier

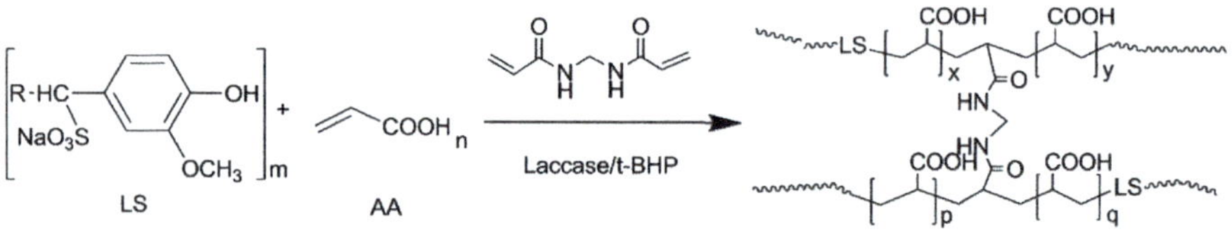

Fig. 6 Synthesis of LS-g-AA hydrogel from lignosulfonate and acrylamide catalyzed by laccase. Reprinted with permission [39]. Copyright 2016 Elsevier

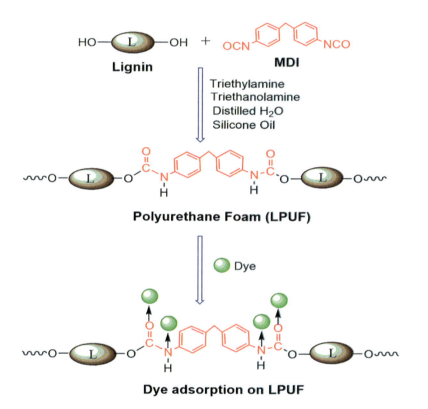

Fig. 7 Schematics of LPUF synthesis and application. Reprinted with permission [40]. Copyright 2016 Royal Society of Chemistry

be described well by the pseudo-second-order equation and Langmuir equation, respectively. Moreover, the LPUF could be used for 20 regeneration cycles with a cumulative adsorption capacity of 1.33 g/g to MG (Fig. 7).

Tang et al. [42] have reported their studies on the preparation of a lignin sulfonate polymer by a simple emulsion polymerization method and the adsorption properties of the lignin sulfonate polymer towards malachite green (MG) dyes (as shown in Fig. 8). The obtained lignin sulfonate polymer showed an effective

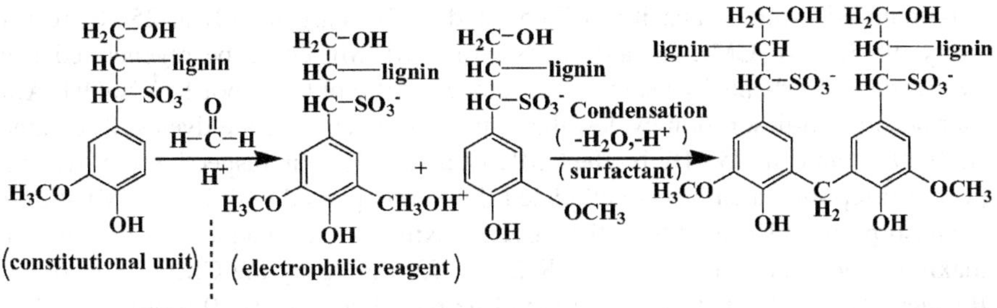

Fig. 8 Schematic diagram of the lignin sulfonate polymer. Reprinted with permission [43]. Copyright 2016 Royal Society of Chemistry

adsorption of MG with a maximum adsorption capacity of 60.2 mg/g for MG according to the Langmuir equation. A Lignosulfonate-based mesoporous material was further prepared by grafting of acrylic acid and acrylamide onto the backbone of lignosulfonate for adsorbing MG from the water. This synthesized material was mesoporous confirmed by using N_2 adsorption/desorption curve. The BET surface area was 118 m^2/g and the mesopores are in an average diameter of 3.8 nm. Due to the presence of pores in the mesoporous material, it showed an enhanced adsorption capacity, 150.376 mg/g for MG.

Song et al. [44] prepared a hydrogel consisted of acylated hemicelluloses, acrylic acid, and sodium lignosulfonate by using initiator ammonium persulfate and N,N,N',N'-tetramethylethane-1,2-diamine. The honeycomb-like morphology was observed in the prepared hydrogel. The adsorption kinetics of methylene blue (MB) by the hydrogel was fitted well with pseudo-second-order kinetics and the isotherm was fitted well with the Langmuir isotherm model, respectively. The adsorption capacity of MG by the hydrogel could reach to 2691 mg/g. Even after a further cycle, the hydrogel exhibited an approximately 80% adsorption efficiency for MG, and accordingly, it was proposed to be a promising material for dye removal from wastewater [44].

Adebayo et al. prepared carboxyl-methyl lignin CML from the acid hydrolysis lignin from sugarcane bagasse (as shown in Fig. 9) [46]. The CMLs was further bound with Al(III) (CML-Al) and Mn(II) (CML-Mn) for the removal of Procion Blue MX-R (PB) in aqueous solutions. The experiment optimum pH and contact time were 2.0 and 5 h, respectively. The CML-Al and CML-Mn showed a

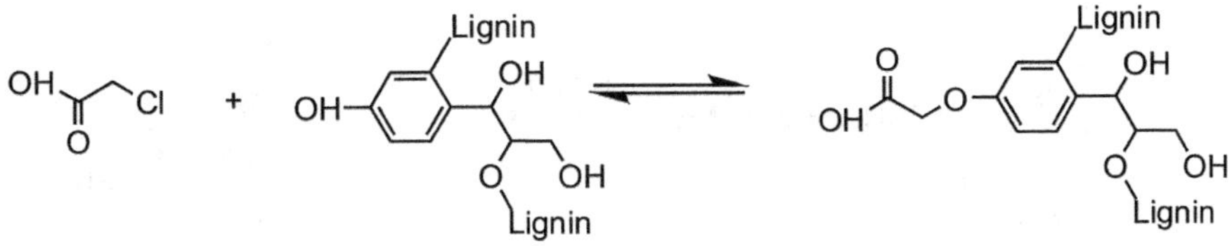

Fig. 9 Schematic of Carboxyl-methyl lignin (CML) from lignin and monochloroacetic acid. Reprinted with permission [46]. Copyright 2014 Elsevier

maximum adsorption capacity of 73.52 and 55.16 mg/g for PB at 25 °C, respectively. 98.33% of CML-Al and 98.08% of CML-Mn could be regenerated from dye-loaded adsorbents by using 50% acetone + 50% of 0.05 mol L^{-1} NaOH. After four adsorption-desorption cycles, the removal efficiency of the dyes still remained ca. 93.97% and ca. 75.91% by the CML-Al and CML-Mn, respectively. Silva et al. [47] also reported a carboxy-methylated lignin complexed with Fe^{3+} (CML-Fe) for the adsorption of Brilliant Red 2BE (BR-2) textile dye from aqueous solutions. The maximum adsorption capacity of BR-2 was 73.6 mg/g by the CML-Fe adsorbent. Besides, the dye-loaded adsorbent could be recycled with 0.050 mol/L NaOH.

3 Modified Lignin for Heavy Metals Adsorption

Heavy metal ions are more toxic than dyes and cover a wider region of pollutants during the past decades. Modification of lignin by introducing the desired properties in physical, chemical and mechanic properties is a key issue to achieve proper characters of lignin including the hydrophilicity, hydrophobicity and adsorption ability. The most common strategies in the modification of lignin for heavy metal ion adsorption include amination, methylolation, alkylation, carboxylation, acylation, sulfonation/sulfomethylation, phosphorylation, and copolymerization etc.

Ge et al. [49] prepared an aminated lignin-based adsorbent from alkaline lignin grafted by methylamine and formaldehyde, as shown in Fig. 10. Kinetic adsorption suggested the aminated lignin could adsorb Pb(II) in water quickly and the adsorption process could be described well by a second-order model. The obtained adsorbent presented a maximum adsorption amount of 60.5 mg/g for Pb(II) that was 4.2 fold of the original alkaline lignin. Ge et al. [50] further investigated the influence of numbers of the carbon in an alkyl, from C2 (ethyl) to C18 (octadecyl), on the adsorption capacity of lignin for the lead ion. The results indicated that the carbon number had a strong influence on the adsorption of Pb(II). A suitable carbon number of alkyl (C4) helped the adsorption of Pb(II), due to the electron donating ability of the alkyl groups. Recently, Huang et al. [51] reported a modified enzymatic hydrolysis lignin-containing nitrogen and sulfur moieties, which showed a high adsorption capacity of 180 mg/g for Hg(II) at 25 °C. The adsorption kinetics

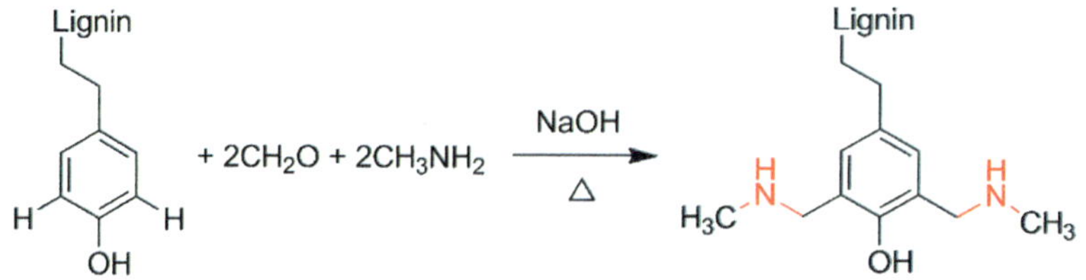

Fig. 10 Synthetic diagram of a Mannich base from lignin. Reprinted with permission [49]. Copyright 2015 Elsevier

could be fitted well with the pseudo-second-order equation, while the adsorption isotherms could be fitted well with the Freundlich equation. Besides, the increasing value of the constant (n) with temperature indicated the adsorption became more favourable at a higher temperature.

Carboxyl is a common component of lignin. The frequency of the group can be adjusted by chemical modifications as well as the hydrophilicity and polyelectrolyte characters. Quintana et al. [52] investigated the oxidized lignins from sulfuric acid pretreated cane bagasse, soda pulping bagasse, eucalypt Kraft lignin and commercial Kraft lignin. The results indicated the oxidized lignins showed higher adsorption capacities than the original lignins, due to the higher contents of carboxyl groups in the oxidized lignins. Peternele et al. [53] reported a functionalized formic lignin from sugarcane bagasse for the adsorption of Pb(II) and Cd(II). Batch adsorption equilibrium could be described well by the Langmuir equation. The oxidative modification of a wheat straw organosolv lignin has been investigated by Dizhbite et al. [54] to introducing –COOH and –OH groups. The oxidation was conducted under a polyoxometalate $H_3[PMo_{12}O_{40}]$ and O_2 or H_2O_2. The oxidative modification did not damage the lignin skeleton, while the carboxyl and hydroxyl group's contents in lignin increased distinctly. As expected, the oxidized lignin showed high adsorption amounts of 35.9 and 155.4 mg/g toward Cd(II) and Pb(II), respectively, at pH 5, 20 °C.

Sulfonate functional groups can be introduced into lignin via sulfomethylation and sulfonation, which has been reported as an effective way to improve the hydrophilicity of lignin [55, 56]. Li et al. [29] prepared a modified lignin-containing both amino and sulfonic groups (as shown in Fig. 11). It was found that the modified lignin could adsorb heavy metals effectively even at low pH values. The dynamic adsorption and equilibrium adsorption could be fitted well by the pseudo-second-order equation and D–R equation, respectively. Xu et al. [57] developed a mesoporous lignin-based biosorbent (MLBB) from rice straw. The MLBB had an excellent adsorption performance for Pb(II) with a maximum capacity of 952 ± 31 mg/g at 20 °C, which was due to the large surface area (186 m^2/g), plenty of mesopores (d_p = 5.5 nm) and high content of sulfonic groups (S: 2.51 ± 0.01%)].

Dithiocarbamated lignin is the most reported modified lignin due to its good adsorption ability toward heavy metal ions. Ge et al. [58] developed a lignin-based

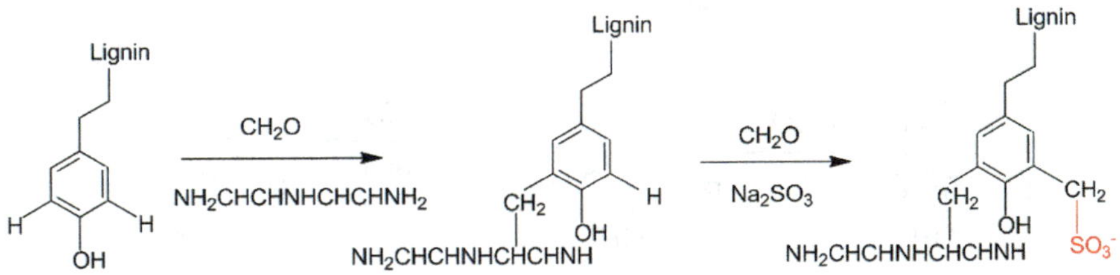

Fig. 11 Synthetic diagram of a modified lignin-containing amino and sulfonic groups. Reprinted with permission [29]. Copyright 2014 Elsevier

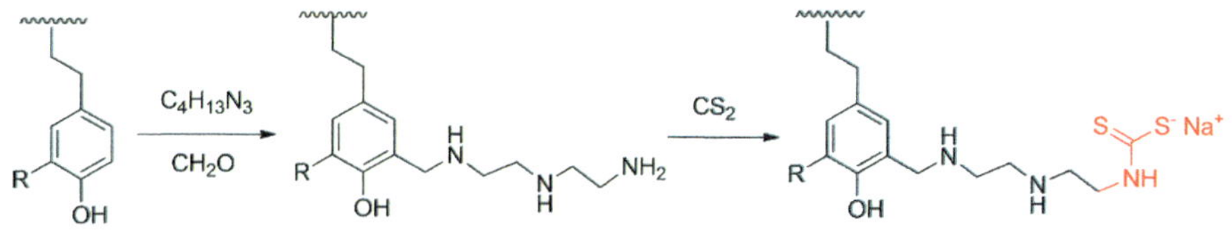

Fig. 12 Synthetic diagram of dithiocarbamated lignin. Reprinted with permission [58]. Copyright 2016 Royal Society of Chemistry

dithiocarbamate (LDTC), as shown in Fig. 12. The developed LDTC showed a high adsorption amount of 175.9 and 103.4 mg/g toward copper and lead ions, respectively. Li et al. [15] developed a porous lignin-containing a large number of mesopores and functional groups. The surface area of the modified lignin increase 11 fold of lignin to 22.3 m^2/g. Accordingly, the modified lignin had a high adsorption capacity of 188 mg/g to Pb(II), 13 fold of the lignin and 7 fold of activated carbon. Ge et al. [27] further prepared a dithiocarbamate modified lignin from organosolv lignin. The modified lignin showed a high adsorption amount of 210 mg/g to Hg(II). The adsorption kinetics could be fitted well by a pseudo-second-order equation, and the adsorption equilibrium could be fitted well by the Freundlich equation. Li et al. [59] prepared a lignin xanthate resin (LXR) with xanthate functional groups (–CSS$^-$) (as shown in Fig. 13). The adsorption capacity of Pb(II) was 64.9 mg/g at pH 5.0, 30 °C.

Ge et al. [26] prepared a new kind of lignin microspheres (LMS) through an inverse suspension copolymerization method. The LMS was in diameter of 348 μm with plenty of amine groups (total N: 7.5 mmol/g). LMS showed an adsorption capacity of 33.9 mg/g for lead ions at pH 6.0, 25 °C, and the adsorption kinetics could be fitted well by the pseudo-second-order equation. Liang et al. [60] prepared a lignin-based resin (LBR) from sodium lignosulfonate and glucose. The maximum adsorption capacity of Cr(VI) by the LBR was 57.68 mg/g. The adsorption equilibrium could be fitted well by the Freundlich equation. The calculated thermodynamic parameters of the adsorption (ΔG, ΔH and ΔS) indicated that the adsorption of Cr(VI) by the LBR is a spontaneous and endothermic process. Parajuli et al. [61] developed a crosslinked lignocatechol gel from catechol and

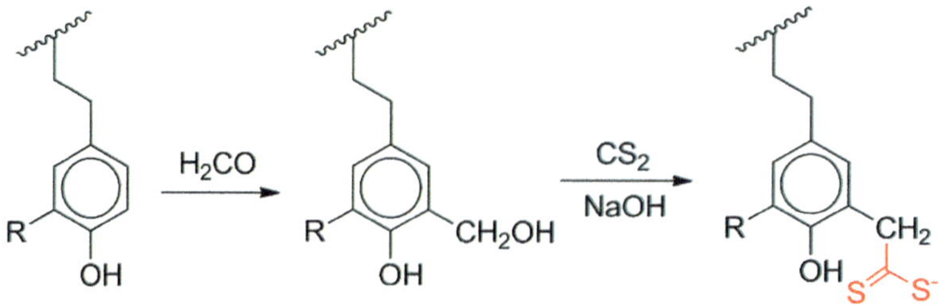

Fig. 13 Synthetic diagram of the lignin xanthate resin (LXR). Reprinted with permission [59]. Copyright 2015 Elsevier

wood lignin. The gel showed a saturated adsorption amount of 37.05 mg/g toward Pb(II) at pH 5.2. The adsorption process was mainly due to the cation exchange mechanism.

Lignin composites are now considered as a talented candidate to traditional adsorbents for cleanup of heavy metal ions in water. Qin et al. [62] synthesized a composite from poly (ethylene imine) and lignin. The composite presented a high adsorption amount of 98.0 mg/g to Cu(II), 78.0 mg/g to Zn(II) and 67.0 mg/g to Ni(II) at pH 6.0, 25 °C. In addition, the composite had good recyclability stability within 5 adsorption-desorption cycles. Klapiszewskia et al. [63] prepared new TiO_2/lignin and TiO_2-SiO_2/lignin hybrids. The adsorption capacity of Pb(II) by the TiO_2/lignin and TiO_2-SiO_2/lignin was 35.7 and 59.9 mg/g, respectively, at pH 5.0, 20 °C. Kinetic analysis revealed that the adsorption followed by the pseudo-second-order equation that meant a chemical interaction occurred during the adsorption process. Equilibrium adsorption was well described well by the Langmuir equation that meant a monolayer coverage of the adsorbates on the homogeneous surface of the adsorbents. Li et al. [64] prepared a composite lignin sphere from sodium alginate and epichlorohydrin. The obtained sphere showed an excellent removal efficiency (95.6 ± 3.5%, C_0 = 25.0 mg/L) to Pb(II). Li et al. [65] reported a new nano-composite composed of lignin and carbon nanotubes (L-CNTs). The as-prepared nano-composite not only showed a good water-dispersibility and environmentally friendliness but also presented an excellent adsorption ability to Pb(II) with a maximum adsorption amount of 235 mg/g. The nano-composite with a lignin layer has advanced adsorption ability, low in cost and environmentally friendliness, and therefore is a talented alternative for wastewater treatment. Klapiszewski et al. [66] synthesized a 'green' adsorbent with a high surface area of 223 m^2/g from the commercial silica Syloid®244 and Kraft lignin for the removal of nickel(II) and cadmium(II). The results indicated the adsorbents had maximum sorption capacities of 77.11 mg/g for Ni(II) and 84.66 mg/g for Cd(II), respectively. The kinetics adsorption data could be described well by a pseudo-second-order equation while the adsorption isotherms fitted well with Langmuir equation. Yao et al. [67] synthesized a composite composed of bentonite/sodium lignosulfonate with acrylamide and maleic anhydride (BLPAMA). Results showed that the adsorption of Pb(II) by the BLPAMA was correlated with pH values but not with the temperatures. It showed a maximum adsorption capacity of 314.8 mg/g for Pb(II) at pH 5.0, 25 °C.

The above mentioned modified lignin adsorbents for adsorption of heavy metals are tabulated in Table 2, from which it can be seen that the adsorption capacity was much dependent on the chemical methods for lignin modification and heavy metals species as well as the temperature and pH values. According to the analysis of the published literature, the research on the development of advanced lignin-based adsorbents, especially lignin-based nano-composites, is greatly needed in the coming decades.

Table 2 Comparison of adsorption capacity of heavy metals with modified lignins

Modified lignin	Heavy metal	T (°C)/pH	Q_m (mg/g)	Reference
Mannich base from lignin	Pb(II)	25 ± 1/6.0	60.5	[49]
EHL-NS	Hg(II)	25/6.0	180	[51]
CMLSCB	Pb(II)	30/6.0	122.9	[68]
BL	Pb(II) Cd(II)	20/5.0	35.9 155.4	[54]
ASL	Cu(II) Pb(II)	25 ± 0.5/6.0	6.3 49.6	[29]
MLBB	Pb(II)	20/7.0	952 ± 31	[57]
LDTC	Cu(II) Pb(II)	25 ± 0.5/6.0	175.9 103.4	[58]
SFPL	Pb(II)	25 ± 0.5/5.0	188	[15]
Dithiocarbamate functionalized organosolv lignin	Hg(II)	25 ± 0.5/5.0	210	[27]
LMS	Pb(II)	25/6.0	33.9	[26]
LBR	Cr(VI)	50/2.0	57.68	[60]
Crosslinked lignocatechol gel	Pb(II)	25/5.2	37.05	[61]
Poly (ethylene imine) anchored lignin	Cu(II) Zn(II) Ni(II)	25/6.0	98.0 78.0 67.0	[62]
TiO$_2$/lignin TiO$_2$-SiO$_2$/lignin	Pb(II)	20/5.0	35.7 59.9	[63]
PLS	Pb(II)	30.2/5.0	31.8	[64]
Siliceous lignin	Pb(II)	70/2.0	–	[69]
L-CNT	Pb(II)	25/6.3	235	[65]
Silica/lignin	Ni(II) Cd(II)	25/3.0 9.0/–	77.11 84.66	[66]
BLPAMA	Pb(II)	25/5.5	314.8	[67]

4 Modified Lignin for Other Pollutants Adsorption

Except for the above discussed main pollutants (dyes and heavy metals), some other pollutants by lignin-based adsorbents are also reported in literature during the past decades. Although there are few papers on the topic, the removal of some other toxic and hazardous pollutants is also very important. Saad et al. [70] investigated two lignins, including alkaline lignin and organosolv lignin, for the adsorption of 2,4-dinitroanisole (DNAN) from the water. The adsorption of DNAN on both lignins could be described well with pseudo-second-order equation. The organosolv lignin showed a maximum adsorption amount of 7.5 mg/g to DNAN while the alkaline lignin showed a maximum adsorption amount of 8.5 mg/g to DNAN. The adsorption equilibrium for either alkali or organosolv lignin could be described well by the Freundlich model.

Chen et al. [71] reported spherical lignin beads as an adsorbent to adsorb L-lysine in water. The results indicated the beads showed a maximum adsorption capacity of 67.11 mg/g to L-lysine. Dynamic adsorption was fitted well by the pseudo-first-order equation, and the adsorption of L-lysine on the beads was initially determined by film diffusion, and then by intra-particle diffusion. Adsorption equilibrium was described well with the Langmuir equation.

Żółtowska-Aksamitowska et al. [72] firstly investigated the use of chitin modified Kraft lignin as an effective sorbent of ibuprofen and acetaminophen. Batch adsorption results indicated the modified lignin showed an adsorption capacity of 400.39 μg/g to ibuprofen and 267.07 μg/g to acetaminophen, respectively. Adsorption isotherms data could be described with the Langmuir equation for ibuprofen, and with the Freundlich equation for acetaminophen, while the adsorption kinetics followed well to the pseudo-second-order equation for both pollutants (R^2 = 0.999). These results indicated that the adsorption belonged to a chemisorption. Furthermore, the used adsorbents could be easily regenerated with ethanol (yield 82.2%) in the case of ibuprofen and methanol (yield 80.8%) in the case of acetaminophen.

Application of the oxidative modification of sulfate kraft lignin with sodium periodate under mild conditions is suggested in order to obtain a sorbent for detoxication of spillage places of rocket fuels based on 1,1-dimethylhydrazine and to purify wastewaters containing this compound [73]. It was found that processing with sodium periodate at 55 °C for 20 h resulted in a more than two-fold increase in the content of carbonyl and quinone groups in lignin and a three-fold increase in the adsorption capacity for 1,1-dimethylhydrazine. The adsorbent can bind 6.7% of 1,1-dimethylhydrazine and substantially surpasses in this parameter other lignin-based adsorbents.

Table 3 listed all these reported modified lignin-based adsorbents for adsorption of some other pollutants in water. As could be seen, the adsorption capability of lignin-based adsorbents for other species is much lower than those for dyes and heavy metals. Therefore, it is in a great need of developing new lignin-based adsorbents with enhanced adsorption capability toward the emerging pollutants in the future.

Table 3 Adsorption capacity of modified lignin for other pollutants in water

Modified lignin	Species	T (°C)/pH	Q_m (mg/g)	Reference
Alkali and organosolv lignin	DNAN	25/5.7	7.5	[70]
Spherical lignin beads	L-lysine	25/9.0	67.11	[71]
Kraft lignin	Ibuprofen acetaminophen	25/6.0	0.40 0.27	[72]
Periodate oxidation of lignin	UDMH	55/5.0	67 ± 1	[73]

5 Outlook and Conclusions

With the depletion of fossil-based resources and increasing environmental concerns, the research and development of new low-cost adsorbents derived from renewable resources have received more and more attention. The utilization of aromatic lignin in replace of the fossil-based carbon will help to the establishment of the sustainable society. In this content, the adsorption of dyes, heavy metals and some other pollutants by various modified lignin materials have been collected and discussed in this study. Based on the above discussions, adsorption performance of these modified lignin materials for dyes, heavy metals, and other pollutants is expected to be amplified in the near future. It should be noted that the cost-effectiveness is firstly important because low manufacturing cost and high adsorption capability of an adsorbent is desired in practical large-scale applications. Lignin is an abundant polymer derived from plant kingdom that can provide CO_2 neutral, cost-effective, environmentally friendly and therefore can be used as building blocks to create "green" adsorbents. Secondly, although modified lignin is efficient to capture different kinds of pollutants through physical or chemical interactions, regeneration by feasible methods should be studied carefully as it is very important for the improvement of economics. The regeneration of lignin-based materials could be carried out by the most common solvent extraction method with EDTA, HCl, HNO–, NaCl, and NaOH solutions et al. Thirdly, experiments should not only stay in the lab. As all known, the industrial wastewater always contains many kinds of contaminants. Although different kinds of modified lignin adsorbents can be obtained with good adsorption capacity and selectivity for the target contaminants, such as methylene blue (MB), toxic metals (Hg), and (Pb, Cu) et al., it is desirable in developing a multipurpose adsorbent from lignin that can be used to adsorb different kinds of contaminants simultaneously. Therefore, specific attention should be focused on the modification of lignin matrix, via hybriding, cross-linking and grafting to develop lignin-based advanced composite, and accordingly broadening the kinds of pollutants for efficient removal and improving the reusability of the modified lignin-based composites.

Acknowledgements Financial support from the National Natural Science Foundation of China (No. 21264002, 21464002), and Guangxi Natural Science Foundation (No. 2015GXNSFBA139215, 2016GXNSFAA380329) is gratefully acknowledged.

References

1. Abe A, Dusek K, Kobayashi S (2010) Biopolymers: lignin, proteins, bioactive nanocomposites. Springer, Berlin
2. Upton BM, Kasko AM (2016) Strategies for the conversion of lignin to high-value polymeric materials: review and perspective. Chem Rev 116:2275–2306

3. Holladay JE, White JF, Bozell JJ, Johnson D (2007) Top value-added chemicals from biomass—volume II—results of screening for potential candidates from biorefinery lignin. In: Pacific Northwest National Laboratory (PNNL), Richland, WA (US), 2007, Medium: ED; Size: PDFN
4. Zakzeski J, Bruijnincx PCA, Jongerius AL, Weckhuysen BM (2010) The catalytic valorization of lignin for the production of renewable chemicals. Chem Rev 110:3552–3599
5. Li Z, Zhang J, Qin L, Ge Y (2018) Enhancing antioxidant performance of lignin by enzymatic treatment with laccase. ACS Sustain Chem Eng 6:2591–2595
6. Calvo-Flores FG, Dobado JA (2010) Lignin as renewable raw material. Chemsuschem 3:1227–1235
7. Thakur VK, Thakur MK, Raghavan P, Kessler MR (2014) Progress in green polymer composites from lignin for multifunctional applications: a review. ACS Sustain Chem Eng 2:1072–1092
8. Rezakazemi M, Maghami M, Mohammadi T (2018) High loaded synthetic hazardous wastewater treatment using lab-scale submerged ceramic membrane bioreactor. Periodica Polytechnica Chem Eng 62:299–304
9. Rezakazemi M, Khajeh A, Mesbah M (2018) Membrane filtration of wastewater from gas and oil production. Environ Chem Lett 16:367–388
10. Rezakazemi M, Dashti A, Riasat Harami H, Hajilari N (2018) Fouling-resistant membranes for water reuse. Environ Chem Lett 1–49
11. Azimi A, Azari A, Rezakazemi M, Ansarpour M (2017) Removal of heavy metals from industrial wastewaters: a review. Chem Bio Eng Rev 4:37–59
12. Shirazian S, Rezakazemi M, Marjani A, Moradi S (2012) Hydrodynamics and mass transfer simulation of wastewater treatment in membrane reactors. Desalination 286:290–295
13. Rezakazemi M, Shirazian S, Ashrafizadeh SN (2012) Simulation of ammonia removal from industrial wastewater streams by means of a hollow-fiber membrane contactor. Desalination 285:383–392
14. Gupta VK (2009) Application of low-cost adsorbents for dye removal—a review. J Environ Manage 90:2313–2342
15. Li Z, Xiao D, Ge Y, Koehler S (2015) Surface-functionalized porous lignin for fast and efficient lead removal from aqueous solution. ACS Appl Mater Interfaces 7:15000–15009
16. Rezakazemi M, Zhang Z (2018) 2.29 desulfurization materials A2. In: Dincer I (ed) Comprehensive energy systems. Elsevier, Oxford, pp 944–979
17. Rezakazemi M, Sadrzadeh M, Matsuura T (2018) Thermally stable polymers for advanced high-performance gas separation membranes. Progr Energy Combust Sci 66:1–41
18. Rezakazemi M, Marjani A, Shirazian S (2018) Organic solvent removal by pervaporation membrane technology: experimental and simulation. Environ Sci Poll Res
19. Rezakazemi M, Ebadi Amooghin A, Montazer-Rahmati MM, Ismail AF, Matsuura T (2014) State-of-the-art membrane based CO_2 separation using mixed matrix membranes (MMMs): an overview on current status and future directions. Prog Polym Sci 39:817–861
20. Foroutan R, Esmaeili H, Abbasi M, Rezakazemi M, Mesbah M (2017) Adsorption behavior of $Cu(II)$ and $Co(II)$ using chemically modified marine algae. Environ Technol 1–9
21. Ge YY, Cui XM, Liao CL, Li ZL (2017) Facile fabrication of green geopolymer/alginate hybrid spheres for efficient removal of $Cu(II)$ in water: batch and column studies. Chem Eng J 311:126–134
22. Saleh TA, Gupta VK (2014) Processing methods, characteristics and adsorption behavior of tire derived carbons: a review. Adv Colloid Interfac 211:93–101
23. Rafatullah M, Sulaiman O, Hashim R, Ahmad A (2010) Adsorption of methylene blue on low-cost adsorbents: a review. J Hazard Mater 177:70–80
24. Ahluwalia SS, Goyal D (2007) Microbial and plant derived biomass for removal of heavy metals from wastewater. Bioresource Technol 98:2243–2257
25. Deng S, Ting Y-P (2005) Characterization of PEI-modified biomass and biosorption of $Cu(II)$, $Pb(II)$ and $Ni(II)$. Water Res 39:2167–2177

26. Ge YY, Qin L, Li ZL (2016) Lignin microspheres: an effective and recyclable natural polymer-based adsorbent for lead ion removal. Mater Des 95:141–147
27. Ge Y, Wu S, Qin L, Li Z (2016) Conversion of organosolv lignin into an efficient mercury ion adsorbent by a microwave-assisted method. J Taiwan Inst Chem Eng 63:500–505
28. Wawrzkiewicz M, Bartczak P, Jesionowski T (2017) Enhanced removal of hazardous dye form aqueous solutions and real textile wastewater using bifunctional chitin/lignin biosorbent. Int J Biol Macromol 99:754–764
29. Ge Y, Li Z, Kong Y, Song Q, Wang K (2014) Heavy metal ions retention by bi-functionalized lignin: synthesis, applications, and adsorption mechanisms. J Ind Eng Chem 20:4429–4436
30. Ge Y, Li Z (2018) Application of lignin and its derivatives in adsorption of heavy metal ions in water: a review. ACS Sustain Chem Eng 6:7181–7192
31. Laurichesse S, Avérous L (2014) Chemical modification of lignins: towards biobased polymers. Prog Polym Sci 39:1266–1290
32. Nair V, Panigrahy A, Vinu R (2014) Development of novel chitosan–lignin composites for adsorption of dyes and metal ions from wastewater. Chem Eng J 254:491–502
33. Albadarin AB, Collins MN, Naushad M, Shirazian S, Walker G, Mangwandi C (2017) Activated lignin-chitosan extruded blends for efficient adsorption of methylene blue. Chem Eng J 307:264–272
34. Lou T, Cui G, Xun J, Wang X, Feng N, Zhang J (2018) Synthesis of a terpolymer based on chitosan and lignin as an effective flocculant for dye removal. Coll Surf A 537:149–154
35. Li YL, Wu M, Wang B, Wu YY, Ma MG, Zhang XM (2016) Synthesis of magnetic lignin-based hollow microspheres: a highly adsorptive and reusable adsorbent derived from renewable resources. ACS Sustain Chem Eng 4:5523–5532
36. Couch RL, Price JT, Fatehi P (2016) Production of flocculant from thermomechanical pulping lignin via nitric acid treatment. ACS Sustain Chem Eng 4:1954–1962
37. Xu WJ, Zhang WS, Li Y, Li W (2016) Synthesis of acrylic-lignosulfonate resin for crystal violet removal from aqueous solution. Korean J Chem Eng 33:2659–2667
38. Wang Y, Xiong Y, Wang J, Zhang X (2017) Ultrasonic-assisted fabrication of montmorillonite-lignin hybrid hydrogel: highly efficient swelling behaviors and super-sorbent for dye removal from wastewater. Coll Surf A 520:903–913
39. Yu C, Wang F, Zhang C, Fu S, Lucia LA (2016) The synthesis and absorption dynamics of a lignin-based hydrogel for remediation of cationic dye-contaminated effluent. React Funct Polym 106:137–142
40. Kumari S, Chauhan GS, Monga S, Kaushik A, Ahn J-H (2016) New lignin-based polyurethane foam for wastewater treatment. RSC Adv 6:77768–77776
41. Feng Q, Cheng H, Chen F, Zhou X, Wang P, Xie Y (2016) Investigation of cationic dye adsorption from water onto acetic acid lignin. J Wood Chem Technol 36:173–181
42. Tang Y, Hu T, Zeng Y, Zhou Q, Peng Y (2015) Effective adsorption of cationic dyes by lignin sulfonate polymer based on simple emulsion polymerization: isotherm and kinetic studies. RSC Adv 5:3757–3766
43. Tang Y, Zeng Y, Hu T, Zhou Q, Peng Y (2016) Preparation of lignin sulfonate-based mesoporous materials for adsorbing malachite green from aqueous solution. J Environ Chem Eng 4:2900–2910
44. Song X, Chen F, Liu S (2016) A lignin-containing hemicellulose-based hydrogel and its adsorption behavior. BioResources 11:6378–6392
45. Luo H, Ren S, Ma Y, Fang G, Jiang G (2015) Preparation and properties of kraft lignin-N-isopropyl acrylamide hydrogel. BioResources 10:3507–3519
46. Adebayo MA, Prola LDT, Lima EC, Puchana-Rosero MJ, Cataluña R, Saucier C, Umpierres CS, Vaghetti JCP, da Silva LG, Ruggiero R (2014) Adsorption of Procion Blue MX-R dye from aqueous solutions by lignin chemically modified with aluminium and manganese. J Haz Mater 268:43–50
47. da Silva LG, Ruggiero R, Gontijo PdM, Pinto RB, Royer B, Lima EC, Fernandes THM, Calvete T (2011) Adsorption of Brilliant Red 2BE dye from water solutions by a chemically modified sugarcane bagasse lignin. Chem Eng J 168:620–628

48. Guo K, Gao B, Li R, Wang W, Yue Q, Wang Y (2018) Flocculation performance of lignin-based flocculant during reactive blue dye removal: comparison with commercial flocculants. Environ Sci Pollut R 25:2083–2095
49. Ge Y, Song Q, Li Z (2015) A Mannich base biosorbent derived from alkaline lignin for lead removal from aqueous solution. J Ind Eng Chem 23:228–234
50. Li Z, Xiao D, Kong Y, Ge Y (2015) Enhancing lead adsorption capacity by controlling the chain length of alkyl amine grafted lignin. BioResources 10:2425–2432
51. Huang W-X, Zhang Y-H, Ge Y-Y, Qin L, Li Z-L (2017) Soft nitrogen and sulfur incorporated into enzymatic hydrolysis lignin as an environmentally friendly antioxidant and mercury adsorbent. BioResources 12:7341–7348
52. Quintana GC, Rocha GJM, Goncalves AR, Velasquez JA (2008) Evaluation of heavy metal removal by oxidised lignins in acid media from various sources. BioResources 3:1092–1102
53. Peternele WS, Winkler-Hechenleitner AA, Pineda EAG (1999) Adsorption of Cd(II) and Pb(II) onto functionalized formic lignin from sugar cane bagasse. Bioresource Technol 68:95–100
54. Dizhbite T, Jashina L, Dobele G, Andersone A, Evtuguin D, Bikovens O, Telysheva G (2013) Polyoxometalate (POM)-aided modification of lignin from wheat straw biorefinery. Holzforschung 67:539–547
55. Li Z, Pang Y, Ge Y, Qiu X (2011) Evaluation of steric repulsive force in the aqueous dispersion system of dimethomorph powder with lignosulfonates via X-ray photoelectron spectroscopy. J Phys Chem C 115:24865–24870
56. Li Z, Ge Y (2011) Extraction of lignin from sugar cane bagasse and its modification into a high performance dispersant for pesticide formulations. J Brazil Chem Soc 22:1866–1871
57. Xu F, Zhu TT, Rao QQ, Shui SW, Li WW, He HB, Yao RS (2017) Fabrication of mesoporous lignin-based biosorbent from rice straw and its application for heavy-metal-ion removal. J Environ Sci 53:132–140
58. Ge Y, Xiao D, Li Z, Cui X (2014) Dithiocarbamate functionalized lignin for efficient removal of metallic ions and the usage of the metal-loaded bio-sorbents as potential free radical scavengers. J Mater Chem A 2:2136–2145
59. Li Z, Kong Y, Ge Y (2015) Synthesis of porous lignin xanthate resin for Pb2 + removal from aqueous solution. Chem Eng J 270:229–234
60. Liang F-B, Song Y-L, Huang C-P, Zhang J, Chen B-H (2013) Adsorption of hexavalent chromium on a lignin-based resin: equilibrium, thermodynamics, and kinetics. J Environ Chem Eng 1:1301–1308
61. Parajuli D, Inoue K, Ohto K, Oshima T, Murota A, Funaoka M, Makino K (2005) Adsorption of heavy metals on crosslinked lignocatechol: a modified lignin gel. React Funct Polym 62:129–139
62. Qin L, Ge Y, Deng B, Li Z (2017) Poly(ethylene imine) anchored lignin composite for heavy metals capturing in water. J Taiwan Inst Chem Eng 71:84–90
63. Klapiszewski L, Siwinska-Stefanska K, Kolodynska D (2017) Preparation and characterization of novel TiO2/lignin and TiO2-SiO2/lignin hybrids and their use as functional biosorbents for Pb(II). Chem Eng J 314:169–181
64. Li Z, Ge Y, Wan L (2015) Fabrication of a green porous lignin-based sphere for the removal of lead ions from aqueous media. J Haz Mater 285:77–83
65. Li Z, Chen J, Ge Y (2017) Removal of lead ion and oil droplet from aqueous solution by lignin-grafted carbon nanotubes. Chem Eng J 308:809–817
66. Klapiszewski L, Bartczak P, Wysokowski M, Jankowska M, Kabat K, Jesionowski T (2015) Silica conjugated with kraft lignin and its use as a novel 'green' sorbent for hazardous metal ions removal. Chem Eng J 260:684–693
67. Yao Q, Xie J, Liu J, Kang H, Liu Y (2014) Adsorption of lead ions using a modified lignin hydrogel. J Polym Res 21:465
68. Peternele WS, Winkler-Hechenleitner AA, Gómez Pineda EA (1999) Adsorption of Cd(II) and Pb(II) onto functionalized formic lignin from sugar cane bagasse. Bioresour Technol 68:95–100

69. Cui J, Sun H, Wang X, Sun J, Niu M, Wen Z (2015) Preparation of siliceous lignin microparticles from wheat husks with a facile method. Ind Crop Prod 74:689–696
70. Saad R, Radovic-Hrapovic Z, Ahvazi B, Thiboutot S, Ampleman G, Hawari J (2012) Sorption of 2,4-dinitroanisole (DNAN) on lignin. J Environ Sci 24:808–813
71. Chen GF, Liu MH (2012) Adsorption of L-lysine from aqueous solution by spherical lignin beads: kinetics and equilibrium studies. BioResources 7:298–314
72. Żółtowska-Aksamitowska S, Bartczak P, Zembrzuska J, Jesionowski T (2018) Removal of hazardous non-steroidal anti-inflammatory drugs from aqueous solutions by biosorbent based on chitin and lignin. Sci Total Environ 612:1223–1233
73. Kozhevnikov AY, Ul'yanovskaya SL, Semushina MP, Pokryshkin SA, Ladesov AV, Pikovskoi II, Kosyakov DS (2017) Modification of sulfate lignin with sodium periodate to obtain sorbent of 1,1-dimethylhydrazine. Russ J Appl Chem + 90:516–521

Synthesis and Characterization and Application of Chitin and Chitosan-Based Eco-friendly Polymer Composites

Aneela Sabir, Faizah Altaf and Muhammad Shafiq

1 Introduction

1.1 History of Chitosan

The chitosan history begins from the 19th century, by Rouget 1859 who discussed the deacetylated parent chitin polymer (the second most plenteous carbohydrate) in nature first time. With the passage of time, a considerable amount of work has been done on chitosan especially during last 20 years, and its potential for different bio applications [1].

Chitosan is obtained from natural sources that are the external skeleton of crustaceans, fungi, and insects and has to be biocompatible and decomposable. It's being a copolymer contains N-acetyl-2-amino-2-deoxy-d-glucopyranose and 2-amino-2-deoxy-d-glucopyranose, the monomers are joined together by (1 → 4) glycosidic bonds. The carbon and hydrogen is a major constituent of chitosan is very alike to cellulose, in which 1,4 is linked with d-glucosamine and with varying number of N-acetylation, the difference is only that the hydroxyl group is replaced acetylamino group at the C_2 position. It is a nearly synthetic derivative of aminopolysaccharide which has peculiar structures, dimensional characteristics, extremely planned practicality and a large vary of usage in medicinal speciality and alternative manufacturing areas [2]. The structure of chitosan is given in Fig. 1.

The removal of the acetyl group from chitin to produce chitosan needs a reaction with highly strong NaOH solution (water or alcohol based) with maintaining safe

---

A. Sabir (✉) · M. Shafiq
Department of Polymer Engineering and Technology, University of the Punjab, Lahore 54590, Pakistan
e-mail: aneela.pet.ceet@pu.edu.pk

F. Altaf
Department of Environmental Sciences, Fatima Jinnah Women University, Rawalpindi 46000, Pakistan

© Springer Nature Switzerland AG 2019
Inamuddin et al. (eds.), *Sustainable Polymer Composites and Nanocomposites*,
https://doi.org/10.1007/978-3-030-05399-4_46

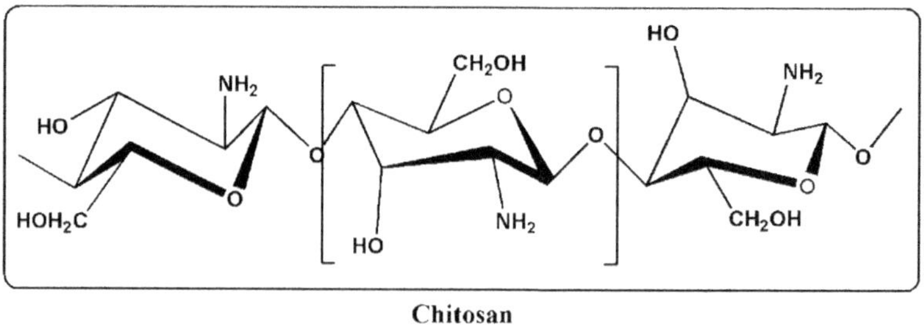

Fig. 1 Structure of chitosan

conditions that ensure the reaction mixture does not interact with oxygen and for this purpose reaction mixture is either purged with nitrogen or by adding $NaBH_4$ so to control unwanted depolymerization and production of reactive species. Chitosan has become attention-grabbing not solely because of its synthesis from plenteous natural resources however as a result of it's highly well-matched and efficient biomaterial that is quite able to be applied. As chitin is poorly soluble in water-based and organic solvents, whereas chitosan as an anthropogenic creation of chitin is highly appropriate for the application in the biomedical field. The best features of magnificent biocompatibility and marvellous bio-decomposition that protect ecosystem protection and less poisonousness with multipurpose medical accomplishments like antibacterial activity and low immunogenicity providing plentiful openings for more development [3].

Chitosan has the most structure changes being diagrammatic by the perspective quantities of N-acetyl-d-glucosamine and d-glucosamine remaining, give definite chemical structure variations. This distinction within formula provides rise to many groups of chitosan that are notable. Chitosan may have various range number of deacetylation (40–98%) and molecular mass (5–104 and 2–106 Da). The degree of deacetylation and polymerization actually describe the molecular mass of the polymer. They are two necessary factors indicating the employment of chitosan for numerous solicitations. Some other parameters may alter the physical properties of CS including the order of the amino and acetamido groups and also the quality of crustacean shells (crabs, etc.) which were wastage of food industry in past but now are commercially used for chitin and CS production [4] (Fig. 2).

The purification of chitin and its method can have an effect on quality of chitosan. The crystallinity and polymorphism of chitosan are highly dependent on the beginning of the polymerization and extraction in thorough process. An aliphatic and straight chain structure having the incorporated chitosan in its blend act as a viscosity enhancing agent that work in acidic conditions and behave like false plastic material possess lower viscosity as shear rate increases. The viscosity rises by increasing the concentration of chitosan. Viscosity can have a strong effect on enhancing the biological properties like wound-recovery and osteogenesis improvement and biodegradation by lysozyme [5].

The chitosan solubilizes in solutions containing acid (below pH 6.0) because of amine groups (pKa = 6.3) quaternization which enables chitosan to become

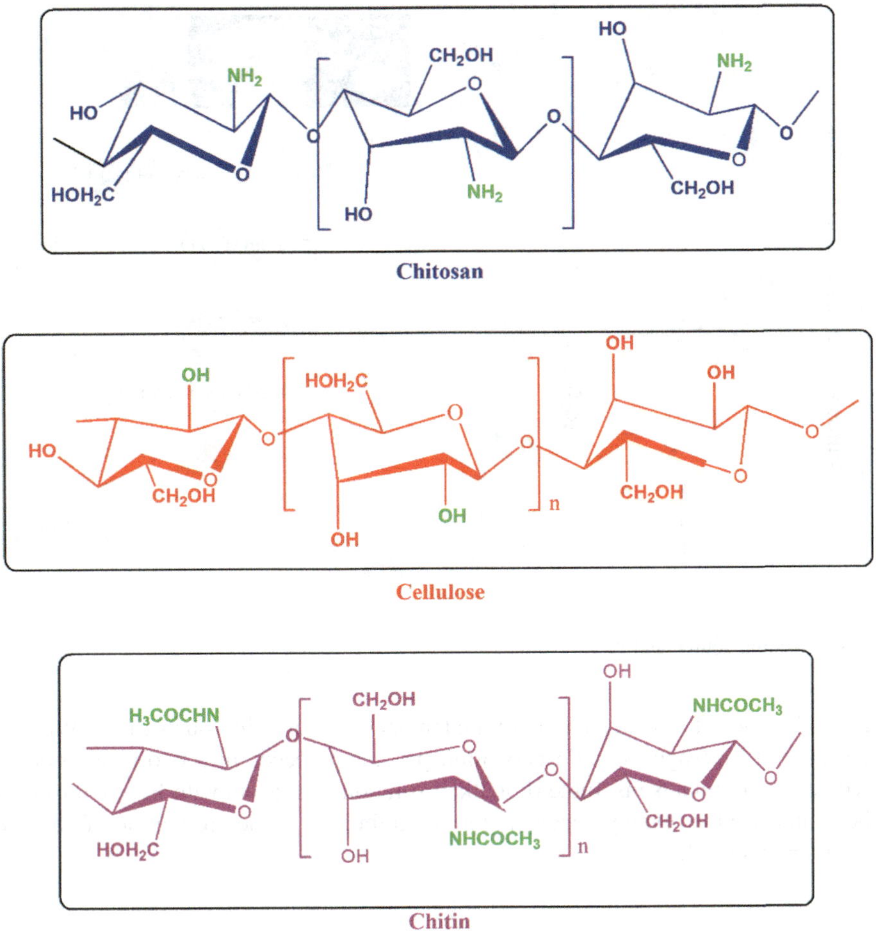

Fig. 2 The molecular structure of chitosan, cellulose and chitin

aqua-soluble. The existence of the –NH$_2$ showed that pH considerably changes the partially ionic state and characteristics of chitosan. When pH is low, protonation of amines takes place causing a positively charged species making it soluble. However at higher pH (higher than 6), chitosan's amines get deprotonated and charge loss occurs and makes insoluble ion electrolyte [6].

2 Production

Chitin is a basic precursor of chitosan production using basically two different approaches, i.e. equally distributed deacetylation and heterogeneous deacetylation. In the equally distributed deacetylation procedure, the chitin is allowed to dissolve

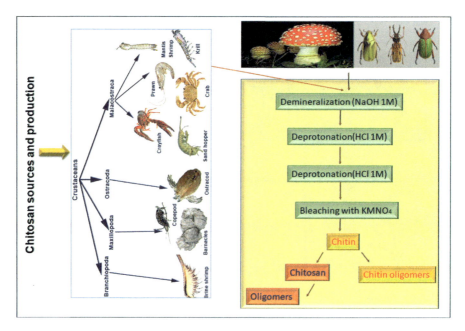

Fig. 3 Chitin, source and its application

in an alkaline solution at an appropriate temperature with vigorous stirring. While in the case of heterogeneous deacetylation process proceeds in a two-phase system, where chitin remains insoluble (hot alkali solution). Chitosan with FA (0–65%) can be synthesized by evenly deacetylation of chitin [5, 7]. The production of chitosan is given in Fig. 3.

3 Chitosan Oligomers

Owing to higher molecular mass and highly viscous nature of chitosan, it is possible to subject chitosan polymer for depolymerization (chitonolysis) results in the production of chitosan of low-molecular-weight i.e. oligomers and monomers also. These chitosan oligomers showed excellent solubility and have been used in numerous applications. There is a variety of chemical, physical and enzymatic techniques that are used to synthesize chitosan having smaller chains of repeating units (3–9) [8].

The acid hydrolysis synthesis restricts the use of chemical methods along with conventional heat procedure. It has some drawbacks including highly expensive, lower final product yield an acid residue. Degradation by acids is not a simple process; the breakage in the presence of water undergoes and results in the random generation of repeating units, D-glucosamine on increasing reaction time.

So, hydrolysis using concentrated HCl has usually been modified. First by employing 35% HCl at a temperature of 80 °C for a shorter period of time. This process produced chitosan oligomers having 1–15 and 20–40 chains. The use of nitric acid (HNO_3) for oligomer production is choosy, rapid, and easily manageable with good stoichiometric products. Hydrogen peroxide (H_2O_2) may also be convenient to breakdown chain of monomers and generate hydroxyl radicals [9].

Another process to synthesize chitosan small chains using hot H_3PO_4 has been described in the literature [10]. The obtained yields were sated as 10–20% with DP 6–8. Moreover, two kinds of chitosan oligomers (DP 7.3 and 16.8) were also produced through chitosan consistent hydrolysis using 85% H_3PO_4 at ambient temperature for four weeks. Fluoride-based hydrolysis of chitosan using dried HF appeared to be more suitable way than traditional chemical polymer breakage in term of good products yield. However, this procedure also has practical restrictions as required an additional processing for defluorination. The polymer decomposition can also be obtained effectively using microwave assisted technology supported by the adding alkali carbonates to synthesize l-molar mass chitosan within a short period of time [11].

The production of chitosan oligomers also carried out using enzymes. Enzymatic synthesis cause chitosan depolymerized easily with a variety of hydrolase including hemicellulases, lipases, cellulases, and lysozyme, papain amylases, pectinase, pronase etc. The enzyme-based polymer degradation is favourable for the production of the chitosan. The reaction rate can be managed by means of reaction time, pH, temperature and recombinant approaches and of physical ways as sonication and electromagnetic radiation [12].

4 Modifications of Chitosan

There are a variety of materials and method that have been used to modify chitosan to enhance biofunctionality and other desired properties for utilization in different applications. Surface modification methods include blending with different other compound and derivatives such as coverings, reaction with an oxygen based compound or by employing surfactant. Moreover, the fabrication of chitosan to prepare stable, porous bioscaffold, functionalized surface, lyophilization is also carried out. Blending using numerous preservatives may change its biomedical compatibility. It is necessary to evaluate the biomedical compatibility of numerous chitosan that is considered so they can be used for wound healing and other related treatments. The other benefits associated with these include easy processability to form membranes, scaffolds, nanofibrils, gels, nanofibres, beads, nanoparticles, microparticles and sponge-like forms (Fig. 2). Due to these properties and biocompatibility, widely employed in wound healing and tissue engineering, and in drugs and gene delivery [13].

Cross-linking of chitosan is another approach to enhance the performance. In crosslinked polymeric crosslinkers interconnect the chains of polymers that in turn help in the development of a 3D framework. The properties of crosslinked polymers depend on viscosity and molar ratio of crosslinking with a corresponding polymer chain. A precise number of crosslinks per chain is needed to allow the development of a framework. The structures of chitosan formed are: (a) self-crosslinking of chitosan; (b) composite network of polymer (c) nearly intercalating network; and (d) cationic/anionic crosslinking. Nature of crosslinking agent decides the type of bonding and interactions (weather it bond covalently or not) [14].

5 Derivatives of Chitosan

The chemical modification of chitosan can be carried out to anchor different functional groups including primary amine and primary and secondary OH groups (Fig. 4).

5.1 Quaternized Chitosan and N-Alkyl Chitosan

Quaternized chitosan can be termed as methylated chitosan. It is obtained using methyl iodide. Usually, the reaction between methyl iodide and chitosan is carried out in a basic environment, one of the most forthright way of quaternizing chitosan [15]. Proteins and peptide-based medicine are extremely adsorbed by an excellent sorbent which is chitosan. The working principle of chitosan for aiding the para-cellular transportation of water-loving drugs was recommended to be a mixture of bio adhesion and a passing broadening of the strengthen joint in the thin film mediated by H^+ functionalized chitosan in its open coiled configuration [16]. Chitosan and its metal derivatives facilitate the paracellular transportation of hydrophilic drugs by combining effect of bioadhesion and a transient broadening of the tight connections in the membrane arbitrated by protonated chitosan in its uncoiled configuration. However at 7.4 pH, chitosan and salts unable to further enhance the permeation rate because of dissolution issues. This character suggests that only chitosan can be an excellent sorbent in its original PKa which matches in the intestinal. This is the only reason that chitosan and its derivatives cannot be targeted by amino-linked protein-based drugs into the human intestine part that is jejunum to the colon [17].

N,N,N-trimethyl chitosan chloride (TMC) which is quaternarily derived are more stable in aqueous solution as compare to chitosan at higher pH. Such derivate of chitosan has used to enhance sorption for medicine. In literature tested for fluorescein-isothiocyanate dextran (FD4, MW4400) and mannitol, etc. The chitosan that is trimethylated also possesses mucous adhesive properties dependent on a number of quaternization in general mucous adhesive characteristics increases by the enhanced degree of quaternization. It is because of the enchantment of positive role of chitosan which increases its bonding with anionic mucin resulting in mucous adhesion [18].

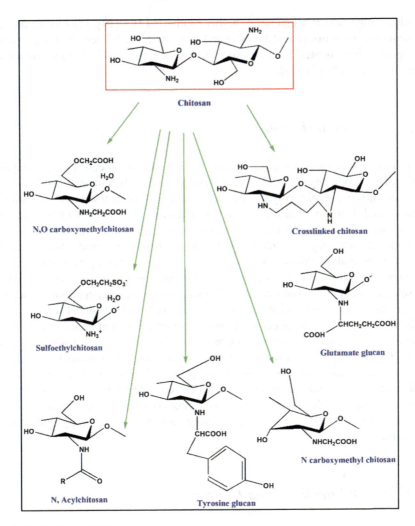

Fig. 4 Derivatives of chitosan

5.2 Hydroxyalkyl Chitosan

Hydroxyl (OH^-) functionalized chitosan is prepared by precursors such as ethylene oxide and propylene oxide (epoxide) reacting chitosan, glycidol. Basically, in an epoxide, the reaction occurred mainly at the groups like NH_2 or alcohol OH, giving *N*-hydroxyalkyl/*O*-hydroxyalkyl chitosan or maybe both. The ratio of *O*/*N*-substitution depends on types of catalyst (NaOH/HCl) used and reaction temperature. In the absence of a catalyst, *N*-hydroxypropylation is produced, while the use of acid catalysts results in predominantly *N*- but very less *O*-alkylated chitosan product. By using basic catalysts *O*-alkylation is dominant with the ability to produce oligomers at temperature higher (above 40 °C) [19]. Peng et al. prepared hydroxypropyl chitosan and investigated it as antibacterial activity [20]. Whereas Dang et al. investigated it for its probable use as an injectable carrier (temperature sensitive) for cells [21–23].

5.3 Carboxyalkyl Chitosan

The introduction of acidic groups in polymer backbone is called as carboxyalkylation. By introducing –COOH at the –NH_2 groups attached to chitosan, both acidic n basic mixed polyelectrolytes (that also contain both cationic and anionic) are synthesized. If the amount is changed then the exchange of the –COOH containing the group, one can get a molecular chain with different charge densities which provide an easy way of controlling PKa-based behaviour. Both, O-carboxyalkyl and N-carboxyalkyl chitosan have been synthesized by employing various reacting circumstances using single halogenated carboxylic acid to obtain the N contrasted O selection [24]. N-Carboxymethyl chitosan possesses some peculiarities including water solubility, remarkable biological, physical and chemical characteristics (highly dense, film and aqueous dynamic volume, gelation characteristics) also, those of which enables it to be optable in food products and make-over products. CH_3COOH modified chitosan also used in preparing of various hydrogel (porous, pH-sensitive, cross-linked hydrogels) for protein drug delivery systems as N,N-Dicarboxymethyl chitosan exhibited excellent chelating capabilities and its chelation with calcium phosphate suitable for osteogenesis thus encouraging bone calcination. O-CH_3COOH chitosan also shows antimicrobial activity and adhesive properties [25].

5.4 Sugar Functionalized Chitosan

Sugar-modified chitosan was first time reported by Hall and Yalpani. They prepared chitosan that is bounded to sugar reductive N-alkylation with $NaBH_3CN$ and aldehydic sugar derivative. At first, the sugar modified chitosan had been examined for rheological studies; but later on, this type of functionalization has been utilized to induce cell-specified sugars into chitosan. The synthesis of sugar bound chitosan, D- and L-fucose, and their interactions with lectin and cells was also reported [26]. Stredanska et al. prepared a chitosan derivative of lactose which is useful in the repair of articular cartilage. Kaneko et al. accomplished chemoenzymatic method for introduction of large carbohydrate as amylose on chitosan. The amylose-grafted chitosan also has been prepared it does not have a solubility in any solvent, (aqueous CH_3COOH and C_2H_6OS) which usually dissolve chitosan and amylose both and hence applied various specific fields [27, 28].

5.5 Cyclodextrin Linked Chitosan

Chitosan that contains cyclodextrin (CD) chains are produced with an intent to join extraordinary attributes of chitosan with the capability of cyclodextrin to shape complexes that are non-covalent in nature with various temporary particles changing their physical-chemical characteristics for the enhanced timely medicinal delivery framework, beauty care products, and systematic science and analytical chemistry. There are diverse means to connect the interface of chitosan to cyclodextrin. A researcher named Sakairi and his colleagues arranged a-CD-connected chitosan utilizing 2-*O*-formylmethyla-CD by reductive *N*-alkylation and affirmed the permanent-temporary particle complex of with p-$C_6H_5NO_2$. Another group of researchers, Auzely-Velty and Rinaudo additionally announced nearly same method of chitosan preparation containing cyclodextrin chain by amination in reducing conditions with the investigations of development of addition complexes with 4-tert-butyl benzoic acid [29]. The CD-chitosan derivatives arranged comparably with cyclodextrin single aldehyde has been assessed for mucous adhesion by a similar group. Chen and Wang acquired cyclodextrin linked chitosan utilizing tosylated b-cyclodextrin and further assessed the capability of b-CD for the release of I-131 in vivo and enhanced solubility [30, 31].

5.6 N-Acyl Chitosan

Anhydride and acyl chlorides are used to synthesize *N*-Acyl derivatives of chitosan. By and large, acylation reactions result carried out in an aqueous solution of CH_3COOH/CH_3OH mediums, pyridine/chloroform, pyridine, trichloroacetic acid/dichloroethane, ethanol/methanol mixture, methanol/formamide or DMA–LiCl. Because of genuinely unique reactivity of the two groups (hydroxyl and the amino) at the monomer unit of chitosan, acylation can be carried out in control manner at the predictable sites, i.e. on either $–NH_2$, –OH, or on the two groups. One of the techniques for obtaining acylated chitosan derivative incorporates the solid state thermal breakdown of its acyl-ammonium derivative. This technique was utilized to obtain amides of chitosan got from acids, for example, trifluoroacetic, acrylic, acidic, methacrylic and myristic [32].

N-Acylated chitosan with long polymer chain (C6–C16) chlorides expanded its water repelling character (hydrophobic self-gathering) and caused vital improvements in its basic properties. It was shown in enhanced mechanical characteristics of drugs synthesized utilizing these subsidiaries. The discharge attributes of the drugs showed that discharge is managed by dispersion or by swelling took after by dissemination; depend upon both the acylated chain length and the level of acylation. The acylation can be accomplished region selectively at the amino site by utilizing as trityl assemble at the basic OH group. This approach was utilized to get ready *N*-haloacyl 6-0-triphenylmethyl chitosan which can be additionally replaced by amines as tributylamine, pyridine, imidazole, triethylamine, *N*-chlorobetainyl chloride. The derivatives of betaine have two notable points of attention over the base chitosan: (i) their water solubility at physical pH, and (ii) they have a perpetual cationic charge on the polysaccharide backbone [33].

5.7 O-Acyl Chitosan

There are two main benefits of introducing of water repelling species into chitosan by ester linkage; the first one is that this water repelling species makes chitosan soluble in organic solvent and the second benefit refers to hydrolysis of ester linkage by lipase-like enzymes. Additionally, glycosidases enzyme helps in the breakdown of glycoside linkage of chitosan. Such properties make *O*-acyl derivatives of chitosan best materials for biodegradable coatings. It has been reported that *O,O*-didecanoylchitosan, *O*-succinyl chitosan was synthesized with *N*-phthaloylchitosan as an intermediate. In this procedure several steps are necessary to avoid phthaloylation of –NH_2 group, *O*-acylation, and at the end of the synthetic process the protected amino group is removed by N_2H_4. In recent years, a procedure is reported in which *O*-acylation of chitosan is carried out by methyl sulfonate [32].

5.8 Thiolated Chitosan

The chitosan is modified by adding –NH_2 group with –SH prompts the arrangement of –SH functionalized chitosan are water-loving larger molecules displaying non bonded –SH group clusters at the polymer spine. Up until now 4 kinds of –SH functionalized chitosan have been prepared: conjugated species as chitosan–4-thiobutylamidine, chitosan–cysteine, and chitosan–thioethylamidine conjugate. Numerous characteristics of chitosan are enhanced by particular stagnancy of –SH bunches assigning it to the auspicious new classification of –SH monomers utilized as a part of specific for the in the spreadable organization of water loving large molecules [34].

5.8.1 Mucous Adhesion Properties

Another promising property of chitosan is mucous adhesion because of charged connections among the cationic polymeric –NH_2 group and adversely ionic $C_{11}H_{19}NO_9$ and $C_6H_6O_3S$ of the mucus. These mucous adhesive characteristics of chitosan can enhance by the immobilization of thiol bunch on the polymer. The improvement of mucous adhesion can be clarified by the S^{2-} bonds development with cysteine enriched subdomains of subdomains sugar bonded proteins, which are more forceful than non-covalent bonds. This hypothesis was reinforced by the tensile strength results obtained with tablets of –SH functionalized chitosan, which exhibited a cationic connection among the level of adjustment with –SH containing moieties and the adhesive characteristics of the polymer [35].

5.8.2 Increase in Permeable Properties

The saturation of marker that is present between intestinal mucous layers of cells can be improved 1-3-overlap using thiolated chitosan as compare to virgin chitosan. Chitosan has the pervasion upgrading capacities with increment in the intercellular passage of absorption, which is vital for the transportation of water-loving substances (peptides and antisense oligonucleotides used for therapy) through the membrane. The working principle beneath this diffusion upgrading impact is by all accounts in light of the cations of the polymer, which cooperate with the cell layer bringing about an auxiliary revamping of tight intersection related proteins [36].

The pervasion upgrading impact of chitosan can be emphatically enhanced by the stagnancy of –SH groups. The take-up of light named bacitracin, for example, was enhanced 1.6-fold using 0.5% of chitosan–cysteine conjugate rather than virgin chitosan. The pervasion enhancing the impact of –SH functionalized chitosan has been examined with penetration mediator glutathione which shows that chitosan–TBA/GSH is a possibly profitable instrument for repressing the ATPase action of P-gp (P-glycoprotein) in the digestive tract [35].

5.8.3 Cohesive Properties

The chitosan with reduced –SH works on the chitosan chain permit –SH modified chitosan to develop inter and intra-molecular atomic S^{2-} bonds bringing about cross-linking of the polymeric chains. Subsequently –SH modified chitosan show, other than their solid mucous adhesive and saturation improving characteristics and incredible durable quality. This quality gives a robust attachment and strength of transporting networks being founded on –SH modified chitosan and can ensure a delayed managed discharge of installed therapy agent. –SH modified chitosan present in situ gelling capabilities because of –SH groups oxidation at physiological pH-values, which bring about the development of disulfide bonds (inter and intra-molecular). To use –SH modified chitosan in nasal, buccal, vaginal, and ocular mucous treatment, it needed to be crosslinked and in situ gelling at pH ranging between 5 and 6 [37].

5.9 Sulfate Modified Chitosan

Sulfate derivatives of a chitosan exhibit quite significant group of chitosan-based materials that can show a wide range of bio-based activities. Sulfonation is carried out using various reagents including oleum, concentrated sulfuric acid, sulfur trioxide, sulfur trioxide/trimethylamine, chlorosulfonic acid–sulfuric acid, sulfur trioxide/pyridine, sulfur trioxide/sulfur dioxide, tetrahydrofuran, and formic acid over wide ranges of temperature or with help of microwave irradiation. On chitosan, the additional sulfa group can be further substituted to form sulfanilamide derivatives of chitosan followed by the reaction among $-NH_2$ or $-OH$ (C6 position) groups [38].

Such types of chitosan sulfates possess anti-coagulant and iron agglutination control activities because of the chemical structural resemblance to heparin. In addition to that chitosan, sulfates have shown other biological activities such as antioxidant, anti-sclerotic, anti-viral, anti-HIV, anti-bacterial, and enzyme control activities. Sulfation of chitosan results in the conversion of some of the $-NH_2$ groups to negative ion centers and the polymer with improved multi electrolyte characteristics which can be used for generating potential drug transporters microcapsules or micelles form. N-Alkyl-O-sulfated chitosan show amphiphilic character as it consists alkyl substituted polymer chains having water repelling nature and SO_4^{2-} groups with water-loving nature. Both waters loving and repelling polymer converted into micelles which physically entrap drugs that are water-insoluble as taxol in substantial concentration. Another very promising quality of chitosan sulfates is sorption that is used for recovering metal ions [39].

5.10 Phosphorylated Chitosan

Phosphorous acid and formaldehyde are allowed to react with chitosan either in consecutive phases or at the same time in an acidic solution in water give condition to form a single or double bond of nitro-phosphonic-methylene chitosan. The framework of phosphorylated chitosan materials and a number of rearrangements reliant on reaction conditions, reactant ratio, and most significantly reaction time. H_3PO_4 modified chitosan is prepared in the P_2O_5—the CH_4O_3S system also. Using H_3PO_4 modified chitosan; the novel gel beads mixed with multi electrolyte have produced for ibuprofen drug as model medication by utilizing a gel that is ionotropic with P_3O_{10} for managed drug conveyance framework management during oral intake by keeping away from the medicine discharge in the exceptionally acidic gastric liquid area of the stomach. The pre-modified chitosan can likewise be stretched out with P comprising groups for instance the $-COOH$ group of CH_3COO^- chitosan created to respond with $-NH_2$ of phosphatidylethanolamine managing both acidic and basic polymer. This polymer was explored for its practicality as a conveyance transporter for the transfection of hydrophobic model drug ketoprofen by shaping dots on ionic cross-linking by sodium tripolyphosphate [40].

5.11 Enzymatic Modification of Chitosan

To modify chitosan another approach can be done by using enzymes and this method is thought-provoking because of its particularity and ecological benefits in comparison with chemo functionalization. Enzymes present the likelihood of removing the threats related to chemical substances in term of to health and safety. Enzymatic grafting of phenol species on the chitosan was first studied by Payne et al. to analyze aqueous solubility in a basic environment (Scheme 17). Quercetin and rutin quinones were effectively synthetically connected to low molecular weight chitosan. The quercetin-adjusted chitosan demonstrated an improvement of plastic, cancer prevention agent and antimicrobial properties and in addition to that also showed thermal degradability [41].

5.12 Graft Copolymers of Chitosan

Chitosan physio-chemical properties can be changed by graft copolymerization which is an attractive system for expansion of their application. The characteristics of the side chains, as well as atomic structure, length, and number comprehensively manage the attributes of the subsequent unite copolymers. Up till now, numerous research projects were carried out to study the effects of these parameters on the connecting factors and the features of polymer grafted chitosan [42].

5.12.1 Grafting Co-polymerization by Radical Production

Various types of polymeric species such as poly (vinylic and acrylic) artificial precursors are the most often grafted on polysaccharides. Radical polymerization is commonly used to prepare copolymers. In the archetypal method of free radicals production in which during first step free radical attached at backbone structure of polymeric chain followed by these free radicals react as large initiators for small repeating agents. The grafting % age and grafting efficacy are mainly achieved by initiator type precursor and initiator amount, reaction time and temperature [43].

5.12.2 Polycondensation to Form Co-polymerization

Grafting co-polymer of polysaccharides is not often prepared by condensation polymerization typically due to the vulnerability of the sugar chain backbone to elevated temperature and punitive conditions of the characteristic polycondensation reactions. The condensation polymerization process has successfully produce lactic acid, LA grafted chitosan of D, L-lactic acid in nonappearance of a catalyst which generates a pH-sensitive hydrogel. The polylactide condensation has also been attained by application of catalyst 4-dimethylaminopyridine in which polylactide was attached through the OH^{-1} group's phthaloyl chitosan. $-NH_2$ group is linked to chitosan by condensation through carbodiimide process on the surface of PLA to increase the stickiness, strength and cell compatibility of human endothelial cells [42].

5.12.3 Coupling to Copolymerize via Oxidation

Polyaniline is polymerized on the chitosan to form conductive polymers by a well-known process called oxidative coupling. Copolymerization is not generally utilized for getting grafted copolymers of a long chain of sugar for the most part because of the vulnerability of the saccharide at elevated temperature and extreme conditions required for performing polycondensation reaction. However, the pH-sensitive hydrogel was formed by successful graft copolymerization of lactic acid (LA) over chitosan by condensation without catalyst which results in a hydrogel that is pH-sensitive. Another method is used to connect—an NH_2 group of chitosan onto the polylactic acid surface which is carbodiimide procedure to make good cell compatibility grip and bolster the expansion of human endothelial cells [44].

5.13 Film-Forming Properties of CS

The inherent film forming ability and antimicrobial character enable chitosan to make potential use in the packaging industry. However, chitosan forms rigid films so required plasticizers in order to lower frictional forces (hydrogen bonds or ionic forces) among polymer chains, for the enhancement of mechanical properties. The induction of polyols for the fabrication of film may overcome this problem and allows the film to be mechanically stable for the required time period. Chitosan films also offer a well-established application that is food packaging which reduces oxygen permeability and water vapour permeability that results in the increased storage of food product by giving it longer shelf life [45] (Fig. 5).

5.14 Chitosan and Proteins

The combination of chitosan with other hydrocolloids results in improvements of functional features of chitosan-based films. Chitosan and pectin coated membranes have been formed by the interface of the chitosan (positive ion groups) with pectin (negative ion groups). This combination results in lowering of water vapour transmission rates (WVTRs) has been observed. The physical and mechanical characteristics of these biobased membranes were improved by joining various

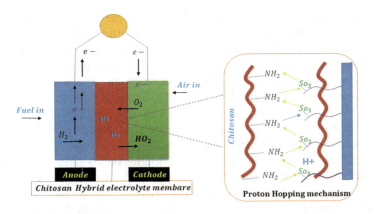

Fig. 5 Cs based membrane in fuel cell application

types of proteins including soy protein, collagen, gelatin and milk proteins with polysaccharides (e.g. alginates, chitosan, starches and cellulose) [46]. Gelatin and chitosan combination results in fabrication of homogeneous membrane because of to the better miscibility among both biopolymers results into the improvement of characteristics of the thus formed mixed membranes as compared polymers films. The betterment in mechanical perspectives occurred due to development of electrostatic forces among chitosan ammonium groups and the gelatin carboxylate groups. However incomplete miscibility among polymer and protein results brittle structure as in the case of chitosan/soy protein blended membranes and phase separation occurs. The brittles are increases upon increasing protein contents. Such type of studies were aimed to develop edible films with anti-bacterial properties [47].

5.15 Chitosan and Starch Blends

Food packaging requires good mechanical properties as a basic necessity that must be edible coatings and films to enhance manual or machinery based food handling or pharmaceutical products. Tapioca and rice are two main materials that are used as starch sources that are used to form blends with chitosan and produce edible food coatings, the consequential films possess dynamic properties better than one single polymer. In South America famous crop is tapioca. The membranes produced by this food product have good physical properties like tasteless, odourless, colourless and oxygen impermeability. The only drawback associated with these membranes is brittleness that results in insufficient mechanical strength. The blending of tapioca and chitosan with the plasticizing agent of glycerol was first done by Chillo et al. During their research they investigated mechanical, dynamic mechanical and ostensible viscosity of the film-forming solution [48].

The mechanical strength is critical for eatable plastic films and covering to enhance mechanical treatment of food item or medicinal stuff. Starch from custard and rice has been combined with chitosan to utilized mix qualities than from the single polymer alone. Custard edible films show suitable physical attributes since they have no colour and taste and also impermeable to oxygen. Be that as it may, films demonstrate weakness with insufficient mechanical strength Chillo et al. have prepared (CS) and custard starch films with and glycerol (plasticizer). The mechanical properties, water vapour porousness and shade of the mix films have been examined and improvement was found [49].

The effect of biodegradable mixture of CS ratios on TS films was investigated. The results showed that improvement in TS of TS of prepared occurred by the maximum induction of chitosan was done 1:1 ratio of chitosan and starch was used. The enhancement in TS of the films upon increase of rice starch and chitosan ratios (2:1–0.5:1) is attributed to the development of hydrogen bonding between molecules of chitosan (NH_2) and the hydroxyl group of rice starch (OH). Amylopectin and amylose are attractive raw materials may use as obstacles in packaging films. They utilized for the preparation of bio-decompose able films to replace plastics partially or entirely as they are less expensive and renewable as well as better mechanical strength. To decrease the water vapour permeability of the films starch was blended with different proteins which also increases tensile strength [50].

5.16 Edible Membranes of Gelatin and Chitosan

The edible coating and membranes made up of the composite are fabricated to obtain combine benefits of every constituent. While proteins and polysaccharides are used as a strong polymer matrix, lipids give the best barrier toward vapours of water. As chitosan and gelatin both possess water loving character with better attraction as well as matching, they produce membranes of composite materials with improved characteristics. Chitosan/gelatin composite has been utilized widely for the synthesis of frameworks in biomedical applications. Rivero et al. prepare gelatin/chitosan, composite biodegradable films and investigate membrane barrier properties and structural power and to analyze their micro to the nanostructure. The composite has a showed even and regular upper layer as examined by Scanning Emission Microscopy and X-ray analysis. The influence molecular mass of chitosan along with the level of deacetylation on the physical and chemical characteristics of the prepared membrane was determined. Results exposed that bonding among chitosan and gelatin were stronger in films composed of larger molecular mass chitosan or greater level of deacetylation as compare to films made up of low molar mass or deacetylation level [51].

5.17 Composite of Chitosan, Carrageenan and Alginate

Membranes based upon alginate are impenetrable into lipids at the same time as other water-loving long chains of sugar molecules which also possess water vapour permeability. Though, a sacrificing agent used in the membrane is alginate gel film in which membrane lost its own water content prior to the loss of food water content. Polyelectrolyte produced by mixing of chitosan with both carrageenans and alginates which were utilized to get microstructure capsules of cell embodiment and gadgets for the managed arrival of medicine or different materials. It appears there is great probability to examine this collaboration to create consumable membranes from these substances which could be of implausible esteem [52].

5.18 Chitosan and Clay Natural Polymers

Natural polymers are hydrophilic in nature so they have the low mechanical strength and high moister barrier. Many different types of methods have been discovered to improve such kind of problems associated with biodegradable packaging films made up of chitosan [53].

These may include induction of plasticizing agents (glycerol) to increase the flexibility in the final product. In other methods silicates nanoparticles (e.g. sodium montmorillonite MMT) are added into chitosan to enhance barrier and mechanical characteristics of end-user product. MMT is a layered silicate having increased thickness and surface area and thickness enables its use for strengthening uses. The literature reported the synthesis and characterization of MMT/chitosan-based composites and membranes. Chitosan and MMT interact with each other and develop homogenous film. In an ion exchange reaction of nanocomposites based upon MMT/Chitosan occur among chitosan and sodium MMT. The chitosan offered high attraction toward montmorillonite host. Powder X-ray diffraction and Thermogravimetric analysis (TGA) and analysis showed that due to electrostatic interaction among cationic chitosan molecules and anionic silicate layers the thermal stability of chitosan improved significantly. In addition to that prepared nanocomposites also showed a synergistic effect on antibacterial properties against S. aureus and E. colcertain bacterial species [54].

The antibacterial property enhances markedly on increasing amount of MMT. The biodegradation rate of nanocomposites film of chitosan/montmorillonite is also the most prevalent than that of the virgin Chitosan polymer. These results showed that MMT/CS nanocomposites have potential antimicrobial applications especially for those comprising 0.1% of MMT. Chitosan/montmorillonite bio-based composite microparticles showed more temperatures of thermal degradation as compared to pristine particles of chitosan. The glycerol and MMT were added into chitosan and the dual effect is examined by a group of researchers. The structural

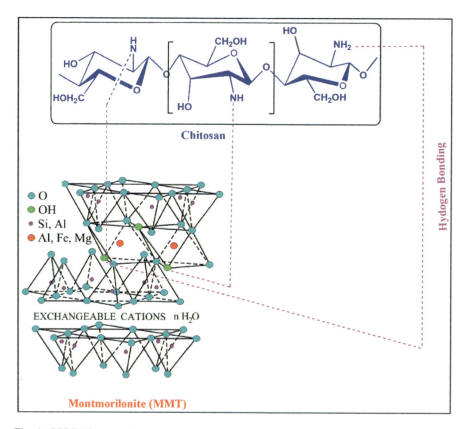

Fig. 6 CS/MMT composite membrane

strength of glycerol-based nanocomposites was increased by increasing clay loading. This is because of syngertic impacts of both plasticizing agents and clays [55] (Fig. 6).

Hydrogen bonding framework was modified by glycerol inside the material which permits the good combination of matrix with filler, which in turn simplify strain transformation among the supporting matrix phase to enhance mechanical strength. 30% water vapour incorporation decreases by glycerol phase addition. Hydrogen bonding of montmorillonite and chitosan was significantly reduced by glycerol that blocks flocculation and breaks the montmorillonite stacks that are randomly arranged in space. The membranes of chitosan prepared without glycerol show 50% reduced permeability because of flocculation and alignment of montmorillonite stacks. Because of these frameworks material has good capability to use the membrane in the field of packaging. Glycerol adjusts the hydrogen bonding inside the substance and allows good cooperation amongst filler and matrix, subsequently encouraging the pressure to the support stage and enhancing its powerful properties [55].

5.19 Properties of Gas Permeability of Edible Coatings

The number of many eatable membrane coverings for fruits for example cellulose, zein, soy protein, chitosan and casein. They possess many required properties such as no taste and odour along with transparency. Though, oxygen permeability can be enhanced by antibacterial activity which was managed by adding potassium sorbate in starch-based membranes used for sweet potato coverings. But oxygen impermeability can better be achieved by composite membranes of chitosan and starch. 15% addition of chitosan, the starch of sweet potato source showed incredibly lower oxygen permeability. There are a few conceivable edible coatings for natural products, for example, soy protein, cellulose, casein, zein, and chitosan. These were picked since they have the attractive qualities of for the most part being transparent, odourless, and tasteless [56].

5.20 Antimicrobial Applications

Cations present in chitosan makes material an antimicrobial agent. The advances in the field of antimicrobial material need characteristics found in chitosan. It is a non-toxic polymer with dynamic antimicrobial properties that may be used as an excellent matrix for eatable coating membranes. The chitosan bonding or chelation with endotoxins related to the bacterial type of Gram-negative reduced the toxicity of such bacteria. Due to this good chelating property, EDTA like external chelating agents is not necessary. This property made chitosan good inhibitor of a wide range of microbes such as viruses, bacteria and fungi etc. [57].

5.21 Anti-inflammatory Applications

Chitosan has a range of encouraging biomedical applications and right now, is considered as another inventive material in wound healing, a hemostatic agent, antimicrobial, lipid binding effects as demonstrated by the substantial number of reports throughout the most recent couple of years. Chitosan films that are planned for wound administration may incite absence of pain by giving a good, charming and relaxing impact when connected uncovered skin or injured part. Chitosan possess fantastic agony helps when it was connected as a covering agent to open injuries, for example, burnt skin and scraped spots, skin grafted regions and skin sores [58].

Because of anti inflammatory impacts of chitosan, it provides benefits for the management of delayed swelling and sore at the injury site. Chitosan which is water solubility strangles the discharge and possession of pro-inflammatory cytokines and induction of NO_2 to prepare astrocytes which is common nerve cells in CNS and is effectively associated with inflammatory events related to the cytokine. In addition,

N-acetylglucosamine is anti inflammatory drug and is prepared in the human body from glucose. Chito-oligosaccharides (atomic weight of 5 kDa) demonstrated preferred hostile to inflammatory specialists over nonsteroidal against swelling. Chitosan applies anti-selling impacts by restraining protein possession and weakening the pre-inflammatory cytokines [59].

5.22 Biomedical Applications of Chitosan

Chitosan is used in different health care disciplines and hygienic applications, such as carrier, wound treatment, gene distribution, and bone tissue manufacturing. Chitosan is used to prepare artificial kidney membrane, it is applied in hypocholesterolemic agents, and supports for immobilized enzymes and drug delivery systems, absorbable sutures. Chitosan has many benefits due to its nontoxicity and biodegradability without damaging the environment. It slowly breaks down to harmful products that are absorbed completely in the body as it is biocompatible [60] (Fig. 7).

5.23 Chitosan-Based Composite Scaffolds in Wound Healings

The biocompatibility of composite nanofibers of chitosan/sericin was manufactured by electrical spinning having a great texture with a small diameter between 200 and

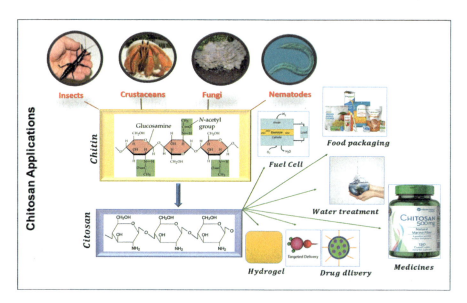

Fig. 7 Applications of chitosan

400 nm. As these composite membranes have property to block the activity of gram positive and negative bacteria so can be used as wound dressing membranes. They also have cell multiplication property so heal the wounds in short time in approximate 24 h and more than 90% at all test exhibiting the benign and biocompatible nanofibres toward cells [61].

AgS based alginate/chitosan membranes are good for wound healing as silver has sensitizing properties while chitosan and alginate heal the wound by covering. The particular sulfonation of oxide ion and additionally chitosan may create a powerful anti-retroviral agent that shows a substantially higher inhibition impact on the contamination of AIDS infection. The managed $C_{10}H_9AgN_4O_2S$ discharge was appeared with incorporated chitosan layers with water vapour vanishing, satisfactory swelling capacity, cytocompatibility and delayed antibacterial action [62].

The iron oxide, chitosan and gelatin-based nanofiber composite membranes show enhanced antibacterial properties along with the mechanical support that possesses a material with more promising wound dressing abilities. The freezing-thawing method was used to develop chitosan/PVA/chitosan/MMT nanocomposite as a biocompatible wound dressing. Enhanced mechanical characteristics, as well as other features including good swelling behavior biocompatibility and antibacterial activity, made it required a candidate for wound dressing applications [63].

The sponge-like dressings in light of glutamate chitosan (large molar mass) and sericin was produced for the cure of older skin sores. The sericin measurements improved skin healing is appropriate to apply a defensive impact on oxidative harm to human fibroblasts. In addition, the upgraded bandage can enhance fibroblast expansion, that is, to enhance injury recuperating. Several studies conducted the research utilization of $scCO_2$ as an environmentally benign media to actuate chitosan porosity frameworks. Sponges of chitosan with cross-linking stacked of anti-toxin medicines the norfloxacin were set up by dissolvable vanishing system [64].

6 Introduction to Chitin

Chitin is obtained from the exoskeleton of crustacea, creepy crawlies, and a few fungi [65]. The fundamental business wellsprings of chitin are the shrimps, krills, lobsters, and crab waste shells. On the planet, a millions of tons of chitin are reaped every year and subsequently, this biopolymer provides an inexpensive and accessible source [5, 66] (Fig. 8).

In chitin, the degree of acetylation (DA) is ordinarily 0.90 showing the occurrence of amino groups. The N-acetylation, i.e. the proportion of 2-acetamido-2-deoxy-d-glucopyranose to 2-amino-2-deoxy-d-glucopyranose basic units strikingly affects its solubility and structural properties [67] (Fig. 9).

The polymorphic types of chitin vary in the arrangement and polarities of neighbouring chains in progressive polymer chain [68]. For the most part, the individual chains expect a basically straight texture, which experiences one full coil

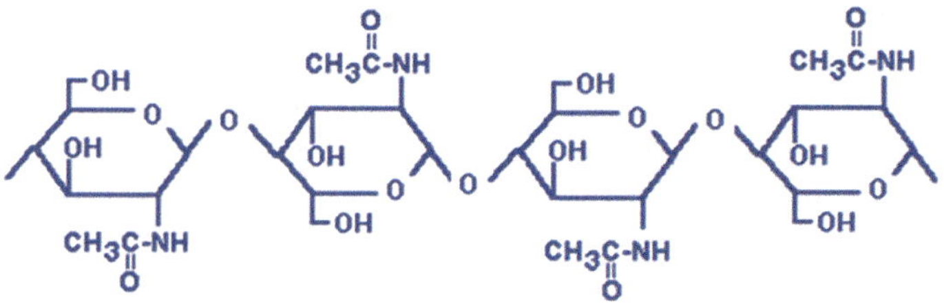

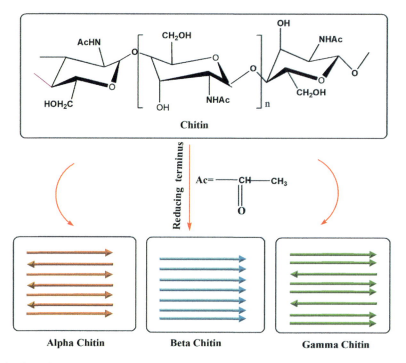

Fig. 8 The molecular structure of chitin

Fig. 9 The polymorphic types of chitin

each having diameter 10.0–10.49 A' along the axis of the chain. Since the chain contain chiral glycosidic units, and all units are associated by an oxygen atom that connections carbon one of one glycosidic unit to carbon four of a neighbouring unit, a particular "left" and "right" arrangement can be allotted to every polymeric chain. The most widely recognized chitin allomorph is known as A-conformation [69], where unit cell possess orthorhombic form while individual polymer chains show antiparallel design. In this manner, contiguous chains are situated in inverse ways. Another less regular allomorph termed as b-conformation relates to a monoclinic unit cell having polymer chains arranged parallel form by weaker intermolecular bonding. A third shape, c-chitin, have two parallel chains parallel form in a relationship with chain arranged in an antiparallel way [70]. This may be reflected as a variation of a shape. XRD and NMR studies of these distinct allomorphic forms have been shown to be different [71, 72].

7 Chemical Modifications of Chitin

Chitin is a synthetic biopolymer with excellent stability. a-chitin in view of its insoluble nature it is seldom exposed to chemical reaction with the exception of the synthesis of chitosan following by deacetylation. b-Chitin has generally more reactive in nature [73]. As indicated by Noishiki et al. b-chitin can be changed into thermodynamically more stable a-chitin by treating with NaOH (20%) followed by resining with DI water. Chitin has been handled in an assortment of approaches to get altered physicochemical characteristics [74]. Vincendon prepared chitin solution in phosphoric acid (H_2SO_4) at ambient temperature, he observed that viscosity and molar mass were decreased with time but the degree of acetylation remain unchanged. The chitin is first allowed to disperse in NaOH (concentrated solution) and kept at room temperature for 3–5 h; the obtained basic chitin is then poured into ice at 0 °C. This system permitted the synthesis of chitin membrane with great mechanical strength and transparency. The prepared chitin is indistinct and by maintaining specific circumstances, it can be broken down in the presence of water. This wonder is explained by the fact of to the lowering of molecular weight under basic environment and deacetylation to some extent [75–77].

It is cleared that to obtain water dissolvable chitin, the level of deacetylation must be almost near to 50% and, likely, that the acetyl group must be consistently dispersed along the chain to keep pressing of chains causing from the interruption of the second configuration in the strong alkaline medium. The impact of this disruption was inspected and it was demonstrated that the dispersion, irregular or blockwise arrangement, is imperative in regulating structural characteristics. Deacetylation around 51%, of an exceedingly deacetylated chitin within the sight of acidic anhydride, produce a water-dissolvable derivative [78].

The chitin derivatives reported in literature are phosphoryl chitin carboxymethyl chitin [79], hydroxyalkyl chitin [13], N-and O-sulfated chitin, hydroxybutyl chitin [80], fluorinated chitin [81], (diethyl amino)ethylchitin [82], mercapto chitin and chitin carbamates [83]. Chitin can be blended with regular or manufactured polymers; and can be cross-linked using various crosslinking agents (TEO, epichlorohydrin, glutaraldehyde, and so forth.) to modify and obtain desired properties [84].

8 Chitin Fiber Formation

Chitin sutures have noteworthy characteristics over different other fibers for biomedical applications. One study showed that chitin fibers have practically identical characteristics to as those lactide and collagen fibers have. The linear chain framework of chitin is enables it to form fibers and film alike to cellulose fibers. The existence of the microfibrils in chitin that are typically engrained in a matrix of protein showed the possibility of chitin spinning into fibers. The polyamide-type structure ought to be separated to empower solubilization of chitin into a dissolvable form. This needs either softening or disintegration in apt solvents. Melting turning is discounted as chitin deteriorates before softening. Regarding this many endeavours have been carried out to control disintegration of chitin and turning of chitin and chitosan into fiber shape. The arrangement of chitin fibers for the creation of suture with absorbing capacity, dressings, and decomposable substrates for the development of human skin cells filaments has been accounted so for [85, 86] (Fig. 10).

9 Preparation of Blends with Other Fibers/Polymers

The induction of chitin fibers into various synthetic polymers, composites or blend results in wide range of interesting characteristics. Young and Eichhorn discussed the process chitin fiber formation in detail. In the synthesis of blend composed of alginate and chitin, the spinning of their mixture solution was carried out by a spinneret with a coagulating bath having aqueous $CaCl_2$ and ethanol. The durable interface developed via the intermolecular hydrogen bonding was utilized to certify excellent miscibility. The good dry tensile strength and elongation to break were gained with 30 wt% of chitin content. While the tensile strength and elongation to break lowered by further increasing chitin content (water-soluble). Moreover, the addition of chitin into the blended fiber can enhance the water-holding characteristics of the blended fiber comparable to pristine alginate fiber [13, 87].

The treatment of chitin fibers with silver nitrate solution were results in enhancement of great antibacterial activity to Staphylococcus aureus. Noteworthy change in properties have been accounted for mixes of chitin/CS fibers with different natural and synthetics fibers to obtain chitin–glycosaminoglycans, chitin–silk fibroin, chitin–cellulose, chitin–cellulose–silk fibroin, and chitin–cellulose–silk fibroin, CS–tropocollagen, and chitin–natural rubber blends. The incorporation of chitin fibers in poly(lactic acid) polymer indicated reasonable mechanical properties and maintenance for settling destructive bone cracks, yet likely had lacking solidness for applications, for example, bone plates for settling cortical bone breaks [88].

Extraordinary properties could be worked by suitable compound alteration to create a progression of chemically modified fibers, for example, *N*-acylCSs,

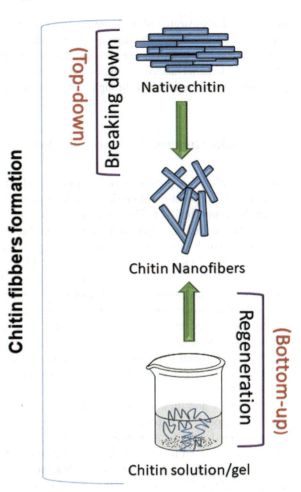

Fig. 10 Chitin fiber formation

N-arylidene-and N-alkylidene CSs, N-acetylCS, chitin–tropocollagen and CS–transition metal complexes. The crystallinity and surface charge thickness of the deacetylated chitin can be expanded on treatment with hydrochloric acid treatment to enhance the fiber properties. It ought to be noticed that East and Qin utilized warmth treatment for getting regenerated by reaction (N-acetylation) between CS and acetic acid. The best properties for rigidity (4 g/d) and modulus (100 g/d) for chitin were accounted for by the blended ester of chitin or CS acetic acid derivation/format polymer [89].

The utilization of chitin whisker course might be useful getting ready high-quality fibers. A further change in fiber properties could be accomplished with the utilization of spinning fiber from lyotropic fluid crystalline solution. Fiber spinning from fluid crystalline solution has huge benefits for expanded quality and different properties. Irradiation of chitin fibre-strengthened poly (caprolactone) composite demonstrated 45% change in rigidity and pliable modulus as for those of the untreated examples. Polymers, for example, polyvinylpyrrolidone, methylcellulose, and sulfite cellulose are accounted for to be utilized to change the properties of chitin filaments added to the spinning solution. Encourage change in fiber properties could be affected through suitable synthetic alterations [90].

10 Chitin General Characterization

The name 'chitin' is gotten from the Greek word 'chiton', which means a layer of mail [91]. The utilization of chitin was first depicted by the French scientific expert, Henri Braconnot in 1811. The structure of chitin $(C_8H_{13}O_5N)n$ shows resemblance with cellulose structure, yet with 2-acetamido-2-deoxy-β-D-glucose (NAG) monomer units, which are joined to each other by means of β(1 → 4) linkages. The material type of chitin is generally a white and hard nitrogenous polysaccharide which is inelastic. It has also been considered to be the main cause of beach contamination in coastal areas. The widespread presence of chitin in the biosphere and its insolubility prompted the possibility that chitin ought to survive in fossils. There have for sure been reports of fossilized chitinous materials, e.g., in Pogonophora and in insect wings preserved inside amber. The immunogenicity of chitin (disregarding nearness of nitrogen in its structure) is uncommonly low. Chitin is a very insoluble material that takes after cellulose in its low dissolvability and chemical non-reactivity.

Chitin now and then is thought to be a cellulose derivative; be that as it may, it doesn't happen in cellulose creating creatures. There is no, by and large, acknowledged classification concerning the level of N-deacetylation of chitin and its derivatives. Chitin has a high level of nitrogen (6.89%) contrasted with artificially substituted cellulose derivatives that must be set up with a lower nitrogen content (1.25%). The vast majority of the normally happening polysaccharides, e.g., alginic, dextran, cellulose, pectin, corrosive, agar, agarose and carrageenan are nonpartisan or acidic in nature, while chitosan is an example of a profoundly fundamental polysaccharide. Other remarkable properties of chitin composite include improvement of capacity to frame films, biocompatibility, biodegradability, non-poisonous quality, atomic adsorption properties, and so on. Despite a few reports demonstrating the synthesis of functionalized chitosan derivatives with chemical modification of the amino acid, not very many of these have worthy dissolvability by and large natural solvents, or binary dissolvable frameworks. Some synthetically adjusted chitin and chitosan derivatives having enhanced dissolvability all in all natural solvents have been accounted for.

11 Chemical Structure and Properties

The individual sugar units in chitin structure are turned 180° concerning each other, and pairs develop the disaccharide N,N'-diacetylchitobiose [(GlcNAc)$_2$] [75, 92, 93]. The single polymer chains can be depicted as helices, in which each sugar unit is transformed as for its neighbours. Such a structure prompts high strength as the inflexible strips are associated with O3-H $\rightarrow$ O5 and O6-H $\rightarrow$ O7 hydrogen bonds. Chitin additionally has three distinctive crystalline allomorphs: the α-, β- and γ-shapes depending upon the orientation of microfibrils [94].

The commonest type of chitin is α-chitin. Its unit cell is made out of two N,N'-diacetylchitobiose units shaping two chains in an antiparallel course of action. Along these lines, nearby polymer chains keep running in inverse ways, held together by O6-H $\rightarrow$ O6 hydrogen bonds, and the chains are held in sheets by O7 $\rightarrow$ H-N hydrogen bonds [95]. This gives a factual blend of –CH$_2$OH orientation identical to a large portion of the oxygen molecules on every residue, having the capacity to develop intra and intramolecular hydrogen bonds. This outcome in two unique kinds of amide group; all are engaged in the development of interchain C=O $\rightarrow$ H-N bonds, while 50% of the amide group additionally fact as acceptors for O6-H $\rightarrow$ O=C intramolecular hydrogen bonds. Development of these intermolecular hydrogen bonds prompts a significant stable structure [96].

β-chitin is a less regular type of chitin, where the unit cell is a N,N'-diacetylchitobiose unit, giving a polymer balanced out as a rigid and inflexible ribbon, by O3 $\rightarrow$ O5 intramolecular similar as α-chitin, H-bonds [97]. The chains in this structure are joined in sheets by C=O $\rightarrow$ H-N H-bonds among the amide group and by the –CH$_2$OH side chains, which prompts development of intersheet H-bonds to the carbonyl oxygens on the neighbouring chains (O6-H $\rightarrow$ O7). This gives a structure of parallel poly-N-acetylglucosamine chains having no intersheet H-bonds. The parallel game plan of polymer chains in β-chitin takes into consideration more adaptability than the antiparallel gameplan found in α-chitin, yet the resultant polymer still has massive quality [98].

γ-Chitin is the third allomorph, having blended parallel and antiparallel configuration. It is found in mushrooms [99]. Chitin is constantly found to be cross-linked to other auxiliary segments except for the β-chitin found in diatoms. Chitin is discovered covalently clung to glucans in the fungal cell wall, either directly attached, as in Candida albicans (1) or by means of peptides. Also, in insects and other invertebrates, chitin is constantly connected with particular proteins, with both covalent and noncovalent holding. This affiliation implies it creates the observed ordered framework. There are likewise differing degrees of mineralization, for example, calcification, and sclerotization, including associations with phenolic and lipids [95]. In organisms (fungi and invertebrates) there have been studied different degrees of deacetylation, providing a continuum of structure among chitin (fully acetylated) and chitosan (fully deacetylated) [100]. Although either acids or alkalis can be utilized to deacetylate chitin, the fact that glycosidic bonds are more vulnerable to acid which would damage the chain, the alkali deacetylation procedure is used more often [101].

N-deacetylation of chitin can be carried out by or homogeneous or heterogeneous reaction mixture [94]. The difference amongst chitin and chitosan with various degrees of deacetylation isn't strict. Setting a couple of special cases aside, chitin founds naturally related with other basic polymers like proteins or glucans, which frequently contribute over half of the mass in chitin-containing tissue [102]. Chitin can be N-deacetylated to such a degree, to the point that it gets noticeable solubility in dilute acidic and formic acids. The acetylated units of chitin prevail and the degree of acetylation is normally 0.90, while chitosan is a partially or fully N-deacetylated derivative possess a degree of deacetylation of greater than 0.65. Numerous analytical techniques have been utilized to find out degree of deacetylation such as IR spectroscopy, gel permeation chromatography, pyrolysis gas chromatography, and UV-vis spectrophotometry, solid-state NMR, H NMR spectroscopy, thermal analysis, acid hydrolysis, separation spectrometry methods, various titration schemes, HPLC and infrared spectroscopy [91].

12 Chitin Biosynthesis

The process of biosynthesis and cellular processing of chitin is very complicated, multi-faceted and interconnected sequence of occasions which begins intracellularly and ends up in inclusion of chitin in external supra-macromolecular framework such as cuticles, arthropod and fungi cell walls [103, 104].

The whole process consists of different distinct steps:

1. Successive biotransformation of sugars (specifically glucose or trehalose). This step comprises biochemical reactions such as amination, phosphorylation, and development of the enzymic substrate.
2. Chitin synthase (CS) prepares the chains. The CS enzyme is a part of a protein/carbohydrate cluster which is narrowly topologically packed molecules. Such types of organization confirm the amalgamation of nascent chitin polymers into a crystalline fibril.
3. The configuration of chitin molecules which have long chains.
4. Polymer translocation over the plasma membrane.
5. Crystallization and development of microfibrils by inter-chain hydrogen bonding.
6. Association with arthropod cuticular proteins or with other carbohydrates in fungal cell walls.

Chitosomes are cytoplasmic microvesicles. These microvesicles have been distinguished by electron microscopy utilizing fungal frameworks [105, 106]. The plenitude of chitosomes at the hyphal tip infers their critical part in CS trafficking to pre-decided areas. Chitosomes begin from organelles, for example, endoplasmic reticulum and Golgi. Chitosomes vesicles having zymogenic chitosan bunch. After fusion of the chitosomes with the plasma membrane, the chitosan units become activated through proteolytic reactions [107].

After in the long run combination of the chitosomes with the plasma film, the chitosan units become triggered via proteolytic reactions. Chitosan addition into plasma layers includes the intervention of focusing on and recognition proteins. Chitosome-like structures likewise have been accounted for in without cell framework derived from insects [108]; be that as it may, it has not been elucidated whether they are associated with in vivo chitin development. Also, chitosome-like vesicles have not yet been depicted in intact insect epidermal cells. Diverse types of chitin are combined with the activity of the catalyst, chitin synthase UDP-*N*-acetyl-D-glucosamine [109] chitin4-β-*N*-acetylglucosaminyl-transferase. Chitin synthase utilizes UDP-*N*-acetylglucosamine (UDPGlcNAc) as the enacted sugar donor to create the chitin polymer. Candy and Kilby first proposed a chitin blend biosynthetic pathway in insects. The proposed process began with glucose and finished with UDP-GlcNAc. Jaworski et al. utilizing cell extricates from the southern armyworm Spodoptera eridania, at last, settled the entire pathway from UDP-GlcNAc to chitin. Results obtained from many subsequent studies conducted with synthesis from different insects reinforced the molecular mechanism of this pathway [110] (Fig. 11).

13 Industrial Processing of Chitin

The detachment of chitin from crustaceans, for example, crayfish, crab, shrimp, and different creatures such as fungi is a tedious procedure [111] It needs 17–72 h including 1–24 h of HCl treatment and NaOH processing of 16–48 h. This time taking chitin isolation process needs more energy expenditure, and consequently builds the cost of generation. Similar to crabs and shrimps, barnacles likewise have a place with the Crustacea family. Their shell structures are less crystalline and they are accounted for to contain a greater number of minerals than different individuals from the crustacean family [112]. These minerals are for the most part made up of calcite and calcium phosphate. It has been demonstrated that these two mineral materials in the carapace can be effortlessly expelled from the carapace structure by utilizing HCl. In such animals, the protein just has some bonds with the chitin so it can be expelled from effortlessly due to low-crystalline shell structure of the barnacle species. In this manner, chitin isolation from barnacle species would be a nearly speedy process. *Chelonibia patulais* (a barnacle species in the subphylum Crustacea), lives episodically on animals (turtles, crabs, whales and molluscs) or on rocks at the seashore where there is shallow water is present. In a method of the segregation of chitin from C. spatula shells, their demineralization is carried out in 1 M HCl for 10 min and then allowed to deproteinized in 2 M NaOH for 20 min. The finishing of the entire procedure takes just 30 min. It begins with HCl dribbling solution (1 M) over 10 g of the dust acquired from ground C. spatula shells along

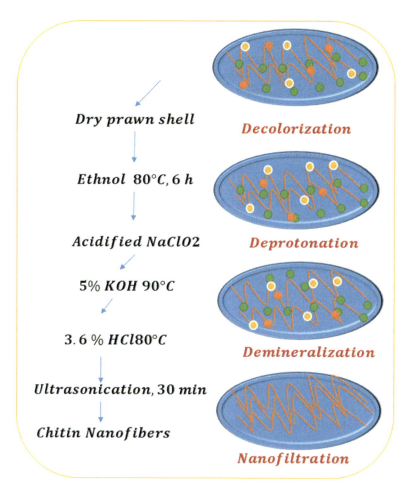

Fig. 11 Development of chitin nanofibers

and allowed to stir at room temperature for 10 min. In the event that the HCl is included too rapidly, an energetic effervescence happens that may prompt flooding. The sample at that point ought to be flushed with refined water until the point when neutral pH esteem is gotten. In an investigation, subsequent to drying the examples in an oven, 376 mg of material stayed toward the finish of the procedure. Considering that most of the first mass comprised of minerals, these have been expelled by methods of HCl.

Proteins are also present in shells and are removed by a deproteinization procedure by refluxing with a base for 20 min. This process yields 311 mg of dry chitin. Getting chitin from shrimp shells is related with food industries, for example, shrimp handling, while the creation of chitosan–glucan from parasitic mycelia is related with fermentation, for example, that such as that producing citric acid from

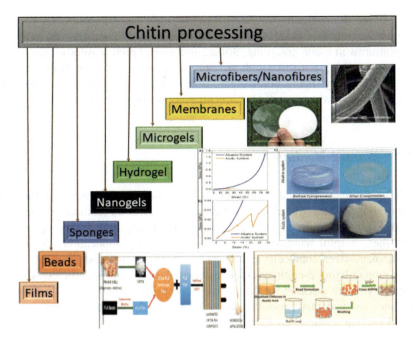

Fig. 12 Chitin processing

Aspergillus niger. For the most part, crustacean shells processing includes the expulsion of proteins and after that disintegration of calcium carbonate, which is available in the shells in a higher amount. The conventional method for obtaining chitosan from these sources additionally includes deacetylation in 40% sodium hydroxide at 120 °C for 1–3 h. This treatment gives 70% deacetylated chitosan [91] (Fig. 12).

As of late, a "green transformation" of agroindustrial waste by utilizing biological activity of Cunninghamella elegans strains and Rhizopus arrhizus has been accounted for the production of chitin and chitosan. Such industrial sources have noteworthy points of interest such as avoiding allergic reactions in individuals susceptible to shellfish antigens and reduction in time and cost of production [113].

14 Chitin Biomedical and Nanomedical Applications

14.1 Tissue Engineering

The principal motivations behind tissue building can be arranged as to repair, supplant, keep up, or upgrade the capacity of a specific tissue or organ [114]. Chitin-based materials, which can be manufactured into tubular structures, can be effectively connected in tissue designing of nerves and veins as a format for cells.

Chitin-based frameworks are flexible items and can be optimized for some, regenerative purposes [115]. Chitin has been effectively connected to create polymer platforms in tissue designing. Some fundamental prerequisites to plan polymer platforms are high porosity (with apt pore measure dispersion); biodegradability, structural integrity, being non-dangerous to cells; biocompatibility; collaborating with the cells to advance cell attachment; empowering cell work.

14.2 Wound Healing

Madhumathi et al. created α-chitin/nanosilver composite frameworks for wound healing usage. These frameworks were known to have an antibacterial movement toward S. aureus and E. coli, and also blood-clotting capacity. Such properties have made them helpful nanostructures for wound healing applications. So also, the β-chitin/nanosilver composite platforms has been created and examined for this application utilizing β-chitin hydrogel composed of silver nanoparticles. Also, these frameworks were assessed for their cell grip properties utilizing Vero cells, and the outcomes showed that nanosilver consolidated chitin platforms were perfect for wound healing applications [22] (Fig. 13).

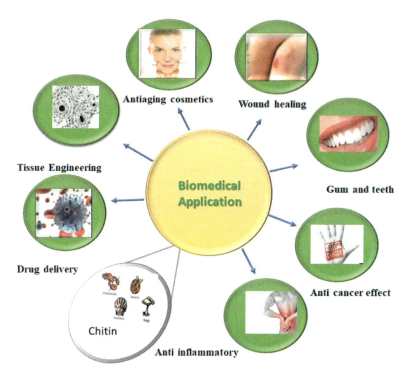

Fig. 13 Chitin applications

14.3 Drug Delivery

Carboxymethyl chitin (CMC) has been mostly used for delivery of many different kinds of drugs. CMC nanoparticles were synthesized via a cross-linking with CaCl2 and $FeCl_3$. A spherical morphology was observed in SEM images of CMC nanoparticles, with diameters ranging from 200 to 250 nm. 5-fluorouracil (5-FU) drug-loaded nanoparticles also exhibited alike morphological feature. In one study the anticancer drug 5-FU was laden into CMC nanoparticles using emulsion cross-linking technique, and these were found to have a sustained and controlled drug-release profile at pH near to neutral. An anti-HIV drug delivery application was also reported by Dev et al. [116] using poly(lactic acid) (PLA)/CS nanoparticles [116]. In addition, lamivudine (antiretroviral drug) was laden into the PLA/CS nanoparticles. In this case absorption spectrophotometry was used to evaluate the encapsulating efficiency and the in vitro drug release performance of CS nanoparticles along with drug [117, 118].

Water-dissolvable carboxymethyl chitin (CMC) was utilized to deliver drugs. CMC nanoparticles were prepared by cross-linking $CaCl_2$ and $FeCl_3$. A circular morphology was seen in SEM photos of CMC nanoparticles, having diameter running from 200 to 250 nm. 5-fluorouracil (5-FU) loaded stacked nanoparticles likewise demonstrated comparable morphology. MTT examine comes about demonstrated that they were non-poisonous to typical fibroblast L929 mouse cells. The hydrophobic anticancer medication 5-FU was stacked into CMC nanoparticles by means of an emulsion cross-connecting strategy, and these were found to have a controlled and managed sedate discharge profile at pH-6.8. A hostile to HIV drug delivery application was likewise revealed by Dev et al. [116] utilizing poly (lactic acid) (PLA)/CS nanoparticles. What's more, lamivudine (a hydrophilic antiretroviral tranquillize) was stacked into the PLA/CS nanoparticles. For this situation absorption spectrophotometry was utilized to assess the effectiveness and the in vitro drug discharge behaviour of PLA/CS nanoparticles with loaded drugs.

14.4 Cancer Diagnosis

The new advance FA-based carboxymethyl chitosan (CMCS) facilitated to manganese doped zinc sulfide (ZnS:Mn) quantum dots (FA-CMCS-ZnS:Mn) composite nanoparticles were created by Mathew et al. [119]. This multifunctional framework could be utilized for focusing on, controlled drug release and tumour cell imaging. The chosen anticancer medication was 5-FU, which is utilized for breast tumour treatment. L929 cells were utilized to affirm the non-harmfulness of FA-CMCS-ZnS:Mn nanoparticles. Moreover, the MCF-7 breast cancer cell line was utilized to consider imaging, particularly focusing on and cytotoxicity of the drug stacked nanoparticles. The in vitro imaging of malignancy cells with the nanoparticles was examined utilizing fluorescence microscopy [120].

14.5 Chitin-Based Dressings

British Textile Technology Group (BTTG) protected a system for preparing chitin-based fibrous dressing [121]. In this strategy, the chitin/chitosan strands were acquired from microorganisms (rather than shrimp shells) and were not created by the conventional fibre-spinning method. The strategy can be outlined as follows:

1. The arrangement of mycelia of micro-fungi from Mucor mucedo culture developing in a supplemented medium.
2. Cleaning and deproteinizing of the mycelial matt with NaOH to chitin/chitosan precipitates.
3. Blanching and further cleaning.
4. Arrangement and scattering of the filaments utilizing the paper-production hardware.
5. Filtration and wet-laid synthesis of a fiber matt; extra mechanical quality furnished by blending with different filaments [122].

14.6 Antiaging Cosmetics

Morganti et al. prepared block copolymer nanoparticles (BPN) made out of phosphatidylcholine and linoleic acid nanocomposite along with hyaluronan and chitin nanofibrils (PHHYCN). The nanoconstructs were utilized to encapsulate items including cholesterol, melatonin, caffeine, creatine, vitamin E and C, glycine, amino acids, and arginine. The thought was to utilize these nanocarriers for skin revival, as all the individual components had demonstrated some action in such manner. The skin treated with the dynamic chitin nanofibrils with BPN was appeared to be milder softer as well as more hydrated following one month of treatment. Both fine wrinkles and wrinkle lines were diminished not long after the initial 15 days of treatment with injectable dynamic chitin nanofibril containing BPN, too less occurrence of telangiectasia, hence over-all face appearance was particularly ameliorated during the reversion period [123].

15 Conclusion

Chitosan is derivative of chitin is obtained from natural sources that are the external skeleton of crustaceans, fungi, and insects and has to be biocompatible and decomposable. The chitosan history begins in the 19th century, by Rouget 1859 who discussed the deacetylated parent chitin polymer (the second most plenteous carbohydrate) in nature first time. The surface functionalization of chitosan can be done using different enzymes so-called enzymatic modification. There is a variety of chitosan derivatives which have been prepared and utilized in the different field of life such as quaternized chitosan and N-alkyl chitosan N-Acyl chitosan, O-Acyl chitosan and Thiolated chitosan which is used in edible coatings to prevent gas permeability.

The crystallinity and polymorphism of chitosan are highly dependent on the beginning of the polymerization and extraction in thorough process. The inherent film forming ability and antimicrobial character enable chitosan to make potential use in the packaging industry. Chitosan is also utilized in direct methanol fuel cell barrier of its excellent methanol permeation property.

The utilization of chitin was first depicted by the French scientific expert, Henri Braconnot in 1811. The structure of chitin $(C_8H_{13}O_5N)n$ shows resemblance with cellulose structure, yet with 2-acetamido-2-deoxy-β-D-glucose (NAG) monomer units, which are joined to each other by means of $β(1 → 4)$ linkages. Chitin additionally has three distinctive crystalline allomorphs: α-, β-and γ-shapes depend upon the orientation of microfibrils. There is a variety of field where chitin is and its derivatives are being used among than most important ones are Tissue engineering and Antiaging cosmetics.

References

1. de Britto D, Celi Goy R, Campana Filho SP, Assis OB (2011) Quaternary salts of chitosan: history, antimicrobial features, and prospects. Int J Carbohydr Chem
2. Dash M, Chiellini F, Ottenbrite R, Chiellini E (2011) Chitosan—a versatile semi-synthetic polymer in biomedical applications. Prog Polym Sci 36(8):981–1014
3. Bu X, Pei J, Zhang F, Liu H, Zhou Z, Zhen X et al (2018) The hydration mechanism and hydrogen bonding structure of 6-carboxylate chitooligosaccharides superabsorbent material prepared by laccase/TEMPO oxidation system. Carbohydr Polym
4. Ahmed S, Ikram S (2017) Chitosan: derivatives, composites and applications. Wiley
5. Arrouze F, Essahli M, Rhazi M, Desbrieres J, Tolaimate A (2017) Chitin and chitosan: study of the possibilities of their production by valorization of the waste of crustaceans and cephalopods rejected in Essaouira. J Mat Environ Sci: Journal of Materials and Environmental Science 8(7):2251–2258
6. Hattori H, Tsujimoto H, Hase K, Ishihara M (2017) Characterization of a water-soluble chitosan derivative and its potential for submucosal injection in endoscopic techniques. Carbohyd Polym 175:592–600
7. Hamed I, Özogul F, Regenstein JM (2016) Industrial applications of crustacean by-products (chitin, chitosan, and chitooligosaccharides): a review. Trends Food Sci Technol 48:40–50
8. Feng Y, Kopplin G, Sato K, Draget KI, Vårum KM (2017) Alginate gels with a combination of calcium and chitosan oligomer mixtures as crosslinkers. Carbohyd Polym 156:490–497
9. Gokara M, Kimavath GB, Podile AR, Subramanyam R (2015) Differential interactions and structural stability of chitosan oligomers with human serum albumin and α-1-glycoprotein. J Biomol Struct Dyn 33(1):196–210
10. Ji X, Li B, Yuan B, Guo M (2017) Preparation and characterizations of a chitosan-based medium-density fiberboard adhesive with high bonding strength and water resistance. Carbohyd Polym 176:273–280
11. Cheon JY, Lee HM, Park WH (2018) Formation of silver nanoparticles using fluorescence properties of chitosan oligomers. Mar Drugs 16(1):11

12. Naqvi S, Moerschbacher BM (2017) The cell factory approach toward biotechnological production of high-value chitosan oligomers and their derivatives: an update. Crit Rev Biotechnol 37(1):11–25
13. Pillai C, Paul W, Sharma CP (2009) Chitin and chitosan polymers: chemistry, solubility and fiber formation. Prog Polym Sci 34(7):641–678
14. Ahmed S, Ikram S (2016) Chitosan based scaffolds and their applications in wound healing. Achievements Life Sci 10(1):27–37
15. Thanou M, Florea B, Geldof M, Junginger H, Borchard G (2002) Quaternized chitosan oligomers as novel gene delivery vectors in epithelial cell lines. Biomaterials 23(1):153–159
16. Liu B, Wang D, Yu G, Meng X (2013) Adsorption of heavy metal ions, dyes and proteins by chitosan composites and derivatives—a review. J Ocean Univer China 12(3):500–508
17. Prashanth KH, Tharanathan R (2007) Chitin/chitosan: modifications and their unlimited application potential—an overview. Trends Food Sci Technol 18(3):117–131
18. Polnok A, Borchard G, Verhoef J, Sarisuta N, Junginger H (2004) Influence of methylation process on the degree of quaternization of N-trimethyl chitosan chloride. Eur J Pharm Biopharm 57(1):77–83
19. LogithKumar R, KeshavNarayan A, Dhivya S, Chawla A, Saravanan S, Selvamurugan N (2016) A review of chitosan and its derivatives in bone tissue engineering. Carbohyd Polym 151:172–188
20. Peng Y, Han B, Liu W, Xu X (2005) Preparation and antimicrobial activity of hydroxypropyl chitosan. Carbohyd Res 340(11):1846–1851
21. Araldi SJ, Tudryn GJ, Hart CE, Carlton AJ (2017) Chemically modified mycological materials having absorbent properties: Google patents
22. Jayakumar R, Chennazhi K, Muzzarelli R, Tamura H, Nair S, Selvamurugan N (2010) Chitosan conjugated DNA nanoparticles in gene therapy. Carbohyd Polym 79(1):1–8
23. Krause T, Baumeister J, Weber D, Lang G, Beyer A, Florig E et al (2005) Hair treatment compositions containing N-hydroxy-alkyl-O-benzyl chitosans and methods of using same: Google patents
24. Karp J, Joshi N, He X, Bhagchandani S (2017) Self assembled gels for controlled delivery of encapsulated agents to cartilage: Google patents
25. Yin T, Zhang Y, Liu Y, Chen Q, Fu Y, Liang J, Huo M (2018) The efficiency and mechanism of N-octyl-O, N-carboxymethyl chitosan-based micelles to enhance the oral absorption of silybin. Int J Pharm 536(1):231–240
26. Sashiwa H, Aiba S-I (2004) Chemically modified chitin and chitosan as biomaterials. Prog Polym Sci 29(9):887–908
27. Chtchigrovsky M, Primo A, Gonzalez P, Molvinger K, Robitzer M, Quignard F, Taran F (2009) Functionalized chitosan as a green, recyclable, biopolymer-supported catalyst for the [3 + 2] Huisgen cycloaddition. Angew Chem 121(32):6030–6034
28. Srbová J, Slováková M, Křípalová Z, Žárská M, Špačková M, Stránská D, Bílková Z (2016) Covalent biofunctionalization of chitosan nanofibers with trypsin for high enzyme stability. React Funct Polym 104:38–44
29. Auzély-Velty R, Rinaudo M (2002) New supramolecular assemblies of a cyclodextrin-grafted chitosan through specific complexation. Macromolecules 35(21):7955–7962
30. Martel B, Devassine M, Crini G, Weltrowski M, Bourdonneau M, Morcellet M (2001) Preparation and sorption properties of a β-cyclodextrin-linked chitosan derivative. J Polym Sci Part A: Polym Chem 39(1):169–176
31. Wang J, Chen C (2014) Chitosan-based biosorbents: modification and application for biosorption of heavy metals and radionuclides. Biores Technol 160:129–141
32. Badawy ME, Rabea EI, Rogge TM, Stevens CV, Smagghe G, Steurbaut W, Höfte M (2004) Synthesis and fungicidal activity of new N,O-acyl chitosan derivatives. Biomacromolecules 5(2):589–595
33. Sun T, Zhu Y, Xie J, Yin X (2011) Antioxidant activity of N-acyl chitosan oligosaccharide with same substituting degree. Bioorg Med Chem Lett 21(2):798–800

34. Zahir-Jouzdani F, Mahbod M, Soleimani M, Vakhshiteh F, Arefian E, Shahosseini S, Atyabi F (2018) Chitosan and thiolated chitosan: novel therapeutic approach for preventing corneal haze after chemical injuries. Carbohyd Polym 179:42–49
35. Ways TM, Lau WM, Khutoryanskiy VV (2018) Chitosan and its derivatives for application in mucoadhesive drug delivery systems. Polymers 10(3):267
36. Chaffanel F, Charron-Bourgoin F, Soligot C, Kebouchi M, Bertin S, Payot S et al (2018) Surface proteins involved in the adhesion of *Streptococcus salivarius* to human intestinal epithelial cells. Appl Microbiol Biotechnol, 1–15
37. Leitner V, Marschütz M, Bernkop-Schnürch A (2003) Mucoadhesive and cohesive properties of poly (acrylic acid)-cysteine conjugates with regard to their molecular mass. Eur J Pharm Sci 18(1):89–96
38. Yuan N-Y, Tsai R-Y, Ho M-H, Wang D-M, Lai J-Y, Hsieh H-J (2008) Fabrication and characterization of chondroitin sulfate-modified chitosan membranes for biomedical applications. Desalination 234(1–3):166–174
39. Zhang C, Ping Q, Zhang H, Shen J (2003) Preparation of N-alkyl-O-sulfate chitosan derivatives and micellar solubilization of taxol. Carbohyd Polym 54(2):137–141
40. Shanmugam A, Kathiresan K, Nayak L (2016) Preparation, characterization and antibacterial activity of chitosan and phosphorylated chitosan from cuttlebone of Sepia kobiensis (Hoyle, 1885). Biotechnol Rep 9:25–30
41. Karaki N, Aljawish A, Humeau C, Muniglia L, Jasniewski J (2016) Enzymatic modification of polysaccharides: mechanisms, properties, and potential applications: a review. Enzyme Microb Technol 90:1–18
42. Thakur VK, Thakur MK (2014) Recent advances in graft copolymerization and applications of chitosan: a review. ACS Sustain Chem Eng 2(12):2637–2652
43. Zhou T, Zhu Y, Li X, Liu X, Yeung KW, Wu S, Chu PK (2016) Surface functionalization of biomaterials by radical polymerization. Prog Mater Sci 83:191–235
44. Carreira A, Gonçalves F, Mendonça P, Gil M, Coelho J (2010) Temperature and pH responsive polymers based on chitosan: applications and new graft copolymerization strategies based on living radical polymerization. Carbohyd Polym 80(3):618–630
45. Kim KM, Son JH, Kim SK, Weller CL, Hanna MA (2006) Properties of chitosan films as a function of pH and solvent type. J Food Sci 71(3)
46. Twu Y-K, Huang H-I, Chang S-Y, Wang S-L (2003) Preparation and sorption activity of chitosan/cellulose blend beads. Carbohyd Polym 54(4):425–430
47. Xu Y, Du Y (2003) Effect of molecular structure of chitosan on protein delivery properties of chitosan nanoparticles. Int J Pharm 250(1):215–226
48. Xu Y, Kim KM, Hanna MA, Nag D (2005) Chitosan–starch composite film: preparation and characterization. Ind Crops Prod 21(2):185–192
49. Chillo S, Flores S, Mastromatteo M, Conte A, Gerschenson L, Del Nobile MA (2008) Influence of glycerol and chitosan on tapioca starch-based edible film properties. J Food Eng 88(2):159–168
50. Vásconez MB, Flores SK, Campos CA, Alvarado J, Gerschenson LN (2009) Antimicrobial activity and physical properties of chitosan–tapioca starch based edible films and coatings. Food Res Int 42(7):762–769
51. Nagahama H, Maeda H, Kashiki T, Jayakumar R, Furuike T, Tamura H (2009) Preparation and characterization of novel chitosan/gelatin membranes using chitosan hydrogel. Carbohyd Polym 76(2):255–260
52. Cheng L, Bulmer C, Margaritis A (2015) Characterization of novel composite alginate chitosan-carrageenan nanoparticles for encapsulation of BSA as a model drug delivery system. Curr Drug Deliv 12(3):351–357
53. Darder M, Colilla M, Ruiz-Hitzky E (2005) Chitosan–clay nanocomposites: application as electrochemical sensors. Appl Clay Sci 28(1–4):199–208
54. Günister E, Pestreli D, Ünlü CH, Atıcı O, Güngör N (2007) Synthesis and characterization of chitosan-MMT biocomposite systems. Carbohyd Polym 67(3):358–365

55. Hsu S-H, Wang M-C, Lin J-J (2012) Biocompatibility and antimicrobial evaluation of montmorillonite/chitosan nanocomposites. Appl Clay Sci 56:53–62
56. Mohammadi R, Mohammadifar MA, Rouhi M, Kariminejad M, Mortazavian AM, Sadeghi E, Hasanvand S (2018) Physico-mechanical and structural properties of eggshell membrane gelatin-chitosan blend edible films. Int J Biol Macromol 107:406–412
57. Hai TAP, Sugimoto R (2018) Surface modification of chitin and chitosan with poly (3-hexylthiophene) via oxidative polymerization. Appl Surf Sci 434:188–197
58. Santos-Moriano P, Fernandez-Arrojo L, Mengibar M, Belmonte-Reche E, Peñalver P, Acosta F, Fernández-Lobato M (2018) Enzymatic production of fully deacetylated chitooligosaccharides and their neuroprotective and anti-inflammatory properties. Biocatal Biotransform 36(1):57–67
59. Vasconcelos DP, Costa M, Neves N, Teixeira JH, Vasconcelos DM, Santos SG et al (2018) The use of chitosan porous 3D scaffolds embedded with resolvin D1 to improve in vivo bone healing. J Biomed Mat Res Part A
60. Singh G, Manohar M, Arya SK, Siddiqui WA, Stenström TA (2017) Potential biomedical applications of chitosan–and chitosan-based nanomaterials. Chitosan Deriv Compos Appl, 385–408
61. Cremar L, Gutierrez J, Martinez J, Materon L, Gilkerson R, Xu F, Lozano K (2018) Development of antimicrobial chitosan based nanofiber dressings for wound healing applications. Nanomed J 5(1):6–14
62. Heidari F, Bahrololoom ME, Vashaee D, Tayebi L (2015) In situ preparation of iron oxide nanoparticles in natural hydroxyapatite/chitosan matrix for bone tissue engineering application. Ceram Int 41(2):3094–3100
63. Jayakumar R, Prabaharan M, Kumar PS, Nair S, Tamura H (2011) Biomaterials based on chitin and chitosan in wound dressing applications. Biotechnol Adv 29(3):322–337
64. Choi YS, Lee S, Hong SR, Lee Y, Song K, Park M (2001) Studies on gelatin-based sponges. Part III: a comparative study of cross-linked gelatin/alginate, gelatin/hyaluronate and chitosan/hyaluronate sponges and their application as a wound dressing in full-thickness skin defect of rat. J Mat Sci: Materials in Medicine 12(1):67–73
65. Srinivasan H, Kanayairam V, Ravichandran R (2018) Chitin and chitosan preparation from shrimp shells *Penaeus monodon* and its human ovarian cancer cell line, PA-1. Int J Biol Macromol 107:662–667
66. Abdelmalek BE, Sila A, Haddar A, Bougatef A, Ayadi MA (2017) β-Chitin and chitosan from squid gladius: biological activities of chitosan and its application as clarifying agent for apple juice. Int J Biol Macromol 104:953–962
67. Kabalak M, Aracagök YD, Torun M (2017) Extraction and physicochemical properties of chitins from four different insect species
68. Sudha PN, Saranya M, Gomathi T, Gokila S, Aisverya S, Venkatesan J, Anil S (2017) Perspectives of chitin- and chitosan-based scaffolds dressing in regenerative medicine. Chitosan Deriv Comp Appl, 253–269
69. Yu Z, Lau D (2017) Flexibility of backbone fibrils in α-chitin crystals with different degree of acetylation. Carbohyd Polym 174:941–947
70. Akpan E, Gbenebor O, Adeosun S (2018) Synthesis and characterisation of chitin from periwinkle (*Tympanotonus fusatus* (L.)) and snail (*Lissachatina fulica* (Bowdich)) shells. Int J Biol Macromol 106:1080–1088
71. Gbenebor OP, Akpan EI, Adeosun SO (2017) Thermal, structural and acetylation behavior of snail and periwinkle shells chitin. Prog Biomat 6(3):97–111
72. Kaya M, Bağrıaçık N, Seyyar O, Baran T (2015) Comparison of chitin structures derived from three common wasp species (Vespa crabro Linnaeus, 1758, *Vespa orientalis* Linnaeus, 1771 and *Vespula germanica* (Fabricius, 1793)). Arch Insect Biochem Physiol 89(4):204–217
73. Silva SS, Mano JF, Reis RL (2017) Ionic liquids in the processing and chemical modification of chitin and chitosan for biomedical applications. Green Chem 19(5):1208–1220

74. Isono Y, Noishiki Y (2018) Method for manufacturing water-insoluble molded article and water-insoluble molded article: Google patents
75. Roy JC, Salaün F, Giraud S, Ferri A, Chen G, Guan J (2017) Solubility of chitin: solvents, solution behaviors and their related mechanisms. Solubility of Polysaccharides, InTech
76. Tachaboonyakiat W (2017) Antimicrobial applications of chitosan. Chitosan based biomaterials, vol 2. Elsevier, pp 245–274
77. Vincendon M (1997) Regenerated chitin from phosphoric acid solutions. Carbohyd Polym 32(3–4):233–237
78. Jothimani B, Sureshkumar S, Venkatachalapathy B (2017) Hydrophobic structural modification of chitosan and its impact on nanoparticle synthesis—a physicochemical study. Carbohyd Polym 173:714–720
79. Jayakumar R, Menon D, Manzoor K, Nair S, Tamura H (2010) Biomedical applications of chitin and chitosan based nanomaterials—a short review. Carbohyd Polym 82(2):227–232
80. Gulati K, Meher MK, Poluri KM (2017) Glycosaminoglycan-based resorbable polymer composites in tissue refurbishment. Regenerative Med 12(4):431–457
81. Cao N, Lyu Q, Li J, Wang Y, Yang B, Szunerits S, Boukherroub R (2017) Facile synthesis of fluorinated polydopamine/chitosan/reduced graphene oxide composite aerogel for efficient oil/water separation. Chem Eng J 326:17–28
82. Yu C, Kecen X, Xiaosai Q (2018) Grafting modification of chitosan. Biopolymer grafting. Elsevier, pp 295–364
83. Badawy ME, Rabea EI (2017) Chitosan and its modifications as biologically active compounds in different applications. Adv Physicochem Properties Biopolym (Part 2), 1
84. Olicón-Hernández DR, Uribe-Alvarez C, Uribe-Carvajal S, Pardo JP, Guerra-Sánchez G (2017) Response of ustilago maydis against the stress caused by three polycationic chitin derivatives. Molecules 22(12):1745
85. Swatloski RP, Barber PS, Opichka T, Bonner JR, Gurau G, Griggs CS, Rogers RD (2017) Process for electrospinning chitin fibers from chitinous biomass solution: Google patents
86. Zou H, Lin B, Xu C, Lin M, Zhan W (2018) Preparation and characterization of individual chitin nanofibers with high stability from chitin gels by low-intensity ultrasonication for antibacterial finishing. Cellulose 25(2):999–1010
87. Kong K, Davies RJ, McDonald MA, Young RJ, Wilding MA, Ibbett RN, Eichhorn SJ (2007) Influence of domain orientation on the mechanical properties of regenerated cellulose fibers. Biomacromology 8(2):624–630
88. Rinaudo M (2006) Chitin and chitosan: properties and applications. Prog Polym Sci 31(7):603–632
89. Khor E, Lim LY (2003) Implantable applications of chitin and chitosan. Biomaterials 24(13):2339–2349
90. Khor E (2014) Chitin: fulfilling a biomaterials promise. Elsevier
91. Kumar MNR (2000) A review of chitin and chitosan applications. React Funct Polym 46(1):1–27
92. Beier S, Bertilsson S (2013) Bacterial chitin degradation—mechanisms and ecophysiological strategies. Front Microbiol 4:149
93. Kumirska J, Weinhold MX, Thöming J, Stepnowski P (2011) Biomedical activity of chitin/chitosan based materials—influence of physicochemical properties apart from molecular weight and degree of N-acetylation. Polymers 3(4):1875–1901
94. Younes I, Rinaudo M (2015) Chitin and chitosan preparation from marine sources. Structure, properties and applications. Mar Drugs 13(3):1133–1174
95. Friedman AJ, Phan J, Schairer DO, Champer J, Qin M, Pirouz A, Modlin RL (2013) Antimicrobial and anti-inflammatory activity of chitosan–alginate nanoparticles: a targeted therapy for cutaneous pathogens. J Invest Dermatol 133(5):1231–1239
96. Gooday GW (1990) The ecology of chitin degradation. Advances in microbial ecology. Springer, pp 387–430
97. Badwan AA, Rashid I, Al Omari MM, Darras FH (2015) Chitin and chitosan as direct compression excipients in pharmaceutical applications. Mar Drugs 13(3):1519–1547

98. Yen M-T, Yang J-H, Mau J-L (2009) Physicochemical characterization of chitin and chitosan from crab shells. Carbohyd Polym 75(1):15–21
99. Ospina Álvarez SP, Ramírez Cadavid DA, Escobar Sierra DM, Ossa Orozco CP, Rojas Vahos DF, Zapata Ocampo P, Atehortúa L (2014) Comparison of extraction methods of chitin from Ganoderma lucidum mushroom obtained in submerged culture. BioMed Res Int
100. Yang T-L (2011) Chitin-based materials in tissue engineering: applications in soft tissue and epithelial organ. Int J Mol Sci 12(3):1936–1963
101. Hajji S, Younes I, Ghorbel-Bellaaj O, Hajji R, Rinaudo M, Nasri M, Jellouli K (2014) Structural differences between chitin and chitosan extracted from three different marine sources. Int J Biol Macromol 65:298–306
102. Xu Q, Wang C-H, Wayne Pack D (2010) Polymeric carriers for gene delivery: chitosan and poly (amidoamine) dendrimers. Curr Pharm Des 16(21):2350–2368
103. Chen Q, Zhang J-W, Chen L-L, Yang J, Yang X-L, Ling Y, Yang Q (2017) Design and synthesis of chitin synthase inhibitors as potent fungicides. Chin Chem Lett 28(6):1232–1237
104. Tang B, Yang M, Shen Q, Xu Y, Wang H, Wang S (2017) Suppressing the activity of trehalase with validamycin disrupts the trehalose and chitin biosynthesis pathways in the rice brown planthopper, *Nilaparvata lugens*. Pestic Biochem Physiol 137:81–90
105. Ruiz-Herrera J, Lopez-Romero E, Bartnicki-Garcia S (1977) Properties of chitin synthetase in isolated chitosomes from yeast cells of *Mucor rouxii*. J Biol Chem 252(10):3338–3343
106. Wang P, Bi S, Wu F, Xu P, Shen X, Zhao Q (2017) Differentially expressed genes in the head of the 2nd instar pre-molting larvae of the nm2 mutant of the silkworm, *Bombyx mori*. PloS One 12(7):e0180160
107. Cohen E (2001) Chitin synthesis and inhibition: a revisit. Pest Manag Sci 57(10):946–950
108. Yang M, Wang Y, Jiang F, Song T, Wang H, Liu Q, Kang L (2016) miR-71 and miR-263 jointly regulate target genes chitin synthase and chitinase to control locust molting. PLoS Genet 12(8):e1006257
109. Bowen A, Chen-Wu J, Momany M, Young R, Szaniszlo P, Robbins P (1992) Classification of fungal chitin synthases. Proc Natl Acad Sci 89(2):519–523
110. Chen Q, Jin S, Zhang L, Shen Q, Wei P, Wei Z et al (2017) Regulatory functions of trehalose-6-phosphate synthase in the chitin biosynthesis pathway in *Tribolium castaneum* (Coleoptera: Tenebrionidae) revealed by RNA interference. Bull Entomol Res, 1–12
111. Kaya M, Sargin I, Tozak KÖ, Baran T, Erdogan S, Sezen G (2013) Chitin extraction and characterization from *Daphnia magna* resting eggs. Int J Biol Macromol 61:459–464
112. Kaya M, Karaarslan M, Baran T, Can E, Ekemen G, Bitim B, Duman F (2014) The quick extraction of chitin from an epizoic crustacean species (*Chelonibia patula*). Nat Prod Res 28 (23):2186–2190
113. Philibert T, Lee BH, Fabien N (2017) Current status and new perspectives on chitin and chitosan as functional biopolymers. Appl Biochem Biotechnol 181(4):1314–1337
114. Jayakumar R, Nair S, Furuike T, Tamura H (2010) Perspectives of chitin and chitosan nanofibrous scaffolds in tissue engineering. Tissue Engineering, Intech
115. Madihally SV, Matthew HW (1999) Porous chitosan scaffolds for tissue engineering. Biomaterials 20(12):1133–1142
116. Dev A, Binulal N, Anitha A, Nair S, Furuike T, Tamura H, Jayakumar R (2010) Preparation of poly (lactic acid)/chitosan nanoparticles for anti-HIV drug delivery applications. Carbohyd Polym 80(3):833–838
117. Mourya V, Inamdar NN, Tiwari A (2010) Carboxymethyl chitosan and its applications. Adv Mat Lett 1(1):11–33
118. Huang Y, Yao M, Zheng X, Liang X, Su X, Zhang Y et al (2015) Effects of chitin whiskers on physical properties and osteoblast culture of alginate based nanocomposite hydrogels. Biomacromolecules 16(11):3499–3507
119. Mathew ME, Mohan JC, Manzoor K, Nair S, Tamura H, Jayakumar R (2010) Folate conjugated carboxymethyl chitosan–manganese doped zinc sulphide nanoparticles for targeted drug delivery and imaging of cancer cells. Carbohyd Polym 80(2):442–448

120. Wu S, Huang Z, Yue J, Liu D, Wang T, Ezanno P, Pan H (2015) The efficient hemostatic effect of Antarctic krill chitosan is related to its hydration property. Carbohyd Polym 132:295–303
121. Komi DEA, Sharma L, Cruz CSD (2017) Chitin and its effects on inflammatory and immune responses. Clin Rev Allergy Immunol, 1–11
122. Elieh-Ali-Komi D, Hamblin MR (2016) Chitin and chitosan: production and application of versatile biomedical nanomaterials. Int J Adv Res 4(3):411
123. Morganti P, Palombo P, Palombo M, Fabrizi G, Cardillo A, Svolacchia F, Mezzana P (2012) A phosphatidylcholine hyaluronic acid chitin–nanofibrils complex for a fast skin remodeling and a rejuvenating look. Clin Cosmet Invest Dermatol 5:213

Nanocomposites for Environmental Pollution Remediation

Anjali Bajpai, Maya Sharma and Laxmi Gond

List of abbreviation

BENT	Bentonite
CHT	Chitosan
CNFs	Carbon nanofibers
CR	Congo red
CV	Crystal violet
DB	Disperse blue
DEA	Diethanolamine
FG	Fast green
GL	Gelatine
HAL	Halloysite
LBL	Layer by layer
MB	Methylene blue
MG	Malachite green
MMT	Montmorillonite
MO	Methyl orange
MWCNT	Multiwall carbon nanotubes
NPs	Nanoparticles
OVU	Organovermiculite
PA6	Polyamides 6
PAL	Palygorskite
PFNC	Polymer-functionalized nanocomposites
PLSN	Polymer-layered silicate nanocomposites
POPs	Persistent organic pollutants
RB	Rose Bengal

A. Bajpai (✉) · M. Sharma · L. Gond
Department of Chemistry, Government Science College,
A Centre for Excellence in Science Education, Pachpedi,
Jabalpur 482001, India
e-mail: abs_112@rediffmail.com

© Springer Nature Switzerland AG 2019
Inamuddin et al. (eds.), *Sustainable Polymer Composites and Nanocomposites*,
https://doi.org/10.1007/978-3-030-05399-4_47

RhB	Rhodamine B
RR	Remazol red
SEP	Sepiolite
TEA	Triethanolamine

1 Introduction

Latter half of the past century has witnessed the great industrial revolution, which brought about significant changes in lifestyle. However, this was on the cost of anthropogenic deterioration of the environment, since several irrational industrial and agricultural activities, cause surface water quality degradation. It is the top priority in the twenty-first century to reduce/eliminate heavy metal ions and other contaminants from wastewater to improve water quality [1–3].

1.1 Heavy Metal Pollution

Among all the pollutants present in wastewater, heavy metals are most serious environmental and health threats. The industries, which contribute significantly to emanate potentially toxic metals in surface and ground waters, seriously aggravate the water pollution. The mining and metallurgical industries play a major role in the release of metals loaded effluents in the river network or indirectly by disposing of solid waste containing residual metals in the environment. These mines tailing dumps weather over time, accelerating the mobility of metal ions which are then continuously released into the surface and groundwater. The toxic heavy metal ions can migrate from sediments into the water and pose threat to flora and fauna through bioaccumulation [1–3].

Emissions from mines, metal wastes, gasoline, paints, fertilizers, manure, sewage sludge, pesticides, irrigation, coal combustion, spillage of petrochemicals, atmospheric deposition contaminate the soil with heavy metals. Heavy metals are elements with atomic number greater than 20 and density at least five times greater than that of water. Pb, Cr, As, Zn, Cd, Cu, Hg and Ni ions are commonly found as contaminants [2, 4]. Pb has raised special concerns as it adversely affects the human health and because of its frequent release in wastewaters through effluents coming from smelters and refineries, battery, steel, printing and glass industries [4].

Most of the metals do not undergo microbial or chemical decomposition; hence soil becomes their major sink [5]. However, their chemical forms change (speciation) and their bioavailability become possible. Toxic metals present in soil severely inhibit biodegradation of organic contaminants [6]. Humans and the ecosystem are at risk through direct ingestion or contact with contaminated soil/drinking groundwater, reduction in agricultural land. The food chain is contaminated through food deteriorated by phytotoxicity. Consequently, the soil ecosystems require characterization, remediation and risk assessment for appropriate protection and restoration [7].

Techniques available for remediation of contaminated sites are immobilization, soil washing and phytoremediation [8]. Despite cost-effectiveness and eco-friendly nature, field applications of these technologies are employed in developed countries only. These techniques are not availed in most of the developing countries due to inadequate awareness of inherent advantages and principles of operation. Development of technologies to remediate contaminated sites is attracting the interest of the scientific community.

1.2 Organic Pollutants

Domestic sewage, urban run-off, industrial effluents, agricultural wastewater generate organic pollution. Textile, food processing, pulp/paper making, agriculture and aquaculture industries are the sources of organic pollutants. The decomposition of organic pollutants consumes dissolved oxygen at a rate much greater than its replenishment, thus oxygen depletion adversely affects the stream biota. Organic pollutants present as suspended solids obstruct the light for photosynthesis to aquatic biota, when settled down they affect the habitat of invertebrates residing on the river bed. Common organic pollutants are pesticides, fertilizers, hydrocarbons, phenols, plasticizers, biphenyls, detergents, oils, greases, pharmaceuticals, proteins and carbohydrates. Persistent organic pollutants (POPs) are organic compounds and mixtures such as polychlorinated biphenyls, polychlorinated dibenzo-p-dioxins and dibenzofurans, organochlorine pesticides, such as hexachlorobenzene and dichloro-diphenyl-trichloroethane, dibenzo-p-dioxins and dibenzo-p-furans [9]. POPs have long-range transportability and are toxic and bioaccumulate in animals.

1.2.1 Water Pollution by Synthetic Organic Dyes

Over 100,000 dyes are commercially available and more than 7,00,000 tonnes are produced annually. Synthetic dyes are water soluble and found in trace quantities in industrial effluents. Many of them are also toxic and/or carcinogenic [10]. Inefficient processes of dyeing textile fibres can cause the colourants being released together with the effluents. The presence of dyes in very low concentrations (even 10 ppm) in water imparts a colour, making it undesirable for use [11]. The reactive dyes that are discharged into the water bodies are not biodegradable, hence, are toxic to aquatic life. Usually, removal of hazardous materials is not considered unless an environmental issue is recognized. Removal of dyes from wastewaters is difficult because of their inert properties and their low concentration in wastewater [12, 13]. These problems and their solution have been comprehensively reviewed taking into account the advantages and disadvantages of purification techniques such as nanotechnology, ultrasound, microwave, catalysis, biosorption, enzymatic treatments, advanced oxidation processes, etc. [14].

1.3 Methods for the Remediation of Pollutants

Environmental pollution has raised worldwide concern. Remediation of natural and anthropogenic contaminants from the water and air is the need of hour especially heavy metal species which affect the health [15]. Several techniques are employed to reduce the concentration of metal ion in wastewater [2]. A brief outline of some of the methods and their salient features can be given as under:

A. **Methods for physical separation from soil**

The efficiency of physical separation depends on soil characteristics. Following methods have been deemed reliable in this context:

- Flotation
- Hydrodynamic classification
- Mechanical screening
- Magnetic separation
- Gravity concentration
- Attrition scrubbing
- Electrostatic separation

B. **Methods of wastewater remediation**

Some of the popular methods for wastewater remediation include:

i. Chemical Precipitation

- Precipitation of heavy metals as hydroxide, sulphide, carbonate and phosphate, which are insoluble
- Generation of micro fine particles
- Removal of sludge by increasing particle size by chemical precipitants, coagulants, and flocculants

ii. Coagulation and Flocculation

- Electrostatic interaction between pollutants and coagulant/flocculants
- Flocculation of discrete particles into larger ones
- Removal or separation of coagulated-flocculated particles by filtration, straining or floatation

iii. Electrochemical treatments of heavy metals

- Precipitation by forming coagulants through electrolytic oxidation
- Precipitation in a weakly acidic or neutralized catholyte as hydroxides

- Electro-deposition, electrocoagulation, electro-flotation and electro-oxidation
- Precipitation by forming coagulants through electrolytic oxidation
- Precipitation in a weakly acidic or neutralized catholyte as hydroxides
- Electro-deposition, electrocoagulation, electro-flotation and electro-oxidation

iv. Ion exchange

- Widely applicable in industries for water treatment
- Convenient operation utilizes low-cost materials
- Special ion exchanger contains cations or anions
- Synthetic organic ion exchange resins are used frequently
- Applicable only for metal ion solutions of low concentration
- *pH* of the aqueous phase affects the sensitivity

v. Membrane filtration

- Removal of suspended solids, organic and inorganic contaminants is dependent on the size of the particles retainable
- Various membrane filtration techniques such as ultrafiltration, nanofiltration and reverse osmosis employed for heavy metal remediation

vi. Electrodialysis

- Ionized species passed through an ion exchange membrane by application of an electric potential
- Thin sheets of plastic materials with anionic or cationic characteristics are used as membranes
- Anions and cations migrate toward the anode and the cathode respectively

vii. Biodegradation

- Microorganisms catabolise organic material with aeration and agitation and settle down solid
- Activated sludge containing bacteria is perpetually re-circulated back to the aeration tank to enhance organic decomposition
- Activated sludge, trickling filters, stabilization tanks are in wide use

viii. Adsorption

- Common mechanism applies to organic and inorganic pollutants removal
- Liquid-solid intermolecular attractive forces induce some of the solute molecules from the solution to be stuck at the solid surface when an adsorbate solution comes in contact with an adsorbent (a solid with a highly porous surface)
- All sorts of valencies/attractive forces of the constituent atoms of the adsorbate material are satisfied in the bulk. However, atoms on the surface of the adsorbent are exposed and hence attract adsorbates. The exact nature of the bonding depends on the species. Adsorption process may be physical sorption (van der Waals interactions) or chemisorption (covalent bonding). Electrostatic attraction is also possible.

Traditional methods for treatment of contaminated water include adsorption [16], chemical precipitation [17], ion-exchange [18], reverse osmosis [19], membrane technologies [20], solvent extraction [21, 22], electrochemical treatment [17], biological treatment [23, 24], phytoremediation [8] and flotation [25]. The disadvantages of these methods to restrain their application are significant capital, energy, and operational costs; the addition of chemicals, generation of hazardous wastes and poor efficiency for wastewaters with a low concentration of heavy metals [26]. Ion exchange and reverse osmosis are more attractive methods owing to the recovery of valuable pollutants, however, these are economically unfavourable due to relatively high investment and operational cost. Adsorption is of special interest due to its economy, versatility, and efficiency, especially for the remediation of trace amounts of pollutants [27].

Developing countries have high population density and poor availability of funds, hence low cost, and sustainable remedial options are required [28]. A critical review of the subject presented the best available remedial strategies for heavy metals usually found in contaminated soils along with their sources, potential biohazards, and chemistry.

2 Adsorption: An Advantageous Process for Pollution Remediation

Simplicity and cost efficiency are the attractive features of adsorption. Further, the use of eco-friendly techniques is possible, which requires minimum skill for implementation. Hence, several high-efficiency adsorbents have been developed, such as activated carbon, quartz, sand, bentonite, zeolite and oxide materials, for heavy metal ion remediation [29]. Heavy metal species emanating from industrial activities are multifarious and complex, therefore, diverse adsorbents have been explored [16, 29].

Environmentally friendly adsorbents are efficient and economic for water decontamination. Adsorbent materials, such as chitosan, clays, zeolites and activated carbons, serve as low-cost adsorbents [29]. However, they have weak affinity for metal ions. Functional groups have therefore been introduced in these materials to enhance their adsorption capacity. Still, the existing adsorbents exhibit efficiency for specific heavy metal. This limitation is attributed to hard and soft acids and bases characteristics [30].

The design of adsorbents with abundant and accessible chelating sites with high affinity is a key challenge. A host of sorbents were applied for the removal of metal ions from wastewater by covalent grafting of coordination groups on the surface of porous materials, such as silica gel and clay [31]. However, small and irregular pore structures allow only a fraction of the surface-bound ligands to bind a particular metal ion [2]. Biosorption has gained interest in recent years due to its efficiency in extenuating heavy metals, particularly those present in low concentration. Biosorption based on biomass-derived sorbents is a cost-effective and sustainable alternative to water treatment [32, 33].

2.1 Biosorption

Biosorption is a potentially attractive technology for removal and/or recovery of toxic/rare heavy metals from industrial effluents by use of low-cost biosorbents in a metabolically-free manner [32, 33]. Biosorption involves complex mechanisms such as coordination, chelation, ion exchange, physical adsorption and/or ion entrapment in inter- and intra-fibrillar capillaries and spaces of the polysaccharide network, etc. [32]. The characteristics of the biosorbents, the physicochemical properties of the heavy metals and the microenvironment of the contact solution influence the operating mechanisms. Biosorption is a passive process involving the affinity between sorbate and biomass sorbent. It is advantageous because of its: (i) low cost, (ii) high efficiency, (iii) potential metal recovery, (iv) easy operation, (v) high sustainability and (vi) reduced sludge handling. Its disadvantages include: (i) limited pilot and industrial scale studies, (ii) possible increments in cost (as wastes become commodities), (iii) process mechanism of high complexity and (iv) disposal of biosorbents at the end of life [34, 35].

Biopolymers as biosorbents

Biopolymers are composed of polysaccharides, fatty acids and proteins, with a variety of abundant functional groups as active sites. Extensive research has been devoted to exuberant natural polymers or agriculture waste products to be employed as biosorbents, e.g. fungi, yeasts, bacteria, algae, chitin and chitosan. Several low-cost and renewable biological materials have been explored viz.: macroalgae, agricultural remainders, industrial waste, stuff, sludge, raw plants, etc. [36].

Biosorption experiments are typically performed by batch processing [37]. Biosorption of metal ion using agricultural or vegetable biomass, however, has been demonstrated to be amenable to continuous processing [38]. Column mode of operation is a continuous process and of greater practical importance due to its simplicity and low-cost of operation and easy scale up from laboratory to industrial scale [37].

Hydrogels have attracted wide attention as adsorbents for remediation of heavy metal ions, ammonia and dyes [39]. Polysaccharide-based hydrogels are attractive as adsorbents because they are biodegradable and biocompatible and have been successfully applied for the organic and inorganic pollutant remediation [39, 40]. A wide range of polysaccharides based hydrogels including gums, collagen, cellulose, alginate, carrageenan and chitosan have also been successfully applied for this purpose [41]. Hybrid hydrogel composite of gelatine (GL) and clinoptilolite synthesized by free radical grafting and were utilized for metal ions remediation from the mine effluents [42].

Chitin, an acetylamino polysaccharide, the second most abundant natural resource, next to cellulose, is found in the exoskeletons of crabs, arthropods and cell walls of some fungi. Chitin and its deacetylated modification, chitosan are excellent metal ligands [43]. Chitin is amorphous and intractable, hence limits its use due to slight inertness and difficulty in processing. Cellulose/chitin beads exhibited greater adsorption capacity for heavy metals than that of pure chitin flakes [44, 45]. Thus, only some functional group content does not decide adsorption abilities of biosorbents, some other factors or their combination are decisive.

The bio-sorbents are preferred because of their unique properties such as biodegradability, low-cost and abundance [46]. However, low adsorption capacity and slow rate seriously restrict their applications. Further, being hydrophilic, they are prone to microbial attack in aqueous environments. Hence, they are being modified for improvement in their properties [47]. Some of the biosorbents are immobilized (cross-linked) in a synthetic polymer matrix [48] and/or grafted onto an inorganic support material such as silica [49] to yield particles/beads with the desired mechanical strength and chemical resistance, to ensure recyclability for subsequent cycles. Nevertheless, cross-linking decreases the adsorption efficiency of biosorbents [48, 50]. Furthermore, the cross-linking agents such as glutaric dialdehyde and ethylene glycoldiglycidyl ether are physiologically toxic and the crosslinked beads are non-biodegradable. For the development of new polymeric materials with desirable properties, polymer blending is a simple process [51]. Cellulose and clay are arousing the attention as reinforcements of various polymeric blends because they are abundant in nature, have high mechanical strength, good chemical resistance, economic and above all environmentally friendly [52, 53]. An article critically reviewed the physical and chemical treatment methods to improve the fibre-matrix adhesion and their characterization [54].

In water decontamination technology, nanostructured adsorbents are emerging as the cynosure of all eyes [55]. However, the nanoparticles suffer some drawbacks, such as a decrease in surface area due to agglomeration, difficulty in recyclability and environmental dangers caused by particle-elapsing [55]. Hybrid nanocomposites are preferred for providing high specific surface area and better adsorption properties cost-effectively by immobilizing nanoparticles onto large solid substrates [56].

3 Nanocomposites

Current research and development are being focussed on polymer matrix based nanocomposites. On nanoscale, the dimensions of the particle, platelet or fibre are in 1–100 nm. Owing to the interesting observations involving exfoliated clay, carbon nanotubes, carbon nanofibers, exfoliated graphite (graphene), nanocrystalline metals and a variety of nanoscale inorganic filler or fibre modifications, interest in polymer matrix based nanocomposites is gaining impetus. The primary area of interest is the reinforcement aspects of nanocomposites however, barrier properties, flammability resistance, electrical/electronic properties, membrane properties, polymer blend compatibilization are important for various potential applications. An important consideration is the synergistic advantage of nanoscale dimensions relative to larger scale modification. The thickness of individual clay layers is much smaller than the wavelength of visible light, hence in well-exfoliated clay/polymer nanocomposites, optical properties of the polymer are not significantly affected rendering it optically clear. To optimize the properties of the resultant nanocomposite, it is desired to interpret the property changes as the particle (or fibre) dimensions reduce to the nanoscale level [57].

3.1 Bio-nanocomposites

Nanostructured hybrid organic-inorganic composites are invoking concern of academia and industry. Bio-nanocomposites exhibit multidimensional properties like biocompatibility, antimicrobial activity and biodegradability [58].

A review described bio-nanocomposites based on different polysaccharides functionalized by different nanofillers such as montmorillonite (MMT), Ag, SiO_2, TiO_2 and ZnO. These were used in regenerative medicine, drug delivery, tissue engineering, electronics and food packaging. The recent advancements were discussed in the light of technical and scientific issues [59].

Cellulose a hydrophilic substrate is a good choice to fabricate hybrid nanocomposites due to its origin from renewable resources and long chains capable of self-assembly [60]. The hydroxyl groups on the surface of cellulose provide numerous active sites for the proliferation of inorganic oxides. Immobilization of nano-TiO_2 onto cellulose matrix was done by various methods such as sol-gel [61], hydrothermal [62], layer by layer (LBL) assembly [63] and hydrolysis treatment [63]. Hydrothermal and hydrolysis methods require heating to attain certain temperature, microwave irradiation provides fast heating rates and molecular homogeneity, quick reaction kinetics, rapid crystallization and simplified procedure [64].

3.1.1 Bionanocomposites for Pollution Remediation

Some of the recent significant applications of bionanocomposites for metal ion remediation and adsorption of organic dyes are summarized in Tables 1 and 2, respectively.

Table 1 Bionanocomposites for metal ion remediation

Nanocomposite	Pollutant	Q_{max}^a	Remarks	Ref.
Chitosan coated iron-oxide nanocomposites	As(III)	267.2 mg/g	Only 13% loss in initial adsorption capacity after 5 repeated adsorption cycles	[94]
Chitosan film loaded with silver nanoparticles (CS-AgNPs)	Al(III), Cd(II), Cu(II), Co(II), Fe(III), Ni(II), Pb(II), Zn(II).		Feasible material for solid-phase extraction of metal pollutants from surface waters	[95]
Chitosan-rectorite nanospheres immobilized on polystyrene fibrous mats CS-REC	Cu(II)	98.87 mg/g	The addition of REC containing Ca^{2+} could also improve the metal adsorption because of cation exchange	[96]
Multiwall carbon nanotubes (MWCNTs)/chitosan nanocomposite	Cu(II), Zn(II), Cd(II), Ni(II)	Cu(II)-100% Zn(II)-99% Cd(II)-98% Ni(II)-100%	Greater efficiency for the target metal ion and suitable to be used in different environmental applications	[97]
Organo vermiculite (OVU) Chitosan (CHT)	Cd(II)	10% CHT-95.9 mg/g 50% CHT-97.1% 65% CHT-141 mg/g	The adsorption capacity was increased by raising the solution pH, which most likely favours a chelation mechanism between the chitosan chains and the Cd(II) cations	[98]
Chitin/magnetite/ multiwalled carbon nanotubes magnetic nanocomposite	Cr(VI)	90.7%	Potential and promising adsorbent for environmental remediation	[99]

(continued)

Table 1 (continued)

Nanocomposite	Pollutant	Q^a_{max}	Remarks	Ref.
Chitosan and carbon nanofibers (CNFs)-supported iron (Fe)-oxide nanoparticles (NPs)	Cr(VI)	80.0 mg/g	Regenerative and reused four times efficiently	[100]
Zero-valent copper-chitosan nanocomposites	Cr(VI)	99%	Excellent adsorption behaviour	[101]
Dead yeast biomass and titania nanoparticles	Cr(VI)	162.07 mg/g	Efficiency (99.92%)	[102]
Fe_3O_4 coated glycine doped polypyrrole magnetic nanocomposite (Fe_3O_4@gly-PPy NC)	Cr(VI)	238–303 mg/g	PPy moiety reduced Cr(VI) to Cr(III) Max. adsorption in 25–45 °C range and pH 2	[103]
Chitosan and Alginate nanocomposites	Cr(VI)	108.8 mg/g	Optimum adsorption at pH 5.0	[104]
Polyamides 6 (PA6)/Chitosan@Fe_xO_ycomposite nanofibers	Cr(VI)	64.3225 mg/g	High Cr(VI) removal efficiency of the nanofibrous membrane by electrospinning and pyrolytic reaction	[105]
Magnetic cellulose nanocomposite beads	Pb(II)	2.86 mg/g	Adsorption processes spontaneous, endothermic, controlled by chemical mechanisms and material reusable	[106]
MnO_2/chitosan nanocomposites	Pb(II)	Noncrosslinked-90.56%, crosslinked 59.93%	Potential adsorbent for the removal of lead ions	[107]
Nanocomposite based on nano silica filled xanthan gum grafted with polyacrylamide (XG-g-PAM/SiO_2)	Pb(II)	537.63 mg/g	Efficient adsorbent for the treatment of battery industry wastewater due to higher	[108]

(continued)

Table 1 (continued)

Nanocomposite	Pollutant	Q_{max}^a	Remarks	Ref.
Pectin zirconium(IV) selenotungstophosphate (Pc/ZSWP)	Cu(II), Th(IV)	Cu(II) 2.48 mg/g Th(IV)- 4.37 mg/g	hydrodynamic radius/volume High ion exchange capacity	[109]
Starch/SnO$_2$ nanocomposite	Hg(II)	192 mg/g	Regenerative capacity (94%) was maintained till fourth cycle	[110]

$^aQ_{max}$ = maximum adsorption capacity

3.2 Clay-Based Nanocomposites

Clays and their minerals are abundant and cheap materials successfully used for decades as adsorbents for removing toxic heavy metals from aqueous solutions. Clays and their minerals, composed a large family of adsorbents both in natural and modified forms, effectively remove most of the chemical contaminants from aqueous solution.

Layered silicate clay minerals are extensively used nano-reinforcements due to easy availability, low cost, and environmentally benign properties. Very low levels of nanofiller (5 wt%) matches the reinforcement efficiency of conventional composites with 40–50% of classical fillers, because of the high aspect ratio and high surface area of the nanolayered clays. Extensive studies on the preparation and properties of biodegradable polymer-clay nanocomposites were conducted on polylactide, polyhydroxy butyrate, thermoplastic starch, polybutylene succinate and polycaprolactone. Clay nanoparticles enhance mechanical properties, thermal stability, barrier properties and control the biodegradation rate of the biopolymers [65].

A detailed review compiled thorough research during 2006–2016 and highlighted the key conclusions of adsorption studies which used clay minerals as adsorbents. The structure, classification, and chemical composition of various clay minerals were outlined and analysis of their adsorption behaviour was described. The author suggested the need to modify and develop a synthetic method of novel materials and for application as an adsorbent for different pollutants [66].

Halloysite (HAL) possesses disordered aluminium silicate structure, number of adsorption sites and unparalleled nanotubular morphology, which makes it a promising adsorbent. HAL is especially attractive due to its low production cost, inertness for the environment. HAL can be modified by multistep intercalation, interlayer grafting reactions or both. HAL-based hybrids can be designed with adsorption functionalities targeted towards individual contaminants (e.g., polar/apolar or positively/negatively charged) by appropriate selection of organic molecules. Matusik summarized the studies on the use of pure and modified HAL for the remediation of chosen inorganic and organic pollutants [67].

Table 2 Bionanocomposites for organic dye adsorption

Nanocomposite	Organic dye	Q^a_{max}	Remarks	Ref.
Chitosan cenospheres (10:3) nanocomposite	Orange 25 (DO) and Disperse Blue (DB)	97.30% for DO, 94.22% for DB	Crosslinking agent glutaraldehyde	[111]
Chitosan/carbon black fiber	p-nitrophenol, congo red (CR) and methyl orange (MO) dyes	% reduction of CR and MO-99	Inhibited growth of E. coli by 47.8%	[112]
Chitosan-copper (CS-Cu) nanocomposite	Rhodamine B (RhB) and CR	Under light RhB-70% CR-93%	Excellent photocatalytic performer and antimicrobial agent	[113]
Chitosan-g-poly (acrylamide)/ZnS (ChPA/ZS) nanocomposite	MO and CR	75% of CR and 69% of MO was degraded after 4 h	MW synthesis Removal by photocatalytic degradation	[114]
Chitosan–polyvinyl alcohol (CHT-PVA) polymer matrix	Bisphenol A (BPA)	97.6%	Optimal pH for maximum adsorption 6.0, regenerative and reusable within five cycles	[115]
Chitosan-SnO$_2$ nanocomposites	MO and RhB	Photocatalytic degradation at 365 nm	High crystallinity, high surface area, and small particle size enable superior photocatalytic degradation	[116]
Chitosan-TiO$_2$ nanocomposite	RhB and CR	Photocatalytic degradation RhB-65% CR-57%	Excellent antimicrobial agent and antifungal activity	[117]
Composite of polyaniline, starch, polypyrrole, chitosan/aniline and chitosan/pyrrole using peanut waste	Crystal Violet (CV)	100.6 mg/g	Good desorption properties, recycling ability	[118]
Fe$_3$O$_4$ MNPs[b] and gum xanthan based hydrogels nanocomposites	Malachite green (MG)	497.15 mg/g	Nanocomposite can be repeatedly used for the adsorption	[119]
Gelatin-Zr(IV) phosphate nanocomposite (GT/ZPNC)	Methylene blue (MB) and fast green (FG)	87.81% MB and 89.91% FG	Photocatalytical degradation of pollutants and antimicrobial efficiency against E. coli	[120]

(continued)

Table 2 (continued)

Nanocomposite	Organic dye	Q_{max}^a	Remarks	Ref.
Guar gum/Al_2O_3 nanocomposite	MG	~80–90% degradation of MG under solar irradiation	Effective photocatalyst	[121]
Guar gum–cerium (IV) tungstate nanocomposite (GG/CTNC)	MB		Potential adsorbent MB from aqueous system	[122]
Gum ghatti-based biodegradable hydrogel	MB and MV	98% of MB and 95% of MV from aqueous solution	Polymer degraded fully within 50 days in soil compost	[123]
Hybrid TiO_2/microcrystalline cellulose	MB	Photocatalytic degradation of 40–90% of MB	Catalytic adsorption of MB from aqueous solutions	[124]
MWCNTS/chitin/magnetite (MCM) nanocomposite	Rose Bengal (RB)		Adsorption of RB occurred in different steps	[125]
Gum ghatti/TiO_2 nanoparticles-based hydrogel nanocomposite	MB	1305.5 mg/g	More efficient in adsorbing cationic dyes than anionic dyes	[126]
TiO_2-impregnated chitosan nano-grafts	Remazol red RB-133 (RR RB-133)	116.3 mg/g	Photoactive under sunlight-irradiation, five adsorption/sunlight-assisted self-cleaning photo regeneration cycles	[127]
Alginate/carboxymethyl cellulose TiO_2 nanoparticles (TiO_2-NPs) and graphene oxide (GO)	CR	degradation efficiency 98%	Efficient visible light responsive photocatalyst, degradation efficiency retained up to seven consecutive cycles	[128]

[b]MNP = magnetic nanoparticles

Clay minerals and their modified derivatives constitute a large family of adsorbents. A review presented literature on adsorption of diverse contaminants such as hydrophobic organic materials, dyes, heavy metal ions, oxyanions, radioactive nuclides, etc. Different mechanisms, such as surface adsorption, partition, ion exchange, surface precipitation and structural incorporation are involved in the uptake of these contaminants [68]. Organoclays are synthesized in the laboratory, and are smart organic-rich clay materials, for geo-environmental applications because of strong sorption affinity for nonpolar organic molecules. Organic cations displace naturally occurring inorganic cations (e.g., Na^+, Ca^{2+}) in the clay. By selecting a suitable organic cation that is exchanged into the clay interlayer, sorption can be managed to be linear or non-linear, competitive or non-competitive. Most commonly used clay is MMT for organoclays due to its high cation exchange capacity and quaternary ammonium cations as the organic phase. An overview of properties, synthesis, applications, synthesis techniques, modifications with several chemical compounds such as quaternary alkylammonium salts and biomolecules such as enzymes was given in a review, particularly focused on bentonite (BENT) and polymer nanocomposites [69]. Zhao et al. [70] exhaustively reviewed organoclays with reference to the fundamental behaviour and geochemical properties.

Photocatalysis emerged as an active field of research endeavour for environmental pollution abatement. This approach is more effective in combining complementary components with the inorganic semiconductor active material. Three such types of nanocomposite assemblies were discussed on the basis of combining semiconductor (usually transition metal oxides) with clays, carbon and metals [71].

3.3 Polymer-Layered Silicate Nanocomposites (PLSN)

Reinforcing fillers for polymer nanocomposites are layered silicates due to their economy and abundance for several years [72].

Polymer-layered silicate nanocomposites (PLSNs) are much superior to conventional micro-, macrocomposites due to unexpected properties [73]. The nanometric scale provides good reinforcement at less than 10% low filler loading as compared to conventional fillers which require more than 30% [74]. The mechanical properties, thermal resistance, solvent resistance and ionic conductivity are improved and gas permeability is reduced on reinforcement with clay [75].

Polymer-functionalized nanocomposites (PFNCs) retain the inherent remarkable surface properties of nanoparticles, functional groups provide specific bindings to target pollutants, and the polymer support materials provide high stability and processability. A review discussed synthesis, characterization and adsorption performance of PFNCs [76].

3.3.1 Structure of PLSN

Layered silicates are built of two structural units. In simplest 1:1 structures (e.g. kaolinite) a silica tetrahedral sheet is fused to an aluminium octahedron by sharing the oxygen atoms [77]. MMT is a smectite with 2:1 phyllosilicate structure, stacked layers made of two silica tetrahedrons fused to an edge-shared octahedral sheet of alumina (Fig. 1). The thickness of layer is ~ 1 nm and the lateral dimensions vary from 300 Å to several microns, giving an aspect ratio greater than 1000. The adjacent layers (interlayer or gallery) are held together by van der Waals forces; hence water and other polar molecules can be accommodated causing expansion of

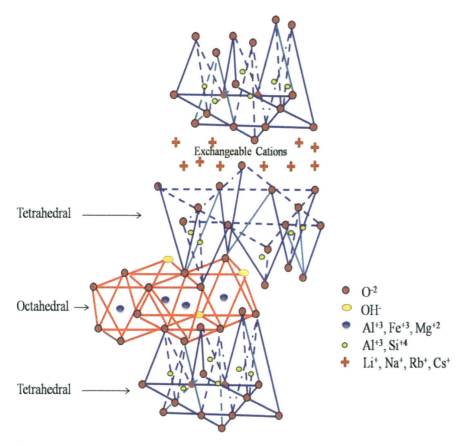

Fig. 1 The structure of 2:1 layered silicates

the lattice. Negative charges are generated by isomorphic substitution within the layers which are counter balanced by hydrated sodium or calcium ions present in the interlayer [78].

The extent of dispersion of the individual silicate platelets within the polymer matrix contributes to the enhancement of substantial properties of nanocomposites. Intercalated and exfoliated morphological structures are proposed for the layered silicate nanocomposites. The exfoliated layers impart the greatest reinforcement in nanocomposites. However, only hydrophilic polymers can be incorporated in clay. Hence, the silicate surface is modified by exchanging the interlaminar cations with organic cations to form organically modified montmorillonite. This results in better compatibility of polymer and silicate layers [79]. Organic cationic surfactants, like primary, secondary, tertiary and quaternary alkylammonium or alkylphosphonium cations are used to improve the surface properties to become compatible with the hydrophobic polymer matrix, moreover, their long aliphatic tails result in a greater interlayer spacing [80].

3.3.2 Methods of Preparation of PLSN

Clay/polymer nanocomposites [81] have been prepared by several methods, widely applied methods are described as under:

(i) Solution method: clay and polymer components are dissolved in a polar organic solvent, where polymers remain in the extended state in the inter gallery space. An intercalated nanocomposite is formed on solvent evaporation [82].
(ii) Interlamellar method or in situ polymerization: clay is completely dispersed in the polar monomer solution than the addition of curing agent forms an exfoliated composite.
(iii) Melt intercalation method: molten thermoplastic is included by blending with silicate to optimize the interaction. The temperature should be kept below the decomposition temperature of the clay modifier; trial and error are practised with different compatibilizers.

Schematic presentation of various types of nanocomposites with probable polymer/layered silicate structures is shown in Fig. 2.

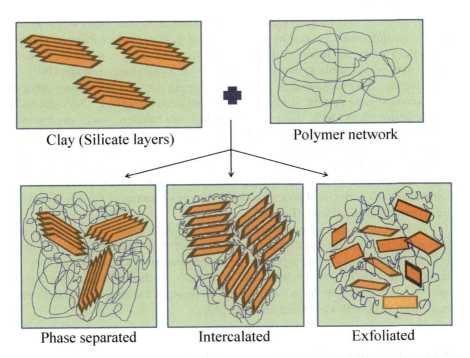

Fig. 2 Probable polymer/layered silicate structures in various types of nanocomposites

3.3.3 Challenges in Use of PLSN

Complete dispersion of the filler in the polymer matrix to prepare true nanocomposite is a scientific and technical challenge. Pandey et al. reviewed the durability of nanocomposites along with the technical problems and their feasible solutions [83]. The preparation, processing, properties, characterization, crystallization behaviours, melt rheology, and possible future applications of nanocomposites of biodegradable polymers and layered silicate have been reviewed [84, 85]. Nanotechnology increases tensile strength, flexural modulus, heat resistance and dyeability of polymeric materials [86]. Choudalakis and Gotsis reviewed the permeability and mechanisms for the transport of gas molecules and the procedures for the measurement of permeability and diffusivity in polymer nanocomposites [87].

A review examined and summarized preparation, properties, and applications of magnetic nanoparticle/clay mineral nanocomposites. Their potentials and applications in electromagnetic devices, magnetorheological fluids/ferrofluids, magnetic adsorbents, catalysts and biomaterials and the existing problems and challenges were discussed. Suggestions for future studies were given such as: to emphasize decreasing polydispersity, increasing functionalities, uncovering the preparation-modification-structure-magnetism-activity relationships and advancing the practical applications [88].

The potential toxicological effects of unmodified or modified clay minerals and their nanocomposites are gaining impetus. There is scarce information on toxicity biomarkers such as immune modulatory effects or alteration of the genetic expression, which needs systematic in vitro–in vivo extrapolation studies [89].

Currently, food packaging is the fast developing application utilizing clays of each phyllosilicate groups: kaolinite, MMT and sepiolite (SEP). Clays generally induce cytotoxicity with different underlying mechanisms, such as necrosis/apoptosis, oxidative stress or genotoxicity as suggested by in vitro studies. However, most of in vivo experiments performed in rodents did not show toxicity even at 5000 mg/kg doses. Pulmonary exposure is the most frequent in humans; clays are usually mixed with other minerals, which induce pneumoconiosis intrinsically. However, oral toxicity is not high; a strict control of the concentrations can provide beneficial uses (Fig. 3).

Clay minerals form highly stable suspensions in water; hence their rational use is important. The volume and treatment complexity of water contaminated with them is increased because of wide application. Current treatment techniques are either not economically viable, not environmentally friendly, or both. Flocculating agents such as polyelectrolytes have the potential to separate the above-mentioned minerals from industrial wastewater effluents [90].

3.3.4 PLSN with Bio-based Polymer Matrices

Favourable characteristics of clay nanocomposites with bio-based polymer matrices are improved *Young's modulus*, reduced gas permeability, and improved fire

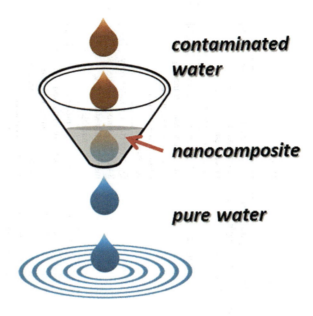

Fig. 3 Schematic presentation of application of nanocomposites for water purification

retardancy. They can be applied as melt-processed thermoplastic mouldings, packaging films and coatings [91].

Different types of nanocomposites are possible depending on the inorganic filler used. In PLSN the reinforcing phase has only one nano level dimension. Intercalated nanocomposites consist of a regular array of the polymer into the clay galleries with fixed interlayer spacings, and in exfoliated nanocomposites, 1 nm thick layers are separately dispersed in the polymeric matrix to form a monolithic structure at the microscale.

MMT is most commonly employed in the synthesis of nanocomposites. It is obtained from bentonite (BENT) [92].

Structural, morphological and textural features of natural microfibrous clay minerals Sepiolite (SEP) and palygorskite (PAL) make them useful for the preparation of nanocomposite materials. SEP- and PAL-based bionanocomposites have diverse applications as bioplastics, membranes, biomedicine, drug delivery systems and additives of vaccines, tissue engineering, sensor devices, bioreactors and as a source of supported graphene [93].

3.4 Clay-Based Nanocomposites for Pollution Remediation

Some of the recent researches involving the application of nanocomposites comprising of clay and organic components (biopolymers, organic compounds and or synthetic polymers) are summarized in Table 3.

Table 3 Clay based nanocomposites for metal ion and organic contaminant remediation

Nanocomposite	Inorganic component	Organic component	Pollutant	Q_{max}	Comments	Ref.
Attapulgite/poly(acrylic acid-co-acrylamide)	ATP	P(AA-co-AM)	Pb(II) and Cu(II)	35.94 mg/g for Pb(II) and 30.35 mg/g for Cu(II)	Exhibit selective adsorption, adsorption desorption process enhanced their desired reusability	[129]
Organo-bentonite/Fe_3O_4 (OB-Fe_3O_4 PSA)	BENT	Poly(sodium acrylate) (PSA)	Th(IV)	6.55 mmol/g	Adsorption capacity was still higher than 3.6 mmol/g after five consecutive adsorption–desorption processes	[130]
Halloysite/alginate nanocomposite beads	HAL	Alginate	Pb(II)	325 mg/g	Q_{max} for Hal nanotubes only 84 mg/g. Nanocomposite mechanically stronger, easily separable	[131]
Nanotubularhalloysite	HAL	Amino alcohols	Pb(II), Cd(II), Zn(II), and Cu(II)	Pb(II)—64.1 (HD), 58.8 (HT); Cd(II)—41.8 (HD), 41.7 (HT); Zn(II)—52.9 (HD), 42.9 (HT); Cu(II)—64.1 (HD), 75.2 (HT) *HD is HAL grafted with diethanolamine and HT is HAL grafted with triethanolamine	No swelling properties, diffusion of metal ion and complex binding with the grafted group. The material was non-recyclable	[132]
Polypyrrole-coated halloysite nanotube nanocomposite (PPy-HNTs NC)	HAL nanotube	Ppy	Cr(VI)	149.25 mg/g	Optimum adsorption at pH 2.0 at 25 °C, some of the Cr(VI) reduced to Cr(III)	[133]

(continued)

Table 3 (continued)

Nanocomposite	Inorganic component	Organic component	Pollutant	Q_{max}	Comments	Ref.
Kaolinite supported bimetallic Fe/Ni nanoparticles (K–Fe/Ni)	Kaolinite, bimetallic Fe/Ni nanoparticles		Pb (II) and NO_3^-	86.3% of Pb(II) and 73.6% of NO_3^-	Reactivity maintained throughout 15 days and removal efficiency maintained till 3 successive cycles	[134]
Di- and triethanolamine grafted kaolinites	Kaolinites	Diethanolamine (DEA) and triethanolamine (TEA)	Cd(II), Zn(II), Pb (II) and Cu(II)		Grafted DEA or TEA donating N and O to form complexes with metals	[135]
PPy-OMMT NC	MMT	Polypyrrole	Cr(VI)	112.3, 119.34, 176.2 and 209.6 mg/g at 292, 298, 308 and 318 K, respectively	The selective adsorption of Cr(VI) was demonstrated in binary adsorption systems with co-existing ions	[136]
PPy-OMMT	MMT	Polypyrrole (ppy)	Cr(VI)	Fixed bed adsorption column studies	High bed mass, low inlet Cr (VI) concentration and flow rate enhanced removal efficiency	[137]
Fe-Ni bimetallic nanoparticles (Fe-Ni NPs) and Nanocomposites, (MMT) clay	MMT		Cr(VI)		Fe-Ni NPs exhibited a complete reduction of Cr (VI) occurs via Cr(II) to Cr (0) below pH 4 due to the generation of reactive H species	[138]
Fe_3O_4/MMT nanocomposite (Fe_3O_4/ MMT NC)	MMT		Pb(II), Cu(II), Ni (II)	89.72%(Pb(II)), 94.89%(Cu(II)), 76.15% (Ni(II))	Response surface methodology was utilized in designing the experiment	[139]

(continued)

Table 3 (continued)

Nanocomposite	Inorganic component	Organic component	Pollutant	Q_{max}	Comments	Ref.
Ammonium-pillared montmorillonite-$CoFe_2O_4$	MMT	Calcium alginate beads	Cs(I)	86.46 mg/g	High selectivity for Cs^+ in the presence of Na^+ and could be rapidly separated from the mixed solution under an external magnetic field	[140]
ZnO/MMT nanocomposite	MMT		Pb(II) and Cu(II)		Stability and durability of the nanocomposite allowed it to be utilized for at least three times in adsorptive removal	[141]
Chitosan/clay nanocomposite (CCN)	Nanoclay (Cloisite 10A)	Chitosan	Cr(VI)	357.14 mg/g	Adsorption at wide pH range, pH 3 most suitable	[142]
Chitosan-nanoclay composite (CNC)	Organoclay (Cloisite 20A)	Chitosan	Cr(VI)	128.43 mg/g	Maximum adsorption at pH 2	[143]
Chitosan–palygorskite (CP) composites. Mass ratios 1:1 (CP1) 1:2 (CP2) and 2:1 (C2P)	PALY	Chitosan	Pb(II)	201.5, 154.5, 147.1, 27.7 and 9.3 mg g^{-1} for CP1, C2P, CP2, chitosan and palygorskite	The surface properties and specific interaction between chitosan and palygorskite in the composites-governing factors in metal ion adsorption	[144]
Polysaccharide–fibrous clay bionanocomposites	Sepiolite and PALY	Polysaccharide (starch, alginate or chitosan)	Heavy metal ions		Films exhibit improved mechanical properties, good water resistance, biocompatibility and biodegradability	[145]

(continued)

Table 3 (continued)

Nanocomposite	Inorganic component	Organic component	Pollutant	Q_{max}	Comments	Ref.
FeO-Fe$_3$O$_4$ beads		Polyvinyl alcohol (PVA)/ sodium alginate (SA)	Cr(VI)	76.3–100%	Beads containing 0.075 wt % FeO and 0.30 wt% Fe$_3$O$_4$ was mechanically stable and highly efficient in removing Cr(VI) from 99.3 to 76.3% with increasing pH from 3.0 to 11.0	[146]
Clay mineral polymer nanocomposites (CPN)	MMT	Block-(diblock or triblock) copolymers	Toluene, ethylbenzene and xylene, Cu(II), Pb (II) and Cr$_2$O$_7^{2-}$	Cu(II)-6.11 mg/g, Pb (II)-10 mg/g, Cr$_2$O$_7^{2-}$ -36.9 mg/g	Dealt with the effect on interlayer space of Mt during reaction	[147]
Xanthan gum/ Methionine-bentonite (XG/Meth-bent) nanocomposite	BENT	Xanthan gum	Congo red dye	530.549 mg/g	Desorption with regeneration (up to fifth cycle) was best observed by NaOH	[148]
Bentonite/g-C3N4 and Ag$_3$PO$_4$	BENT	Graphitic carbon nitride (g-C$_3$N$_4$)	Rhodamine B	Photocatalytic degradation	The composite containing 20 wt% Ag$_3$PO$_4$ exhibited the highest photocatalytic activity in decolorizing RhB	[149]
biomass-derived carbon@montmorillonite	MMT	Glucose biomass as a carbonaceous source	Methylene blue (MB)	194.2 mg/g	Carbon–clay nanocomposites could be explored as a new type of the highly efficient and green adsorbent	[150]
CdSe-MMT nanocomposites	CdSe-MMT		Indigo carmine	Photocatalytic degradation	Adsorptive removal Indigo Carmine (IC) on the composite surface occurs via photocatalytic degradation under visible light	[141]

(continued)

Table 3 (continued)

Nanocomposite	Inorganic component	Organic component	Pollutant	Q_{max}	Comments	Ref.
Arquad® 2HT-75 organobentonite	BENT	Arquad® 2HT-75 (alkyl ammonium surfactant)	Phenol and p-nitrophenol	Phenol-2.1 mg/g and p-nitrophenol-6.0 mg/g	Surface having positive zeta potential, effective against ionizable organic contaminants	[151]
Chitosan-Bentonite (Cts-Bent)	BENT	Chitosan	Highly efficient inactivation of bacteria		Ag and ZnO nanoparticles in the chitosan matrix work as a disinfectant in wastewater treatment	[152]
Copper-complexed clay/polyacrylic acid	Acid-treated BENT, natural PAL	Polyacrylic acid	NH_3 gas	Composite with 75% clay-75.1 mg/g, with 66% clay-80.0 mg/g	Efficiently adsorbed NH_3 gas from the contaminated air and proved to be an effective and cheaper method of air filtering	[153]
Chitosan immobilized Bentonite (CIB)	BENT	Chitosan	Remediation of distilleries (vinasse) wastewater	83% COD reduction and 78% colour removal	Immobilization of chitosan with CTAB strongly affects the sorption capacity of the clay mineral	[154]
Tubular halloysite	HAL	Organosilane and organophosphoric acid			Regulate the physical (solubility, dispersion, hydrophilicity/hydrophobicity, rheology, etc.) and the chemical (reactivity, biotoxicity, electrochemistry, etc.) controlled release and pollution remediation	[155]

(continued)

Table 3 (continued)

Nanocomposite	Inorganic component	Organic component	Pollutant	Q_{max}	Comments	Ref.
TiO_2 and Fe_2O_3-halloysite	HAL		Aniline, 2-chloro- and 2,6-dichloroaniline	Photocatalytic degradation	Photocatalytic active under UV radiation	[156]
MMT–CuO–Chitosan, MMT–CuO–Gum ghatti, and MMT–CuO poly lactic acid	MMT	Gum ghatti (GG), Chitosan (Ch), poly lactic acid (PLA)	Monocrotophos, and organophosphate insecticide	83.99% on MMT–CuO–PLA, 71.6% on MMT–CuO–Ch, 62.1% on MMT–CuO–GG, 40.8% on MMT–CuO		[157]
Montmorillonites intercalated with polymeric Fe/Al	MMT	Hexdecyl-trimethyl-ammonium bromide (HDTMA)	Phenol	Physical adsorption	Al/HDTMA and MMT/HDTMA possess a good affinity for phenol, in comparison to polymeric Al/Fe modified- and starting montmorillonites	[158]
Organoclay (PC-VER)	Modifying vermiculite (VER)	Phosphatidylcholine (PC)	Oxytetracycline (OTC) and ciprofloxacin (CIP)		PC-VER could serve as a low-cost, suitable and eco-friendly material for adsorption of antibiotics	[159]
Nanocomposites of ferrocenyl surfactants and clay	MMT		In situ remediations of many organic compounds in the environment		The crystal lattice iron redox reactions mechanism proposed for catalysis	[160]
TiO_2–sepiolite nanocomposites	SEP		β-naphthol	Photocatalytic degradation	High water vapour uptake properties and degradation ability make it a useful applicant in energy and environment	[161]

References

1. Vorosmarty CJ et al (2010) Global threats to human water security and river biodiversity. Nature 467:555–561. https://doi.org/10.1038/nature09440
2. Fu FL, Wang Q (2011) Removal of heavy metal ions from wastewaters: a review. J Environ Manage 92:407–418. https://doi.org/10.1016/j.jenvman.2010.11.011
3. Shannon MA et al (2008) Science and technology for water purification in the coming decades. Nature 452:301–310. https://doi.org/10.1038/nature06599
4. Ngah WSW, Hanafiah M (2008) Removal of heavy metal ions from wastewater by chemically modified plant wastes as adsorbents: a review. Bioresour Technol 99:3935–3948. https://doi.org/10.1016/j.biortech.2007.06.011
5. Kirpichtchikova TA et al (2006) Speciation and solubility of heavy metals in contaminated soil using X-ray microfluorescence, exafs spectroscopy, chemical extraction, and thermodynamic modeling. Geochim Cosmochim Acta 70:2163–2190. https://doi.org/10.1016/j.gca.2006.02.006
6. Maslin P, Maier RM (2000) Rhamnolipid-enhanced mineralization of phenanthrene in organic-metal co-contaminated soils. Bioremediat J 4:295–308. https://doi.org/10.1080/10889860091114266
7. Kabata-Pendias A (2017) Trace elements in soils and plants, 3rd edn. CRC Press, Boca Raton, FL, USA
8. Salt DE et al (1995) Phytoremediation—a novel strategy for the removal of toxic metals from the environment using plants. Bio-Technology 13:468–474. https://doi.org/10.1038/nbt0595-468
9. Jones KC, de Voogt P (1999) Persistent organic pollutants (pops): state of the science. Environ Poll 100:209–221 https://doi.org/10.1016/s0269-7491(99)00098-6
10. Forgacs E, Cserhati T, Oros G (2004) Removal of synthetic dyes from wastewaters: a review. Environ Int 30:953–971. https://doi.org/10.1016/j.envint.2004.02.001
11. Malik R, Ramteke DS, Wate SR (2007) Adsorption of malachite green on groundnut shell waste based powdered activated carbon. Waste Manage 27:1129–1138. https://doi.org/10.1016/j.wasman.2006.06.009
12. Salleh MAM et al (2011) Cationic and anionic dye adsorption by agricultural solid wastes: a comprehensive review. Desalination 280:1–13. https://doi.org/10.1016/j.desal.2011.07.019
13. Ngah WSW, Teong LC, Hanafiah M (2011) Adsorption of dyes and heavy metal ions by chitosan composites: a review. Carbohydr Polym 83:1446–1456. https://doi.org/10.1016/j.carbpol.2010.11.004
14. Green chemistry for dyes removal from waste water: research trends and applications (2015). Wiley, Hoboken, NJ, USA
15. Alsbaiee A et al (2016) Rapid removal of organic micropollutants from water by a porous beta-cyclodextrin polymer. Nature 529:U146–U190. https://doi.org/10.1038/nature16185
16. Crini G (2006) Non-conventional low-cost adsorbents for dye removal: a review. Bioresour Technol 97:1061–1085. https://doi.org/10.1016/j.biortech.2005.05.001
17. Ku Y, Jung IL (2001) Photocatalytic reduction of Cr(VI) in aqueous solutions by uv irradiation with the presence of titanium dioxide. Water Res 35:135–142. https://doi.org/10.1016/s0043-1354(00)00098-1
18. Luo T, Abdu S, Wessling M (2018) Selectivity of ion exchange membranes: a review. J Membrane Sci 555:429–454. https://doi.org/10.1016/j.memsci.2018.03.051
19. Missimer TM, Maliva RG (2018) Environmental issues in seawater reverse osmosis desalination: intakes and outfalls. Desalination 434:198–215. https://doi.org/10.1016/j.desal.2017.07.012
20. Alkhudhiri A, Darwish N, Hilal N (2012) Membrane distillation: a comprehensive review. Desalination 287:2–18. https://doi.org/10.1016/j.desal.2011.08.027
21. Knez Z et al (2014) Industrial applications of supercritical fluids: A review. Energy 77:235–243. https://doi.org/10.1016/j.energy.2014.07.044

22. Djas M, Henczka M (2018) Reactive extraction of carboxylic acids using organic solvents and supercritical fluids: a review. Sep Purif Technol 201:106–119. https://doi.org/10.1016/j.seppur.2018.02.010
23. Banat IM et al (1996) Microbial decolorization of textile-dye-containing effluents: a review. Bioresour Technol 58:217–227. https://doi.org/10.1016/s0960-8524(96)00113-7
24. Aksu Z (2005) Application of biosorption for the removal of organic pollutants: a review. Process Biochem 40:997–1026. https://doi.org/10.1016/j.procbio.2004.04.008
25. Prakash R, Majumder SK, Singh A (2018) Flotation technique: its mechanisms and design parameters. Chem Eng Process 127:249–270. https://doi.org/10.1016/j.cep.2018.03.029
26. Gundogdu A et al (2009) Biosorption of pb(ii) ions from aqueous solution by pine bark (pinus brutia ten.). Chem Eng J 153:62–69. https://doi.org/10.1016/j.cej.2009.06.017
27. Khan S et al (2008) Health risks of heavy metals in contaminated soils and food crops irrigated with wastewater in Beijing, China. Environ Pollut 152:686–692. https://doi.org/10.1016/j.envpol.2007.06.056
28. Li J et al (2018) Synthesis of highly porous inorganic adsorbents derived from metal-organic frameworks and their application in efficient elimination of mercury(II). J Colloid Interface Sci 517:61–71. https://doi.org/10.1016/j.jcis.2018.01.112
29. Gupta VK, Suhas (2009) Application of low-cost adsorbents for dye removal—a review. J Environ Manage 90:2313–2342. https://doi.org/10.1016/j.jenvman.2008.11.017
30. Pearson RG (1963) Hard and soft acids and bases. J Am Chem Soc 85:3533. https://doi.org/10.1021/ja00905a001
31. Mercier L, Detellier C (1995) Preparation, characterization and applications as heavy-metals sorbents of covalently grafted thiol functionalities on the interlamellar surface of montmorillonite. Environ Sci Technol 29:1318–1323. https://doi.org/10.1021/es00005a026
32. Volesky B, Holan ZR (1995) Biosorption of heavy-metals. Biotechnol Progr 11:235–250. https://doi.org/10.1021/bp00033a001
33. Wang JL, Chen C (2009) Biosorbents for heavy metals removal and their future. Biotechnol Adv 27:195–226. https://doi.org/10.1016/j.biotechadv.2008.11.002
34. Volesky B (2001) Detoxification of metal-bearing effluents: biosorption for the next century. Hydrometallurgy 59:203–216. https://doi.org/10.1016/s0304-386x(00)00160-2
35. Fomina M, Gadd GM (2014) Biosorption: current perspectives on concept, definition and application. Bioresour Technol 160:3–14. https://doi.org/10.1016/j.biortech.2013.12.102
36. Vijayaraghavan K, Balasubramanian R (2015) Is biosorption suitable for decontamination of metal-bearing wastewaters? A critical review on the state-of-the-art of biosorption processes and future directions. J Environ Manage 160:283–296. https://doi.org/10.1016/j.jenvman.2015.06.030
37. Long YC et al (2014) Packed bed column studies on lead(II) removal from industrial wastewater by modified agaricus bisporus. Bioresour Technol 152:457–463. https://doi.org/10.1016/j.biortech.2013.11.039
38. Ungureanu G et al (2017) Biosorption of antimony oxyanions by brown seaweeds: batch and column studies. J Environ Chem Eng 5:3463–3471. https://doi.org/10.1016/j.jece.2017.07.005
39. Zhang JP, Wang AQ (2015) Polysaccharide-based composite hydrogels for removal of pollutants from water. In: Dragan ES (ed) Advanced separations by specialized sorbents, vol 108. Chromatographic science series. CRC Press-Taylor & Francis Group, Boca Raton, pp 89–126
40. Guilherme MR et al (2015) Superabsorbent hydrogels based on polysaccharides for application in agriculture as soil conditioner and nutrient carrier: a review. Eur Polym J 72:365–385. https://doi.org/10.1016/j.eurpolymj.2015.04.017
41. Fosso-Kankeu E et al (2010) A comprehensive study of physical and physiological parameters that affect bio-sorption of metal pollutants from aqueous solutions. Phys Chem Earth 35:672–678. https://doi.org/10.1016/j.pce.2010.07.008
42. Fosso-Kankeu E et al (2017) Thermodynamic properties and adsorption behaviour of hydrogel nanocomposites for cadmium removal from mine effluents. J Ind Eng Chem 48:151–161. https://doi.org/10.1016/j.jiec.2016.12.033

43. Sag Y, Aktay Y (2000) Mass transfer and equilibrium studies for the sorption of chromium ions onto chitin. Process Biochem 36:157–173. https://doi.org/10.1016/s0032-9592(00)00200-4
44. Zhou D et al (2004) Cellulose/chitin beads for adsorption of heavy metals in aqueous solution. Water Res 38:2643–2650. https://doi.org/10.1016/j.watres.2004.03.026
45. Dao Z et al (2004) Development of a fixed-bed column with cellulose/chitin beads to remove heavy-metal ions. J Appl Polym Sci 94:684–691. https://doi.org/10.1002/app.20946
46. Morosanu I et al (2017) Biosorption of lead ions from aqueous effluents by rapeseed biomass. N Biotechnol 39:110–124. https://doi.org/10.1016/j.nbt.2016.08.002
47. Kumar DPJ, Raj V (2017) A review on the modification of polysaccharide through graft copolymerization for various potential applications. Open Med Chem J 11:109–126. https://doi.org/10.2174/1874104501711010109
48. Ngah WSW, Endud CS, Mayanar R (2002) Removal of copper(ii) ions from aqueous solution onto chitosan and cross-linked chitosan beads. React Funct Polym 50:181–190
49. Rangsayatorn N et al (2004) Cadmium biosorption by cells of spirulina platensis tistr 8217 immobilized in alginate and silica gel. Environ Int 30:57–63. https://doi.org/10.1016/s0160-4120(03)00146-6
50. Lee ST et al (2001) Equilibrium and kinetic studies of copper(ii) ion uptake by chitosan-tripolyphosphate chelating resin. Polymer 42:1879–1892. https://doi.org/10.1016/s0032-3861(00)00402-x
51. Sawatari C, Kondo T (1999) Interchain hydrogen bonds in blend films of poly(vinyl alcohol) and its derivatives with poly(ethylene oxide). Macromolecules 32:1949–1955. https://doi.org/10.1021/ma980900o
52. Ollier R, Perez CJ, Alvarez V (2012) Effect of relative humidity on the mechanical properties of micro and nanocomposites of polyvinyl alcohol. In: Armas AF (ed) 11th international congress on metallurgy & materials sam/conamet 2011, vol 1. Procedia materials science. Elsevier Science Bv, Amsterdam, pp 499–505. https://doi.org/10.1016/j.mspro.2012.06.067
53. Gemeiner P et al (1998) Cellulose as a (bio)affinity carrier: properties, design and applications. J Chromatogr B-Anal Technol Biomed Life Sci 715:245–271. https://doi.org/10.1016/s0378-4347(98)00047-4
54. George J, Sreekala MS, Thomas S (2001) A review on interface modification and characterization of natural fiber reinforced plastic composites. Polym Eng Sci 41:1471–1485. https://doi.org/10.1002/pen.10846
55. Zhang WX (2003) Nanoscale iron particles for environmental remediation: an overview. J Nanopart Res 5:323–332. https://doi.org/10.1023/a:1025520116015
56. Bet-moushoul E et al (2016) TiO_2 nanocomposite based polymeric membranes: a review on performance improvement for various applications in chemical engineering processes. Chem Eng J 283:29–46. https://doi.org/10.1016/j.cej.2015.06.124
57. Paul DR, Robeson LM (2008) Polymer nanotechnology: nanocomposites. Polymer 49:3187–3204. https://doi.org/10.1016/j.polymer.2008.04.017
58. Sharma M, Bajpai A (2018) Superabsorbent nanocomposite from sugarcane bagasse, chitin and clay: synthesis, characterization and swelling behaviour. Carbohydr Polym 193:281–288. https://doi.org/10.1016/j.carbpol.2018.04.006
59. Zafar R et al (2016) Polysaccharide based bionanocomposites, properties and applications: a review. Int J Biol Macromol 92:1012–1024. https://doi.org/10.1016/j.ijbiomac.2016.07.102
60. Luo Y, Huang JG (2015) Hierarchical-structured anatase-titania/cellulose composite sheet with high photocatalytic performance and antibacterial activity. Chem Eur J 21:2568–2575. https://doi.org/10.1002/chem.201405066
61. Huang JG, Kunitake T (2003) Nano-precision replication of natural cellulosic substances by metal oxides. J Am Chem Soc 125:11834–11835. https://doi.org/10.1021/ja037419k
62. Luo Y, Xu JB, Huang JG (2014) Hierarchical nanofibrous anatase-titania-cellulose composite and its photocatalytic property. Cryst Eng Comm 16:464–471. https://doi.org/10.1039/c3ce41906a

63. Li H, Fu SY, Peng LC (2013) Surface modification of cellulose fibers by layer-by-layer self-assembly of lignosulfonates and TiO_2 nanoparticles: effect on photocatalytic abilities and paper properties. Fiber Polym 14:1794–1802. https://doi.org/10.1007/s12221-013-1794-8
64. Ding KL et al (2007) Facile synthesis of high quality TiO_2 nanocrystals in ionic liquid via a microwave-assisted process. J Am Chem Soc 129:6362–6363. https://doi.org/10.1021/ja070809c
65. Nayak PL et al (2008) Nanocomposites from polycaprolactone (PCL)/soy protein isolate (SPI) blend with organoclay. Polym Plast Technol Eng 47:600–605. https://doi.org/10.1080/03602550802059402
66. Uddin MK (2017) A review on the adsorption of heavy metals by clay minerals, with special focus on the past decade. Chem Eng J 308:438–462. https://doi.org/10.1016/j.cej.2016.09.029
67. Matusik J (2016) Chapter 23–halloysite for adsorption and pollution remediation. In: Yuan P, Thill A, Bergaya F (eds) Developments in clay science, vol 7. Elsevier, pp 606–627. https://doi.org/10.1016/b978-0-08-100293-3.00023-6
68. Zhu RL et al (2016) Adsorbents based on montmorillonite for contaminant removal from water: a review. Appl Clay Sci 123:239–258. https://doi.org/10.1016/j.clay.2015.12.024
69. de Paiva LB, Morales AR, Diaz FRV (2008) Organoclays: properties, preparation and applications. Appl Clay Sci 42:8–24. https://doi.org/10.1016/j.clay.2008.02.006
70. Zhao Q et al (2017) Review of the fundamental geochemical and physical behaviors of organoclays in barrier applications. Appl Clay Sci 142:2–20. https://doi.org/10.1016/j.clay.2016.11.024
71. Rajeshwar K, Chanmanee W (2012) Bioinspired photocatalyst assemblies for environmental remediation. Electrochim Acta 84:96–102. https://doi.org/10.1016/j.electacta.2012.04.072
72. Pavlidou S, Papaspyrides CD (2008) A review on polymer-layered silicate nanocomposites. Prog Polym Sci 33:1119–1198. https://doi.org/10.1016/j.progpolymsci.2008.07.008
73. Manias E et al (2001) Polypropylene/montmorillonite nanocomposites. Review of the synthetic routes and materials properties. Chem Mater 13:3516–3523. https://doi.org/10.1021/cm0110627
74. Teh PL et al (2004) On the potential of organoclay with respect to conventional fillers (carbon black, silica) for epoxidized natural rubber compatibilized natural rubber vulcanizates. J Appl Polym Sci 94:2438–2445. https://doi.org/10.1002/app.21188
75. Alexandre M, Dubois P (2000) Polymer-layered silicate nanocomposites: preparation, properties and uses of a new class of materials. Mater Sci Eng R-Reports 28:1–63. https://doi.org/10.1016/s0927-796x(00)00012-7
76. Lofrano G et al (2016) Polymer functionalized nanocomposites for metals removal from water and wastewater: an overview. Water Res 92:22–37. https://doi.org/10.1016/j.watres.2016.01.033
77. Miranda-Trevino JC, Coles CA (2003) Kaolinite properties, structure and influence of metal retention on ph. Appl Clay Sci 23:133–139. https://doi.org/10.1016/s0169-1317(03)00095-4
78. Leroux F, Besse JP (2001) Polymer interleaved layered double hydroxide: a new emerging class of nanocomposites. Chem Mater 13:3507–3515. https://doi.org/10.1021/cm0110268
79. Mousa A, Karger-Kocsis J (2001) Rheological and thermodynamical behavior of styrene/butadiene rubber-organoclay nanocomposites. Macromol Mater Eng 286:260–266. https://doi.org/10.1002/1439-2054(20010401)286:4%3c260:aid-mame260%3e3.0.co;2-x
80. Kiliaris P, Papaspyrides CD (2010) Polymer/layered silicate (clay) nanocomposites: an overview of flame retardancy. Prog Polym Sci 35:902–958. https://doi.org/10.1016/j.progpolymsci.2010.03.001
81. Gao FG (2004) Clay/polymer composites: the story. Mater Today 7:50–55. https://doi.org/10.1016/s1369-7021(04)00509-7
82. Deka BK, Maji TK (2011) Effect of TiO_2 and nanoclay on the properties of wood polymer nanocomposite. Compos A 42:2117–2125. https://doi.org/10.1016/j.compositesa.2011.09.023
83. Pandey JK et al (2005) An overview on the degradability of polymer nanocomposites. Polym Degrad Stab 88:234–250. https://doi.org/10.1016/j.polymdegradstab.2004.09.013

84. Ray SS, Okamoto M (2003) Polymer/layered silicate nanocomposites: a review from preparation to processing. Prog Polym Sci 28:1539–1641. https://doi.org/10.1016/j.progpolymsci.2003.08.002
85. Ray SS, Bousmina M (2005) Biodegradable polymers and their layered silicate nano composites: in greening the 21st century materials world. Prog Mater Sci 50:962–1079. https://doi.org/10.1016/j.pmatsci.2005.05.002
86. Ataeefard M, Moradian S (2012) Investigation the effect of various loads of organically modified montmorillonite on dyeing properties of polypropylene nanocomposites. J Appl Polym Sci 125:E214–E223. https://doi.org/10.1002/app.34812
87. Choudalakis G, Gotsis AD (2009) Permeability of polymer/clay nanocomposites: a review. Eur Polym J 45:967–984. https://doi.org/10.1016/j.eurpolymj.2009.01.027
88. Chen L et al (2016) Functional magnetic nanoparticle/clay mineral nanocomposites: preparation, magnetism and versatile applications. Appl Clay Sci 127:143–163. https://doi.org/10.1016/j.clay.2016.04.009
89. Maisanaba S et al (2015) Toxicological evaluation of clay minerals and derived nanocomposites: a review. Environ Res 138:233–254. https://doi.org/10.1016/j.envres.2014.12.024
90. Shaikh SMR et al (2017) Influence of polyelectrolytes and other polymer complexes on the flocculation and rheological behaviors of clay minerals: a comprehensive review. Sep Purif Technol 187:137–161. https://doi.org/10.1016/j.seppur.2017.06.050
91. Liu AD, Berglund LA (2012) Clay nanopaper composites of nacre-like structure based on montmorrilonite and cellulose nanofibers-improvements due to chitosan addition. Carbohydr Polym 87:53–60. https://doi.org/10.1016/j.carbpol.2011.07.019
92. Paranhos CM et al (2007) Microstructure and free volume evaluation of poly(vinyl alcohol) nanocomposite hydrogels. Eur Polym J 43:4882–4890. https://doi.org/10.1016/j.eurpolymj.2007.10.001
93. Ruiz-Hitzky E et al (2013) Fibrous clays based bionanocomposites. Prog Polym Sci 38:1392–1414. https://doi.org/10.1016/j.progpolymsci.2013.05.004
94. Neeraj G et al (2016) Adsorptive potential of dispersible chitosan coated iron-oxide nanocomposites toward the elimination of arsenic from aqueous solution. Process Saf Environ Prot 104:185–195. https://doi.org/10.1016/j.psep.2016.09.006
95. Djerahov L et al (2016) Chitosan film loaded with silver nanoparticles-sorbent for solid phase extraction of Al(III), Cd(II), Cu(II), Co(II), Fe(III), Ni(II), Pb(II) and Zn(II). Carbohydr Polym 147:45–52. https://doi.org/10.1016/j.carbpol.2016.03.080
96. Tu H et al (2017) Chitosan-rectorite nanospheres immobilized on polystyrene fibrous mats via alternate electrospinning/electrospraying techniques for copper ions adsorption. Appl Surf Sci 426:545–553. https://doi.org/10.1016/j.apsusc.2017.07.159
97. Salam MA, Makki MSI, Abdelaal MYA (2011) Preparation and characterization of multi-walled carbon nanotubes/chitosan nanocomposite and its application for the removal of heavy metals from aqueous solution. J Alloys Compd 509:2582–2587. https://doi.org/10.1016/j.jallcom.2010.11.094
98. Padilla-Ortega E et al (2016) Ultrasound assisted preparation of chitosan-vermiculite bionanocomposite foams for cadmium uptake. Appl Clay Sci 130:40–49. https://doi.org/10.1016/j.clay.2015.11.024
99. Salam MA (2017) Preparation and characterization of chitin/magnetite/multiwalled carbon nanotubes magnetic nanocomposite for toxic hexavalent chromium removal from solution. J Mol Liq 233:197–202. https://doi.org/10.1016/j.molliq.2017.03.023
100. Khare P et al (2016) Microchannel-embedded metal-carbon-polymer nanocomposite as a novel support for chitosan for efficient removal of hexavalent chromium from water under dynamic conditions. Chem Eng J 293:44–54. https://doi.org/10.1016/j.cej.2016.02.049
101. Wu SJ, Liou TH, Mi FL (2009) Synthesis of zero-valent copper-chitosan nanocomposites and their application for treatment of hexavalent chromium. Bioresour Technol 100:4348–4353. https://doi.org/10.1016/j.biortech.2009.04.013

102. Choudhury PR et al (2017) Removal of Cr (VI) by synthesized titania embedded dead yeast nanocomposite: optimization and modeling by response surface methodology. J Environ Chem Eng 5:214–221. https://doi.org/10.1016/j.jece.2016.11.041
103. Ballav N et al (2014) Synthesis, characterization of Fe_3O_4@glycine doped polypyrrole magnetic nanocomposites and their potential performance to remove toxic Cr(VI). J Ind Eng Chem 20:4085–4093. https://doi.org/10.1016/j.jiec.2014.01.007
104. Gokila S et al (2017) Removal of the heavy metal ion chromium(VI) using chitosan and alginate nanocomposites. Int J Biol Macromol 104:1459–1468. https://doi.org/10.1016/j.ijbiomac.2017.05.117
105. Li CJ et al (2014) Preparation of polyamides 6 (PA6)/chitosan@fexoy composite nanofibers by electrospinning and pyrolysis and their Cr(VI)-removal performance. Catal Today 224:94–103. https://doi.org/10.1016/j.cattod.2013.11.034
106. Luo XG et al (2016) Adsorptive removal of lead from water by the effective and reusable magnetic cellulose nanocomposite beads entrapping activated bentonite. Carbohydr Polym 151:640–648. https://doi.org/10.1016/j.carbpol.2016.06.003
107. Mallakpour S, Madani M (2016) Functionalized-MnO_2/chitosan nanocomposites: a promising adsorbent for the removal of lead ions. Carbohydr Polym 147:53–59. https://doi.org/10.1016/j.carbpol.2016.03.076
108. Ghorai S et al (2012) Novel biodegradable nanocomposite based on XG-g-PAM/SiO_2: application of an efficient adsorbent for Pb^{2+} ions from aqueous solution. Bioresour Technol 119:181–190. https://doi.org/10.1016/j.biortech.2012.05.063
109. Sharma G, Pathania D, Naushad M (2014) Preparation, characterization and antimicrobial activity of biopolymer based nanocomposite ion exchanger pectin zirconium(IV) selenotungstophosphate: application for removal of toxic metals. J Ind Eng Chem 20:4482–4490. https://doi.org/10.1016/j.jiec.2014.02.020
110. Naushad M et al (2016) Synthesis and characterization of a new starch/SnO_2 nanocomposite for efficient adsorption of toxic Hg^{2+} metal ion. Chem Eng J 300:306–316. https://doi.org/10.1016/j.cej.2016.04.084
111. Markandeya et al (2017) Statistical optimization of process parameters for removal of dyes from wastewater on chitosan cenospheres nanocomposite using response surface methodology. J Clean Prod 149:597–606. https://doi.org/10.1016/j.jclepro.2017.02.078
112. Ali F et al (2017) Bactericidal and catalytic performance of green nanocomposite based on chitosan/carbon black fiber supported monometallic and bimetallic nanoparticles. Chemosphere 188:588–598. https://doi.org/10.1016/j.chemosphere.2017.08.118
113. Nithya A et al (2017) A potential photocatalytic, antimicrobial and anticancer activity of chitosan-copper nanocomposite. Int J Biol Macromol 104:1774–1782. https://doi.org/10.1016/j.ijbiomac.2017.03.006
114. Pathania D et al (2016) Photocatalytic degradation of highly toxic dyes using chitosan-g-poly (acrylamide)/ZnS in presence of solar irradiation. J Photochem Photobiol, A 329:61–68. https://doi.org/10.1016/j.jphotochem.2016.06.019
115. Simsek EB et al (2017) Carbon fiber embedded chitosan/PVA composites for decontamination of endocrine disruptor bisphenol-A from water. J Taiwan Inst Chem Eng 70:291–301. https://doi.org/10.1016/j.jtice.2016.11.008
116. Gupta VK et al (2017) Degradation of azo dyes under different wavelengths of uv light with chitosan-SnO_2 nanocomposites. J Mol Liq 232:423–430. https://doi.org/10.1016/j.molliq.2017.02.095
117. Karthikeyan KT, Nithya A, Jothivenkatachalam K (2017) Photocatalytic and antimicrobial activities of chitosan-TiO_2 nanocomposite. Int J Biol Macromol 104:1762–1773. https://doi.org/10.1016/j.ijbiomac.2017.03.121
118. Tahir N et al (2017) Biopolymers composites with peanut hull waste biomass and application for crystal violet adsorption. Int J Biol Macromol 4:210–220. https://doi.org/10.1016/j.ijbiomac.2016.10.013

119. Mittal H et al (2014) Fe_3O_4 mnps and gum xanthan based hydrogels nanocomposites for the efficient capture of malachite green from aqueous solution. Chem Eng J 255:471–482. https://doi.org/10.1016/j.cej.2014.04.098
120. Thakur M et al (2017) Efficient photocatalytic degradation of toxic dyes from aqueous environment using gelatin-Zr(IV) phosphate nanocomposite and its antimicrobial activity. Colloids Surf B: Biointerfaces 157:456–463. https://doi.org/10.1016/j.colsurfb.2017.06.018
121. Pathania D et al (2016) Novel guar gum/Al_2O_3 nanocomposite as an effective photocatalyst for the degradation of malachite green dye. Int J Biol Macromol 87:366–374. https://doi.org/10.1016/j.ijbiomac.2016.02.073
122. Gupta VK et al (2014) Adsorptional removal of methylene blue by guar gum-cerium (iv) tungstate hybrid cationic exchanger. Carbohydr Polym 101:684–691. https://doi.org/10.1016/j.carbpol.2013.09.092
123. Mittal H, Maity A, Ray SS (2015) Effective removal of cationic dyes from aqueous solution using gum ghatti-based biodegradable hydrogel. Int J Biol Macromol 79:8–20. https://doi.org/10.1016/j.ijbiomac.2015.04.045
124. Virkutyte J, Jegatheesan V, Varma RS (2012) Visible light activated TiO_2/microcrystalline cellulose nanocatalyst to destroy organic contaminants in water. Bioresour Technol 113:288–293. https://doi.org/10.1016/j.biortech.2011.12.090
125. Salam MA, El-Shishtawy RM, Obaid AY (2014) Synthesis of magnetic multi-walled carbon nanotubes/magnetite/chitin magnetic nanocomposite for the removal of rose bengal from real and model solution. J Ind Eng Chem 20:3559–3567. https://doi.org/10.1016/j.jiec.2013.12.049
126. Mittal H, Ray SS (2016) A study on the adsorption of methylene blue onto gum ghatti/tio2 nanoparticles-based hydrogel nanocomposite. Int J Biol Macromol 88:66–80. https://doi.org/10.1016/j.ijbiomac.2016.03.032
127. Essawy AA, Sayyah SM, El-Nggar AM (2017) Wastewater remediation by TiO_2-impregnated chitosan nano-grafts exhibited dual functionality: high adsorptivity and solar-assisted self cleaning. J Photochem Photobiol B: Biol 173:170–180. https://doi.org/10.1016/j.jphotobiol.2017.05.044
128. Thomas M et al (2017) Self-organized graphene oxide and TiO_2 nanoparticles incorporated alginate/carboxymethyl cellulose nanocomposites with efficient photocatalytic activity under direct sunlight. J Photochem Photobiol, A 346:113–125. https://doi.org/10.1016/j.jphotochem.2017.05.037
129. Liu P et al (2015) Synthesis of covalently crosslinked attapulgite/poly (acrylic acid-co-acrylamide) nanocomposite hydrogels and their evaluation as adsorbent for heavy metal ions. J Ind Eng Chem 23:188–193. https://doi.org/10.1016/j.jiec.2014.08.014
130. Wu LS et al (2013) Organo-bentonite-Fe_3O_4 poly(sodium acrylate) magnetic superabsorbent nanocomposite: synthesis, characterization, and thorium(IV) adsorption. Appl Clay Sci 83–84:405–414. https://doi.org/10.1016/j.clay.2013.07.012
131. Chiew CSC et al (2016) Halloysite/alginate nanocomposite beads: kinetics, equilibrium and mechanism for lead adsorption. Appl Clay Sci 119:301–310. https://doi.org/10.1016/j.clay.2015.10.032
132. Matusik J, Wscislo A (2014) Enhanced heavy metal adsorption on functionalized nanotubular halloysite interlayer grafted with aminoalcohols. Appl Clay Sci 100:50–59. https://doi.org/10.1016/j.clay.2014.06.034
133. Ballav N et al (2014) Polypyrrole-coated halloysite nanotube clay nanocomposite: synthesis, characterization and Cr(VI) adsorption behaviour. Appl Clay Sci 102:60–70. https://doi.org/10.1016/j.clay.2014.10.008
134. Shi LN et al (2014) Functional kaolinite supported fe/ni nanoparticles for simultaneous catalytic remediation of mixed contaminants (lead and nitrate) from wastewater. J Colloid Interface Sci 428:302–307. https://doi.org/10.1016/j.jcis.2014.04.059
135. Koteja A, Matusik J (2015) Di-and triethanolamine grafted kaolinites of different structural order as adsorbents of heavy metals. J Colloid Interface Sci 455:83–92. https://doi.org/10.1016/j.jcis.2015.05.027

136. Setshedi KZ et al (2013) Exfoliated polypyrrole-organically modified montmorillonite clay nanocomposite as a potential adsorbent for Cr(VI) removal. Chem Eng J 222:186–197. https://doi.org/10.1016/j.cej.2013.02.061
137. Setshedi KZ et al (2014) Breakthrough studies for Cr(VI) sorption from aqueous solution using exfoliated polypyrrole-organically modified montmorillonite clay nanocomposite. J Ind Eng Chem 20:2208–2216. https://doi.org/10.1016/j.jiec.2013.09.052
138. Kadu BS et al (2011) Efficiency and recycling capability of montmorillonite supported Fe-Ni bimetallic nanocomposites towards hexavalent chromium remediation. Appl Catal B 104:407–414. https://doi.org/10.1016/j.apcatb.2011.02.011
139. Kalantari K et al (2015) Rapid and high capacity adsorption of heavy metals by Fe_3O_4/montmorillonite nanocomposite using response surface methodology: preparation, characterization, optimization, equilibrium isotherms, and adsorption kinetics study. J Taiwan Inst Chem Eng 49:192–198. https://doi.org/10.1016/j.jtice.2014.10.025
140. Zheng XM et al (2017) Ammonium-pillared montmorillonite-co-Fe_2O_4 composite caged in calcium alginate beads for the removal of Cs^+ from wastewater. Carbohydr Polym 167:306–316. https://doi.org/10.1016/j.carbpol.2017.03.059
141. Chikate RC, Kadu BS (2014) Improved photocatalytic activity of cdse-nanocomposites: effect of montmorillonite support towards efficient removal of indigo carmine. Spectrochim Acta A Mol Biomol Spectrosc 124:138–147. https://doi.org/10.1016/j.saa.2013.12.099
142. Pandey S, Mishra SB (2011) Organic-inorganic hybrid of chitosan/organoclay bionanocomposites for hexavalent chromium uptake. J Colloid Interface Sci 361:509–520. https://doi.org/10.1016/j.jcis.2011.05.031
143. Kahraman HT (2017) Development of an adsorbent via chitosan nano-organoclay assembly to remove hexavalent chromium from wastewater. Int J Biol Macromol 94:202–209. https://doi.org/10.1016/j.ijbiomac.2016.09.111
144. Rusmin R et al (2015) Structural evolution of chitosan-palygorskite composites and removal of aqueous lead by composite beads. Appl Surf Sci 353:363–375. https://doi.org/10.1016/j.apsusc.2015.06.124
145. Alcantara ACS et al (2014) Polysaccharide-fibrous clay bionanocomposites. Appl Clay Sci 96:2–8. https://doi.org/10.1016/j.clay.2014.02.018
146. Lv XS et al (2013) Fe-0-Fe_3O_4 nanocomposites embedded polyvinyl alcohol/sodium alginate beads for chromium (vi) removal. J Hazard Mat 262:748–758. https://doi.org/10.1016/j.jhazmat.2013.09.036
147. Chen HH et al (2017) Feasible preparation and characterization of tunable novel montmorillonite/block-copolymers based composites as potential dual adsorbent candidates. Appl Clay Sci 137:192–202. https://doi.org/10.1016/j.clay.2016.12.028
148. Ahmad R, Mirza A (2017) Green synthesis of xanthan gum/methionine-bentonite nanocomposite for sequestering toxic anionic dye. Surf Interfaces 8:65–72. https://doi.org/10.1016/j.surfin.2017.05.001
149. Ma JF et al (2016) Nanocomposite of exfoliated bentonite/g-C_3N_4/Ag_3PO_4 for enhanced visible-light photocatalytic decomposition of rhodamine B. Chemosphere 162:269–276. https://doi.org/10.1016/j.chemosphere.2016.07.089
150. Ai LH, Li LL (2013) Efficient removal of organic dyes from aqueous solution with ecofriendly biomass-derived carbon@montmorillonite nanocomposites by one-step hydrothermal process. Chem Eng J 223:688–695. https://doi.org/10.1016/j.cej.2013.03.015
151. Sarkar B et al (2011) Structural characterisation of arquad (R) 2HT-75 organobentonites: surface charge characteristics and environmental application. J Hazard Mat 195:155–161. https://doi.org/10.1016/j.jhazmat.2011.08.016
152. Motshekga SC, Ray SS (2017) Highly efficient inactivation of bacteria found in drinking water using chitosan-bentonite composites: modelling and breakthrough curve analysis. Water Res 111:213–223. https://doi.org/10.1016/j.watres.2017.01.003
153. Liu EM et al (2016) Copper-complexed clay/poly-acrylic acid composites: extremely efficient adsorbents of ammonia gas. Appl Clay Sci 121:154–161. https://doi.org/10.1016/j.clay.2015.12.012

154. El-Dib FI et al (2016) Remediation of distilleries wastewater using chitosan immobilized bentonite and bentonite based organoclays. Int J Biol Macromol 86:750–755. https://doi.org/10.1016/j.ijbiomac.2016.01.108
155. Tan DYP (2016) Surface modifications of halloysite. In: Yuan P (ed) Developments in clay science. Elsevier, pp 167–201
156. Szczepanik B et al (2017) Synthesis, characterization and photocatalytic activity of TiO_2-halloysite and Fe_2O_3-halloysite nanocomposites for photodegradation of chloroanilines in water. Appl Clay Sci 149:118–126. https://doi.org/10.1016/j.clay.2017.08.016
157. Sahithya K, Das D, Das N (2016) Adsorptive removal of monocrotophos from aqueous solution using biopolymer modified montmorillonite-CuO composites: equilibrium, kinetic and thermodynamic studies. Process Saf Environ Prot 99:43–54. https://doi.org/10.1016/j.psep.2015.10.009
158. Jiang JQ, Cooper C, Ouki S (2002) Comparison of modified montmorillonite adsorbents—part i: preparation, characterization and phenol adsorption. Chemosphere 47:711–716. https://doi.org/10.1016/s0045-6535(02)00011-5
159. Liu S et al (2017) Preparation and characterization of organo-vermiculite based on phosphatidylcholine and adsorption of two typical antibiotics. Appl Clay Sci 137:160–167. https://doi.org/10.1016/j.clay.2016.12.002
160. Swearingen C, Macha S, Fitch A (2003) Leached ferrocenes at clay surfaces: potential applications for environmental catalysis. J Mol Catal A: Chem 199:149–160. https://doi.org/10.1016/s1381-1169(03)00031-1
161. Karamanis D et al (2011) Water vapor adsorption and photocatalytic pollutant degradation with TiO_2-sepiolite nanocomposites. Appl Clay Sci 53:181–187. https://doi.org/10.1016/j.clay.2010.12.012

Correction to: Extraction of Nano Cellulose Fibres and Their Eco-friendly Polymer Composite

Bashiru Kayode Sodipo and Folahan Abdul Wahab Taiwo Owolabi

Correction to:
Chapter "Extraction of Nano Cellulose Fibres and Their Eco-friendly Polymer Composite" in: Inamuddin et al. (eds.), *Sustainable Polymer Composites and Nanocomposites*, https://doi.org/10.1007/978-3-030-05399-4_8

The original version of this chapter was inadvertently published with the incorrect author sequence and corresponding author. This has now been corrected.

The updated version of this chapter can be found at
https://doi.org/10.1007/978-3-030-05399-4_8

© Springer Nature Switzerland AG 2019
Inamuddin et al. (eds.), *Sustainable Polymer Composites and Nanocomposites*,
https://doi.org/10.1007/978-3-030-05399-4_48